UCSMP The University of Chicago School Mathematics Project

Geometry

Authors

John Benson
Ray Klein
Matthew J. Miller
Catherine Capuzzi-Feuerstein
Michael Fletcher
George Marino
Nancy Norem Powell
Natalie Jakucyn
Zalman Usiskin

Director of Evaluation

Denisse R. Thompson

UChicago**Solutions**

Authors

3rd EDITION AUTHORS

John Benson *Mathematics Teacher*
Evanston Township H.S., Evanston, IL

Ray Klein *Mathematics Teacher (retired)*
Glenbard West High School, Glen Ellyn, IL

Matthew J. Miller *Mathematics Teacher*
Washington High School, Cedar Rapids, IA

Catherine Capuzzi-Feuerstein *Mathematics Teacher*
Westfield High School, Westfield, NJ

Michael Fletcher *Mathematics Teacher*
W.P. Davidson High School, Mobile, AL

George Marino *Mathematics Teacher (retired)*
Proviso West High School, Hillside, IL

Nancy Norem Powell *Mathematics Teacher*
Bloomington High School, Bloomington, IL

Natalie Jakucyn *Mathematics Teacher*
Glenbrook South High School, Glenview, IL

Zalman Usiskin *Professor of Education*
The University of Chicago

AUTHORS OF EARLIER EDITIONS

Arthur Coxford *(deceased)*
Professor of Mathematics Education
University of Michigan, Ann Arbor, MI

Virginia Highstone *Mathematics Teacher*
York High School, Elmhurst, IL

Daniel B. Hirschhorn
Assistant Professor of Mathematics
Illinois State University, Normal, IL

Hester Lewellen
Assistant Professor of Mathematics
Baldwin-Wallace College, Berea, OH

Nicholas Oppong
Assistant Professor of Mathematics Education
University of Georgia, Athens, GA

Richard DiBianca
UCSMP

Merilee Maeir
UCSMP

http://ucsmp.uchicago.edu/secondary/overview

UChicago**Solutions**

Send all inquiries to:
UChicagoSolutions
6030 S. Ellis Avenue, 2nd Floor
Chicago, IL 60637

ISBN 978-1-943237-03-6
ISBN 1-943237-03-4

1 2 3 4 5 6 7 8 9 RRDW 21 20 19 18 17 16 15

UCSMP Evaluation, Editorial, and Production

Director of Evaluation
Denisse R. Thompson
Professor of Mathematics Education
University of South Florida, Tampa, FL

Coordinator of School Relations: Carol Siegel

Executive Managing Editor: Clare Froemel

Production Coordinator: Benjamin R. Balskus

Editorial Staff: Catherine Ballway (lead),
Asaf Hadari, Matthew McCrea, Alex Tomasik,
Scott Neff, Kathlyn Nguyen Ngo, Daniel Boutwell

Evaluation Consultant
Sharon L. Senk, *Professor of Mathematics*
Michigan State University, East Lansing, MI

Evaluation Assistants: Sophia Zhang, Gladys Mitchell,
Julian Owens, Shravani Pasupneti, Zhuo Zhang

Production Assistants: Rachel Huddleston,
Nurit Kirshenbaum, Alex Liu, Sarah Mahoney,
Gretchen Neidhardt, S.L. Schieffer, Yaya Tang,
Yayan Zhang, Don Reneau, Elizabeth Olin,
Loren Santow

Technology Assistant: Luke I. Sandberg

Since the first two editions of *Geometry* were published, millions of students and thousands of teachers have used the materials. Prior to the publication of this third edition, the following teachers and schools participated in evaluations of the trial version during 2006–2007.

Mark Brooks
Andover Central High School
Andover, Kansas

Stefanie Geeve
Glenbard West High School
Glen Ellyn, Illinois

Michael Buescher
Hathaway Brown School
Shaker Heights, Ohio

Ryan Nelson
Jenison High School
Jenison, Michigan

Mike Hendricks
Kewaskum High School
Kewaskum, Wisconsin

Katy Rodriguez
Nolan Catholic High School
Ft. Worth, Texas

Jacquelyn Boswell
Oxford High School
Oxford, Michigan

Erik Hanson
Valley View Middle School
Edina, Minnesota

Michael Fletcher
W.P. Davidson High School
Mobile, Alabama

Jay Loyd
Greenwood High School
Greenwood, Arkansas

Christopher M. Davidson
Oxford High School
Oxford, Michigan

Nancy Zenere
St. Rita of Cascia High School
Chicago, Illinois

The following schools participated in field studies in 1993–1994, 1988–1989, 1987–1988, or 1986–1987 as part of the first edition or the second edition research.

Collins High School
Corliss High School
Taft High School
Hyde Park Career Academy
Chicago, IL

Glenbrook South High School
Glenview, IL

Lake Park High School
Roselle, IL

Rich South High School
Richton Park, IL

Lyons Township High School
La Grange, IL

Olympia High School
Stanford, IL

Niles Township High School West
Skokie, IL

Chaparral High School
Scottsdale, AZ

Marietta High School
Marietta, GA

Fayette County High School
Fayetteville, GA

Sandy Creek High School
Tyrone, GA

Fruitport High School
Fruitport, MI

*Cincinnati Academy of
    Mathematics and Science*
Walnut Hills High School
Aiken High School
Cincinnati, OH

Taylor Allderice High School
Carrick High School
Pittsburgh, PA

Irvine High School
Irvine, CA

Mendocino High School
Mendocino, CA

North Hunterdon High School
Annandale, NJ

Franklin High School
Somerset, NJ

Arsenal Technical High School
Indianapolis, IN

Lakeridge High School
Lake Oswego, OR

Ashland High School
Ashland, OR

Astoria High School
Astoria, OR

Wando High School
Mt. Pleasant, SC

Olympic High School
Silverdale, WA

Lincoln High School
Wisconsin Rapids, WI

UCSMP The University of Chicago School Mathematics Project

The University of Chicago School Mathematics Project (UCSMP) is a long-term project designed to improve school mathematics in Grades Pre-K–12. UCSMP began in 1983 with a 6-year grant from the Amoco Foundation. Additional funding has come from the National Science Foundation, the Ford Motor Company, the Carnegie Corporation of New York, the Stuart Foundation, the General Electric Foundation, GTE, Citicorp/Citibank, the Exxon Educational Foundation, the Illinois Board of Higher Education, the Chicago Public Schools, from royalties, and from publishers of UCSMP materials.

From 1983 to 1987, the director of UCSMP was Paul Sally, Professor of Mathematics. Since 1987, the director has been Zalman Usiskin, Professor of Education.

UCSMP *Geometry*

The text *Geometry* has been developed by the Secondary Component of the project, and constitutes the core of the fourth year in a seven-year middle and high school mathematics curriculum. The names of the seven texts around which these years are built are:

- *Pre-Transition Mathematics*
- *Transition Mathematics*
- *Algebra*
- *Geometry*
- *Advanced Algebra*
- *Functions, Statistics, and Trigonometry*
- *Precalculus and Discrete Mathematics*

Why a Third Edition?

Since the second edition, there has been a general increase in the performance of students coming into middle and high schools due to a combination of increased expectations and the availability of improved curricular materials for K–8 grades. The UCSMP third edition is more ambitious and takes advantage of the increased knowledge students bring to the classroom.

These increased expectations and the increased levels of testing that have gone along with those expectations are requiring a broad-based, reality-oriented, easy-to-comprehend approach to mathematics. UCSMP third edition materials were written to better accommodate these factors.

The writing of the third edition of UCSMP is also motivated by the recent advances in technology both inside and outside the classroom, coupled with the widespread availability of computers with Internet connections at schools and at homes.

Yet another factor for the continued importance of the UCSMP curriculum is the increase in the number of students taking a full course in algebra before ninth grade. With the UCSMP curriculum, these students will have four years of mathematics beyond algebra before calculus and other college-level mathematics. UCSMP is the only curriculum so designed. Thousands of schools have used the first and second editions and have noted success in student achievement and in teaching practices. Research from these schools shows that the UCSMP materials really work. Many of these schools have made suggestions for additional improvements in future editions of the UCSMP materials. We have attempted to utilize all of these ideas in the development of the third edition.

UCSMP *Geometry*–Third Edition

All lessons have been reviewed and examined afresh for this edition but the overall structure and mathematical prerequisites are the same as in previous editions. The previous editions of UCSMP courses introduced many features that are retained and sometimes enhanced in this third edition. There is **wider scope**, including significant amounts of algebra employed to motivate, justify, extend, and otherwise enhance the geometry. The coordinate and transformation approaches are particularly important because coordinates connect geometry with algebra, and transformations are functions which allow all figures to be considered as geometric. These two features enable this text to be particularly beneficial in any further study of algebra and functions. A **real-world orientation** has guided both the selection of content and the approaches allowed the student in working out exercises and problems, because being able to do mathematics is of little use to an individual unless he or she can apply that content. This edition contains lessons applying geometry to music and architecture. We ask students to **read mathematics**, because students must read to understand mathematics in later courses and must learn to read technical matter in the world at large. The use of **new and powerful technology** is integrated throughout, with *graphing calculators* and *dynamic geometry systems* assumed for activities found throughout the materials.

Four dimensions of understanding are emphasized: skill in drawing, visualizing, and following algorithms; understanding of properties, mathematical relationships, and proofs; using geometric ideas in realistic situations; and representing geometric concepts with coordinates, networks, or other diagrams. We call this the SPUR approach: **S**kills, **P**roperties, **U**ses, **R**epresentations.

The **lessons** include prose designed to show why the content is important and to explain how ideas are related to each other, and fully-developed examples that often show multiple worked-out solutions and checks. Each lesson includes a question set that begins with **Covering the Ideas** questions that demonstrate the student's knowledge of the overall concepts of the lesson. **Applying the Mathematics** questions go beyond lesson examples with an emphasis on real-world problem solving. **Review** questions relate to previous lessons in the course or to content from earlier courses. **Exploration** questions ask students to explore ideas related to the lesson, and frequently have many possible answers.

The **book organization** is designed to maximize the acquisition of both skills and concepts. The daily review feature allows students several nights to learn and practice important concepts and skills. At the end of each chapter, a carefully focused Self-Test and a Chapter Review, each keyed to objectives in all the dimensions of understanding, are used to solidify performance of skills and concepts from the chapter so that they may be applied later with confidence. To increase retention, important ideas are reviewed in later chapters.

New instructional features for this edition include: a "Big Idea", highlighting the key concept of each lesson; Mental Math questions at the beginning of lessons to sharpen "in your head" skills; activities in virtually every lesson to develop concepts and skills; Guided Examples that provide partially completed solutions to encourage active learning; and Quiz Yourself stopping points to periodically check understanding as you read.

Comments about these materials are welcomed. Please address them to:

The University of Chicago School Mathematics Project
http://ucsmp.uchicago.edu
ucsmp@uchicago.edu
773-702-1130

▷ Contents

Getting Started   1

Chapter 1	4

Points and Lines

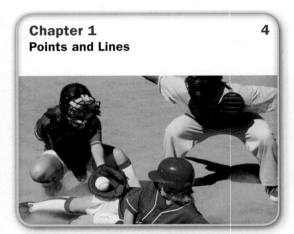

	Chapter 2	58

The Language and Logic of Geometry

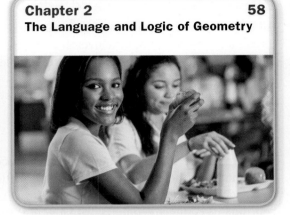

1-1	Points and Lines as Locations	6
1-2	Ordered Pairs as Points	13
1-3	Other Types of Geometry	19
1-4	Undefined Terms and First Definitions	27
1-5	Postulates for Points and Lines in Euclidean Geometry	32
1-6	Betweenness and Distance	38
1-7	Using a Dynamic Geometry System (DGS)	44
▷	Projects	51
▷	Summary and Vocabulary	53
▷	Self-Test	54
▷	Chapter Review	55

2-1	The Need for Definitions	60
2-2	Conditional Statements	66
2-3	Converses	72
2-4	Good Definitions	77
2-5	Unions and Intersections of Figures	83
2-6	Polygons	89
2-7	Conjectures	95
▷	Projects	101
▷	Summary and Vocabulary	103
▷	Self-Test	104
▷	Chapter Review	106

Chapter 3   110
Angles and Lines

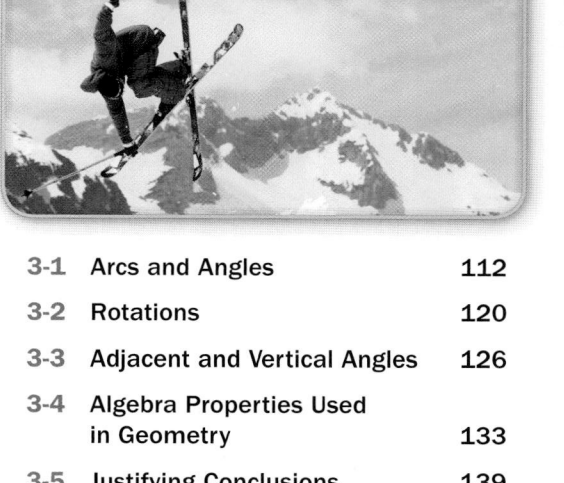

3-1	Arcs and Angles	112
3-2	Rotations	120
3-3	Adjacent and Vertical Angles	126
3-4	Algebra Properties Used in Geometry	133
3-5	Justifying Conclusions	139
3-6	Parallel Lines	145
3-7	Size Transformations	151
3-8	Perpendicular Lines	159
3-9	The Perpendicular Bisector	165
	Projects	172
	Summary and Vocabulary	174
	Self-Test	176
	Chapter Review	178

Chapter 4   182
Congruence Transformations

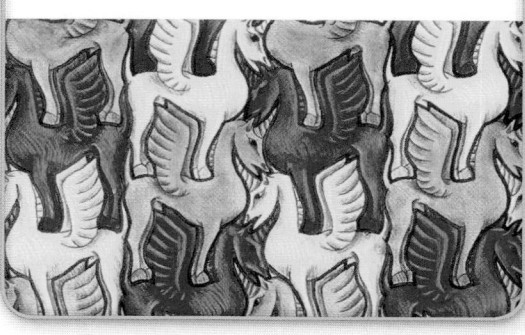
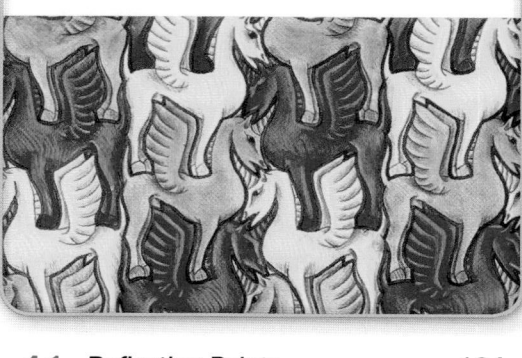

4-1	Reflecting Points	184
4-2	Reflecting Figures	189
4-3	Miniature Golf and Billiards	196
4-4	Composing Reflections over Parallel Lines	202
4-5	Composing Reflections over Intersecting Lines	210
4-6	Translations as Vectors	216
4-7	Isometries	223
4-8	Transformations and Music	230
	Projects	241
	Summary and Vocabulary	242
	Self-Test	244
	Chapter Review	246

Chapter 5   250
Proofs Using Congruence

5-1	When Are Figures Congruent?	252
5-2	Corresponding Parts of Congruent Figures	257
5-3	One-Step Congruence Proofs	263
5-4	Proofs Using Transitivity	269
5-5	Proofs Using Reflections	277
5-6	Auxiliary Figures and Uniqueness	282
5-7	Sums of Angle Measures in Polygons	288
	Projects	296
	Summary and Vocabulary	298
	Self-Test	300
	Chapter Review	302

Chapter 6   306
Polygons and Symmetry

6-1	Reflection Symmetry	308
6-2	Isosceles Triangles	316
6-3	Angles Inscribed in Circles	324
6-4	Types of Quadrilaterals	331
6-5	Properties of Kites	339
6-6	Properties of Trapezoids	345
6-7	Rotation Symmetry	351
6-8	Regular Polygons	356
6-9	Frieze Patterns	363
	Projects	368
	Summary and Vocabulary	369
	Self-Test	371
	Chapter Review	373

Chapter 7   378
Applications of Congruent Triangles

7-1	Drawing Triangles	380
7-2	Triangle Congruence Theorems	386
7-3	Using Triangle Congruence Theorems	393
7-4	Overlapping Triangles	400
7-5	The SSA Condition and HL Congruence	406
7-6	Tessellations	413
7-7	Properties of Parallelograms	419
7-8	Sufficient Conditions for Parallelograms	426
7-9	Diagonals of Quadrilaterals	431
7-10	Proving That Constructions Are Valid	436
	Projects	440
	Summary and Vocabulary	442
	Self-Test	443
	Chapter Review	445

Chapter 8   450
Lengths and Areas

8-1	Perimeter	452
8-2	Fundamental Properties of Area	457
8-3	Areas of Irregular Figures	463
8-4	Areas of Triangles	468
8-5	Areas of Quadrilaterals	474
8-6	The Pythagorean Theorem	480
8-7	Special Right Triangles	487
8-8	Arc Length and Circumference	494
8-9	The Area of a Circle	500
	Projects	506
	Summary and Vocabulary	508
	Self-Test	509
	Chapter Review	511

Chapter 9   **516**
Three-Dimensional Figures

9-1	Points, Lines, and Planes in Space	518
9-2	Prisms and Cylinders	525
9-3	Pyramids and Cones	532
9-4	Drawing in Perspective	538
9-5	Views of Solids and Surfaces	544
9-6	Spheres and Sections	550
9-7	Reflections in Space	559
9-8	Making Polyhedra and Other Surfaces	564
9-9	Surface Areas of Prisms and Cylinders	570
9-10	Surface Areas of Pyramids and Cones	577
	Projects	583
	Summary and Vocabulary	585
	Self-Test	587
	Chapter Review	589

Chapter 10   **594**
Formulas for Volume

10-1	Fundamental Properties of Volume	596
10-2	Multiplication, Area, and Volume	603
10-3	Volumes of Prisms and Cylinders	609
10-4	Volumes of Pyramids and Cones	615
10-5	Organizing and Remembering Formulas	623
10-6	The Volume of a Sphere	628
10-7	The Surface Area of a Sphere	633
	Projects	638
	Summary and Vocabulary	639
	Self-Test	640
	Chapter Review	642

Chapter 11   **646**
Indirect Proofs and Coordinate Proofs

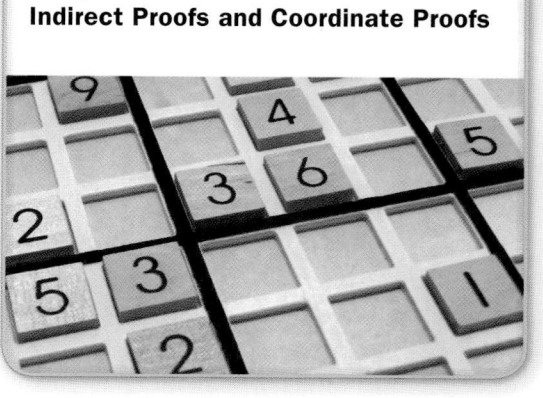

11-1	Ruling Out Possibilities	648
11-2	The Logic of Making Conclusions	654
11-3	Indirect Proof	663
11-4	Proofs with Coordinates	670
11-5	The Pythagorean Distance Formula	676
11-6	Equations of Circles	682
11-7	Means and Midpoints	688
11-8	Theorems Involving Midpoints	694
11-9	Three-Dimensional Coordinates	700
	Projects	708
	Summary and Vocabulary	709
	Self-Test	711
	Chapter Review	713

Chapter 12   **716**
Similarity

12-1	Size Transformations Revisited	718
12-2	Review of Ratios and Proportions	726
12-3	Similar Figures	731
12-4	The Fundamental Theorem of Similarity	738
12-5	Can There Be Giants?	745
12-6	The SSS Similarity Theorem	750
12-7	The AA and SAS Triangle Similarity Theorems	756
	Projects	764
	Summary and Vocabulary	765
	Self-Test	766
	Chapter Review	768

Chapter 13   772
Similar Triangles and Trigonometry

13-1	The Side-Splitting Theorems	774
13-2	The Angle Bisector Theorem	781
13-3	Geometric Means in Right Triangles	786
13-4	The Golden Ratio	793
13-5	The Tangent of an Angle	800
13-6	The Sine and Cosine Ratios	807
	Projects	814
	Summary and Vocabulary	816
	Self-Test	818
	Chapter Review	819

Chapter 14   822
Further Work with Circles

14-1	Chord Length and Arc Measure	824
14-2	Regular Polygons and Schedules	831
14-3	Angles Formed by Chords or Secants	837
14-4	Tangents to Circles and Spheres	844
14-5	Angles Formed by Tangents and a General Theorem	852
14-6	Three Circles Associated with a Triangle	859
14-7	Lengths of Chords, Secants, and Tangents	866
14-8	The Isoperimetric Inequality	872
14-9	The Isoperimetric Inequality in Three Dimensions	879
	Projects	886
	Summary and Vocabulary	889
	Self-Test	891
	Chapter Review	893

Postulates	S1
Theorems	S4
Formulas	S11
Selected Answers	S15

Glossary	S75
Index	S89
Photo Credits	S110
Symbols	S112

To the Student: Getting Started

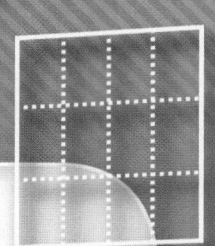

Welcome to *Geometry*! We hope you enjoy this book and find it useful; it was written for you.

This year, you will be drawing, constructing, measuring, visualizing, comparing, transforming, and classifying geometric figures, and applying these ideas in a wide variety of situations both inside and outside of mathematics. These few pages contain some information to help you get started.

Studying Mathematics

A goal of this book is to help you learn mathematics on your own, so that you will be able to understand the mathematics in newspapers, magazines, on television, at work, and in school. The authors, who are experienced teachers, have tried to write as though they were having a conversation with you about geometry. Everything in the book is written with the expectation that with guidance from a teacher, you will read it, learn from it, and enjoy that experience.

One thing we *know* is that not all students learn the same way. You have to find a way that works best for you. Here are a few suggestions.

Looking at an example is often necessary, but it is rarely enough. Most of us don't really learn something until we have done it, usually many times. You can watch an athlete, actor, juggler or dancer perform, but you cannot replicate their performance without trying, failing, and improving through organized attempts. Learning geometry is easier than learning to juggle five objects, but the process of learning is similar, and an important thing is to try and not give up.

Reading the Lessons

Carefully reading each lesson is a good way to start. As you know, reading math is different from reading email, a comic book, a novel, or a newspaper. For one thing, it takes longer. You have to slow down and be more careful. We want you to read *and understand* what came before the words that you are reading. So, at times, you will see QY (Quiz Yourself) questions to help you know whether you understand. The answers to QY questions are found at the end of each lesson. Usually there is no QY question, so after you read a sentence, ask yourself if you understood it. You might have to look up a word. You might have to study a diagram. You may wish to read ahead to see if the idea is explained later in the lesson.

Guided examples are in most lessons to help you work through problems. You should copy these examples into a notebook so that you can refer to them. Try to finish each guided example by yourself. Then check what you wrote with the answers in the back of this book.

You might have to ask someone for help. Many students have found study groups to be helpful. A good idea is to get the phone number or the email address of a few students in your class, so you can share thoughts about what you are learning. And perhaps most importantly, ask your teacher for help with ideas that are still not clear to you after you have worked on them.

Working on Questions

There are questions in every lesson. An important part of learning geometry is to wonder, or speculate about what might happen if something is true about a figure. It can be useful and enjoyable to ask "what if" questions. If you can develop the habit of asking yourself these questions, you may find that geometry becomes rather exciting, like a mystery, a puzzle, or a video game. For example, suppose you learn a fact about a square. You might ask yourself if that fact is also true about something other than a square. It is the intent of these questions for you to do a bit of thinking before you continue to read. Even if we do not ask the question, you will benefit greatly from asking them yourself.

It is likely that there will be some questions that stump you. The most important thing is that you do not give up. You should expect some difficult questions. Learning takes place when a person tackles a question she or he cannot answer at first, and after a while and some work, the person figures it out.

Tools Needed for This Book

Geometry is the study of visual patterns. These patterns are found in physical objects. These patterns are also found in representations of numbers and other ideas whose origin is not visual. Good drawings can help you to learn geometry. Poor drawings sometimes hide patterns in geometry just as poor computation can hide patterns in arithmetic. Thus, for this course, you should have good drawing equipment.

In addition to the notebook paper, sharpened pencil, and erasers that you should always have, you will find it helpful to have some drawing equipment.

- **Ruler** (marked in both centimeters and inches, clear plastic)
- **Protractor** (circular or semicircular, clear plastic)
- **Compass** (that tightens with a screw, holds regular pencils)
- **Plain (unlined) paper**
- **Graph paper**
- **Tracing (or patty) paper** (see-through paper that can be easily folded)
- **Graphing calculator** (many have DGS built in)
- **Dynamic Graphing System (DGS)** (see Lesson 1-7)

All these things may be familiar to you except the DGS. A DGS allows a person to make thousands of drawings in the time it takes to make two or three by hand. That not only speeds up drawing, but makes it possible to the user to investigate more complicated figures much more easily. DGS software is on many graphing calculators and is available for all of today's computers.

Getting Acquainted with This Book

It is always helpful to spend some time getting acquainted with your textbook. The questions that follow are designed to help you become familiar with UCSMP *Geometry*.

We hope you join the hundreds of thousands of students who have enjoyed this book. We wish you much success.

Questions

COVERING THE IDEAS

1. What is geometry?

2. Name four things you will experience by the end of the year in geometry.

3. List four items of drawing equipment besides paper and pencil that you will need for your work in *Geometry*.

4. In what lesson can you learn about a DGS?

5. What things other than words are important to understand when reading about geometry?

KNOWING YOUR TEXTBOOK

In 6–14, answer the questions by looking at the Table of Contents, the lessons and chapters of the textbook, or the end-of-book material.

6. In what lesson is mathematics and music discussed?

7. Look at several lessons.

    a. What are the four categories of questions at the end of each lesson?

    b. What word is formed by the first letters of these categories?

8. Suppose you have just finished the questions in Lesson 2-5.

    a. On what page can you find answers to check your work?

    b. What answers are given?

9. On what page can you find the vocabulary for Chapter 1?

10. What should you do after taking a Self-Test at the end of a chapter?

11. What differentiates the Self-Test answers from the answers to the Questions in each lesson?

12. This book has some Appendices. What do they cover?

13. Use the Index to find out where in this book you can find information on the golden ratio.

14. What is in the Glossary?

Chapter

1 Points and Lines

▶ **Contents**

1-1 Points and Lines as Locations

1-2 Ordered Pairs as Points

1-3 Other Types of Geometry

1-4 Undefined Terms and First Definitions

1-5 Postulates for Points and Lines in Euclidean Geometry

1-6 Betweenness and Distance

1-7 Using a Dynamic Geometry System (DGS)

Geometry is foremost the study of figures and visual patterns. All of these figures are based on points and lines. Different meanings of points and lines lead to different types of geometry. *Euclidean synthetic geometry* helps us to describe objects and their locations in the physical world. *Euclidean plane coordinate geometry* enables graphs on coordinate grids to be used in studying figures. *Discrete geometry* involves points and lines, such as those used to create the pictures you see on cell phones, televisions, and computer screens. Graph theory helps in solving problems such as routing airplanes between large airports.

These different types of geometry are like different games played with the same kinds of equipment. You have seen situations like these before. Baseball and softball are both played with bats and balls. The field looks very much the same, with four bases that are the vertices of a square. In both games, a pitcher throws a ball toward home plate and a batter tries to hit the ball. Pitches are either balls or strikes or are hit by the batter. When the batter hits the ball, the batter tries to run around the bases to score runs, or the batter might be called out.

However, the rules of baseball and the rules of softball are different. A softball is larger.

The distance between the bases in softball is shorter. The pitcher pitches underhand in softball and overhand in baseball. Baseball and softball are *two different types of games* played with bats and balls.

Similarly, different rules for points and lines lead to the different types of geometry studied in this chapter. In geometry, the basic rules are called *postulates*. In this chapter, you will encounter the postulates about points and lines in Euclidean geometry. These postulates provide the basis for the work with Euclidean synthetic geometry and Euclidean plane coordinate geometry that you will see in the rest of this book.

Lesson
1-1

Points and Lines as Locations

Vocabulary
coordinate

coordinatized

distance between two points
 on a coordinatized line

important

▶ **BIG IDEA** Objects and locations in the real world are often represented by points and lines so that they can be studied.

> Of a point there is little to write
> It neither has length, width, nor height
> But on inquisition
> It does have position
> Like a star in the dark of the night
> (George Marino)

Mental Math

How many segments are drawn in this figure?

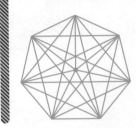

From earliest times, people have used language and pictures to describe shapes of natural things and objects they made. Thousands of years ago, as soon as people began to have crops and livestock, questions of ownership of land arose, and people needed to be able to determine how much land a person owned. Through these and other activities, a large number of today's geometrical principles arose, such as "The area of a rectangle equals its length times its width" and "The shortest path between two points is along a line."

From the years 650 to 250 BCE, Greek mathematicians logically organized the knowledge they knew about shape and size. They called this body of knowledge "geometria," from the words *geo*, meaning "earth" and *metria,* meaning "measure."

Geometry, as Seen by the Greeks

The Greek mathematicians of that period made significant advances in number theory and geometry. They considered points not as actual physical dots, but as idealized dots with no size. For them, a point represented the exact location of this idealized dot.

When two different points are exact locations, then there is exactly one line containing them. This line contains the shortest path connecting the two points. In Lesson 1-5, we will explain why this description of points and lines is the one used in what is called *Euclidean synthetic geometry.*

Euclidean geometry helps to describe and work with objects in the everyday world. Here is a typical example of a problem that can be solved using this geometry.

A Problem: Connecting to a Pipeline

The Alaska oil pipeline is a collection of connected cylinders 4 feet (about 120 cm) in diameter and about 800 miles (1300 km) long. Members of the U.S. Congress have debated whether to expand oil drilling into the Arctic National Wildlife Refuge. If the new oil wells are built, they will need to supply oil to the existing pipelines at pumping stations. It is very expensive to build new pumping stations on the existing pipeline, so oil companies try to connect two new wells to the pipeline in one place whenever possible. They need to know where they should tap into the pipeline so that the small pipelines from the two oil wells to the connecting point use the least amount of materials. The least expensive solution to their problem can be found using geometry.

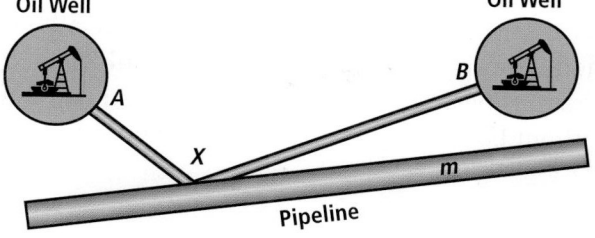

Simplifying the Situation

Each oil well takes up some space, and the pipeline is a cylinder. But because the distances are so great, we can think of each oil well as a *point* and the cylinder as a *line*. Now the problem looks like the diagram below. We call the fixed points A and B, and the point on the pipeline whose location is unknown we call X. The line we call m. The distance between A and X is written AX.

There are 32 distinct herds of caribou in Alaska. This photo shows part of the Porcupine caribou herd migrating through the Arctic National Wildlife Refuge.

Source: Alaska Department of Fish and Game

Activity

MATERIALS Tracing paper

This activity simulates the idea of finding the best place for the connection. The goal of this activity is to locate a point X on the line m that minimizes the sum $AX + BX$.

Step 1 Trace the picture at the right. X_1 is a possible location of X. Measure to find $AX_1 + BX_1$.

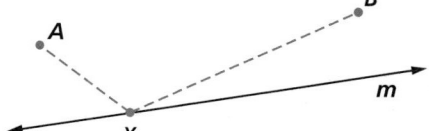

Step 2 Locate a second point X_2, on line m so that $AX_2 + BX_2$ is less than your answer to Step 1.

Step 3 Experiment to locate a point X so that $AX + BX$ is as small as possible.

Step 4 Suppose you wished to communicate your solution to the oil company. How would you describe to them where they should put the pumping station?

Points as Exact Locations

The oil wells are huge, but to connect them to the pipeline, it is only necessary to think of each well as a point. When points are used to model these sorts of situations, we think of each of them as an *exact location* having no size or shape. We think of lines as extending without bound in two opposite directions and as having no thickness.

> **Euclidean Synthetic Geometry**
>
> **Description of a point**
> A point is an exact location.
>
> **Description of a line**
> A line is a set of points extending in both directions containing the shortest path between any two points on it.

Number Lines

In the Activity, you were asked to explain how you would describe the position of the connection to the oil company. One way to describe the position is to choose any point along the line and call it the zero point. Moving 2 miles on the line "away from" that point is ambiguous, because it is not clear in which direction one should travel. To address this problem, choose one direction to be positive and one direction to be negative. Every distance from the zero point can be labeled with exactly one number, its **coordinate.** When this is done, the line is said to be **coordinatized.** On the number line at the right, point O has coordinate 0, point A has coordinate 3, and point B has coordinate –2.

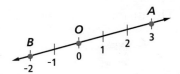

Distance on a Number Line

The distance between two points can be calculated if you know their coordinates. Recall from algebra that the absolute value of a number n is written $|n|$. When n is positive, it equals its absolute value. For instance, $|6.2| = 6.2$. The absolute value of a negative number is the opposite of that number: $|-2500| = 2500$. The absolute value of 0 is 0: $|0| = 0$.

The absolute value of the coordinate of a point is its distance from the zero point on the number line. An algebraic relationship between absolute value and square roots is: for all x, $|x| = \sqrt{x^2}$. For instance, $|-2500| = \sqrt{(-2500)^2} = \sqrt{6,250,000} = 2500$, and $|6.2| = \sqrt{6.2^2} = 6.2$.

> **Definition of Distance between Two Points on a Coordinatized Line**
>
> The **distance between two points on a coordinatized line** is the absolute value of the difference of their coordinates. In symbols, the distance between two points with coordinates a and b is $|a - b|$.

Since $|a - b| = \sqrt{(a - b)^2}$, the distance between two points with coordinates a and b can also be found by calculating $\sqrt{(a - b)^2}$. The square root form of the distance may be easier to work with in some situations than the absolute value form.

In an example, when we want to emphasize what you should write down, we use print that looks like this. The other text in the solution is what you might be thinking in order to arrive at the solution.

GUIDED

Example

A Guided Example is an example in which some, but not all, of the work is shown. You should try to complete the example before reading on. Answers to Guided Examples are in the Selected Answers section at the back of the book.

Find AB on the number line on the previous page.

Solution 1 The coordinate of A is __?__. The coordinate of B is __?__. Find their difference by subtracting. Then take the absolute value of the difference.

$$AB = |3 - (-2)| = |\underline{\ ?\ }| = \underline{\ ?\ }$$

Solution 2 The coordinate of A is __?__. The coordinate of B is __?__.

$$AB = \sqrt{(3 - (-2))^2} = \sqrt{\underline{\ ?\ }} = \underline{\ ?\ }$$

Check Count the units to verify that there are __?__ units between A and B.

CAUTION: When A and B are points, AB always means the distance from A to B, not their product. You cannot multiply points.

Because of properties of absolute value, if you switch the order of the points, the distance is unchanged.

$$BA = |-2 - 3| = |-5| = 5.$$

For any two points A and B, $AB = BA$.

See Quiz Yourself at the right.

Quiz Yourself (QY) questions are designed to help you follow the reading. You should try to answer each Quiz Yourself question before reading on. The answer to the Quiz Yourself is found at the end of the lesson.

Tape measures (when stretched) and rulers resemble coordinatized lines. For example, if you wish to buy draperies or blinds for a window, you need the dimensions of the window. Suppose a tape measure crosses the top of a window at the 6-inch mark and the bottom at the 81-inch mark. Then the window is $|6 - 81|$ or $|-75|$ or 75 inches tall. When you connect two locations with a ruler, the edge of the ruler lies on a line. Whenever you use a ruler, you are applying the definition of distance. Many highways in the United States are coordinatized using mile markers. The zero point is at the point where the interstate crosses a state line. Website programmers use these distances to calculate the road distance between locations on a map.

The Point of Discussing Points

Geometry is built on clear and consistent rules. Before anyone can start thinking of geometry in a formal sense, everyone must agree on common terms. In this lesson, points stand for exact locations. In the next few lessons, you will see other descriptions of a point that serve as the basis for other types of geometry. As in baseball and softball, the descriptions we select determine the results we get.

▶ **QUIZ YOURSELF**

Find the distance between two points on a number line with coordinates −100 and −300.

Questions

COVERING THE IDEAS

These questions cover the content of the lesson. If you cannot answer a Covering the Ideas question, you should go back to the reading for help in obtaining an answer.

1. The word "geometry" comes from the word *geometria*. In what language is that word and what does it mean?

2. What is a mathematical way to represent a location?

3. How is the period at the end of a sentence different from the way a point is described in this lesson?

4. For what value or values of b is it true that $|b| = -b$?

5. For what value or values of c is it true that $\sqrt{c^2} = \sqrt{(-c)^2}$?

6. Let a and b be coordinates of points on a number line. What is the distance between a and b?

In 7–9, give the distance between two points with the given coordinates.

7. –7 and 7   8. –10.3 and –12.1   9. x and y

10. What is the symbolic way to describe the distance between points C and D on a number line?

11. A customer in a grocery store gives a cashier $16.00. The cashier replies "you are off by 63 cents." How much did the customer actually owe?

12. A carpenter has used her tape measure so much that the end has worn away. She can no longer see the first couple of inches. When she measures a board, she places the 3-inch mark at one end of the board. At the other end of the board, the ruler reads 52 inches. How long is the board?

APPLYING THE MATHEMATICS

These questions extend the content of the lesson. You should study the examples and explanations if you cannot answer the question. For some questions, you can check your answers with the ones in the Selected Answers section at the back of this book.

13. When a new subdivision is built, the power company provides a connection from the power plant for the new houses to share. On a new street with three new houses, draw an idealized model for this situation. Explain how your idealized model is different from the actual situation.

14. On a trip, you are k miles from your destination at 4 P.M. and y miles from your destination 2 hours later.
    a. How far have you traveled in the two hours?
    b. What was your average speed for the two hours?

15. Use the number line at the right.
    a. Calculate AB, BC, and AC.
    b. True or false. $AB + BC = AC$.
    c. The point D has coordinate $\sqrt{700}$. Trace the number line and locate D on it.
    d. Find the exact value of DB.
    e. Estimate DA to the nearest tenth of a unit.

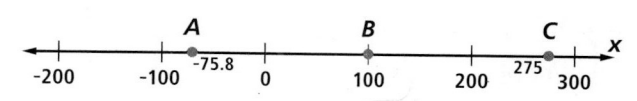

16. Use the following road mileage chart for four cities in Alabama. How much farther is the drive from Birmingham to Mobile if you stop in Auburn?

	Auburn	Birmingham	Huntsville	Mobile
Auburn	0	110	211	221
Birmingham	110	0	101	258
Huntsville	211	101	0	356
Mobile	221	258	356	0

In 17 and 18, draw a number line and coordinatize it so that the numbers from –10 to 10 show.

17. The center of a circle with a radius of 4 is placed on a number line at the point with coordinate 1.7. What are the two points where the circle intersects the line?

18. a. A circle is drawn with center –2.7. One of the points at which the circle intersects the line has coordinate 1.4. What is the other point of intersection?

    b. Generalize the result in Part a.

REVIEW

Every lesson contains review questions to practice ideas you have studied earlier.

19. In this sequence of dots, one row is added to each term to get the next term. **(Previous Course)**

    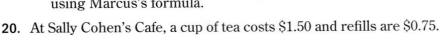

    a. Find the number of dots in the 5th and 6th terms.

    b. Marcus realized the number of dots in the nth term is $\frac{n(n+1)}{2}$. Confirm your answer to Part a by using Marcus's formula.

20. At Sally Cohen's Cafe, a cup of tea costs $1.50 and refills are $0.75.

    a. Make a table of the cost of a cup of tea with up to five refills.

    b. Plot these points on a coordinate grid. Be sure to label the axes. **(Previous Course)**

21. Solve the system. $\begin{cases} 2x + y = 1 \\ -x + 2y = 7 \end{cases}$ **(Previous Course)**

EXPLORATION

These questions ask you to explore topics related to the lesson. Sometimes you will need to use references found in a library or on the Internet.

22. Model a part of your neighborhood in the following manner: Let a point represent your home and another point represent your school. Use lines to represent the different paths to school from your home. Make your drawing reasonably accurate with respect to scale.

23. On page 5, differences between baseball and softball are discussed. Many sports have different rules depending on who is playing. What differences are there between the basketballs and basketball courts used in men's and women's basketball? Is the difference the same at the college level and professional level?

Lesson 1-2 Ordered Pairs as Points

Vocabulary

plane coordinate geometry

horizontal line

vertical line

oblique line

standard form of an equation of a line

slope-intercept form of an equation of a line

y-intercept

slope

▶ **BIG IDEA** The ordered pairs of real numbers and the lines that you graphed in algebra are points and lines in Euclidean plane coordinate geometry.

Lines are *one-dimensional,* while the pipeline problem of Lesson 1-1 takes place in two dimensions. *Two-dimensional objects* lie in *planes.* We think of a plane as a flat surface, like a table top or a tennis court without any boundaries or edges. Floors, roads, and even the surface of some regions on Earth are flat like a plane.

Mental Math

Give the coordinates of 6 points on the line with equation $x + 4y = -10$.

What Is Coordinate Geometry?

Number lines show us that a point on a 1-dimensional line can be located by a single number. Around the year 1630, Pierre de Fermat and René Descartes realized that it is useful to identify a location in a plane by an *ordered pair* of real numbers. At the right, the three points (0, 0), (4, 1), and (–2.1, –1.5) are graphed. The study of geometric figures using points as ordered pairs of real numbers is called **plane coordinate geometry.**

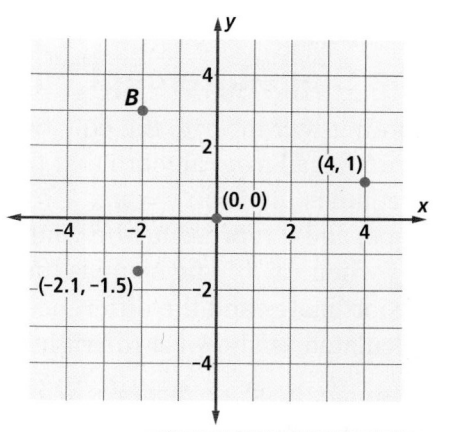

🛑 See Quiz Yourself 1 at the right.

In algebra you graphed equations of lines. These graphs consisted of all the ordered pairs that solved a linear equation. Since we can view points as ordered pairs, we can view lines as sets of ordered pairs. This type of geometry is called *Euclidean plane coordinate geometry.*

▶ **QUIZ YOURSELF 1**

The point *B* above has integer coordinates. What are they?

The Standard Form of an Equation of a Line

There are many ways to write an equation of a line. One form that occurs frequently is the **standard form**, $Ax + By = C$. In the standard form, the values of A, B, and C determine the slope and location of the line. When $A = 0$, the equation of the line is of the form $By = C$, and the line is **horizontal.** When $B = 0$, the equation of the line is of the form $Ax = C$, and the line is **vertical**. When neither A nor B is zero, the line is **oblique** (neither horizontal nor vertical).

Euclidean Plane Coordinate Geometry

Description of a point
A point is an ordered pair of real numbers.

Description of a line
A line is the set of ordered pairs of real numbers (x, y) satisfying an equation of the form $Ax + By = C$, where A and B are not both zero.

Horizontal Line
$y = 1.5$

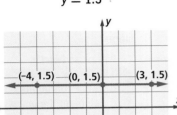

Vertical Line
$x = -3$

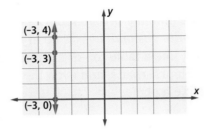

Oblique Line
$x + y = 3$

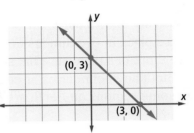

The Slope-Intercept Form of an Equation of a Line

Another way to write the equation of a line is the slope-intercept form. Recall from algebra that the **slope-intercept form** of an equation of a line is the form $y = mx + b$. In this form, m represents the slope, and b represents the y-intercept. The **y-intercept** is the value of y when $x = 0$. The **slope** is the ratio of the difference between y-coordinates and the difference between x-coordinates. It can be calculated as shown at the right.

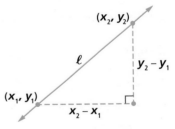

Slope $= m = \dfrac{y_2 - y_1}{x_2 - x_1}$

Examine the slope formula. When $y_2 = y_1$, the line is horizontal and its slope is 0. When $x_2 = x_1$, the line is vertical and its slope is undefined.

Example 1
Graph the line with equation $y = 2x - 2$ and check your answer.

Solution 1 When $x = 0, y = -2$.
When $x = 1, y = 2 \cdot 1 - 2 = 0$.

The two points on the graph are $(0, -2)$ and $(1, 0)$.
Plot these points and connect them.

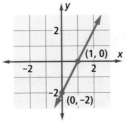

Check It looks like the line passes through the point with coordinates $(2, 2)$. $2 \cdot 2 - 2 = 2$, so we probably have a good graph of the equation.

Solution 2 Start at the y-intercept, -2, and advance according to the slope, 2 (up 2, right 1). The graphs are the same, so the slope checks.

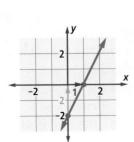

Example 2

Write an equation of the line that contains the points (3, 1) and (–2, 11).

Solution To write the equation of the line, we need to find the values of m and b in the equation $y = mx + b$.

$$m = \frac{y_2 - y_1}{x_2 - x_1} = \frac{11 - 1}{-2 - 3} = \frac{10}{-5} = -2$$

You now know that this particular line is of the form $y = -2x + b$. To find the value of b, substitute the coordinates of one of the points for x and y. We choose the point (3, 1).

$$y = -2x + b$$
$$1 = -2(3) + b$$
$$1 = -6 + b$$
$$7 = b$$

Substitute the value of b into the equation $y = -2x + b$. **The equation is $y = -2x + 7$.**

STOP **See Quiz Yourself 2 at the right.**

▶ QUIZ YOURSELF 2

How can you tell that the answer to Example 2 is correct?

GUIDED

Example 3

Write an equation for the line containing the points (4, 9) and (–2, 6).

Solution First find the slope using the formula

$$m = \frac{y_2 - y_1}{x_2 - x_1} = \frac{6 - ?}{-2 - ?} = \underline{\quad ? \quad}.$$

Now you know that this line has an equation of the form $y = \frac{1}{2}x + b$.

Substitute the coordinates of one of the given points to find b.

$$y = \frac{1}{2}x + b$$
$$\underline{\quad ? \quad} = \frac{1}{2}(\underline{\ ? \ }) + b$$
$$b = \underline{\quad ? \quad}$$

So an equation of the line that contains the points (4, 9) and (–2, 6) is $y = \underline{\ ? \ }x + \underline{\ ? \ }$.

A line that is in standard form can be "converted" into slope-intercept form. If B does not equal zero in the standard from, it is possible to solve for y. The resulting equation is $y = \frac{Ax}{B} + \frac{C}{B}$. If we let $m = -\frac{A}{B}$ and $b = \frac{C}{B}$, then we have the form $y = mx + b$. Converting from standard form to slope-intercept form is useful if you are trying to use a graphing utililty to graph the equation of a line.

Example 4

Convert the equation $3x - 2y = 6$ into slope-intercept form and graph it.

Solution

$3x - 2y = 6$	Given
$-2y = -3x + 6$	Add –3x to both sides.
$y = \frac{3}{2}x - 3$	Divide both sides by –2.

Start at the y-intercept, –3, and advance according to the slope: up 3, right 2. Plot the point (2, 0) and draw the line from it to (0, –3).

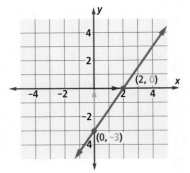

Ramps, braces, slides, and many other structures are often modeled by lines. With an equation for a line, you can determine locations of points on the structure.

GUIDED

Example 5

The maintenance crew at the Splash 'n' Slide water park have found that their main attraction, the Big Kahuna Water Slide, needs some additional vertical bracing. There already exists a 10-foot vertical brace near the bottom and a 46-foot vertical brace near the top that is 72 feet from that bottom brace. Two more vertical braces are needed that are equally spaced between these two braces. Model the slide with a line, letting the bottom of the slide be the point (0, 10).

a. What are the possible coordinates for the top of the slide?

b. Find an equation for the line modeling the slide.

c. How long should each of the additional braces be?

Solution

a. Draw a picture.

Since the top of the bottom brace is at (0, 10) and the 46-foot tall brace is __?__ feet away, the coordinates for the top of the slide are (__?__, __?__).

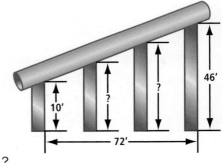

b. From Part a, the line modeling the slide contains (0, 10) and (__?__, __?__). Its slope is $\frac{? - 10}{? - 0} = $ __?__. The y-intercept of this line is __?__. Therefore, the equation of the line is $y = $ __?__$x + $ __?__.

c. The horizontal distance between the original two braces is 72 feet, so two equally spaced braces between them would have to be located at __?__ feet and __?__ feet away from the shorter brace.

Substituting these values for x in the equation results in a vertical length of __?__ for the shorter of the two new braces, and a vertical length of __?__ for the longer of the two.

Questions

COVERING THE IDEAS

1. **a.** What is the idea behind a *plane* in geometry?
   b. Give several examples of real-world objects that resemble planes.

2. In a coordinate plane, how can the location of points be described?

3. What is the slope of the graph of $y = 3x - 1$?

4. Write an equation for the line through (6, 8) and (2, –10).

In 5–9, classify the line with the given equation as *horizontal*, *vertical*, or *oblique*.

5. $x = 89$ 6. $x + y = 89$ 7. $y = x + 89$ 8. $y = 89$ 9. $x = 0$

In 10–13, convert the equation to slope-intercept form and sketch a graph.

10. $y = 2x - 5$ 11. $2x + 3y = 6$ 12. $x = 3$ 13. $3x + y = -1$

APPLYING THE MATHEMATICS

14. **a.** Write an equation for the vertical line that passes through (0, 5).
    b. Write an equation for the horizontal line that passes through (0, 5).
    c. Write an equation in slope-intercept form for an oblique line that passes through (0, 5).

15. **a.** Graph the line with equation $y = 3x + 2$.
    b. Graph the line with equation $y = 8$.
    c. Find the point of intersection of the two lines.

16. Fire escapes zigzag down the sides of many tall buildings. Each set of stairs of the fire escape in the building pictured here goes up about 12 feet and over about 6 feet. Model the bottom fire escape stairs by a line beginning at (0, 30). (The bottom of the fire escape is 30 feet off the ground.)
    a. What point stands for the top of the bottom fire escape stairs?
    b. Give an equation for the line modeling the bottom fire escape stairs.

17. A wheelchair ramp usually can be no steeper than the ramp pictured at the right, by the Americans with Disabilities Act (ADA) specifications.

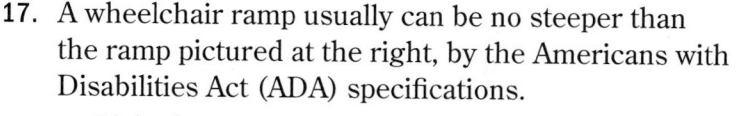

2 ft

24 ft

   a. If the left point of this ramp is at (–12, 0), where is the right point?
   b. What is the slope of this ramp?
   c. What is an equation for this ramp?
   d. How high is the ramp at its middle?

18. **a.** Graph the line with equation $4x + 3y = 6$.
    b. Graph the line with equation $x = 3$.
    c. Find the point of intersection of the two lines.

REVIEW

19. Find BC, where B and C are the points marked below. (**Lesson 1-1**)

20. What are the coordinates of all the points at distance 3 from the point 2.12 on the number line? (**Lesson 1-1**)

21. Evaluate the expression $\sqrt{a^2 + b^2}$ when $a = -8$ and $b = -15$. (**Previous Course**)

QUIZ YOURSELF ANSWERS

1. $(-2, 3)$

2. Answers vary. Sample: Substitute 3 for x and 1 for y in $y = -2x + 7$ to see if the point $(3, 1)$ is on the line.

EXPLORATION

22. When driving in a city, taxicabs are restricted to city streets, and so distances traveled are measured in the number of blocks the taxi has to travel to get from one place to another. In the map of part of Chicago at the right, each north-south block is one taxicab block, and each east-west block is half a taxicab block. Thus, the distance from point A, at W. Byron St. and N. Sacramento Ave., to point B, at W. Irving Park Rd. and N. Francisco Ave., is 2 taxicab blocks. Find all other street intersections that are 2 taxicab blocks from point A. To make your task easier, put a grid on the map, with A being at the origin, so B is at $(1, 1)$. Then you can name the other intersections by their coordinates. On what shape do all the points seem to lie?

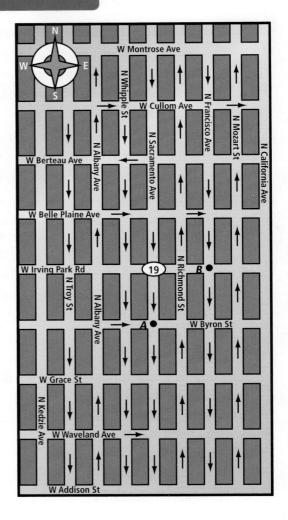

Lesson 1-3
Other Types of Geometry

Vocabulary

pixels

discrete point

discrete line

discrete geometry

graph theory

arc

network

node

vertices, vertex

transversable network

even node, odd node

▶ **BIG IDEA** Points and lines in discrete geometry and in graph theory have different properties than points and lines in Euclidean geometry.

In Lesson 1-1, a point was described as an exact location. In Lesson 1-2, a point was described as a location in the plane identified by an ordered pair of real numbers. In this lesson, you will examine two other common descriptions of points. This is important because different descriptions of points and lines serve as the foundation for different types of geometry.

Mental Math

If you take any of the 3 paths from *A* to *B*, any of the 4 paths from *B* to *C*, and any of the 5 paths from *C* to *D*, how many different routes are possible from *A* to *D*?

Georges Seurat's *Sunday Afternoon on the Island of Grande Jatte* measures $6\frac{3}{4}$ feet by 10 feet.

Dots and Points

Look at the Seurat painting. In both the original of this painting and the picture reproduced here, points are represented by dots. These dots have both length and width.

Dots come in many different shapes, colors, and sizes. Consider, for example, the image you see when you look at a computer screen or digital camera. The image you see consists of thousands of tiny dots called **pixels** that are combined to create the image you see on the screen. These dots are so small that it is very hard to distinguish one from another.

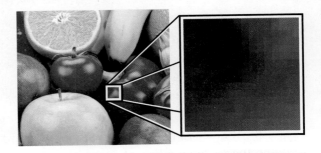

On the other hand, consider a scoreboard. The writing on the scoreboard is made up of larger dots created by light bulbs.

Discrete Geometry

When a point is described as a dot of some size, a line is made up of points that could have space between them. In other words, when using these descriptions, between two points on a line there is not necessarily another point. Points are called *discrete points* and the lines are called *discrete lines*. The study of **discrete points** and **discrete lines** is one type of geometry called **discrete geometry.**

Discrete Geometry	
Description of a point	**Description of a line**
A point is a dot.	A line is a set of dots in a row.

One difference between Euclidean synthetic geometry and discrete geometry occurs when you consider crossing lines. In Euclidean geometry, when two lines cross, they are guaranteed to intersect at a point that is on each line, because there are no spaces or gaps in the line. For instance, in the upper figure at the right, lines m and n cross at point A, which is on both lines. This is not necessarily true in discrete geometry because the points on a line have space in between them. The lower diagram at the right shows two discrete lines that cross and have a point in common.

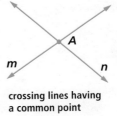

crossing lines having a common point

crossing lines having a common point

 See Quiz Yourself at the right.

In both Euclidean and discrete geometry, points represent locations. However this is not true in *graph theory,* where points simply represent items to be connected and lines are the connectors. This will become clearer after you work through the following very famous problem.

> ▶ QUIZ YOURSELF
>
> Draw an example of two discrete lines crossing in which the two lines do not share a common point.

The Königsberg Bridge Problem

Through the city of Kaliningrad, Russia, flows the Pregol'a River, and in this part of the river are two islands. In the 1700s this city was part of East Prussia and was known as Königsberg. Seven bridges connected the islands to each other and to the shores, and it was common on Sundays for people to take walks over the bridges. These walks and bridges led to a problem. Can you walk over each bridge exactly once?

The drawing at the right is based on one that first appeared in an article by the great mathematician Leonhard Euler (OY ler). The islands are *A* and *D;* the bridges are *a, b, c, d, e, f,* and *g;* the shores of the river are *B* and *C.*

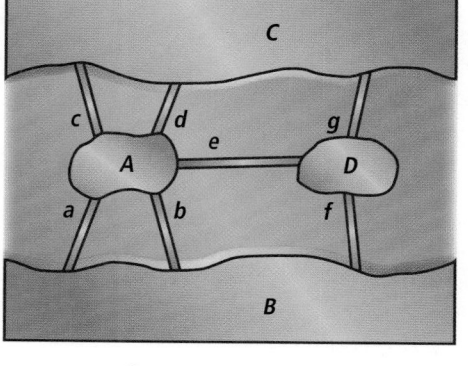

The Königsberg Bridge Network

To solve the Königsberg Bridge Problem, Euler redrew the map with the islands *A* and *D* as very small, and lengthened the bridges. This doesn't change the problem.

Then he realized that the shores *B* and *C* could be small. This again distorts the picture but it doesn't change the problem.

Finally—and this was the big step—he thought of the land areas *A, B, C,* and *D* as points and the bridges *a* through *g* as **arcs** connecting them. The result, shown at the right, is a **network** of points and arcs. In this network there is a path (though not necessarily direct) from any point to any other point. Networks are sometimes called *graphs*. This is why the geometry of networks is called **graph theory**. In a network, the *only* points are the endpoints of arcs. These endpoints have no size and are called **nodes** or **vertices**. (The singular of "vertices" is vertex.)

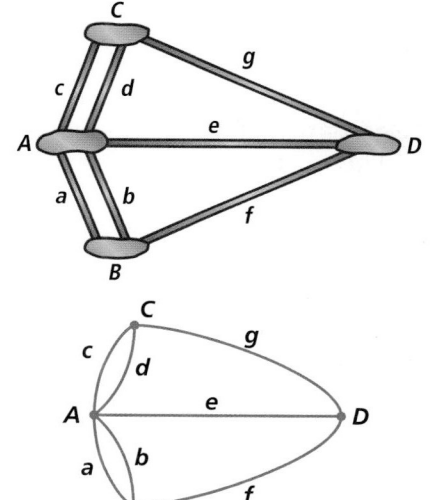

Activity 1

MATERIALS Tracing paper

Königsberg Bridge Problem: Trace the Königsberg Bridge network and use your drawing to find a way to walk across all seven bridges of Königsberg so that each of the seven bridges (*a, b, c, d, e, f,* and *g*) is crossed exactly once. (For instance, if you start at *D* and walk on *f, b,* and *a,* then you are at point *B,* and you are stuck because you have already walked on all the bridges from *B.*) Try to explain any difficulties you may encounter.

Arcs and Nodes

Euler was able to rephrase the Königsberg Bridge Problem to become: "Without lifting a pencil off the paper, can one trace over all the arcs of this network exactly once?" If the answer is yes, this kind of network is called a **traversable network**. Euler noticed that the number of arcs at each node provided a clue as to whether or not a network was traversable. As shown below, nodes in a network can have different numbers of arcs.

Leonhard Euler

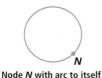

Node *N* with arc to itself

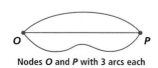

Nodes *O* and *P* with 3 arcs each

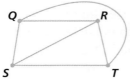

Four nodes with 3 arcs each
(Arcs *QT* and *SR* do not intersect.)

There are no points in the middle of an arc, but between two nodes there can be many different arcs connecting them.

> **Graph Theory**
>
> **Description of a point**
> A point is a node of a network.
>
> **Description of a line**
> A line is an arc connecting either two nodes or one node to itself.

In the Königsberg Bridge Problem, there are 5 arcs at vertex *A* (*c, d, e, b,* and *a*) and 3 arcs at each of vertices *B, C,* and *D.* Before we show you Euler's solution to the Königsberg Bridge Problem, here are three more networks to consider. In these networks, the arcs are drawn as segments.

> ## Example 1
>
> How many arcs are at each node of networks I, II, and III?
>
> I
>
>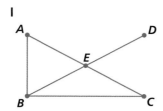
>
> II
>
>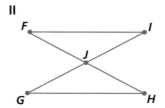
>
> III
>
>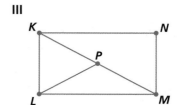

Solution

Network I: There are 4 arcs at node E, 3 at B, 2 at A and C, and 1 at D.

Network II: There are 4 arcs at node J and 2 arcs at F, G, H, and I.

Network III: There are 3 arcs at P, K, L, and M, and 2 arcs at N.

If the number of arcs at a node is even, the node is called an **even node.** Otherwise it is an **odd node.** In Example 1, *A, C, E, F, G, H, I, J,* and *N* are even nodes while *B, D, K, L, M,* and *P* are odd nodes.

Activity 2

Look again at the networks in Example 1.

1. Where must you start and end in order to traverse network I?

2. Where must you start and end in order to traverse network II?

3. Is it possible to traverse network III?

Euler's Solution to the Königsberg Bridge Problem

Euler realized that when a path goes through a node, it uses two arcs: one to the node and one away from it. This led him to realize that when a network has an odd node, it must be the starting or finishing point for a traversable path. In Activity 2, you should have found that Network I is traversable only if you start at *B* or *D.* If you started at *B,* then you ended at *D* and vice versa.

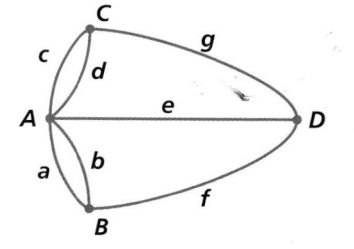

Euler realized that all four nodes in the Königsberg network are odd. Since a traversable network can only have one starting point and one finishing point, the Königsberg network is not traversable. Whenever a network has more than two odd nodes, it is not traversable.

Example 2

Look back at the networks from Example 1. Use Euler's reasoning to explain why networks I and II are traversable while network III is not.

Solution Network I is traversable because there are exactly two odd nodes (B and D). Network II is traversable because there are no odd nodes. Network III is not traversable because there are 4 odd nodes.

Graph theory and problems like the Königsberg Bridge Problem have many applications.

Graph theory is used in computer networking, city planning, security, airline flight scheduling, and many other areas. For instance, a curator of a museum would be interested in finding the minimum number of security cameras needed to provide security for all of the exhibits in the building. A traveling salesman might want to minimize the distance on his route in order to maximize sales. A restaurant manager might want to know the most efficient way of assigning tables to wait staff to maximize efficiency. All of these questions could be addressed using graph theory.

You have now seen examples from four different kinds of geometry: Euclidean synthetic geometry, Euclidean plane coordinate geometry, discrete geometry, and graph theory. These are presented for you as a way of showing that the word "geometry" can have different meanings depending upon how you describe the simplest components of that geometry—the point and the line. Euclidean geometry is by far the most useful and most common of these geometries, so the two types of Euclidean geometry are the ones we will examine most in the remainder of this book.

Questions

COVERING THE IDEAS

1. What type of geometry best describes the lines you see in each of the following diagrams?

   a.  b.   c.

2. How are points described in Euclidean plane coordinate geometry?

3. How many lines contain any two points in Euclidean plane coordinate geometry?

4. How are points described in discrete geometry?

5. Is it possible for two lines in Euclidean synthetic geometry to cross without having a point in common?

6. How is a point described in graph theory?

7. In graph theory, what does it mean for a network to be *traversable*?

8. What is an *odd node*?

9. When you graph a line on a graphing calculator, what kind of geometry best describes the line that the calculator creates?

10. How do you know if a network is traversable?

11. Below you see a "square" in each of the three types of geometry described in this lesson: Euclidean synthetic geometry, graph theory, and discrete geometry. Name the type of geometry represented by each "square".

a.
b.
c.

In 12–14, tell whether the network is traversable.

12.
13.
14.

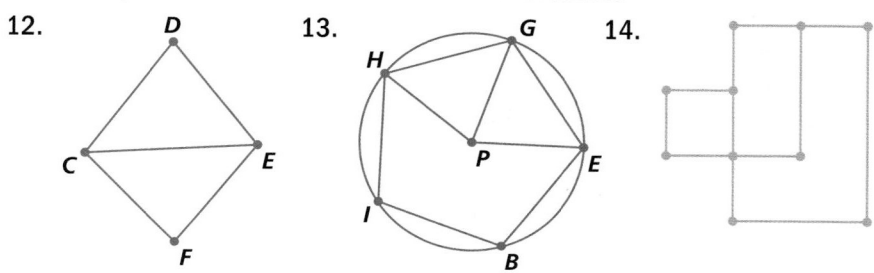

APPLYING THE MATHEMATICS

15. Draw a network with 2 nodes and 4 arcs. Is this network traversable?

16. A map of Frytown, Iowa is shown at the right. A snowplow driver wants to know if it is possible to plow all of the roads in Frytown without passing over the same road twice.

    a. If you think of this map as a network in graph theory, what do the arcs represent and what do the nodes represent?

    b. Is this network traversable? If so, where can the driver start and stop?

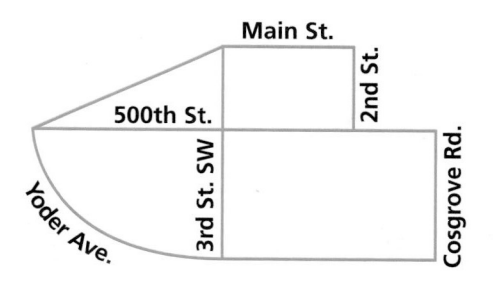

17. Refer to the map of Königsberg from the lesson. Euler proved that it was impossible to travel over all of the bridges without crossing over at least one bridge twice. Let us suppose, however, that you travel back in time and build one additional bridge anywhere that you choose. Where could you build the bridge so that it becomes possible to walk over every bridge without crossing over any bridge twice? Are there multiple correct answers to this problem?

18. Create a network that has 4 nodes and 7 arcs and is possible to traverse. Use arrows to show the path that you used to traverse it.

19. Create a network that has 4 nodes and 7 arcs but is impossible to traverse. Explain why your network is impossible to traverse.

20. Consider the capital letters of the English alphabet. Find 3 letters that cannot be drawn without picking up your pencil or retracing a line and find 3 letters that can. Explain your choices using the ideas from graph theory.

REVIEW

In 21-23, graph the line with the given equation, and classify it as horizontal, vertical, or oblique. (Lesson 1-2)

21. $20y = 5$ 22. $x + 7y = 4$ 23. $x + y = 5 + y$

24. Selena's hair is 10 cm long and growing at a rate of 2 cm per month. (Lesson 1-2, Previous Course)
   a. Graph a line representing Selena's hair length over time.
   b. Write an equation of the line in slope-intercept form.

25. What does it mean for a line to be *coordinatized*? (Lesson 1-1)

26. Suppose Megan, Namiko, and Omar are standing in a straight line. Megan and Omar are 20 feet apart and Namiko is 7 ft from Omar. What are the possible distances between Namiko and Megan? (Lesson 1-1)

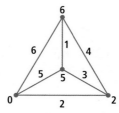

EXPLORATION

27. a. Explain why it is impossible to trace the drawing of a box shown at the right without ever lifting up your pencil and without tracing the same path twice.
   b. What is the fewest number of times you need to pick up your pencil in order to trace the box?

28. There are many puzzles using networks. Here is one: Start with a network that has *n* arcs. (The networks at the right have 4 nodes and 6 arcs.) Name each *node* with a different number from 0 to *n*. Then number each *arc* by the difference of the nodes it connects. For example, in the graceful network at the right, the arc connecting nodes 5 and 2 is named 3 because $5 - 2 = 3$. The goal is to name the nodes in such a way that the *n* arcs are numbered with all the integers from 1 to *n*. Such a network is called a *graceful* network.

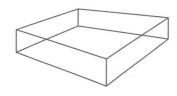

graceful
(arcs are numbered from 1 to 6)

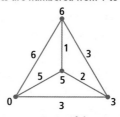

not graceful
(two arcs are numbered 3)

Number the nodes in these three networks to make them graceful.

a.   b.   c.

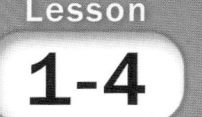

Lesson 1-4

Undefined Terms and First Definitions

Vocabulary

circularity

undefined terms

figure

space

collinear

plane figure

coplanar

one-dimensional

two-dimensional

three-dimensional

▶ **BIG IDEA** Because it is impossible to define all words and phrases in terms of simpler words and phrases, some words must be left undefined.

T-ball is a simple form of baseball or softball. Watching a child's first T-ball game, you might notice players running in the "wrong" direction, throwing to the "wrong" person, advancing to the "wrong" base, standing in the "wrong" place for a given position, and so on. How do we know that these children are "wrong?" T-ball is played according to a set of rules and regulations. In order to be a good T-ball player, players need to learn the rules and regulations set forth by their league. These rules include bats, bases, and the T-ball.

If the rules do not cover all aspects of the game, are ambiguous, or are inconsistent, then knowing the rules would not result in an organized game. The rule book must be sufficient and consistent. That is, the rules must address every possible situation and the rules must not contradict each other. Baseball, softball, and T-ball have different rules even though the rules use many of the same terms. They are like the different types of geometry.

Learning the rules of a game requires understanding the terms that are used. For instance, in T-ball, what is a base? But accurate definitions are not just used in games and sports. Law, economics, science, and labor relations, in addition to mathematics, are some of the fields in which precise definitions are necessary. For example, disputes may occur because individuals do not know or agree on the meanings of "overtime" or "freedom" or "force." However, it is impossible to define every term. Some terms need to be left undefined.

Undefined Terms

In the previous lessons, you have seen the word "point" used in many contexts. A point may be a dot, a location, an ordered pair, or a node in a network. This can cause confusion. Imagine a T-ball game in which everyone was interpreting the rules differently! Like the T-ball rule book, geometry definitions must be clear. That is, every geometric term must have a definition that is descriptive enough to distinguish the term from every other term.

Mental Math

What is the maximum number of different lines that can pass through at least 2 of 6 given points of a plane?

Formulating a *good definition* for every word is difficult without some building blocks. Suppose you were asked to define "point," and defined it as a *spot*.

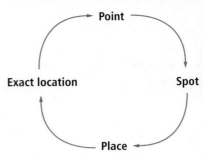

What about the word "spot?"   A spot is a *place*.
What about the word "place?"   A place is an *exact location*.

Trying to define this term, you might find the best description of an exact location to be a *point*.

You have returned to the original word that you were trying to define. This "circling back" is called **circularity.** (Circularity means that you have circled back to any word previously defined, not necessarily the original one.) To avoid circularity, certain basic geometric terms are forced to be *undefined*. These terms are the building blocks of the defined terms. They provide an initial reference upon which other terms can be defined. In this book, following a tradition begun by the German mathematician David Hilbert about 100 years ago, we choose *point, line,* and *plane* as **undefined terms**.

Also not defined in this book are very common English words—articles, prepositions, and conjunctions. Likewise, we assume that you are familiar with many words used in algebra or arithmetic, such as "equation," "number," "equals," "is less than," and so on. Other words, such as *if* and *then,* are the basic terms of logic, and are left undefined, too.

> **Undefined Terms (in this book)**
>
> Geometric terms: point, line, plane

Definitions Using Undefined Terms

Definitions are made using undefined terms. Here are definitions that apply to all of the types of geometries that you have studied:

A **figure** is a set of *points*.

Space is the set of all *points*.

Three or more *points* are **collinear** if there is a *line* that contains all of them.

A **plane figure** is a set of *points* that are all in one *plane*.

Four or more *points* are **coplanar** if there is a *plane* that contains all of them.

 See Quiz Yourself 1 at the right.

> ▶ **QUIZ YOURSELF 1**
>
> Identify three common English words in the definitions at the left that are undefined but not mentioned earlier on this page.

GUIDED

Example 1

A photo of a set of stairs is at the right. Use the definition of collinear and coplanar to name three sets of three collinear points and three sets of four coplanar points.

Solution Points are collinear if they lie on the same line. One set of three collinear points is A, E, I. Two others are ___?___ and ___?___.

Points are coplanar if they lie on the same plane. Each step is part of a plane. Other parts of the staircase are parts of planes also. This provides many opportunities to choose coplanar points from this diagram. One set of four coplanar points is on the bottom step: A, E ,F, B. Two other sets of four coplanar points are ___?___ and ___?___.

When all points in a set are collinear, then the space, the type of geometry, and the figures in it are all called **one-dimensional**. A good example would be the points on a line. When the points in a set are not all collinear, but are all coplanar, then the space and the geometry are called **two-dimensional**. The top of a table is an example of a two-dimensional space. Most real objects do not lie in a single plane. They are **three-dimensional** figures. A house is a three-dimensional figure.

(STOP) **See Quiz Yourself 2 at the right.**

> ▶ **QUIZ YOURSELF 2**
>
> Is the set of stairs in the previous example 1-dimensional, 2-dimensional, or 3-dimensional?

GUIDED

Example 2

Classify each of the following objects as 1-dimensional, 2-dimensional, or 3-dimensional.

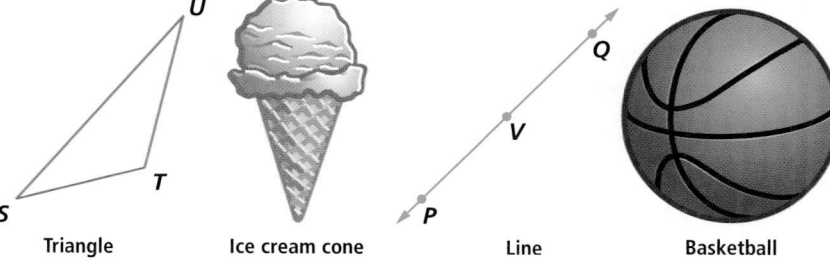

| Triangle | Ice cream cone | Line | Basketball |

Solution Not all points of the triangle are collinear, but they are all coplanar. The triangle is ___?___-dimensional.

The ice cream cone is made of many points that do not lie in one plane. The ice cream cone is ___?___-dimensional.

All points are collinear. The line is ___?___-dimensional.

The points on the basketball are not collinear or coplanar. The basketball is ___?___-dimensional.

If a word or phrase is not defined, it could mean anything. Point might mean "elephant"! So, to clarify that we are talking about a particular kind of point or line, we need to assume that points and lines have certain properties. For instance, if we assume that between two points, there is always another point, you know we are not talking about networks. Some of the assumptions about points and lines for this book are given in the next lesson. These assumptions and undefined terms serve as "rules" for our "game" of geometry.

Questions

COVERING THE IDEAS

1. Why are undefined terms necessary?

2. What are three undefined geometric terms used in this book?

3. The following are definitions of the words "money" and "currency." Why are these definitions considered circular?
   Money: The official currency, coins, and negotiable paper notes issued by a government.
   Currency: Money in any form when in actual use as a medium of exchange, especially circulating paper money.

4. The set of all points is ___?___.

5. Name at least 3 other sets of coplanar points that are not mentioned in Example 1.

6. Consider the cereal box pictured at the right.
   a. Name two sets of three collinear points.
   b. Name two sets of four coplanar points.

APPLYING THE MATHEMATICS

7. Look up the word "point" in a dictionary. List two definitions.

8. Find a definition of the word "whirlpool" in a dictionary. Then find a definition of a key word in whirlpool's definition. Continue this process until circularity occurs.

9. Consider the following set of rules for a card game with a standard deck of cards.
   Rule 1: Every even-numbered card is worth 3 points.
   Rule 2: Every odd-numbered card is worth 2 points.
   Rule 3: Jacks, Kings, and Queens are each worth 4 points.
   Rule 4: All the cards are dealt.
   Rule 5: The player with the highest point value wins.
   a. Explain why the rules for a game of cards are not sufficient.
   b. Add a rule to make the set of rules sufficient.

10. Consider the picture of a top at the right.
    a. Name three points that are collinear.
    b. Name four points that are not collinear, but are coplanar.

11. Wrapping a present requires several steps.
    Step 1 Measure and cut the wrapping paper.
    Step 2 Fold the paper to cover the box and tape it closed.
    Step 3 Give the wrapped gift to your friend.
    Identify which steps of this process are 1-, 2-, and 3-dimensional.

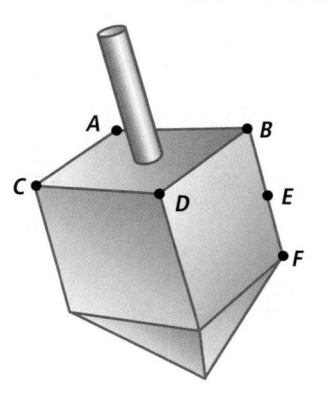

REVIEW

12. *The New York City Bridge Problem.* At the right is a drawing of the five boroughs of New York City, Randall's Island, and New Jersey. (No water separates Brooklyn and Queens.) Bridges and tunnels connect the regions. Draw a network like the Königsberg bridge network to represent New York and determine if you could take a driving tour of New York going over each bridge and through each tunnel exactly once. **(Lesson 1-3)**

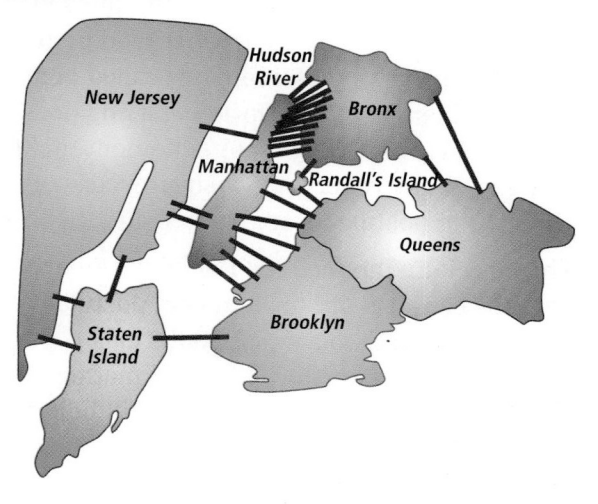

13. Name two differences between Euclidean synthetic geometry and discrete geometry. **(Lesson 1-3)**

14. Find an equation for the horizontal line passing through (1, –3). **(Lesson 1-2)**

15. Write the slope-intercept form of the equation $x + 3y = 4$. **(Lesson 1-2)**

16. Write an equation for the line through (0, 0) and (11, 15). **(Lesson 1-2)**

EXPLORATION

17. a. A point can be thought of as a zero-dimensional figure. If you were to "pull" this point in one direction a certain fixed distance so that it sweeps out a path, this new figure would be a segment. This segment is a 1-dimensional figure. If you "pull" this segment perpendicularly in one direction that same fixed distance, you will sweep out a square. If you continued this process of "pulling" the figure in one direction perpendicularly the same fixed distance, what 3-dimensional figure will you sweep out?

    b. Could you continue this process one more time? Research the term *hypercube*.

QUIZ YOURSELF ANSWERS

1. Answers may vary. Sample: is, called, that

2. 3-dimensional

1-5

Postulates for Points and Lines in Euclidean Geometry

Vocabulary

postulate

theorem

Euclidean geometry

parallel lines

> ▶ **BIG IDEA** The Point-Line-Plane Postulate for Euclidean geometry includes rules that the undefined terms *point, line,* and *plane* must obey and is the basis for deduction of properties of figures.

Different descriptions of points and lines lead to very different types of geometry. To make clear what kind of geometry we are studying, we begin by making assumptions called **postulates.** Postulates explain undefined terms and serve as a starting point for logically deducing, or proving, other statements. A statement that follows from postulates, definitions, and other previously proved statements is called a **theorem.** Notice that postulates are chosen and theorems are proven.

The Greek mathematician Euclid wrote one of the most famous (and perhaps earliest) organizations of postulates and theorems of all time, around 300 BCE, in a set of books called *Elements.* The mathematics in Euclid's *Elements* has inspired mathematical research for over 23 centuries, and the books were still being used as a geometry text into the twentieth century. This is why most of the world calls the geometry you study this year **Euclidean geometry.** The postulates in this book were picked to fit this geometry. The postulates for Euclidean geometry fit two notions of point described in earlier lessons: as an idealized location and as an ordered pair in the plane.

Mental Math

Write down four sentences using the word "line" in which it has different meanings.

The Point-Line-Plane Postulate of Euclidean Geometry

Our postulates for Euclidean geometry are introduced in Chapters 1–4 and 8–11. A complete list is on page S2. The first postulate for Euclidean geometry is called the *Point-Line-Plane Postulate.* It includes three assumptions about points, lines, and planes. The first assumption of the Point-Line-Plane postulate is sometimes restated as "two points determine a line."

Point-Line-Plane Postulate

a. *Unique Line Assumption*
   Through any two points there is exactly one line. If the two points are in a plane, the line containing them is in the plane.

At the right, line ℓ is the unique line through points A and B. $\overleftrightarrow{AB}$, $\overleftrightarrow{BA}$, and ℓ are three different names for this line. The symbol $\overleftrightarrow{AB}$ is read "line AB."

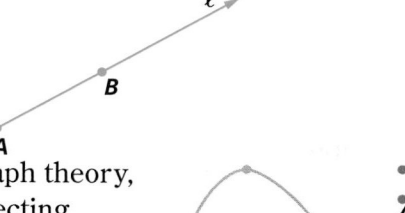

The Unique Line Assumption does not apply to graph theory, where there can be more than one line (arc) connecting two points (nodes). It also does not apply to lines in discrete geometry because two points might not be part of a line.

You have already seen examples of this assumption in action in algebra and in Example 2 of Lesson 1-2, where you found the equation of a line passing through any two given points.

The second part of the Point-Line-Plane postulate assures that lines in Euclidean geometry contain infinitely many points. This postulate shows that lines are *not* made up of dots placed next to each other.

Point-Line-Plane Postulate

b. *Number Line Assumption*
Every line is a set of points that can be put into a one-to-one correspondence with the real numbers, with any point on it corresponding to 0 and any other point corresponding to 1.

Therefore, any line in Euclidean geometry can be made into a number line. From this we see that any line in Euclidean geometry has infinitely many points. The Number Line Assumption does not apply to graph theory or discrete geometry, where we can have a finite number of points on a line.

STOP See Quiz Yourself 1 at the right.

▶ **QUIZ YOURSELF 1**

How do we know that every line in Euclidean geometry has infinitely many points?

The third part of the Point-Line-Plane postulate assures that Euclidean geometry can study 1-, 2-, and 3-dimensional figures.

Point-Line-Plane Postulate

c. *Dimension Assumption*
(1) There are at least two points in space.
(2) Given a line in a plane, there is at least one point in the plane that is not on the line.
(3) Given a plane in space, there is at least one point in space that is not in the plane.

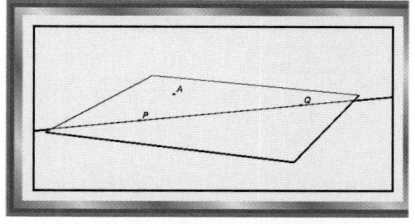

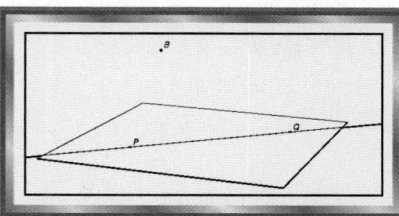

STOP **See Quiz Yourself 2 at the right.**

Intersecting and Parallel Lines

Using the Point-Line-Plane postulate, we can deduce other properties of points and lines.

A reasonable question to consider is, "In how many points can two different lines intersect?"

In algebra, if you graphed the system $\begin{cases} x + 2y = 5 \\ 2x + y = 4 \end{cases}$ on the coordinate plane, you would find that $(1, 2)$ is the only point of intersection of these two lines. You also know that sometimes lines do not intersect and sometimes two equations describe the same line.

We can show that two different lines cannot intersect at two different points. Suppose lines m and n intersect at point P. If there were another point Q where the lines intersected, then both m and n would pass through P and Q. However, the Unique Line Assumption says that only one line can pass through the points P and Q, so the two lines can intersect at only one point.

> **► QUIZ YOURSELF 2**
>
> In the coordinate plane, give an example of a point that does not lie on the line $y = x + 2$.

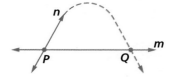

Line Intersection Theorem	
Two different lines intersect in at most one point.	

You saw examples of the Line Intersection Theorem in algebra when you solved systems of linear equations. Recall that three common ways to do this are substitution, linear combinations, and graphing.

GUIDED

Example 1

Find the coordinates of the intersection of the two lines whose equations are $\begin{cases} 3x - 4y = -17 \\ 5x + 6y = -3 \end{cases}$.

Solution Because both equations are in standard form, this problem lends itself to the linear combination method.

$$\begin{cases} 3(3x - 4y) = (-17)3 \\ 2(5x + 6y) = (-3)2 \end{cases} \rightarrow \begin{cases} \underline{\quad ? \quad} \\ 10x + 12y = -6 \end{cases} \rightarrow 19x = -57 \rightarrow \underline{\quad ? \quad}$$

Now substitute -3 in for x into one of the two original equations and solve for y.

$$3(-3) - 4y = -17$$
$$-9 - 4y = \underline{\quad ? \quad}$$
$$-4y = -8$$
$$y = \underline{\quad ? \quad}$$

The coordinates of the intersection point are ($\underline{\ ?\ }$, $\underline{\ ?\ }$).

Example 2

Find the coordinates of the intersection of the two lines whose equations are given by $\begin{cases} y = 3x - 7 \\ y = -2x + 13 \end{cases}$.

Solution Because both equations are solved for y, this problem lends itself to graphing. We choose to use graphing technology to graph the lines and find the point of intersection.

The point of intersection of the graphs seems to be (4, 5). The point is verified by substitution of its coordinates in both equations:

$$5 = 3 \cdot 4 - 7 \quad \text{and} \quad 5 = -2 \cdot 4 + 13$$

 See Quiz Yourself 3 at the right.

When coplanar lines do not intersect in exactly one point, they are called parallel. For instance, the rails of railroad tracks are parallel to each other.

Definition of Parallel Lines

Two coplanar lines m and n are called **parallel lines**, written $m \parallel n$, if and only if they have no points in common or they are identical.

 See Quiz Yourself 4 at the right.

Example 3

In the coordinate plane, are the lines $\begin{cases} y = x + 2 \\ y = x + 4 \end{cases}$ parallel?

Solution If you substitute $x + 2$ for y in the second equation, you get $x + 2 = x + 4$. Subtracting x from both sides, you get $2 = 4$. This is never true, so these two lines never intersect. Since $x + 2 \neq x + 4$, the lines are parallel.

▶ **QUIZ YOURSELF 3**

Do Example 2 again using the substitution method. Start by substituting $3x - 7$ for y in the second equation.

▶ **QUIZ YOURSELF 4**

According to this definition, is a line parallel to itself?

Questions

COVERING THE IDEAS

1. What is Euclid's *Elements*?

2. **a.** Draw two points, A and B, and the line they determine.
   b. Name the line you drew in Part a in three ways.

3. Name two types of geometry for which the postulates of Euclidean geometry are appropriate.

4. Name two types of geometry for which the postulates of Euclidean geometry are inappropriate.

5. For the line at the right, which postulate allows you to make 0 the coordinate of B and 1 the coordinate of A?

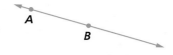

6. **Fill in the Blank** If two coplanar lines do not intersect, then they are ___?___.

7. What is the difference between a *postulate* and a *theorem?*

8. Given the system of equations $\begin{cases} y = 3x - 5 \\ x + 2y = 5 \end{cases}$, find all the points in which the corresponding lines intersect.

9. Find the point of intersection of the lines with equations $-2x + 3y = 10$ and $8x - 4y = -8$.

10. Explain why if m and n are two coplanar lines that are not parallel, then they intersect in exactly one point.

APPLYING THE MATHEMATICS

11. Draw a line m and a point P not on m. Suppose your paper is a portion of a plane. Describe where you would have to place your pencil to indicate the location of a point in three-dimensional space that is not in that plane.

12. The Unique Line Assumption states that two points determine a line. Given the points with coordinates $(-7, 8)$ and $(15, 11)$, find an equation for the line that passes through these points.

13. Show that both $(4, 0)$ and $(1, 2)$ satisfy both equations of the system $\begin{cases} 2x + 3y = 8 \\ 4x + 6y = 16 \end{cases}$. Why does this not violate the Line Intersection Theorem?

14. Graph the equations in Question 13 using a graphing utility. Explain the result.

15. **True or False** In Euclidean geometry:
    a. A line contains infinitely many points.
    b. A plane contains infinitely many points.
    c. For line ℓ in plane M, all points in M are also on ℓ.
    d. If two coplanar lines intersect, then they have a point in common.
    e. A line has thickness.
    f. There are lines with only two points.

16. a. Graph the lines with equations $y = 10x$ and $y = 5(7 + 2x)$ on the same set of axes.
    b. Are the lines parallel?

17. From the Number Line Assumption, you deduced that there are infinitely many points in a line. Using the Postulates and Theorem in this lesson, give an argument to deduce the following theorem: There are infinitely many lines in a plane.

REVIEW

18. **True or False** In geometry we must define every term we use. (**Lesson 1-4**)

19. **True or False** Any set of three points is collinear in Euclidean geometry. (**Lesson 1-4**)

20. **True or False** A sphere is an example of a three-dimensional figure. (**Lesson 1-4**)

21. The network pictured at the right is not traversable. Add an arc, so that the network becomes traversable. (**Lesson 1-3**)

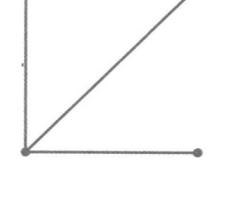

22. Marina travels among three islands. Island *A* has three bridges to Island *B*, Island *B* has two bridges to Island *C*, and one bridge connects Islands *A* and *C*.
    a. Draw a picture of this network.
    b. Is it possible for Marina to start on one island, cross each bridge exactly once, and end up where she started? (**Lesson 1-3**)

23. Before typing her letter, Tamara set a 0.75-inch left margin and a 1-inch right margin on an 8.5-inch-wide document. What is the new width of the typing area? (**Lesson 1-1**)

EXPLORATION

24. This lesson refers to Euclid's *Elements*. Research this work and write a paragraph about it, including information about how it was structured, what new ideas it introduced, and how it influences mathematicians and other scientists.

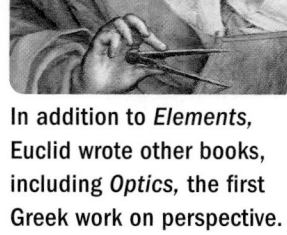

In addition to *Elements*, Euclid wrote other books, including *Optics*, the first Greek work on perspective.

Lesson
1-6

Betweenness and Distance

Vocabulary

between

segment, line segment

endpoints

ray

opposite rays

▶ **BIG IDEA**   The distance between two points is measured along the coordinatized line that contains them and obeys rules that are summarized in the Distance Postulate.

If you were to ask people the question "Is Chicago between Los Angeles and New York?", many people would answer "Yes." To a geography teacher, this might seem like a reasonable answer. However, in geometry, we use a more precise definition of the word *between*. To define precisely what is meant for a *point* to be between two other points, we start by considering numbers.

A *number* is between two others if it is greater than one of them and less than the other. For example, 4 is between –2 and 7.3, because 4 > –2 and 4 < 7.3. The number –2 is not considered to be between itself and 7.3.

A *point* is **between** two other points if it is on the same line and its coordinate is between their coordinates. For example, point U above, with coordinate 4, is between the other two points A and B. Note that point A is not considered to be between A and B because the number –2 is not between –2 and 7.3. If point E is between A and B, then the double inequality $-2 < x < 7.3$ describes the possible values of the coordinate x of point E.

🛑 **See Quiz Yourself at the right.**

Activity

• P

MATERIALS Tracing paper

Find and then trace the two points P and Q on this page onto another sheet of paper. Draw the shortest path from P to Q.

Q•

Mental Math

a. A 45-minute train ride ends at 12:13 P.M. When did it start?

b. A train ride begins at 10:42 A.M. and ends at 11:08 P.M. of the same day. How long was it?

c. A 122-minute train ride leaves at 3:51 P.M. Due to construction there is an anticipated delay of between 5 and 13 minutes. What is the earliest and latest time the train is expected to arrive?

▶ **QUIZ YOURSELF**

One day Cesar's temperature ranged between 100.4°F and 96.3°F. Describe his possible temperatures t during that day with a double inequality.

The **segment** (or **line segment**) with **endpoints** A and B, denoted $\overline{AB}$, is the set consisting of the distinct points A and B and all points between A and B.

On a number line, the graph of two numbers, such as –3 and 7.4, and *all* points having coordinates between them, is a segment. The points for –3 and 7.4 are the endpoints of the segment.

The symbol $\overline{AB}$ is read "segment AB." On this number line, $\overline{AB}$ consists of all points on the line whose coordinates satisfy the inequality $-3 \le x \le 7.4$.

Rays

A *ray* is like a laser beam. A laser beam starts at a point and, if not blocked, continues forever in a particular direction. In geometry, a ray consists of an endpoint and all points of a line on one side of that endpoint. The idea of betweenness helps to give a precise definition of ray.

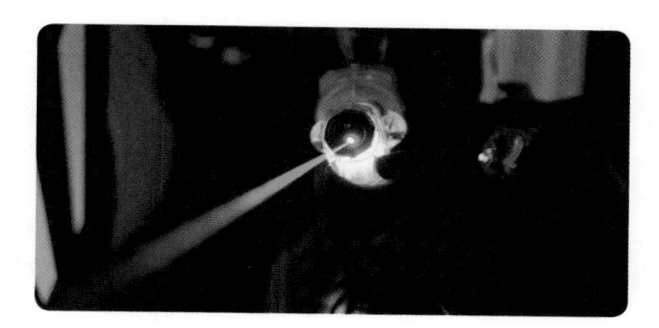

Definition of Ray

The **ray** with endpoint A and containing a second point B, denoted $\overrightarrow{AB}$, consists of the points on $\overline{AB}$ and all points for which B is between the point and A.

A ray is named by its endpoint and any other point on the ray. The symbol $\overrightarrow{AB}$, read "ray AB," is written with the left letter representing the endpoint of the ray and the right letter representing any other point on the ray.

The graphs of the simplest inequalities in algebra are rays. For example, if A has coordinate 2 and B has coordinate 8, then $\overrightarrow{AB}$ consists of all points with coordinates $x \ge 2$.

If C has coordinate 1, then $\overrightarrow{AC}$ points in the opposite direction of $\overrightarrow{AB}$. $\overrightarrow{AC}$ consists of all points with coordinates $x \le 2$. $\overrightarrow{AB}$ and $\overrightarrow{AC}$ are called *opposite rays*.

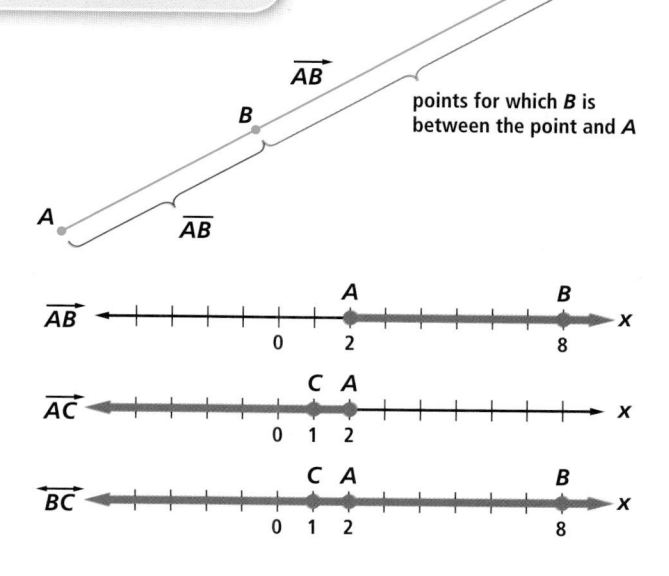

Definition of Opposite Rays

$\overrightarrow{AB}$ and $\overrightarrow{AC}$ are **opposite rays** if and only if A is between B and C.

Assumptions about Distance

On the first page of this lesson, you were asked to draw the "shortest" path from P to Q. In order to have the idea of "shortest" in our geometry, we need some assumptions about distance. The first two parts of the Distance Postulate repeat information from Lesson 1-1.

Distance Postulate

a. *Uniqueness Property* On a coordinatized line, there is a unique distance between two points.

b. *Distance Formula* If two points on a line have coordinates x and y, the distance between them is $|x - y|$.

There is a simple property relating betweenness and length. Suppose you are traveling from X to Z along $\overline{XZ}$. Let Y be between X and Z, as shown at the right.

To go from X to Z, you first travel on $\overline{XY}$, then on $\overline{YZ}$. Clearly, the length of $\overline{XY}$ plus the length of $\overline{YZ}$ is the length of the whole segment $\overline{XZ}$. In symbols, $XY + YZ = XZ$. We call this the *Additive Property of Distance*.

Distance Postulate

c. *Additive Property* If B is on $\overline{AC}$, then $AB + BC = AC$.

Example

In the figure at the right, E is a point between points G and F. If $GE = 9$ and $GF = 27$, what is EF?

Solution Sketch and mark the picture. Then use the Additive Property of Distance.

$$GE + EF = GF$$
$$9 + EF = 27$$
$$EF = 27 - 9 = 18$$

Caution: Be careful to distinguish the following symbols. They look similar, but their meanings are quite different.

$\overleftrightarrow{AB}$ is the *line* determined by points A and B.

$\overrightarrow{AB}$ is the *ray* with endpoint A containing B.

$\overrightarrow{BA}$ is the *ray* with endpoint B containing A.

$\overline{AB}$ is the *segment* with endpoints A and B.

AB is the *distance* between A and B, or the length of $\overline{AB}$.

Of the five symbols above, the first four represent sets of points and the last one represents a number.

Questions

COVERING THE IDEAS

1. Look at a map of the United States. In the geometric sense, is Chicago between Los Angeles and New York?

2. Define *ray*.

3. **Fill in the Blank** A laser beam is like the geometric figure called a(n) ___?___.

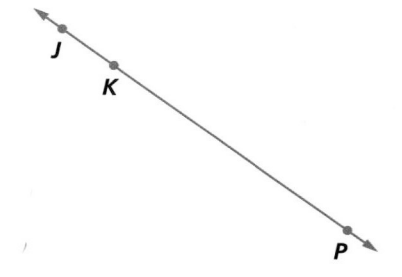

4. a. On a number line, graph the set of numbers that satisfy $w \geq 3$.
   b. What figure is the graph?

5. a. On a number line, graph the set of numbers that satisfy $2.4 \geq d \geq 0.6$.
   b. What figure is the graph?

6. Suppose points A and C have coordinates –4 and 17, respectively.
   a. If point B is between A and C, give the range of coordinates x for B.
   b. If D is on $\overrightarrow{CA}$, give the range of coordinates x for D.

7. If point H is between points G and K, then what is the relationship between the lengths HK, HG, and GK?

8. a. In the figure at the right, K is between J and P. If $JK = 13$ and $JP = 57$, what is KP?
   b. Name a pair of opposite rays in the figure.

9. **True or False**
   a. $\overleftrightarrow{KM} = \overleftrightarrow{MK}$
   b. $KM = MK$
   c. $\overrightarrow{KM} = \overrightarrow{MK}$
   d. $\overline{KM} = \overline{MK}$

10. **Matching** Match each symbol with the correct description.
    a. PQ (i)   segment
    b. $\overline{PQ}$ (ii) length
    c. $\overrightarrow{PQ}$ (iii) line
    d. $\overleftrightarrow{PQ}$ (iv) ray

APPLYING THE MATHEMATICS

11. Points A, B, C, and D are collinear. In each of the parts below, draw a figure for which the given statement is true.
    a. $\overrightarrow{BC}$ and $\overrightarrow{BD}$ are not opposite rays.
    b. $\overline{BC}$ is contained in $\overrightarrow{AD}$.
    c. $\overline{BC}$ is not contained in $\overrightarrow{AD}$.

12. Ana, Beth, Cody, and Destiny are all friends who have just completed a 5K race. Here is what we know about their results:

    Ana and Beth finished 5 minutes apart.
    Cody was 2 minutes slower than Beth.
    Destiny was 2 minutes slower than Ana.
    Ana finished the race before Cody.
    Destiny had a finishing time of 25 minutes.

    What are the finishing times of Ana, Beth, and Cody?

In 13 and 14, use the figure below. Points, E, L, M, Q, and T are collinear. Explain why each of the following statements is *false*. Rewrite the statement to make it true.

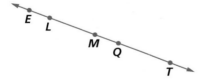

13. $MQ > QM$ 14. $EM = LM$

15. In 2007, Yao Ming was the tallest player in the National Basketball Association. His height was 7 feet 6 inches. This could be considered his length. On his body, what are the endpoints of the segment whose length is his height?

Yao Ming

In 16–18, use the figure below. Points J, D, V, and N are collinear.

16. If $JV = 16$ and $JN = 21$, find NV.

17. If $JV = 20$, $JD = 11$, and $DN = 13$, find VN.

18. Suppose $JD = 2x + 3y$, $DV = 7 - y$, $VN = 9x - 2y$, and $JN = 32$. Solve for x.

19. Explain why the graph of $3 < x < 14$ on a number line is not a segment.

20. A family from Indianapolis, Indiana, is planning a vacation to see relatives in Springfield, Missouri. They check a road atlas and find that the distance between these two cities is 428 miles, which they decide is too long for a one-day drive. St. Louis, Missouri is "between" these two cities, so the family decides to drive to St. Louis on the first day of their vacation. The family checks a road map and finds that their first day's drive will be a distance of 242 miles, and the drive from St. Louis to Springfield on the second day will be 217 miles. Explain the apparent discrepancy in their findings.

REVIEW

21. True or False In Euclidean geometry:
  a. Through any three points there is exactly one line.
  b. Every plane has infinitely many points.
  c. Any line is parallel to itself. (**Lesson 1-5**)

22. Use the assumptions of Euclidean geometry to explain why, in a plane, there is a set of points that are not collinear. (**Lessons 1-4, 1-5**)

23. Give an example to show that the Unique Line Assumption is not true in discrete geometry. (**Lessons 1-3, 1-5**)

In 24 and 25, solve the given inequality. (**Previous course**)

24. $4x + 7 < 5x + 2$

25. $\frac{a}{6} - \frac{a}{2} \geq \frac{1}{8}$

EXPLORATION

26. Suppose A, B, and C are points in the plane.
  a. Give a simple description of the following set: the set consisting of the points A, B, and C, and all the points that are between A and B or between B and C or between C and A.
  b. Does your answer depend on whether A, B, and C are collinear?

QUIZ YOURSELF ANSWER

$96.3 < t < 100.4$

Lesson
1-7
Using a Dynamic Geometry System (DGS)

Vocabulary

dynamic geometry system (DGS)

window

menu

> ▶ **BIG IDEA** The Triangle Inequality Postulate guarantees that the shortest distance between two points is the length of the line segment they determine.

Drawing a picture is often a great way to start solving a problem in geometry. The picture gives your mind a frame of reference to organize your thinking. The kind of picture depends on what the problem is and what you know about it. This is not unlike what happens in other situations. Sometimes a rough diagram is enough, a photograph might be better, or you may need a movie to understand the situation. A quick sketch might tell you where the doors to a building are located, a photograph would be better than a sketch for identifying a person, and a movie would be the best for seeing how an expert figure skater executes a particular move.

Mental Math

If you cut a square cake with 8 slices parallel to its sides, what is the largest number of pieces you can form?

If you are investigating an unfamiliar situation and your carefully drawn picture does not give you enough information, it might be helpful to have a movable drawing that has the same properties that your problem describes.

In recent years, this idea of moving a drawing has become possible using a dynamic drawing system, either on your computer or on your calculator. The key word here is *dynamic,* meaning "marked by continuous change." A **dynamic geometry system** (referred to from here on as a **DGS**), enables you to look at hundreds of well drawn, carefully measured drawings in a short period of time. This enables you to look for patterns and notice relationships. Finding patterns and noticing relationships in figures is a large part of what it means to do geometry.

The activities in this lesson will help you understand how to use a DGS. You will discover an interesting and important fact about Geometry as you explore. Have fun. Don't be afraid to try things and see what happens. That is all part of the reason a DGS is useful.

Using a DGS

The area in which a DGS can draw is called the **window.** Outside the window is a **menu** of shapes and options. The screen will also likely show you a menu of *tools* or a *toolbox* that the DGS can use to change the drawings. A screen from one such piece of software is pictured below with Activity 1.

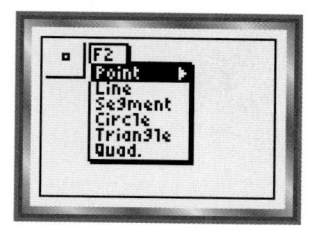

Activity 1

MATERIALS DGS

Step 1 Use a DGS to construct a segment and label its endpoints *A* and *B*.

Step 2 Construct another point on $\overline{AB}$ and label it *C*.

Step 3 Measure the lengths *AB, AC,* and *CB*. Learn how you can change the number of decimal places to which these measures are given. Give the lengths to 2 decimal places.

Step 4 Drag point *C* along the segment $\overline{AB}$. Compute *AC + CB*. Compare this measure to that of *AB*. Do your data verify the Additive Property of Distance mentioned in Lesson 1-6? Explain why it does or why it does not.

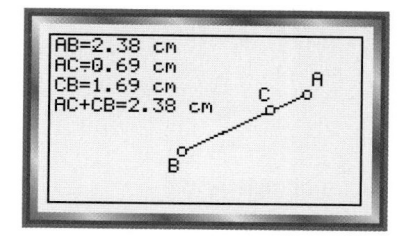

Step 5 Drag one of the endpoints of the segment to various positions on the screen. What happens to the measurements on the screen as you move the endpoint? Is it still true that *AC + CB = AB*?

In Activity 1, you should have verified by measurement that the Additive Property of Distance holds true when *C* is on $\overline{AB}$. But what happens when *C* is not on $\overline{AB}$?

Dynamic Capabilities

The dynamic capabilities of a DGS can be employed to generate many instances of a general pattern quickly.

Activity 2

MATERIALS DGS

Step 1 Construct points A, B, and C on the screen.

Step 2 Measure CA and CB. Then find the sum $AC + BC$. Make this sum accurate to as many places as your DGS allows.

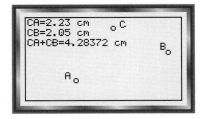

Step 3 Drag the point C to other locations, so that the sum $AC + BC$ is as small as possible. Note the location.

Step 4 Find two other locations of C that make the sum $AC + BC$ as small as the first location you found. Where are these other locations?

In Activity 2, you should have found that the location of point C that minimizes the sum is anywhere between point A and point B. From Lesson 1-6, this means that C must be on $\overline{AB}$.

Activity 3

MATERIALS DGS

Step 1 Start with a clear screen on your DGS. Construct a triangle ABC on your screen.

Step 2 Measure the lengths AB, AC, and BC. Calculate the sum $AC + BC$.

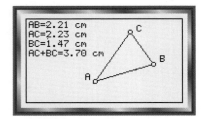

For Steps 3–8, answer the question: Which is greater, $AC + BC$ or AB?

Step 3 Move vertex C far away from $\overline{AB}$.

Step 4 Move vertex C very close to $\overline{AB}$.

Step 5 Move vertex C onto $\overline{AB}$.

Step 6 Move vertex C below $\overline{AB}$.

Step 7 Move vertex B to a different location.

Step 8 Move vertex A to a different location.

In Activity 3, you should have found that when C is not on $\overline{AB}$, then $AC + BC > AB$. If C is on $\overline{AB}$, then we are in the same situation as Activity 2. This result, which we take to be a postulate, is called the Triangle Inequality.

> ### Triangle Inequality Postulate
>
> The sum of the lengths of any two sides of a triangle is greater than the length of the third side.

For any triangle ABC such as the one drawn at the right, this relationship means three inequalities are true.

$$AB + BC > AC$$
$$BC + AC > AB$$
$$AB + AC > BC$$

By taking the Triangle Inequality as a postulate, we are assuming in our geometry that the shortest path between two points is along the segment that joins them.

$AB = 7$
$AC = 5$
$AB + AC = 12$
$BC = 8$

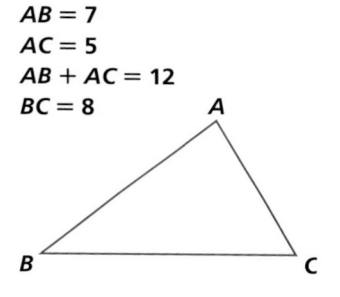

Example

Can 8, 16, and 7 be the lengths of the sides of a triangle? Why or why not?

Solution 1 Because of the Triangle Inequality, the sum of the lengths of any two sides of a triangle is greater than the length of the third side.

$8 + 16 > 7$ and $7 + 16 > 8$, but $8 + 7$ is not greater than 16. So, 8, 16, and 7 cannot be the lengths of the sides of a triangle.

Solution 2 No. If there were a triangle PQR with sides of lengths 8, 16, and 7, then it would be a shorter route to travel from Q to P and then to R than it would be to travel along $\overline{QR}$. This would violate the Triangle Inequality Postulate.

Questions

1. What are some benefits and some disadvantages of drawing figures in geometry by hand? What are some of the limitations of drawing on a calculator or computer?

2. What is a DGS?

3. When using a DGS, what does one call the area in which one can draw?

4. In Activity 1, what procedure did you use on your DGS to draw a segment?

5. In measuring segments, the DGS you used gave lengths to how many decimal places?

6. In Activity 1, Step 5, when you dragged one endpoint of the segment, what happened to the other endpoint?

7. In Activity 3, Steps 3–6, you dragged vertex C. Which length remained constant during this activity?

8. State the Triangle Inequality Postulate.

9. Consider a triangle HGK. Give three inequalities satisfied by the lengths of the sides of this triangle.

In 10 and 11, tell whether the numbers can be lengths of the three sides of a triangle.

10. **a.** 5, 5, 9   **b.** 5, 5, 10   **c.** 5, 5, 11

11. **a.** 11, 21, 13   **b.** 13, 21, 7   **c.** 14, 22, 8

12. Quincy is using a DGS and constructs four segments $\overline{AB}$, $\overline{BC}$, $\overline{CD}$, and $\overline{AD}$. If he drags point C, which segments also move? Which segments stay the same?

13. Maria was doing Activity 1 and found measurements such that $AC = 1.13$ cm, $BC = 1.59$ cm, and $AB = 2.73$ cm. These measurements seemed to contradict her previous finding because $AC + BC$ does not equal AB, so she asked her teacher what she could have possibly done wrong. Give an explanation that her teacher might have given to her.

14. Carl was doing Activity 2. He dragged point C so that it was right on top of point B. Is the Additive Property of the Distance Postulate still true in this case?

15. Points *D*, *E*, and *F* are on a line. What can be said about the position of these points if each of the following is true?

   a. $DE + EF = DF$ b. $FD + ED = FE$ c. $DE - FE = FD$

16. Points *J*, *K*, and *L* are on a plane. For what positions of these three points is each of the following true?

   a. $KJ + KL > JL$ b. $KL + JL = JK$ c. $KL + LJ < KJ$

17. Can $\sqrt{40}$, $\sqrt{60}$, and $\sqrt{140}$ be the lengths of the sides of a triangle?

18. Two sides of a triangle have lengths 3 and 8. Give all possible values of *x*, the length of the third side.

19. A stick that is 10 cm long is broken into three pieces, all of which have integer lengths. These pieces are placed end-to-end to make a triangle.

   a. What is the shortest possible length of one of the sides of such a triangle?

   b. How many different such triangles can be made?

   c. What is the longest possible length of one of the sides of such a triangle?

20. **Fill in the Blanks** Jessica's school is 45 miles from the center of Boston, Massachusetts, and 38 miles from the center of Providence, Rhode Island. With this information, you know that the cities are between __?__ and __?__ miles apart.

21. A road map states the driving distance between Anchorage, Alaska, and Vancouver, British Columbia, as 2290 miles. The driving distance between Anchorage and Seattle, Washington, is given as 2758 miles, and the driving distance between Vancouver and Seattle is given as 142 miles. Choose one of the following and explain your choice.

   A This information violates the Triangle Inequality Postulate, so one of the numbers is in error.

   B The distances could be correct because the roads connecting the cities are not straight lines.

   C If the distances were given in kilometers instead of miles, the Triangle Inequality Postulate would not be violated.

22. If a triangle has sides of length $(2x - 1)$ cm, $(3x + 2)$ cm, and $(12 - x)$ cm, find all possible values of *x*.

23. On a DGS, construct a circle and three points on it. Construct a triangle so that your figure looks similar to the one at the right. If you drag the circle, what happens to the triangle? If you drag the triangle, what happens to the circle?

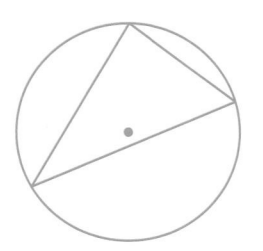

24. In the figure at the right, Ms. Álvarez needs to drive from her house to the grocery store. She is in a hurry, so she drives along Elm Street and then turns on 3rd Avenue and proceeds to the store.

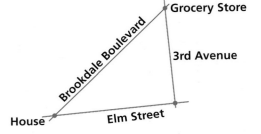

   a. Is there is a shorter route she could have taken?
   b. Is it possible that her route is faster? Write a short explanation why this might be true.

25. a. On a DGS, construct a figure similar to the one at the right.
   b. Describe a way for your DGS to calculate the perimeter of your figure.

26. a. On a DGS, construct a circle.
   b. Describe a way for your DGS to calculate the area of the circle.

REVIEW

27. The solutions to the double inequality $-6 \leq 2x \leq 0$ describe a segment on the number line. What are the coordinates of the segment's endpoints? (**Lesson 1-6**)

In 28–31, describe the graph of all solutions to the inequality on a number line. (**Lesson 1-6**)

28. $x + 2 \leq x + 3$

29. $3x + 2 \leq x + 4$

30. $1 \leq z \leq 5$

31. $4n > 2$

32. Explain why there are infinitely many points in every segment. (**Lessons 1-5, 1-6**)

33. In the pipeline problem from Lesson 1-1, what is the answer to the problem if the two oil wells are on opposite sides of the pipeline? (**Lesson 1-1**)

34. Draw the set of solutions to $y \leq x + 4$ on the coordinate plane. (**Previous Course**)

EXPLORATION

35. Use a DGS to draw one of the figures below.

   a.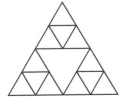

   b.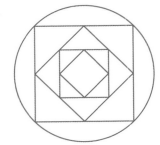

Chapter 1 Projects

A project presents an opportunity for you to extend your knowledge of a topic related to the material of this chapter. You should allow more time for a project than you do for typical homework questions.

1 Finding the Shortest Route

Many road maps have inserts that look like the network shown below. Some even have miles and/or times on them. Travelers have a quick way of looking at times or distances between major cities. Ambulance or fire truck drivers could use these networks to plan the best route to their emergencies. Dijkstra's algorithm can be used to find the shortest path through a network.

a. Find out what Dijkstra's algorithm is and use it to find the shortest distance between point *A* and point *J*.

b. Find a road map and use Dijkstra's algorithm to find the shortest distance and the shortest time between two cities. Are the paths the same? Is there more than one path with the same time/distance?

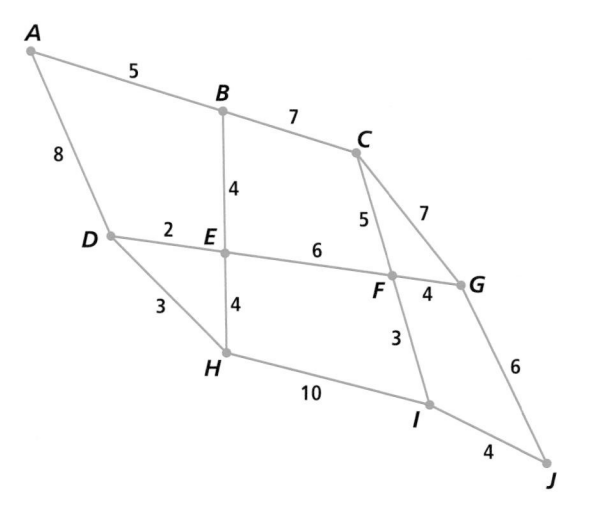

2 Using a DGS to Draw

In order to illustrate their dynamic capabilities, many DGS have animation capabilities. Draw the figure of a person on a DGS, and make him dance using the animation capabilities. Try to include as many features as you can in this person, as long as they move logically. For instance, if you give a person fingers, they should stay attached to his hand, if you give him irises, they should stay inside the eyes.

3 Circular Definitions in English

Suppose someone wished to write an English dictionary that had no circular definitions in it. Do you think such a thing is possible? If so, what are some things that person could do in order to remove the circular definitions? If not, why not? Prepare a debate with a friend discussing this topic.

4 String Art

Who says you can't create curves with points and line segments? Line designs form a basis for mathematical understanding of geometric shapes and relationships of points and segments. Attractive and sophisticated line designs can be produced and created using only a ruler, compass, protractor, pencil, and paper. Curves can be formed using *envelopes of lines*. Computers with DGS can be used to imitate this procedure. Create a line design and use your design to make it into string art!

5 Hamiltonian Circuits

In Lesson 1-3 you saw one special kind of path that may or may not exist in a network, and a method to determine if a given network did indeed have such a path. Another kind of special path is a *Hamiltonian circuit*.

a. Find out what a Hamiltonian circuit is. Draw seven networks that have a Hamiltonian circuit, and seven that do not.

b. Try to find methods to determine if a given network has a Hamiltonian circuit. Compare the methods you find with those of your classmates, to see which methods are most efficient.

c. Are any of the methods that you or your classmates found as fast as the one given in Lesson 1-3? Find out why this might be so.

Chapter 1 Summary and Vocabulary

This page lists new terms and phrases for this chapter. You should be able to give a general description and specific example of each. For starred terms (*), you should be able to give a good definition.

● Geometry is the study of visual patterns. The basic building blocks of these patterns are **points** and **lines.** Many conceptions of points and lines are studied in this chapter. In **discrete geometry,** points are dots and lines are collections of dots in a row. It is possible that between two dots, there are no other dots. In **graph theory,** points are nodes in networks and lines are arcs joining the nodes. Lines have only two points on them, and many lines can connect two points. In **Euclidean synthetic geometry,** points are locations and lines are the shortest paths between the locations. In **Euclidean plane coordinate geometry,** points are ordered pairs and lines are sets of ordered pairs (x, y) satisfying $Ax + By = C$.

● Because of circular reasoning, it is impossible to define all terms in any system. The terms left undefined in geometry are *point, line,* and *plane.* Therefore, any of the conceptions of point above might be possible. In this book we choose the assumptions that are in the **Point-Line-Plane Postulate** indicating which properties points and lines satisfy. These assumptions form a starting point for deducing properties of figures in Euclidean geometry. They apply also to Euclidean plane coordinate geometry.

● The properties of points and lines as ordered pairs and as locations are the same, so what you learned in algebra about points and lines can be used in geometry. For example, every nonvertical line is the set of ordered pairs (x, y) satisfying an equation of the form $y = mx + b$, where m is its slope and b its y-intercept. The intersection point of two nonparallel lines can be found by solving a system of equations.

Postulates, Theorems, and Properties

Point-Line-Plane Postulate (p. 32)	Distance Postulate (p. 40)
Unique Line Assumption (p. 32)	Uniqueness Property (p. 40)
Number Line Assumption (p. 33)	Distance Formula (p. 40)
Dimension Assumption (p. 33)	Additive Property (p. 40)
Line Intersection Theorem (p. 34)	Triangle Inequality Postulate (p. 47)

Vocabulary

1-1
coordinate
coordinatized
*distance between two points on a coordinatized line

1-2
plane coordinate geometry
horizontal, vertical, oblique line
standard form, slope-intercept form of an equation for a line
y-intercept, slope

1-3
pixels
discrete point, discrete line
discrete geometry
graph theory
arc, network, node
vertices, vertex
traversable network
even node, odd node

1-4
circularity
undefined terms
*figure, plane figure
*space
collinear, coplanar
one-dimensional
two-dimensional
three-dimensional

1-5
*postulate, *theorem
Euclidean geometry
*parallel lines

1-6
between
*segment, line segment
endpoints
*ray, *opposite rays

1-7
dynamic geometry system (DGS)
window, menu

Chapter 1 — Self-Test

1. At the beginning of a trip, the train conductor announced the total distance traveled would be 650 kilometers. After a while the conductor announces the train is 230 kilometers from its destination. How far has the train traveled?

2. Write an equation in slope-intercept form for the line passing through the points $(2, 2)$ and $(-1, -7)$.

3. Which of the assumptions of Euclidean geometry can be used to explain why Question 2 has only one solution?

4. Points A, B, and C are on a plane. What can be said about their positions if $AC = 2$, $AB = 6$, and $BC = 4$?

In 5-7, determine whether *segment*, *ray*, or *neither segment nor ray* best describes the graph of the given inequality on the number line.

5. $-5 < x \le 2$ 6. $y \ge 4$ 7. $3 \le z \le 6$

In 8 and 9, use the network at the right, which represents cities on a salesman's route and train lines connecting them.

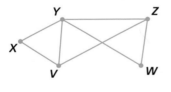

8. Is the network traversable?

9. If the network is traversable, what are the possible start and finish vertices for a traversing path? If not, explain why not.

10. Determine whether these numbers could be the lengths of the three sides of a triangle.
    a. 1, 2, 3   b. $\frac{13}{5}$, 4.37, $\sqrt{37}$ c. $\sqrt{5}$, 6, $\sqrt{10}$

11. A marching band forms the letter I at a game. What type of geometry creates this figure?

12. a. Graph $y = 2x - 4$ and $y = -2x + 4$ on the same coordinate grid.
    b. Do these lines intersect? If so, give their point of intersection.

13. In the number line below, P has coordinate –5 and T has coordinate 7. What is PT?

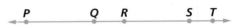

14. Using a DGS, construct points A and B and a circle going through B with center A. Measure AB. Place point C on the circle and measure AC.
    a. Try moving the point C around the circle. What happens to AC and why?
    b. Move B to change the size of the circle. What happens to AB and AC?

15. **True or False** In graph theory, it is not possible for a line to contain only one point.

16. At the right is a drawing of a pyramid. Name four coplanar points.

17. Find the intersection of the lines given by the system $\begin{cases} y = 2x - 8 \\ 2x + 2y = 0 \end{cases}$.

18. Find the coordinates for all the points on a number line whose distance from the point –1 is equal to 3.5.

19. a. Write the equation $3 - 2x = 5 + 3y$ in slope-intercept form.
    b. Graph the set of points satisfying the equation in the coordinate plane.

Chapter 1 Chapter Review

The Chapter Review questions are grouped according to the SPUR Objectives in this chapter.

SKILLS Procedures used to get answers

OBJECTIVE A Analyze networks.
(Lesson 1-3)

In 1–4, refer to the network at the right.

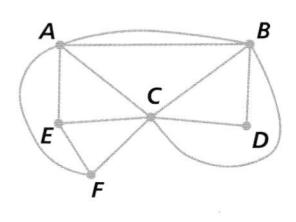

1. a. Which nodes are odd?
   b. Which nodes are even?
2. Is this network traversable? How do you know?
3. Which nodes are connected by more than one arc?
4. a. Give an example of two arcs that do not intersect.
   b. Give an example of two arcs whose intersection is one point.
   c. Give an example of two arcs whose intersection is two points.
5. If a network has exactly two odd nodes, A and B, what can you say about a traversable path in this network?
6. **True or False** Any network with an odd number of nodes is traversable.

OBJECTIVE B Use a DGS with points and distances. (Lesson 1-7)

In 7–9, use a DGS to draw three points: A, B, and C, and connect them with segments.

7. When you drag point C, which segments move as well? Which segment(s) do not move?

8. Display the distance BC. Where would you drag the point C so that $BC = 0$?
9. Construct another point, D. Have the DGS display $DA + DB + DC$. Now drag the point D. Does this quantity seem to be greater when D is outside or inside the triangle formed by A, B and C?

PROPERTIES Principles behind the mathematics

OBJECTIVE C Recognize and use geometric notation for one-dimensional ideas. (Lessons 1-5, 1-6)

In 10–13, use the diagram of the line $\overleftrightarrow{OP}$ below.

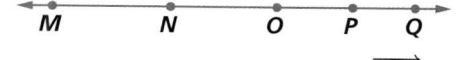

10. Give another name for the ray $\overrightarrow{ON}$.
11. Is $\overrightarrow{NO}$ the ray opposite to $\overrightarrow{ON}$?
12. Give a name for the set of points that is in both $\overrightarrow{NO}$ and $\overrightarrow{ON}$.
13. Name a point on $\overrightarrow{MQ}$ that is not between M and Q.
14. Let ℓ be the line given by $y = 2x + 6$ and m be the line given by $2y - 4x = 1$. Is it true that $\ell \parallel m$? Why or why not?

OBJECTIVE D Given a property of points and lines, tell whether it is true for each of the four descriptions of points. (Lessons 1-1, 1-2, 1-3, 1-5)

15. **True or False** In discrete geometry, between two points on a line there is a third point.
16. **True or False** Points in the coordinate plane have size.

17. Is it true in graph theory that between any two points there is at most one line? Why or why not?

18. Is it true in Euclidean geometry that through any three points, there is exactly one line? Explain your answer.

OBJECTIVE E Recognize the use of undefined terms and postulates. (Lessons 1-4, 1-5)

19. Write down the undefined geometric terms in this sentence: If three points in a plane are collinear, that is, if there is a line that contains all three of them, then one of the points is between the other two.

20. When defining terms, what does "circularity" mean?

In 21 and 22, consider the following paragraph: Given a line ℓ in a plane, there is a point P that is not on that line. Let A be a point on the line ℓ. The lines ℓ and $\overleftrightarrow{PA}$ intersect at point A. No other point on ℓ can be on $\overleftrightarrow{PA}$.

21. Which assumption of Euclidean geometry was used in the first sentence?

22. What theorems or postulates were used in the sentence "No other point in ℓ can be on $\overleftrightarrow{PA}$"?

OBJECTIVE F Apply properties of betweenness and the Triangle Inequality Postulate. (Lessons 1-6, 1-7)

23. Suppose $AB = 4$, $AC = 3$, and $BC = 7$. Which of the three points A, B, or C is between the other two?

24. Suppose $OP = PQ = QO = 2$. Is it possible for the three points O, P, and Q to be collinear?

25. Can 4, 7, and 10 be the lengths of the three sides of a triangle?

26. Point M is between L and N. $LN = 6$, $LM = 2$. Find MN.

USES Applications of mathematics in real-world situations

OBJECTIVE G Apply the definition of distance to real situations. (Lesson 1-6)

27. Lauren is at mile 237 on the highway and plans to exit at mile x.
   a. How many miles from her exit is she?
   b. If she knows she has 63 miles left before she exits, give two possibilities for the mile marker at her exit.

In 28–30, use the following data: Luna took a trip in her truck. When she left home, the odometer read 15,153 miles. When Luna returned home from her trip, the odometer read 15,343 miles.

28. How many miles had Luna traveled?

29. What did the odometer read at the halfway point of her trip?

30. If Luna made the entire trip by truck, is it possible for her trip to include a visit to a location that was 150 miles from her home? Why or why not?

OBJECTIVE H Use the various descriptions of points and lines to model real-world situations. (Lessons 1-1, 1-2, 1-3, 1-4)

31. a. The diagram below shows the floor plan for a museum. Use a network to represent the diagram; the nodes should be the different rooms and the arcs should be the doors connecting them.

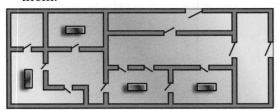

   b. At the end of the day, the curator wishes to walk through the museum, and shut every door without retracing his steps. Can this be done? Explain your answer.

32. A group of seven travelers is sharing a room in a youth hostel. Maurice, Beau, and Amber speak English. Beau, Maurice, and Gabrielle speak French. Heidi and Nicolas speak German. Nicolas, Amber, and Pablo speak Spanish.

   a. Use a network representing who can speak directly to whom.

   b. If Heidi wishes to ask Beau a question, into how many languages will her message have to be translated for Beau to understand it?

   c. If Nicolas leaves the room, will Beau be able to communicate with Heidi?

33. Nodin's car breaks down in the desert. Luckily he has a global positioning system (GPS), that tells him that he is 7 miles south and 8 miles west of Moab, Utah.

   a. Model this situation in the coordinate plane.

   b. Suppose Nodin wishes to walk to Moab in a straight line. Find an equation for this line.

34. The diagram at the right represents the positions of street lights in a small area of a city.

   a. What do you think the discrete lines represent?

   b. What do you think the intersections of the lines represent?

REPRESENTATIONS Pictures, graphs, or objects that illustrate concepts

OBJECTIVE I Determine distance on a number line. (Lessons 1-1, 1-6)

35. Use the number line to determine TW.

S	T	U		V	W	X
-41	-36	-31		-21	-16	-11

In 36 and 37, give the distance between two points with the given coordinates.

36. -12 and 36

37. x and $-y$

38. If C and D are on the number line, $C = 2$, and $CD = 3$, what could the coordinate of D be?

39. If $A = 2$, $B = 10$, and $AC = CB = 4$, what is the coordinate of the point C?

OBJECTIVE J Graph points and lines in the coordinate plane. (Lesson 1-2)

40. Graph the set of points satisfying $3y - 2 = 4$.

41. Graph $y = x$ and $x = 3$ on the same coordinate axes.

In 42 and 43, classify the line given by the equation as horizontal, vertical, or oblique.

42. $y = -2x + 7$

43. $x + y = y + 2$

OBJECTIVE K Write and graph equations of lines. (Lesson 1-2)

44. Write an equation for the line through the points $(-1, 4)$ and $(2, 5)$.

45. Write the equation $5x - 2y = -6$ in slope-intercept form.

46. Draw a graph for the equation $5x - 3y = -15$.

47. Write an equation for the vertical line through the point $(-4, -4)$.

OBJECTIVE L Find the point of intersection of two lines. (Lesson 1-5)

In 48 and 49, find the point of intersection of each pair of lines.

48. $\begin{cases} 3x - 4y = 2 \\ 3x + 8y = -1 \end{cases}$

49. $\begin{cases} y = 2x - 4 \\ 4x - 5y = -16 \end{cases}$

In 50 and 51, what is the number of solutions to the given system?

50. $\begin{cases} y = x - 1 \\ y = 3x + 5 \end{cases}$

51. $\begin{cases} y = 2x + 5 \\ y = 2x + 7 \end{cases}$

Chapter

2 The Language and Logic of Geometry

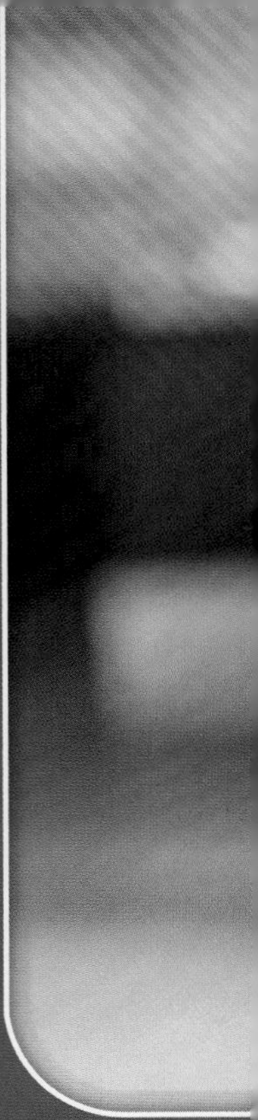

Contents

2-1 The Need for Definitions

2-2 Conditional Statements

2-3 Converses

2-4 Good Definitions

2-5 Unions and Intersections of Figures

2-6 Polygons

2-7 Conjectures

In 2005, Americans drank an average of approximately 60 gallons of soda. Since 1975 the average school-age child, from age 6 to age 17, has drunk more carbonated soft drinks than milk. Type 2 diabetes, historically considered an adult disease, is increasingly being reported among young American children. Many people feel that soft drinks are a factor in this increase.

By 2006, more than 27 states and the District of Columbia had responded to these data by enforcing "junk food" laws that required schools to restrict access to junk food in vending machines and cafeterias.

A debate has ensued over the *definition* of junk food. The United States Department of Agriculture (USDA) is responsible for, among other things, writing the dietary guidelines for the United States. The USDA limits its definition of junk food to sodas, water ices, chewing gum, and candies made mostly of sugar. The Federal Trade Commission (FTC) is responsible for protecting the public against unfair business practices. In order to place any controls on the advertising of junk food, the FTC would first have to define "junk food," but there are "no clear standards for doing this." Further, some states have formulated their own definitions of junk food. For example, New Jersey includes all items listing sugar as the first ingredient as junk food.

Some people have wanted a federal definition of "junk food" that would be the standard across the nation. As a result, some members of Congress have drafted a bill, called the Child Nutrition Promotion and School Lunch Protection Act, with the aim of standardizing the federal definition of "junk food" so that all schools across the United States would ban the same foods from their vending machines and cafeterias.

Careful definitions are also one of the features of mathematical reasoning. In this chapter, the language and logic of mathematics are applied to simple geometric figures, to some of the other mathematics you know, and to everyday situations.

Lesson 2-1

The Need for Definitions

Vocabulary

convex set

nonconvex set

▶ **BIG IDEA** Definitions of mathematical terms are explicitly stated both for clarity and because it is important in communication that people use the same meaning for a term.

You may wonder why a discussion on junk food was included in a geometry book. The reason is that it points out the need for careful definitions. Careful definitions are necessary in almost every field and are found throughout mathematics. When ideas are not carefully defined, people may not agree with what is written about those ideas. For example, astronomers recently clarified the definition of the word *planet*. Their reason for rethinking the definition centered around Pluto. Pluto had been called a planet since its discovery in 1930. However, other astronomical bodies revolving around our Sun were discovered that have characteristics similar to that of Pluto, and at least one of these bodies is larger than Pluto. In August 2006, the International Astronomical Union reclassified Pluto as a *dwarf planet,* separated from the eight *classical planets.*

There have been many legal cases that focus on the precise definition of a word. In the 1980s, Procter & Gamble won a $125,000,000 settlement based on its patented definition of the term "dual-textured cookie." There are terms that we use in everyday life such as *love, success,* and *tyranny* that take on different meanings based on the parties involved. Do you think everyone in your class would agree on the definition of *love,* of *planet,* of *triangle,* or of *circle*?

Mental Math

Suppose 4 laps around a track is a distance of one mile.

a. How many laps equal a half mile?

b. How many miles is one lap?

c. How many laps equal 3.5 miles?

Clyde Tombaugh could not afford to go to college immediately after high school, but that did not stop him from discovering Pluto at age 24.

Source: Academy of Achievement

GUIDED

Example 1

What *rectangles* appear to be in the figure at the right? Name your choices before reading on. Explain why you made your choices.

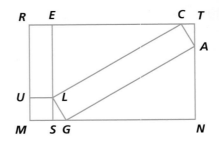

Solution In previous courses, you learned that a rectangle is a four-sided figure with four right angles. So, one strategy is to start by naming all the four-sided figures, then analyze each one to determine if it has four right angles. Figures are named by stating their vertices in order. You may start at any vertex and proceed clockwise or counterclockwise.

The four-sided figures with one vertex at R are RCLU, RTNM, ___?___, and ___?___.

The other four-sided figures are ___?___, ___?___, ___?___, and ___?___.

The four-sided figures that appear to be rectangles are ___?___.

STOP **QY1**

It is impossible to decide which figures in Guided Example 1 are rectangles without a careful definition of a rectangle. Some people might argue that *RELU* is too long to be a rectangle. Others might argue that *ULSM* is a square and so it cannot be a rectangle. Some might say *LGAC* is on a slant, and hence cannot be a rectangle.

Our view, which agrees with the common mathematical definition, is that *RELU, ULSM,* and *LGAC* each appear to be rectangles because each appears to have four right angles. All squares are rectangles. A rectangle may be tall. A tilted rectangle is still a rectangle. Of the four-sided figures listed, only *RCLU* and *ULGM* do not appear to be rectangles because they do not appear to contain four right angles.

We are careful to say that these figures *appear* to be rectangles because we cannot be sure that the angles are right angles by just looking at them. Right angles cannot be assumed from a diagram. When we want to assert that an angle is a right angle, we put a small square by its vertex. The symbol ⌐ in a diagram indicates that an angle is a right angle.

Defining Geometric Terms

You have seen and used many geometric terms in previous courses. This has helped you gain some geometric intuition that will be very valuable to you in this course. However, in this book, geometry terms are carefully defined. Although your instincts will generally lead you in the right direction, learning the precise definition of a term is essential to the logical development of geometry.

▶ **QY1**

Is *RTMN* another name for *RTNM*? Explain.

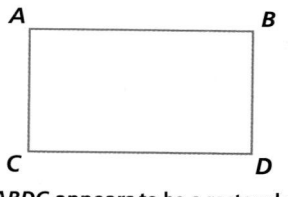

ABDC **appears** to be a rectangle

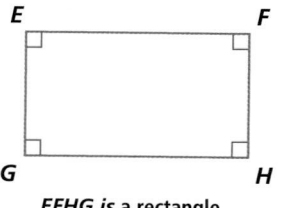

EFHG **is** a rectangle.

Six figures are drawn below. All of the figures share certain characteristics. They all have straight sides, and each one encloses a region. These are characteristics of *polygons*. Polygons are not limited to three or four sides. At the right are two different seven-sided polygons, called *heptagons*.

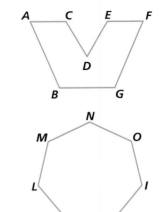

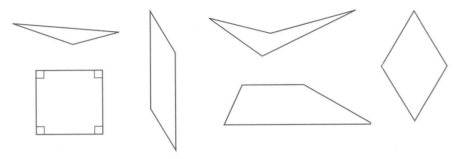

 QY2

Yet these drawings do not tell us exactly what a polygon is. We have not defined the word *polygon* yet!

▶ QY2

Name each of the two heptagons above in two ways.

Activity 1

Which figures below do you think appear to be polygons? Why?

A   B  C   D   E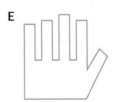

Your answers may be different from those of others. But, in this chapter, you will learn vocabulary for the careful definition of *polygon* that will be used throughout this book. These definitions will help you answer questions like the one in Activity 1. They will also serve as a building block for terms such as *triangle, quadrilateral,* and *pentagon*.

Convex Sets

One way to categorize polygons is by using the terms *convex* and *nonconvex*. The term *convex* distinguishes between sets of points that have "dents" and those that do not. Sets of points that do not have dents in them are called *convex sets*.

▶ **READING MATH**

The word *convex* has the same meaning when it describes lenses. Lenses that are not convex are called *concave*.

convex lens

concave lens

Definition of Convex Set

A **convex set** is a set of points in which every segment that connects points of the set lies entirely in the set.

A set that is not convex is called, quite appropriately, a **nonconvex set**. Because a figure is just a set of points, we can apply the definition of a convex set to figures.

Example 2

Determine if the figures at the right are convex or nonconvex and justify your answer.

Solution Figure I is ___?___ because all segments connecting points in the figure lie in the figure.

Figure II is nonconvex because ___?___.

Figure III is ___?___ because ___?___.

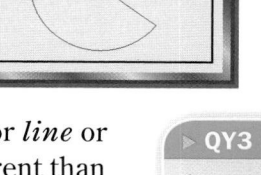

Figure I   Figure II   Figure III

STOP QY3

Activity 2

MATERIALS DGS

Construct a nonconvex figure similar to the one at the right. Describe the region where point A can be dragged so that the new figure will be convex.

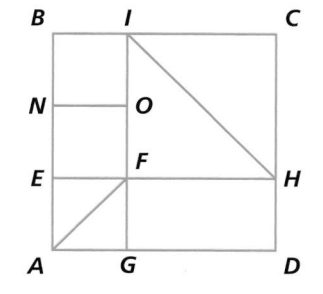

Caution: Many words used in mathematics, such as *angle* or *line* or *plane,* have non-mathematical meanings that are quite different than their mathematical meaning. The definitions of terms we give in this book apply only to the mathematical meanings of the terms.

▶ QY3

A teacher places a star ☆ on your paper. Is the star a convex or nonconvex region?

Questions

COVERING THE IDEAS

1. Write your own definition of *junk food*.

2. How much did Procter & Gamble win in its settlement about its patented definition of the term "dual-textured cookie"?

3. Give a reason why Pluto was reclassified as a *dwarf planet*.

4. Give an example of a word that might take on different meanings for different people.

5. **True or False** Right angles can be assumed from a diagram.

6. Consider the figure at the right.
   a. List all the 4-sided figures.
   b. Which of the 4-sided figures appear to be *squares*?

7. Draw a figure that has some, but not all, properties of a rectangle. Explain why your drawing is not a rectangle.

8. In Activity 1, the block letter K is considered as a figure. Draw two other block letters, one you would classify as a polygon and one you would not. Explain why you chose your figures.

9. Why are careful definitions needed in geometry?

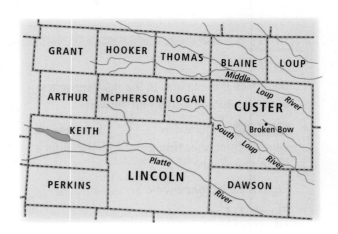

In 10 and 11, use the portion of the map of the state of Nebraska at the right.

10. Name a county that is convex.

11. Name a county that is nonconvex.

APPLYING THE MATHEMATICS

12. Draw a nonconvex 10-sided figure.

13. **Multiple Choice** The word *midpoint* is carefully defined in Lesson 2-4. But before reading that lesson, in which of the following pictures do you think point J is the midpoint of $\overline{MC}$?

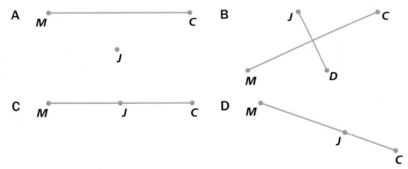

14. During lunch, a hungry student stares at a friend about to chomp down on a tasty looking carrot. "Here," says his friend, breaking the carrot, "have half." Do you think each piece should be called a "half carrot"? Why or why not?

15. Which of the following figures appears to be a circle? For each figure you think to be a circle, explain why.

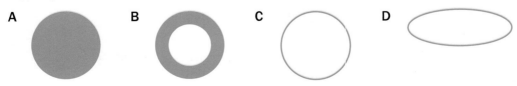

16. In parts of New York City, there are avenues and streets. The avenues run north/south and the streets run east/west. The length of a city avenue block is $\frac{1}{20}$ mile. The length of a city street block is $\frac{1}{5}$ mile. If Jasmin walks 2 blocks north, then 2 blocks east, then 2 blocks south, then 2 blocks west, she will end up where she started.

    a. Do you think the path she traveled should be called a rectangle? Why or why not?

    b. Do you think it should be called a square? Why or why not?

17. When the points (2, 3), (2, 5), (4, 5), (4, 3), and (2, 3) are connected in order, a square is formed. Give an example of a list of points that you think do not form a polygon when connected in order.

18. A cube is drawn at the right. The points G, E, O, and M are connected in order. They form a four-sided figure $GEOM$ in which each angle is a right angle. Should $GEOM$ be called a rectangle? Explain your answer.

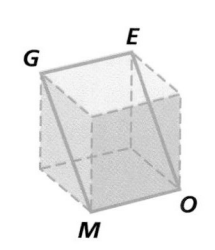

REVIEW

19. In the figure at the right, is it possible to have $AD = 10$, $AB = 5$, $BC = 1$, and $CD = 8$? If the answer is no, explain why not. (**Lesson 1-7**)

20. Triangle ABC has sides whose lengths are all integers. Suppose $AB = 4$ and $BC = 6$. What are the possible values for AC? (**Lesson 1-7**)

21. On a number line, point A has coordinate 0 and point B has coordinate 1. (**Lesson 1-6**)

    a. Use an inequality to describe $\overrightarrow{AB}$.

    b. Use an inequality to describe the ray opposite to $\overrightarrow{AB}$.

22. What is the definition of *parallel lines*? (**Lesson 1-5**)

23. Solve the inequality $2z + 3 < 4$. (**Previous Course**)

EXPLORATION

24. a. Use a reliable source to find definitions for the following terms as they are used in astronomy: *asteroid, planetoid, centaur, trans-Neptunian,* and *solar system.*

    b. Use the definitions to complete the chart with the above terms from most general to most specific:

    c. Where would Pluto lie in this chart?

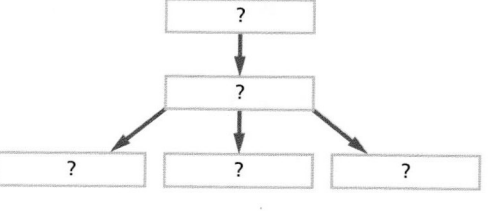

Lesson 2-2
Conditional Statements

Vocabulary

statement

compound statement

conditional statement

antecedent

consequent

instance of a conditional

counterexample to a
    conditional

▶ **BIG IDEA** Conditional statements, those which can be put in the form "If ..., then ...," are the basis of logical thinking in mathematics.

Conditional Statements

A **statement** is a sentence that is either true or false and not both. "My dog has fleas" is a statement because either the dog has fleas or it doesn't. A **compound statement** is a sentence that combines two or more statements with some type of *connective* such as *and, or,* or *if ... then.* "My dog has fleas and fleas make dogs itch" is an example of a compound statement. If a compound statement can be written in *if ... then* form, then the statement is called a **conditional statement**. Below are several examples of conditional statements.

> If the light is red, then you must stop.
>
> If $x = y$, then $x^2 = y^2$.
>
> If you live in Springfield, then you live in Illinois.
>
> A figure is a polygon if it is a triangle.
>
> You can get your driver's license if you pass the test.
>
> If I move my queen here, then you will take it with your rook.

The two statements that make up a conditional statement have special names. The statement that follows "if" is called the **antecedent**, while the other statement, which often follows "then" is called the **consequent**. (Alternate names for antecedent and consequent are *hypothesis* and *conclusion,* respectively.)

> If <u>the light is red</u>, then <u>you must stop</u>.
>   *antecedent*   *consequent*

Both the antecedent and consequent alone would be complete sentences if they began with a capital letter and ended with a period. Notice that this statement can be written as follows without losing any meaning.

> <u>You must stop</u> when <u>the light is red</u>.
>   *consequent*   *antecedent*

The queen has not always been the strongest chess piece, nor has it always been a queen. When chess first spread to Europe, the piece next to the king was the *ferz* and was the second-weakest piece on the board.

Example 1

Write the antecedent and consequent of the following conditional:
If two unique lines intersect, then they intersect at exactly one point.

Solution The antecedent follows the word *if*. Antecedent:
Two unique lines intersect. The consequent follows the word *then*.
Consequent: They intersect at exactly one point.

 QY1

Rewriting Sentences as Conditionals

There are sentences that have the same meaning as a conditional,
but do not contain the words *if* or *then*. Some that you have seen in
this book are:

> Given any two points, there is exactly one line containing them.
>
> The slope of the line containing points (x_1, y_1) and (x_2, y_2) is $\frac{y_2 - y_1}{x_2 - x_1}$.

These statements do not contain *if* or *then,* but they can easily be
rewritten as conditional statements:

> If there are two points, then there is exactly one line
> containing them.
>
> If the points (x_1, y_1) and (x_2, y_2) are on a line, then the slope
> of that line is $\frac{y_2 - y_1}{x_2 - x_1}$.

In addition, sometimes statements follow the pattern *All A are B,*
or *Every A is a B.* These statements can be rewritten as the
conditional *If something is an A, then it is a B.* In geometry, often
the "something" is "a figure."

STOP QY2

Abbreviating Conditionals

Sometimes it is easier to abbreviate a conditional by replacing the
statements with letters. We will always use lower case letters to do
this. Consider the statement, *If I move my queen here, you will
take it with your rook.* You can let $q = $ *I move my queen here*
and $r = $ *you will take it with your rook.* Now you can rewrite the
statement as *If q, then r.* A still shorter way of writing this is $q \Rightarrow r.$
The symbol "$\Rightarrow$" means "implies" and takes the place of "if ... then."
Thus, the statement "$q \Rightarrow r$" is read "q implies r."

STOP QY3

▸ **QY1**

Write the antecedent
and consequent of the
statement: *You can get
your driver's license if you
pass the test.*

▸ **QY2**

Rewrite the following
statement as a
conditional: *All segments
are convex sets.*

▸ **QY3**

Let $g = $ *you understand
geometry,* and let
$m = $ *you are a budding
mathematician.* Rewrite
the statement $g \Rightarrow m$
in English.

Instances of Conditionals

Consider this statement: *If you live in Springfield, then you live in Illinois.* Suppose Oscar lives in Springfield, Illinois. Oscar lives in Springfield, so the antecedent is true. He also lives in Illinois, so the consequent is true. Because Oscar is an example of someone who makes both the antecedent and the consequent true, we call him an *instance of the conditional.*

> ### Definition of Instance of a Conditional
>
> An **instance of a conditional** is a specific case in which both the antecedent (*if* part) and the consequent (*then* part) of the conditional are true.

Truth and Falsity of Conditionals

A conditional statement is *true* if, for every possible case in which the antecedent is true, the consequent is also true. It is often hard to show that a conditional is true because you must show that the conditional holds for *all* cases in which the antecedent is true.

> ### Example 2
> Oscar says that because he lives in Springfield, Illinois, the above conditional statement, *If you live in Springfield, then you live in Illinois,* is true. Is Oscar correct? Justify your answer.
>
> **Solution** No. Oscar is one instance of the conditional. To prove the conditional true, he would have to show that everyone from a place called Springfield is also from Illinois.

To prove that a conditional statement is *false,* all you need is one instance in which the antecedent is true, but the consequent is false. This instance is a *counterexample to the conditional.*

> ### Definition of Counterexample to a Conditional
>
> A **counterexample to a conditional** is a specific case for which the antecedent (*if* part) of the conditional is true and its consequent (*then* part) is false.

In Example 2, one person from Springfield, Massachusetts, (or any of the 33 cities named Springfield in the United States alone that are not in Illinois) would serve as a counterexample to the conditional statement.

According to the 2000 U.S. Census, about 152,000 of the people who lived in a place called Springfield lived in Springfield, Massachusetts.

That person lives in Springfield, so the antecedent is true, but the person does not live in Illinois, so the consequent is false. Notice that to prove a conditional statement is true requires you to show that it always works in all cases, but proving it false requires only one counterexample!

STOP QY4

Venn Diagrams

A helpful way to analyze a conditional statement is to model it using a *Venn diagram*. Venn diagrams show relationships among sets.

At the right is a Venn diagram that shows the relationship between squirrels and things that cannot drive cars. It shows that the set of squirrels is a *subset* of the set of things that cannot drive cars because the set of squirrels is completely contained within the set of things that cannot drive cars. From this diagram we can determine that the statement *If you are a squirrel, then you cannot drive a car* is true. This is because all squirrels are within the inner circle, which also means they are within the outer circle. However, the statement *If you cannot drive a car, then you are a squirrel* is not true (obviously). Using the Venn diagram, we can see this because if an item is in the outer circle, it is not necessarily in the inner circle.

STOP QY5

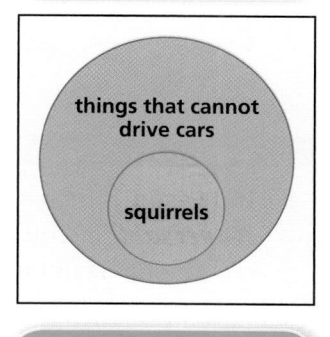

Example 3

Use the Venn diagram at the right to create one true conditional statement and one false conditional statement about fictitious "wibbles" and "squibbles."

Solution

True Statement: If it is a squibble, then it is a wibble.
False Statement: If it is a wibble, then it is a squibble.

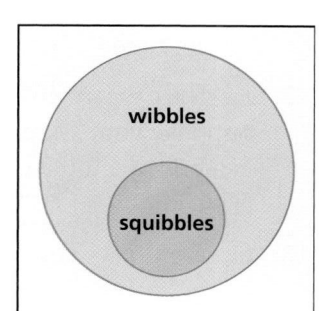

Questions

COVERING THE IDEAS

1. **Fill in the Blank** An if-then statement is commonly called a ___?___ statement.

2. **Fill in the Blanks** The clause following *if* in a conditional statement is called the ___?___, while the clause following *then* is called the ___?___.

3. Let w = *today is Wednesday* and t = *tomorrow is Thursday*.
   a. Rewrite $w \Rightarrow t$ in English.
   b. What is the antecedent of the conditional?
   c. What is the consequent?

4. What must you do to prove that a conditional statement is true?

5. What is a *counterexample*?

6. Rewrite this statement as a conditional: *When you add two even numbers, the sum is another even number.*

7. Use the Venn diagram at the right to write a true conditional statement.

APPLYING THE MATHEMATICS

8. In Lesson 1-3, there is the following statement: *A network is traversable when it has two or zero odd nodes.* Write this as a conditional statement.

9. Rewrite the Unique Line Assumption Postulate from Lesson 1-5 as a conditional statement.

10. Consider the following statement: *If $x^2 = 16$, then $x = 4$.* Provide a counterexample to this conditional.

11. Consider this conditional: *If today is the 30th day of the month, then tomorrow is the 1st day of the next month.*
    a. One instance of this conditional is June 30. Give all other instances of this conditional.
    b. Provide a counterexample that shows why this conditional is not true.

12. A *prime* is any number whose only factors are 1 and itself. Suppose your friend says that if n is an even integer, then $n^2 + 1$ is a prime. Find a counterexample that shows that your friend is wrong.

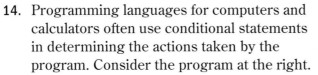

13. Create three true conditional statements about the fictitious animals shown in the Venn diagram at the right.

14. Programming languages for computers and calculators often use conditional statements in determining the actions taken by the program. Consider the program at the right.

```
Display "Are you in a geometry class?"
Display "Yes/No"
Prompt x
If x = "Yes": Display "Good for you!"
Else: Display "I'm sorry!"
```

   a. In the program, what happens if you answer the question with a "Yes"?
   b. What happens if you answer the question in any other manner?

REVIEW

15. Characterize the shaded set of points as convex or nonconvex. (**Lesson 2-1**)

a.

b.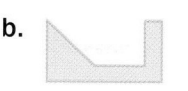

c.

16. Is a ray a convex figure? (**Lesson 2-1**)

17. Do you think a ray should be considered a polygon? Why or why not? (**Lesson 2-1**)

18. In how many points do the lines with equations $2y + 3x = 4$ and $y = -1.5x + 7$ intersect? (**Lesson 1-5**)

19. Describe the set {–1, 0, 1, 2, 3, . . .} in set-builder notation. (**Previous Course**)

EXPLORATION

20. Consider the following statement: *If n is a positive integer, $n^2 + n + 41$ is a prime number.*

    a. Choose two different values for n to see if you get a prime number.

    b. Does your answer for part a prove that the conditional statement is true?

    c. Explain why $n = 41$ is a counterexample.

    d. Find another counterexample.

21. A Rube Goldberg machine is a machine or contraption that performs a very simple task in a very complex way. Below is a diagram of one such machine.

    a. Write a conditional statement that summarizes the actions of the machine.

    b. What are the antecedent and the consequent of the conditional statement that you wrote?

QY ANSWERS

1. The antecedent is *you pass the test.* The consequent is *you can get your driver's license.*

2. If a figure is a segment, then it is a convex set.

3. If you understand geometry, then you are a budding mathematician.

4. You go outside when it is not raining and you do not get wet.

5. Answers vary. Sample: A baby is not a squirrel and a baby cannot drive a car.

Lesson

2-3

Converses

Vocabulary

converse

▶ **BIG IDEA** Every conditional statement has a converse, found by switching its antecedent and consequent, and the converse may or may not be true even if the conditional is true.

Here is a true conditional.

If you own a Chihuahua, then you own a dog.

Let $c = $ *You own a Chihuahua*, and let $d = $ *You own a dog*. Then the conditional can be written in symbols as $c \Rightarrow d$.

This conditional is true because every Chihuahua is a dog. Thus, by satisfying the antecedent (you own a Chihuahua), you automatically satisfy the consequent (you own a dog). Now examine the conditional statement shown below.

If you own a dog, then you own a Chihuahua.

This conditional can be written symbolically as $d \Rightarrow c$.

 QY1

The conditional statements $c \Rightarrow d$ and $d \Rightarrow c$ are called *converses*. In the converse of a statement, the antecedent and the consequent switch places.

Definition of Converse
The **converse** of $p \Rightarrow q$ is $q \Rightarrow p$.

Notice that both the original statement and its converse are conditional statements, and that they are converses of one another.

 QY2

Mental Math

a. If there are exactly 34 states in the U.S. with a city or town called "Springfield," how many states do not have a town called "Springfield"?

b. What percentage of states in the U.S. have a city or town called "Springfield"?

▶ **QY1**

Provide a counterexample to prove that $d \Rightarrow c$ is false.

▶ **QY2**

Write the converse of the statement: *If you live in Kentucky, then you live in the United States.*

Example 1

Consider this conditional statement: *If $x = 4$, then $x + 2 = 6$.*

a. Write the converse of the statement.

b. Is the statement true?

c. Is the converse true?

Solution

a. The antecedent follows *if* and the consequent follows *then*. We switch these to make the converse. If $x + 2 = 6$, then $x = 4$.

b. Yes, it is true, since $4 + 2 = 6$.

c. Yes, the converse is true. Solving $x + 2 = 6$ gives $x = 4$.

The Truth or Falsity of Converses

As you can see in the Chihuahua and equation examples, knowing that a statement $p \Rightarrow q$ is true does not tell you whether the converse $q \Rightarrow p$ is true or false. Any combination is possible.

GUIDED

Example 2

Let $s = $ *A figure is a square.* Let $r = $ *A figure is a rectangle.* The Venn diagram at the right shows the relationship between the figures that make s true and the figures that make r true.

a. Write $s \Rightarrow r$ in words. Is $s \Rightarrow r$ true according to the Venn diagram?

b. Write the converse of $s \Rightarrow r$ in words. Is it true according to the Venn diagram?

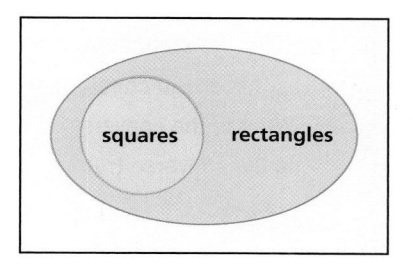

Solution

a. If a figure is a ___?___, then it is a ___?___. This statement is ___?___ according to the Venn diagram.

b. If a figure is a ___?___, then it is a ___?___. This statement is ___?___ because there are many rectangles that are not squares.

Example 3

Let p be the statement $|x| < 5$. Let q be the statement $|x| \leq 4$.

a. Draw the graph of each inequality on a separate number line.

b. Write $p \Rightarrow q$.

c. Is $p \Rightarrow q$ true? Explain your reasoning.

d. Write the converse of your statement from Part b.

e. Is the converse true? Explain why or provide a counterexample.

Solution

a. $|x| < 5$ means the distance from x to 0 is less than 5. The graph for $|x| < 5$ is:

$$-5 \ -4 \ -3 \ -2 \ -1 \ 0 \ 1 \ 2 \ 3 \ 4 \ 5 \quad x$$

(Continued on next page)

$|x| \leq 4$ means the distance from x to 0 is less than or equal to 4. The graph for $|x| \leq 4$ is:

b. The statement $p \Rightarrow q$ is "If $|x| < 5$, then $|x| \leq 4$."

c. Any number satisfying $4 < x < 5$ or $-5 < x < -4$ serves as a counterexample to p $\Rightarrow$ q. A few of the infinitely many counterexamples that are possible are -4.9, 4.2, and 4.81. So p $\Rightarrow$ q is not true.

d. The converse is "If $|x| \leq 4$, then $|x| < 5$."

e. The graphs show that when a number is between -4 and 4, it must also be between -5 and 5. So the converse is true.

Example 4

Consider the statement: *If $n = 3$, then $n^2 - 5n + 6 = 0$.*

a. Is this a true conditional statement?

b. What is the converse?

c. Is the converse true?

Solution

a. Remember that a conditional is true if every time the antecedent is true, so is the consequent. $n = 3$ is true only when n is 3. The original statement is true because $3^2 - 5 \cdot 3 + 6 = 0$.

b. The converse is "If $n^2 - 5n + 6 = 0$, then n = 3."

c. Factor the quadratic trinomial. $n^2 - 5n + 6 = 0$.
$$(n - 3)(n - 2) = 0$$
Now find the values of n that make the product equal to zero by solving two linear equations.
$$n - 3 = 0 \text{ or } n - 2 = 0$$
$$\text{So } n = 3 \text{ or } n = 2.$$
Both 2 and 3 satisfy this equation, so there are two values of n that make the antecedent true. Thus, just because the antecedent ($n^2 - 5n + 6 = 0$) is true, it does not mean that n must be 3. There is a value of n for which $n^2 - 5n + 6 = 0$ that is not 3, so the converse is not true.

Questions

COVERING THE IDEAS

1. Define *converse*.

2. What is the converse of the statement $d \Rightarrow c$?

In 3 and 4, the original statement is true.

    a. Write the converse.

    b. If the converse is true, explain why. If it is false, provide a counterexample.

3. If $x > 4$, then $x > 2$.

4. If $AB + BC = AC$, then B is on $\overline{AC}$.

5. **a.** Graph the solutions to the inequality $|x| \leq 3\frac{1}{3}$ on a number line.

    b. Write a true statement using $|x| \leq 3\frac{1}{3}$ as the antecedent.

    c. Write a true statement using $|x| \leq 3\frac{1}{3}$ as the consequent.

6. Erin claims that if two squares have equal perimeters, then they have equal areas.

    a. Is she correct?   **b.** Is the converse also true?

7. Consider this statement: *If $x^2 = 121$, then $x = 11$.*

    a. Is the statement true?

    b. Write the converse.

    c. Is the converse true?

APPLYING THE MATHEMATICS

In 8 and 9, a statement is given.

    a. Identify the antecedent and the consequent.

    b. Write the converse of each statement.

8. It is impossible to traverse a network if it has three odd nodes.

9. When two different lines intersect, they do so at one point.

10. Let m be the statement $x > -1$. Let n be the statement $x > -2$.

    a. Explain why it is true that $m \Rightarrow n$.

    b. Write the converse of $m \Rightarrow n$ in words, and then provide a counterexample to show why the converse is false.

11. Consider as true this statement: *All fleepers are beepers.*

    a. Draw a Venn diagram that could represent this relationship.

    b. Create a true conditional statement about fleepers and beepers.

    c. Write the converse of your conditional statement. In order for the converse to be true, what would have to be true about fleepers and beepers?

12. Decide if the statement is true or false. If the statement is false, provide a counterexample.

    a. If $x > y$, then $x + 2 > y + 2$.   **b.** If $x > y$, then $2x > 2y$.

    c. If $x > y$, then $-2x > -2y$.   **d.** If $x > y$, then $x^2 > y^2$.

13. Is the converse of the following statement true?

    If C is between A and B, then C is on $\overrightarrow{AB}$.

14. Create a true conditional statement for which *a network has two odd nodes* is the antecedent, and whose converse is not a true statement.

15. Assume for this problem that the following statement is true:
If you work hard, then you will become wealthy.
Suppose Nathan is wealthy. Can we assume that he worked hard? Why or why not?

REVIEW

16. Write the antecedent and the consequent in René Descartes's famous statement, "I think, therefore I am." **(Lesson 2-2)**

17. Is it possible to find a counterexample to the statement below? Explain your answer.

    If x is a number such that x < 5 and x > 7, then x = 13. **(Lesson 2-2)**

18. Give a counterexample to the statement *Every two lines are parallel.* **(Lesson 2-2)**

19. **a.** Can the numbers 1, 1, and 3 be the lengths of three sides of a triangle?

    b. Can the numbers 1, 1, and 3 be the lengths of three out of the four sides of a 4-sided figure? **(Lessons 2-1, 1-7)**

20. A triangle has vertices $(0, 4)$, $(3, 0)$, and $(3, 4)$. The three sides of the triangle lie on 3 lines. Find equations for these lines. **(Lesson 1-2)**

René Descartes (1596-1650)

EXPLORATION

21. For this problem, assume that all of these statements are true:

    If I am late for work, then I will be fired.
    If I oversleep, then I will miss the bus.
    If my alarm does not go off, then I will oversleep.
    If I miss the bus, then I will be late for work.

    a. Create a Venn diagram that displays the relationship between all of the antecedents and consequents.

    b. Create a conditional statement that links the first event to the last event.

QY ANSWERS

1. Answers very. Sample: Any person who owns a Golden Retriever but not a Chihuahua would be such an example.

2. If you live in the United States, then you live in Kentucky.

Lesson 2-4

Good Definitions

Vocabulary

midpoint

if and only if

biconditional

circle

radius, radii

diameter

▶ **BIG IDEA** Good definitions are clear, accurate, short, and contain only undefined terms or terms that have already been defined.

Properties of a Good Definition

In this book, you have seen good definitions for several words, including *convex* and *segment*. A good definition satisfies three properties.

    I   It uses only words either commonly understood, defined earlier, or purposely undefined.

    II It accurately describes what is being defined.

    III It includes no more information than is necessary.

In this lesson, you will see other definitions and examine definitions in more detail.

Mental Math

a. How many different vertical lines pass through the origin?

b. How many different horizontal lines pass through the origin?

c. How many different oblique lines pass through the origin?

A Good Definition for the Midpoint of a Segment

Consider the term *midpoint*. We want *midpoint* to refer to a special point on the segment that divides it into two parts of equal length. In a flight over water, if there is trouble before an airplane reaches the midpoint of its flight path, the plane goes back to its point of origin. If there is trouble after it passes that point, the plane goes on to its destination. If the pilot and the navigator have different ideas of what *midpoint* means, there will be confusion. Here is a good definition of midpoint.

Definition of Midpoint

The **midpoint** of a segment $\overline{AB}$ is the point M on $\overline{AB}$ with $AM = MB$.

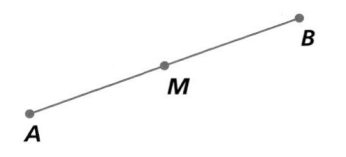

Notice how this definition satisfies the properties of a good definition. The definition uses only words that are commonly understood (*is, the, on, equal*), defined earlier (*segment, measure*), or purposely undefined (*point*). The definition is clear. The type of item being defined is specified; it is a point. In the figure at the right, $AC = CB$ as noted by the red tick marks. But C is not on $\overline{AB}$, so even though $AC = CB$, C cannot be the midpoint. The definition also identifies how the midpoint is different from other points on the segment. The midpoint is different because it is the only point that divides the segment into two parts of equal length. Finally, there is not extra information like "$AM = \frac{1}{2}AB$" in the definition. Writing "M is on $\overline{AB}$ and $AM = MB$" is sufficient to identify what is meant by *midpoint*.

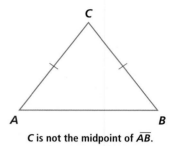

C is not the midpoint of $\overline{AB}$.

GUIDED

Example 1

Why is each of these not a good definition of midpoint?

a. The midpoint of $\overline{AB}$ is the point between A and B.

b. The midpoint is the ideal location to minimize commuting to A and B.

c. The midpoint of $\overline{AB}$ is a point on $\overline{AB}$ that makes $AM = MB$, $AB = 2AM$, and $AB = 2MB$.

Solution

a. The statement is not clear because it does not distinguish the midpoint from other points that lie between A and B. (It violates Property __?__.)

b. The statement uses many terms neither previously defined nor understood. (It violates __?__.)

c. The statement contains too much information. (It violates __?__.)

Definitions as Biconditionals

A definition usually gives a short name ("the midpoint of $\overline{AB}$") for something with a longer name ("the point M on $\overline{AB}$ with $AM = MB$"). The two names are completely interchangeable, so a good definition can be written as a conditional and its converse, both of which are true. Thus, a good definition can be expressed symbolically as:

$$p \Longrightarrow q \text{ and } q \Longrightarrow p.$$

This is abbreviated $p \Longleftrightarrow q$, read "p **if and only if** q." Because $p \Longleftrightarrow q$ is the combination of two conditionals, it is called a **biconditional**. For instance:

Let p be statement "D is between A and B."
Let q be the statement "$AD + DB = AB$."

Then $p \Longleftrightarrow q$ is the biconditional "D is between A and B if and only if $AD + DB = AB$." This is the geometric definition of *between*.

Every good definition can be written as a true biconditional. Below are the two true conditionals that make up the definition of midpoint. The first conditional goes from the defined *term* to tell you its *characteristics*.

$$\underline{\text{If } M \text{ is the midpoint of } \overline{AB}}, \text{ then } \underline{M \text{ is on } \overline{AB} \text{ and } AM = MB}.$$
$$\qquad\quad \text{term} \qquad\qquad \Longrightarrow \qquad \text{characteristics}$$

The second conditional in a good definition starts with the characteristics that allow you to use the defined term. It is the converse of the first conditional.

$$\underline{\text{If } M \text{ is on } \overline{AB} \text{ and } AM = MB}, \text{ then } \underline{M \text{ is the midpoint of } \overline{AB}}.$$
$$\qquad \text{characteristics} \qquad\qquad \Longrightarrow \qquad\quad \text{term}$$

Written as a biconditional, the definition of midpoint is as follows:

M is the midpoint of $\overline{AB} \Longleftrightarrow M$ is on $\overline{AB}$ and $AM = MB$.
M is the midpoint of $\overline{AB}$ if and only if M is on $\overline{AB}$ and $AM = MB$.

A Good Definition of a Circle

We use the "if-and-only-if" form of a definition when we want to stress the two directions of a definition. However, most of the time we use the word *is* or *means* between the word being defined and its characteristics because it is shorter.

Consider a familiar figure, the circle. Here is a good definition of circle.

Definition of Circle, Radius

A **circle** is the set of all points in a plane at a certain positive distance, its **radius**, from a certain point, its center.

The definition of a circle can be reworded to describe each point on it. The circle with center C and radius r is the set of all points P in a plane with $PC = r$. A circle with center C is often called "circle C," which can be written as "$\odot C$."

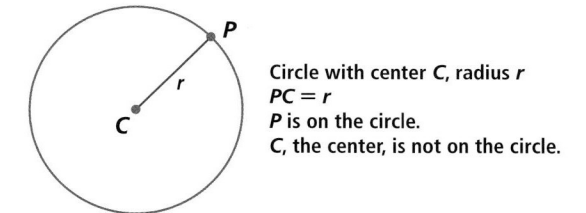

Circle with center C, radius r
$PC = r$
P is on the circle.
C, the center, is not on the circle.

> **READING MATH**
>
> As in English, terms in mathematics may have more than one meaning. The term *radius* can mean either the *distance* from the center to a point on a circle or any segment connecting the center with a point on the circle. A circle has only one radius (the distance) but infinitely many **radii** (the segments). Likewise, the term **diameter** can mean either any segment connecting two points on a circle and containing its center or the length of that segment, which is twice the radius.

Example 2

a. Write the conditional part of the definition of circle in the direction *term* ⟹ *characteristics.*

b. Write the conditional part of the definition of circle in the direction *characteristics* ⟹ *term.*

c. Write the definition of a circle as a biconditional.

Solution

a. The *term* ⟹ *characteristics* direction is a conditional with the defined term *circle* in the antecedent. If a figure is a circle, then it is the set of points in a plane at a certain distance from a certain point.

b. The *characteristics* ⟹ *term* direction is a conditional with the defined term *circle* in the consequent. If the figure is the set of all points in a plane at a certain distance from a certain point, then it is a circle.

c. Connect the term and the characteristics with the words *if and only if.* A figure is a circle if and only if it is the set of all points in a plane at a certain distance from a certain point.

Questions

COVERING THE IDEAS

1. List the properties of a good definition.

2. a. Define: *midpoint of a segment.*
   b. What two terms that are defined earlier are contained in this definition?

In 3–5, fill in the blank.

3. To combine $a \Rightarrow b$ and $b \Rightarrow a$, we write a __?__ b.

4. The symbol ⟺ is read __?__ .

5. If P is a point on circle C with radius 5, then $PC =$ __?__ .

6. Write the definition of *midpoint* as a biconditional.

7. What two meanings does the word *diameter* have?

APPLYING THE MATHEMATICS

8. **Fill in the Blanks** Suppose P is the midpoint of $\overline{MN}$ and $MN = 12$. Draw a diagram of this situation, and fill in each blank with a number.

   a. $MP =$ __?__
   b. $PN =$ __?__
   c. $MP + PN =$ __?__
   d. $MN =$ __?__
   e. $\dfrac{MP}{PN} =$ __?__
   f. $\frac{1}{2}MN =$ __?__
   g. $MN =$ __?__ $\cdot MP$

9. Why is each of these *not* a good definition of *circle*?

   a. A circle is the set of points in a plane at a certain distance from a certain point and it goes around the center.

   b. A circle is the plane section of a sphere.

   c. In a circle, radii are equal.

   d. A circle is the set of points around a certain point.

In 10 and 11, one of the conditionals of a previous definition is given. Tell whether the statement goes in the direction *term ⟹ characteristics* or *characteristics ⟹ term*.

10. If two coplanar lines are parallel, then they have no points in common or are identical.

11. If S is the set of all possible points, then S is space.

12. Let p be the statement *your cell phone is charged.* Let q be the statement *you can receive calls.*

   a. Write out $p \Longleftrightarrow q$ in words.

   b. Explain why $p \Longleftrightarrow q$ is false.

13. Write the definition of *convex set* from Lesson 2-1 as a biconditional.

14. **Multiple Choice** The shaded portion of the figure at the right is the interior of a circle with center P and radius r. Which of these is a good definition of *interior of a circle*?

   A The interior of a circle with center P is the set of points inside circle P.

   B The interior of the circle with center P and radius r is the set of points whose distance from P is less than r.

   C The interior of the circle with center P is the set of points that are neither on nor outside circle P.

15. Is the center of a circle on the circle?

16. In common English, "bisect" means to cut in half. In the figure, D is the midpoint of $\overline{AB}$. The tick marks show $AD = DB$. The point D, or any ray, line, or segment not in $\overline{AB}$ containing D, is called a *bisector* of $\overline{AB}$. Write a good mathematical definition of *bisector of a segment*.

REVIEW

17. a. Write the converse of the statement *If x and y are both even, then x + y is even.*

   b. Is the converse true? (**Lesson 2-3**)

18. Give an example of a conditional that is false and whose converse is false as well. (**Lesson 2-3**)

19. Identify the antecedent and the consequent in the following statement: *If it is true that if you buy 100 lottery tickets then you are sure to win, then I will buy 100 lottery tickets.* (**Lesson 2-2**)

20. The statement *There exists an integer between 1 and 7 that is greater than 10* is false. Is it possible to show that it is false using a counterexample? (**Lesson 2-2**)

21. Consider the statement *Every convex figure is a polygon.* (**Lessons 2-2, 2-1**)

    a. Rewrite the statement in if-then form.

    b. Use a counterexample to show that it is false.

22. Find all solutions to the system $\begin{cases} |x-2| = 1 \\ |x-0| = 1 \end{cases}$. (**Lesson 1-1, Previous Course**)

EXPLORATION

23. On a number line, let A be the point -1 and B be the point 3.

    a. Show that 1 is the midpoint of $\overline{AB}$. Notice that $1 = \frac{-1+3}{2}$.

    b. Use the same process to find the midpoint of $\overline{CD}$ where C is the point 3.5 and D is the point 5.5. Use your knowledge of averages to explain why this method works.

24. *Webster's Second New Revised Dictionary* includes this definition:

    > dodo *n.* 1. An extinct flightless bird, *Raphus cucullatus,* once found on the island of Mauritius.

    a. Write the definition in biconditional form.

    b. What words in the definition are exact synonyms for dodo? Why do you think that the dictionary does not define dodo using only those words?

    c. Is this a good definition by the standard in this book? Do you think the standards of *Webster's Dictionary* are different from ours? Why or why not?

The name *dodo* may have originally come from the Dutch word *dodoor* which means "sluggard" or from the Portuguese word *doudo* which means "foolish" or "simple."

Source: www.birds.mu

Unions and Intersections of Figures

Vocabulary

union of two sets

intersection of two sets

angle

null set (empty set)

▶ **BIG IDEA** Given two figures, you can form their *union* by including all points in either of them, or their *intersection* by including only those points in both of them.

Points, lines, segments, and rays are the building blocks of more complex figures. The two most common ways of combining figures, or any other sets, is to take their *union* or their *intersection*.

Mental Math

Evaluate $3x - 4y$ when $(x, y) =$

a. $(7, 0)$.

b. $(0, 13)$.

c. $(7, 13)$.

d. $(13, 7)$.

Definition of Union, Intersection of Two Sets

The **union of two sets** A and B, written $A \cup B$, is the set of elements that are either in A or in B, or in both A and B.

The **intersection of two sets** A and B, written $A \cap B$, is the set of elements that are in both A and B.

We sometimes use Venn diagrams to visualize intersections and unions. The purple shaded region at the left is the intersection of sets A and B, and the purple shaded region at the right is their union.

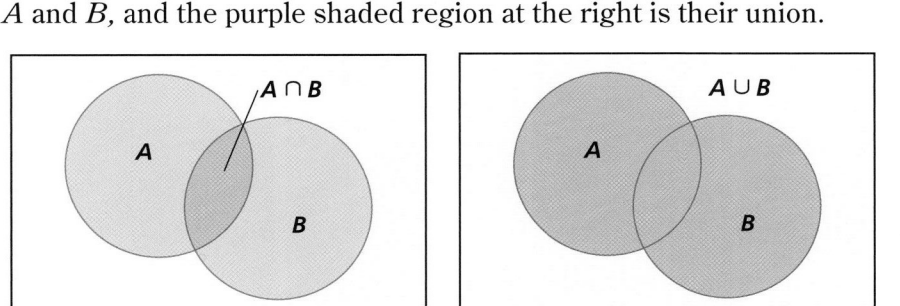

You have already seen intersections and unions in this book. The intersection of two nonparallel lines is the point P that is on both lines. The union of two rays with the same endpoint is an **angle.**

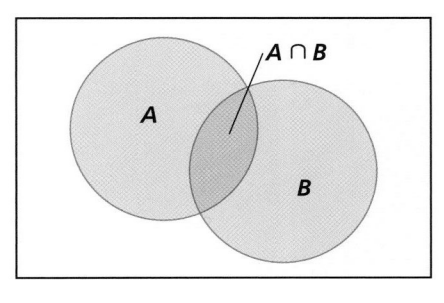

$m \cap n = \{P\}$

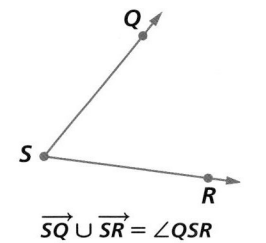

$\overrightarrow{SQ} \cup \overrightarrow{SR} = \angle QSR$

▶ **QY1**

Name the elements of $\{3, 4, 5\} \cap \{4, 8, 3\}$.

STOP **QY1**

Example 1

Given $\overleftrightarrow{MR}$ with point N between M and R.

a. What is $\overrightarrow{MR} \cap \overrightarrow{NM}$?

b. What is $\overrightarrow{NR} \cup \overrightarrow{MN}$?

Solution

a. $\overrightarrow{MR} \cap \overrightarrow{NM}$ is the set of points on both $\overrightarrow{MR}$ and $\overrightarrow{NM}$. It may help you to draw each ray separately.

The points that satisfy both of those conditions make up the segment with endpoints M and N, $\overline{MN}$. So, $\overrightarrow{MR} \cap \overrightarrow{NM} = \overline{MN}$.

b. $\overrightarrow{NR} \cup \overrightarrow{MN}$ consists of the points either on $\overrightarrow{NR}$, on $\overrightarrow{MN}$, or on both. Again draw a picture. These points make up the ray $\overrightarrow{MR}$, which can also be named $\overrightarrow{MN}$. So, $\overrightarrow{NR} \cup \overrightarrow{MN}, = \overrightarrow{MR}$.

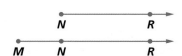

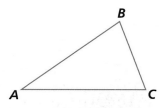

Unions and Intersections in Coordinate Geometry

When you solve a system of equations or inequalities, you are finding the intersection of solution sets. For example, the solution to the system $\begin{cases} y = 2x + 3 \\ y = \frac{1}{2}x - 1 \end{cases}$ is $\left(-\frac{8}{3}, -\frac{7}{3}\right)$. That means that the intersection of the graphs of $y = 2x + 3$ and $y = \frac{1}{2}x - 1$ is the point $\left(-\frac{8}{3}, -\frac{7}{3}\right)$.

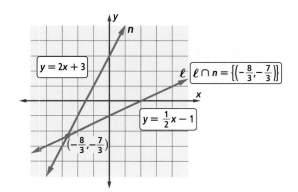

Unions and Intersections in Polygons

It is convenient to describe geometric shapes using the ideas of intersection and union. For example, at the right,

$$\triangle ABC = \overline{AB} \cup \overline{BC} \cup \overline{CA}.$$

Often we speak of the intersection of two lines, line segments, or rays. When a set of points contains just one point, we usually drop the set braces { }. In $\triangle ABC$, for example, you can write $\overline{AB} \cap \overline{BC} = \{B\}$, or $\overline{AB} \cap \overline{BC} = B$. In words, segment $\overline{AB}$ intersects segment $\overline{BC}$ at point B.

The set containing no elements is called the **null set** (or **empty set**). This set is written { } or Ø. If figures have no points in common, then their intersection is the null set. In symbols, the situation at the right is written as $\overrightarrow{FL} \cap \overline{MN} = \{\ \}$, or $\overrightarrow{FL} \cap \overline{MN} = \emptyset$.

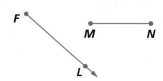

 QY2

▶ QY2

a. Draw a convex, four-sided polygon *ABCD*.

b. Describe the following in words: (polygon *ABCD*) $\cup \overline{AC} \cup \overline{BD}$.

c. Describe the following in words: (polygon *ABCD*) $\cap (\overline{AC} \cup \overline{BD})$.

Unions and Intersections in the Real World

Both unions and intersections of figures occur in practical situations. Below is a map of some of the stops of the subway system for Washington, D.C., and some of its surrounding communities. The system consists of five routes, each named by a color: Red Line, Blue Line, Green Line, Yellow Line, and Orange Line.

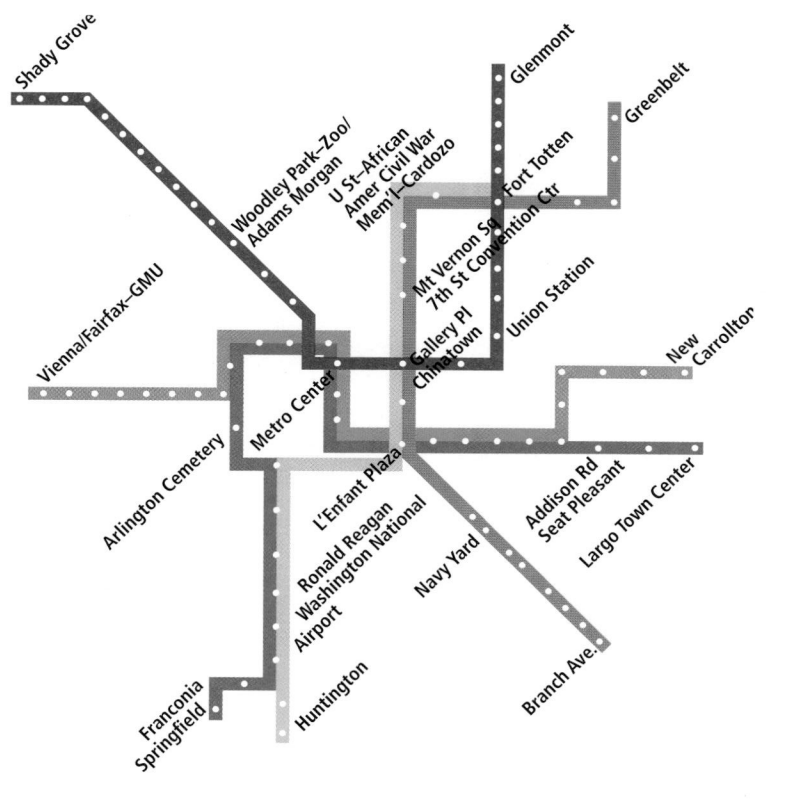

Example 2

Refer to the map of the Washington, D.C., subway system above.

Let R = Red Line, B = Blue Line, G = Green Line, O = Orange Line, and Y = Yellow Line.

a. What does $R \cup B \cup G \cup O \cup Y$ mean?

b. What does $O \cap G$ mean?

Solution

a. $R \cup B \cup G \cup O \cup Y$ consists of all lines, so $R \cup B \cup G \cup O \cup Y$ means this entire part of the transit system.

b. $O \cap G$ is the station that the two lines share. According to the map, $O \cap G$ is the L'Enfant Plaza station.

Questions

1. a. Draw a Venn diagram for two sets, G and H, and shade the union of G and H.

    b. Draw a Venn diagram for two sets, J and K, and shade the intersection of J and K.

In 2–4, describe $A \cup B$ and $A \cap B$.

2. $A = \{-2, 0, 3, 7, 12, 18\}$, $B = \{0, 1, 2, 3, 4, 5\}$

3. $A = \{(x, y): y = 2x - 1\}$, $B = \{(x, y): y = 3x - 1\}$

4. $A =$ the graph of $4x - 3y = 12$; $B =$ the graph of $2x + 5y = -7$.

In 5 and 6, refer to Example 1.

5. Explain why $\overrightarrow{NM} \cap \overrightarrow{NR} = N$.

6. Explain why $\overrightarrow{NM} \cup \overleftrightarrow{MR} = \overleftrightarrow{MR}$.

7. Describe $\odot P \cap \odot R$, as shown in the figure at the right.

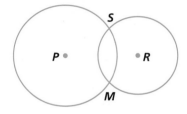

In 8–10, consider the figure at the right.

8. Describe $AFMT \cap \triangle LMF$.

9. Describe $\overline{FM} \cup \overline{ML} \cup \overline{LF}$.

10. Describe $\overline{ML} \cap \overline{AT}$.

In 11 and 12, refer to Example 2.

11. Choose the correct word from those in parentheses: The Washington, D.C., subway system is the (*union, intersection*) of its five lines.

12. What is $R \cap G$?

13. Let $A =$ the set of real numbers x with $x < 40$, $B =$ the set of real numbers x with $20 < x$.

    a. Describe $A \cup B$.

    b. Describe $A \cap B$.

14. Let $B =$ the set of people who are citizens of Tokyo, Japan, and let $C =$ the set of people who are citizens of Japan.

    a. Draw a Venn diagram of $B \cap C$.

    b. Describe the set $B \cap C$ in words.

    c. Draw a Venn diagram of $B \cup C$.

    d. Describe the set $B \cup C$ in words.

Most Japanese men and women wear kimonos only for special occasions. However, some women wear them every day.

In 15 and 16, name all segments in the figure described by the two triangles.

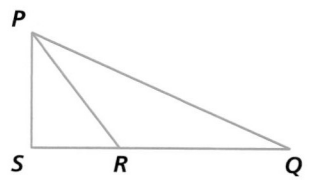

15. $\triangle PQR \cap \triangle PQS$

16. $\triangle PQR \cup \triangle PQS$

17. Let A = set of all ordered pairs (x, y) with $2x + 5y \geq 10$, and let B = the set of all ordered pairs (x, y) with $3x - 4y < 24$.
   a. Draw a graph of $A \cap B$.
   b. Draw a graph of $A \cup B$.

18. m and n are two lines in a plane. $m \cap n = \varnothing$. Draw a picture that could represent m and n. Describe the relationship between the two lines you have drawn.

19. Suppose that, on a coed volleyball team, F = the set of females rostered and M = the set of males rostered.
   a. Describe the set $F \cup M$ in words.
   b. What is $F \cap M$? Explain your answer.

REVIEW

20. Why is each of the following not a good mathematical definition of *speed limit*? (**Lesson 2-4**)
   a. A speed limit tells how fast you must go.
   b. A speed limit gives the highest speed you can go. If you go faster the police may give you a speeding ticket. On some roads it may be dangerous to go faster than the speed limit.
   c. A speed limit is the maximum legal speed to avoid prosecution under the state's legislative sanctions.

21. Consider the statement *If Q is the set of elements in both set A and set B, then Q is the intersection of sets A and B.* Is this the (term $\Rightarrow$ characteristics) or the (characteristics $\Rightarrow$ term) direction of the definition of intersection? (**Lesson 2-4**)

22. Write the two conditionals that make up the biconditional *P is on circle O with radius r if and only if PO = r.* (**Lessons 2-4, 2-2**)

23. Give the antecedent and consequent of this statement: *When you work more than 40 hours per week, you will receive bonus pay for overtime.* (**Lesson 2-4**)

24. a. Write the converse of the statement *If B is on $\overrightarrow{AC}$, but not between A and C, then AC = AB − BC.* (**Lesson 2-3**)
   b. Is the statement true?
   c. Is its converse true?

25. Solve for x and check: $x = 2(180 - x)$. (**Previous Course**)

26. Graph all possibilities for q on a number line: $22 + q > 180$. **(Previous Course)**

EXPLORATION

27. This puzzle (from *Mathematical Puzzles of Sam Loyd,* edited by Martin Gardner) combines some ideas of union and intersection:

"It is told that three neighbors who shared a small park, as shown in the sketch, had a falling out over each other's chickens. The owner of the large house at the back, complaining that his neighbor's chickens annoyed him, built an enclosed pathway from his door to the gate at the bottom of the picture. The man on the right built a path to the gate on the left, and the man on the left built a path to the gate on the right. None of the paths intersected. Draw the three paths correctly."

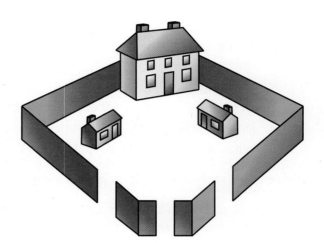

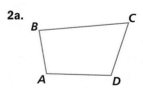

Lesson

2-6

Polygons

Vocabulary

polygon

sides

vertex, vertices

consecutive (adjacent) vertices

consectutive (adjacent) sides

diagonal

n-gon

triangle

quadrilateral

pentagon

hexagon

heptagon

octagon

nonagon

decagon

polygonal region

convex polygon

equilateral triangle

isosceles triangle

scalene triangle

hierarchy

▶ **BIG IDEA** Some unions of segments in the plane are polygons; the number of segments classifies a polygon as a *triangle, quadrilateral, pentagon,* and so on.

Even a figure as simple as a triangle is hard to define. Using the language from the last lesson, Activity 1 starts us in the direction of a good definition.

Activity 1

In 1–5, use figures that are unions of segments and consider the following conditions:

I   There are three or more segments.

II The figure lies entirely in a plane.

III Each segment intersects exactly two others in the figure.

IV Each segment intersects other segments only at its endpoints.

If you think any of the figures are impossible to draw, explain why you think this is so.

1. Draw a figure satisfying I, II, III, and IV.
2. Draw a figure satisfying I, II, and IV, but not III.
3. Draw a figure satisfying II, III, and IV, but not I.
4. Draw a figure satisfying I, III, and IV, but not II.
5. Draw a figure satisfying I, II, and III, but not IV.

We could use the following as the definition of a triangle: *A triangle is the union of three segments.* The results from Activity 1 should convince you that this would not be a good definition of *triangle* because there are unions of three segments that do not look like figures we want to be triangles.

Mental Math

Find the point of intersection of

a. $x = 2$ and $y = -17$.

b. $x = a$ and $y = b$.

c. $x = m$ and $y = -x$.

Unions of three segments that are not triangles

a.   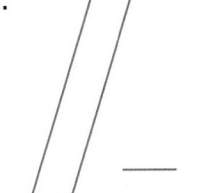   b.   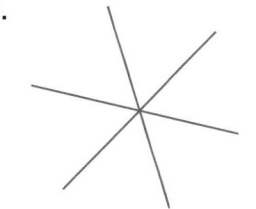   c.   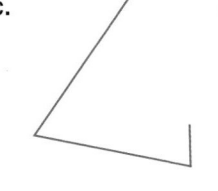   d.   ✕   e.

You can see that a triangle is not just any union of three segments. Each segment must intersect the others, and the intersections must be at endpoints. These criteria help to get a good definition not only of *triangle,* but also of the more general term *polygon.*

> ### Definition of Polygon
>
> Let P_1, P_2, . . . , and P_n be n distinct points in the same plane. The union of the n segments $\overline{P_1P_2} \cup \overline{P_2P_3} \cup \ldots \cup \overline{P_{n-1}P_n} \cup \overline{P_nP_1}$ is a **polygon** if and only if each segment intersects exactly two others and if no segments sharing an endpoint are collinear.

Activity 1 shows that this definition does not contain more information than is necessary and all parts of the definition are needed.

Activity 2

Work in groups and examine Figures a, b, c, d, and e on the previous page. Explain why each of the unions of segments is not a polygon by stating which part of the definition it violates.

Parts of Polygons

To describe polygons, terminology is needed. The segments that make up a polygon are its **sides.** The endpoints of the sides are the **vertices** of the polygon. (The singular of vertices is **vertex.**) A polygon can be named by giving its vertices in order clockwise or counterclockwise. Many names are possible; two names for the polygon shown at the right are *POLYGN* and *GYLOPN.*

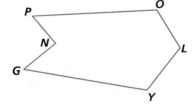

POLYGN = $\overline{PO} \cup \overline{OL} \cup \overline{LY} \cup \overline{YG} \cup \overline{GN} \cup \overline{NP}$

🛑 QY

Consecutive (or **adjacent**) **vertices** are endpoints of a side. For instance, G and Y are consecutive vertices of *POLYGN.* Note that our definition of polygon does not allow three consecutive vertices to be collinear. **Consecutive** (or **adjacent**) **sides** are sides which share an endpoint, such as $\overline{PO}$ and $\overline{OL}$. A **diagonal** of a polygon is a segment connecting nonconsecutive vertices. For example, in the polygon *POLYGN* above at the right, $\overline{NY}$ and $\overline{PG}$ are diagonals. Diagonals do not have to be drawn to exist.

> ▶ QY
>
> What is wrong with using *PGNYLO* as a name for the polygon above?

Names of Polygons

A polygon with n sides is called an ***n*-gon**. When n is small, the polygons have special names: a **triangle** has 3 sides, a **quadrilateral** has 4, a **pentagon** has 5, a **hexagon** has 6, a **heptagon** 7, an **octagon** 8, a **nonagon** 9, a **decagon** 10, and a **dodecagon** 12.

Polygonal Regions

From its definition, every polygon lies entirely in one plane. It separates the plane into two other sets—its *interior* and its *exterior*. The union of a polygon and its interior is a **polygonal region**.

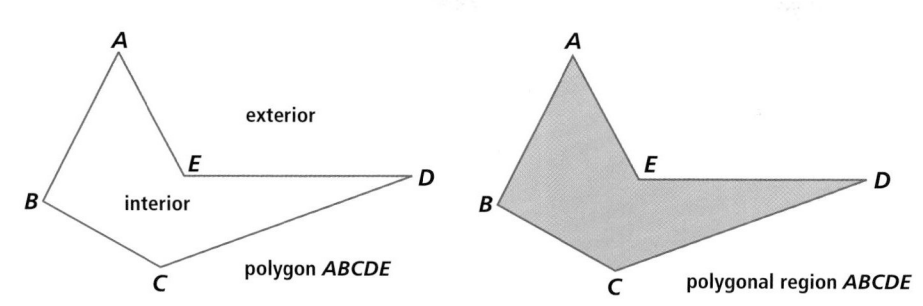

polygon *ABCDE*

polygonal region *ABCDE*

A polygon is called a **convex polygon** if and only if its corresponding polygonal region is convex. Many commonly used polygons, such as squares and parallelograms, are convex.

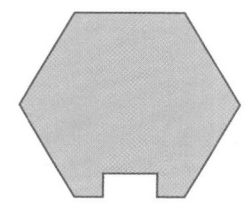

convex hexagon

nonconvex decagon

Types of Triangles

Triangles with special characteristics are given specific names. When the lengths of the sides of a triangle are considered, three possible triangles occur. An **equilateral triangle** has all three sides of equal length. An **isosceles triangle** has two (or more) sides of equal length. (An equilateral triangle is also isosceles.) A triangle with no sides of the same length is called a **scalene triangle.**

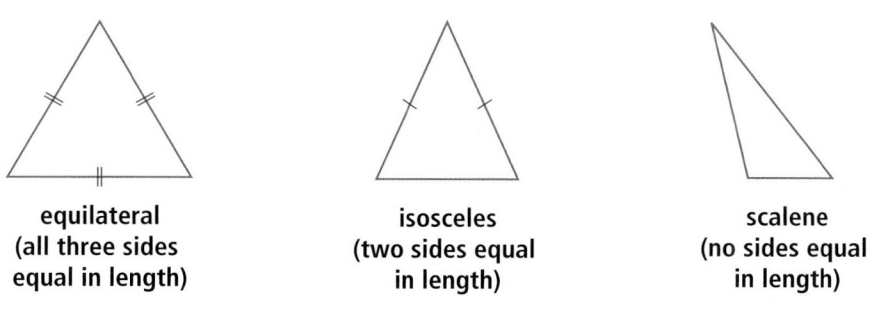

equilateral
(all three sides
equal in length)

isosceles
(two sides equal
in length)

scalene
(no sides equal
in length)

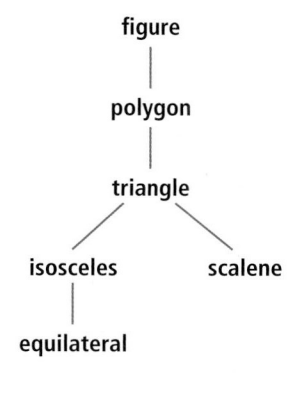

The classification of triangles by sides is shown at the right in a **hierarchy**. Each name includes all the shapes below it to which it is connected. Thus, every isosceles triangle is a triangle, a polygon, and a figure, but an isosceles triangle is not a scalene triangle.

Example

According to the hierarchy, an equilateral triangle is a special type of what figures?

Solution Find *equilateral triangle* at the lower left and go up the hierarchy. An equilateral triangle is an isosceles triangle, a triangle, a polygon, and a figure.

Questions

COVERING THE IDEAS

1. Why is each figure not a polygon?

   a. 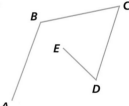 b.  c.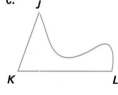

2. a. How many vertices does a heptagon have?

   b. Is a nonagon a polygon? Why or why not?

   c. How many angles does an *n*-gon have?

3. Use polygon *ABCDEF* at the right.

   a. Name its vertices.

   b. Name its sides.

   c. Name a diagonal.

   d. Name a pair of consecutive sides.

   e. Name a pair of adjacent vertices.

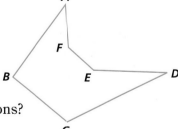

4. How do polygons differ from polygonal regions?

5. Draw a convex decagonal region.

In 6 and 7, characterize the polygonal region as convex or nonconvex.

6.    7.

8. Use polygon *KLMNO* at the right.

    a. Name a diagonal that is entirely in the interior of the polygon.

    b. Name a diagonal that is entirely in the exterior of the polygon.

    c. Name a diagonal that is not an answer to Part a or b.

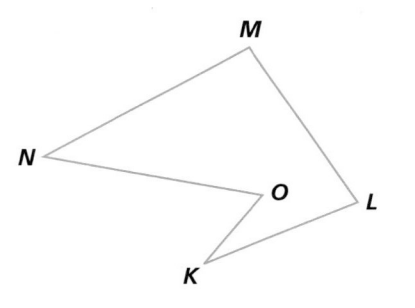

In 9–11, classify the polygon by the number of its sides and by whether or not it is convex.

9.    10.    11.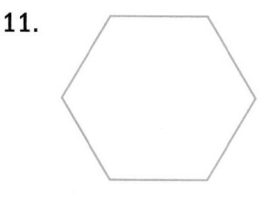

12. Draw a convex 13-gon.

13. According to the hierarchy given in the lesson, a scalene triangle is a special type of what figures?

APPLYING THE MATHEMATICS

14. Explain why no triangle has diagonals.

In 15–17, the map of one of the states of the United States is shown. The boundary is very much like a polygon. Give the name of the polygon and of the state.

15. 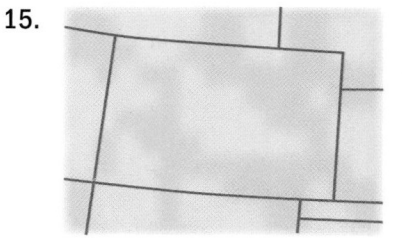   16.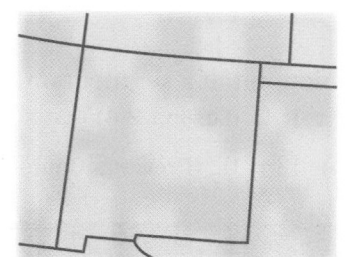

17.

18. Consider the quadrilateral *MATH* with vertices at $M = (1, 5)$, $A = (5, 4)$, $T = (4, -2)$, and $H = (3, 2)$. Is this quadrilateral convex or nonconvex?

19. Consider a coordinate plane and the graphs of the inequalities $y \geq 0$, $x \leq 4$, and $y \leq x$.

   a. Describe the figure formed by the intersection of these sets of points.

   b. Write the coordinates of the vertices of this figure.

20. a. Trace the hexagon *FLOWER* shown at the right and draw all of its diagonals.

   b. How many distinct diagonals are there?

21. Name a polygon whose number of sides is equal to its number of diagonals.

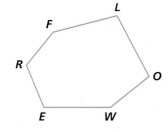

REVIEW

22. Draw two triangular regions whose intersection is a quadrilateral region. (**Lesson 2-5**)

23. Suppose you know that $A \cup B = A \cap B$. What can you say about A and B? (**Lesson 2-5**)

24. Describe the intersection of a circle and a radius of that circle. (**Lessons 2-5, 2-4**)

25. a. Solve the inequalities $2x + 1 \leq 5$ and $4x + 10 \geq 0$.

   b. Find the intersection of their solution sets and graph it on a number line. (**Lesson 2-5, Previous Course**)

26. According to the definition of a circle, is a point a circle? Why or why not? (**Lesson 2-4**)

27. Find the coordinate of the midpoint of a segment on a number line with endpoints at x and $x + 4$. (**Lesson 2-4**)

28. Solve the equation $x^2 - 17x + 42 = 0$. (**Previous Course**)

EXPLORATION

29. Some names for polygons are no longer used or are used only rarely. For each of the following names of polygons, guess how many sides the polygon has, and then check your guesses by looking in a large dictionary or online.

   a. duodecagon
   b. enneagon
   c. pentadecagon
   d. tetragon
   e. trigon
   f. undecagon

QY ANSWER

PGNYLO is not a possible name because $\overline{PG}$ is not a side of the polygon.

Lesson 2-7

Conjectures

Vocabulary

conjecture

▶ **BIG IDEA** By studying figures, you may see properties that you think are true, but only a proof can establish their truth.

In the following activity you are asked to make some educated guesses based on the appearance of the figures.

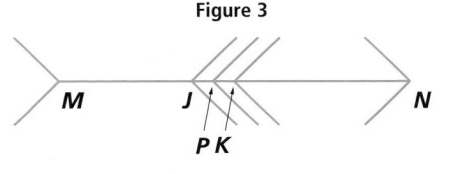

Mental Math

Name a commonly found

a. 3-sided figure.

b. 4-sided figure.

c. 6-sided figure.

d. 8-sided figure.

Activity 1

Consider each of the figures below. Using only your eyes, make an educated guess to answer the question about each figure.

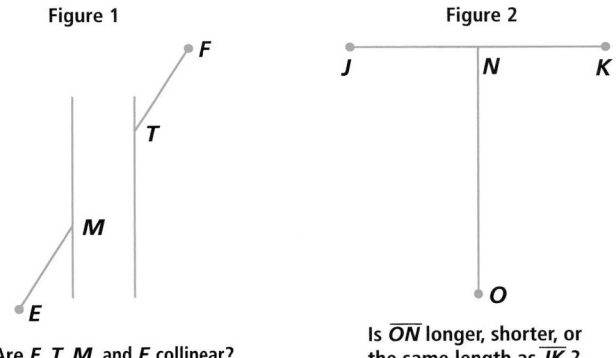

Figure 1

Are *F*, *T*, *M*, and *E* collinear?

Figure 2

Is $\overline{ON}$ longer, shorter, or the same length as $\overline{JK}$?

Figure 3

Is *J*, *P*, or *K* the midpoint of $\overline{MN}$?

Conjectures

In mathematics, an educated guess or opinion is called a **conjecture.**

People use conjectures frequently in ordinary conversation as well as in mathematics:

> If carbon emissions are not reduced, then global warming will continue.

> If inflation rates increase, then the federal interest rate will increase.

 QY1

▶ **QY1**

Use a ruler to determine if your conjectures in Activity 1 were correct.

Chalk artists use *trompe l'oeii* ("trick the eye") in their sidewalk art. The trompe l'oeil aspect of the figures in Activity 1 may have caused you to guess incorrectly.

Activity 2

MATERIALS: DGS (optional)

Step 1 Draw a circle.

Step 2 Put three points on the circle and label them *A*, *B*, and *C*.

Step 3 Draw $\overline{AB}$. Draw $\overline{BC}$.

Step 4 Measure ∠*ABC*.

Step 5 Drag point *B* and explore the possible measures of ∠*ABC*.

Step 6 Move points *A* and *C* to new positions.

Step 7 Measure ∠*ABC* again.

Step 8 Drag point *B* and explore the possible measures of ∠*ABC*.

Repeat Steps 6–8 as many times as you want. Based on your investigation, make a conjecture.

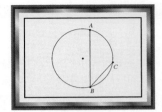

The Truth of a Conjecture

Most conjectures that are generalizations can be written in if-then form. Consequently, they are to be proved or disproved as you would any other conditional. To tell whether a conjecture is true or false, mathematicians often start by examining specific cases in which the antecedent is true. For conjectures about geometric figures, this means drawings are made and explored as you did in Activity 2.

 QY2

In Activity 2, you were asked to experiment with different positions of point *B* on the circle. This is a good way to start because if one counterexample is found, the conjecture is not true. When instances of the conjecture are found to be true, then there is evidence that the conjecture is true. But finding examples in which the conjecture is true is not enough to say it is always true. In mathematics, for a conjecture to be accepted in all cases, it must be *proved* for all those cases.

 QY3

Notice the difference between *proof* and *disproof*. If you find one figure in which the conjecture is false, you can conclude the conjecture is false. You have *disproved* the conjecture. However, even if the conjecture is true for every figure you have checked, this still does not prove the conjecture. It just shows that the conjecture is reasonable.

> ▶ **QY2**
>
> Write your conjecture from Activity 2 as a conditional.

> ▶ **QY3**
>
> Suppose a supercomputer generated 10 million true instances of this conjecture: *If a number is even, then it can be expressed as the sum of two primes.* For instance, $100 = 53 + 47$. Does this show that the conjecture is true?

Later in this course you will learn how to prove statements in geometry. Throughout the book, you will see proofs of geometric theorems and you will be asked to write your own proofs for other statements.

Activity 3

MATERIALS: DGS (optional)

Step 1 Consider the figures below. Fill the blanks with the number of regions created in Figures 3, 4 and 5.

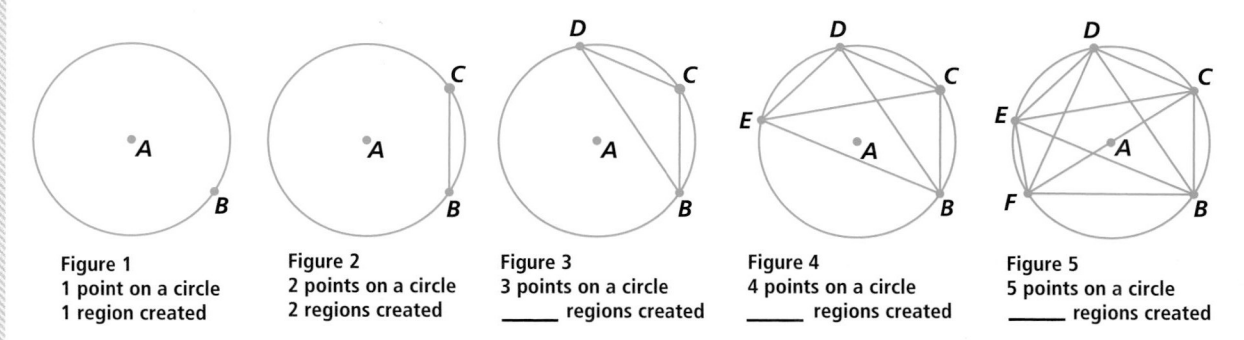

Figure 1
1 point on a circle
1 region created

Figure 2
2 points on a circle
2 regions created

Figure 3
3 points on a circle
_____ regions created

Figure 4
4 points on a circle
_____ regions created

Figure 5
5 points on a circle
_____ regions created

Step 2 Look for a numerical pattern.

Step 3 Make a conjecture about the maximum number of regions created by connecting each of 6 points in a circle to every other point.

Step 4 Test your conjecture by creating a circle with 6 points, connecting each point to every other point and then counting the number of regions created.

Step 5 Does your conjecture seem to be true, or have you found a counterexample to your conjecture?

A DGS can be very helpful in testing conjectures.

Using a DGS to Test a Conjecture

In Activity 4 on the next page, the ability to create one quadrilateral and add its diagonals, have the DGS measure the diagonals, and then drag a vertex or two to create the next quadrilateral will make your task very quick. If you would like to keep many examples on the screen to test other conjectures, copy and paste a duplicate picture and drag it to make your next example. This will also make it easy to examine more examples quickly. Once comfortable with your DGS, you may choose to test conjectures that you have about situations that are not presented in this book as DGS activities by creating the situation yourself.

Activity 4

MATERIALS: DGS (optional)

Consider the following statement:

The diagonals of a quadrilateral are equal in length.

Do you think this is true for any quadrilateral? Examine the figures below and make a conjecture. Draw additional quadrilaterals to test your conjecture.

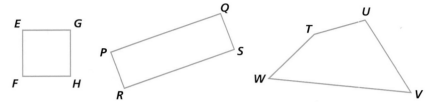

The quadrilaterals drawn above may have influenced you to believe that the diagonals of all quadrilaterals would be equal in length. Sometimes data seem to support a conjecture that is not actually true for all cases. The first two quadrilaterals are rectangles. They are not a good representation of "all quadrilaterals." Quadrilateral *UTWV* was purposefully drawn with *UW = TV*. If you use a DGS, you could consider figures that have any combination of many different characteristics.

When a conjecture is not true, as in Activity 4, you may try to *refine* it. This means you change the statement slightly so that the conjecture is true. For example, in Activity 4, if the diagonals of all quadrilaterals are not equal in length, is there any type of quadrilateral for which the diagonals are always equal in length?

 QY4

> ▶ QY4
>
> Based on your explorations in Activity 4, write a refined conjecture that may be correct.

Questions

COVERING THE IDEAS

1. What is a conjecture?

2. **Fill in the Blank** To show that a conjecture is true for all cases, it must be ___?___.

3. What is needed to show that a conjecture is false?

4. What does *refining a conjecture* mean?

In 5 and 6, sketch a counterexample to the statement.

5. If a quadrilateral has its vertices on a circle, then the measures of the opposite angles are equal.

6. If a quadrilateral has its vertices on a circle, then the lengths of the opposite sides are equal.

APPLYING THE MATHEMATICS

7. Anisa makes the conjecture "There are no white tigers." How can she prove this conjecture?

In 8–10, refine the conjecture to make it true.

8. The product of two integers is a positive integer.

9. The sum of two whole numbers is an even whole number.

10. If a and b are the lengths of two sides of a triangle, then $a + b > c$.

Multiple Choice In 11–13, a conjecture is stated. Choose the answer that best indicates your feeling about the statement. Draw several pictures to help you choose.

A   The conjecture is true and in my mind needs no proof.

B   The conjecture may be true, but I need a proof or a similar argument before I would believe it.

C   The conjecture is probably not true, but I would be sure only if I had a counterexample.

D   The conjecture is definitely false. No argument is needed to convince me that it is false.

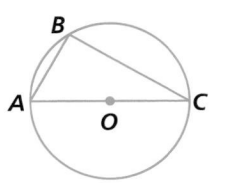

This is not a white tiger.

11. If the midpoints of two sides of a triangle are joined, the segment is half the length of the third side.

12. If the midpoints of four sides of a quadrilateral are joined, then the resulting figure is a rectangle.

13. In the figure at the right, if $\overline{AC}$ is a diameter of $\odot O$, then $\angle ABC$ is a right angle no matter where point B is on the circle O.

14. Perform the following steps using a DGS to create $\triangle DEF$.

Step 1 Construct points D and E.

Step 2 Construct a circle with center D and radius $\overline{DE}$.

Step 3 Construct a circle with center E and radius $\overline{ED}$.

Step 4 Name one of the intersection points of the two circles F.

What is the relationship between the side lengths of $\triangle DEF$?

15. Perform the following steps using a DGS.

Step 1 Draw $\triangle ABC$.

Step 2 Connect the midpoints of the sides to form $\triangle DEF$.

Step 3 Use your DGS to calculate Area($\triangle ABC$), Area($\triangle DEF$), and $\frac{\text{Area} \triangle ABC}{\text{Area} \triangle DEF}$.

Step 4 Move points A, B, and C around the screen.

Complete this conjecture: If the midpoints D, E, and F of $\triangle ABC$ are connected, then ___?___.

REVIEW

16. Trace the figure at the right. Draw a single diagonal to divide the given figure into two convex quadrilaterals. (**Lessons 2-6, 2-5**)

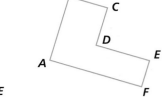

17. Consider the pentagon at the right. Give two possible names for the pentagon in terms of its vertices. (**Lesson 2-6**)

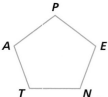

18. Let $A = \{1, 2, 3\}$, $B = \{2, 3, 4\}$, $C = \{3, 4, 5\}$. What is $A \cap (B \cup C)$? (**Lesson 2-5**)

19 What is wrong with the following definition?
A segment is the intersection of two rays. (**Lessons 2-5, 2-4**)

20. Find all values of x for which 1, 1, and x can be the lengths of the three sides of a triangle. (**Lesson 1-7**)

EXPLORATION

21. Pappus was a Greek geometer who lived in the third century CE Pappus discovered a number of very interesting theorems. One theorem made a conclusion based on the following situation:

Step 1 Create a line with points A, B, and C on it.

Step 2 Create a second line with points X, Y, and Z on it.

Step 3 Draw $\overleftrightarrow{AY}$, $\overleftrightarrow{BX}$, $\overleftrightarrow{BZ}$, $\overleftrightarrow{CY}$, $\overleftrightarrow{AZ}$, and $\overleftrightarrow{CX}$.

Step 4 Name $T = \overleftrightarrow{AY} \cap \overleftrightarrow{BX}$.

Step 5 Name $H = \overleftrightarrow{BZ} \cap \overleftrightarrow{CY}$.

Step 6 Name $E = \overleftrightarrow{AZ} \cap \overleftrightarrow{CX}$.

Step 7 Make a conjecture about the relationship between points T, H, and E.

Step 8 Use the tools in your DGS to support your conclusion. This may require you to find the length of segments, measures of angles and/or draw points, lines or segments.

Step 9 Use your DGS to create many examples of the above situation by dragging points A, B, C, X, Y, and Z. Does your conjecture still hold?

Step 10 Complete the conditional statement with a logical consequent:
If given two sets of collinear points $\{A, B, C\}$ and $\{X, Y, Z\}$, with $\overleftrightarrow{AY} \cap \overleftrightarrow{BX} = T$, $\overleftrightarrow{BZ} \cap \overleftrightarrow{CY} = H$, and $\overleftrightarrow{AZ} \cap \overleftrightarrow{CX} = E$, then ___?___.

QY ANSWERS

1. Answers vary. Sample: Yes, my conjectures were correct.

2. If an angle is formed with three points on a circle and the vertex moves to different places on the circle, then there are two possible angle measures for $\angle ABC$, depending on the location of B. The measures add up to 180°.

3. No. There is evidence that the conjecture is true, but it has not been proven.

4. Answers vary. Sample: The diagonals of a rectangle are equal in length.

Chapter 2 Projects

1 Venn Problem Solving

Venn diagrams can be used to organize information in order to solve problems. Solve this problem and write another problem for a classmate to solve.

A survey of 100 seniors at Artsbe Goode High School reported 82 students are taking an art class, 33 are in the band, 27 are in the choir, and 42 students not in any music program. If everyone in a music program is also taking an art course, how many students are participating in art, band, and choir?

2 Polygons Are a Matter of Opinion

There are many different definitions of polygons and of polygonal regions.

a. Go to the library and find several high school geometry textbooks. See how these books define polygons and polygonal regions. Find several more definitions using the Internet. Collect at least five different definitions.

b. Do all of the definitions you collected refer to the same things? Try to find examples of shapes that are included by some of the definitions, but not by others.

3 Tangram Shapes

Find a set of tangrams.

a. Use the tangrams to form the house below. Draw and label your diagram, clearly showing how you fit the tangram pieces together.

b. Use the tangrams to form the capital letter E below. Follow the directions in Part a.

c. Form at least three other shapes with the tangrams, and follow the directions in Part a. Present your diagrams to the class.

4 Intersection, Union, Addition, and Multiplication

For any three sets A, B, and C, it is true that $A \cap (B \cup C) = (A \cap B) \cup (A \cap C)$. If you replace $\cup$ with $+$ and $\cap$ with $\cdot$, you get a familiar property from algebra. This is no coincidence.

a. Draw a Venn diagram illustrating the above statement.

b. Which property from algebra does the statement resemble?

c. List as many different ways as you can in which union is similar to addition. Explain your reasoning.

d. List as many different ways as you can in which intersection is similar to multiplication. Explain your reasoning.

5 Convex or Nonconvex?

At first glance, a basketball may appear to be convex. Upon further inspection, you can see that the ball has ridges and dimples that make it nonconvex. List other real-world objects that appear convex but are not and explain why they are not. Can you think of a real-world object that is convex?

Chapter 2

Summary and Vocabulary

In this chapter, the language and logic of geometry are discussed and applied to segments, circles, and polygons. A language is needed in geometry to name figures and to agree on their characteristics. This requires good definitions. A good definition is an accurate description of an idea that involves only words defined earlier, words commonly understood, or words purposely undefined.

  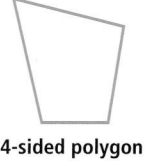  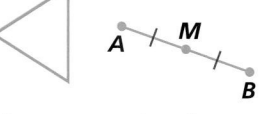

| circle | hexagon | 4-sided polygon | not a polygon | not a polygon |

The logic of mathematics is based on the truth or falsity of conditional statements. A statement is a sentence that is either true or false. Every conditional statement has an antecedent, its *if* part, and a consequent, its *then* part. If its antecedent is p and its consequent is q, the conditional is $p \Rightarrow q$.

The logic in this chapter dealt with three questions:

(1) How can you tell if $p \Rightarrow q$ is false?
You need a counterexample. A counterexample is a situation for which p is true and q is false.

(2) How are $p \Rightarrow q$ and $q \Rightarrow p$ related?
$q \Rightarrow p$ is the converse of $p \Rightarrow q$. The truth or falsity of one does not tell you anything about the truth or falsity of the other.

(3) What happens when both $p \Rightarrow q$ and $p \Rightarrow q$ are true?
The statement $p \Rightarrow q$ and its converse $q \Rightarrow p$ together form the biconditional $p \Leftrightarrow q$. We then say "p if and only if q." Every definition can be reworded as an if-and-only-if statement and can be separated into two conditionals.

A conjecture is an educated guess or opinion. A useful way to investigate the truth of some conjectures is to use a dynamic geometry system (DGS). A DGS allows the examination of many instances of a conjecture in a matter of seconds.

Vocabulary

2-1
*convex set, nonconvex set

2-2
statement
compound statement
conditional statement
*antecedent, consequent
instance of a conditional
counterexample to a
   conditional

2-3
*converse

2-4
*midpoint
*if and only if, biconditional
*circle
*radius, radii, diameter

2-5
*union, intersection of
   two sets
*angle
null set (empty set)

2-6
*polygon
sides
vertex, vertices
consecutive (adjacent)
   vertices
consecutive (adjacent)
   sides
diagonal
*n-gon
*triangle, quadrilateral
*pentagon, hexagon
*heptagon, octagon
*nonagon, decagon
polygonal region
*convex polygon
*equilateral triangle
*isosceles triangle
*scalene triangle
hierarchy

2-7
conjecture

Chapter 2 Self-Test

Take this test as you would take a test in class. You will need a calculator. Then use the Selected Answers section in the back of the book to check your work.

1. Which property of a good definition is violated if this definition were given in this chapter? *A tetrahedron is a polyhedron with four faces.*

2. **Matching** Match each term with the most appropriate drawing:
   a. convex hexagon
   b. nonconvex pentagon
   c. convex quadrilateral

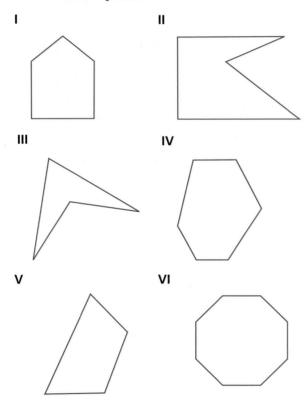

I II

III IV

V VI

3. Identify the antecedent and consequent of the following conditional: *A triangle is acute if a triangle has three acute angles.*

4. Write the conditional in Question 3 using the symbol ⇒.

5. Rewrite the following statement as a conditional. *All polyhedrons have at least four faces.*

6. Rewrite the following definition as a biconditional. *An isosceles triangle is a triangle with at least two sides having the same length.*

7. Consider the diagram below. Describe the given set.

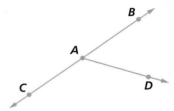

a. $\overrightarrow{AB} \cup \overrightarrow{AC}$ b. $\overrightarrow{AB} \cup \overrightarrow{AD}$
c. $\overrightarrow{BC} \cap \overrightarrow{CB}$ d. $\overrightarrow{AD} \cap \overleftrightarrow{CB}$

8. Let *p* be the statement *△MNR is equilateral.* Let *q* be the statement *△MNR has three angles with equal measure.* Write *p* ⇔ *q* in words.

9. Miranda is 15 and has 16-year-old friends who all drive cars. She makes the conjecture "If you are 16 years old, you drive a car," and claims it is true because everyone she knows who is 16 drives a car. Is Miranda right? If not, how would you prove her wrong?

10. Mr. Numkena said, "If you do not use a pencil on this test, I will give you a failing grade." Mr. Numkena always tells the truth. Is the converse of Mr. Numkena 's statement true? Explain why or why not.

11. Consider the conditional, "If $x < 1$, then $\frac{1}{x} > 1$."

    a. Give an instance of the conditional.

    b. Give two counterexamples to the conditional.

    c. Add a phrase to the conditional so it becomes true.

12. Use the definition of polygon to explain why each figure is not a polygon.

    a. 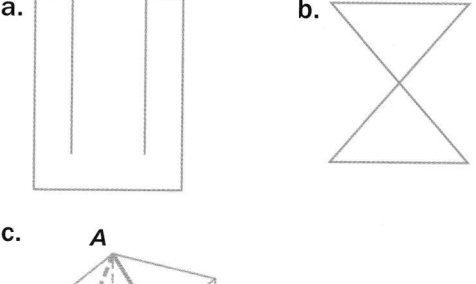 b.

    c.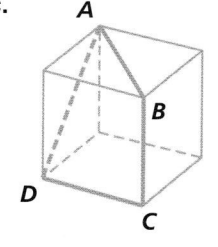

13. Mrs. Hawkins set up a tasting station at a grocery store with a sign that said, "Please try some chips!" Diamond walked up to the display and took two chips. Her friend Darius took the rest of the bowl. They each believed they took "some" chips. What could Mrs. Hawkins have written to clear up any confusion?

14. Explain why $x = -5 \Rightarrow x^2 = 25$ is true, and why $x^2 = 25 \Leftrightarrow x = -5$ is false.

15. What name is given to the union of Mexico, the United States, and Canada?

16. President James Garfield is the only President of the United States of America who has created an original proof of the Pythagorean Theorem. As such, he is the only member of the intersection of what two sets? Draw a Venn diagram representing the situation.

17. An advertisement said, "If your test score does not improve by at least ten points after you take this class, we will refund your money." Rachel signed up for the course, took the test for the second time, got a score of 610, and got her money back. What, if anything, can we conclude about Rachel's score the first time she took the test?

18. Draw a hierarchy to classify each of the following: isosceles triangle, figure, triangle, polygon, scalene triangle.

19. a. Draw a convex pentagon.

    b. Draw all of the diagonals of the pentagon.

Chapter 2 Chapter Review

SKILLS
PROPERTIES
USES
REPRESENTATIONS

SKILLS Procedures used to get answers

OBJECTIVE A Distinguish between convex and nonconvex figures. (Lessons 2-1, 2-6)

In 1–4, characterize each figure as convex or nonconvex.

1.

2.

3.

4.

5. Draw a convex pentagonal region.

6. Draw a nonconvex octagonal region.

OBJECTIVE B Identify the antecedent and the consequent of a conditional statement. (Lesson 2-2)

In 7–11, identify the antecedent and the consequent in each conditional statement.

7. If two different lines are parallel, then they do not intersect.

8. A figure is convex if whenever it contains two points, it contains the segment connecting them.

9. When $x = 4$, then $x > -2$.

10. The solution to the equation $ax + b = 0$ is $x = \frac{-b}{a}$ if $a \neq 0$.

11. Assuming that $(a + b = c + b)$, $a = c$.

OBJECTIVE C Use and interpret the symbols $\Rightarrow$ and $\Leftrightarrow$, and write conditionals and biconditionals. (Lessons 2-2, 2-4)

In 12 and 13, $p = (AB + BC > AC)$ and $q = (C$ is not between A and $B)$. Write down the given statement in words.

12. $p \Rightarrow q$

13. $p \Leftrightarrow q$

In 14–18, write the statement using $\Rightarrow$ or $\Leftrightarrow$ appropriately.

14. A polygon has 8 sides if and only if it is an octagon.

15. If a person is over the age of 25, then the person is an adult.

16. Every quadrilateral is a polygon.

17. When A is between B and C, $BA + AC = BC$.

18. Every triangle is convex.

19. Separate the following biconditional into its two conditionals: *A point is on a circle if and only if its distance from the center of the circle is equal to the radius of the circle.*

OBJECTIVE D Draw and identify polygons. (Lesson 2-6)

20. How many sides does a decagon have?

21. In the pentagon *ABCDE,* how many diagonals are there?

22. What does it mean for a polygon to be convex?

23. Draw a quadrilateral. Find two points that illustrate the following statement: *The exterior of a polygon is not convex.*

In 24 and 25, a type of polygon is given.

    a. Draw the polygon and name its vertices.

    b. Give one correct name for the polygon and one incorrect name in terms of its vertices.

    c. Name 3 sides, neither of which is adjacent to the other two.

24. a nonconvex hexagon

25. a convex heptagon

26. **Multiple Choice** Which of the following is a nonagon?

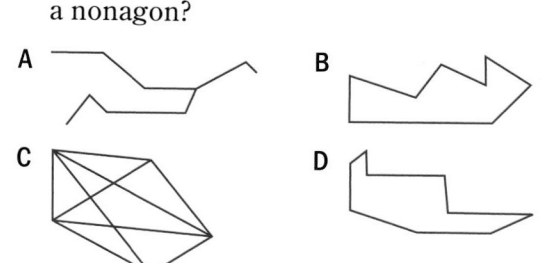

A   B

C   D

PROPERTIES The principles behind the mathematics

OBJECTIVE E Write the converse of a conditional. (Lesson 2-3)

In 27 and 28, a conditional is given.

    a. Write the converse of the statement.

    b. Say if the converse is true or false.

27. For any real numbers a, b, and c, if $a + c = b + c$, then $a = b$.

28. If you are 16 years old, then you cannot remember the 1930s.

True or False In 29–31, tell whether each statement is true or false and explain your answer.

29. If a statement is true then its converse is true.

30. If the converse of a statement is true, then the statement is true.

31. A statement is true if and only if its converse is true.

OBJECTIVE F Know and apply the properties of a good definition. (Lesson 2-4)

In 32–34, explain why the given definition is not a good definition.

32. A triangle consists of three points and every point between two of them and every point on the segments connecting any two of them.

33. A circle is a set of points that are placed around a point in a circle.

34. Point C on $\overline{AB}$ is the midpoint of $\overline{AB}$ if it looks like $AC = CB$.

35. Consider the definition: Two circles are called *intersecting circles* if and only if they intersect in exactly two points.

    a. Is this a good definition?

    b. Write this definition as two conditionals.

OBJECTIVE G Evaluate conditionals and conjectures. (Lessons 2-2, 2-7)

In 36–38, a conjecture is given.

    a. Draw an instance of the conjecture.

    b. Draw a counterexample to the conjecture.

36. If a figure is the union of three segments, then it is convex.

37. The union of two convex polygonal regions is convex.

38. The diagonals of a quadrilateral intersect in a single point.

In 39 and 40, refer to the conjecture: *Any order of the vertices of a polygon gives a correct name for that polygon.* Explain if the given figure is an instance of or a counterexample to the conjecture.

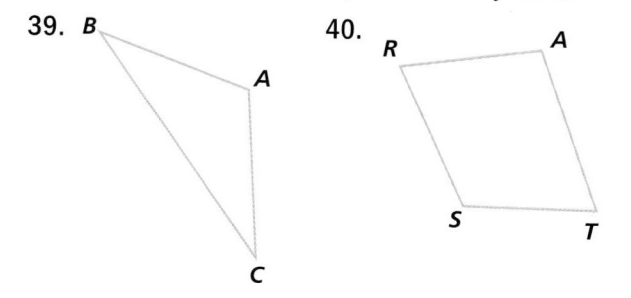

39. B   40.

OBJECTIVE H Use and interpret union and intersection of two sets. (Lesson 2-5)

In 41–43, refer to the figure below and describe the given set.

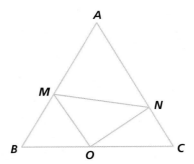

41. $\triangle ABC \cap \triangle MNO$ **42.** $\triangle NOC \cap \triangle ONM$

43. $(\triangle NOC \cap \triangle ONM) \cup (\triangle OMB \cap \triangle OMN)$
$\cup (\triangle AMN \cap \triangle MNO)$

44. Draw a picture of the following situations.

   a. Two triangles whose intersection is one point

   b. Two triangles whose intersection is two points

   c. Two triangles whose intersection is three points.

45. Give an example of two polygonal regions whose union is not convex.

USES Applications of mathematics in real situations

OBJECTIVE I Apply properties of conditionals to real situations.
(Lessons 2-2, 2-3)

46. A shampoo ad shows a model with beautiful hair using the shampoo.

   a. What conditional statement do the creators of the ad wish you to believe?

   b. Does the fact that the model has beautiful hair prove the statement?

   c. Give an example of something that could show that the statement is false.

47. Julian's friend told him that if you practice long enough, you can teach yourself how to juggle. Julian tried for days, and still can not juggle. Has Julian shown his friend's statement to be false? Why, or why not?

48. Explain why the converse of the following statement may not be true: *If you are the fastest runner in the world, then you can win an Olympic gold medal.*

OBJECTIVE J Apply unions and intersections to real-life situations.
(Lesson 2-5)

49. Let O be the set of people who were elected to be President of the United States exactly one time. Let T be the set of people who were elected to be the President of the United States exactly twice.

   a. What is the intersection of O and T?

   b. Let P be the set of all the presidents of the United States. Are there any elements of P that are not in the union of O and T?

50. A group of friends has a study session every Friday, and every week a different member is responsible for bringing food. Let V be the set of friends that are vegetarians. Let S be the set of friends that do not like spicy food.

   a. What does the set $V \cup S$ describe?

   b. What does the set $V \cap S$ describe?

   c. One week Tierra brought a spicy meat dish. Which set describes the people who probably did not like the food that week?

   d. Attempting to learn from Tierra's mistake, Kaitlyn brought a meat dish and a spicy vegetable dish. Which set describes the people who probably did not like the food that week?

OBJECTIVE K Apply the need for good definitions to real situations. (Lessons 2-1, 2-4)

51. The international system of units defines a foot to be the distance light travels in vacuum in $\frac{3048}{2,997,924,580,000}$ second. It is believed that originally, the measurement "foot" was taken to be the length of a grown man's foot (with his shoe on). What are the problems in using the old definition of a foot? How are these problems addressed by the new definition?

52. In the United States a person is not considered to be a medical doctor unless the person is licensed to practice medicine. Why is it important to carefully define exactly who is a medical doctor?

OBJECTIVE L Evaluate conjectures about real situations using conditionals. (Lesson 2-7)

53. One version of Murphy's Law states that whatever line you are waiting in will move slower than the other lines.
    a. Write this version of Murphy's Law as a conditional statement.
    b. What could happen to disprove this conjecture?

54. Carla believes that the tooth fairy exists. She decides to prove her conjecture by showing that the conditional statement *If I put a tooth under my pillow at night, there will be a dollar there in the morning* is true.
    a. What would be necessary to show that Carla's conditional statement is true?
    b. If Carla's conditional statement were proved to be true, would this show that her conjecture is true?

55. Give an instance and a counterexample to the conjecture: *Slow and steady wins the race.*

56. A chess player conjectures that his move will cause him to win in three turns. Which does the chess player probably hope that his opponent's next moves provide: an instance or a counterexample to his conjecture?

REPRESENTATIONS Pictures, graphs, or objects that illustrate concepts

OBJECTIVE M Draw hierarchies of triangles and polygons. (Lesson 2-6)

57. Draw the hierarchy relating the following: square, quadrilateral, polygon, convex quadrilateral, nonconvex quadrilateral, rectangle.

58. Draw the hierarchy relating the following: scalene triangle, figure, isosceles triangle, hexagon.

OBJECTIVE N Use and interpret Venn diagrams. (Lesson 2-2, 2-5)

59. A health club offers two classes. There are 24 participants in the yoga class and 16 participants in the aerobics class with a total of 33 people registered for the classes. Use a Venn diagram to find how many people are registered for both classes, and shade the corresponding region on your diagram.

60. Draw a Venn diagram relating I, the set of isosceles triangles; E, the set of equilateral triangles; and S, the set of scalene triangles.

61. Create a Venn diagram relating dodecagons, nonagons, and polygons.

62. Let $M = \{x: x \geq 4\}$ and $N = \{x: x > 7\}$.
    a. Create a Venn diagram relating M and N.
    b. What is $M \cap N$?

Chapter

3

Angles and Lines

Contents

3-1   Arcs and Angles

3-2   Rotations

3-3   Adjacent Angles and Vertical Angles

3-4   Algebra Properties Used in Geometry

3-5   Justifying Conclusions

3-6   Parallel Lines

3-7   Size Transformations

3-8   Perpendicular Lines

3-9   The Perpendicular Bisector

In watching gymnastics, figure skating, skateboarding, or other competitions, spectators are often awestruck by the number of twists and turns that athletes do. One of the most difficult ice skating jumps is the triple axel, which requires $3\frac{1}{2}$ revolutions. In gymnastics, athletes rotate themselves about uneven bars gaining momentum for landings that often are preceded by various types of flips. In diving, one of the most difficult dives off a 3-meter springboard is the reverse $3\frac{1}{2}$ somersault with a half twist. American Freestyle aerial skier Jeret Peterson's signature "hurricane" consists of five twists and three flips.

At the 2005 National Figure Skating Championships, Kimmie Meissner became only the second female U.S. figure skater to land a triple axel in competition.

110

Mathematically, each twist and flip is a rotation about an imaginary axis through the athlete's body. In this chapter you will learn more about rotations, as well as the angles that create them.

Angles play an important role in other sports, too. Soccer players decide where to plant their nonkicking foot based on the angle at which they'd like to kick the ball. Tennis players often strategically try to place the ball far from the net to limit the angle at which the ball is returned.

In bodybuilding, performing exercises that extend muscles from various angles has a profound effect on muscle growth. Good bowlers know that rolling the ball straight down the lane does not create the most strikes.

Angles are found in all sorts of geometric figures. They occur whenever lines intersect. They are found in all polygons. There are simple relationships between angles, parallel lines, and perpendicular lines, and surprising connections with the slopes of lines you studied in algebra. Angles are also formed by curves and planes, but in this chapter we concentrate on the basic angles of the plane—those formed by lines and those found in polygons.

Lesson

3-1   Arcs and Angles

Vocabulary

arc

endpoints of an arc

measure of an arc

semicircle

minor arc

major arc

angle

sides of an angle

vertex of an angle

central angle

zero angle

acute angle

right angle

obtuse angle

straight angle

interior of an angle

exterior of an angle

▶ **BIG IDEA**   Every angle has a measure that is a number in the interval from 0 to 180, the number 180 being half the measure in degrees of a full circle.

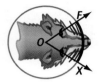 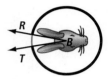

Above are the top views of the *fields of vision* of a fox (at the left) and a rabbit (at the right) as the rabbit is running from left to right away from the fox.

In nature, prey, such as rabbits, tend to have greater fields of vision than their predators, possibly so that they can see danger coming. A rabbit will turn its head slightly while running to keep the fox in its field of vision. Predators tend to have narrower fields of vision but are better able to focus.

In the diagram above, the pairs of rays $\overrightarrow{OF}$ and $\overrightarrow{OX}$ and $\overrightarrow{BR}$ and $\overrightarrow{BT}$ represent the boundaries of each animal's field of vision. The rabbit can see almost all the way round to the back of its head, much more than the fox. The darker part of the circle shows how much the animals can see.

Mental Math

Suppose an airline allows passengers to check a maximum of 50 lb of luggage. What is the maximum allowable weight of a passenger's second bag, if his or her first bag weighs

a. 10 lb?

b. x lb?

c. $a + b$ lb

Activity 1

Step 1 Hold two objects at the same height in front of your face with your arms outstretched.

Step 2 Without moving your head or rotating your eyes, move these two objects at relatively the same speed in opposite directions around an imaginary circle, with your body being the center of this circle. Try to keep both objects in your field of vision. At what point can you no longer see the objects?

Step 3 Draw a circle and show your field of vision as done above for the fox and rabbit.

The Number of Degrees in an Arc

To measure a field of vision, we measure an *arc* of a circle. An **arc** of a circle is a set of points made up of two points on a circle, called the **endpoints of the arc**, and all points on the circle between those two points. Suppose you start at point X on the edge of the field of vision of the fox and walk on the circle until you reach point F. The part of the circle you have walked is arc XF, written $\overset{\frown}{XF}$.

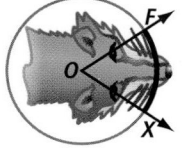

The rabbit's field of vision is almost the entire circle. If the endpoint R were rotated clockwise towards T, the field of view of the rabbit would be decreased, making the arc smaller. The **measure of an arc** is the amount of a circle that is traced by that arc, and the measure of an arc $\overset{\frown}{RT}$ is denoted by $m\overset{\frown}{RT}$. Arcs are usually measured in degrees, denoted °.

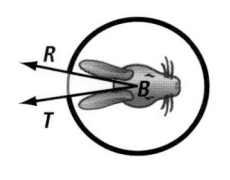

The arc measure of the entire circle is considered to be 360°. The division of the circle into 360° comes from the Babylonians, who lived in what is Iraq today. They knew that the year was about 365 days long, and that stars moved about $\frac{1}{365}$ of the way around an imaginary circle in the sky each day. By splitting the circle into 360 parts, each part of the circle was approximately the same as the change in the position of a star from one night to the next.

Minor Arcs, Semicircles, and Major Arcs

Two points on a circle determine two arcs. If the points are the endpoints of a diameter, as are points B and C in $\odot A$ at the right, then each arc is a **semicircle** whose degree measure is 180°. Otherwise, the two points determine arcs of different measures, one less than 180° and the other greater than 180°. An arc with measure less than 180° is called a **minor arc**, while an arc with measure greater than 180° is called a **major arc**.

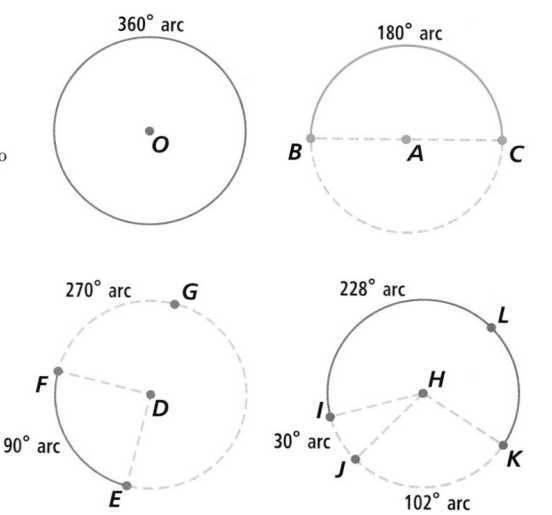

To distinguish minor arcs from major arcs, we always name major arcs with three letters. (Sometimes we name minor arcs with three letters if there might be confusion or if we do not know the measures.) In $\odot D$ at the right, $\overset{\frown}{EF}$ is the minor arc and $m\overset{\frown}{EF} = 90°$, while $\overset{\frown}{FGE}$ is the major arc and $m\overset{\frown}{FGE} = 270°$.

In general, if $x°$ is the measure of one of the two arcs determined by points A and B on a circle, then $(360 - x)°$ is the measure of the other arc.

Of course, you can split a circle into more than two arcs. You have seen this in circle graphs. On the previous page, ⊙H is split into three arcs.

 QY1

Angles

Imagine yourself at the center of circle D looking at point E. Your line of sight is the ray $\overrightarrow{DE}$. If you turn right to face point F, your line of sight will be the ray $\overrightarrow{DF}$. You will have turned from one side of an *angle* to the other.

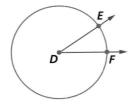

▶ QY1

Name the two arcs determined by points *I* and *K* in ⊙*H* shown on the previous page, and give the measure of each arc.

> **Definition of Angle**
>
> An **angle** is the union of two rays that have the same endpoint.

The rays are the **sides of the angle**. The intersection of the rays is the **vertex of the angle.**

The angle above at the right may be named ∠EDF or ∠FDE (the vertex is named by the middle letter) or, if there is no confusion, ∠D. ∠D is the union of rays $\overrightarrow{DE}$ and $\overrightarrow{DF}$.

Caution: Notice that the angle symbol ∠ and the symbol < for "is less than" look quite a bit alike, but they are not the same.

When the vertex of an angle is at the center of a circle, then we call the angle a **central angle** of the circle. In the figure above, because D is the center of the circle, ∠D is a central angle.

▶ QY2

In the figure below, give two other names for ∠*BAM*.

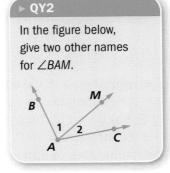

 QY2

Pictured at the right are two special angles. ∠STA is a *straight angle*, and ∠ZRO is a *zero angle*. Straight and zero angles are special because some different properties apply to them. If you are asked to identify angles in a given figure or problem, you should ignore straight and zero angles unless they are specifically mentioned.

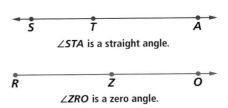

∠*STA* is a straight angle.

∠*ZRO* is a zero angle.

The Measure of an Angle

A *protractor* is an instrument for measuring angles. By looking at this picture of a protractor measuring ∠BVA, you can see that the outside of the protractor is a semicircular arc. Scales along the semicircle go from 0° to 180°, one in the clockwise direction and the other in the counterclockwise direction.

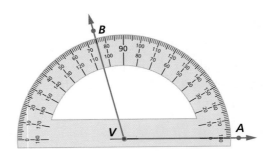

To measure a nonstraight angle, place the center of the semicircle (at the bottom center of the protractor) over the vertex of the angle you want to measure. The measure of the angle $\angle BVA$ equals the measure of the minor arc determined by where rays $\overrightarrow{VB}$ and $\overrightarrow{VA}$ intersect the circle. By reading the bottom scale on the protractor from right to left, $m\angle BVA = 105°$. Notice the similarity to measuring distance with a ruler. If x is the number where the ray $\overrightarrow{VB}$ intersects the protractor and y the number where $\overrightarrow{VA}$ intersects the protractor, then $m\angle BVA = |x - y|$.

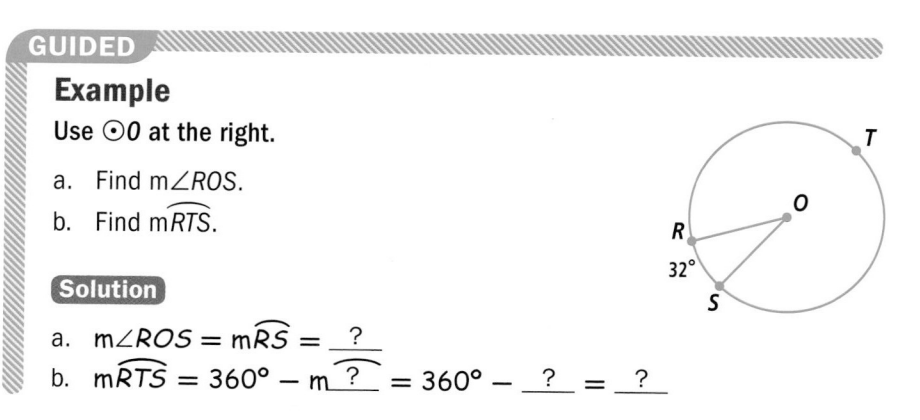

GUIDED

Example

Use $\odot O$ at the right.

a. Find $m\angle ROS$.

b. Find $m\widehat{RTS}$.

Solution

a. $m\angle ROS = m\widehat{RS} = \underline{\ ?\ }$

b. $m\widehat{RTS} = 360° - m\underline{\ ?\ } = 360° - \underline{\ ?\ } = \underline{\ ?\ }$

From this example, we see that the measure of a minor arc is equal to the measure of the corresponding central angle, while the measure of a major arc is equal to 360 minus the measure of the corresponding central angle.

The Angle Measure Postulate

We summarize the basic assumptions about angles in the *Angle Measure Postulate*. The first two assumptions of the Angle Measure Postulate ensure that every angle can be measured and an angle can be created with any given measure from 0° to 180°.

A straight central angle will divide its circle in half; thus, the measure of a straight angle is $\frac{360°}{2} = 180°$. A zero angle has no arc between its rays, so a zero angle has measure 0°. The *Zero Angle Assumption* and the *Straight Angle Assumptions* of the Angle Measure Postulate formalize these ideas.

Angle Measure Postulate

a. *Unique Measure Assumption*
   Every angle has a unique measure from 0° to 180°.

b. *Unique Angle Assumption*
   Given any ray $\overrightarrow{VB}$ and a real number r between 0 and 180, there is a unique angle *BVA* on each side of $\overrightarrow{VB}$ such that m∠*BVA* = r°.

c. *Straight Angle Assumption*
   If $\overrightarrow{VA}$ and $\overrightarrow{VB}$ are opposite rays, then m∠*BVA* = 180°.

d. *Zero Angle Assumption*
   If $\overrightarrow{VA}$ and $\overrightarrow{VB}$ are the same ray, then m∠*BVA* = 0°.

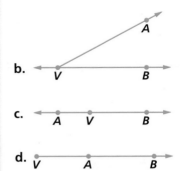

In Lesson 3-3, you will see a fifth assumption of the Angle Measure Postulate.

The Angle Measure Postulate tells us that every angle has a unique measure from 0° to 180°. In this book, angles are always measured in degrees, so we frequently omit the degree symbol ° when measuring angles. In this range, angles are classified into one of five types.

Definitions of Types of Angles

If *m* is the measure of an angle, then the angle is:
   a **zero angle** if and only if $m = 0$;
   an **acute angle** if and only if $0 < m < 90$;
   a **right angle** if and only $m = 90$;
   an **obtuse angle** if and only if $90 < m < 180$;
   a **straight angle** if and only if $m = 180$.

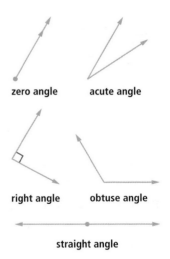

zero angle   acute angle

right angle   obtuse angle

straight angle

Interior and Exterior Angles

Every angle except a zero angle separates the plane into two nonempty sets other than the angle itself. If the angle is not a straight angle, exactly one of these sets is convex. The convex set is called the **interior of the angle.** The nonconvex set is the **exterior of the angle.** For a straight angle, both sets are convex and either set may be considered its interior.

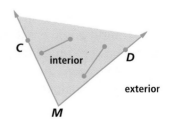

Assumptions from Drawings

Geometry is filled with diagrams. But you must be careful when viewing them. Diagrams can be misleading! For example, an angle that looks like its measure is 90° might actually have measure 90.01°. There are limits to what information you can use from a drawing.

From a figure you *can* assume	From a figure you *cannot* assume
collinearity and betweenness of points drawn on the same line	collinearity of three or more points that are not drawn on lines
intersections of lines at a given point	parallel lines
points in the interior of an angle, on an angle, or in the exterior of an angle	exact measures of angles or segments

Activity 2

Examine the figure at the right. Line *t* intersects lines *m* and *n*. Tell whether you can assume the statement from the figure.

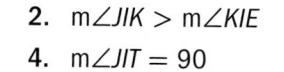

1. *T* is on *n*.
2. m∠JIK > m∠KIE
3. *m* ∥ *n*
4. m∠JIT = 90
5. *T* is between *A* and *E*.
6. *A*, *I*, and *E* are collinear.
7. *J*, *K*, and *L* are collinear.

Questions

COVERING THE IDEAS

In 1–3, draw a diagram that fits each description.

1. a zero angle named ∠*NUL*
2. a straight angle named ∠*FLT*
3. an obtuse angle named ∠*NMC*

In 4–6, consider the angle drawn at the right.

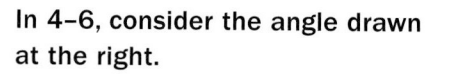

4. Give five names for this angle.
5. Name a point in its interior.
6. Name a point in its exterior.

In 7–11, decide whether the statement can be assumed from the diagram at the right.

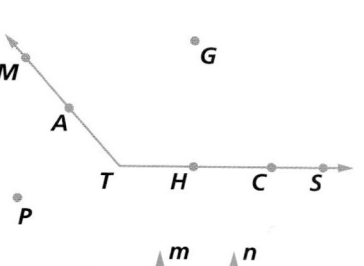

7. *F*, *D*, *C* are collinear.
8. *m* ∥ *n*
9. *C* is between *D* and *G*.
10. $\overrightarrow{AD}$ intersects $\overrightarrow{FG}$ at *D*.
11. *E* is in the interior of ∠*ADF*.

12. Use the diagram of circle *A* at the right.

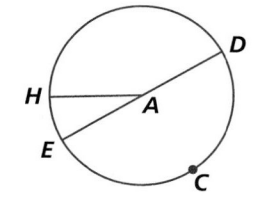

   a. Name all minor arcs.
   b. Name two major arcs.

APPLYING THE MATHEMATICS

13. Determine if each of the following situations creates an *acute*, an *obtuse*, a *right*, or a *straight* angle.

   a. the angle, formed by most flip-open cell phones when the phone is opened

   b. the angle that most tree trunks make with the ground

   c. the largest angle between the blades of a pair of scissors

   d. the angle formed by the hands of a clock at 7:45

14. Phillip Umpkin cut the pie at the right into 10 equal pieces by slicing it through its center.

   a. What is the measure of the arc formed by each piece?

   b. If $\frac{1}{5}$ of the pie has been eaten, what is the measure of the arc created by the remaining pieces?

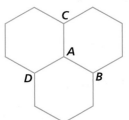

15. In the diagram of a portion of a tile floor at the right, the three angles with vertex A have the same measure.

   a. Name these angles.

   b. What is their measure?

16. What is the measure of the angle formed by the hands of the clock at 1 P.M.?

17. The points of a compass are labeled clockwise in degrees like a protractor. Airport runways are numbered with their compass direction, except that the final zero is dropped. The runway heading north is labeled 36 at the bottom, for 360°. North-south and east-west runways are labeled at the right.

   a. The Orchard Airport Expansion Project planners want to add a runway labeled 14. Copy the diagram at the right and draw runway 14.

   b. What number would be at the other end of the runway numbered 14?

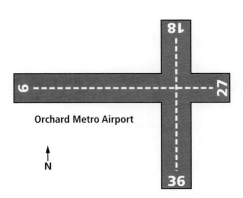

Orchard Metro Airport

18. The field of view for humans is about 210°. Draw an arc that represents a person's field of vision.

REVIEW

19. Consider the conjecture: If two circles intersect, then they intersect at two points. (**Lesson 2-7**)

   a. Draw an instance of the conjecture.

   b. Draw a counterexample to the conjecture.

   c. Draw two circles that are neither an instance nor a counterexample to the conjecture.

20. In the figure at the right, the diagonal in the hexagon cuts it into two quadrilaterals. What other combinations of polygons can be created by drawing a diagonal in a hexagon? (**Lesson 2-6**)

21. Describe the segment $\overline{AB}$ in the figure below as the intersection of two rays. (**Lesson 2-5**)

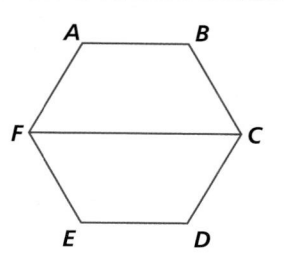

22. Consider this conditional statement: *If a definition is a good definition, then it accurately describes what is being defined.*
    a. Is this statement true? (**Lesson 2-4**)
    b. Is the converse of this statement true? Why or why not? (**Lesson 2-3**)

23. **Fill in the Blanks** If two sides of a triangle have lengths 10″ and 17″, then the third side can have any length between __?__ and __?__. (**Lesson 1-7**)

24. In graph theory, a *complete graph* on *n* vertices is a network with *n* nodes and arcs connecting each node to every other node. (**Lesson 1-3**)

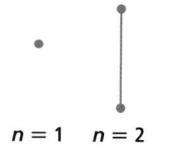   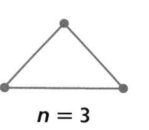   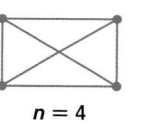   

$n = 1$ $n = 2$ $n = 3$ $n = 4$ $n = 5$

    a. Draw a complete graph for $n = 6$.
    b. How many arcs does a complete graph on 6 nodes have?

25. A 10-foot ladder is leaned against a wall as shown at the right. Think of the ladder as a line segment. (**Lesson 1-2**)
    a. If the bottom left point of the ladder is (0, 0), where is the top right point?
    b. What is the slope of the ladder?
    c. What is an equation for the line containing this ladder?

EXPLORATION

26. Degrees can be divided into minutes and seconds.
    a. From a dictionary or other source, find out how many minutes are in one degree.
    b. How many seconds are in one degree?
    c. The width of the moon covers an angle of about 30 minutes in the sky. How many moons placed next to each other would extend from one point on the horizon to the point on the opposite side of the horizon?

QY ANSWERS

1. $m\widehat{IK} = 30° + 102° = 132°$; $m\widehat{ILK} = 228°$

2. $\angle MAB$ and $\angle 1$

Arcs and Angles   **119**

Lesson

3-2 Rotations

Vocabulary

preimage

image

▶ **BIG IDEA** Every rotation has a point that is fixed that is its center and a magnitude, either positive or negative, that is the measure of an arc clockwise or counterclockwise.

Mental Math

A rectangle has sides of length s ft and 2s ft.

a. What is the perimeter of the rectangle?

b. What is the area of the rectangle?

Rotations

We turn frequently to see objects or to hear better or to move from one place to another. Hands on clocks turn, combination locks often involve turning, and wheels turn. The mathematical model for a turn is a *rotation*.

In Lesson 3-1, you saw that a fox's field of view is limited. How far will it have to turn its head to see something directly behind it? The fox can turn its head clockwise or counterclockwise. In mathematics, *counterclockwise* is the positive direction and *clockwise* is the negative direction.

Example 1

In the figure at the right, the dashed ray $\overrightarrow{OH}$ represents the direction the fox is facing. If the fox's field of vision is 85°, how much of a turn is necessary so that the fox can see something directly behind it?

Solution Point X, the edge of the fox's field of vision, has to move along the circle until it is behind the fox, 180° from the green dashed line. The fox can already see 42.5° to its right, so if it turns clockwise it needs to turn its head 180 − 42.5 = 137.5. This turn is clockwise, so the rotation has a negative magnitude. The fox has to turn −137.5°.

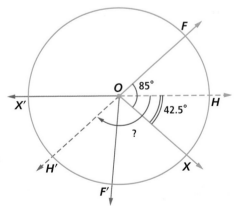

Activity

MATERIALS Protractor, tracing paper
Here is an algorithm for rotating △PQO 100° about G.

Step 1 Trace △PQO and point G onto tracing paper.

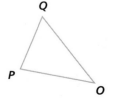

Step 2 On a second piece of paper, trace △PQO on the previous page again, but name the vertices P', Q' and O'.

Step 3 On the first sheet, draw $\overrightarrow{PG}$.

Step 4 Use a protractor to draw ∠PGM so that m∠PGM = 100°. (Make sure you draw $\overrightarrow{GM}$ at least as long as $\overrightarrow{GP}$.)

Step 5 Place the second piece of paper on top of the first, lining up the two papers on point G.

Step 6 Holding point G, rotate the top piece of paper so that point P' lies on $\overrightarrow{GM}$. These two sheets display the image of △PQO when rotated 100° about G as shown at the right.

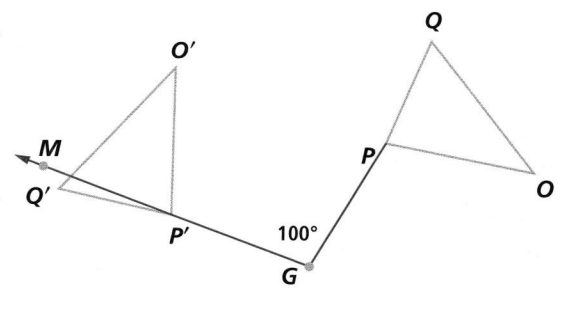

Step 7 Draw $\overline{GO}$, $\overline{GO'}$, $\overline{GQ}$, and $\overline{GQ'}$. Measure ∠OGO'. Measure ∠QGQ'. How do the measures of these angles compare to m∠PGP'?

In the Activity, the original figure △*PQO* is the **preimage**. The new figure △*P'Q'O'* is the **image** of △*PQO* under a rotation of 100°. Rotations always have a *center* whose image is itself. In the Activity, the center is *G*. Because the turn in the activity is counterclockwise, the *magnitude of the rotation* is positive, 100°. So, the magnitude of a rotation can be either positive or negative depending on the direction of the rotation. We denote rotations with a capital *R* and a subscript that states the center and the magnitude of the rotation. In the Activity, the rotation image of △*PQO* 100° about *G* can be denoted: $R_{G,100}$ (△*PQO*). When the center is obvious, or when there is only one center, we may assume the center is known and write R_{100}(△*PQO*).

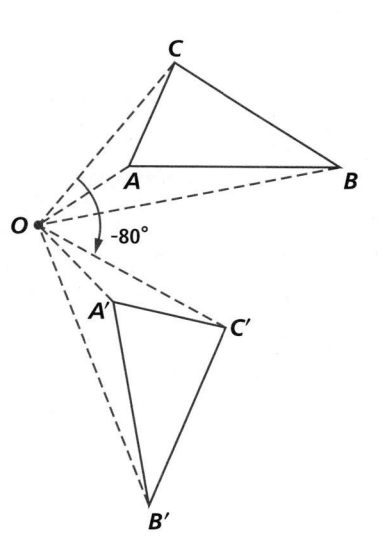

In the figure at the right, △*A'B'C'* is the image of preimage △*ABC*. The magnitude of this rotation is –80° because the rotation is in the clockwise direction. It is denoted R_{-80}(△*ABC*).

Rotations can also have magnitudes outside the range of –180° to 180°. For example, rotating 750° means rotating two 360° counterclockwise revolutions (2 · 360° = 720°) plus an additional 30°. So a rotation of 750° yields the same image as a rotation of 30°. Any rotation can be converted to a rotation in the range of –180° to 180°, by adding or subtracting a multiple of 360°. A figure skater who has rotated $1\frac{1}{2}$ revolutions rotated $1\frac{1}{2}$ · 360°.

🛑 **QY1**

> ▶ **QY1**
>
> Convert R_{-400} to a rotation with a magnitude in the range of –180° to 180°.

Example 2

Pentagon *P'E'N'T'A'* is the image of *PENTA* under a rotation about the center *G*. What is the magnitude of the rotation?

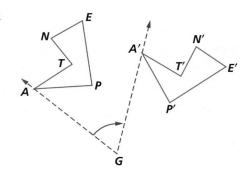

Solution Draw $\overrightarrow{GA}$ and $\overrightarrow{GA}'$ and measure $\angle AGA'$. $m\angle AGA' =$ 65°. Because the rotation is clockwise, the magnitude of the rotation is negative. Hence, the magnitude of the rotation is –65°.

A DGS can perform a rotation rather quickly. Find out the commands that your DGS uses to perform rotations. Generally, you must select the figure to be rotated, the center of the rotation, and identify the magnitude.

Example 3

Use a DGS to rotate △*ABC* about point *P* with a magnitude of 40°.

Solution Construct circle *O* and identify points *S* and *T* on it. Measure ∠*TOS*. Select △*ABC*, the point *P*, and the measure of ∠*TOS* and use the rotation command. If the image is not labeled, then label the corresponding points.

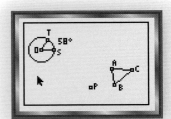
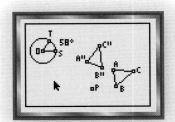

Drag point *T* around the circle until m∠*TOS* = 40°. △*A'B'C'* will rotate about *P* until $R_{P,\,40}(\triangle ABC) = \triangle A'B'C'$.

 QY2

> **QY2**
>
> What happens when you move point *T* to point *S*?

Rotations can be useful in picturing repeating patterns. For instance, on a piano keyboard, the notes repeat themselves as shown below. The three marked keys form a C-major chord.

When the 12 different possible notes are placed on a circle as shown at the right, the notes of a C-major chord are represented by the vertices of triangle *CEG*. To find the notes of the other major chords, you can rotate the triangle clockwise about the center *O* of the circle.

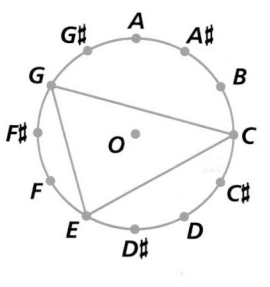

Example 4

Name the notes of a D♯ (D-sharp) major chord.

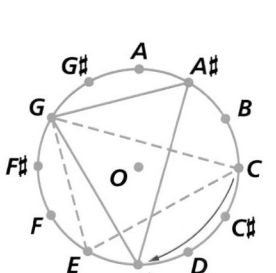

Solution Rotate △*CEG* so that the image of C is D♯. This is a rotation of 3 places in the clockwise direction. The notes are D♯, G and A♯.

STOP QY3

▶ QY3

What is the magnitude of the rotation in Example 4?

Questions

COVERING THE IDEAS

1. What happens if you rotate an object 360° about a point?

2. Trace the figure at the right and draw $R_{D, 110}$ (△*ABC*).

3. In the figure at the right, *H′J′I′G′* represents a flag that is the rotation image of *HJIG* about *K*. What is the magnitude of this rotation?

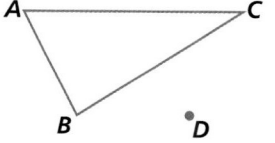

4. Copy point *P* and △*XYZ* onto a piece of tracing paper. Use another sheet of tracing paper to rotate △*XYZ* counterclockwise about point *P* by m∠*ABC*.

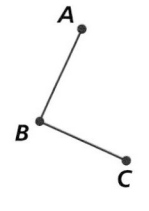

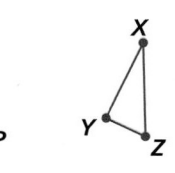

5. Name all the notes of a D-major chord.

6. An Olympic discus thrower may do $1\frac{1}{2}$ rotations before releasing the discus. Through how many degrees does the athlete turn?

7. In the Activity, how do m∠OGO′, m∠QGQ′ and m∠PGP′ compare?

APPLYING THE MATHEMATICS

8. Refer to this quotation by tennis player Maria Sharapova after she won the U.S. Open in 2006: "I figured I lost the last four times against Justine [Henin] so I thought I would just flip everything 360° and do the total opposite." What do you think she meant to say? Explain your answer.

9. Trace the diagram at the right and rotate △ABC 60° about B.

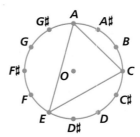

10. At the right is a circle with an A-minor chord represented by △ACE.
   a. Find the notes of a G♯-minor chord.
   b. What is the magnitude of the rotation you needed to perform to answer Part a?

11. In ice skating, a triple axel consists of $3\frac{1}{2}$ full rotations in the air. In the 2006 U.S. Figure Skating Championships, American pairs skaters Rena Inoue and John Baldwin Jr. became the first pair to complete a triple axel in competition. What is the total number of degrees that a skater rotates in a triple axel?

12. Refer to the graph at the right. Suppose $\overline{BA}$ is rotated 270° about the origin. Let $\overline{B'A'}$ be the image of $\overline{BA}$.
   a. What are the coordinates of B' and A'?
   b. Write an equation for the line containing points B' and A'.

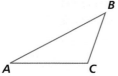

13. A point P is placed on a circle as shown at the right. P is rotated about O a magnitude w. A segment is drawn from P to its image P'. P' is then rotated about O at the same magnitude w. A segment is drawn from P' to its image P''. Continue this process until reaching the original point P.
   a. What magnitude w would result in a triangle whose sides have the same length?
   b. What magnitude w would result in a pentagon whose sides have the same length?

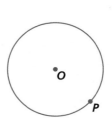

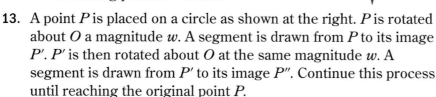

REVIEW

14. Refer to ⊙A at the right with diameter $\overline{BC}$. **(Lesson 3-1)**
    a. Name the straight angle.
    b. Name two right angles.
    c. Name an acute angle and give its measure.
    d. Name two obtuse angles and give their measure.

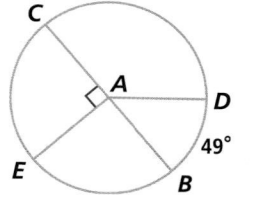

15. What fraction of a circle is determined by a 45° central angle? **(Lesson 3-1)**

16. Refine the following conjecture: Every odd integer between 1 and 10 is a prime number. **(Lesson 2-7)**

17. In the coordinate plane, let I be the set of points whose x-coordinate is an integer and J be the set of points whose y-coordinate is an integer. **(Lesson 2-5)**
    a. Describe the set $I \cap J$.
    b. Describe the set $I \cup J$.

18. Let r = You forgot your locker combination and s = You cannot open your locker. **(Lessons 2-3, 2-2)**
    a. Write $r \Rightarrow s$ in words.
    b. Write the converse of $r \Rightarrow s$ in words. Is it true?

EXPLORATION

19. In an accurate circle graph, the ratios of the measures of the central angles should be the same as the ratios of the quantities graphed. Find the total population of students in each grade in your school. Make an accurate circle graph of the data.

20. The design below was created by rotating a polygon 60° about a given point. The image of the polygon was then rotated 60° about the same point. This process was repeated to get the entire design. Create a new design in which a polygon is rotated 45° about a fixed point. Once your design is complete, create different designs by manipulating your original polygon. Print three different designs based on your 45° design.

Lesson

3-3

Adjacent Angles and Vertical Angles

Vocabulary

adjacent angles

angle bisector

complementary angles

supplementary angles

complements

supplements

linear pair

vertical angles

▶ **BIG IDEA**   A basic property of angles is that the measure of an angle formed by the outside rays of adjacent angles is the sum of the measures of the adjacent angles.

Adjacent Angles

Some pairs of angles have special names depending on their positions relative to each other. *Adjacent angles* are created when a ray is drawn in the interior of an angle. In the figure at the right, $\overrightarrow{OB}$ is in the interior of $\angle COA$ and is the common side of $\angle COB$ and $\angle BOA$.

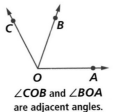

∠*COB* and ∠*BOA* are adjacent angles.

Mental Math

In the figure below, name:

a. a zero angle.

b. an acute angle.

c. an obtuse angle.

d. a straight angle.

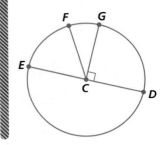

Definition of Adjacent Angles

Two nonstraight and nonzero angles are **adjacent angles** if and only if a common side is in the interior of the angle formed by the noncommon sides.

Identification of adjacent angles can be tricky.

GUIDED

Example 1

Tell why the indicated angles of the figures below are not adjacent.

1. ∠*G* and ∠*E*

2. ∠7 and ∠8

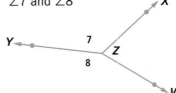

Solution

1. The sides of ∠*G* are __?__ and __?__. The sides of ∠*E* are __?__ and __?__. Thus, ∠*G* and ∠*E* do NOT have a common side.

2. The common side of ∠7 and ∠8 is __?__. __?__ is not in the interior of ∠*XZW*.

If two adjacent angles are equal in measure, then their common side is an *angle bisector*.

Definition of Angle Bisector

$\overrightarrow{VR}$ is an **angle bisector** of $\angle PVQ$ if and only if $\overrightarrow{VR}$ (except point V) is in the interior of $\angle PVQ$, and $m\angle PVR = m\angle RVQ$.

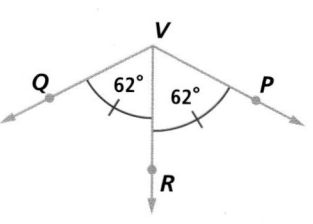

$\overline{VR}$ bisects $\angle PVQ$.

When $\overrightarrow{VR}$ is the bisector of $\angle PVQ$, we say that $\overrightarrow{VR}$ bisects $\angle PVQ$.

Notice that the definition of angle bisector is similar in form to the definition of midpoint. The tick marks on the arcs in the angles mean that the measures of the two angles are equal.

Adjacent angles are important because their measures can be added to obtain the measure of the largest angle they form. For instance, in the figure above at the right, $m\angle QVP = 62 + 62 = 124$. This is the last assumption we state about angles. It completes the Angle Measure Postulate, which was first introduced in Lesson 3-1.

Angle Measure Postulate

e. *Angle Addition Assumption*
   If angles *AVC* and *CVB* are adjacent angles, then
   $m\angle AVC + m\angle CVB = m\angle AVB$.

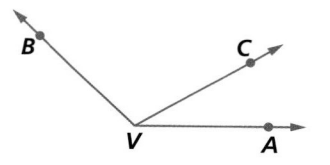

GUIDED

Example 2

From the drawing at the right, determine:

a. $m\angle DAE$.   b. $m\angle BAE$.

Solution

a. $\overrightarrow{AC}$ is in the interior of $\angle DAE$. Using the Angle Addition Assumption,

$$m\angle DAE = m\angle EAC + m\angle \underline{\ ?\ }$$
$$= \underline{\ ?\ } + 35$$
$$= \underline{\ ?\ }$$

b. Think of A as the center of a circle. The measures of the angles about A add to 360. So,

$$m\angle BAE = 360 - (\underline{\ ?\ } + \underline{\ ?\ } + \underline{\ ?\ })$$
$$= 360 - \underline{\ ?\ }$$
$$= \underline{\ ?\ }$$

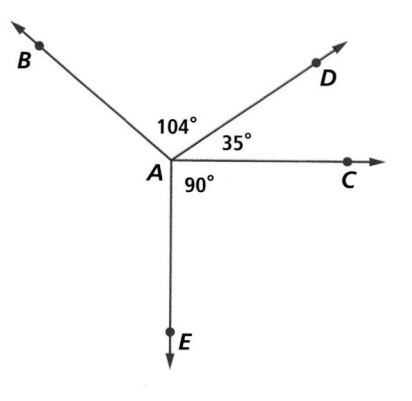

Complementary and Supplementary Angles

The Angle Addition Assumption allows you to add the measures of adjacent angles. However, we also give a special name to angle pairs whose measures add to 90 or 180, whether or not they are adjacent.

Definitions of Complementary Angles and Supplementary Angles

If the measures of two angles are r and s, then the angles are:
complementary angles if and only if $r + s = 90$;
supplementary angles if and only if $r + s = 180$.

Notice that the letter C (for complementary) comes before the letter S (for supplementary) as 90 comes before 180. This may help you remember which word goes with which number.

In the figure at the right, $\angle A$ and $\angle B$ are complementary angles. You can also say $\angle A$ and $\angle B$ are **complements**, or $\angle A$ is a complement of $\angle B$. In the same figure, $\angle A$ and $\angle C$ are supplementary angles, $\angle A$ and $\angle C$ are **supplements**, and $\angle A$ is a supplement of $\angle C$. Notice that complementary and supplementary refer only to the measures of the angles, not to where they are drawn. An angle on the floor and an angle on the moon are complements if the sum of their measures is 90.

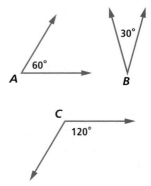

If the measure of an angle is x, then the measure of any angle complementary to it is $90 - x$ and the measure of an angle supplementary to it is $180 - x$. As a consequence, there is the following theorem.

Equal Angle Measures Theorem

If two angles have the same measure, their complements have the same measure.

If two angles have the same measure, their supplements have the same measure.

Example 3

Suppose you are given two angles A and B in which $m\angle A = m\angle B$. If the measure of a complement of $\angle A$ equals $5x + 20$ and the measure of a complement of $\angle B$ equals $3x + 40$, what is the measure of a supplement of $\angle B$?

Solution Because $m\angle A = m\angle B$, their complements are equal in measure. Use this information to write an equation and solve for x.

$$5x + 20 = 3x + 40$$
$$2x = 20$$
$$x = 10$$

Substituting 10 for x in 3x + 40 gives 3(10) + 40 = 70. This is the measure of a complement of ∠B. Thus, m∠B = 90 − 70 = 20. Because m∠B = 20, the measure of a supplement of ∠B is 160.

Linear Pairs

Certain adjacent angles form *linear pairs*. In the figure at the right, ∠1 and ∠2 are a linear pair.

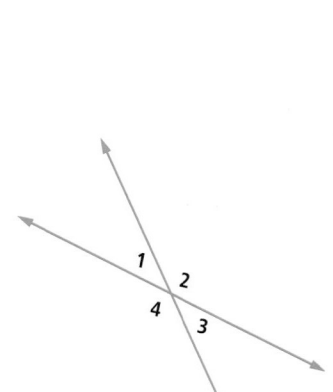

▶ QY

Refer to the figure at the left.

a. If m∠1 = 52, what is m∠2?

b. If m∠1 = q, what is m∠2?

Definition of Linear Pair

Two adjacent angles are a **linear pair** if and only if their noncommon sides are opposite rays.

Linear pairs have a special property. When ∠1 and ∠2 are a linear pair, $\overrightarrow{VB}$ and $\overrightarrow{VA}$ are opposite rays (definition of linear pair). As a result, m∠AVB = 180 (Straight Angle Assumption). Because of this, m∠1 + m∠2 = 180 (Angle Addition Assumption). Consequently, ∠1 and ∠2 are supplementary (definition of supplementary angles). This argument proves the following theorem.

Linear Pair Theorem

If two angles form a linear pair, then they are supplementary.

🛑 QY

Vertical Angles

When two lines intersect, four angles are formed. Each nonadjacent pair is a pair of *vertical angles*.

Definition of Vertical Angles

Two nonstraight angles are **vertical angles** if and only if the union of their sides is two lines.

The word vertical here does not mean "straight up and down." In the figure at the right, ∠1 and ∠3 are one pair of vertical angles. ∠2 and ∠4 are another pair of vertical angles. An important theorem stems from this figure.

Vertical Angles Theorem

If two angles are vertical angles, then their measures are equal.

Using properties from algebra, the following five steps give a *proof* of the *Vertical Angles Theorem*.

1. ∠1 and ∠3 are vertical angles because that is given.

2. ∠1 and ∠2 are a linear pair, and ∠2 and ∠3 are a linear pair based on the given diagram.

3. So m∠1 + m∠2 = 180, and m∠2 + m∠3 = 180 because the two angles forming a linear pair are supplementary (Linear Pair Theorem).

4. m∠1 + m∠2 = m∠2 + m∠3 by the Transitive Property of Equality.

5. Thus, m∠1 = m∠3 by the Addition Property of Equality (Add –m∠2 to each side.)

Questions

COVERING THE IDEAS

In 1–4, fill in the blanks.

1. The measures of two supplementary angles add to __?__.

2. The measures of two complementary angles add to __?__.

3. The measures of two vertical angles are __?__.

4. Two angles form a __?__ __?__ if and only if their noncommon sides are opposite rays.

5. Using the diagram at the right, tell whether each pair of angles is adjacent. If the angles are not adjacent, explain why they are not.

    a. ∠AOB and ∠COD   b. ∠AOC and ∠BOC

    c. ∠COB and ∠COD   d. ∠EOD and ∠EOA

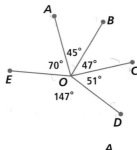

6. a. In the figure at the right, ∠AOB and ∠BOC are adjacent angles. Write an equation relating m∠AOB, m∠BOC, and m∠AOC.

    b. If m∠AOB = 23 and m∠AOC = 103, find m∠BOC.

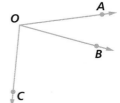

7. Suppose ∠7 and ∠8 are supplementary and adjacent angles, with m∠7 = 155.

    a. Sketch a possible situation.   b. Find m∠8.

In 8–10, use the figure at the right.

8. **Fill in the Blank** ∠1 and ∠4 are __?__ angles.

9. If m∠1 = 35, find the measures of ∠2, ∠3, ∠4, and ∠5.

10. If m∠1 = y, find the measures of ∠2, ∠3, ∠4, and ∠5.

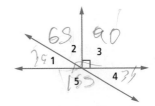

11. Suppose that an angle has five times the measure of a complement to it. Find the measure of the angle.

12. In the figure at the right, $\overrightarrow{CF}$ bisects $\angle DCE$. $m\angle DCE = 124$ and $m\angle FCE = 6x + 4$. Find x.

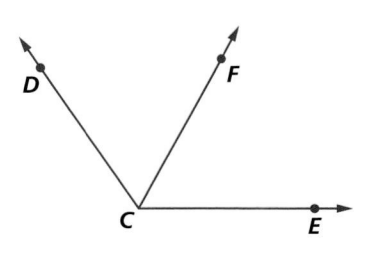

13. Is the statement always, *sometimes but not always*, or *never* true?
    a. The angles formed when an angle is bisected are adjacent angles.
    b. The angles formed when an obtuse angle is bisected are obtuse.

14. $\angle A$ has a measure of 39.2°. What is the difference between the measure of a supplement to $\angle A$ and the measure of the complement to $\angle A$?

APPLYING THE MATHEMATICS

15. Draw a diagram in which $\angle AOB$ is adjacent to $\angle BOD$, $\angle AOB$ is adjacent to $\angle COA$, and $\angle COD$ is not adjacent to $\angle AOC$ or to $\angle BOD$.

In 16 and 17, sketch a possible drawing of $\angle 1$ and $\angle 2$.

16. $\angle 1$ and $\angle 2$ are a linear pair in which $\angle 1$ is acute.

17. $\angle 1$ and $\angle 2$ are adjacent and complementary, and $m\angle 1 = 2m\angle 2$.

18. a. What is the measure of an angle with the same measure as its complement?
    b. What is the measure of an angle with the same measure as its supplement?

19. An angle's measure is 9 more than twice the measure of a supplement to it. Find the measures of the angle and its supplement.

20. The measure t of an angle is less than the measure of its supplement. Find all possible values of t.

21. Suppose an angle measure is x and its supplement is y. Graph the set of all possible ordered pairs (x, y).

22. Given an angle on a DGS, explain what you would do to construct a supplement of that angle.

23. The figure at the right shows the upper left portion of a compass. The ray creating the NW direction bisects the angle formed by the N and W directions. Similarly the rays creating the NNW and WNW directions are bisectors. If you were to go from W to NNW, how many degrees would you turn?

24. Find $m\angle AOC$ in the diagram at the right.

$(x + 6)°$ $(3x - 30)°$

25. Every type of grain has a unique angle of repose. The angle of repose is the angle that a pile of that particular grain makes with the ground (shown in the diagram) when it is allowed to fall into a pile from a grain elevator. Old McDonald would like to measure the angle of repose of this pile of corn, but he cannot get into the pile of corn to measure the angle. Write a sentence or two explaining how he could do this using the concept of linear pairs of angles.

26. a. Write the converse of the Linear Pair Theorem.
    b. Explain in words and with a drawing why the converse is not true.

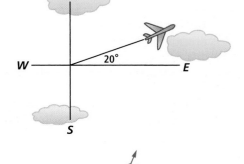

27. The plane in the diagram at the right is heading in a direction 20° north of east, written 20° N of E. How many degrees would it have to turn in order to head due north?

28. In the diagram at the right, $\overrightarrow{DG}$ bisects $\angle EDF$. Find m$\angle GDF$ if m$\angle EDG = 20v + 16$ and m$\angle GDF = 14v + 19$.

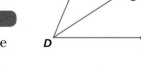

REVIEW

29. B is the image of point A under a rotation of 45° about C. Name two magnitudes for a rotation about C that maps B onto A. (Lesson 3-2)

30. Find the magnitude of a rotation in the coordinate plane about the origin under which the image of $(x, 0)$ is $(-x, 0)$. (Lesson 3-2)

31. Your protractor does not have a notch for the angle 37.432°. Which postulate assures that an angle with this measure exists? (Lesson 3-1)

32. Refer to the diagram at the right. Is the union of $ABCDE$ and $EDHGF$ a polygon? Why or why not? (Lessons 2-6, 2-5)

33. Let A be the set of positive integers. Let B be the set of even positive integers. Identify the members of set C so that $B \cap C = \{\ \}$ and $B \cup C = A$. (Lesson 2-5)

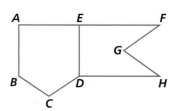

EXPLORATION

34. The words *compliment* and *complement* sound alike but mean different things. They are called *homonyms*.
    a. What is the meaning of *compliment*?
    b. Find at least two other mathematical terms that have homonyms.

QY ANSWERS

a. 128

b. $180 - q$

Lesson 3-4

Algebra Properties Used in Geometry

> ▶ **BIG IDEA** The properties of operations of real numbers that you used in arithmetic and algebra can be applied in geometry.

In Chapter 1, we discussed how the rules of a game affect the outcome. When you find measures in figures, you combine arithmetic, algebra, and geometry. Consequently, the properties of numbers and operations in arithmetic and algebra are important in geometry. In this lesson you are asked to recall some properties from arithmetic and algebra that you will need in this course.

Types of Properties

There are three types of properties in mathematics:

- *Defining properties* are those assigned to a word in its definition.
- *Assumed properties* are postulates.
- *Deduced properties* are theorems.

Below are properties of real numbers, but they are true also for all algebraic expressions and variables that stand for real numbers.

Describing Equality

The first set of properties describes basic properties of the undefined equal sign. The names of these properties begin with the letters R, S, and T. That can help you remember their names.

Postulates of Equality

For any real numbers a, b, and c:

Reflexive Property of Equality: $a = a$
Symmetric Property of Equality: If $a = b$, then $b = a$.
Transitive Property of Equality: If $a = b$ and $b = c$, then $a = c$.

Mental Math

Decide whether the rules of each game are an instance, a counterexample, or neither of this conditional: *If you have a higher score, you win.*

a. basketball

b. chess

c. golf

The Reflexive Property of Equality states that every number is equal to itself. This may seem unnecessary to write, but it is important when you want to note that a segment's length does not change when it is considered as a side of two different triangles. The Symmetric Property of Equality allows you to say things like $x = 3$ when your work said $3 = x$. The Transitive Property of Equality enables you to conclude that if two measures are each equal to a third measure, then the first two measures are equal.

Equality and the Operations of Addition and Multiplication

The second set of properties shows how equality is related to the operations of addition and multiplication. You have used these properties when solving equations.

> ### Postulates of Equality and Operations
>
> For any real numbers a, b, and c:
>
> *Addition Property of Equality:* If $a = b$, then $a + c = b + c$.
> *Multiplication Property of Equality:* If $a = b$, then $ac = bc$.

Example

Show the postulates of equality and the operations used in solving $50x - 8 = 120$.

Solution We indicate what was done by giving steps at the left and writing properties on the right.

$50x - 8 = 120$	*Given*
$50x = 128$	*Addition Property of Equality*
	(If $a = b$, then $a + c = b + c$. Here, $c = 8$.)
$x = \frac{128}{50}$	*Multiplication Property of Equality*
	(If $a = b$, then $ac = bc$. Here $c = \frac{1}{50}$.)
$x = 2.56$	*Rewrite the fraction as a decimal.*

These properties are also used in geometry. Suppose that $m\angle AVB = m\angle CVD$ in the figure at the right. By adding $m\angle BVC$ to both sides of the equation, you can conclude that $m\angle AVC = m\angle BVD$.

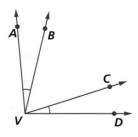

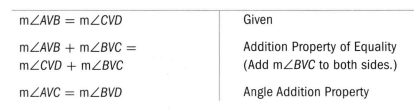

$m\angle AVB = m\angle CVD$	Given
$m\angle AVB + m\angle BVC = m\angle CVD + m\angle BVC$	Addition Property of Equality (Add $m\angle BVC$ to both sides.)
$m\angle AVC = m\angle BVD$	Angle Addition Property

Properties of Inequality

Now we ask: If the equal sign = is replaced by the inequality sign < in the preceding postulates, are the resulting properties true?

Is there a Reflexive Property of Inequality? No, because –5 < –5 is not true.

Is there a Symmetric Property of Inequality? If –1 < 72, is 72 < –1? No.

Is there a Transitive Property of Inequality? Given –4 < 9 and 9 < 13, is it true that –4 < 13? Yes. This would be the case whatever numbers we used.

There is an Addition Property of Inequality. For example, if $7y + 2 < 0$, then –2 can be added to both sides to get $7y < -2$.

The Multiplication Property of Inequality is slightly more complicated. Consider the true inequality –6 < 10. If you multiply both sides by a positive number such as $\frac{1}{2}$, you get –3 < 5 and the direction of the inequality does not change. But if you multiply both sides of –6 < 10 by a negative number such as –2, the direction of the inequality must change: 12 > –20.

🛑 QY

These properties of inequality are formally stated below in the third set of postulates.

▸ QY

What happens if you multiply both sides of the inequality –6 < 10 by 0?

Postulates of Inequality and Operations

For any real numbers a, b, and c:

Transitive Property of Inequality: If $a < b$ and $b < c$, then $a < c$.
Addition Property of Inequality: If $a < b$, then $a + c < b + c$.
Multiplication Property of Inequality:
 If $a < b$ and $c > 0$, then $ac < bc$.
 If $a < b$ and $c < 0$, then $ac > bc$.

These postulates are useful in geometric situations.

If $m\angle P < m\angle Q$ and $m\angle Q < m\angle R$, then the Transitive Property of Inequality allows you to conclude that $m\angle P < m\angle R$.

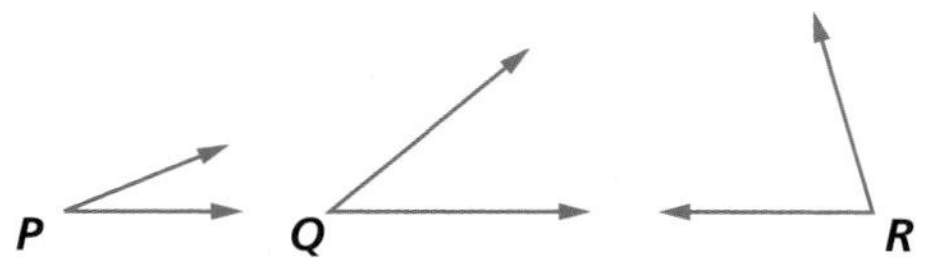

In the figure at the right, $m\angle DBC = x$, $m\angle HEF = y$, and $m\angle ABD = m\angle GEH = z$. If $x < y$, then by the Addition Property of Inequality, $x + z < y + z$. Then $m\angle ABC < m\angle GEF$.

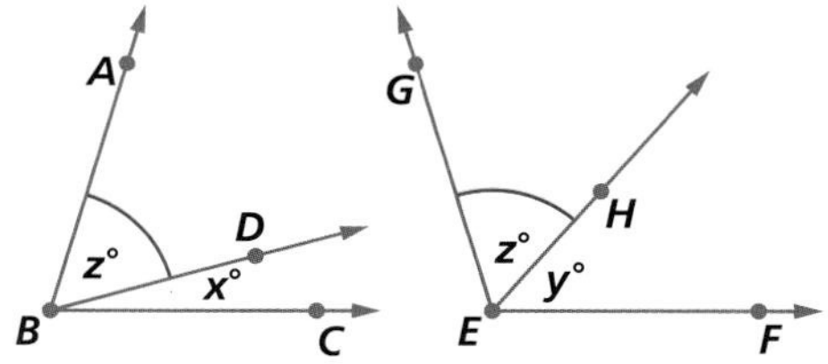

Equations and Inequalities

A fourth set of properties applies to both equations and inequalities. The mathematics of these two properties should be familiar to you, but the formal statements of them may not be.

Postulates of Equality and Inequality

For any real numbers a, b, and c:

Equation to Inequality Property:
   If a and b are positive numbers and $a + b = c$, then $c > a$ and $c > b$.

Substitution Property:
   If $a = b$, then a may be substituted for b in any expression.

The Equation to Inequality Property is sometimes informally stated as "The whole is greater than any of its parts." For example, in the figure at the right, the length of the whole segment is greater than any part of it. That is, $XY + YZ = XZ$, so $XZ > XY$ and $XZ > YZ$.

You have used the Substitution Property many times. Suppose you know $2x + y = 18$ and you know $y = 3x$. You may substitute $3x$ for y and get $2x + 3x = 18$.

Other Assumed Properties of Operations

You are familiar with other properties of numbers and operations. Among these are the Commutative and Associative Properties of Addition and Multiplication; the properties of 0, 1, and –1 in addition and multiplication; and the properties of opposites and reciprocals. These properties are so well-known that we usually use them without explicitly stating them. But one property of this type is so important that, when it is used, we often name it. It is the *Distributive Property of Multiplication over Addition*: For any real numbers a, b, and c: $a(b + c) = ab + ac$. For example, $3(x + -y) = 3x + -3y$ and $2A + 3A = (2 + 3)A = 5A$. You will see the Distributive Property applied often in the discussions of area and volume later in this book.

Questions

COVERING THE IDEAS

1. What undefined term do the Postulates of Equality describe?

2. Name the three assumed properties of equality.

3. Because $\frac{1}{2} = 0.5$ and $0.5 = 50\%$, what can be concluded using the Transitive Property of Equality?

4. **a.** Solve $7x + 17 = 31$.

   b. What properties did you use in finding your solution?

5. **a.** Solve $7x + 17 < 31$.

   b. What properties did you use in finding your solution?

6. Given $m\angle APB = m\angle CPD$ in the figure at the right. What can you conclude using the Addition Property of Equality and the Angle Addition Postulate?

7. Suppose $m\angle T > m\angle Q$. What can you conclude about $-4m\angle T$ and $-4m\angle Q$?

8. In the figure at the right, $DE + EF = DF$. What can you conclude using the Equation to Inequality Property?

9. Suppose an angle has measure $90 - x$, and $x = 23$.

   a. What does the Substitution Property allow you to conclude?

   b. What is the measure of the angle?

APPLYING THE MATHEMATICS

10. **Multiple Choice** According to the Distributive Property, if A, B, and C are points, then $2(AB + BC) = $ __?__.

    A $2AB + 2BC$ **B** $4B(A + C)$ **C** $2AB + BC$ **D** $4AB + 4BC$

In 11–13, a situation is given. What postulate mentioned in this lesson explains why the situation is true?

11. Mrs. Vasquez earns more at her job than her husband earns at his job. If they receive raises of the same amount, then Mrs. Vasquez will still earn more.

12. It's the same distance from my house to school as it is from your house to school. And it is of course the same distance from school to my house as it is from my house to school. So it is the same distance from school to my house as it is from your house to school.

13. The grocery bill for eight steaks and two slabs of ribs came to $65.89, so the eight steaks must have cost less than $65.89.

14. Steps in solving the equation $1024 - 51x = 89$ are shown below. What postulate was used to obtain each equation?

    a. $1024 = 89 + 51x$ **b.** $935 = 51x$

    c. $\frac{935}{51} = x$ **d.** $x = \frac{935}{51}$

15. **Multiple Choice** Suppose that in a triangle ACE, $AC = CE$. Which of these postulates justifies $CE = AC$?

    A Symmetric Property of Equality

    B Addition Property of Equality

    C Transitive Property of Equality

16. The symbol ≠ means "is not equal to."
    a. Is this relation reflexive?
    b. Is this relation symmetric?
    c. Is this relation transitive?
    d. **True or False** For any real numbers a, b, and c: if $a \neq b$, then $a + c \neq b + c$.
    e. **True or False** For any real numbers a, b, and c: if $a \neq b$, then $ac \neq bc$.

17. Consider the figure at the right.

   a. If you know $PQ = RS$, write an argument using the properties of this lesson and the definition of betweenness to show that $PR = QS$.
   b. If "=" in Part a were changed to "<," how would this change the problem and solution in Part a?

REVIEW

18. Use the figure at the right to find the value of x. **(Lesson 3-3)**

19. a. State the converse of the Vertical Angles Theorem.
       (Lessons 3-3, 2-3)
    b. Explain why this converse is not true. **(Lesson 3-3)**

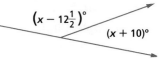

20. In the diagram at the right, the upper star is the image of the lower star under a rotation about the center D. Trace the diagram and estimate the magnitude of the rotation. **(Lesson 3-2)**

21. $A = (0, 5)$, $B = (5, 5)$, and $C = (5, 0)$. Find the coordinates of the images of A, B, and C under a rotation of $-90°$ about the origin. **(Lesson 3-2)**

22. Trace Circle O at the right and draw a semicircle that contains five of the named points. **(Lesson 3-1)**

23. The field of vision for humans is about $210°$. Two people are sitting back-to-back at the center of a large circle. Is there any point on the circle that neither of them can see? Explain your answer. **(Lesson 3-1)**

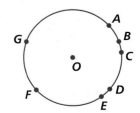

EXPLORATION

24. Using the Equation to Inequality Property and the picture at the right, write as many inequality statements as you can.

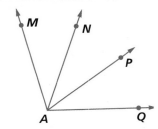

QY ANSWER

The inequality becomes false because 0 is not less than 0.

Lesson
3-5 Justifying Conclusions

Vocabulary

proof

▶ **BIG IDEA** The basis of mathematical reasoning is the making of conclusions justified by definitions, postulates, or theorems.

Simplifying expressions, solving equations, and applying algebraic techniques all represent a series of justified steps. If the given information p and the conclusion q following it are important enough, then the statement $p \Rightarrow q$ is labeled as a theorem. Showing q follows from p *proves the conditional* $p \Rightarrow q$.

When you write a proof, you need to indicate the given information p, the statement q that follows from the conditional, and the proof showing how q follows from p.

Mental Math

Find an equation of a line through (2, –3) that is

a. horizontal.

b. vertical.

c. oblique.

Definition of Proof

A **proof** of a conditional is a sequence of justified conclusions starting with the antecedent and ending with the consequent.

Every time you have written down how you solved an equation, you have proved a conditional. When you solve $4x - 12 = 10$ and find that $x = 5.5$, your steps are the conclusions in the proof of if $4x - 12 = 10$, then $x = 5.5$. When you justify the steps, you have written a proof!

Example 1 shows a proof using justifications from algebra.

Example 1

Given: $4x - 12 = 10$

Prove: $x = 5.5$

Proof

Conclusions	Justifications
1. $4x - 12 = 10$	1. Given
2. $\quad 4x = 22$	2. Addition Property of Equality (Add 12 to both sides.)
3. $\quad x = 5.5$	3. Multiplication Property of Equality (Multiply both sides by $\frac{1}{4}$.)

Placing the statements in columns is a way to organize your proof. The two columns remind you that each statement must have a justification. Here is an example with a justification from geometry.

Example 2

Given: P and Q are points on a circle with center O.

Prove: $OP = OQ$

Proof

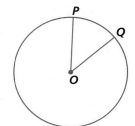

Conclusions	Justifications
1. P and Q are points on a circle with center O.	1. Given
2. $OP = OQ$	2. Definition of a circle (A circle is the set of points in a plane at a given distance from a given point.)

 QY

▶ **QY**

If you know that $m\angle ABC = 46$, what is the justification for the conclusion that $\angle ABC$ is acute?

Proof Outside Mathematics

The idea of proof is also used outside mathematics. Investigations use the following assumptions:

(1) No two people have exactly the same fingerprints.

(2) The fingerprints of a person cannot be altered.

From these two assumptions, we can prove: If fingerprints F that are the same as those that a person P has now are found at a place, then person P must have been at that place.

Given Fingerprints F that are the same as P's fingerprints now were found at a place.

Prove P was at the place.

Proof

Conclusions	Justifications
1. Fingerprints F were found at a place.	1. Given
2. P's fingerprints are F.	2. Given
3. P's fingerprints were F at the time the fingerprints were left at the place.	3. Assumption (2)
4. P was at the place.	4. Assumption (1)

The justification "Given" means that this information is the antecedent of what is to be proved. A justification can also be a definition (as in Example 2), a postulate (as in Example 1), or a previously proved theorem. Here are two of the geometry theorems that have already been proved in this book:

- Linear Pair Theorem: If two angles form a linear pair, then they are supplementary.

- Vertical Angles Theorem: If two angles are vertical angles, then they have the same measure.

Notice how one of these theorems is used in the following proof and how we use one of the things that you can assume from a figure.

GUIDED

Example 3

Given: the figure at the right

Prove: $m\angle 1 + m\angle 2 = 180$

Proof

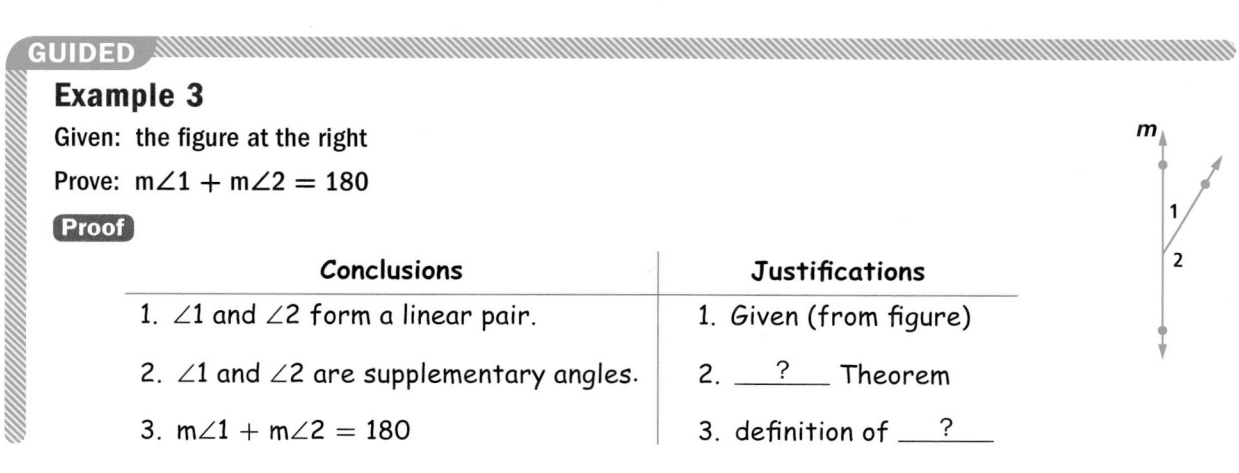

Conclusions	Justifications
1. $\angle 1$ and $\angle 2$ form a linear pair.	1. Given (from figure)
2. $\angle 1$ and $\angle 2$ are supplementary angles.	2. ___?___ Theorem
3. $m\angle 1 + m\angle 2 = 180$	3. definition of ___?___

Why Are There Proofs?

The basic reason for proofs in mathematics is:

- A proof establishes, without question, the truth of a statement from the postulates and definitions in the system.

There are other reasons for proofs. Here are four of them.

- Sometimes people disagree. What is obvious to one person may not be obvious to another person.

- Unexpected results can be verified. For instance, in Chapter 1 you learned Euler's Theorem about traversable networks.

- If a statement cannot be proved after many people have worked on it for a long time, it is quite possible that either (1) the statement cannot be proved or disproved from what you know (definitions, theorems, postulates) or (2) the statement is not true even though it looks true.

- Proofs can help you organize mathematical ideas and make connections among them.

To write justifications, you obviously need to be familiar with definitions, postulates, and theorems. When you write a proof, you can only use the justifications that have been presented before the statement you are proving. As you go through this book, the supply of definitions and theorems you can use as justifications will grow. This will give you yet another reason for proof. You will be able to prove some wonderful and useful properties of geometric figures.

Questions

COVERING THE IDEAS

In 1–3, state the justification for the conclusion.

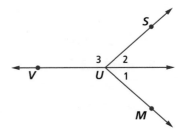

1. Given: Angles 1 and 2 as shown at the right.
   Conclusion: m∠*SUM* = m∠1 + m∠2.

2. Given: Angles 2 and 3 as shown at the right.
   Conclusion: Angles 2 and 3 are supplementary.

3. Given: m∠3 = 135.
   Conclusion: ∠3 is an obtuse angle.

4. Define *proof.*

5. What three kinds of statements can be used as justifications in a proof?

6. List four reasons why there are proofs in mathematics.

APPLYING THE MATHEMATICS

7. Fill in the missing parts of this proof.

   Given ∠1 and ∠2 are vertical angles.

   Prove ___?___

   Proof

Conclusions	Justifications
a. ∠1 and ∠2 are vertical angles.	a. ___?___
b. ___?___	b. Vertical Angles Theorem (If two angles are vertical angles, then their measures are equal.)

8. Write the conditional statement of the form $p \Rightarrow q$ for the proof in Question 7.

9. Refer to the proof using fingerprints on page 140. If "fingerprints" is replaced by DNA, does the proof still make sense? Why or why not?

10. Given m∠3 = m∠4, a student concluded that ∠3 and ∠4 are vertical angles, using the definition of vertical angles as the justification. What is wrong with this reasoning?

In 11–13, state the justification for the conclusion.

11. Given: W is between A and C.
    Conclusion: $CW + AW = AC$

12. Given: Line m contains points A and B.
    Line n contains points A and B.
    Conclusion: $m = n$

13. Given: $CW + AW = AC$, $CW = 50$, and $AC = 65$
    Conclusion: $50 + AW = 65$

Multiple Choice In 14 and 15, use the figure at the right. Which is the correct justification for the conclusion?

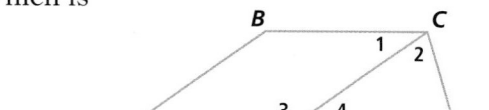

14. Given: $m\angle 1 = 36$; $m\angle 2 = 70$
    Conclusion: $m\angle BCD = 106$

    A Angle Addition Property   **B** Linear Pair Theorem
    C definition of obtuse angle   **D** Vertical Angles Theorem

15. Given: $\angle 3$ and $\angle 4$ form a linear pair.
    Conclusion: $\angle 3$ and $\angle 4$ are supplementary angles.

    A Angle Addition Property
    B Linear Pair Theorem
    C If two angles are supplementary, then the sum of their measures is 180.
    D If the sum of the measure of two angles is 180, then they are supplementary angles.

16. Write the justifications for each step in this proof of the theorem:
    If two angles are supplementary and one is obtuse, then the other is acute.

    Given Angles 1 and 2 are supplementary. $\angle 1$ is obtuse.

    Prove $\angle 2$ is acute.

    Proof

Conclusions	Justifications
a. Angles 1 and 2 are supplementary. $\angle 1$ is obtuse.	a. ___?___
b. $m\angle 1 + m\angle 2 = 180$	b. ___?___
c. $m\angle 1 > 90$	c. ___?___
d. $m\angle 1 + m\angle 2 > 90 + m\angle 2$	d. ___?___
e. $180 > 90 + m\angle 2$	e. ___?___
f. $90 > m\angle 2$	f. ___?___
g. $\angle 2$ is acute	g. ___?___

REVIEW

17. $\angle ABC$ and $\angle DBC$ are complementary angles, and m$\angle ABC -$ m$\angle DBC = -50$. Find m$\angle ABC$. **(Lesson 3-3)**

18. A race is conducted on a circular track. The winner is the first runner to complete 3 laps. How many degrees will the winner's body have rotated while completing three laps around the track? **(Lesson 3-2)**

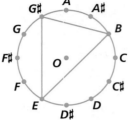

19. The chord represented in the figure at the right is EG♯B, which is called the E major chord. Name the notes in an F♯ major chord. **(Lesson 3-2)**

20. Use the diagram at the right to find the value of x. **(Lesson 3-2)**

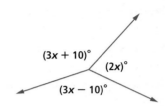

21. Classify each statement below as a statement you *can* conclude from the figure or a statement you *cannot* conclude from the figure at the right. **(Lesson 3-1)**

    a. $\overleftrightarrow{BC} \perp \overleftrightarrow{FG}$ b. $FC > FD$ c. $FD > DC$

22. Consider the conditional: *If a figure is made up of eight segments, it is an octagon.* **(Lessons 2-6, 2-2)**

    a. Draw an instance of the conditional.

    b. Draw a counterexample to the conditional.

23. In New York City, out of 100 people surveyed, 62 said they watched Yankees games and 49 said they watched Mets games. Each person surveyed watched at least one team.
    Let $Y =$ the set of people who watched Yankees games and $M =$ the set of people who watched Mets games. **(Lesson 2-5)**

    a. How many people are in $Y \cup M$?

    b. How many people are in $Y \cap M$?

24. Tell whether the numbers can be the lengths of the sides of a triangle. **(Lesson 1-7)**

    a. 4.5, 5.5, 10   b. 7, 7, 7

25. In the figure at the right, if you place point D so that $\overrightarrow{CD}$ bisects $\angle ACB$, what is an equation for the line $\overleftrightarrow{CD}$? **(Lesson 1-2)**

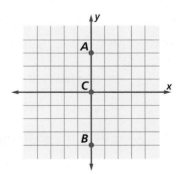

EXPLORATION

26. Felicia's father claimed that the sum of any two odd integers is an even integer. Write an argument justifying this claim.

Lesson
3-6
Parallel Lines

Vocabulary

transversal

corresponding angles

▶ **BIG IDEA** When a line intersects two or more parallel lines, the angles that are formed have equal measures.

Lines have *tilt*. The tilt of a line can be measured by the angle it makes with some line of reference. It is often convenient to make this reference line horizontal. A road might be described as having a *grade* of 7°.

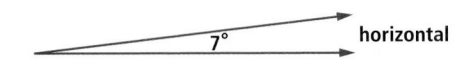

Mental Math

What is the measure of

a. the complement of a 36° angle?

b. the supplement of a 36° angle?

c. the complement of a $p°$ angle?

d. the supplement of a $p°$ angle?

Corresponding Angles

Consider the angles formed when two lines m and n are intersected by a third line ℓ, called a **transversal**. We say that the transversal "cuts" the lines. Eight angles are formed: four by m and the transversal, four by n and the transversal. Any pair of angles in similar locations with respect to the transversal and each line is called a pair of **corresponding angles**. In the drawing, angles 1 and 5 are corresponding angles because both angles are to the left of the transversal and above lines m and line n, respectively. The pairs of angles 2 and 6, 3 and 7, and 4 and 8 are also corresponding angles.

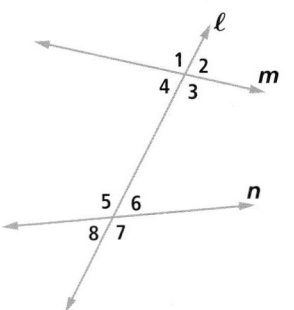

Recall the definition of parallel lines: Two coplanar lines are *parallel* if and only if they are the same line or they do not intersect. In symbols, $(\ell \parallel m) \Leftrightarrow (\ell = m \text{ or } \ell \cap m = \varnothing)$. We say that segments or rays are parallel if the lines containing them are parallel.

Activity 1

MATERIALS DGS

Step 1 Start with a clear DGS screen. Construct two nonparallel lines that do not intersect on the screen.

Step 2 Construct a transversal and its points of intersection with the two lines in Step 1.

(continued on next page)

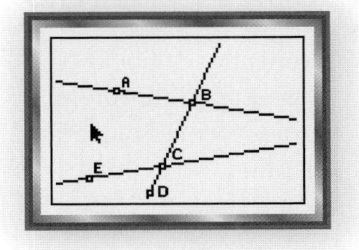

Step 3 Label the points A, B, C, D, and E as in the figure at the bottom of previous page.

Step 4 Measure the corresponding angles ∠ABC and ∠ECD.

Step 5 Drag the line $\overleftrightarrow{AB}$, the line $\overleftrightarrow{EC}$, or the transversal $\overleftrightarrow{DC}$ to change the tilt of each of these. Try to drag the lines so that measured angles have equal measurement.

Step 6 Make a conjecture about the measures of the angles ∠ABC and ∠ECD and the relationship between lines $\overleftrightarrow{AB}$ and $\overleftrightarrow{EC}$.

The results of Activity 1 might make you think that if two lines have the same tilt, that is, if they make corresponding angles with equal measures, then the lines are parallel. Yet, on Earth, two north-south streets make the same angle with any east-west street, but they would intersect at the North Pole (and South Pole!) if extended. Assumptions about parallel lines are needed to ensure that the geometry you are studying is not the geometry of the curved surface of Earth. We make the following assumption.

Corresponding Angles Postulate

Suppose two coplanar lines are cut by a transversal.

a. If two corresponding angles have the same measure, then the lines are parallel.
Abbreviation: (measures of corr. ∠s =) ⇒ (∥ lines)

b. If the lines are parallel, then corresponding angles have the same measure.
Abbreviation: (∥ lines) ⇒ (measures of corr. ∠s =)

Segments or rays are considered to be parallel if the lines containing them are parallel. For instance, by Part a of the Corresponding Angles Postulate, if segments are drawn that, together with line ℓ, create angles with equal measures, then the segments are parallel. Palm fronds show many parallel segments as shown at the right.

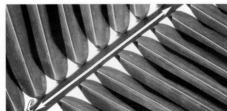

In the figure at the right, the small red arrows on lines m and n indicate that m and n are parallel. By Part b of the Corresponding Angles Postulate, if m ∥ n, then m∠1 = m∠2.

Properties of linear pairs and vertical angles can be used with the Corresponding Angles Postulate to determine the measures of angles formed by parallel lines. In the solution to Example 1, we mention the geometric justifications.

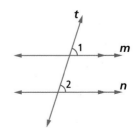

Example 1

In the figure at the right, $m \parallel n$. Find $m\angle 7$.

> **Solution** $m\angle 8 = 85$ because ($\parallel$ lines) $\Rightarrow$ (measures of corr. $\angle$s =).
> Because $m\angle 7 + m\angle 8 = 180$ by the Linear Pair Theorem, then by
> substitution, $m\angle 7 + 85 = 180$. So $m\angle 7 = 95$.

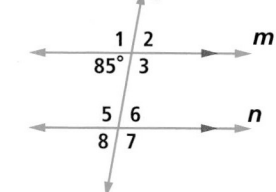

Slopes of Lines

In the coordinate plane, the tilt of a nonvertical line is indicated by
the slope of the line. You learned about slope in algebra and slope
was defined in Lesson 1-2 of this book.
Some other ideas about slope are
reviewed here.

Recall that the slope of a line is
the change in y-values divided by
corresponding change in x-values. It
tells how many units the line goes up or
down for every unit the line goes to the
right. The slope of every horizontal
line is equal to zero, while the slope of a
vertical line is undefined.

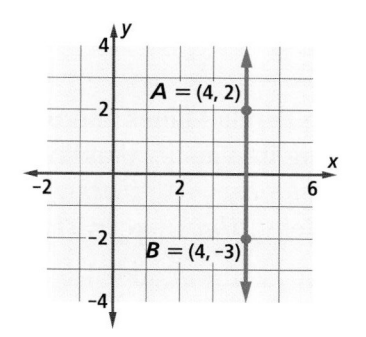

slope of $\overleftrightarrow{PQ} = \frac{-1 - (-1)}{1 - 4} = \frac{0}{-3} = 0$ slope of $\overleftrightarrow{AB} = \frac{-3 - 2}{4 - 4} = \frac{-5}{0}$; undefined

The slope of a line can be found from an equation for that line.

Example 2

Find the slope of the line with equation $2x - 5y = -10$.

> **Solution** Solve the equation for y and put it into slope-intercept form. If
> $2x - 5y = -10$, then $-5y = -2x - 10$ and $y = \frac{2}{5}x + 2$. From this, it can
> be seen that **the slope is $\frac{2}{5}$.**

Slopes of Parallel Lines

Activity 2

MATERIALS DGS

Step 1 Start with a clear DGS screen and show the coordinate axes.

Step 2 Construct two parallel lines on the screen.

Step 3 Measure the slope of each of these lines.

Step 4 Drag the lines so as to change their slopes. Make a conjecture as to
the relationship between parallel lines and their slopes.

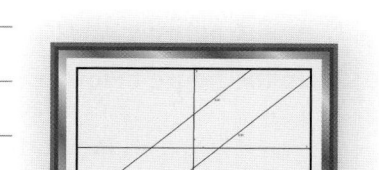

The results of Activity 2 should convince you that regardless of the tilt of nonvertical parallel lines, their slopes will always be equal. This illustrates the following theorem. A general proof requires quite a bit of algebra and is omitted.

Parallel Lines and Slopes Theorem

Two nonvertical lines are parallel if and only if they have the same slope.

Thus, to determine whether nonvertical lines are parallel, you only have to know their slopes. All vertical lines, of course, are parallel to each other.

 QY1

The Parallel Lines and Slopes Theorem is a biconditional. When using it to justify that lines are parallel, you can abbreviate it: ($=$ *slopes*) $\Rightarrow$ ($\parallel$ *lines*). When using it to justify equal slopes, you can write: ($\parallel$ *lines*) $\Rightarrow$ ($=$ *slopes*).

Suppose $\ell \parallel m$ and also $m \parallel n$. Because ($\parallel$ *lines*) $\Rightarrow$ ($=$ *slopes*), ℓ and m must have the same slopes, and so do m and n. By the Transitive Property of Equality, ℓ and n have the same slopes. Then, because ($=$ *slopes*) $\Rightarrow$ ($\parallel$ *lines*), ℓ and n are parallel. This sequence of justified conclusions proves a simple theorem.

Transitivity of Parallelism Theorem

If line ℓ is parallel to line m and line m is parallel to line n, then line ℓ is parallel to line n.

In other words, if two lines are each parallel to a third line, then they are parallel to each other.

 QY2

Questions

COVERING THE IDEAS

1. In the figure at the right, which line is a transversal?

2. In the figure below, which of the lines is a transversal?

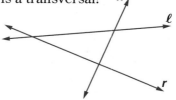

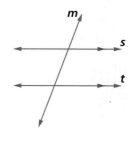

> ▶ **QY1**
>
> Why does the Parallel Lines and Slopes Theorem stipulate that the two lines are nonvertical?

> ▶ **QY2**
>
> Does the Transitivity of Parallelism Theorem apply when the lines in question are vertical? Why or why not?

In 3–6, refer to the figure at the right.

3. ∠2 and __?__ are corresponding angles.

4. ∠8 and __?__ are corresponding angles.

5. If m∠1 = m∠5, then what must be true about lines *m* and *n*?

6. If *m* ∥ *n*, name all sets of angles with equal measures.

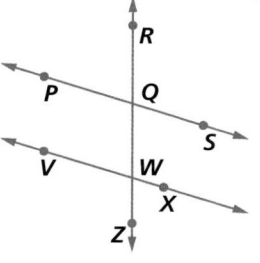

In 7 and 8, use the figure at the right.

7. If $\overleftrightarrow{PQ} \parallel \overleftrightarrow{VW}$, and m∠*PQR* = 70, find m∠*QWV*.

8. **True or False** If m∠*PQR* = m∠*VWZ*, then $\overleftrightarrow{PQ} \parallel \overleftrightarrow{VW}$.

9. **True or False** The Corresponding Angles Postulate is true on the surface of Earth.

10. Find the slope of the line through points (2, 5) and (–3, 8).

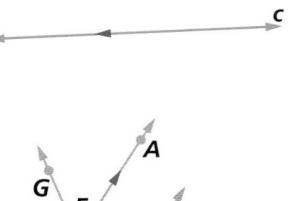

In 11 and 12,

    a. Find the slope of the line with the given equation.

    b. Find the slope of a line parallel to the line with the given equation.

11. $4y = -2x + 7$
12. $-3x + 2y = 8 - 4x$

13. For which lines is slope undefined?

14. In the figure at the right, which lines do you know are parallel?

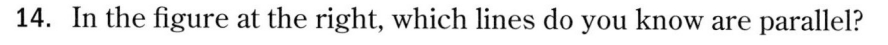

15. Suppose you write *When two lines are parallel to the same line, they are parallel to each other* as a justification for a step. What is a name of the justification?

16. Given the figure at the right with m∠*GEA* = 56 and with $\overleftrightarrow{AD} \parallel \overrightarrow{ST}$, find each measure.

    a. m∠*GED*   b. m∠*DEC*   c. m∠*AEC*

    d. m∠*ECT*   e. m∠*ECS*   f. m∠*PCT*

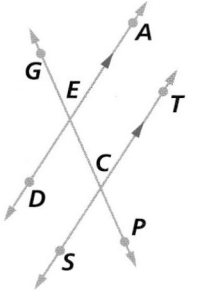

APPLYING THE MATHEMATICS

17. Are all horizontal lines parallel? Justify your answer.

18. Given the lines with equations *x* = *b* and *x* = *c*. Are the graphs of these two lines parallel? Justify your answer.

19. A quadrilateral in which both pairs of opposite sides are parallel is a *parallelogram*. Use slope to show that the figure connecting the points with the coordinates shown at the right is a parallelogram.

20. Determine whether the lines with equations $3x - 6 = 4y + 11$ and $4x + 11 = -3y + 6$ are parallel.

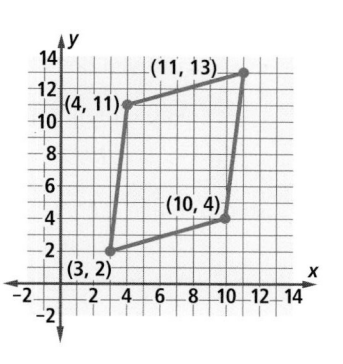

In 21 and 22, use the figure at the right.

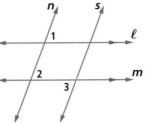

21. Which lines are parallel if m∠1 = m∠2?

22. Which lines are parallel if m∠2 = m∠3?

23. Using the figure from Question 16, prove that m∠AEG = m∠PCS. State all of the justifications you use.

24. Consider the figure at the right with $\overleftrightarrow{PR} \parallel \overleftrightarrow{ST}$, m∠VRP = 5x, m∠RST = 105, and m∠PTW = 3x. Find:

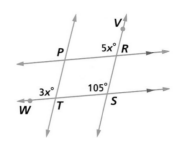

    a. x.   **b.** m∠PTS.   **c.** m∠RPT.

REVIEW

25. In the figure at the right $AB = AC$. Which property of algebra allows you to conclude that $AB + BC = AC + BC$? **(Lesson 3-4)**

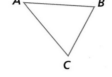

26. Use the figure at the right to justify the following conclusion: m∠AEC = m∠DEB. **(Lesson 3-3)**

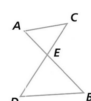

27. ∠ABC is a straight angle. $\overrightarrow{BD}$ bisects ∠ABC. $\overrightarrow{BE}$ bisects ∠DBC. **(Lesson 3-3)**

    a. Draw a picture of this situation.

    b. If m∠EBC = 3x + 5, find x.

28. The point $A = (0, 2)$ is rotated 90° about the origin to point B. **(Lesson 3-2)**

    a. What are the coordinates of B?

    b. Write an equation for the line containing A and B.

EXPLORATION

29. Use a DGS to complete the following activity.

    Step 1 Construct a triangle △ABC on a clear screen.

    Step 2 Construct a point D on side $\overline{AB}$.

    Step 3 Construct a line parallel to side $\overline{BC}$ through point D.

    Step 4 Label the point of intersection of the parallel line and side $\overline{AC}$ as point E.

    Step 5 Measure the distances AD, AB, AE, and AC.

    Step 6 Make a conjecture about a relationship among these measures.

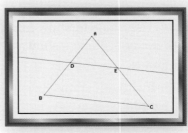

Lesson
3-7
Size Transformations

Vocabulary

concurrent

point of concurrency

transformation

mapping, maps

size change, size
  transformation

▶ **BIG IDEA** By multiplying the coordinates of points on a figure by a fixed nonzero number k, you can create an image of the figure that looks very much like the original figure.

Suppose you are given a picture and you would like to enlarge it or shrink it without distorting the picture. How could you accomplish such a task? The activity below shows one way to do this.

Activity 1

MATERIALS Ruler, protractor, graph paper

Step 1 Make a copy of the sailboat below. This sailboat is the preimage.

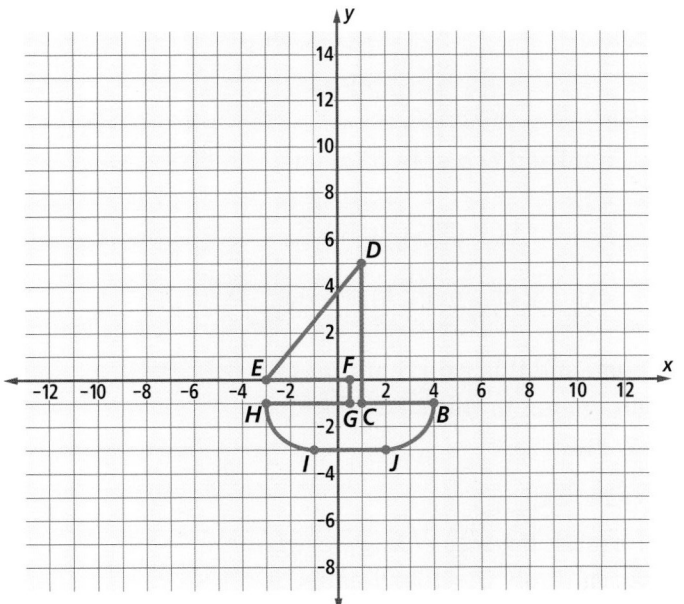

Mental Math

Classify the triangle as equilateral, isosceles, or scalene.

a.

b.

c.

Step 2

   a. On the same grid, form a new picture by multiplying each coordinate of the named points by 3. Label these new points E', D', and so on. For example, because $B = (4, -1)$, $B' = (3 \cdot 4, 3 \cdot (-1)) = (12, -3)$. The new picture is called the *image*.

   b. Describe the shape of the image.

(continued on next page)

Size Transformations **151**

Step 3 The lines $\overleftrightarrow{BB'}$, $\overleftrightarrow{CC'}$, $\overleftrightarrow{DD'}$, and so on, through points and their images are **concurrent**. This means that they have a point in common. This point is called the **point of concurrency**. What is the point of concurrency of these three lines?

In Steps 4–6, answer the question. Support your answer with an example from the diagram.

Step 4 How are the lengths in the preimage related to the lengths in the image?

Step 5 How are the measures of the angles in the preimage related to the measures of the angles in the image?

Step 6 How are the slopes of segments in the preimage related to slopes of the corresponding segments in the image?

You should have found in this Activity that you created a figure whose lengths were 3 times as long, but otherwise all of the other qualities of the shape of the sailboat remained the same. In other words, the figure looked exactly the same, just bigger.

The correspondence in the Activity between the preimage and the image is an example of a *transformation*.

Definition of Transformation

A **transformation** is a correspondence between two sets of points *A* and *B* such that:
1. each point in set *A* corresponds to exactly one point in set *B*, and
2. each point in set *B* corresponds to exactly one point in set *A*.

A transformation is often called a **mapping**, and it is said that the transformation **maps** a preimage onto an image. In this case, the correspondence can be described by the rule

$$(x, y) \rightarrow (3x, 3y).$$

For obvious reasons, we call this transformation a *size change* or *size transformation* of magnitude 3. You can multiply the coordinates of all points on the plane by any positive number (not just 3) to produce an image under a size change. In this size change, the origin $(0, 0)$ is on the line containing any point and its image. We call $(0, 0)$ the *center* of the size change.

Definition of Size change (Size transformation)

When $k \neq 0$, the transformation under which the image of (x, y) is the point (kx, ky) is the **size change** (**size transformation**) of magnitude k and center $(0, 0)$.

STOP **QY1**

The size change of magnitude k and center $(0, 0)$ is denoted by the symbol S_k. The Activity at the beginning of this lesson pictures a sailboat and its image under the size change S_3. The image of a point P under S_k is written as $S_k(P)$. For instance, in the Activity, $S_3(J) = S_3(2, -3) = (3 \cdot 2, 3 \cdot -3) = (6, -9)$.

▶ QY1

In the definition of *size change*, is it possible for k to be zero? If so, what is the image of point (x, y) under this size transformation?

Example 1

If $W = (6, -24)$, find $S_{\frac{2}{3}}(W)$.

Solution $S_{\frac{2}{3}}(6, -24) = \left(\frac{2}{3}(6), \frac{2}{3}(-24)\right) = (4, -16)$

Size changes have many nice properties. In the Activity, you should have noticed that segments and their images are parallel.

GUIDED

Example 2

In the figure of the Activity, show that $\overline{DE}$ and its image $\overline{D'E'}$ are parallel.

Solution $D = (1, 5)$, so $D' = S_3(D) = (\underline{\ ?\ }, \underline{\ ?\ })$.
$E = (-3, 0)$, so $E' = S_3(\underline{\ ?\ }) = (\underline{\ ?\ }, \underline{\ ?\ })$.
The slope of $\overleftrightarrow{DE}$ is $\underline{\ ?\ }$.
The slope of $\overleftrightarrow{D'E'}$ is $\underline{\ ?\ }$.
Because the slopes are $\underline{\ ?\ }$, the lines are parallel.

By generalizing Guided Example 2, we are able to prove that this property holds for any points and any size change S_k.

S_k Theorem 1: Parallel Property

Under a size change S_k, the line through any two preimage points is parallel to the line through their images.

To prove this theorem, we need to consider any points and any size change. So, we need variables.

Proof

Conclusions	Justifications
1. Let $A = (x_1, y_1)$ and $B = (x_2, y_2)$ be any two points in the plane.	1. Given
2. Let $A' = S_k(A) = (kx_1, ky_1)$ and $B' = S_k(B) = (kx_2, ky_2)$.	2. definition of S_k
3. slope of $\overleftrightarrow{AB} = \frac{y_2 - y_1}{x_2 - x_1}$	3. definition of slope
4. slope of $\overleftrightarrow{A'B'} = \frac{ky_2 - ky_1}{kx_2 - kx_1}$	4. definition of slope
5. $\frac{ky_2 - ky_1}{kx_2 - kx_1} = \frac{k(y_2 - y_1)}{k(x_2 - x_1)} = \frac{y_2 - y_1}{x_2 - x_1}$	5. Distributive Property of Multiplication over Addition; Equal Fractions Property
6. $\overleftrightarrow{AB} \parallel \overleftrightarrow{A'B'}$	6. Parallel Lines and Slopes Theorem (If two lines have the same slope, then they are parallel.)

The S_k Parallel Property helps to explain why figures look so much like their images.

Collinearity

Recall from algebra that when three or more points lie on the same line, then the slopes between any two of them are equal. From the S_k Parallel Property, we can show that the image of a line is a line. We say that S_k *preserves collinearity*. A proof requires the Uniqueness of Parallels Theorem that you will study in Chapter 5.

S_k Theorem 2: Collinearity Is Preserved

Under S_k, the images of collinear points are collinear.

Angle Measure

In the Activity at the beginning of this lesson you probably noticed that the angles in the image and the angles in the preimage had the same measure. This should make sense because if they didn't, the figures wouldn't look the same. We now turn our attention to demonstrating why this is true.

S_k Theorem 3: Angle Measure Is Preserved

Under S_k, an angle and its image have the same measure.

Given $S_k(\angle ABC) = \angle A'B'C'$

Prove $m\angle ABC = m\angle A'B'C'$

Proof

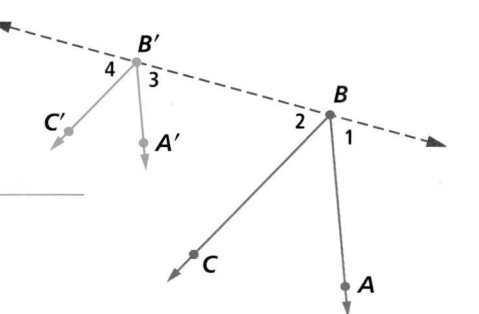

Conclusions	Justifications
1. $A' = S_k(A)$, $B' = S_k(B)$, $C' = S_k(C)$	1. Given
2. $BC \parallel B'C'$ and $AB \parallel A'B'$	2. S_k Theorem 1
3. $m\angle 1 = m\angle 3$, $m\angle 2 = m\angle 4$	3. Corresponding Angles Postulate (($\parallel$ lines) $\Rightarrow$ (measures of corresponding $\angle$s $=$))
4. $m\angle 1 + m\angle 2 + m\angle ABC = 180$, $m\angle 3 + m\angle 4 + m\angle A'B'C' = 180$	4. Angle Addition Assumption and definition of straight angle
5. $m\angle 1 + m\angle 2 + m\angle ABC = m\angle 3 + m\angle 4 + m\angle A'B'C'$	5. Transitive Property of Equality
6. $m\angle 1 + m\angle 2 + m\angle ABC = m\angle 1 + m\angle 2 + m\angle A'B'C'$	6. Substitution Property of Equality (Substitute $m\angle 1$ for $m\angle 3$ and $m\angle 2$ for $m\angle 4$ in the right side of equation.)
7. $m\angle ABC = m\angle A'B'C'$	7. Addition Property of Equality (Subtract $m\angle 1 + \angle 2$ from both sides of the equation.)

The proofs of the two parts of the S_k Theorem do not work for vertical lines because each of the proofs relies on the concept of slope and vertical lines have undefined slope. The proofs of these three theorems for the special case of vertical lines need to be done in a slightly modified manner.

Length

The transformation S_k has other properties. One important property that you should have noticed in the Activity is that the lengths in the image of the sailboat were 3 times the lengths in the original sailboat. Multiplication of lengths is indeed true for all S_k, but the proof cannot be done now because we have not developed the geometry yet to prove it. However, you can show that some specific size transformations do not preserve length.

GUIDED

Example 3

Prove that, $S_{\frac{1}{2}}$ does not preserve length.

Solution You only need to find two points for which the length is not preserved. We pick points A and B for which we know the distance between them.

Let $A = (0, 4)$ and $B = (8, 4)$.

Step 1 Using the formula for $S_{\frac{1}{2}}$ we know that $A' = (\underline{\ ?\ }, \underline{\ ?\ })$ and $B' = (\underline{\ ?\ }, \underline{\ ?\ })$

Step 2 $AB = \underline{\ ?\ }$, and $A'B' = \underline{\ ?\ }$.

Step 3 Because $AB \neq A'B'$, S_k does not preserve length.

 QY2

> **QY2**
>
> Explain why the solution in Example 3 will not work for S_1.

Questions

COVERING THE IDEAS

In 1–5, refer to the Activity in this lesson.

1. The image is found under a transformation S_k. What is the value of k?

2. **Fill in the Blanks** The slope of $\overline{EB} = \underline{\ ?\ }$, and the slope of $\overline{E'B'} = \underline{\ ?\ }$.

3. **Fill in the Blanks** $BC = \underline{\ ?\ }$ and $B'C' = \underline{\ ?\ }$.

4. Give an argument that shows that the line through F and F' also contains $(0, 0)$.

5. In the Activity you found that the slope of $\overline{ED}$ and the slope of $\overline{E'D'}$ were equal. What property in this lesson generalizes this finding?

6. Let $G = (-2, -2)$, $H = (0, 3)$, $I = (6, 3)$, and $J = (8, -2)$.
   a. What kind of quadrilateral is $GHIJ$?
   b. Find the image of $GHIJ$ under the transformation $S_{1.5}$.
   c. What kind of quadrilateral is $G'H'I'J'$? Explain how you know.
   d. Verify that $\overline{GH}$ and $\overline{G'H'}$ are parallel.
   e. How do HI and $H'I'$ compare?

7. If $S_k(10, -25) = (4, -10)$, then what is k?

8. **True or False** If two lines on a preimage are perpendicular, then the images of those lines will be perpendicular under any transformation S_k.

9. What is the image of $(8, 20)$ under the transformation $S_{\frac{3}{4}}$?

10. Define *transformation*.

APPLYING THE MATHEMATICS

11. Refer again to the Activity in this lesson. For what value of k would S_k map the larger sailboat back onto the smaller sailboat?

12. **a.** $\triangle LMN$ is shown at the right. Graph its image under $S_{0.5}$. Label the new points L', M', and N'.
    b. Find the slopes of $\overleftrightarrow{MN}$ and $\overleftrightarrow{M'N'}$ to verify that they are equal.
    c. Find an equation for $\overleftrightarrow{NN'}$.
    d. Find an equation for $\overleftrightarrow{MM'}$.
    e. From Parts c and d, determine the point of intersection of $\overleftrightarrow{MM'}$ and $\overleftrightarrow{NN'}$.

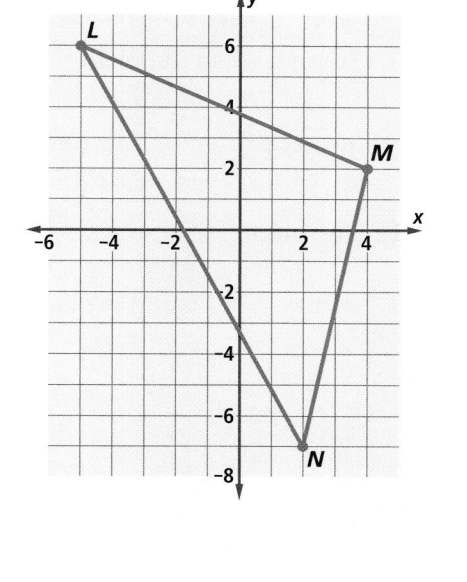

13. Complete the steps below to show that any line drawn through a point and its image under S_k (when $k \neq 0$) contains the origin.
    a. Let (a, b) be any point on the plane except the origin. Let $S_k(a, b) = (ka, kb)$, where $k \neq 1$. An equation for the line through (a, b) and (ka, kb) is _____?_____.
    b. Show that $(0, 0)$ is on the line you found in part a.

14. Is there a value of k so that an image under S_k is smaller than its preimage? If so, give such a value. If not, explain why not.

15. In the diagram at the right, $S_k(E) = E'$, $S_k(F) = F'$, and $S_k(G) = G'$. The coordinate grid has purposely been omitted. Trace a copy of the preimage and image shown.
    a. Locate the origin of the coordinate system that was used.
    b. What is the value of k in the S_k that was used? How do you know?

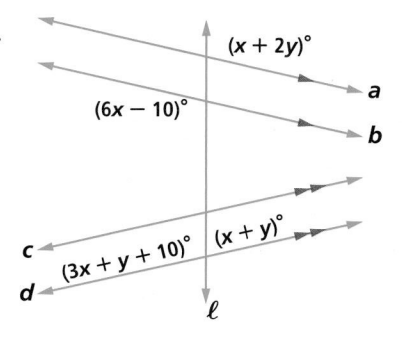

REVIEW

16. The line ℓ passes through the points $(0, 5)$ and $(-1, 4)$. Find an equation for a line k parallel to ℓ and containing the point $(7, 7)$. **(Lessons 3-6, 1-2)**

17. In the figure at the right, $a \parallel b$ and $c \parallel d$. Find the values of x and y. **(Lesson 3-6)**

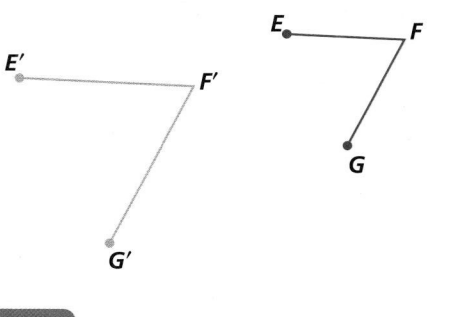

Size Transformations **157**

18. Use the diagram at the right, in which $a \parallel b$, to justify the conclusion, "$\angle GDE$ is obtuse." (**Lessons 3-6, 3-5**)

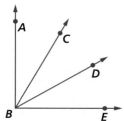

19. C is the midpoint of $\overline{AB}$, as shown below.

Suppose AB is an integer, $AB < 8$, and $AC > 3$. What is AB? Explain your reasoning. (**Lesson 3-4**)

20. **True or False** Refer to the diagram at the right. (**Lesson 3-3**)

    a. $\angle CBD$ and $\angle EBD$ are adjacent.

    b. $\angle DBE$ and $\angle EBA$ are adjacent.

    c. $\angle ABC$ and $\angle DBE$ are adjacent.

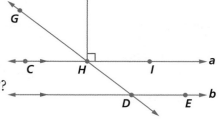

21. The angle formed by adjacent sides of a piece of paper is $90°$. One piece of paper was placed on another as shown in the figure below. What is the measure of $\angle 1$? (**Lesson 3-3**)

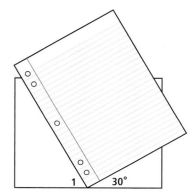

22. A Ferris wheel completes one full turn every five minutes. Tyra spent 25 minutes on the Ferris wheel. How many degrees did the wheel rotate during this time? (**Lesson 3-2**)

At 212 feet tall, the Texas Star is the largest Ferris wheel in North America.

EXPLORATION

23. a. Repeat all questions for the Activity in the lesson but by multiplying the coordinates by –3 instead of 3.

    b. How do the answers to part a compare with the answers found in the Activity?

24. Cut out a cartoon character in a newspaper. Trace the character onto a coordinate grid so that the character surrounds the origin. Then apply $S_{2.5}$ to this character and describe the image.

QY ANSWERS

1. No (The size change is only defined for $k \neq 0$.)

2. Answers may vary. Sample: S_1 results in points that coincide with their images, so length is preserved.

Lesson
3-8
Perpendicular Lines

Vocabulary

perpendicular

opposite reciprocals

▶ **BIG IDEA**   The lines containing the rays of a right angle are perpendicular; all lines perpendicular to the same line are parallel.

Mental Math

A National Hockey League team plays an 82-game regular season and 10 playoff games.

a. How many games has the team played at the midpoint of its regular season?

b. How many games has the team played at the midpoint of its season, including playoffs?

c. Write an expression for the number of games a team has played at the midpoint of its season, including playoffs, if it played n playoff games.

Perhaps the most striking features of Winchester Cathedral, pictured above, are the tall vertical spires. The spires were built during the Perpendicular Gothic period in England, in the 14th century. The designers of these buildings emphasized vertical lines. All of these lines are perpendicular to the ground, but not all perpendicular lines are horizontal and vertical. In this lesson, you will explore the properties of perpendicular lines and their relationship to parallel lines.

Definition of Perpendicular

Two segments, rays, or lines are **perpendicular** if and only if the lines containing them form a 90° angle.

At the right, $\angle ABC$ is a right angle, so m and n are perpendicular. The symbol $\perp$ is read "is perpendicular to." You can write $m \perp n$, $\overleftrightarrow{AB} \perp \overleftrightarrow{BC}$, or $m \perp \overleftrightarrow{BC}$ to indicate perpendicularity. You cannot assume drawn lines are perpendicular unless they are marked or you are told they are perpendicular.

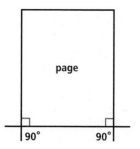

When intersecting lines form one right angle, the Vertical Angles and Linear Pair Theorems force the other three angles to be right angles.

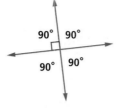

Pairs of Perpendicular Lines

The two side edges of this page are each perpendicular to the bottom edge. If the edges are extended, corresponding 90° angles appear. By the Corresponding Angles Postulate, the side edges are parallel. This argument proves a useful theorem.

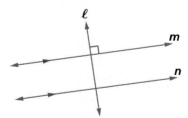

> **Two Perpendiculars Theorem**
>
> If two coplanar lines ℓ and m are each perpendicular to the same line, then they are parallel to each other.

In symbols, if $\ell \perp n$ and $m \perp n$, then $\ell \parallel m$. Because parallel lines are not perpendicular, you can see that the relation "is perpendicular to" *does not* satisfy the Transitive Property.

Suppose you are given the lines as shown at the right with one perpendicular relation ($\ell \perp m$) and one parallel relation ($m \parallel n$) among the lines. Then there are 90° angles where ℓ intersects m. Because ($\parallel$ lines) $\Rightarrow$ (corr. $\angle$s =), there are also 90° angles where ℓ intersects n. So $\ell \perp n$. This simple argument proves that if $\ell \perp m$ and $m \parallel n$, then $\ell \perp n$. In words, this relationship is stated as follows.

> **Perpendicular to Parallels Theorem**
>
> In a plane, if a line is perpendicular to one of two parallel lines, then it is also perpendicular to the other.

Slopes of Perpendicular Lines

You saw in the last lesson that the slopes of parallel lines are always equal. The Activity below shows a relationship between the slopes of perpendicular lines.

Activity 1

MATERIALS DGS

Step 1 Display the coordinate axes.

Step 2 Construct $\overleftrightarrow{AB} \perp \overleftrightarrow{BC}$.

Step 3 Display the slopes of $\overleftrightarrow{AB}$ and $\overleftrightarrow{BC}$.

At this point, your sketch might look like the one shown at the right.

1. As you drag the lines to create other lines with different slopes, what is the relationship between the signs of the slopes of the perpendicular lines?

2. What do you notice about the magnitudes of the slopes?

3. Display the product of the slopes. Drag the lines around the screen. What do you notice?

4. Drag your lines until one of them is vertical. What happens to the product and why?

From the Activity, you should have drawn the conclusion below.

Perpendicular Lines and Slopes Theorem

Two nonvertical lines are perpendicular if and only if the product of their slopes is –1.

This conclusion allows us to use the slope of one line to calculate the slope of any line perpendicular to it.

Example 1

Find the slope of a line perpendicular to $y = -\frac{2}{3}x + 7$.

Solution The slope of the perpendicular line and the slope of the given line must have a product of –1. The slope of the original line is $-\frac{2}{3}$. Let $m =$ the slope of the perpendicular line. We now know that

$$-\frac{2}{3} m = -1.$$
$$\text{So } m = -\frac{3}{2} \cdot -1 = \frac{3}{2}.$$

Notice in the above example that the original line has a slope of $-\frac{2}{3}$ and the line perpendicular to it has a slope of $\frac{3}{2}$. These are known as *opposite reciprocals*. **Opposite reciprocals** are two numbers whose product is –1.

GUIDED

Example 2

Write an equation of the line ℓ that passes through the point $(2, -3)$ and is perpendicular to the line $y = 4x - 1$.

Solution The slope of the line ℓ is the opposite reciprocal of 4, so it is __?__.

At this point, you know the equation of this line is of the form:

$$y = \underline{\ ?\ } x + b.$$

Now we substitute $x = 2$ and $y = -3$ and solve for b.

$$-3 = \underline{\ ?\ }(2) + b$$
$$-3 = \underline{\ ?\ } + b$$
$$\underline{\ ?\ } = b$$

So, an equation of the line ℓ is: $y = \underline{\ ?\ }$.

Summary of Parallel and Perpendicular Lines

In this and the preceding lesson, you have seen how parallel and perpendicular lines can be determined either through angle measures or by slopes. The table below relates parallelism and perpendicularity to the conceptions of points and lines in Euclidean and coordinate geometry that you saw in Chapter 1.

Idea	Euclidean synthetic geometry	Euclidean plane coordinate geometry
point	location	ordered pair (x, y)
line	$\ell \longleftrightarrow \ \ \bullet \ P \longrightarrow$	$Ax + By = C$
measure tilt of line	angle measure	slope
parallel lines	corresponding angles equal in measure	slopes equal
perpendicular lines	lines form 90° angles	product of slopes = –1

Questions

COVERING THE IDEAS

1. Define *perpendicular lines*.

2. Why might the term "Perpendicular Gothic" be confusing to someone studying lines that intersect at right angles?

3. Trace line ℓ at the right. Draw a line through point P perpendicular to ℓ.

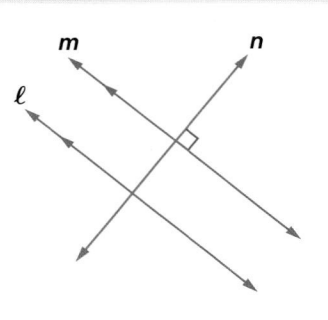

4. In the diagram at the right, $\ell \parallel m$ and $m \perp n$. What theorem justifies the conclusion that $\ell \perp n$?

5. In the drawing below, $\overline{AD} \perp \overline{AB}$ and $\overline{BC} \perp \overline{AB}$. What theorem justifies the conclusion that $\overline{AD} \parallel \overline{BC}$?

6. A line has slope 2. What is the slope of a line perpendicular to it?

7. Write the equation of the line through $(-4, 6)$ that is perpendicular to the line with equation $y = \frac{2}{3}x + 1$.

8. An oblique line has slope x. What is the slope of a line perpendicular to it?

APPLYING THE MATHEMATICS

9. Graph the equations $y = 3x + 5$ and $y = 3x + 10$. Find an equation of a line that is perpendicular to both lines and that intersects the first line in the first quadrant and the second line in the second quadrant.

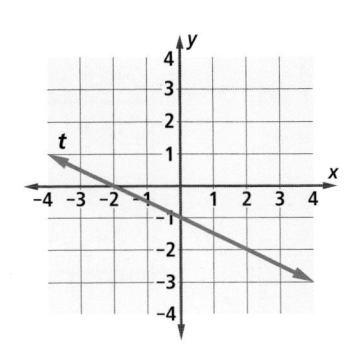

10. Suppose a line s is perpendicular to line t at the right. What is the slope of s?

In 11 and 12, refer to the figure at the right.

11. In the figure, $\overrightarrow{AD} \perp \overrightarrow{AC}$. Compute the following angle measures.
    a. $m\angle DAC$ b. $m\angle DAE$ c. $m\angle DAE + m\angle BAC$

12. If $m\angle DAC$ remains unchanged, and the angle rotates about point A while still remaining above $\overleftrightarrow{EB}$, describe the sum $m\angle DAE + m\angle BAC$.

In 13 and 14, use the figure at the right. Given $\overline{QR} \parallel \overline{SU}$, $\overline{QT} \parallel \overline{PU}$, and $\overline{QT} \perp \overline{TU}$ as indicated.

13. **Matching** Choose the correct justification for each conclusion.
    a. $m\angle P = m\angle TQS$
    b. $m\angle RQP = m\angle S$
    c. $\overline{TU} \perp \overline{UR}$
    d. $m\angle RQT = 90°$

    i. $(\parallel \text{ lines}) \Rightarrow \text{corr. } \angle s =)$
    ii. $(\text{corr. } \angle s =) \Rightarrow \parallel \text{ lines})$
    iii. $(\ell \perp n \text{ and } m \perp n) \Rightarrow (\ell \parallel m)$
    iv. $(\ell \perp m \text{ and } m \parallel n) \Rightarrow (\ell \perp n)$
    v. $(\ell \perp n) \Rightarrow (90° \text{ angle})$
    vi. $(90° \text{ angle}) \Rightarrow (\ell \perp m)$

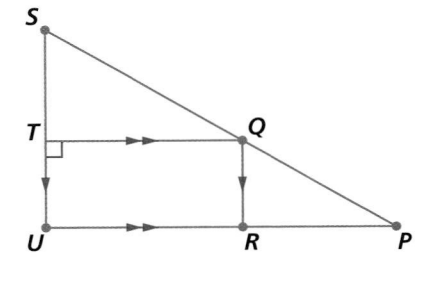

14. If $m\angle PQR = 60°$, give the measure of each angle.
    a. $\angle S$ b. $\angle TQR$ c. $\angle TQS$ d. $\angle P$

15. Write the two conditional statements making up the Perpendicular Lines and Slopes Theorem.

16. Find k if the line through (5, 11) and (-3, 7) is parallel to the line through (-2, k) and (2, 5).

17. Consider this statement: If $\overline{AB} \perp \overline{BC}$ and $\overline{BC} \perp \overline{CD}$, then $\overline{AB} \perp \overline{CD}$. Draw an instance of this statement in a plane, or tell why the drawing is impossible.

REVIEW

18. Let $P = (x, y)$. Given that $S_k(P) = (2x, 2y)$, find k. (Lesson 3-7)

19. Refer to the diagram below. Is there some value of k so that $S_k(ABCD) = AEFG$? If so, what is it? If not, why not? (Lesson 3-7)

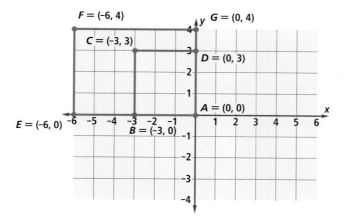

20. In the diagram at the right, showing $\odot O$, D is the midpoint of $\overline{OE}$. Justify the conclusion $DE = OF$. (Lessons 3-5, 3-4)

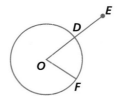

21. Does the relation $\leq$ have the reflexive property? The symmetric property? The transitive property? (Lesson 3-4)

22. The measure of $\angle A$ is twice the measure of its complement. What is m$\angle A$? (Lesson 3-3)

EXPLORATION

23. A person is at a place on Earth where the following is possible. The person goes north 1 mile, then turns east for 1 mile, then turns south for 1 mile and winds up at the starting place.
    a. Name one place where the person might have started.
    b. Find a second place where the person might have started.

Lesson 3-9

The Perpendicular Bisector

▶ **BIG IDEA** The construction of the perpendicular bisector of a segment is a basis for many other constructions.

Here are two similar problems.

- An airplane is having engine trouble. There are two airports close by. By looking at a map, how can you tell which airport is closer?

- There are two cell phone towers. If a cell phone user is moving from one tower, to the other, when will the signal switch from using one tower to using the other?

Each problem can be viewed as involving two points (the airports or the towers) and asking: What points are closer to one of the points than to the other?

In the following activity, you will be looking for points that are the same distance from two points A and B. Such points are said to be **equidistant** from A and B.

Mental Math

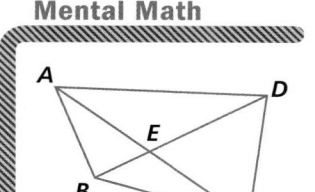

a. Name two pairs of supplementary angles.

b. Name two pairs of vertical angles.

Activity 1

MATERIALS ruler, plain paper

Step 1 On a piece of paper, draw 2 points. Label them A and B.

Step 2 You know the midpoint M of $\overline{AB}$ is equidistant from A and B. Draw the midpoint.

Step 3 Measure to find at least three other points equidistant from A and B.

The points you drew in Activity 1 should be collinear. The line they lie on is called the *perpendicular bisector* of the segment $\overline{AB}$. A **bisector of a segment** $\overline{AB}$ is its midpoint, or any line, ray, or segment that intersects the segment only at its midpoint and is not contained in $\overleftrightarrow{AB}$. Every segment can have many bisectors, but in a given plane, only one line is both a bisector and perpendicular to the segment. This line is the **perpendicular bisector**, or ⊥ bisector, of the segment.

some of the many bisectors of $\overline{XY}$

ℓ is the ⊥ bisector of $\overline{AB}$.

The Perpendicular Bisector **165**

Constructing the Perpendicular Bisector of a Segment

A *geometric construction* is a precise way of drawing that uses specific tools and follows specific rules. From the time of the ancient Greeks, only two tools have been permitted in making a construction: the *unmarked straightedge* and the *compass*. The unmarked straightedge can draw a line through two points A and B. As you know, we write that line as $\overleftrightarrow{AB}$. The compass can draw a circle with a given point A as center and containing a second given point B. If $AB = r$, we name that circle as $\odot(A, r)$ or $\odot(A, AB)$, because AB is the length of the radius $\overline{AB}$.

Look at this finished construction of the perpendicular bisector of $\overline{AB}$ using a straightedge and compass as shown at the right. What do you think was there to start with? What do you think was constructed?

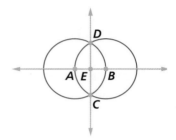

We write the construction using the words Given (for what we start with), Name (for naming new points), and Construct (for identifying what is constructed). With some DGS software, all of these are called CONSTRUCT. New points can only be intersections of constructed figures or points that exist because of postulates.

Activity 2

MATERIALS DGS (or compass and straightedge)
Follow the steps of this construction of the points that are equidistant from airports at A and B.

Step 1 Given points A and B. (Put them in the same relative position as shown at the right.)

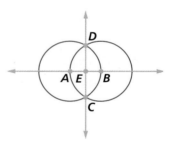

Step 2 Construct $\overline{AB}$.

Step 3 Construct $\odot(A, AB)$ and $\odot(B, AB)$.

Step 4 Name $\odot(A, AB) \cap \odot(B, AB) = \{C, D\}$ (Name the two points of intersection C and D.)

Step 5 Construct $\overleftrightarrow{CD}$. A bonus in constructing $\overleftrightarrow{CD}$ is that the midpoint of $\overline{BA}$ has also been constructed. You can name it E in one more step.

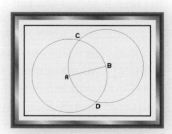

Step 6 Name $\overleftrightarrow{CD} \cap \overline{AB} = \{E\}$.

Now examine the picture you have. When $\overleftrightarrow{CD}$ is the perpendicular bisector of $\overline{AB}$, then all points on $\overleftrightarrow{CD}$ are equidistant from A and B. The points on the same side of $\overleftrightarrow{CD}$ as A are closer to A. The points on the same side of $\overleftrightarrow{CD}$ as B are closer to B.

How a DGS screen might look after Step 4

Voronoi Diagrams

In the real world, there are more than two airports and more than two cell phone towers. The figure at the right started with 3 points X, Y, and Z. Which points in the plane are closer to X than to Y or Z? Which are closest to Y? Which are closest to Z? By drawing perpendicular bisectors of the segments $\overline{XY}$, $\overline{XZ}$, and $\overline{YZ}$, you can outline the regions of the plane closest to each point. The diagram that results is called a **Voronoi diagram,** named after the Russian mathematician Georgy Voronoi (1868–1908).

Constructing the Perpendicular to a Line Through a Point

In the construction of the perpendicular through a given point to a given line, we consider the construction of the perpendicular bisector as just one step, as is the case on most DGS.

Activity 3

MATERIALS Compass, straightedge (or DGS)

Follow the steps to construct the perpendicular to a line $\overleftrightarrow{AB}$ through point C.

Step 1 Given $\overleftrightarrow{AB}$ and point C. (Draw $\overleftrightarrow{AB}$ and point C on your paper.)

Step 2 Construct $\odot (C, CA)$.
Because CA is the radius of $\odot (C, CA)$, that circle will intersect $\overleftrightarrow{AB}$ at A. If it is the case that there is no other point of intersection, then construct $\odot (C, CB)$ and switch the positions of A and B in Step 3.

Step 3 Name $\odot (C, CA) \cap \overleftrightarrow{AB} = \{A, D\}$.

Step 4 Construct the perpendicular bisector of $\overline{AD}$ with the following steps:
  a. Construct $\odot (A, AD)$.
  b. Construct $\odot (D, AD)$.
  c. Construct $\odot (A, AD) \cap \odot (D, AD) = \{E, F\}$.
  d. Construct $\overleftrightarrow{EF}$.

Your final diagram should look like the one at the right. Not only is $\overleftrightarrow{EF}$ the perpendicular bisector of $\overleftrightarrow{AD}$, but $\overleftrightarrow{EF}$ also contains point C. How do we know this will always be the case? We cannot answer this question now, for we need some of the ideas you will study in the next two chapters.

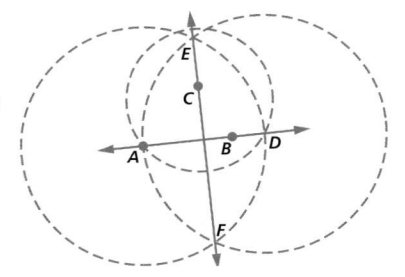

Air traffic controllers manage the movements of aircraft along civil airways, including the coordination of arrivals and departures.

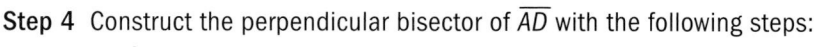

Using the Perpendicular Bisector to Create a Design

Activity 4

MATERIALS DGS

Step 1 Given points A and B.

Step 2 Construct $\overleftrightarrow{AB}$.

Step 3 Name a third point C on $\overleftrightarrow{AB}$ and construct a line perpendicular to $\overleftrightarrow{AB}$ through point C.

Step 4 Name a point D on this perpendicular line but not on $\overleftrightarrow{AB}$.

Step 5 Construct $\odot (C, CD)$.

Step 6 Name $\odot (C, CD) \cap \overleftrightarrow{CD} = \{D, E\}$

After Step 6, your circle may look something like that at the right. Notice that $\overleftrightarrow{AB}$ is the perpendicular bisector of $\overline{DE}$. This is so because they are perpendicular and since $\overline{CD}$ and $\overline{CE}$ are radii of the same circle, $CD = CE$.

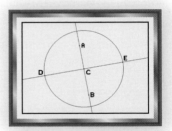

Step 7 Hide the circle. Drag point D along $\overleftrightarrow{DE}$ and notice the effect on point E. Notice also that $\overleftrightarrow{DE}$ can be moved by dragging point C along $\overleftrightarrow{AB}$.

Step 8 Use the Trace tool to trace the path of points D and E as point D is dragged and as the point C is dragged. Try to construct a picture like the one at the right.

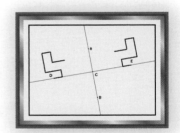

Many interesting pictures can be generated using the constructions in Activity 4. Be creative and make some interesting figures of your own. How are these figures different? What is the role of the perpendicular bisector in all of this? These questions and others relating to them are answered in Chapter 4.

Questions

COVERING THE IDEAS

1. Cell phone towers are at S and T. Copy the points and identify all the points that are closer to S than to T.

2. Define *bisector of a segment*.

3. Define *perpendicular bisector of a segment*.

In 4–5, refer to the first figure in Activity 2.

4. What do you drag to get a vertical trace?

5. What do you drag to get a horizontal trace?

$\overset{\bullet}{S}$

$\overset{\bullet}{T}$

Some cell phone towers are disguised, like this one is in Phoenix, Arizona.

In 6 and 7, use the figure at the right. *D* is the midpoint of $\overline{GF}$.

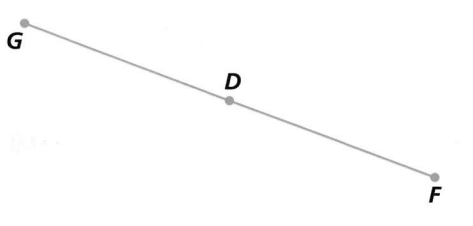

6. Draw a line perpendicular to $\overline{GF}$ that does not go through point *D*.

7. Draw a line that bisects $\overline{GF}$ but is not perpendicular to $\overline{GF}$.

8. Who or what is Voronoi?

9. Trace points *A*, *B*, and *C*. Draw a Voronoi diagram and identify all the points that are closest to each point.

10. In a construction, what tools are allowed?

11. What three types of commands are there in a DGS construction?

12. What figure is described as $\odot (A, AP)$?

13. **a.** Do the following construction.

   Step 1 Given points *A* and *B*.

   Step 2 Construct $\odot (A, AB)$.

   Step 3 Construct $\overleftrightarrow{AB}$.

   Step 4 Name $\odot (A, AB) \cap \overleftrightarrow{AB} = \{B, C\}$

   Step 5 Construct $\odot (B, BC)$.

   Step 6 Construct $\odot (C, BC)$.

   Step 7 Name $\odot (B, BC) \cap \odot (C, BC) = \{D, E\}$

   Step 8 Construct $\overleftrightarrow{DE}$.

   b. How is point *A* related to points *B* and *C*. Why is this true?

   c. Does the line $\overleftrightarrow{DE}$ seem to contain point *A*?

   d. Measure $\angle CAD$. What seems to be true about lines $\overleftrightarrow{DE}$ and $\overleftrightarrow{AB}$?

14. **a.** Do the following construction on a DGS.

   Step 1 Given points *H* and *M*.

   Step 2 Construct $\overleftrightarrow{HM}$.

   Step 3 Name a point *J* on $\overleftrightarrow{HM}$.

   Step 4 Name a point *G* not on $\overleftrightarrow{HM}$.

   Step 5 Construct $\overline{JG}$. Your screen should look somewhat like the one at the right.

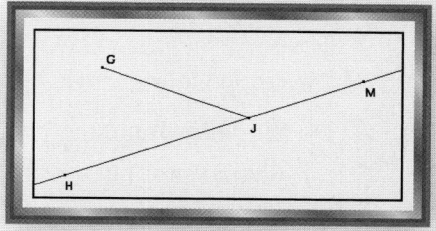

   Step 6 Measure *GJ* to as many places of accuracy as your DGS allows. Drag point *J* and watch as the measure *GJ* changes.

   b. Where should you drag point *J* so as to make *GJ* as small as possible?

APPLYING THE MATHEMATICS

In 15 and 16, trace the 4 points. Then identify all the points in the place that are closest to each of the points.

15. vertices A, B, C, and D of rectangle $ABCD$

16.

$\bullet Y$

X
$\bullet$

W
$\bullet$

$\bullet Z$

17. Refer to the Voronoi diagram with points X, Y, and Z pictured in the lesson.

a. Which points are the same distance from X as from Y?

b. Which points are the same distance from Z as from X?

c. Is any point the same distance from all 3 points? If so, identify it. If not, why not?

18. a. Do the following construction on a DGS.

   Step 1 Given points A, B, and C.

   Step 2 Construct $\triangle ABC$. (Construct 3 segments).

   Step 3 Construct a line parallel to side $\overline{BC}$ that passes through point A.

   Step 4 Construct points on the parallel line such that point A is between these points. Label these points as in the figure at the right.

   b. Find $m\angle EAC$, $m\angle CAB$, $m\angle DAB$, $m\angle BCA$, and $m\angle CBA$.

   c. Find two pairs of angles in your figure that have equal measures.

   d. Drag point A to various locations between points D and E. Is your previous answer to Part c still true? Will it be true for any position of A?

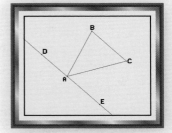

19. a. Do the following construction on a DGS.

   Step 1 Given noncollinear points A, B, and C.

   Step 2 Construct $\overrightarrow{AB}$ and $\overleftrightarrow{AC}$.

   Step 3 Construct a line parallel to $\overleftrightarrow{AC}$ that passes through point B.

   Step 4 Construct a line parallel to $\overleftrightarrow{AB}$ that passes through point C.

   Step 5 Construct the point of intersection of the two lines constructed in Steps 4 and 5 and label that point D as in the figure at the right.

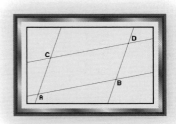

b. Find m∠*BAC*, m∠*ACD*, m∠*CDB*, and m∠*DBA*.

c. Identify pairs of supplementary angles and pairs of congruent angles from those in Part b.

d. Drag lines $\overleftrightarrow{AB}$ and $\overleftrightarrow{AC}$ to change their tilt. Are your answers to Part c still true? Will they always be?

e. Find the sum of the four angle measures that you found in Step 5. What seems to be true about this sum? When you drag lines $\overleftrightarrow{AB}$ and $\overleftrightarrow{AC}$, is your result still true?

REVIEW

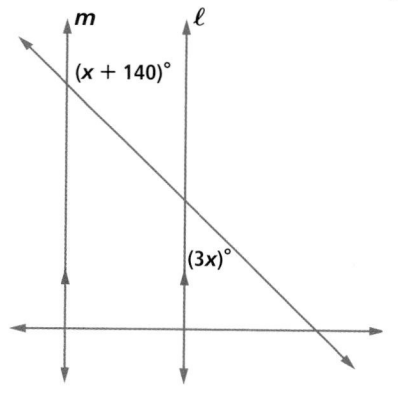

20. In the figure at the right, $\ell \parallel m$, $j \perp m$, and $k \perp \ell$. Prove that $j \parallel k$. (**Lesson 3-8**)

21. Line ℓ has equation $\frac{x - y}{2} = 4$. $m \perp \ell$ and m intersects line ℓ at $(9, 1)$. Find an equation for line m. (**Lesson 3-8**)

22. Is there a point A in the plane that satisfies $S_3(A) = A$? If not, why not? If so, how many such points are there? (**Lesson 3-7**)

23. In the figure at the right, $\ell \parallel m$. Find the value of x. (**Lesson 3-6**)

24. Justify the following conclusion: Two obtuse angles BAC and BAD cannot form a linear pair. (**Lessons 3-5, 3-4**)

25. While sailing, Andy's sail makes a 40° angle with the wind. He then turns the bow (front) of his boat through the wind so that the sail changes from one side to the other, and still makes the same magnitude angle with the wind. (**Lesson 3-1**)

a. Through what angle did the sail rotate?

b. Answer Part a if 40° is replaced by w°.

EXPLORATION

26. Search the internet to find and describe other applications of Voronoi diagrams outside of mathematics.

Chapter 3 Projects

1 Zoom and Size Transformations

Many sites on the Internet give you access to maps and allow you to zoom in on a particular area on the map. Find a map and explain how to zoom. See if you can compute the scale factor for the zoom at that site. Does the scale factor stay constant between levels? Explain how you know.

2 Mazes

Mazes are often made out of parallel and perpendicular lines. Research and find an algorithm, or method, for solving mazes. Draw two difficult mazes and use this method to solve them. Write a short report on the method. Show how rotations, parallel lines, and perpendicular lines are involved in the method.

3 Combination Locks and Rotations

You have probably used a combination lock many times. Opening these locks involves a sequence of rotations.

a. If you are given the combination to open a lock, explain how you could find the magnitude of the rotations necessary to open it.

b. Find out how combination locks work. Write a clear explanation of what you find, include any diagrams you feel are necessary to illustrate your explanation. Be sure to also explain exactly how rotations are involved.

4 Voronoi Diagrams and High Schools

Find a map that includes the area in which you live, and mark on the map the locations of all the public high schools. If necessary, include neighboring areas so multiple high schools will be marked.

a. Draw the Voronoi diagram for these high schools on your map.

b. Find out where the different districts for these high schools are, and draw their borders on the map.

c. What is the relationship between the answers you got in Parts a and b? If the answers are not close to the same, explain why this might be so.

5 Fingerprint Statistics

Collect at least 20 fingerprints of friends and family. The easiest way to record fingerprints is to have people rub an area on a piece of paper with a pencil. Have your participants rub their fingers over the graphite on the paper. Apply a piece of transparent tape to the finger, lift, and place the tape on the sheet of paper.

It is thought that there are four major groups of fingerprints. They are:

Loops: strongly curved patterns, the ends of which enter and exit the finger on the same side (60% of all fingerprints)

Whorls: complete ovals (30% of all fingerprints)

Arches: ridges that run across the finger, arching in the middle (5% of all fingerprints)

Accidentals: combinations of two of the above patterns (5% of all fingerprints)

Classify your fingerprint samples into one of these four groups and see if your collection supports the statistics. Present your data and explain your results.

Chapter 3 Summary and Vocabulary

● Arcs, rotations, and angles are related by the notion of a turn. Arcs are classified as major or minor arcs and have measures between 0° and 360°. In contrast, the magnitude of rotation may be any real number. If its magnitude is positive, the rotation is counterclockwise; if its magnitude is negative, the rotation is clockwise. The measure of an angle, however, is a unique number between 0 and 180. Angles are classified by their measures as zero, acute, right, obtuse, or straight.

● Two angles may be related by their measures. Two angles are complementary if their measures add to 90 and supplementary if their measures add to 180. Two angles may also be related by their positions. If they are vertical angles, their measures are equal. If they form a linear pair, the angles are supplementary.

● A proof of a conditional $p \Rightarrow q$ is a sequence of justified statements leading from p to q. Sometimes the statements made in a proof require using the properties of real numbers. Because of this, we discussed the properties of real numbers and postulates of equality and inequality, and the operations of addition and multiplication.

● Parallel and perpendicular lines are important ideas in geometry. Two lines are parallel if and only if the measures of corresponding angles are equal. Two nonvertical lines are parallel if and only if their slopes are equal. Likewise, two nonvertical lines are perpendicular if and only if their slopes are opposite reciprocals. An important construction is the perpendicular bisector of a segment. You should be able to construct parallel and perpendicular lines by using a variety of tools.

● A transformation is a correspondence between sets of points such that each point in the preimage has exactly one image point, and each point in the image has exactly one preimage point. The specific transformations examined in this chapter include rotations and size transformations. The size transformation with center $(0, 0)$ and magnitude $k \neq 0$ maps (x, y) onto (kx, ky). It preserves collinearity and angle measure and the image of a line ℓ is a line parallel to ℓ.

Vocabulary

Lesson 3-1
arc
endpoints of an arc
measure of an arc
semicircle
minor arc
major arc
*angle
sides of an angle
vertex of an angle
central angle
zero angle
*acute angle
*right angle
*obtuse angle
straight angle
interior of an angle
exterior of an angle

Lesson 3-2
preimage
image

Lesson 3-3
*adjacent angles
*angle bisector
*complementary angles
*supplementary angles
complements
supplements
*linear pair
*vertical angles

Lesson 3-5
*proof

Lesson 3-6
transversal
corresponding angles

Postulates, Theorems, and Properties

Angle Measure Postulate
  Unique Measure Assumption (p. 116)
  Unique Angle Assumption (p. 116)
  Straight Angle Assumption (p. 116)
  Zero Angle Assumption (p. 116)
  Angle Addition Assumption (p. 127)
Equal Angle Measures Theorem (p. 128)
Linear Pair Theorem (p. 129)
Vertical Angles Theorem (p. 129)
Postulates of Equality
  Reflexive Property of Equality (p. 133)
  Symmetric Property of Equality (p. 133)
  Transitive Property of Equality (p. 133)
Postulates of Equality and Operations
  Addition Property of Equality (p. 134)
  Multiplication Property of Equality (p. 134)
Postulates of Inequality and Operations
  Transitive Property of Inequality (p. 135)
  Addition Property of Inequality (p. 135)
  Multiplication Property of Inequality (p. 135)
Postulates of Equality and Inequality
  Equation to Inequality Property (p. 136)
  Substitution Property (p. 136)
Corresponding Angles Postulate (p. 146),
Parallel Lines and Slopes Theorem (p. 148)
Transitivity of Parallelism Theorem (p. 148)
S_k Theorem 1: Parallel Property (p. 153)
S_k Theorem 2: Collinearity Is Preserved (p. 154)
S_k Theorem 3: Angle Measure Is Preserved (p. 155)
Two Perpendiculars Theorem (p. 160)
Perpendicular to Parallels Theorem (p. 160)
Perpendicular Lines and Slopes Theorem (p. 161)

Vocabulary

Lesson 3-7
*concurrent
point of concurrency
*transformation
mapping, maps
size change, size
  transformation

Lesson 3-8
*perpendicular
opposite reciprocals

Lesson 3-9
equidistant
*bisector of a segment
*perpendicular bisector
Voronoi diagram

Chapter 3 Self-Test

Take this test as you would take a test in class. You will need a calculator and tracing paper. Then use the Selected Answers section in the back of the book to check your work.

1. Two angles are complementary. One has measure 81°. What is the measure of the other?

In 2 and 3, refer to the figure at the right.

2. If m∠3 = 27, find m∠4.

3. If m∠3 is 123 less than m∠9, find m∠6.

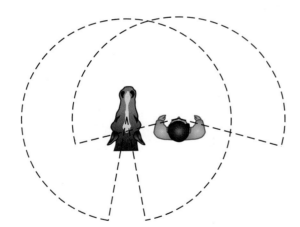

4. Adrian has a 210° field of vision. Adrian's horse has a 340° field of vision. Through what angle on each side of his head does Adrian need to turn his head to cover the same field of vision as his horse?

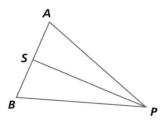

5. In the diagram below, $\overrightarrow{PS}$ bisects ∠APB, m∠APS = 4q + 8, and m∠BPS = 2q + 18. Find m∠APB.

6. ∠1 and ∠2 form a linear pair. m∠1 is 12 more than five times m∠2. Find m∠2.

In 7 and 8, state the algebraic property that justifies the statement.

7. $PQ = RS$ and $SR = MN$, so $PQ = MN$.

8. (m∠ABC = m∠EBF) ⇒ (m∠ABC + m∠CBE = m∠CBE + m∠EBF)

9. What is the measure of the angle formed by the hands of a clock at 10 P.M.?

10. Al was holding a compass that was pointing north when Freda placed a magnet next to the compass equally distant from east and south. The needle moved immediately to the middle of the magnet. What is the magnitude of the rotation of the compass needle?

11. Trace points A and C below. Draw the image of point A under a rotation of 160° about point C. Label the image A'.

In 12 and 13, O is the center of the circle shown at the right and $m\widehat{PMD} = 230°$.

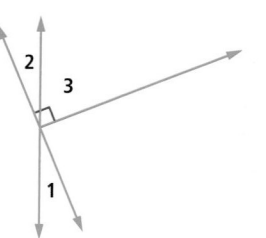

12. Find each measure.

   a. $m\widehat{PD}$ b. $m\angle POM$

   c. $m\widehat{WM}$ d. $m\widehat{MWD}$

13. a. Name three pairs of arcs with the same measure.

   b. Name two pairs of angles with the same measure.

14. Trace the circle and points A and B at the right. Shade all points on or inside the circle that are closer to A than to B.

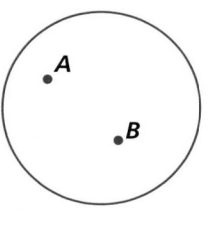

15. Which pairs of lines are parallel in the diagram below? Justify your answer.

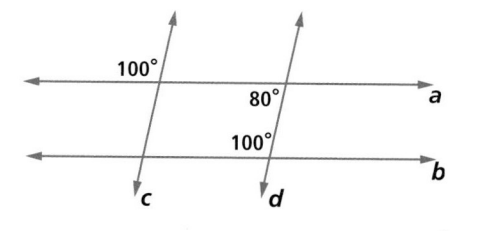

16. In the figure at the right, $a \parallel b$, $b \perp c$, $m\angle 1 = 3x + 57$, and $m\angle 2 = 2x + 2y$. Find y.

17. a. What is the slope of the line with equation $x + 3y = 17$?

   b. What is the slope of a line perpendicular to the line in Part a?

18. In the figure below, R is the midpoint of $\overline{SW}$. Make a conclusion and justify that conclusion.

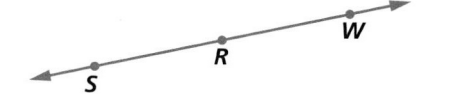

19. Refer to the figure at the right. If $m\angle 2 + m\angle 3 = 90$, why does $m\angle 1 + m\angle 3 = 90$?

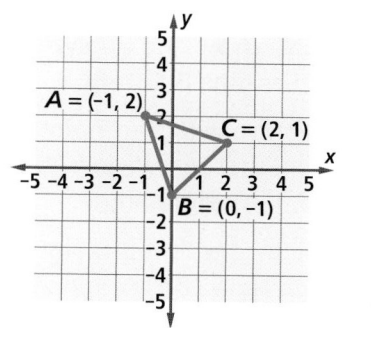

In 20 and 21, consider $\triangle ABC$ below.

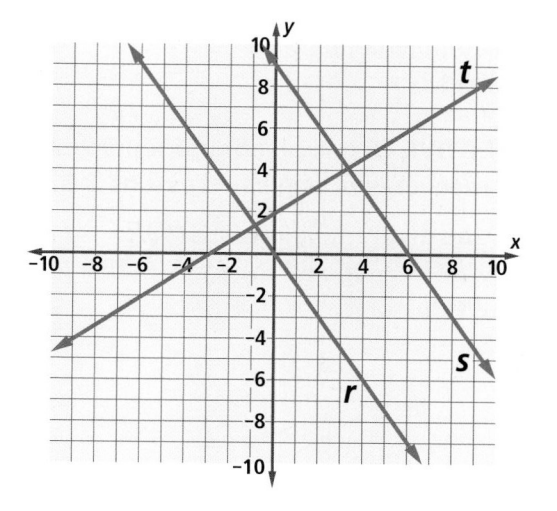

20. Give the coordinates of A', B', and C' when $\triangle A'B'C' = S_3(\triangle ABC)$.

21. **Multiple Choice** In the situation of Question 20, which is true?

   A $m\angle BAC = m\angle B'A'C'$

   B $m\angle BAC = 3m\angle B'A'C'$

   C $3m\angle BAC = m\angle B'A'C'$

   D   None of the above

22. Use the graph below.

   a. $s \parallel r$. Find the slope of s.

   b. $t \perp r$. Find the slope of t.

Chapter 3   Chapter Review

SKILLS Procedures used to get answers

OBJECTIVE A Draw and analyze drawings of angles. (Lessons 3-1, 3-3)

1. Draw a picture with four points A, B, C, and D such that $\angle ABC$ is a straight angle and $\angle ABD$ is a right angle.

2. Draw a picture with four points M, N, O, and P such that $\angle MON$ and $\angle PON$ are supplementary.

In 3–5, refer to the figure at the right.

3. Name two angles that form a linear pair with $\angle AEC$.

4. Name a zero angle.

5. Which angles appear to be obtuse? Which angles appear to be acute?

6. In the figure at the right, $m\angle NOQ = 90$, $m\angle MON = 58$, and $m\angle PON = 32$. Find all the pairs of complementary angles.

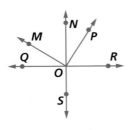

OBJECTIVE B Use algebra to represent and find measures of angles. (Lessons 3-3, 3-4)

7. In the diagram below, $m\angle PQR = 4t + 3$, $m\angle RQS = 2t$, and $m\angle PQS = 150$. Find the value of t.

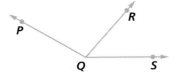

8. Angles $\angle ABC$ and $\angle ABD$ are complementary. Also, $m\angle ABC = 3x - 12$ and $m\angle ABD = 2x + 5$. Find x.

9. Let n be the measure of an obtuse angle. Graph all possible values for n on a number line.

In 10–12, refer to the diagram below.

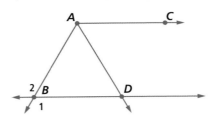

10. If $m\angle 1 = 3y$ and $m\angle 2 = 120$, find y.

11. Suppose that $m\angle ABD = \frac{1}{2}m\angle 2$. Find $m\angle ABD$.

12. If $\overrightarrow{AD}$ bisects $\angle BAC$ and $m\angle BAC = 120$, find $m\angle BAD$.

OBJECTIVE C Determine measures of angles formed by parallel lines, perpendicular lines, and transversals. (Lessons 3-6, 3-8)

In 13 and 14, refer to the diagram below, in which $\ell \parallel m$.

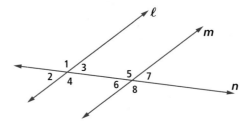

13. If $m\angle 1 = 135$, find the measures of $\angle 2$ through $\angle 8$.

14. If $m\angle 1 = \frac{3}{2}m\angle 2$, find the measures of $\angle 1$ through $\angle 8$.

In 15–17, refer to the diagram at the right.

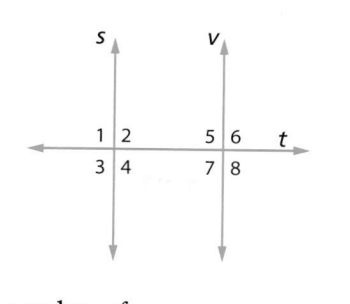

15. If m∠1 = 45 and m∠6 = 135, is s ∥ v?

16. Suppose s ∥ v, m∠5 = 2x, and m∠2 = 3x. Find the value of x.

17. Suppose s ⊥ t. What is m∠5? Explain how you know.

OBJECTIVE D Draw and analyze rotation images. (Lesson 3-2)

In 18–20, refer to the figure below.

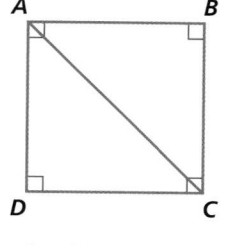

18. Trace the figure and draw its image under a rotation around C of 120°.

19. Where can you place a point O so that the image of ABCD under a rotation around O by 180° is the figure ABCD itself?

20. Trace and draw the rotation image of the figure around E by –40°. What positive magnitude rotation around E has the same image?

In 21 and 22, use the figure below. The 12 points A through L are equally spaced on circle O.

21. If △AJH is rotated –60° about O, what is its image?

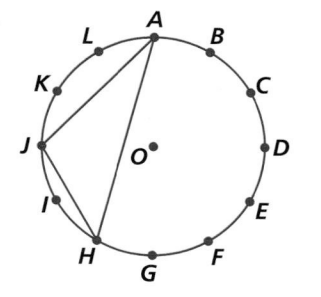

22. Suppose △AJH is the image of a triangle after a rotation of 240°. What was the preimage?

OBJECTIVE E Find the degree measure of arcs and the measures of central angles. (Lesson 3-1)

In 23–26, $\overline{AC}$ and $\overline{DB}$ are diameters of circle O.

23. Name all the central angles in the figure.

24. Suppose m$\widehat{DE}$ = 45. Find m∠EOD and m$\widehat{DAE}$.

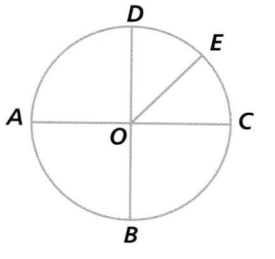

25. **True or False** m$\widehat{ECD}$ = m$\widehat{EBD}$.

26. Suppose $\overrightarrow{OD}$ bisects ∠COA. What is m$\widehat{DA}$?

PROPERTIES Principles behind the mathematics

OBJECTIVE F Recognize and use the postulates of algebra. (Lessons 3-4, 3-5)

27. If m∠1 = m∠2 and m∠1 + m∠2 = m∠3, which postulate of algebra allows you to conclude that m∠3 = 2m∠1?

28. In the figure below, PR = SU, Q is the midpoint of $\overline{PR}$, and T is the midpoint of $\overline{SU}$. Write an argument explaining why PQ = ST.

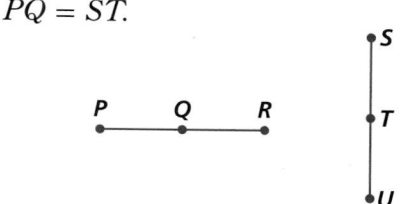

In 29 and 30, refer to the figure below, in which points L, M, N, O are collinear.

29. If LM = MN and MN = NO, what can you deduce about LM and NO?

30. If LM = NO, what can you deduce using the Substitution Property of Equality?

In 31 and 32, give a justification for each conclusion.

31. $\dfrac{5 + 2y}{13} = 22$

$5 + 2y = 286$ a. ___?___

$2y = 281$ b. ___?___

$y = 140.5$ c. ___?___

32. $\dfrac{9 - 2n}{18} = 0.8$

$9 - 2n = 14.4$ a. ___?___

$-2n = 5.4$ b. ___?___

$n = -2.7$ c. ___?___

OBJECTIVE G Give justifications for conclusions involving angles and segments. (Lessons 3-3, 3-5, 3-6, 3-8)

In 33–37, give a justification for each conclusion.

33. Given: $m\angle 2 = 65°$ and $m\angle 6 = 65°$.
Conclusion: $m \parallel n$.

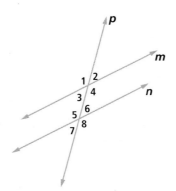

34. Given: $k \parallel \ell$ and $j \perp \ell$.
Conclusion: $j \perp k$.

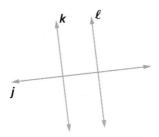

35. Given: $\overleftrightarrow{MB} \cap \overleftrightarrow{TH} = I$ and
$m\angle HIM = 78.6$.
Conclusion: $m\angle BIT = 78.6$.

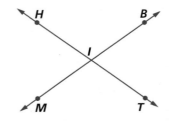

36. Given: M is the midpoint of $\overline{EA}$.
Conclusion: $EM = MA$.

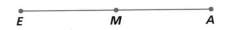

37. Given: $m\angle FOR = 2m\angle FOP$.
Conclusion: $m\angle FOR = 120°$.

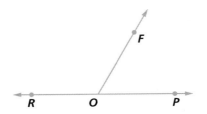

USES Applications of mathematics in real-world situations

OBJECTIVE H Apply angle and arc measures and rotations in real situations. (Lessons 3-1, 3-2)

38. What is the measure of the angle between the hands of a clock showing 5 P.M.?

39. If a circular pie is cut into eight equal slices, what is the degree measure of the arc of 3 pieces of pie?

40. Doing an around-the-world trick with a yo-yo requires a full revolution of the yo-yo. If Tawny completes two-and-a-half around-the-worlds, how many degrees has her yo-yo traveled?

41. If you were facing east and wished to face northwest, by how much must you turn?

OBJECTIVE I Construct and use Voronoi diagrams to solve problems. (Lesson 3-9)

In 42 and 43, the rectangle below represents a college campus. Four students live at points A, B, C, and D. Trace the figure. Locate all places on campus that are:

42. closer to point A than to point C.

43. closer to point B than to any other points.

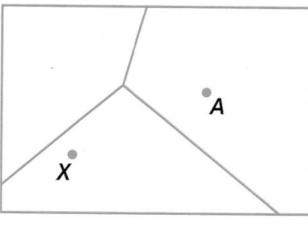

In 44 and 45, use the Voronoi diagram for points A, B, and C shown at the right.

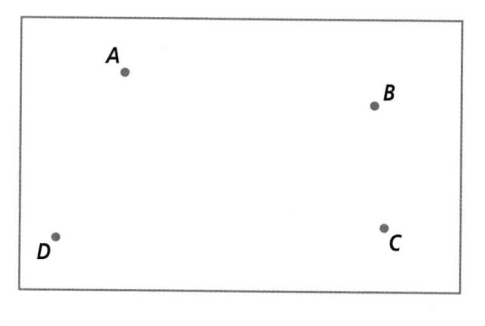

44. Fill in the missing dots B and C in the diagram. Ignore point X.

45. Redraw the Voronoi diagram for points A, B, C, and the fourth point, X.

REPRESENTATIONS Pictures, graphs, or objects that illustrate concepts

OBJECTIVE J Apply size transformations to figures on coordinate axes. (Lesson 3-7)

46. $S_{1.5}(LUV) = \triangle L'U'V'$ as shown below. Graph $\triangle LUV$.

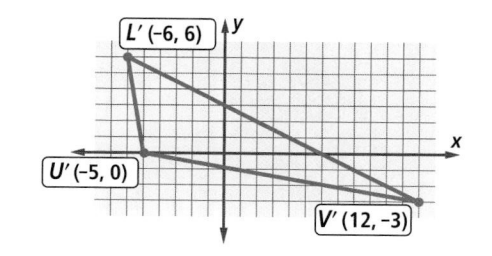

L' (-6, 6)

U' (-5, 0)

V' (12, -3)

In 47 and 48, a drawing for a triangular playlot for the playground at an elementary school is shown below.

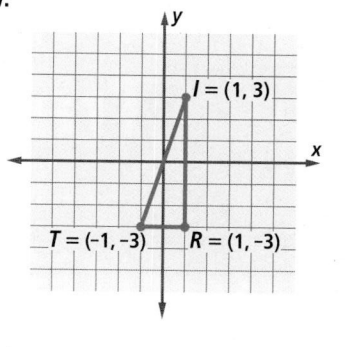

$I = (1, 3)$

$T = (-1, -3)$ $R = (1, -3)$

47. Graph the image of the triangle under S_2.

48. Find the slopes of $\overline{IT}$ and $\overline{I'T'}$. Are the slopes equal?

OBJECTIVE K Determine the slope of a line, and of lines parallel or perpendicular to a given line. (Lessons 3-6, 3-8)

In 49–52, find the slope of the line with the given equation.

49. $y = 2x + 4$

50. $y = 4 - 5.7x$

51. $5y + 4x = 18$

52. $\frac{7}{8}x - \frac{2}{7}y = 4$

53. If the slope of a line is $\frac{5}{3}$, then what is the slope of a line perpendicular to it?

54. Line ℓ has slope 7.4. Maxine says the line with equation $5y + 37x = -15$ is parallel to ℓ. Is she correct? Explain your answer.

In 55 and 56, use the graph below.

55. $n \parallel m$. Find the slope of n.

56. $p \perp m$. Find the slope of p.

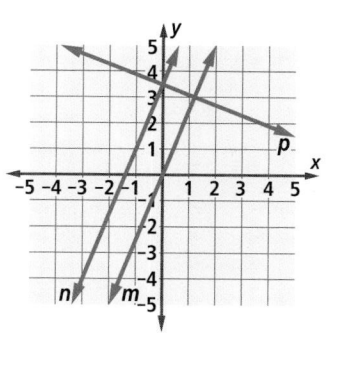

Chapter

4

Congruence Transformations

Contents

4-1 Reflecting Points

4-2 Reflecting Figures

4-3 Miniature Golf and Billiards

4-4 Composing Reflections over Parallel Lines

4-5 Composing Reflections over Intersecting Lines

4-6 Translations as Vectors

4-7 Isometries

4-8 Transformations and Music

The drawing above is part of one of the many *tessellations* drawn by the Dutch artist Maurits C. Escher (1898–1972). This tessellation was designed by piecing together unicorns of the same size and shape. We say that the unicorns are all *congruent* to each other.

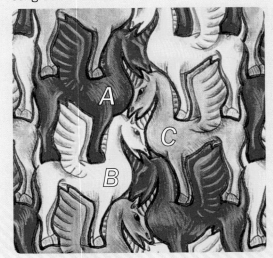

Unicorns A and B are related to each other by a *slide* or *translation.* Each is a translation image of the other. (The translation is vertical.) On the other hand, unicorns A and C have different orientations. However, they are not related by a single *flip* or *reflection.* They are related by another kind of transformation, a *walk* or *glide reflection.*

It is natural to ask for all the possible ways in which the various unicorns could be related to each other. In this chapter, you will learn that there are four types of transformations in the plane that yield congruent figures: reflections (flips), rotations (turns), translations (slides), and glide reflections (walks). These were first identified in 1831 by the French mathematician Michel Chasles (1793–1880).

The first use of transformations dates back to the ancient Greeks around the time of Euclid. However, not until Euler (in 1776) did anyone identify all the kinds of transformations in space that could yield congruent figures. It is interesting that the 3-dimensional analysis of congruence was accomplished before the 2-dimensional. This is probably because the congruent objects seen daily are 3-dimensional.

Studying these various transformations helps a person to become more aware of the movements of objects such as gears (which rotate) and conveyer belts (which slide). More complicated movements, such as those done by robots, can be taken apart into their component moves and analyzed. Transformations also appear in music and help to show some connections between mathematics and music.

Lesson

4-1 Reflecting Points

▶ **BIG IDEA** A reflection over a line is a transformation of the plane in which the line acts like a mirror.

In mirrors, you can see the reflection images of objects that you cannot see directly, such as your eyes. Mirrors produce an image that looks like the original object. Babies are often fascinated with the "other baby" in a mirror.

Mirrors are not the only things that create an image of the object they reflect. Reflection images can be seen in ponds, lakes, puddles, and streams. Water, when still, is a perfect reflecting medium. The picture at the right was taken from under water.

Reflection Images

Activity 1

MATERIALS felt marker, ruler or straightedge

Step 1 Fold a piece of paper in half, and then unfold it. Using a felt marker, draw a closed figure on one half of the paper. Refold the paper along the crease, and trace your figure onto the blank part of the paper. Now unfold the paper and look at your result. Compare it to those that your classmates made.

Step 2 Mark a point on your drawing. Mark a second point at the image of your original point. Draw the segment connecting the two points. Repeat this for several more points on the original figure. Describe the line that is the crease in your paper.

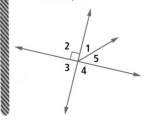
Examine the figure below. Think of the left side as the *preimage*. The *reflection image* on the right side can be drawn by folding over the line *m* and then tracing. Line *m* is called the **reflecting line** (or **line of reflection**).

Line *m* is the perpendicular bisector of the segments connecting corresponding points, such as *E* and *E′*, *L* and *L′*, and *F* and *F′*.

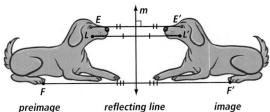

preimage   *reflecting line*   *image*

Definition of Reflection Image

For a point *P* not on a line *m*, the **reflection image** of *P* over line *m* is the point *Q* if and only if *m* is the perpendicular bisector of $\overline{PQ}$. For a point *P* on *m*, the reflection image is *P* itself.

Activity 2

MATERIALS Ruler, protractor, DGS

To draw the reflection image *P′* of point *P* over line *m*, follow this algorithm.

Step 1 Trace line *m* and point *P*.

Step 2 Place your protractor so that its 90° mark and the center of the protractor are on *m*.

Step 3 Slide the protractor along *m* so that the base line goes through *P*.

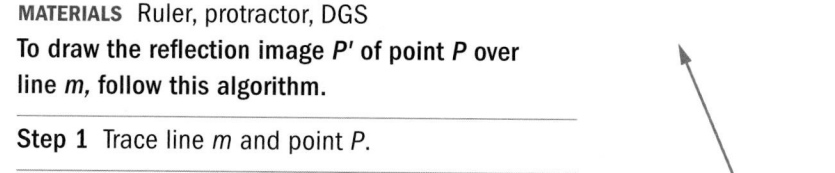

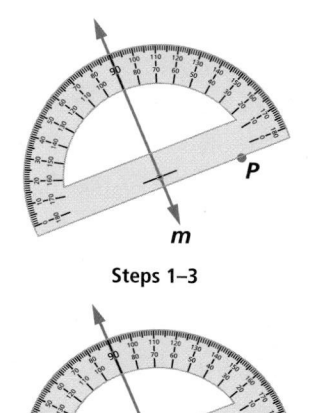

Steps 1–3

Step 4 Measure the distance from *P* to *m*. You may wish to draw the line lightly.

Step 5 Locate *P′* on the other side of *m*, the same distance from *P*.

Step 6 Check your work in two ways.

> **Check 1** If you draw $\overline{PP'}$ it should be perpendicular to and bisected by *m*.

> **Check 2** Fold *P′* over line *m*. It should land on *P*.

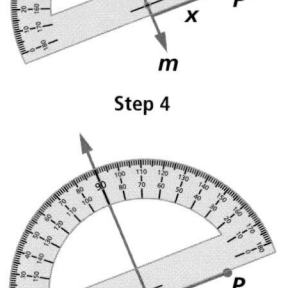

Step 4

Step 7 On a clear DGS screen, construct a line and a point not on the line.

Step 8 Using the reflection tool, reflect the point over the line.

Step 9 Check your construction. Use the DGS to measure one of the angles formed by $\overline{PP'}$ and *m*. Then measure the distances from *P* and *P′* to *m*. The angle should measure 90°, and the distances should be equal.

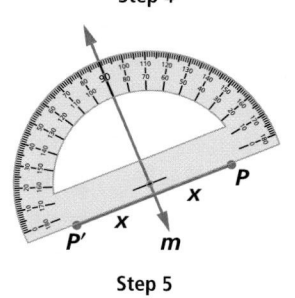

Step 5

Notation for Reflections

A reflection is a type of transformation. We use a lower case letter "*r*" to refer to a reflection. When discussing reflections in general, or when the reflecting line is obvious, we write

$$r(A) = A'.$$

The statement $r(A) = A'$ is read "The reflection image of *A* is *A′*," or "*r* of *A* equals *A′*."

When we want to emphasize the reflecting line *m*, we write

$$r_m(Q) = P.$$

This statement is read, "The reflection image of Q over line m is P," or "r of Q over line m equals P."

The notation above should be familiar to you as it is very similar to what we used for rotations and size transformations in the last chapter. This notation is also used when working with functions in algebra.

 QY

▶ QY

Write using reflection (function) notation: "The reflection image of point Z over the line n is point V."

Example 1

In the figure at the right, the reflection image of A over line m is B. Name B using reflection notation.

Solution You should not write $B = r(A)$ because there is more than one line in the drawing. You should write $B = r_m(A)$.

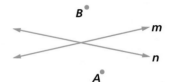

Reflection images can be found easily for points in the coordinate plane if the reflecting line is one of the axes. The image is found using the definition of reflection.

GUIDED

Example 2

Find the reflection image of $(-3, 4)$ over
a. the y-axis.   b. the x-axis.

Solution Draw a coordinate grid and let $P = (-3, 4)$.

a. P is __?__ units from the y-axis.
   The image of P is __?__ units from the y-axis.
   $P' = r_{y\text{-axis}}(-3, 4) = (\underline{\ ?\ }, \underline{\ ?\ })$
b. P is __?__ units from the x-axis.
   The image of P is __?__ units from the x-axis.
   $P' = r_{x\text{-axis}}(-3, 4) = (\underline{\ ?\ }, \underline{\ ?\ })$

Questions

COVERING THE IDEAS

1. **Fill in the Blank** A figure that is to be reflected is called the __?__.

2. Suppose B is the reflection image of A over line m. How are m, A, and B related?

3. **Fill in the Blank** When a point P is on the reflecting line ℓ, then the reflection image of P is __?__.

4. **a.** Draw A', the reflection image of A over line ℓ by using a ruler and protractor.

   b. Find B', the reflection image of point B over ℓ.

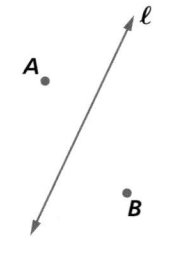

5. **a. True or False** Every reflection is a transformation.

   b. True or False Every transformation is a reflection.

   c. True or False In a reflection over a line m, the reflection image of every point is a different point than the preimage.

6. If Q is a point, write how each expression would be read.

   a. $r(Q)$ **b.** $r_m(Q)$

7. Trace the drawing at the right. Then draw the reflection images of the labeled points over line m.

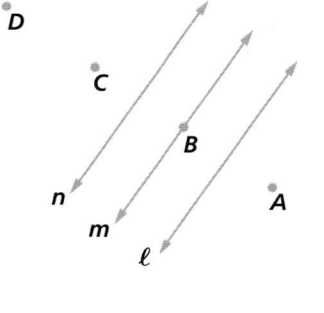

8. In the figure at the right, give the coordinates of each point.

   a. $r_{x\text{-axis}}(P)$ **b.** $r_{y\text{-axis}}(P)$

   c. $r_{y=x}(P)$

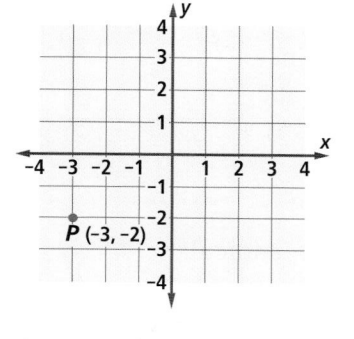

9. Find the image of $(0, 1)$ when reflected over the given line.

   a. the x-axis   **b.** the y-axis

   c. the line $y = x$

APPLYING THE MATHEMATICS

10. Repeat Question 9 for the point (c, d).

11. **a.** In the figure at the right, B, C, and D are three reflection images of point A. Match each image with the correct reflecting line.

    b. Name each image of A using reflection notation.

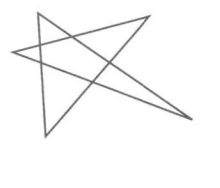

12. In the figure at the right, $r_\ell(P) = T$. Draw ℓ. $T_\bullet$

13. Trace the figures below. Find the line so that one of the figures is the reflection image of the other.

$P_\bullet$

14. **a.** Decipher the message at the right.

    b. Which letter is written incorrectly?

HELP! I'M TRAPPED INSIDE THE PAGE!

15. Trace the figure at the right. Then find its reflection image over line m by folding and tracing.

16. Let $P = (-3, 12)$. If line ℓ has equation $y = 4$, determine the coordinates of $r_\ell(P)$.

REVIEW

17. Refer to the graph at the right.

    a. Is $\ell \parallel m$? Justify your answer.
       (Lesson 3-6)

    b. Is $\ell \perp n$? Justify your answer.
       (Lesson 3-8)

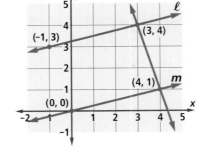

18. **Multiple Choice** Which of the following describes the relationship between the lines with equations $x + 2y = 6$ and $2x - y = 8$? (Lessons 3-8, 3-6)

    A parallel

    B perpendicular

    C neither parallel nor perpendicular

19. In the figure at the right, E is on $\overline{VG}$, $m\angle G = 30$, and $m\angle GEO = 150$. Justify each conclusion.

    a. $m\angle VEO = 30$ (Lesson 3-3)

    b. $\overline{EO} \parallel \overline{GL}$ (Lesson 3-6)

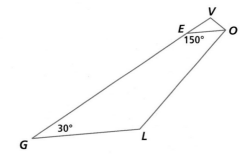

20. Consider the following conditional: *If $|x| = 10$, then $x = 10$.* (Lesson 2-2)

    a. If this conditional is $p \Rightarrow q$, what is p?

    b. Find a counterexample to this conditional.

21. Show that the conditional *If a network is traversable, then it has exactly two odd nodes* is false. (Lessons 2-2, 1-4)

22. a. **Fill in the Blank** If D is on $\overline{XY}$, $XD = 11.2$, and $XY = 26.7$, then $DY = \underline{\quad?\quad}$.

    b. Make a drawing of the situation described in Part a.
       (Lesson 1-6)

EXPLORATION

23. Draw three possible pictures of the antecedent in the sentence below. (Make your pictures look different from each other.) Then write in as many consequents as you think are true.

    If $r_\ell(A) = B$ and $r_\ell(C) = D$, then $\underline{\quad?\quad}$.

QY ANSWER

$r_n(Z) = V$

Lesson
4-2   Reflecting Figures

▶ **BIG IDEA**   The reflection image of a figure can be found by finding the images of enough points to determine the figure.

Activity

MATERIALS DGS

Step 1   Construct polygon *WXYZ*, line ℓ and a point *P* on $\overline{WX}$.

Step 2   Reflect the points *W, X, Y, Z,* and *P* over line ℓ and name them *W′, X′, Y′, Z′,* and *P′* respectively.

Step 3   Connect *W, X, Y,* and *Z* using line segments to create quadrilateral *W′X′Y′Z′*.

Step 4   Drag different parts of your diagram around the screen to answer the questions below.

     a. How do the lengths in the preimage compare to the lengths in the image?

     b. How do the angles in the preimage compare to the angles in the image?

     c. How is the position of *P* related to the position of *P′*?

Leave this DGS screen up and running. You will refer to it again.

Mental Math

Give the coordinates of each image.

a. $r_{x\text{-axis}}$ (4.5, –3.4)

b. $r_{y\text{-axis}}$ (4.5, –3.4)

c. $r_{x\text{-axis}}$ (*a, b*)

d. $r_{y\text{-axis}}$ (*a, b*)

Preservation Properties of Reflections

The Activity illustrates that the reflection image of a polygon is a polygon with sides the same length and angles the same measures as their preimages. Why do the reflected images of complex figures like polygons or the bridge at the right look so much like their preimages? It is because there are properties of figures that hold for both preimages and images. When this occurs, we say the property is *preserved* by reflections. Think about reflections you see in mirrors, draw on paper, or create with a computer. You will notice that each of the five statements on the next page is true. We label them as the *Reflection Postulate*. They are our assumptions about reflections. You may wish to compare them with the properties of size changes from Lesson 3-7.

Reflection Postulate

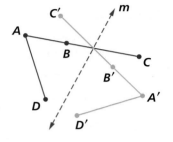

Under a reflection	In other words
a. There is a one-to-one correspondence between points and their images.	Each preimage has a unique (exactly one) image, and each image has a unique preimage.
b. Collinearity is preserved. If three points are collinear, then their images are collinear.	The image of a line is a line.
c. Betweenness is preserved. If B is between A and C, then the image B' is between the images A' and C'.	The image of a line segment is a line segment.
d. Distance is preserved. If $\overline{A'B'}$ is the image of $\overline{AB}$, then $AB = A'B'$.	The distance between two preimage points equals the distance between their image points.
e. Angle measure is preserved. If $\angle B'A'D'$ is the image of $\angle BAD$, then $m\angle B'A'D' = m\angle BAD$.	The measure of the preimage of an angle is equal to the measure of the image of the angle.

The parts of the Reflection Postulate are listed in a logical order. Points must have images before there can be collinearity. Images of collinear points must be collinear before betweenness can be preserved. Betweenness precedes distance and the rays necessary to have angles.

A device for helping you remember the entire postulate is to list the last four properties in alphabetical order (A-B-C-D), as follows: Every reflection is a one-to-one correspondence that preserves Angle measure, Betweenness, Collinearity, and Distance.

 QY1

Drawing Reflection Images of Figures

The **reflection image of a figure** is the set of all the reflection images of points in the figure. While you can reflect a figure point by point, it is impossible to do this for figures with infinitely many points. You need to use a shortcut: simply reflect the points that determine the figure, then connect the points. This is exactly what you did to create the image of polygon $WXYZ$ in the Activity. Because reflections preserve betweenness and collinearity, the remaining points will fall into place. For instance, angles can be determined by a vertex and two points, one on each side. So, if $r(A) = A'$, $r(B) = B'$ and $r(C) = C'$, then the

▶ QY1

Return to the screen of Activity 1 on your DGS. Find one example of each part of the A-B-C-D on this screen.

reflection image of ∠*ABC* is ∠*A'B'C'*. We write *r*(∠*ABC*) = ∠*A'B'C'*. This idea is generalized in the following theorem, which is true for all the transformations you study in this book.

> **Figure Transformation Theorem**
>
> If a figure is determined by certain points, then its transformation image is the corresponding figure determined by the transformation images of those points.

Example 1

Draw the reflection image of pentagon *ABCDE* over line *m*.

 The five points *A, B, C, D,* and *E* determine pentagon *ABCDE*. Find the images of *A, B, C, D,* and *E*. Connect them in order as shown in the figure at the far right.

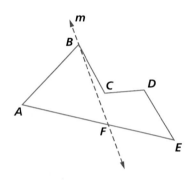

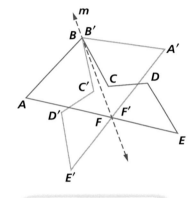

STOP QY2

In Example 1, we write *r*(*ABCDE*) = *A'B'C'D'E'*, keeping preimage and image points in order. Notice that the preimage and image both intersect the reflecting line at *B* and *F*. This is because the image of all points on the reflecting line are the points themselves, and so the image intersects the reflecting line at the same points.

> ▶ **QY2**
>
> How many image points are needed to determine the image of a heptagon?

Orientation of Figures

Although reflections preserve angle measure, betweenness, collinearity, and distance, they do not preserve everything. The image *A'B'C'D'E'* looks reversed from the preimage *ABCDE*. Now we explore that reversal.

When you think of walking along the sides of a polygon, going from vertex to vertex, you are assigning an **orientation** to the polygon. Every polygon has two orientations. The orientation for which the interior is on the *right* as you walk is **clockwise.** When the interior is on the *left,* the orientation is **counterclockwise.**

GUIDED

Example 2

Refer to Example 1 with image *A'B'C'D'E'* and preimage *ABCDE*.
a. Give the orientation of *ABCDE*.   b. Give the orientation of *A'B'C'D'E'*.

Solution

a. As you walk from *A* to *B* and around the polygon, the interior is on the ___?___. So the orientation is ___?___.

b. As you walk from *A'* to *B'* to *C'* to *D'* to *E'* and back to *A'*, the interior is on the ___?___. So the orientation is ___?___.

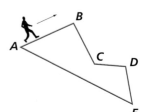

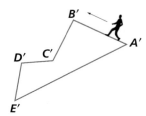

Orientation depends on the order in which the vertices are given. In Guided Example 2, the orientation of $ABCDE$ is clockwise, so the orientation of the image polygon, taking the letters in the corresponding order, is counterclockwise. A figure and its reflection image always have opposite orientation. We take the word "orientation" as undefined, and add the following part to the Reflection Postulate.

Reflection Postulate

Under a reflection	In other words
f. Orientation is reversed. A polygon and its image, with vertices taken in corresponding order, have opposite orientations.	If a preimage has clockwise orientation, its image has counterclockwise orientation, and vice versa.

Think about a pair of gloves or a pair of shoes. The two gloves or shoes appear to have the same shape and size, but they have different orientations. If orientation were not reversed, you would have two shoes for one foot and none for the other!

Questions

COVERING THE IDEAS

1. Can a point have two different reflection images over the same line? Why or why not?

2. When is the reflection image of a point the point itself?

3. △*K'A'Y'* is the reflection image of △*KAY* over line *m* as shown at the right. Answer the following questions. Justify your answers using properties of the Reflection Postulate.

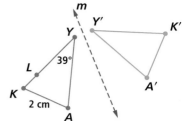

a. **Fill in the Blank** *K'A'* = _?_

b. **Fill in the Blank** m∠*K'Y'A'* = _?_

c. Where would point *L'* lie?

d. Are *K'*, *L'* and *Y'* collinear? Why or why not?

4. Name four properties that reflections preserve.

5. How are the orientations of a figure and its reflection image related?

6. If a polygon *POLYG* is oriented clockwise, what is the orientation of a reflection image *P'O'L'Y'G'*?

7. Trace the drawing at the right.

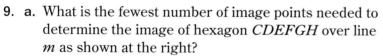
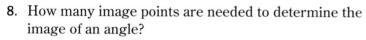
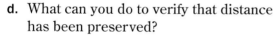

a. Draw $r_k(ABCD)$.

b. Provide three alternate names for *ABCD* that have the same orientation.

c. What is the orientation of *A'B'C'D'*?

d. What can you do to verify that distance has been preserved?

8. How many image points are needed to determine the image of an angle?

9. a. What is the fewest number of image points needed to determine the image of hexagon *CDEFGH* over line *m* as shown at the right?

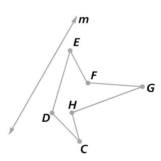

b. If m∠*EDC* = 107, what is the measure of ∠*E'D'C'* (its reflection image)?

c. Trace the figure and draw $r_m(CDEFGH)$.

APPLYING THE MATHEMATICS

10. What is the orientation of the path of a base runner around a baseball diamond?

11. Refer to the drawing of △*ABC* at the right.

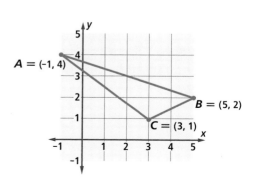

a. What is the slope of $\overleftrightarrow{AB}$?

b. What is the slope of $r_{x\text{-axis}}(\overleftrightarrow{AB})$?

c. What is the slope of $r_{y\text{-axis}}(\overleftrightarrow{AB})$?

d. Make a conjecture based on your answers in Parts a–c.

In 12 and 13, refer to the figure at the right.

Suppose $\odot A$ is reflected about line j.

12. a. If $AF = 108 + 2x$ and $A'F' = 5x$, find the radius of $\odot A$.

    b. What property of reflections did you use to answer Part a?

13. Trace the figure and draw $r_j(\odot A)$.

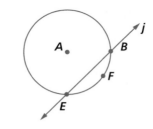

In 14 and 15, trace the drawing. Then draw the reflecting line m so that $\angle XYZ$ is the reflection image of $\angle ABC$ over m.

14.

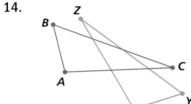

15.

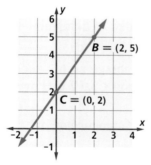

16. In the diagram at the right, $\triangle E'C'D'$ is the refection image of $\triangle ECD$ over line m. $m\angle D = 58^\circ$, $DE = 4$ cm, $m\angle D' = (10x + y)^\circ$, and $D'E' = (x + y)$ cm.

    a. Find x and y.

    b. What property allowed you to set up your equations?

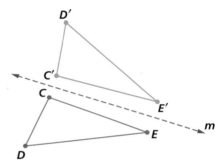

17. Refer to the graph at the right.

    a. What is an equation of $\overleftrightarrow{CB}$?

    b. What is an equation of $r_{y\text{-axis}}(\overleftrightarrow{CB})$?

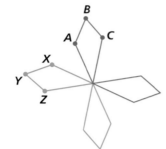

18. **Matching** Assume $r_m(N) = P$, $r_m(M) = Q$, $r_m(O) = T$, and that N, M, and O are not collinear. Choose the correct justification for each conclusion.

    a. $NO = PT$ i. Reflections preserve angle measure.

    b. $m\angle OMN = m\angle TQP$ ii. Reflections preserve betweeness.

    c. $m \perp \overline{NP}$ iii. Reflections preserve collinearity.

    d. $r_m(\triangle NMO) = \triangle PQT$ iv. Reflections preserve distance.

                           v. definition of reflection image

                         vi. Figure Reflection Theorem

19. Brooklyn and Maya begin jogging at the same time. Brooklyn begins at her house at point D and travels clockwise until she reaches home. Maya begins at her house at point C and travels counterclockwise until she reaches home. Assume that the girls run at the same pace and the blocks are of equal length.

    a. Name Brooklyn's path.

    b. Name Maya's path.

    c. Where do Brooklyn and Maya pass each other?

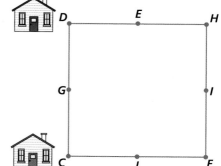

REVIEW

20. Let $P = (a, b)$. Find the coordinates of $P' = r_{y\text{-axis}}(P)$ and $P'' = r_{x\text{-axis}}(P')$. (**Lesson 4-1**)

21. a. Find an equation for the line perpendicular to $y = 3x + 5$ that contains $(4, 7)$. (**Lesson 3-8**)

    b. Find an equation for the line parallel to $y = 3x + 5$ that contains $(4, 7)$. (**Lesson 3-6**)

22. In the figure at the right, $a \parallel b$ and $c \parallel d$. Find y. (**Lesson 3-6**)

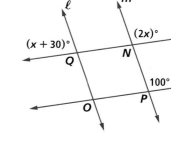

23. In the figure at the right, $w \parallel z$. Is $\ell \parallel m$? Justify your answer. (**Lesson 3-6, 3-5**)

24. Refer to the figure at the right. Draw the rotation image of $ABCD$ about point A with a magnitude of $-45°$. (**Lesson 3-2**)

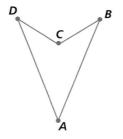

25. Draw two hexagonal regions whose intersection is a convex quadrilateral region. (**Lesson 2-6, 2-5**)

EXPLORATION

26. Many objects come in different orientations. For instance, there are right-handed golf clubs and left-handed golf clubs. Name at least three other objects that come in different orientations.

QY ANSWERS

1. Answers vary. Sample:

   A. $m\angle WXY = m\angle W'X'Y'$

   B. In the preimage, P is between W and X. In the image, P' is between W' and X'.

   C. In the preimage, the points W, P, and X are collinear. In the image, W', P', and X' are collinear.

   D. $WX = W'X'$

2. 7

Lesson
4-3

Miniature Golf and Billiards

Vocabulary

angle of incidence

angle of reflection

▶ **BIG IDEA** By considering reflection images of reflection images, you can determine paths of balls or light beams that bounce off of more than one surface.

When a ball is thrown or rolled without friction or spin against a wall, it bounces off the wall and travels along a ray that is the reflection image of the path the ball would have taken without the wall.

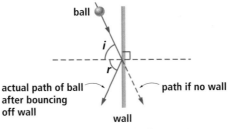

This property is not limited to bouncing balls. It is true of any object traveling without friction that hits a surface, including sound, light, and radio waves. The angles marked *i* and *r* in the figure above are always of equal measure and are referred to in the study of light as the **angle of incidence** (*i*) and **angle of reflection** (*r*).

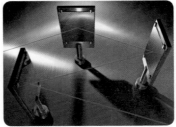

A laser beam being reflected off of three mirrors

Mental Math

Ella's water bottle holds twice as much water as Ryan's. Thurgood's bottle holds $\frac{1}{3}$ as much as Ryan's bottle. Ryan's bottle holds 36 ounces.

a. How much water does Ella's bottle hold?

b. How much water does Thurgood's bottle hold?

c. Together, can their bottles hold a total of a gallon of water?

When you look in a mirror, your eyes see only those light waves that bounce off the mirror in your direction. This is shown in the illustration at the right.

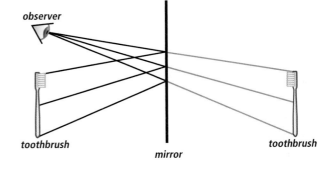

The image appears to be as far "behind" the mirror as the person is in front of it, and the orientation of the image is reversed. All of these results agree with the definition of reflection and the Reflection Postulate.

Miniature Golf

The simple idea of bouncing off a wall has applications in many sports and games, including miniature golf and billiards. In miniature golf, the object is to hit a golf ball into a hole. The hole is often placed so that a direct shot from the tee into the hole is impossible.

At the right is a diagram of a miniature golf hole, as seen from above, with a golf ball at G and the hole at H. The segments represent wooden boards off which the ball can bounce. There is an obstructing board perpendicular to the wall $\overline{AF}$.

In this situation, a good strategy is to bounce, or *carom* (CARE um), the ball off a board, as shown at the right. To find where to aim the ball, start with the hole and work backward. Reflect the hole H over $\overleftrightarrow{AB}$. If you shoot at the image H', the ball will bounce off $\overleftrightarrow{AB}$ at P and go towards the hole.

Billiards

The following discussion of billiards shows how to aim when two or more caroms are needed. Billiards is a game played on a rectangular table with rubber cushions on its sides and no holes. In one version of the game, 3-cushion billiards, the goal is to hit the *cue ball* so that it bounces off three cushions and then hits another ball. To simplify our discussion, we ignore other balls that may be on the table. Pictured at the right is a table with cushions w, x, y, and z.

Suppose you want to shoot cue ball C off x, then y, then z, and finally hit ball B. To set up the shot, consider the walls in *reverse* order. First, reflect B over wall z to get B'. Then take the image B' and reflect it over wall y to get B'' (read "B double prime"). You may have to extend y as shown at the right.

Now reflect the image B'' over wall x to get B''' (read "B triple prime"). Shoot C in the direction of B'''.

Study the diagram at the right to see what happens with the shot. On the way toward B''', the cue ball C bounces off x in the direction of B''. On the way to B'', it bounces off y in the direction of B'. Finally it bounces off z and hits B.

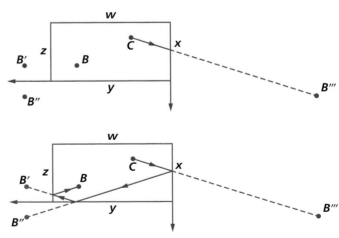

The Oil Well Problem Revisited

In the oil well problem first mentioned in Lesson 1-1, the goal was to find the position of the pumping station X that should be built on the pipeline m so that the sum $AX + BX$ of the distances to oil wells at A and B is as small as possible. This turns out to be a situation similar to miniature golf or billiards.

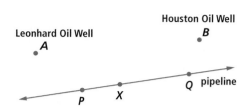

Activity

MATERIALS DGS

Step 1 On a clear DGS screen, construct a line $\overleftrightarrow{PQ}$, a point X on $\overleftrightarrow{PQ}$, and two points A and B on the same side of $\overleftrightarrow{PQ}$.

Step 2 Measure AX, XB, and $AX + XB$.

Step 3 Drag X to a location on $\overleftrightarrow{PQ}$ to minimize the sum of $AX + XB$.

Step 4 Reflect A over $\overleftrightarrow{PQ}$ to get A', the image of A.

Step 5 Construct $\overline{A'B}$ and the point of intersection of this segment and $\overleftrightarrow{PQ}$. What do you notice about the point of intersection and the location of X from Step 3?

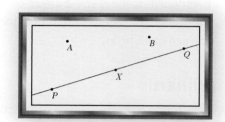

The best position of X can be deduced using properties of reflections. Because A' is the reflection image of A, it follows that $\overline{A'X}$ is the reflection image of $\overline{AX}$. By the Reflection Postulate (distance is preserved), this means the lengths $A'X$ and AX are equal.

You can make the cue ball roll with spin if you hit the ball left of, right of, or down from center.

So, the path A-X-B that represents the sum of the distances of both oil wells to the pumping station has the same length as A'-X-B. ($AX + XB = A'X + XB$).

But, by the Triangle Inequality Postulate, the shortest distance $A'X + XB$ occurs when points A', X, and B are collinear. Therefore, the pumping station should be located at the intersection of the pipeline and the segment connecting the image of A to B to minimize the sum $AX + XB$.

Questions

COVERING THE IDEAS

1. In the figure at the right, a billiard ball (point B) is rolling toward a cushion without spin. Trace the figure and draw the path showing how the ball will bounce off the cushion.

2. Consider the situation of Question 1. If the ball is rolling toward the cushion *with* spin, what phenomenon might not be true?

3. Trace the figure at the right. To shoot a ball from point T to R off the wall x, where should you aim?

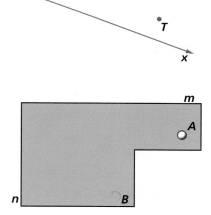

4. Trace the diagram of the miniature golf hole at the right. Draw the path of a shot that would get the ball B into the hole M.

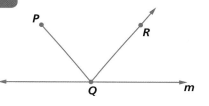

5. Trace the diagram at the right. Draw the path of a shot that would get the golf ball A to bounce off sides m and n and into the hole B.

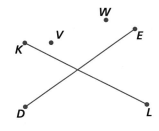

APPLYING THE MATHEMATICS

6. Trace the figure at the right.
   a. A billiard ball P travels along the path shown. Mark the angles that are equal in measure.
   b. Draw a line perpendicular to m through point Q. Label the angle of incidence and the angle of reflection.

7. The figure at the right consists of two intersecting segments $\overline{DE}$ and $\overline{KL}$ and two points V and W. Imagine that a laser beam is to be sent from V, bounced off $\overline{KL}$, and then bounced off $\overline{DE}$ to end up at W.
   a. Trace the figure and draw the path of the laser beam.
   b. Trace the figure again and draw the path of the laser beam if it is to be sent from W, bounced off $\overline{DE}$, and then bounced off $\overline{KL}$ to end up at V.
   c. Are the two paths the same?

In 8 and 9, use the following information: Pocket billiards is played on a table with rubber cushions on its sides and 6 holes called *pockets*. The goal is to hit the cue ball so that it hits an object ball into a pocket. Trace the figure below.

8. Draw the path of a shot that would get the object ball B to bounce off wall y so that it goes into pocket 1 without hitting ball D?

9. Draw the path of a shot that would get the cue ball C to bounce off side x before hitting object ball B.

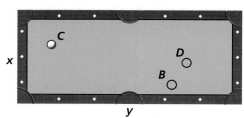

The object ball above is the 9 ball, and the target is the corner pocket.

In 10 and 11, a laser is placed on a coordinate grid at (3, 2) and aimed at (1, 0).

10. The laser beam is reflected off mirrors on the x-axis, the y-axis, the line $y = 10$, and the line $x = 10$.

    a. What are the coordinates of the point where it will first hit the y-axis?

    b. What are the coordinates of the point where it will first hit the line $y = 10$?

    c. What are the coordinates of the point where it will first hit the line $x = 10$?

    d. What are the coordinates of the point where it will hit the x-axis for the second time?

11. There are mirrors at the x-axis, the y-axis, the line $y = 10$, and this time at the line $x = 5$.

    a. What are the coordinates of the point where it will first hit the line $y = 10$?

    b. What are the coordinates of the point where it will first hit the line $x = 5$?

12. In the Activity you reflected point A over $\overleftrightarrow{PQ}$. Repeat the Activity, but this time reflect point B over $\overleftrightarrow{PQ}$. Does the result change?

13. Aunt Annie the Ant needs to travel from the point (–5, 2) to the x-axis and then to the point (4, 4). If Annie wishes to travel the shortest possible distance, where should she touch the x-axis?

14. Jade owns a shop that has two rooms. For security reasons, she wants to be able to see into room B from her position at the cash register in room A. To accomplish this, she places a mirror as shown in the diagram. Copy the drawing at the right and shade the area in room B that Jade can see through the mirror.

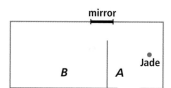

REVIEW

In 15 and 16 use the figure at the right.

15. Trace the figure and draw its reflection image over line ℓ. **(Lesson 4-2)**

16. Which of the named points in the figure coincide with their images? **(Lessons 4-2, 4-1)**

17. Given $r_m(\triangle ABC) = \triangle A'B'C'$, $m\angle ABC = 5x + 3$, and $m\angle A'B'C' = 7x - 7$. Find $m\angle ABC$. **(Lesson 4-2)**

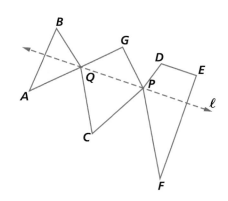

18. Points *A, B, C,* and *D* are vertices of a rectangle. $B = (5, 3)$, $C = (1, 3)$, and $D = (3, 5)$. Find the coordinates of *A*. **(Lesson 3-8)**

19. Use the figure at the right. Justify the conclusion that $x < 40$. **(Lesson 3-6, 3-5)**

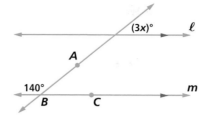

20. In the figure below, $\triangle ABC$ was rotated about *O* to get triangle $\triangle FGD$. What was the magnitude of the rotation? **(Lesson 3-2)**

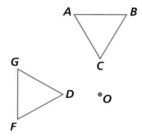

21. Draw a hierarchy for the terms: circle, figure, convex figure, triangle, isosceles triangle, scalene triangle. **(Lesson 2-6)**

EXPLORATION

22. Design a miniature golf hole where it is impossible to get the ball in the hole without first bouncing off two walls.

23. A human's field of vision is about 210°. When driving a car, this is often not enough. For example, when changing lanes, a driver needs to know the positions of the cars behind him. This is why cars have rear-view and side-view mirrors.

    a. Draw a schematic diagram of a driver and car's rear-view and two side-view mirrors.

    b. Draw rays depicting what the driver can see using the mirrors.

    c. With the mirrors, can the driver now see 360°? Explain why or why not.

Lesson
4-4
Composing Reflections over Parallel Lines

▶ **BIG IDEA** The reflection image of a reflection image of a figure over two parallel lines is a translation image of the original figure.

In some of the miniature golf and billiard problems of Lesson 4-3, it was necessary to follow a reflection over one line by a reflection of the image over a second line. The general idea was to apply one transformation and then apply a second transformation to the image of the first. Following one transformation by another in this manner is called *composing* the transformations. In the next three lessons, we analyze what happens when you compose reflections. In this lesson, we examine what happens when two reflections over parallel lines are composed.

Activity 1

MATERIALS tracing paper, straightedge

Step 1 Draw a scalene triangle about postage-stamp size onto tracing paper in the upper left-hand corner. Name it △*ABC*. Draw a line *m* close below it (a line of reflection) such that *m* does not intersect your figure.

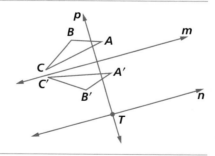

Step 2 Fold on line *m* to reflect your figure. Turn your paper over and trace your figure to create its reflection image △*A'B'C'*.

Step 3 Fold line *m* onto itself to create a perpendicular line to *m* and call it *p*. On *p*, create a point *T* which is located about one-third of the way from *m* to the bottom of the paper.

Step 4 Fold the perpendicular line *p* onto itself through point *T*. You should now have created a line *n* such that *m* ∥ *n*.

Step 5 Fold on your new line *n* and reflect △*A'B'C'* over line *n*. Trace your figure to create its reflection △*A"B"C"*.

Step 6 Unfold your tracing paper.

You have created a figure by performing two reflections over two parallel lines. You will use your work in this activity later in this lesson, so keep your tracing paper triangles near by.

The Language of Composition

In Activity 1, $\triangle ABC$ was reflected over line m to form $\triangle A'B'C'$. $\triangle A'B'C'$ was then reflected over line n to form $\triangle A''B''C''$. This can be expressed using function notation by writing

$$r_m(\triangle ABC) = \triangle A'B'C'$$

and then

$$r_n(\triangle A'B'C') = \triangle A''B''C''.$$

These two equations can be written as one equation by substituting $r_m(\triangle ABC)$ for $\triangle A'B'C'$ in the second equation. This one equation fully describes the relationship.

$$r_n(r_m(\triangle ABC)) = \triangle A''B''C''$$

Using the same notation, you can easily see the relationship between individual parts of the triangles. For example,

$$r_n(r_m(A)) = A''.$$

 QY1

When one transformation is followed by a second, we say that the transformations have been *composed*. In the above situation, the two transformations that are composed are r_n and r_m. The result is called the *composite* of the transformations.

> **QY1**
>
> Express the relationship between $\overline{AB}$ and $\overline{A''B''}$ using function notation.

Definition of Composite

The **composite** of a first transformation S and a second transformation T, denoted $T \circ S$, is the transformation that maps each point P to $T(S(P))$.

$T \circ S$ is read, "T following S." $T(S(P))$ is read "T of S of P." The operation denoted by the small circle $\circ$ is called **composition**. Any transformations can be composed, but for now we are only composing reflections.

In this example $r_n(r_m(A)) = A''$ is equivalent to writing $r_n \circ r_m(A) = A''$. Whichever way you write it, the transformation applied first is written on the right. The reason for this is that with transformations, as in algebra, you must work inside the parentheses first. Thus, in $r_n(r_m(A))$, the reflection r_m is applied before the reflection r_n.

QY2

▸ **QY2**

Rewrite $r_n(r_m(\overline{AB}))$ using the symbol $\circ$.

Translations

Suppose you traced $\triangle ABC$ from Activity 1 onto a second piece of tracing paper. Then you could move the triangle across (or down) the paper from $\triangle ABC$ to $\triangle A''B''C''$. It appears that you could map $\triangle ABC$ onto $\triangle A''B''C''$ by simply sliding $\triangle ABC$ without turning the triangle.

$\triangle A''B''C''$ is an image of $\triangle ABC$ under a *slide* or *translation*. There are several ways to define a translation. We define translation in terms of reflections.

> ### Definition of Translation
>
> A **translation** is the composite of two reflections over parallel lines.

That is, when $m \parallel n$, the transformation $r_n \circ r_m$ is a translation. You may be interested in finding more about the properties of translations and answer questions such as, "How far, and in what direction is the translation?"

Properties of Translations

Because a translation is the composite of two reflections, translations have many of the properties of reflections.

Activity 2

MATERIALS Tracing paper
Use $\triangle ABC$ and $\triangle A''B''C''$ on your tracing paper from Activity 1 to show that the translation preserves angle measure, distance, and orientation.

Step 1 Name the pairs of angles whose measures are equal.

Step 2 Name the pairs of lengths that are equal.

Step 3 Explain why $\triangle ABC$ and $\triangle A''B''C''$ have the same orientation.

Because each reflection preserves angle measure, betweenness, collinearity, and distance, so does the translation. Also, every translation preserves orientation, because the first reflection switches the orientation, and the second switches it back.

Suppose you wanted to completely describe a particular translation. You couldn't just say, "slide the figure" because no one would know how far to slide it. If you said, "slide the figure 5 cm," your instructions would still be incomplete because one would need to know in what direction to slide the figure. Thus, a translation is fully described only when its distance and its direction are known. The distance is the **magnitude of the translation.** It is found by measuring the distance from any point to its image point. The **direction of a translation** is given by the ray from any preimage point through its image point.

Example 1

Find the distance and direction of the translation that mapped △ABC onto △A″B″C″ on your tracing paper.

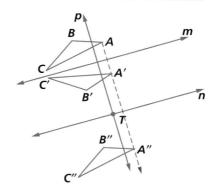

Solution Draw $\overrightarrow{AA}$. The magnitude of the translation is the length AA'', while the direction of the translation is given by $\overrightarrow{AA''}$.

 QY3

Because a translation is defined as the composite of reflections over parallel lines, it should follow that the position of those parallel lines should completely describe the position of the preimage and image. The magnitude and direction of a translation are directly related to the position of the parallel lines that define it in a surprisingly simple way.

Activity 3

MATERIALS tracing paper; ruler

Refer back to your tracing paper from Activity 1.

Step 1 Measure the distance from A to A′.

Step 2 Measure the distance between the two lines along the perpendicular fold-line p that you made in Activity 1. (The distance between two parallel lines is the length of a perpendicular segment joining the two lines.)

Step 3 Compare the measurements from Step 1 and Step 2. AA'' should be twice the distance between the two lines. Is it? Check other points of your triangle to show that these lengths remain related in this way.

Step 4 Fold your tracing paper to make the line perpendicular to m through point A. Does this perpendicular pass through A'' as well?

> ▶ QY3
>
> a. Name two different rays that describe the direction of the translation in Example 1.
>
> b. Name three different segments that you could measure to find the magnitude of the translation in Example 1.

The results in Activity 3 should agree with the following theorem:

> ## Two-Reflection Theorem for Translations
>
> If $m \parallel n$, the translation $r_n \circ r_m$ has magnitude two times the distance between m and n in the direction from m perpendicular to n.

Given $m \parallel n$, $r_m(A) = A'$, $r_n(A') = A''$.

Prove (1) $\overline{AA''} \perp m$ and $\overline{AA''} \perp n$.

      (2) AA'' is twice the distance between m and n.

Proof $\overline{AA'} \perp m$ and $\overline{A'A''} \perp n$ by the definition of reflection image. Because $\overline{AA''} \perp m$, it also must be true that $\overline{AA'} \perp n$ by the Parallel to Perpendiculars Theorem. Through a point not on a line there exists only one line perpendicular to that line, so it must be true that A, A', and A'' are collinear. Thus, $\overline{AA''} \perp m$, so $\overline{AA''} \perp n$ as well.

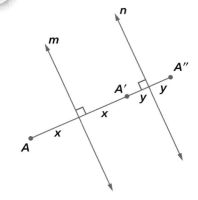

In the figure, the distance between the parallel lines is $x + y$. From the definition of reflection image, $AA' = 2x$ and $A'A'' = 2y$. By the Additive Property of Distance, $AA'' = 2x + 2y$, which is double the distance between the two lines.

A similar argument can be given if A is in a different position relative to m and n. Sometimes the preimage is between the two parallel lines.

Example 2

Let p be the line described by the equation $y = 3$ and q be the line described by the equation $y = 8$. Describe the translation $r_p \circ r_q$ and compare it to the translation $r_q \circ r_p$.

Solution Because the parallel lines are 5 units apart, the magnitude of both $r_p \circ r_q$ and $r_q \circ r_p$ is 10 units. The directions of the translations differ, however. Remember that composed transformations are performed from right to left. Therefore, the translation $r_p \circ r_q$ is directly down because the direction from q to p is down, while the direction of the translation $r_q \circ r_p$ is up because the direction from p to q is up.

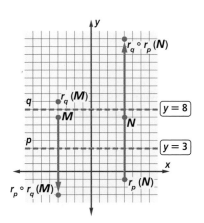

If the parallel lines m and n are known, then you can see the Two-Reflection Theorem for Translations to find the specific magnitude and direction of $r_n \circ r_m$.

GUIDED

Example 3

Using the diagram at the right, prove that it is true that the distance between the preimage A and its translation image A″ is twice the distance between the two lines.

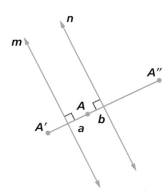

Solution Trace the diagram. Mark the distances as you find them.

The distance between lines m and n is __?__.
The distance from A′ to line m is __?__.
The distance from A″ to line n is __?__.
The total distance from A to A″ is __?__, which is twice the distance from m to n.

Have you ever been in a room that has mirrors on opposite walls? When you look into those mirrors you can see both your face and the back of your head. Suppose the mirrors are 10 feet apart (the width of the room). Using properties of reflections over parallel lines, you can now say that as you face the mirror, the first image of the back of your head appears to be 20 feet away.

Questions

COVERING THE IDEAS

1. **Multiple Choice** Which stands for the transformation that results from first applying r_f and then applying r_s to images under r_f?

   A $r_f \circ r_s$ B $r_s \circ r_f$ C $f \circ s$ D $s \circ f$

In 2–5, use the diagram at the right. Points A to K are equally spaced on a line perpendicular to both m and n. Identify the point by a single letter.

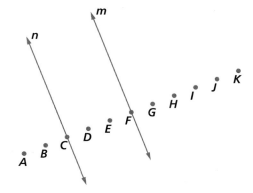

2. a. $r_n(A)$ b. $r_m(r_n(A))$ c. $r_m(r_n(B))$

3. a. $r_m(I)$ b. $r_m \circ r_n(I)$
   c. $r_n(E)$ d. $r_m \circ r_n(E)$

4. a. $r_m(D)$ b. $r_m \circ r_n(D)$ c. $r_n \circ r_m(H)$

5. a. $r_n(C)$ b. $r_m \circ r_n(C)$

6. Define *translation*.

7. Use the diagram at the right to give the coordinate of the image point.

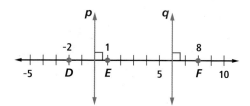

   a. $r_q \circ r_p(E)$ b. $r_p \circ r_q(E)$
   c. $r_p \circ r_q(D)$ d. $r_q \circ r_p(F)$

8. Trace the figure at the right.

   a. Draw $r_x(\triangle LMN)$. Call it $\triangle L^*M^*N^*$.

   b. Draw $r_y \circ r_x(\triangle LMN)$. Call it $\triangle L'M'N'$.

   c. What is the length of $\overline{LL'}$?

   d. Justify your answer to Part c.

   e. What is the measure of $\angle L'M'N'$?

   f. Justify your answer to Part e.

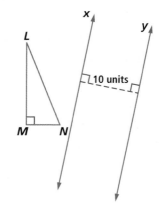

9. Suppose $s \parallel t$. Consider $r_s \circ r_t(\triangle XYZ) = \triangle X''Y''Z''$.

   a. Rewrite the equation without using the ∘ symbol.

   b. Z is reflected first over which line?

   c. What length is guaranteed to be equal to XZ?

   d. What angle will have the same measure as $\angle ZXY$?

   e. If $ZZ'' = 8$ cm, what is the distance between lines s and t?

   f. **Fill in the Blank** $s \perp$ __?__

10. Name five properties of figures that are preserved under translations.

11. In the diagram at the right, $r_n \circ r_m(A) = A''$. Sketch the diagram and use it to provide an argument to show why $AA'' = 2x$, using the properties of reflections. (Do not use the theorem of this lesson.)

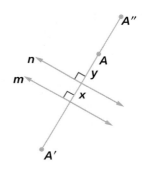

12. Trace the diagram at the right onto a piece of paper. Given $r_\ell(r_k(B)) = B''$, draw ℓ.

13. Suppose the point $A = (2, 3)$ is mapped to the point $A'' = (2, -3)$ under a translation. If the translation is $r_\ell \circ r_m$ and m has equation $y = -2$, what is an equation for ℓ?

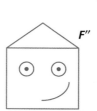

14. Trace F and F''. Find two lines m and n so that $r_n \circ r_m(F) = F''$.

15. On graph paper, draw $\triangle XYZ$ with points $X = (-4, 5)$, $Y = (0, 2)$, and $Z = (-2, -1)$. Shade the interior of the triangle. Sketch the line $x = -1$ and label it m. Sketch the line $x = 5$ and label it n. Find and shade the image $r_n \circ r_m(\triangle XYZ)$. Why is $r_n \circ r_m$ a translation?

16. In a new apartment complex being built, every apartment is the mirror image of the one next door and the one across the hall. On each floor there are 16 apartments on one side of the hall and 16 on the other side. There are three floors of apartments. If one apartment is chosen to be the model apartment that will be shown to prospective tenants, how many apartments have a different orientation than the model apartment?

REVIEW

17. Describe how you could get ball B to hole H in one shot in the miniature golf hole pictured at the right. (**Lesson 4-3**)

18. Refer to the figure at the right.
    a. Draw the shortest path between A and B. (**Lesson 1-7**)
    b. Draw the shortest path between A and B that touches the line ℓ. (**Lesson 4-3**)

19. Is the orientation of $ABCDEF$ below clockwise or counterclockwise? (**Lesson 4-2**)

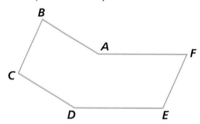

20. Why is the lettering on the ambulance the way it is? (**Lesson 4-1**)

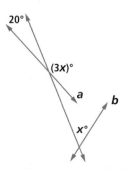

21. Let $A = (0, 4)$ and $B = (0, 8)$. Find an equation for the perpendicular bisector of $\overline{AB}$. (**Lesson 3-8**)

22. a. **True or False** If ℓ and m are lines with slopes k_1 (for ℓ) and k_2 (for m), and $k_1 \cdot k_2 = 0$, then one of the lines is horizontal.
    b. Can one of the lines be vertical? Why or why not?
    (**Lesson 3-8, 3-6**)

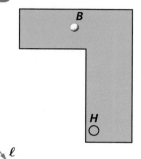

23. Use the figure at the right. Justify the following claim: a is not parallel to b. (**Lesson 3-6**)

EXPLORATION

24. Suppose $\ell \parallel m \parallel n$. Describe $r_m \circ r_n \circ r_\ell$, the composite of three reflections, over these parallel lines. (Hint: Draw some figures and lines ℓ, m, and n, and do the transformation.)

QY ANSWERS

1. $r_n(r_m(\overline{AB})) = \overline{A''B''}$

2. $r_n \circ r_m(\overline{AB})$

3a. $\overrightarrow{BB''}$ and $\overrightarrow{CC''}$

3b. $\overline{AA''}$, $\overline{BB''}$, and $\overline{CC''}$

Lesson

4-5

Composing Reflections over Intersecting Lines

Vocabulary

rotation

> ▶ **BIG IDEA** The reflection image of a reflection image of a figure over two intersecting lines is a rotation image of the original figure.

In Lesson 4-4, you saw that the composite of two reflections over parallel lines was a translation. In this lesson you will examine the composite of two reflections over intersecting lines.

Activity 1

Materials DGS

Step 1 Construct lines j and k that contain a common point O.

Step 2 Construct a quadrilateral $ABCD$. Your DGS screen should look something like the one shown at the right.

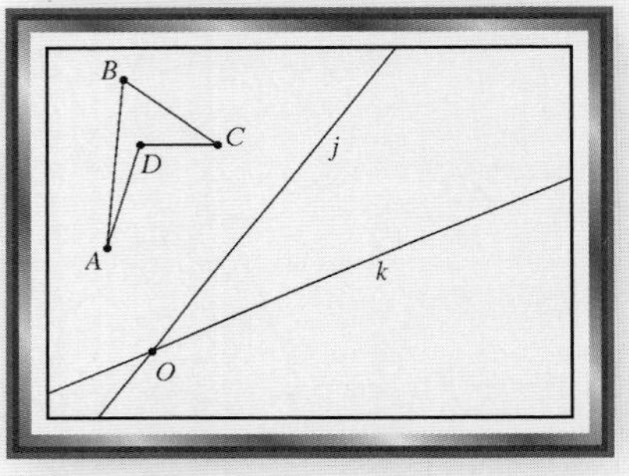

Step 3 Construct $A'B'C'D'$ where $A'B'C'D' = r_k \circ r_j(ABCD)$. You will also have an intermediate image $r_j(ABCD)$ on your screen, but you should hide or erase that.

Step 4 Drag line K around point O to come up with a conjecture about the composition of reflections over intersecting lines. Does there appear to be a single transformation that maps $ABCD$ onto $A'B'C'D'$?

Leave your DGS running; you will refer back to it shortly.

Rotations Defined

You may have noticed in the above activity that the composite of reflections over intersecting lines creates an image that can be obtained by rotating the figure about the intersection of the two lines. As with translations, we define rotations in terms of the reflections so that the properties of rotations can be more easily deduced.

Mental Math

a. Ciara was facing north. After turning counterclockwise, she was facing west. Give a possible magnitude for this rotation.

b. Sharif was facing west. He turned 45° clockwise. What direction was he facing?

c. Monique was facing north. She turned 180° clockwise, and then 90° counterclockwise. Which direction was she facing?

> ### Definition of Rotation
>
> A **rotation** is the composite of two reflections over intersecting lines.

That is, when line j intersects line k, the composite of reflections over j and k is a rotation about the intersection of j and k.

Because each reflection in a rotation $r_j \circ r_k$ preserves angle measure, betweenness, collinearity, and distance, so does the rotation. Since a rotation is a composite of two reflections, orientation is also preserved.

Describing Specific Rotations

Recall that you studied rotations in Lesson 3-2. In that lesson, rotations were described as a *turn* of a given magnitude (the amount of turn) about a center of the rotation (the point about which the figure is turned). Recall that to measure the magnitude of a rotation, you measure the angle formed between any point in the preimage, the center, and the image of that point. Once found, you then need to decide if the magnitude of the rotation is positive (counterclockwise) or negative (clockwise).

Example 1

Describe the rotation that maps △XYZ onto △X'Y'Z'.

Solution The rotation has a magnitude of –102° and a center A.

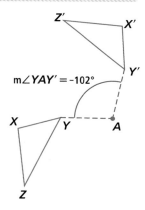

$m\angle YAY' = -102°$

Rotations and Reflections

Activity 2

MATERIALS DGS

Return to your DGS and drag the reflection lines, as well as the quadrilateral, around the screen to see if you can answer the following two questions.

(contitnued on next page)

1. In terms of the intersecting lines, where is the center of the rotation that they describe?

2. Measure the magnitude of the rotation and the angle formed between the two lines (you may have to construct some additional points on those lines to do this). Experiment with your diagram to make a conjecture about the relationship between the measure of the angle formed between the lines and the magnitude of the rotation.

In Activity 2, you may have noticed that the center of the rotation is the point of intersection of the two reflection lines. The magnitude of the rotation, however, is not quite as obvious. The magnitude of the rotation is double the measure of an angle formed by the two lines (in the direction from the first line of reflection to the second line of reflection).

> ## Two-Reflection Theorem for Rotations
>
> If m intersects ℓ, the rotation $r_m \circ r_\ell$ has a center at the point of intersection of m and ℓ and has a magnitude twice the measure of an angle formed by these lines, in the direction from ℓ to m.

Proof Consider any two lines ℓ and m that intersect at point O.

To find the center, notice that point O is on both reflecting lines. So $r_m(r_\ell(O)) = O$. This makes O the center of the rotation.

To find the magnitude, look at the points Q, Q', and Q''. Remember that reflections preserve angle measure. From the reflection over ℓ, $m\angle QOA = m\angle Q'OA = x$. From the reflection over m, $m\angle Q'OB = m\angle Q''OB = y$. The measure of the angle formed between lines ℓ and m is $m\angle AOB$, which has a measure of $x + y$. The angle of rotation is given by $m\angle QOQ''$ and has a measure of $2x + 2y$, which is twice the measure of $\angle AOB$. Thus, the magnitude of rotation is twice the measure of the angle formed by the lines of reflection.

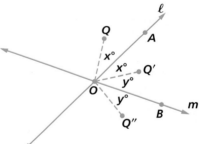

GUIDED

Example 2

Describe the center and measure an angle to determine the magnitude of the rotation $r_q \circ r_e$.

Solution The center of the rotation is the point where the lines intersect, named __?__. An angle between e and q has measure __?__ and from e to q in that direction is counterclockwise. So the magnitude of the rotation is twice that angle, or __?__. If you instead calculate the measure of the acute angle from e to q, the measure is $180° -$ __?__ $=$ __?__.

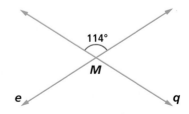

Doubling that gives you __?__, but because that is clockwise, **the magnitude of the rotation is negative, or __?__. Both magnitudes are correct.**

🛑 **QY**

▶ **QY**

Follow Example 2 to describe the rotation $r_e \circ r_q$.

Example 3

$\triangle A''B''C''$ is the image of $\triangle ABC$ under a rotation about point Z. Using a protractor and straightedge, create two lines m and n such that $r_n(r_m(\triangle ABC)) = \triangle A''B''C''$.

Solution Because $m\angle CZC'' = 140$, the magnitude of the rotation is $140°$. (Notice that this is positive since the rotation is counterclockwise.) Because the magnitude of the rotation is $140°$, the angle formed between the lines must be $70°$. Now all you need to do is draw any two lines through Z that meet at a $70°$ angle. Name the lines m and n so that going from m to n is counterclockwise.

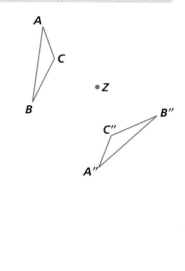

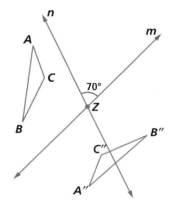

Questions

COVERING THE IDEAS

1. **Fill in the Blanks** A __?__ is a composite of reflections over intersecting lines. The __?__ of the rotation is the point of intersection of the lines.

2. In the figure at the right $r_k(D) = D'$ and $D'' = r_j(D')$.
   a. What is the measure of $\angle DBD''$?
   b. What are the center and magnitude of the rotation that maps D onto D''?
   c. Explain why the statement $r_k \circ r_j(D) = D''$ is not true.

3. Name five properties that rotations preserve.

4. How can you decide whether the composite of reflections over two lines is a rotation or a translation?

5. **Fill in the Blank** To rotate a figure $180°$, you can reflect the figure over two lines where the angle between them is __?__.

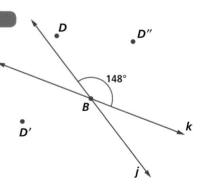

APPLYING THE MATHEMATICS

6. In each of the following diagrams, use properties of rotations and translations to explain why the green figure cannot be the image of the blue under a rotation or a translation.

   a.   b.

   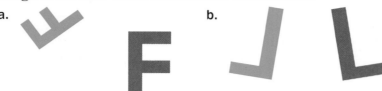

7. In the figure at the right, reflect $\triangle ABC$ over the x-axis, then reflect that image over the y-axis to get $\triangle A''B''C''$.

   a. Give the coordinates of A'', B'', and C''.

   b. Describe the rotation that would map $\triangle ABC$ onto $\triangle A''B''C''$.

8. Draw two lines m and n such that $r_n \circ r_m$ is a rotation of $-60°$.

9. Draw two lines k and ℓ such that $r_k \circ r_\ell$ is a rotation of $60°$.

10. Draw two lines p and q such that $r_q \circ r_p$ is a translation 10 cm to the right.

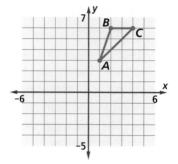

11. The written alphabet of Inuktitut uses symbols to denote sound combinations. One interesting feature of this alphabet is that it consists of 12 basic symbols, each of which is rotated or reflected to denote an additional sound.

    a. Trace the symbol for *va* at the right. Draw the Inuktitut letter *vai* by rotating this figure 90°.

    b. Reflect *va* over a vertical line to draw *vu*.

    c. Rotate *vu* 90° to draw *vi*.

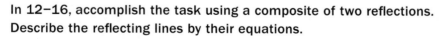

In 12–16, accomplish the task using a composite of two reflections. Describe the reflecting lines by their equations.

   Example: Translate up 4 units.

   Solution: Reflect over $y = -1$ and then reflect over $y = 1$.

12. Give a *different* solution to the above example.

13. Translate left 8 units.

14. Rotate 180° about the origin.

15. Rotate -90° about the origin.

16. Translate left 8 units and then rotate that image 180° about the origin.

The official languages of the territory of Nunavut, Canada, are French, English, Inuktitut, and Inuinnaqtun. The first three langages are shown on the sign above.

REVIEW

17. Why do translations preserve orientation? (Lesson 4-4)

18. Suppose $\ell \parallel m$. Let P be a point. Let $Q = r_\ell \circ r_m(P)$. What is $r_m \circ r_\ell(Q)$? (Lesson 4-4)

19. A transformation T maps $\triangle ABC$ onto $\triangle A'B'C'$ as shown below. Could T be a reflection? Why or why not? (Lesson 4-2)

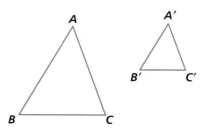

20. Line m is the perpendicular bisector of $\overline{AB}$, as shown at the right. $\overline{AB}$ is reflected over line ℓ. Is m' the perpendicular bisector of $\overline{A'B'}$? How do you know? (Lessons 4-2, 3-9)

21. Find an equation for two different parallel lines that are perpendicular to the line with equation $2x - 4y = 5$. (Lessons 3-8, 3-6)

22. Does the relation "is perpendicular to" have the reflexive property? The symmetric property? The transitive property? Explain your answers. (Lessons 3-8, 3-4)

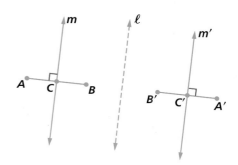

EXPLORATION

23. Use the figure at the right. Suppose $\triangle B'C'D'$ is the image of $\triangle BCD$ under a rotation, but you do not know the center.
   a. Try to find a method for locating the center of this rotation.
   b. Once you find a method for locating a center, use a DGS to create a rotation and then check your method.
   c. Write your method out so that someone could follow your instructions for finding a missing center of rotation.

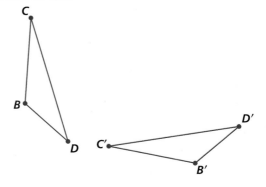

QY ANSWER

The center is the point of intersection of lines e and q. The magnitude is about $132°$ or $-228°$.

Lesson
4-6
Translations as Vectors

Vocabulary

vector

translation vector

initial point

terminal point

horizontal component

vertical component

▶ **BIG IDEA** A translation can be described by a vector that gives its magnitude and direction.

Mental Math

Find an equation for:

a. the horizontal line through (3, 7).

b. a line through (2, 9) parallel to the horizontal line through (3, 7).

c. a line perpendicular to the lines in Parts a and b.

At high school or college football games, a common form of halftime entertainment is the marching band. Marching bands often use the positioning of band members to create letters on the field when seen from above. Suppose band members spell out the letter R on one half of the field. The director might want to translate (slide) the R to a different location on the field. Each band member in the R must receive directions for getting from their beginning location in the first R to their final location in the second R. The directions given to each band member represent what is called a *vector*.

Definition of Vector

A **vector** is a quantity that can be characterized by its direction and magnitude.

Recall that the magnitude and direction of a translation is found by connecting any point with its image. Every arrow in the diagram at the top of the next page represents a translation vector. The length of the arrow is the magnitude of the translation and the direction of the translation is the direction of the arrow. Notice that every arrow is equal in length and parallel to the others.

In general, the *magnitude* of the vector is its length, while the *direction* of the vector is determined by the direction of a ray.

In Lesson 4-4, we defined a translation to be a composite of reflections over parallel lines. In this lesson, we look at translations as individual transformations. Recall that a specific translation is described completely by its magnitude and direction. A *translation vector* communicates exactly that information.

Definition of Translation Vector

A **translation vector** is a vector that gives the length and direction of a particular translation.

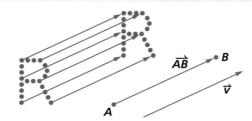

The arrows in the diagram above show two ways of identifying a translation vector. The vector that starts at A and ends at B is denoted $\overrightarrow{AB}$. The point A is called the **initial point** and the point B is called the **terminal point**. A half-arrow or harpoon above a single lower-case letter, as in $\vec{v}$, or a boldface single letter v are other ways to identify a vector.

 QY1

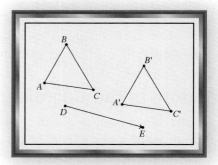

> **QY1**
>
> What are the initial and terminal points of the vector $\overrightarrow{VE}$?

Activity

MATERIALS DGS

Step 1 On a clear DGS screen, draw △ABC.

Step 2 Draw a vector $\overrightarrow{DE}$ that does not intersect the triangle.

Step 3 Translate △ABC by $\overrightarrow{DE}$ to get the image △A′B′C′.

Step 4 Drag point E to various positions on the screen. What happens to the image △A′B′C′? How can you describe the position of this image triangle?

Step 5 Drag point D to various positions on the screen. What happens to the image △A′B′C′? Explain how this compares to what happened in Step 4. Explain why.

Step 6 Drag the entire translation vector $\overrightarrow{DE}$. What happens to the image triangle $A′B′C′$? Explain why.

Drawing Translation Images Given a Vector for the Translation

Suppose you are given a figure to translate, and the translation is described by a vector. To find the image of the figure, you can slide each point of the figure by the distance and direction defined by the vector.

 QY2

As opposed to merely sketching the image, locating the image accurately is a more difficult task. This is shown in the example below.

▶ **QY2**

Trace the diagram in Example 1 below. Slide that same sheet of paper and trace the diagram again to sketch the image of △ABC under the translation by the given vector $\overrightarrow{DE}$.

Example 1

Using the diagram at the right, draw the image of △ABC under the translation by the given vector $\overrightarrow{DE}$.

Solution You can create the image of △ABC by locating the image of each of the vertices A, B, and C and then connecting them with line segments. For the sake of simplicity, we only demonstrate how to translate point A.

Step 1	Draw $\overleftrightarrow{DA}$ as shown.
Step 2	Measure ∠EDA.
Step 3	Draw a line through A parallel to $\overrightarrow{DE}$ by making corresponding angles with the same measure.
Step 4	Create a point A′ on your new line by making DE = AA′.

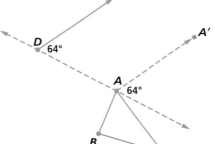

To locate △A′B′C′, you can repeat this process for points B and C and then connect A′, B′, and C′ to complete the triangle.

Vectors in the Coordinate Plane

It is often easier to work with vectors using coordinates. In the diagram at the right, C′D′E′H′ is the image of CDEH under a translation by the vector $\vec{v}$.

You could tell someone how to get from a point on CDEH to a point on C′D′E′H′ by saying "right 5 units and then up 3 units." The vector information is communicated mathematically by writing ⟨5, 3⟩. The symbol ⟨5, 3⟩ describes how to get from the initial point to the terminal point by traveling horizontally and vertically. These horizontal and vertical distances are called the **horizontal component** and **vertical component** of the vector, respectively.

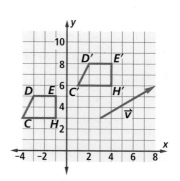

In general, the vector $\langle a, b \rangle$ has horizontal component a and vertical component b. If the initial point of the vector is on the origin, then the vector $\langle a, b \rangle$ displays the coordinates of the terminal point (a, b).

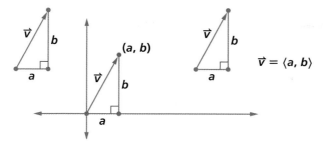

When a figure is translated on the coordinate plane by a vector $\langle a, b \rangle$, the image of the point (x, y) is $(x + a, y + b)$. That is, a is added to the x-coordinate and b is added to the y-coordinate. You may have used this idea to slide figures in algebra.

GUIDED

Example 2

P' is the image of the point $P = (3, -4)$ under a translation by the vector $\langle -5, 8 \rangle$. Find the coordinates of P'.

Solution P' has coordinates $(3 + \underline{\ ?\ }, -4 + \underline{\ ?\ })$. Thus, the coordinates of P' are $(\underline{\ ?\ }, \underline{\ ?\ })$.

Example 3

Suppose a translation maps $Q = (4, 8)$ onto $Q' = (1, 9)$. Find the vector that describes this translation.

Solution The x-coordinate has been shifted 3 units to the left, so the horizontal component is -3. The y-coordinate has been shifted one unit up, so the vertical component is 1. Thus, the translation is described by the vector $\langle -3, 1 \rangle$.

The vectors you have seen in this lesson are a specific type of vector in which the magnitude is a distance. In applications, however, the magnitude of a vector does not have to be a distance. The magnitude of a vector describing the motion of a ship traveling on the water is its speed. The magnitude of a vector representing the force a bowling ball exerts on the ground is its weight. The magnitude of a magnetic field is its pull on an object.

At this point you have now studied three types of transformations in a plane: reflections, rotations, and translations. On the next page is a summary of what determines them.

Type of transformation	Informal term	Specifically determined by	Description in terms of reflections
reflection	flip	line of reflection	reflection over one line
rotation	turn	center and magnitude	composite of two reflections over intersecting lines
translation	slide	vector	composite of two reflections over parallel lines

Questions

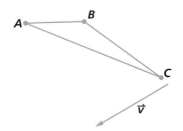

COVERING THE IDEAS

1. **Fill in the Blanks** A vector is a quantity that can be characterized by its ___?___ and ___?___.

2. Trace $\triangle ABC$ at the right. Make a rough sketch of the image of $\triangle ABC$ under the translation with the vector $\vec{v}$.

In 3–5, a translation is described.

    a. Give a coordinate vector for the translation.

    b. Give the image of the point (a, b) under each translation.

3. Translate up 4.

4. Translate 5 to the left.

5. Translate 3.5 right and 7.1 down.

6. Suppose A is translated onto A' by a vector $\overrightarrow{PQ}$. How are the segments $\overline{AA'}$ and $\overline{PQ}$ related?

7. On page 218, what is the vector for the translation that maps $C'D'E'H'$ back onto $CDEH$?

APPLYING THE MATHEMATICS

In 8–10, draw a picture with a preimage, image, and translation vector clearly marked.

8. A slide on a trombone

9. A slide on a zipper

10. A slide on a playground

11. Consider the point $A = (10, 12)$. If A' is the image of A under a translation by the vector $\langle -4, 5 \rangle$, what are the coordinates of A'?

12. A rover is being developed to explore a planet. To move the rover, people will instruct it from Earth. Vectors are used to help direct its travels. The rover began at $(0, 0)$ and units refer to meters. If the rover is now at $(125, 61)$ and you need to get it to $(172, -4)$, what vector could you input into the computer that will move it to the new location?

13. Let c be the horizontal line with equation $y = 3$. Find an equation of the line d such that $r_d \circ r_c$ is the translation with vector $\langle 0, 8 \rangle$.

14. Translate the figure by the given vector $\vec{v}$.

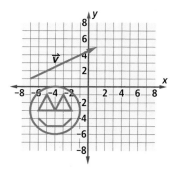

15. Consider the line m described by $y = \frac{1}{2}x + 4$. The points $A = (0, 4)$ and $B = (2, 5)$ are on m.
    a. Find A' and B' under the translation by the vector $\langle 7, 3 \rangle$.
    b. Find an equation for $\overleftrightarrow{A'B'}$.
    c. What is a relationship between m and $\overleftrightarrow{A'B'}$? Explain why this happened for this particular case.
    d. Would the answer to Part c be true if you changed the vector?

REVIEW

16. In the coordinate plane, the line ℓ has equation $y = 4x - 2$ and the line m has equation $5y - 4x = 3$. Is the composite $r_\ell \circ r_m$ a translation or a rotation? How do you know? (Lesson 4-5)

17. a. Trace the diagram at the right, showing $\odot A$, and draw $r_\ell \circ r_m(\triangle ABC)$.
    b. Draw $r_k (\triangle ABC)$. (Lesson 4-5)

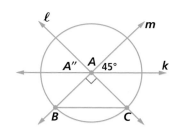

18. The diagram at the right represents a room with a closet in the middle and one wall, $\overline{CD}$, that is a mirror. Trace the diagram. In what direction can a person at point B look in order to see what a person at point A is doing? (**Lesson 4-3**)

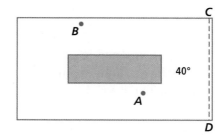

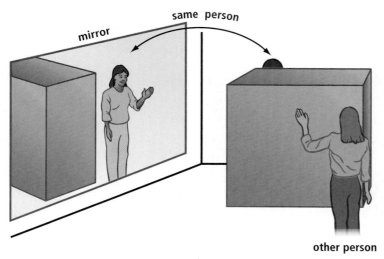

same person

mirror

other person

19. Is the reflection image of an isosceles triangle always an isosceles triangle? How do you know? (**Lesson 4-2**)

20. In the coordinate plane let $A = (0, 0)$ and $B = \left(\frac{1}{2}, \frac{1}{2}\right)$. Give coordinates of two points that are the same distance from A as from B. (**Lesson 3-9**)

EXPLORATION

21. Most DGS have many ways to translate a figure. Use a DGS to find a way to translate a figure without using reflections over parallel lines. Explain at least two different ways you can translate a figure using that DGS.

QY ANSWERS

1. The initial point is V and the terminal point is E.

2.

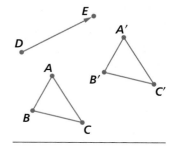

Lesson 4-7

Isometries

▶ **BIG IDEA** There are four types of transformations that preserve distance: reflections, rotations, translations, and glide reflections.

Reflections, translations, and rotations are all examples of *isometries*. An **isometry** is a transformation that can be created by composing reflections. One of the fundamental properties of an isometry is that it preserves distance and collinearity. As a consequence, the image of a figure after an isometry has many of the same properties as the preimage. Thus far, you have seen composites of reflections over one or two lines. A single reflection over one line is, obviously, a reflection. Two lines on a plane are either parallel or intersecting, so all composites of reflections over two lines are rotations or translations. In this lesson you will see what happens when you compose reflections over three lines.

Reflections over Three Lines

When a composite of three reflections is applied to a figure, the orientation must be reversed because there are an odd number of reflections. As a result, we know that the composite of reflections over any three lines cannot be either a translation or a rotation because those transformations preserve orientation.

To examine all cases of composites of reflections over three lines, consider how many points of intersection the lines can have. There are only four possibilities, or *cases*. Three lines can intersect in 0, 1, 2, or 3 points. We consider each case separately.

 QY1

Case 1, 0 points of intersection
In this case, the 3 lines must be parallel. As you can see from the diagram at the right, the composite of reflections over three parallel lines is a reflection over a line parallel to the three existing lines.

Let $T = r_n \circ r_m \circ r_\ell$.
Let p be the line such that $r_m \circ r_\ell = r_n \circ r_p$.
Then $T = r_n \circ (r_m \circ r_\ell) = r_n \circ (r_n \circ r_p) = r_p$.

▶ **READING MATH**

The word *isometric* comes from the Greek root *iso*, which means "equal," and *metric*, from the Greek *metron*, which means "measure" or "length."

▶ **QY1**

Draw four diagrams to show how three lines can intersect in 0, 1, 2, and 3 points.

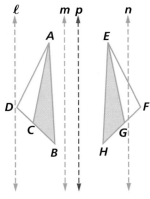

$r_n \circ r_m \circ r_\ell (\triangle ABD) = \triangle EHF = r_p (\triangle ABD)$

Activity 1

MATERIALS DGS (optional)

Draw three parallel lines ℓ, m, and n that are not vertical. Draw a triangle $\triangle PQR$ of your choosing and find $\triangle P'Q'R'$, the image of $\triangle PQR$ under a composition of reflections over the three lines. Verify that $\triangle P'Q'R'$ is the image of $\triangle PQR$ under a reflection.

Case 2, 1 point of intersection

In this case, the three lines are *concurrent*. Recall from Lesson 3-7 that this means that they all have a single point in common. The composite of reflections over three concurrent lines is a reflection over a line that is concurrent with the three existing lines.

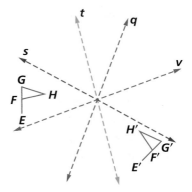

Let $R = r_v \circ r_t \circ r_s$ and let q be the line such that $r_v \circ r_q = r_t \circ r_s$. Then $R = r_v \circ (r_t \circ r_s) = r_v \circ (r_v \circ r_q) = r_q$.

Activity 2

MATERIALS DGS (optional)

Verify the result in Case 2 with a figure and three concurrent lines of your choosing.

Case 3, 2 points of intersection

For this case, two of the lines must be parallel. Let's call the parallel lines m and n and the line that intersects them p, as shown at the right.

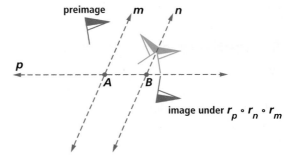

preimage

image under $r_p \circ r_n \circ r_m$

There are six possible orders in which you can do the reflections. No matter what order you choose, however, the composite is a combination of either a translation or a rotation and a reflection. For example: $r_p \circ r_n \circ r_m$ is the translation $(r_n \circ r_m)$ followed by a reflection (r_p). You could also think of this as a reflection (r_m) followed by the rotation $(r_p \circ r_n)$ about point B.

STOP QY2

Case 4, 3 points of intersection

For this case, regardless of the order in which you choose to compose the reflections, the result can always be thought of as a rotation followed by a reflection over a line not concurrent with the first two lines. For example, $r_p \circ r_n \circ r_m$ can be thought of as a rotation about point F followed by a reflection over line p.

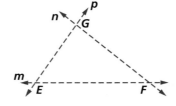

> **QY2**
>
> **Fill in the Blanks**
>
> Describe $r_m \circ r_p \circ r_n$ in two different ways:
>
> $(r_m \circ r_p) \circ r_n$ is a _____?_____ followed by a _____?_____, and
>
> $r_m \circ (r_p \circ r_n)$ is a _____?_____ followed by a _____?_____.

QY3

Cases 3 and 4 cannot be described in terms of a single reflection, rotation, or translation. It turns out that *whenever* an isometry cannot be described by a single reflection, rotation, or translation, it can be described as a single *glide reflection*. A glide reflection is a reflection followed by a translation in a direction parallel to the reflection line.

▶ **QY3**

Using the diagram at the bottom of the previous page, describe $r_m \circ r_p \circ r_n$ in terms of a rotation followed by a reflection.

Definition of Glide Reflection

Let r_m be a reflection and T be a translation with nonzero magnitude and direction parallel to m. Then $G = T \circ r_m$ is a **glide reflection**.

In this definition we use the word *nonzero* because if the translation had a magnitude of zero, then the composite would be a simple reflection. One way to think of a glide reflection is as footprints. If you were to dip your feet in paint and then walk along a sidewalk, then your footprints would be a sequence of glide reflection images. So an informal term for a glide reflection is a *walk*.

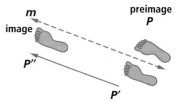

Reflection: r_m
Translation: vector $\overrightarrow{P'P''}$
$\overleftrightarrow{P'P''} \parallel m$

Because a glide reflection is the composite of three individual reflections, glide reflections preserve angle measure, betweenness, collinearity, and distance. A glide reflection reverses orientation, however, because the first reflection reverses orientation, the second reflection brings it back, and the third reflection reverses the orientation once again.

For example, in the diagram at the right, the rightmost "F" shape is the image of the leftmost "F" under the composite $r_n \circ r_m \circ r_p$.

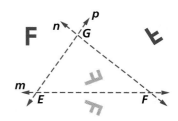

The glide reflection that would achieve this same result is a translation by the vector $\vec{v}$ followed by a reflection over line ℓ, as shown at the right. Notice that it does not matter here which you do first. You could reflect and then translate, or you could translate and then reflect. Both would map the preimage F onto the image F.

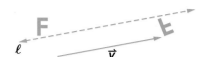

In a later chapter, you will learn how to determine $\vec{v}$ and the location of the glide reflecting line ℓ. For now the important point is that if the composite of reflections over three lines does not result in a reflection, then the composite is a glide reflection.

QY4

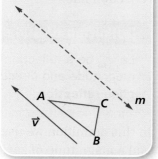

▶ **QY4**

Make a sketch of the glide reflection image of $\triangle ABC$ shown below under the glide reflection G, where $G = T \circ r_m$ and T is described by $\vec{v}$.

Types of Isometries

It is remarkable that the glide reflection completes the types of isometries. That is, a composite of reflections in a plane is always either a reflection, a rotation, a translation, or a glide reflection—no matter how many reflections are composed! It is not difficult to identify which of the four isometries mapped any preimage to its image under an isometry.

Example 1

Identify the isometry that maps one image onto the other.

Solution The hands have the same orientation, so the isometry is a rotation or translation. But it is not a translation. So, the isometry is a rotation.

Example 2

Identify the isometry that maps one image onto the other.

Solution The hands have different orientations, but it is not a reflection because the perpendicular bisectors of segments connecting corresponding points do not coincide. So, the isometry is a glide reflection.

Activity 3

MATERIALS Large sheets of paper

You need at least three people to do this activity.

1. Place your right hand palm down anywhere on a large sheet of paper. Have another person outline your hand. Then place either your right or left palm down somewhere else on the same sheet of paper and have that person outline it. The job of the third person is to identify which of the four isometries maps one handprint onto the other handprint. Do this with different hands of different people until you are sure that you have had one example of each of the four types of isometries.

2. What type of isometries are possible if both people place their right hands on the desk? Explain your answer.

3. What types of isometries are possible if one hand is a right hand and the other is a left hand? Explain your answer.

4. What types of isometries are possible if both hands are left hands? Again explain your answer.

Questions

COVERING THE IDEAS

1. What are the four types of isometries?

2. If one figure is the image of another figure under any of the four isometries, what types of "equal measure" will you notice between the two figures?

3. **Fill in the Blank** When three lines intersect at one point, the composite of reflections over those three lines is a ___?___.

4. **Fill in the Blank** When three lines intersect at two or three points, then the composite of reflections over those lines is a ___?___.

5. Use the diagram at the right.
   a. Name the isometry that maps one of the hands onto the other hand.
   b. Why can't the isometry be a glide reflection?

6. Why is it impossible for any of the hands shown below to be images of one another under an isometry?

7. In the figure at the right, one specific isometry was used to create the blue rectangle from the green rectangle.
   a. If *ABCD* is the image of *EFGH,* what type of isometry was used?
   b. If *ABCD* is the image of *FEHG,* what type of isometry was used?
   c. If *ABCD* is the image of *GHEF,* what type of isometry was used?
   d. The answers to Parts a, b, and c are three different types of isometries. Explain why the fourth type of isometry could not have been used.

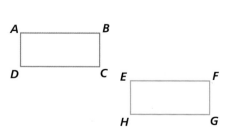

8. Draw two triangles that are images of one another under a glide reflection.

9. Draw three lines, ℓ, m, and n such that $r_n \circ r_m \circ r_\ell$ is equivalent to each of the following, or tell why it is impossible to do so.

   a. reflection   b. translation   c. glide reflection   d. rotation

APPLYING THE MATHEMATICS

10. a. Graph $\overline{AB}$, where $A = (3, 4)$ and $B = (6, 1)$.

    b. Let T be a translation determined by the vector $\langle 5, 0 \rangle$. Graph $G(\overline{AB})$, where $G = T \circ r_{x\text{-axis}}$.

    c. Suppose you reverse the order in the composition. That is, suppose you let $H = r_{x\text{-axis}} \circ T$. Show that $G(\overline{AB})$ is the same segment as $H(\overline{AB})$.

11. The Sitka willow was used by Native Americans in the Pacific Northwest to dry fish, heal wounds, stretch skins, and make baskets. The placement of Sitka willow leaves looks like successive glide reflection images. What part of the plant is the glide-reflecting line?

12. Which of the following "transformations" are like isometries and which are not? Explain your reasoning.

    a. Put air in a tire.   b. Spin a wheel.
    c. Ride your bike a short distance.   d. Blow up a balloon.
    e. Make a yo-yo go up and down.   f. Bounce a ball.

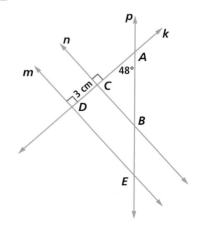

Sitka willow

13. Isometries are often classified as *even isometries* or *odd isometries*. The even isometries are the translations and rotations. The odd isometries are the reflections and the glide reflections.

    a. What do you think is meant by the terms "even" and "odd"?

    b. What do even isometries preserve that odd isometries do not?

REVIEW

14. Suppose T is the translation described by the vector $\langle 0, 7 \rangle$ and S is the translation described by the vector $\langle 0, 6 \rangle$. What vector describes the translation $T \circ S$? What vector describes the translation $S \circ T$? (**Lesson 4-6**)

15. Use the diagram at the right. Describe the transformation that results from each of the following. (**Lessons 4-5, 4-4**)

    a. $r_m \circ r_n$ b. $r_n \circ r_m$
    c. $r_p \circ r_k$ d. $r_k \circ r_m$

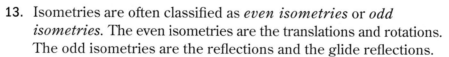

16. In the figure at the right, which of the following is true? (**Lesson 4-5**)

    A. $Q = r_\ell \circ r_m(P)$

    B. $Q = r_m \circ r_\ell(P)$

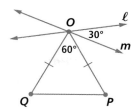

17. Trace the two triangles at the right on a piece of paper. Draw two lines, j and k, so that $r_k \circ r_j(\triangle ABC) = \triangle A''B''C''$. (**Lesson 4-4**)

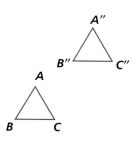

18. Given $\ell \parallel m$ and $j \parallel k$ as shown below, find x and y. (**Lesson 3-6**)

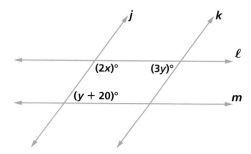

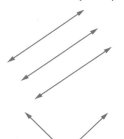
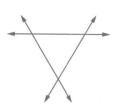

EXPLORATION

19. Explore with a DGS how you might find the vector and line of reflection for a glide reflection.

    a. Do this construction.

       Step 1 Given three noncollinear points A, B, and C.

       Step 2 Construct $\overleftrightarrow{AB}$, $\overleftrightarrow{BC}$, and $\overleftrightarrow{AC}$.

       Step 3 Construct $\triangle XYZ$ and find its image under a composite of reflections over the three lines in any order. Label this image $\triangle X'''Y'''Z'''$.

       Step 4 Hide everything except $\triangle XYZ$, $\triangle X'''Y'''Z'''$, and points A, B, C. If you did everything correctly, a glide reflection will map $\triangle XYZ$ onto $\triangle X'''Y'''Z'''$.

    b. Using your construction, make a conjecture about how to locate the line of reflection for your glide reflection. Test your conjecture by reflecting your image back across the line and then translating it to see if it matches up with your preimage. Once you find a process that works for locating the line of reflection, write down your steps.

Lesson

4-8

Transformations and Music

▶ **BIG IDEA** All of the transformations that you have seen so far in this book—reflections, rotations, translations, glide reflections, and size changes, are found in the sounds and writing of music.

Many people feel that there is a special relationship between mathematics and music that goes beyond the physics of sound. Over 2500 years ago, Pythagoras or his followers discovered the relationship between the lengths of strings and the tones they made. About 200 years ago, the philosopher William Hegel wrote, "Music is architecture translated or transposed from space into time; for in music, besides the deepest feeling, there reigns also a rigorous mathematical intelligence."

You can do mathematics without any musical knowledge, and you can do music with no mathematical knowledge, but to connect them you need to know a little of both. We begin with a discussion of the basics of music.

The Notes of Western Music

A piano keyboard plays the notes that are most commonly used in western music. At the right we label 12 notes (or keys), 7 white and 5 black, of a musical *octave*. Using the common symbol ♯ for "sharp" (the black key to the right of a white key) the 12 notes are named A, A♯, B, C, C♯, D, D♯, E, F, F♯, G, and G♯.

To make a larger keyboard, these 12 notes are repeated. The result is a translation image of the original octave, and the names start all over again. To the right of the black key G♯ is the white key A, then the black key A♯, and so on.

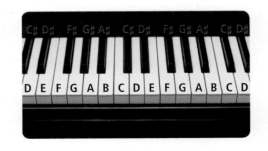

If you take any 12 consecutive notes, you can translate them horizontally again and again and you will form a keyboard. The entire system of notes is based on mathematical translations.

Mental Math

The notes of a C-major chord are represented by the vertices of triangle *CEG*.

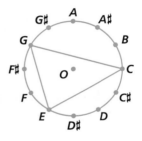

a. What are the notes of an E-major chord?

b. What are the notes of a G#-major chord?

Transposing in Music = 1-Dimensional Translation in Mathematics

If you play the notes G, G, A, F♯, G, and A in this order, you will have played the first six notes of the anthem "America," with the words "My country, 'tis of thee." If you raise the notes by 2, then you would play A, A, B, G♯, A, and B. It will still sound like the first notes of "America," but the pitch will be higher. Pitch is the high or low sound you hear from a note. The original notes are in the *key of G,* while the image notes are in the *key of A.* In music, this process is called *transposing.* Transposing two notes higher on a piano can be described as the horizontal translation where $x + 2$ is the image of x.

The mechanism and strings in upright pianos are perpendicular to the keys.

Music is usually written on *staffs.* A staff is a set of 5 equally spaced parallel lines. As in reading English, you read a staff from left to right. Each line or space between the lines stands for a note. The higher a note is placed on the staff, the higher its pitch. So a staff is like a graph in algebra because each horizontal line stands for a certain pitch. The only exception is that the black keys on the piano are indicated with additional symbols (♯ or ♭) so they will be played instead of the corresponding white keys. The vertical bars on the staff denote the end of a *measure.* Measures represent equal time periods in the music. A commonly used staff, shown at the right, has the note names denoted by capital letters.

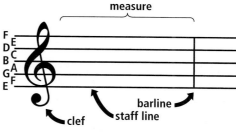

The first line of "America" is written on two staffs below. The second staff shows the same tune transposed 2 notes higher. Notice that the transposed melody is a vertical translation image of the original melody. A horizontal translation on the piano is a vertical translation in written music. We have translated (x, y) to $(x, y + 2)$.

If you transpose 12 notes higher, then you raise the pitch a full octave and the names of the notes are the same: (G, G, A, . . .).

Round in Music =
2-Dimensional Translation in Mathematics

¹Row,	row,	row your	boat	²gent- ly	down the	stream.
C	C	C D	E	E D	E F	G
³Merrily,	merrily,	merrily,	merrily,	⁴life is	but a	dream.
CCC	GGG	EEE	EEE	G F	E D	C

You likely have sung the above round. A *round* in music is a composition in which two or more voices sing the same melody, but at different times, and when each voice finishes the melody, the voice goes back to the beginning. "Row, Row, Row Your Boat" is a round for 4 voices. Next to the words above, we have placed numbers. As the 1st voice sings the round, the numbers 2, 3, and 4 indicate when the 2nd, 3rd, and 4th voices begin.

If the voices are put down so that what is sung at the same time is in the same vertical column, then we see a mathematical pattern. Here we put the notes under a beat (|), because some notes are sung faster than others. The C of the first "Row" is in boldface to show when the notes repeat. The italicized *C G E C* letters each stand for the same note sung three times (in the word "merrily").

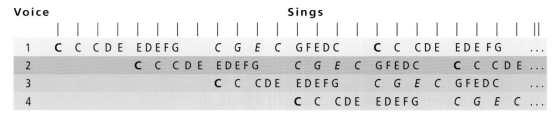

There is a 2-dimensional translation in the above writing. One dimension (vertical) is the change of voice or singer. The other (horizontal) is the time lapse between one voice and another entering the round.

The Round in Musical Notation

Reading written music is like reading a graph of a time series in mathematics. The horizontal axis is time, so anything on the same vertical line is played or sung at the same time. At the top of the next page are the four voices of "Row, Row, Row Your Boat" put into standard musical notation.

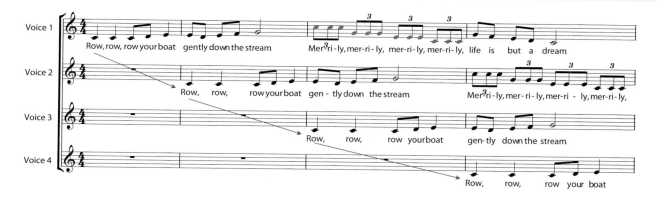

In the music shown here, the filled-in note ($\quarternote$) is called a quarter note and usually has 1 beat. The open note ($\halfnote$) is a half note and has 2 beats. The note with a flag connected to another note ($\eighthnote$) is an eighth note and has $\frac{1}{2}$ beat. The three notes connected to each other ($\eighthnote$) together have 1 beat in all, so each has $\frac{1}{3}$ of a beat. The stem of the note can go up or down ($\quarternote$ or $\quarternote$).

The notes in red (for "Merrily, merrily, merrily, merrily") show a figure and three translation images.

Mer³ri-ly, mer-ri-ly, mer-ri-ly, mer-ri-ly,

The pitch changes, but when you hear the round sung, you hear the similarity of the preimage to its images because the rhythm is the same (and different from anything else in the round).

Changing Tempo in Music = 1-Dimensional Size Change in Mathematics

The *tempo* of a piece of music is how fast it is played or sung. Composers can change the tempo of a piece in two ways. One way is to indicate how many beats there are to the minute. The round "Row, Row, Row Your Boat" might be sung at about 120 beats a minute. There are only 16 beats in the entire round, so a voice would go through the round 7.5 times in a minute. If it were sung at 96 beats a minute, then a voice would go through the round 6 times in the minute, quite a bit slower. Another way to change the tempo is to multiply the values of the notes by 2 or by $\frac{1}{2}$. On the first staff on the next page, the value of each note has been multiplied by 2. Quarter notes became half notes and eighth notes became quarter notes. On the second staff, the value of each note has again been multiplied by 2. Now half notes become whole notes ($\wholenote$) and quarter notes become half notes.

A metronome (MET ra nohm) helps musicians keep time by ticking the selected number of times per minute.

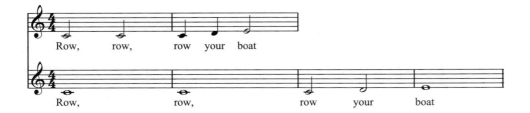

Row, row, row your boat

Row, row, row your boat

Augmentation and Diminution in Music = Expansion and Contraction in Mathematics

Composers will play with a theme or melody by applying the 1-dimensional size change that can be used to change tempo. In classical music, themes are often played slowly for emphasis at the same time that other music is being played. This is called *augmentation* when the notes are held longer and *diminution* when the notes are held shorter. Below is a copy of a part of a piano sonata by Ludwig van Beethoven (1770–1827). The five notes at the beginning (♪♪) are in sixteenths. Following the top staff in a horizontal direction, those five notes are expanded to eighths (in blue), then to quarters (in green), and finally to half notes (in orange), at which time the five notes (in red) are reduced to three. The same occurs for the first five notes of the bottom staff.

Inversion in Music = Horizontal Reflection or Glide Reflection in Mathematics

To *invert* a musical theme means to change it so that it goes down in pitch wherever it went up, or up instead of down. Here we show an inverted image of the first few notes of "My country tis of thee." The image is a glide reflection, with the G line as the reflecting line. If you play these notes, you may realize that many composers start with a theme one way and then approximately invert it to end a piece.

original

My coun-try 'tis of thee

G

inverted

My coun-try 'tis of thee

The Hungarian composer Béla Bartók (1881–1945) sometimes put the theme and its inversion in two parts to be played at the same time.

Here is a fragment of his "Music for Strings, Percussion, and Celesta". The ♭ by a note (called a *flat*) indicates that the note just below it on the piano is to be played. The lightning bolt sign ♮ (called a *natural*) means that the note returns to its original value. The original theme and its inversion are reflection images over a horizontal line that is in the space for the note A.

A celesta (suh LESS tuh) looks like a little piano, but its hammers strike metal plates instead of strings, producing bell-like tones.

Composites of Transformations in Music

The musical fragment below is from *The Art of the Fugue* by Johann Sebastian Bach, one of the greatest composers of all time. It combines many of the ideas that we have mentioned. Voice 3 starts with a theme whose notes are in blue. Four beats later Voice 1 (in red) comes in with the same theme augmented and inverted. Specifically, Voice 1's notes are twice as long as those of Voice 3, and they go in the opposite direction of Voice 3. That is, when Voice 3 goes down, Voice 1 goes up. Then four beats later Voice 2 (in green) comes in. Voice 2 also has an inversion of the original theme, but it is at the original tempo. The object in composing a piece like this is to have it sound right when all the voices are heard together.

Notice how Voice 2 is almost exactly a glide reflection image of Voice 3.

glide reflection (inversion and delay) and size change (augmentation)

glide reflection

original

Retrograde in Music = Vertical Reflection in Mathematics

A *palindrome* is a word or phrase that reads the same backward and forward. Two of the most famous palindromes are: "Madam, I'm Adam." and "A man, a plan, a canal: Panama." Not as famous is this one: "I prefer pi." You can see that a palindrome is reflection-symmetric to a vertical line drawn at the middle of the word (ignoring the shape of the letters). Sometimes you don't even have to ignore the shape of the letters: MOM is reflection-symmetric.

When musicians reverse a theme, it is said to be in *retrograde*. For instance, the "Row, Row, Row Your Boat" melody and its version in retrograde are shown here. The written music of one is a reflection image of the other over a vertical line.

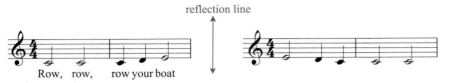

reflection line

Row, row, row your boat

> **▶ READING MATH**
>
> The word *retrograde* starts with the Latin prefix *retro*, which means "backward," so this is a perfect word to describe the reversal of a musical theme. A second related meaning of *retro* is "pertaining to the past."

Sometimes composers will put part of a theme in retrograde along with other music that is being played. Wolfgang Amadeus Mozart (1756–1791) and Franz Joseph Haydn (1732–1809) even wrote entire pieces as a palindrome. Then the written music is reflection-symmetric to a vertical line at the middle of the piece.

Other Musical Transformations

Mozart once wrote a duet for violins that looked on paper almost the same upside-down as right-side up. The violinists could sit with the music on a table between them and each could read the music right-side up! If something reads the same upside-down as right-side up, then it is rotation-symmetric.

Through the centuries to the present day, musicians have taken themes and varied them. Many compositions or parts of compositions are even identified as "Theme and Variations." People who play jazz learn to improvise variations to themes. The variations sometimes use the kinds of transformations identified in this lesson, and others are possible. You can change pitches, or the distances between consecutive notes, or the rhythms, and all of these are mathematical. Musicians also change what instruments play the music in order to achieve different qualities of sound. Some of these transformations are very mathematical and can be programmed into synthesizers and other electronic musical instruments.

Putting It All Together

On the next page we show a composition by Herf Yamaya, composed while he was a student at Glenbrook South High School in Glenview, Illinois. Herf combined many of the types of musical transformations in this piece to be played by the members of a string quartet.

Transformation

Herf Yamaya

Questions

COVERING THE IDEAS

In 1–5, a musical idea is named.
   a. Describe what is meant by the idea.
   b. **Multiple Choice** Choose the mathematical transformation
      that corresponds to the musical modification.
   A reflection   B glide reflection   C translation
   D size change   E rotation

1. change in pitch (transposing)   2. round
3. augmentation   4. inversion   5. retrograde

6. On a piano, the note A near the middle is the 37th note from
   the left side of the piano.
   a. What is the number of the key called C that is to the
      right of the A?
   b. What is the number of the next higher A on the piano?

7. When a piece of music starts on C, to transpose the piece up
   to start on D, the translation $x \rightarrow x + h$ can be used. What is
   the value of h?

8. The music here is of the round "Frère Jacques." What do the
   numbers 1, 2, 3, and 4 indicate?

Fre - re ·Jac - ques, fre - re   Jac - ques,   dor - mez vous,   dor - mez vous?

Son-nez les ma - ti - nes,   son-nez les ma - ti - nes. Ding dong ding.   Ding dong ding.

In 9–13, transformations have been made of the first four notes of
"Frère Jacques."
   a. Name the musical term to describe the transformation.
   b. Describe the geometric transformation that maps the original
      notes onto this image.
   c. Musically, how do the original and the image differ from each
      other?

9.    10.    11.

12.    13.

APPLYING THE MATHEMATICS

In **14** and **15**, use the first seven notes (pictured here) of the song "Nobody Does It Better," the theme song of the James Bond movie "The Spy Who Loved Me" (words by Carol Bayer Sager and music by Marvin Hamlisch).

14. If this melody is transposed so that the first note is a D, write what the other notes would be both as letters and on a musical staff.

15. Write what this theme would be in retrograde.

16. The Mozart piece that can be read upside-down and right-side up looks almost the same after a rotation of 180°. Pieces with a much less common type of rotation symmetry appear the same after a rotation of a smaller angle. Composer Mauricio Kagel proposed an example of such a rotational symmetry in the following example. Measure the angle of rotation used.

17. The two measures at the right, called "Trias Harmonica", by Johann Sebastian Bach, were given as a puzzle for a person to figure out when the eight parts of this round (or canon) would enter. Below is what one person thought Bach meant.

   a. Each part of this solution comes in how many beats after the previous part?

   b. The second measure of the theme is the image of the first measure under what mathematical transformation?

   c. Find other mathematical transformations in the written music.

In 18–21, refer to the composition "Transformation" by Herf Yamaya shown in this lesson. The small numbers 6 and 11 above the second and third lines of the violin I part are measure numbers. There are 16 measures in the piece. The theme is in measures 1 and 2 of the violin I part. Describe what has been done to the theme in the indicated measures.

18. the 3rd and 4th measures of the viola part
19. the 7th and 8th measures of the violin I part
20. the 5th–7th measures of the cello part
21. the 14th–16th measures of the cello part

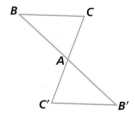

A violin and a viola look similar, but the viola is a little larger.

REVIEW

22. Suppose I and J are isometries. (Lesson 4-7)
    a. Must $I \circ J$ be an isometry?
    b. How do you know?

23. a. Can the composite of two rotations be a glide reflection?
    b. How do you know? (Lesson 4-7)

24. Let ℓ be the graph of $x = 0$. Let T be the translation given by the vector $\langle 2, 3 \rangle$. Let P be the point with coordinates $(4, -5)$. What are the coordinates of $T \circ r_\ell (P)$? (Lesson 4-7)

25. What is the vector for the translation that sends a point in the coordinate plane 3 units to the left and 4 units up? (Lesson 4-6)

26. ℓ and m are two intersecting lines in the plane. In the figure at the right, $\triangle AB'C' = r_\ell \circ r_m(\triangle ABC)$. At what point do ℓ and m intersect? (Lesson 4-5)

27. At the right, A, B, C, and D are equidistant from each other on $\odot O$. (Lessons 3-2, 3-1)
    a. Find m$\angle AOB$.
    b. Find m$\widehat{ABD}$.
    c. **True or False** O is between A and C.

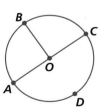

EXPLORATION

28. Find and reproduce the piece that Mozart wrote for two violins that plays nearly the same upside-down as right-side up. Identify the center of the rotation that maps this piece onto itself.

29. Take a piece of music that you like. Write down its theme and then subject the theme to various transformations of your own choosing.

Chapter 4 Projects

1 Reflections and Communications

Fiber optic cables are used to send large amounts of information quickly around the world. Find out how they work and how they use reflections. Illustrate how they work, where they are used, and what the longest fiber optic cables in the world are. Make a small model to demonstrate how they work.

2 Escher Tessellations

M.C. Escher is known for his tessellations, which often illustrate translations (for example, Sky and Water I), rotations (Drawing Hands), and glide reflections (Horseman). There are books and Web sites that teach you how to create your own tessellations. Create your own tessellation(s) that illustrate translations, rotations, and/or glide reflections.

3 Transformations in Music

Peter and the Wolf is a famous composition by Sergei Prokofiev in which a narrator tells a story, accompanied by musical sections that represent each character. Write a (possibly comic) story about musical transformations, and several short tunes to accompany and illustrate the story (they can be in any musical style you like). Together with as many friends as necessary, perform this piece in front of your class.

4 Kaleidoscopes

Sir David Brewster (1781–1868) invented the first kaleidoscope. Find out how kaleidoscopic images use reflections and rotations, and then make your own. Build a kaleidoscope that uses three mirrors and prepare a report for your class on how kaleidoscopes work.

5 Miniature Golf

Design a 9-hole miniature golf course. On each of your holes, show at least two paths that bank off of walls to make a hole in one.

or

Design and build a tabletop model of a playable miniature golf hole. Mark on the walls of the hole at least two places that you should aim the ball to make a hole in one. If you make the walls out of mirrors, you can find/check the paths with the aid of a laser pointer.

or

Work with other students to create a full-size playable miniature golf hole from an 8' × 4' piece of plywood.

Chapter 4 Summary and Vocabulary

○ A transformation is a one-to-one correspondence between sets of points. If the preimage point or figure is F and the transformation is T, we denote the image by $T(F)$. The composite of two transformations, S followed by T, is the transformation $T \circ S$ that maps each point P onto $T \circ S(P)$. Four types of transformations are discussed in this chapter: reflections, translations, rotations, and glide reflections.

○ When point A is not on line m, the reflection image of A over line m, written $r_m(A)$, is the point B such that m is the perpendicular bisector of $\overline{AB}$. If A is on m, then $r_m(A) = A$. Reflections preserve collinearity, betweenness, distance, and angle measure, and switch orientation.

○ Composites of reflections are called **isometries.** A **translation** is a composite of two reflections over parallel lines; its magnitude is twice the distance between the lines and its direction is along the line perpendicular to those lines, from the first line to the second. A **rotation** is a composite of two reflections over intersecting lines; its center is the intersection of the lines and its magnitude is twice the measure of the non-obtuse angle between the lines from the first line to the second. These composites of two reflections have the preservation properties of reflections and also preserve orientation. The hierarchy below shows how the isometries are related to each other.

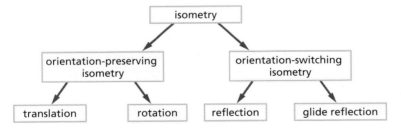

○ A translation can be described by a **vector**. The composite of a translation and a reflection over a line parallel to the direction of the vector is a **glide reflection**. A composite of three reflections is either a single reflection or a glide reflection. Glide reflections have the preservation properties of reflections and reverse orientation.

Vocabulary

4-1
reflecting line
  (line of reflection)
*reflection image

4-2
reflection image of a figure
orientation
clockwise
counterclockwise

4-3
angle of incidence
angle of reflection

4-4
*composite
composition
*translation
magnitude of a translation
direction of a translation

4-5
*rotation

4-6
*vector
translation vector
initial point
terminal point
horizontal component
vertical component

4-7
*isometry
glide reflection

● Reflections have numerous applications. In billiards and miniature golf, using reflections and composites of reflections can help a player make more accurate shots. Reflections are used to determine the shortest path between two points when it is not possible simply to connect the points with a straight line. You saw this in the case of oil pipelines and the same idea is applied in many situations where a minimal distance is desired.

● Transformations also play an important role in understanding physical interactions, such as light and sound. They help explain what composers do with musical themes.

Type of transformation	Specifically determined by	Description in terms of reflections	Informal term
reflection	line of reflection	reflection over one line	flip
rotation	center and magnitude	composite of two reflections over intersecting lines	turn
translation	vector	composite of two reflections over parallel lines	slide
glide	vector and line of reflection	composite of translation and reflection	walk

Postulates, Theorems, and Properties

Figure Transformation Theorem (p. 191)
Reflection Postulate (p. 192)
Two-Reflection Theorem for Translations (p. 206)
Two-Reflection Theorem for Rotations (p. 212)

Chapter

4   Self-Test

Take this test as you would take a test in class. You will need a calculator. Then use the Selected Answers section in the back of the book to check your work.

1. In the figure at the right, $r_n(BRQ) = B'R'Q'$. Trace the figure and draw n.

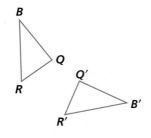

2. Trace the figure below and draw the reflection image of *PENTA* over line *m*.

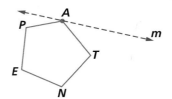

3. $r_\ell(FGLO) = F'G'L'O'$. Trace the figure below and draw the two quadrilaterals *FGLO* and *F'G'L'O'*.

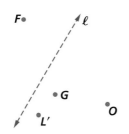

4. Plot $F = (-3, 4)$, $G = (4, 2)$, and $H = (-1, -3)$ on a coordinate plane and plot their reflection images F', G', and H' over the line with equation $y = -1$. Draw $\triangle FGH$ and $\triangle F'G'H'$ and indicate the coordinates of F', G', and H'.

5. Draw two lines so that the composite of reflections over these lines is a rotation of 120°.

6. Draw three lines so that the composite of reflections over those lines is a glide reflection.

True or False In 7–9, it is given that *F*, *M*, and *W* are noncollinear, none of these points is on line ℓ, and $r_\ell(FMW) = F'M'W'$. For each statement below that is true, state a reason; for each that is false, sketch a counterexample.

7. $m\angle FWM = m\angle M'W'F'$

8. $MM' \perp \ell$ 9. $\overrightarrow{FF'} = \overrightarrow{F'F}$

10. In the figure below, $\triangle AEH$ has been reflected over *n*. Its image was then reflected over *m* to create $\triangle MFS$. Which side of $\triangle AEH$ has the same length as $\overline{SM}$?

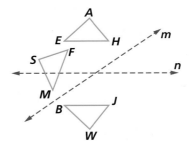

11. $P'V'S'R'$ is the image of *PVSR* under a glide reflection. Line *j* is the glide reflection line. Trace the figure below and draw the translation vector.

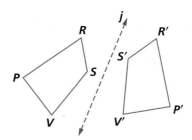

12. Describe the transformation $r_z \circ r_y$, using the figure at the right.

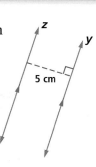

13. Trace the diagram of the miniature golf hole below. Draw a path a ball at B can take to land in the hole at H in one shot.

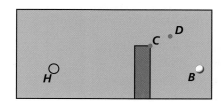

14. Using your tracing from Question 13, draw a path that will get the ball at B into the hole at H in one shot even if a barrier is built from C to D.

15. A laser is placed at (1, 2) and is pointed towards the x-axis. Assume the light reflects off the x-axis and hits the corner (6, 8). Find the coordinates of the point on the x-axis where the light will strike.

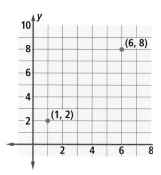

16. In the figure below, there is a mirror at an angle of 45° to the building. Explain why this mirror enables pedestrians to see a car coming out of the driveway.

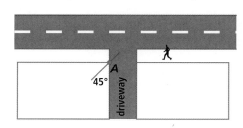

17. Trace the diagram below. In Parts a and b, draw the indicated images of figure H.

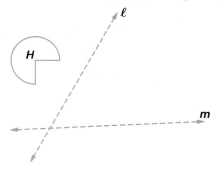

a. $r_m \circ r_\ell(H) = J$ b. $r_\ell \circ r_m(H) = K$

c. How are J and K different?

d. How could you change the position of the lines so that ℓ and m are still distinct lines and $r_m \circ r_\ell(H) = r_\ell \circ r_m(H)$?

18. **Multiple Choice** Which of the following is not an isometry?

A reflection   B rotation

C size change   D translation

E glide reflection

19. Using the diagram below, name the isometry that maps flag I onto flag II.

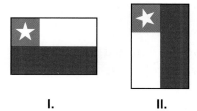

I.   II.

20. **Multiple Choice** Below is a musical theme and its image under a transformation. What is the musical term for this transformation?

A transposition   B inversion

C augmentation   D retrograde

Chapter 4 Chapter Review

SKILLS
PROPERTIES
USES
REPRESENTATIONS

SKILLS Procedures used to get answers

OBJECTIVE A Draw figures by applying the definition of reflection image.
(Lessons 4-1, 4-2)

In 1–4, trace the figure first.

1. Use the figure at the right.
   a. Draw $r_\ell(B)$.
   b. Draw $r_\ell(R)$.

2. Use the figure below.
   a. Draw $r_k(C)$.
   b. Draw $r_m(C)$.

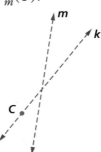

3. Using the figure at the right, draw the reflecting line h for which $r_h(Y) = X$.

4. Using the figure below, draw the reflecting line g for which $r_g(ABCD) = PQRS$.

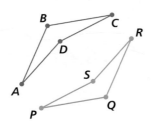

In 5 and 6, draw and label a figure matching the given conditions.

5. $r_q(\triangle SIN) = \triangle COT$
6. $r_s(\triangle RAT) = \triangle CAT$

In 7–10, trace line a at the right and the given figure. Draw the reflection image of the figure over line a.

7. $\triangle GHI$
8. $\triangle MNO$
9. $\overline{JK}$
10. $\overrightarrow{GI}$

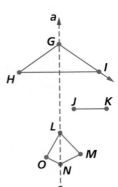

OBJECTIVE B Draw translation and glide-reflection images of figures.
(Lessons 4-6, 4-7)

In 11 and 12, trace the figure at the right, and draw the translation image of *ABCDE* determined by the given vector.

11. $\overrightarrow{FG}$
12. $\overrightarrow{CD}$

13. Trace the figure at the right. If T is the translation determined by vector $\overrightarrow{MN}$, draw $T \circ r_j(PQRS)$.

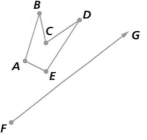

14. Trace the figure below. $\triangle JKL$ is the image of $\triangle GHI$ under a glide reflection. $\overrightarrow{YZ}$ is the translation vector. Draw the glide-reflecting line g.

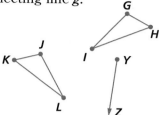

OBJECTIVE C Draw or identify images of figures under composites of two reflections. (Lessons 4-4, 4-5)

In 15 and 16, trace the figure and draw the identified image. Then describe the transformation.

15.

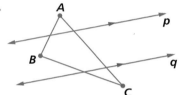

a. $r_p(r_q(\triangle ABC))$ b. $r_q \circ r_p(\triangle ABC)$

16.

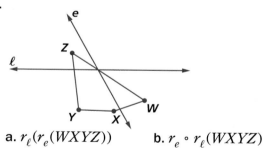

a. $r_\ell(r_e(WXYZ))$ b. $r_e \circ r_\ell(WXYZ)$

PROPERTIES Principles behind the mathematics

OBJECTIVE D Apply properties of reflections to make conclusions, using one or more of the following justifications:

   Definition of reflection
   Reflections preserve distance.
   Reflections preserve angle measure.
   Reflections switch orientation.
   Figure Transformation Theorem

(Lessons 4-1, 4-2)

In 17–20, use the figure below, in which $r_v(\triangle ABC) = \triangle A'B'C'$ and $r_w(\triangle A'B'C') = \triangle A''B''C''$. Justify your answers using the properties of reflections.

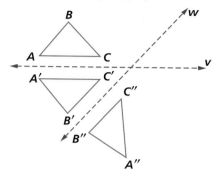

17. Determine the orientation of the following triangles.

   a. $\triangle ABC$ b. $\triangle A'B'C'$ c. $\triangle A''B''C''$

18. **True or False** $r_v(C) = C'$

19. **True or False** $CC' = BB'$

20. **True or False** $m\angle A'B'C' = m\angle A''B''C''$

In 21 and 22, use the figure below, in which $r_{\overleftrightarrow{TV}}(S) = U$.

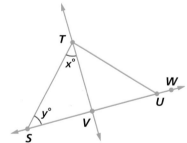

21. Find $m\angle STU$ given $m\angle STV = x$.
22. Find $m\angle TUW$ given $m\angle TSV = y$.

OBJECTIVE E Apply properties of reflections to obtain properties of other isometries. (Lessons 4-4, 4-5, 4-7)

23. **Multiple Choice** A composite of two reflections over intersecting lines is

   A always a reflection.

   B always a rotation.

   C always a translation.

   D either a rotation or a translation.

   E none of the above.

24. **Multiple Choice** Rotations do not preserve:

    A angle measure.   **B** betweenness.

    C collinearity.   **D** distance.

    E Rotations preserve all of the above.

25. Explain why a composite of 5 reflections reverses orientation.

26. What properties do all isometries preserve?

OBJECTIVE F Apply the Two-Reflection Theorems for translations and for rotations. (Lessons 4-4, 4-5, 4-7)

In 27–30, use the diagram below.

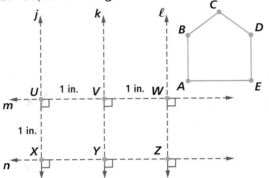

27. Describe the transformation $r_n \circ r_m$.

28. Describe the transformation $r_m \circ r_j$.

29. Write a composite of reflections to rotate $ABCDE$ 180° about Y.

30. Write a composite of reflections to translate $ABCDE$ 2 inches to the left.

USES Real-world applications of mathematics

OBJECTIVE G Determine the isometry that maps one figure onto another. (Lesson 4-7)

In 31–34, name the type of isometry that maps Figure I onto Figure II.

31.

   II.   **I.**

32.

   I.   **II.**

33.

   I.   **II.**

34.

   II.   **I.**

In 35 and 36, use the tessellation below.

35. Which type of isometry maps figure A onto figure B?

36. Which type of isometry maps figure B onto figure C?

OBJECTIVE H Use reflections to find the shortest path from an object to a particular point. (Lesson 4-3)

In 37 and 38, use the diagram to find a path to shoot the ball G into the hole at H that would:

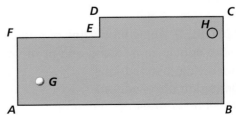

37. bounce off $\overline{AB}$ only.

38. bounce off $\overline{EF}$ and then $\overline{AB}$.

39. Trace the drawing below, and draw the path of a ball that begins at B, bounces off x, y, and z in that order, and then hits A.

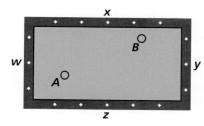

40. Trace the drawing of an oil pipeline $\overleftrightarrow{MN}$ and two oil wells R and S at the right. Find a location for a pumping station P on the pipeline so as to minimize the sum $RP + PS$.

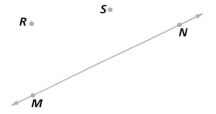

OBJECTIVE I Apply transformations to music. (Lesson 4-8)

41. A melody and its image under a transformation are shown below.

a. Identify the type of geometric transformation exhibited.

b. What musical term or terms describe this transformation?

42. Describe what is meant by *inversion* in music.

43. Suppose you have a piece of music that begins on an E, and you want to lower it to begin on a D. This can be done with a transformation of the form $x \rightarrow x + h$. What is h?

44. **True or False** A rotation of $90°$ represents a transformation in music.

REPRESENTATIONS Pictures, graphs, or objects that illustrate concepts

OBJECTIVE J Find coordinates of reflection and translation images of points over the coordinate axes. (Lessons 4-1, 4-2, 4-6)

45. Find the reflection image of $(8, –2)$ over the x-axis.

46. Find the reflection image of $(–3, –5.5)$ over the y-axis.

47. Find the reflection image of (m, n) over the x-axis.

48. Find the image of $(4, 4)$ when translated by the vector $\langle c, d \rangle$.

49. Find the vector that translates (a, b) to $(0, –1)$.

50. Find the preimage of $(2, 3)$ when translated by the vector $\langle -5, -4 \rangle$.

In 51 and 52, a quadrilateral has vertices $A = (-3, -6)$, $B = (-3, 0)$, $C = (1, 5)$, and $D = (3, -8)$.

51. Graph $ABCD$ and its reflection image over the y-axis.

52. Graph $ABCD$ and its image when translated by the vector $\langle -2, 5 \rangle$.

Chapter

5 Proofs Using Congruence

Contents

5-1   When Are Figures Congruent?

5-2   Corresponding Parts of Congruent Figures

5-3   One-Step Congruence Proofs

5-4   Proofs Using Transitivity

5-5   Proofs Using Reflections

5-6   Auxiliary Figures and Uniqueness

5-7   Sums of Angle Measures in Polygons

Recall that two objects are *congruent* if they are exactly the same size and shape. Congruent objects are everywhere. Teachers duplicate worksheets for students, and businesses photocopy photographs and diagrams. Tool-and-die makers create molds (the dies) for cutting and forging metal so that manufacturers can make identical parts.

Which of the figures below do you think are congruent to Figure A?

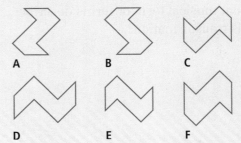

A   B   C

D   E   F

Tiled floors and bricked driveways can form interesting patterns. The congruent figures below fit together nicely to form an attractive pattern. This is possible because they have corresponding parts that are congruent.

People who manufacture dice for games have to make sure the faces of the die are congruent. Congruence helps assure that each face has an identical chance of facing up when the die is randomly tossed.

In this chapter, you will take a close look at some of the properties of congruent figures. You will see how congruent figures are used and use them yourself to deduce properties of angles and segments within particular figures. These properties and results have been key ideas for as long as people have studied geometry, so in this chapter you will be introduced to some of the history of the development of geometry.

Lesson

5-1

When Are Figures Congruent?

▶ **BIG IDEA** Figures are congruent if and only if they are related by an isometry, that is, a reflection, rotation, translation, or glide reflection.

You first dealt with the idea of *congruent figures* when you were very young and copied figures. You may have used tracing paper to make an image of a drawing. With tracing paper you could slide (translate) or turn (rotate) the image. You could even flip (reflect) the paper over or do these movements one after the other (compose them), and you knew that the image you made would still be an exact copy that was congruent to the original. Likewise, you knew you could reverse the process. That is, if you were given a figure and a traced image of it, you could slide, turn, and flip the traced image to coincide with the original figure.

The above processes can be described in the language of mathematics using what you learned in Chapter 4. When isometries are composed, we know that angle measure, betweenness, collinearity, and distance are preserved. All of these properties are deduced from the fact that isometries are composites of reflections. Because reflections preserve these properties, so do all of the rest of the isometries. We call this the A-B-C-D Theorem.

Mental Math

Suppose $A = (3, 11)$ and $\overrightarrow{MN} = \langle -4, 4 \rangle$.

a. If A is the preimage, what is its image under the translation described by $\overrightarrow{MN}$?

b. If A is the image, what is its preimage under the translation described by $\overrightarrow{MN}$?

A-B-C-D Theorem

Every isometry preserves **A**ngle measure, **B**etweenness, **C**ollinearity (lines), and **D**istance (lengths of segments).

The A-B-C-D Theorem is the most important property of isometries. Often, isometries are defined to be transformations that preserve the quantities A-B-C-D. This definition makes sense for transformations of 3-dimensional objects as well.

As a result of the A-B-C-D Theorem, when a figure is the image of another figure under an isometry, the two figures are exactly the same size and shape. In Chapter 4 you also learned that the converse is true, that is, if two images are the same size and shape, exactly one of the four types of isometries maps the preimage onto the image. Using this idea, we can define the term *congruent figures*.

Definition of Congruent figures

Two figures *F* and *G* are **congruent figures**, written *F* ≅ *G*, if and only if *G* is the image of *F* under an isometry.

 QY1

While we are focusing most of our attention on 2-dimensional figures, it is worth noting that congruent shapes appear everywhere around us in the physical world. Some of these are 3-dimensional objects. Light bulbs, for instance, must be congruent to fit properly. The same is true of screws, batteries, coins, soda cans, clothing, and almost anything else for which there are standard sizes and shapes.

Because every isometry is a composite of reflections, we can rephrase the definition in either of the following ways.

> **QY1**
>
> Figure A on page 250 is congruent to Figures B, C, and E. What type of isometry maps Figure A onto each of these figures?

Definition of Congruent figures (alternate form)

Two figures *F* and *G* are **congruent figures** if and only if *G* is the image of *F* under a composite of reflections.

Two figures *F* and *G* are **congruent figures** if and only if *G* is the image of *F* under a rotation, translation, reflection, or glide reflection.

Because of this definition, another term for isometry is **congruence transformation**. When figures are congruent and have the same orientation, we say they are **directly congruent**. Thus, translation or rotation images are directly congruent to their preimages. For instance, two stamps of the same kind are congruent regardless of where they are placed on an envelope. When figures are congruent and have opposite orientation, we say they are **oppositely congruent**. Reflection and glide-reflection images are oppositely congruent to their preimages. You and a mirror image of yourself are oppositely congruent.

At the right, Figures *A* and *B* are directly congruent; Figures *B* and *C* are oppositely congruent.

A   *B*   *C*

Transformations That Do Not Yield Congruent Figures

All isometries produce images that are congruent to their preimages. All isometries are transformations. However, not all transformations are isometries. Transformations that are not isometries do not always produce images congruent to their preimages. On the next page are two transformations that do not produce congruent images.

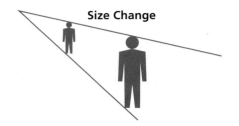

Size Change

Shear

Notice that in each case there are specific distances and angle measures that have not been preserved, so it is easy to see that, in these cases, the image and the preimage are not congruent.

The Equivalence Properties of Congruence

Some of the properties of congruence ($\cong$) are like properties of equality ($=$). Three of these properties are called the *equivalence properties of congruence.* Here we give these properties, as well as justifications that explain why they are true.

> ### Theorem (Equivalence Properties of Congruence)
>
> For any figures F, G, and H:
>
> 1. $F \cong F$ **Reflexive Property of Congruence**
>
> 2. If $F \cong G$, then $G \cong F$.   **Symmetric Property of Congruence**
>
> 3. If $F \cong G$ and $G \cong H$,   **Transitive Property of Congruence**
>   then $F \cong H$.

Proof of 1 Let F be any figure. We need to show that F is congruent to itself. To do that, reflect F over line m and then reflect that image back over m. $r_m \circ r_m(F) = F$, so there is a composite of reflections mapping F onto itself. Thus, by the definition of congruence, $F \cong F$.

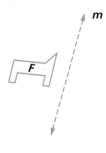

m

F

Proof of 2 If $F \cong G$, then there is a composite of reflections that maps F onto G. To map G back onto F, simply reflect G back over the same lines in reverse order. This means F is the image of G under an isometry, so $G \cong F$.

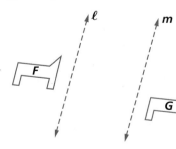

ℓ

m

F

G

STOP QY2

> ▶ QY2
>
> At the left above, $r_m \circ r_\ell(F)G$. Identify the composite of reflections that maps G onto F.

Proof of 3 If $F \cong G$, then there is a composite of reflections that maps F onto G. If $G \cong H$, then there is a composite of reflections that maps G onto H. The composite of these two isometries is another isometry, which is a composite of reflections. This composite of reflections maps F onto H, so $F \cong H$.

STOP QY3

Questions

> ▶ **QY3**
>
> Suppose $r_a \circ r_b \circ r_c(F) = G$ and $r_d \circ r_e \circ r_f(G) = H$. What composite of reflections maps F onto H?

COVERING THE IDEAS

1. Name three places outside of mathematics in which congruent figures may be found.

2. **Fill in the Blank** If $r_m \circ r_n \circ r_\ell(F) = G$, then F __?__ G.

3. Give three phrases that mean the same thing as "isometry."

4. What four properties does the A-B-C-D Theorem say are preserved by isometries?

5. Diagrams in this lesson show a figure and its image under a shear and under a size change. These do not produce congruent figures because they fail to preserve one or more of the A-B-C-D properties. Use those diagrams to answer the following:
   a. Which of these properties are not preserved by the size change?
   b. Which of these properties are not preserved by the shear?

6. **Fill in the Blank** By the Reflexive Property of Congruence, $\overline{AB} \cong$ __?__.

7. **Fill in the Blank** By the Symmetric Property of Congruence, if $\triangle CAT \cong \triangle DOG$, then __?__.

8. **Fill in the Blank** By the Transitive Property of Congruence, if $\angle A \cong \angle B$ and $\angle B \cong \angle C$, then __?__.

9. Which types of isometries create oppositely congruent figures?

APPLYING THE MATHEMATICS

10. a. Plot the points $A = (2, 4)$, $B = (3, 1)$, $C = (4, 4)$ and their images A', B', C' under the transformation T with $T(x, y) = (x + 3, -y)$.
    b. Is T an isometry? Explain your answer.

11. Repeat Question 10 if $T(x, y) = (x + y, x - y)$.

In 12–14, the Figures *F*, *G*, *H*, and *I* below are all congruent to each other. Tell whether the figures are directly or oppositely congruent, and name the type of isometry that would accomplish each mapping.

F   G   H   I

12. **a.** *F* onto *I*
    b. *I* onto *F*

13. **a.** *G* onto *H*
    b. *H* onto *G*

14. **a.** *F* onto *H*
    b. *H* onto *F*

15. Draw a counterexample to show that the statement "Any two squares are congruent" is not true.

In 16–19, tell whether the task uses the idea of congruence.

16. making sure that the replacement glass for a car window fits snugly

17. changing the letter P in Times New Roman font into the letter P in Arial font

18. figuring out that two jigsaw pieces fit together

19. cutting carpet to lay in a room

REVIEW

20. Which of the transformations of music that were shown in Chapter 4 are not isometries? **(Lesson 4-8)**

21. Suppose *ABCD* is a square. Must $\angle ABC$ be a right angle? **(Lesson 3-8)**

22. **a.** Suppose *T* is a translation. Must there be a point *A* with $T(A) = A$? Why or why not? **(Lesson 4-4)**
    b. Suppose *R* is a rotation. Must there be a point *B* with $R(B) = B$? Why or why not? **(Lesson 3-2)**

23. Suppose ℓ is the *x*-axis, *m* is the *y*-axis and *k* is the graph of $y = 3$. **(Lessons 4-5, 4-4)**
    a. Determine $r_\ell \circ r_m \circ r_k(2, 7)$.   **b.** Determine $r_k \circ r_m \circ r_\ell(2, 7)$.

24. The lengths of the three sides of a triangle are 2, 2.5, and *x*. Find all possible values of *x*. **(Lesson 1-7)**

EXPLORATION

25. One type of shear maps every point (x, y) to $(x + ky, y)$, where *k* is any fixed number. On a coordinate system, draw the rectangle whose vertices are $P = (0, 0)$, $Q = (4, 0)$, $R = (4, 3)$ and $S = (0, 3)$. Shear this rectangle using the mapping $(x, y) \rightarrow (x + 2y, y)$ to create $P'Q'R'S'$. Use this instance to discuss what parts of the A-B-C-D Theorem for isometries are also true for shears.

QY ANSWERS

1. A reflection maps A onto B, a glide reflection maps A onto C, and a rotation maps A onto E.

2. $r_\ell \circ r_m$

3. $r_d \circ r_e \circ r_f \circ r_a \circ r_b \circ r_c$

Lesson
5-2
Corresponding Parts of Congruent Figures

Vocabulary

corresponding parts

▶ **BIG IDEA** When figures are congruent, all corresponding parts—angles, segments, regions, and so on—are congruent.

When an AAA battery in your calculator dies, how do you know that a new AAA battery will fit perfectly in your calculator? All AAA batteries are produced to have the same dimensions. They all have the same length and the radius of the top is always the same. In geometric terms,

we say the corresponding parts of all AAA batteries are the same. In this lesson, you will learn how to determine the corresponding parts of congruent figures and how they relate to length and measure.

Mental Math

$CORD = r_{x\text{-axis}}(BELT)$.

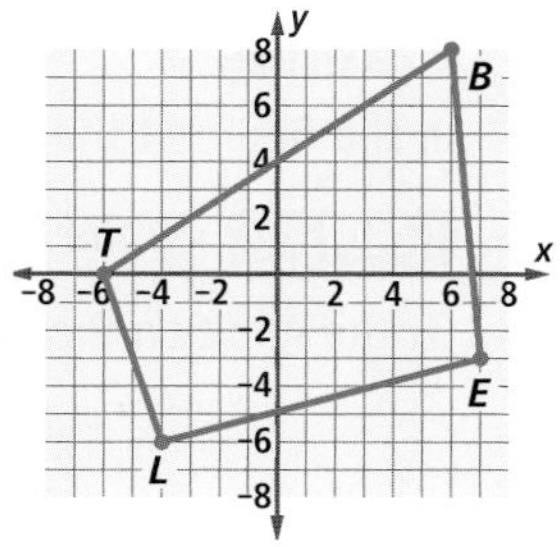

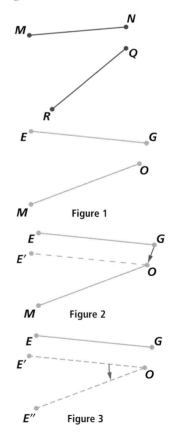

Give the coordinates of:
a. C.
b. O.
c. R.
d. D.

Congruent Segments and Length

If two segments are congruent, then you know that there is an isometry I that maps one onto the other. In the figure at the right $I(\overline{MN}) = \overline{RQ}$.

By the A-B-C-D Theorem (isometries preserve distance), $MN = RQ$. This argument proves the following conditional: *If two segments are congruent, then they have the same length.*

The converse of this statement is: *If two segments have the same length, then they are congruent.* Is this statement true as well? Yes, and the justification is below.

In Figure 1, $EG = MO$. To prove $\overline{EG} \cong \overline{MO}$, we need to show that $\overline{MO}$ is the image of $\overline{EG}$ under an isometry.

Translate $\overline{EG}$ by vector $\overrightarrow{GO}$ to get the image $\overline{E'O}$. (See Figure 2.) Call this translation T.

Rotate $\overline{E'O}$ by m$\angle E'OM$ about center O to get image $\overline{E''O}$. (See Figure 3.) Call this rotation R.

Now, $R \circ T$ is an isometry such that $R \circ T(\overline{EG}) = \overline{E''O}$. Because isometries preserve distance, $E''O = EG$. Thus, using the Transitive Property of Equality, $MO = E''O$. Since E'' lies on $\overleftrightarrow{OM}$, $E'' = M$. Thus, $R \circ T(\overline{EG}) = \overline{MO}$, xand $R \circ T$ is an isometry mapping $\overline{EG}$ onto $\overline{MO}$. Therefore, by the definition of congruence, $\overline{EG} \cong \overline{MO}$.

Figure 1

Figure 2

Figure 3

Because both the conditional and its converse are true, we can state the theorem as a biconditional.

> ## Segment Congruence Theorem
>
> Two segments are congruent if and only if they have the same length.

GUIDED

Example 1

In the figure at the right $\overline{CL} \cong \overline{AR}$. If $CA = 36$ miles, find LR.

Solution By the Segment Congruence Theorem, $CL = \underline{\ ?\ }$.

By the Addition Property of Equality, we can add LA to both sides:

$$CL + LA = \underline{\ ?\ } + LA$$
$$\underline{\ ?\ } = LR$$

Thus, if $CA = 36$ miles, $LR = \underline{\ ?\ }$.

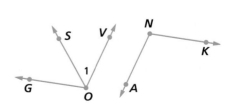

Congruent Angles and Measure

The Segment Congruence Theorem, which connects segments and their lengths, has a counterpart involving angles and angle measure.

> ## Angle Congruence Theorem
>
> Two angles are congruent if and only if they have the same measure.

> ▶ **QY1**
>
> Write the two conditionals that we combined in the Angle Congruence Theorem.

The Angle Congruence Theorem is also a biconditional. In Questions 10 and 11, you are asked for arguments that prove this theorem.

🛑 **QY1**

Example 2

In the figure at the right, $\angle GOV \cong \angle ANK$ and $\overrightarrow{OS}$ bisects $\angle GOV$. If $m\angle GOV = 12x - 2$ and $m\angle ANK = 8x + 34$, find $m\angle 1$.

Solution Because $\angle GOV \cong \angle ANK$,

$$m\angle GOV = m\angle ANK$$
$$12x - 2 = 8x + 34$$
$$4x = 36$$
$$x = 9.$$

By substitution, $m\angle GOV = 12x - 2 = 12(9) - 2 = 106$.
Because $\overrightarrow{OS}$ bisects $\angle GOV$, $m\angle 1 = \frac{106}{2} = 53$.

Corresponding Parts

When you have to replace a battery in a calculator, you know you must insert a new battery that is congruent to the dead battery. Even so, when you insert the new battery in the calculator, it must be in the correct position. The $+$ and $-$ ends of the battery must fit into the correct corresponding places in the calculator.

This same idea is true of congruent figures. It is often not enough to know they are congruent. You must know what parts correspond. This is easy when you know the isometry that mapped one of the figures onto the other.

In the figure at the right, quadrilateral $MAPS$ is the image of quadrilateral $NICE$ under an isometry T, so the two quadrilaterals are congruent. When we write $MAPS \cong NICE$ or $T(MAPS) = NICE$, we *keep the corresponding vertices in order*. This is so that we know which points are the images of which other points. The order tells us that $T(M) = N$, $T(A) = I$, $T(P) = C$, and $T(S) = E$.

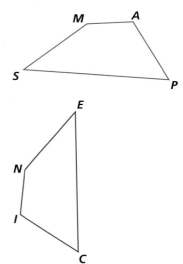

Parts of figures that are images of each other under a transformation are called **corresponding parts**. These parts may be sides, angles, diagonals, arcs, circles, rays, or even complicated figures. If the transformation is an isometry, then any corresponding parts are congruent because of the definition of isometry. This is an important and widely used theorem, so we abbreviate its name.

> ### Corresponding Parts in Congruent Figures (CPCF) Theorem
>
> If two figures are congruent, then any pair of corresponding parts are congruent.

By knowing which points correspond, because every isometry preserves distance, we know that corresponding segments have the same length. For example, in the quadrilaterals above, $MA = NI$ and $MP = NC$. We also know that corresponding angles have the same measure. For instance, $m\angle SPA = m\angle ECI$ and $m\angle SAP = m\angle EIC$ (even though those angles are not drawn).

GUIDED

Example 3

Given: $\triangle TOP \cong \triangle CAN$.

a. List the corresponding angles and sides.

b. Sketch a possible situation and mark the congruent angles and sides.

   (continued on next page)

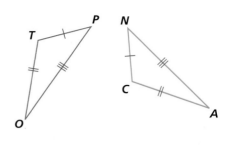

Solution

a. You can use the congruence statement to match the corresponding parts.

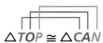

$\triangle TOP \cong \triangle CAN$

So $\angle TOP \cong$ __?__, $\angle OPT \cong$ __?__, and $\angle PTO \cong$ __?__;
and __?__ $\cong \overline{CA}$, __?__ $\cong \overline{AN}$, and __?__ $\cong \overline{CN}$.

b. Two possible triangles are given at the right with tick marks to indicate the congruent segments.

 QY2

> **QY2**
>
> Trace $\triangle TOP$ and $\triangle CAN$ in Part b of the solution, and add tick marks to indicate congruent angles.

Congruence and Equality

Because of the Segment and Angle Congruence Theorems, you can substitute statements of equal measure for some statements of congruence. The following chart lists some substitutions that we often use in this book.

Congruence	Equality
congruent segments	segments of equal length
$\overline{AX} \cong \overline{BY}$	$AX = BY$
congruent angles	angles with the same measure
$\angle ABC \cong \angle DEF$	$m\angle ABC = m\angle DEF$

Questions

COVERING THE IDEAS

For 1–4, fill in the blanks.

1. Congruent segments have the same __?__.
2. If a figure is the image of another under a(n) __?__, then their corresponding parts are congruent.
3. Congruent __?__ have the same measure.
4. Corresponding segments of __?__ figures are the same length.
5. If $\triangle TOP \cong \triangle CAN$ and $m\angle T = 118$, $m\angle A = 27$, $m\angle N = 35$, $CA \approx 3.4$, $TP \approx 2.7$ and $OP \approx 5.2$, find the lengths of the missing sides and the measures of the missing angles.
6. a. Find all the missing measures of the figures shown at the right if $PENTA \cong GRMSW$.
   b. What segment has the same length as $\overline{ET}$?

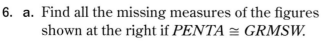
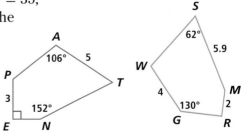

7. Use the tick marks on the figures below to write a congruence statement with the vertices in *correct* order.

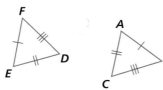

In 8 and 9, use the figure at the right, in which $\overline{GH} \cong \overline{IJ}$.

8. If $GH = 12x$ and $x = 4$, find IJ.

9. Describe a composite of two isometries that maps $\overline{GH}$ onto $\overline{IJ}$.

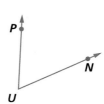

In 10 and 11, you are asked to provide justifications for each conclusion of the Angle Congruence Theorem. Use the diagram at the right.

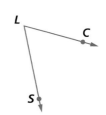

10. Provide the justifications that complete the following proof of the statement: *If two angles are congruent, then they have the same measure.*

Given $\angle CLS \cong \angle PUN$

Prove $m\angle CLS = m\angle PUN$

Proof

Conclusions	Justifications
a. $\angle CLS \cong \angle PUN$	a. ___?___
b. There exists an isometry I for which $I(\angle CLS) = \angle PUN$.	b. ___?___
c. Because $I(\angle CLS) = \angle PUN$, $m\angle CLS = m\angle PUN$.	c. ___?___

11. Consider the converse of the statement from Question 10. The converse is: *If two angles have the same measure, then they are congruent.* So, you need to show that there is an isometry that maps $\angle CLS$ onto $\angle PUN$. This can be done by using a translation followed by a rotation. Answer the following questions regarding the translation and the rotation that accomplish this task.

a. Let T be the translation. What is the translation vector for T?

b. Let $T(\angle CLS) = \angle C'L'S'$. What is the magnitude of the rotation R required to map $\angle C'L'S'$ onto $\angle PUN$?

c. Now that you have shown that $R \circ T(\angle CLS) = \angle PUN$, you know that $\angle PUN$ is the image of $\angle CLS$ under the isometry $R \circ T$. Because of this, you can now conclude that $\angle CLS \cong \angle PUN$. Why?

12. In the diagram at right, $\triangle KLM \cong \triangle KMN$, O is the midpoint of $\overline{KM}$, $OM = 2t + 3$, and $KN = 8t + 8$. Find t.

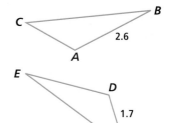

In **13–15**, assume that the two triangles in each figure that appear to be congruent are congruent.

a. Write a congruence statement for each with the vertices in the *correct* order.

b. List, in pairs, all corresponding congruent parts.

13. 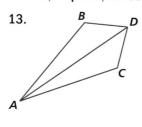 14. 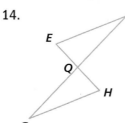 15.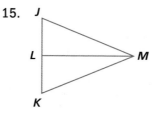

16. In the figure at the right, $\triangle ABC \cong \triangle DEF$.
    a. Find ED.
    b. Use the triangle inequality to find the possible lengths of $\overline{EF}$.

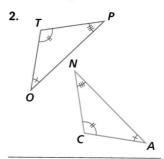

17. Suppose $RUNW \cong ILDS$, with $RU = 8y + x$, $IL = 13x + 20$, $\text{m}\angle NUR = 13y + 2x$, and $\text{m}\angle DLI = 10y + 40$. Find $\text{m}\angle DLI$.

REVIEW

18. For points in the coordinate plane, is the property *A lies above B* preserved by isometries? Explain. **(Lesson 5-1)**

19. When a musical round is written down, what type of isometry is pictured? **(Lesson 4-8)**

20. Suppose ℓ is the perpendicular bisector of $\overline{AB}$. Let S_k be a size transformation. Is $S_k(\ell)$ the perpendicular bisector of $S_k(\overline{AB})$? Explain your answer. **(Lessons 3-9, 3-7)**

21. State the Equation to Inequality Property of real numbers. **(Lesson 3-4)**

22. Let s be the measure of an obtuse angle and t be the measure of an acute angle. Tell whether each statement is true or false. **(Lesson 3-1)**
    a. $s + t < 270$ b. $s < t$ c. $s > t$ d. $s + t > 90$

EXPLORATION

23. How many different noncongruent shapes can you make using exactly five congruent squares if the squares must be attached edge to edge? Two examples are shown at the right.

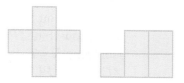

Lesson 5-3

One-Step Congruence Proofs

▶ **BIG IDEA** By giving a justification for a conclusion, you have done a proof that is one step long.

In this lesson, you will work with statements that can be used as justifications in proofs that involve congruence. Most arguments in this lesson are only one step long, so you will need to find only one justification. In the next lesson you will see that by putting a few of these steps together, you can prove some very important theorems.

The fictional detective Sherlock Holmes was famous for making deductions. By observing how a client was dressed, tooth marks on his cane, and nicotine stains on his fingers, the detective was able to deduce how wealthy the person was, that he had a large dog, and that he smoked. Likewise, a doctor can make deductions from a blood test, and a geneticist can make deductions from a DNA sample.

This same idea occurs throughout mathematics. In geometry, you may be given a figure or some other information and be asked to provide conclusions that you can justify.

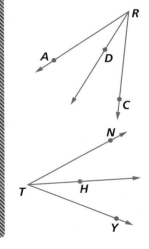

Example 1

S_k is a size transformation of magnitude k. What can you conclude about $\overleftrightarrow{AB}$ and $S_k(\overleftrightarrow{AB})$?

Solution Three theorems are related to size transformations. One says size transformations preserve parallelism, one says they preserve angle measure, and one says they preserve collinearity. Because this problem asks about two lines, the only theorem that seems relevant is the theorem involving parallels. This theorem states that any line will be parallel to its image under a size transformation (S_k Theorem 1). You can conclude that $S_k(\overleftrightarrow{AB}) \parallel \overleftrightarrow{AB}$ because $S_k(\overleftrightarrow{AB})$ is the image of $\overleftrightarrow{AB}$ under a size transformation. To prove this is true, you can write a short proof like the one below.

Conclusions	Justifications
1. S_k is a size transformation.	1. Given
2. $S_k(\overleftrightarrow{AB}) \parallel \overleftrightarrow{AB}$	2. Any line is parallel to its image under a size transformation. (S_k Theorem 1)

Example 2

Given: *A* and *B* are points in a convex set. What can you conclude about $\overline{AB}$?

Solution Recall that a convex set is a set in which every segment that connects points of the set lies entirely within the set. So, from the definition of convex set you can conclude that $\overline{AB}$ must lie entirely within the set. Here is what you might write.

Conclusions	Justifications
1. A and B are points in a convex set.	1. Given
2. $\overline{AB}$ is in the same set as A and B.	2. If a set is convex, then every segment that connects points of the set lies entirely in the set (definition of convex set).

The CPCF theorem from Lesson 5-2 is an often used justification.

GUIDED

Example 3

Given: $\triangle QRS \cong \triangle TUV$

Prove: $QR = TU$

Proof

Conclusions	Justifications
1. $\triangle QRS \cong \triangle TUV$	1. ___?___
2. $\overline{QR} \cong \overline{TU}$	2. ___?___
3. $QR = TU$	3. If two segments are congruent, then they are the same length (___?___ Congruence Theorem).

Definitions are also common justifications.

GUIDED

Example 4

Given: *M* is the midpoint of $\overline{AB}$.

Prove: $AM = MB$

Proof

Conclusions	Justifications
1. ___?___	1. Given
2. ___?___	2. ___?___

Common Justifications for Congruence

Often you know what you need to prove and you need to search for a justification. For instance, someone may tell you that your friend Alicia is at a basketball game. You ask, "How do you know?" The justification may be simple: "She text messaged me that she was there." Or the justification may be a little more complicated, such as, "I saw Alicia walking with Luis. Luis attends all of the school basketball games, and tonight is a basketball game." Both statements justify a conclusion that has been asserted.

Every justification in mathematics is either a postulate, a definition, or a theorem that we have already proved. Sometimes a justification is one direction of a biconditional. To indicate which direction of the biconditional we are using, we often write the correct direction as a simple if-then statement. The following table should help you locate the justification you need. In each case, the name of the property is given first. This is followed by a concise if-then statement suitable for use in a proof as justifications.

When a conclusion is that one figure is congruent to another, you have seen a number of possible justifications.

Table of Justifications	
Some justifications that segments are congruent	**Some justifications that angles are congruent**
Definition of bisector: If a figure is the bisector of a segment, it divides the segment into two congruent segments. (Lesson 3-9)	**Corresponding Angles Postulate:** If lines intersected by a transversal are parallel, then corresponding angles are congruent. (Lesson 3-6)
Definition of midpoint: If a point is the midpoint of a segment, it divides the segment into two congruent segments. (Lesson 2-4)	**Definition of angle bisector:** If a ray bisects an angle, then it divides the angle into two congruent angles. (Lesson 3-3)
CPCF Theorem: If figures are congruent, then corresponding segments are congruent. (Lesson 5-2)	**CPCF Theorem:** If figures are congruent, then corresponding angles are congruent. (Lesson 5-2)
Segment Congruence Theorem: If segments have equal measures, then the segments are congruent. (Lesson 5-2)	**Angle Congruence Theorem:** If the measures of angles are equal, then the angles are congruent. (Lesson 5-2)
Definition of circle: If a figure is a circle, then its radii are congruent. (Lesson 2-4)	**Vertical Angles Theorem:** If angles are vertical angles, then they are congruent. (Lesson 3-3)
Definition of congruence: If a segment is the image of another under an isometry, then the segment and its image are congruent. (Lesson 5-1)	**Definition of congruence:** If an angle is the image of another under an isometry, then the angle and its image are congruent. (Lesson 5-1)

GUIDED

Example 5

Given: $p \parallel q$.

Justify the conclusion that $\angle 4 \cong \angle 6$.

Solution Think: What is given? $p \parallel q$. How are angles $\angle 4$ and $\angle 6$ related to each other? They are corresponding angles.

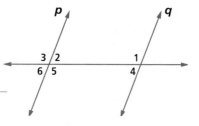

Conclusions	Justifications
1. $p \parallel q$	1. Given

You need a reason that ends with "then angles are congruent." In the table preceding this example, many such statements are listed. Choose the one whose antecedent is about parallel lines. The justification is ___?___.

Now complete line 2.

2. ___?___	2. ___?___

GUIDED

Example 6

In the figure at the right, $\angle AOB$ is the reflection image of $\angle COB$ over $\overleftrightarrow{OB}$. Justify the conclusion that $\angle AOB \cong \angle COB$.

Solution The justification needs to be a conditional that ends with "then the angles are congruent."

The second column of the Table of Justifications lists many such justifications. You need to find one of these whose antecedent has something to do with a reflection.

Find such a reason and either write it out or simply use its name.

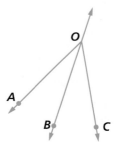

Conclusions	Justifications
1. $r_{\overleftrightarrow{OB}}(\angle AOB) = \angle COB$	1. ___?___
2. $\angle AOB \cong \angle COB$	2. ___?___

Questions

COVERING THE IDEAS

In 1 and 2, make and justify a conclusion from the given information. (There may be more than one possible answer.)

1. In the figure at the right, $r_{\overline{AB}}(\triangle ABC) = \triangle ABD$.

2. In the figure below, $\overleftrightarrow{PQ}$ is the perpendicular bisector of $\overline{ST}$.

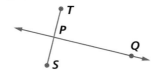

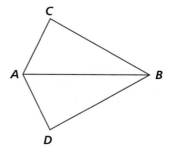

3. What kinds of statements are allowed as justifications in a geometry proof?

4. Write a conditional that might be used to justify $\overline{PQ} \cong \overline{AB}$.

5. Name a theorem that might be used to justify that two angles are congruent.

6. Given: $\triangle PQR$ is the image of $\triangle ABC$ under a rotation of $45°$ about point G.
   a. Draw a possible picture.
   b. Make and justify a conclusion about two segments in the drawing.
   c. Write this as a one-step proof in two-column format.

7. In the figure at the right, the lengths of segments are as indicated. Justify the conclusion $\overline{XZ} \cong \overline{ZY}$.

8. Given: quadrilateral $ABCD \cong$ quadrilateral $PQRS$.
   a. Draw a picture.   b. Make a conclusion.
   c. Justify this conclusion.
   d. Write this as a one-step proof in two-column format.

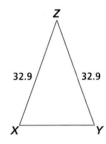

APPLYING THE MATHEMATICS

9. In the figure at the right, parallel lines p and q are intersected by transversal t. Make and justify at least five different conclusions.

10. In the figure at the right, $EA = AB$ and $AB = BC$. Justifications are shown. State the corresponding conclusion.
    a. Segment Congruence Theorem
    b. Transitive Property of Congruence

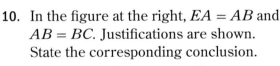

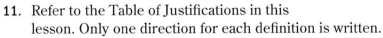

11. Refer to the Table of Justifications in this lesson. Only one direction for each definition is written.
    a. For the definition of midpoint, is this written as term $\Rightarrow$ characteristics or characteristics $\Rightarrow$ term?
    b. Write the definition of midpoint in the other direction from what is written in the table of justifications.

12. In the figure at the right, $\overline{AC}$ and $\overline{BD}$ intersect at the center of $\odot O$. For each conclusion, give a justification.
    a. $\overline{AO} \cong \overline{OC}$ b. O is the midpoint of $\overline{AC}$.

13. T is an isometry and $T(ABCDE) = BCDEA$.
    a. Draw a possible figure.
    b. Give a sequence of conclusions to explain why $\overline{AB} \cong \overline{DE}$.

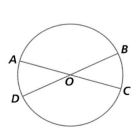

14. Suppose A, B, and C are points and $AB = BC$. Is B is the midpoint of $\overline{AC}$? Why or why not?

REVIEW

15. In the figure at the right, is $\triangle ADE \cong \triangle ABC$? Explain your answer. (**Lesson 5-2**)

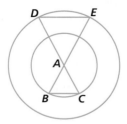

16. Let T be the transformation that maps (x, y) to the point (x^2, y^2). Use the points $(0, 0)$, $(1, 0)$, $(0, -3)$ to show that T is not an isometry. Explain your answer. (**Lesson 5-1**)

17. **Multiple Choice** Suppose you drew a figure on a transparency and wished to produce a figure that was oppositely congruent to it. Which of the following could you do? (**Lesson 5-1**)

A View your figure in a mirror.

B Use a camera to take a picture of your figure.

C Flip the transparency and project it on a wall.

D Turn the transparency 180°.

18. In the figure at the right, $\triangle ABC \cong \triangle A'B'C'$. This translation could have been accomplished by the composition of two reflections over parallel lines. Draw two such parallel lines that do not intersect $\overline{AA'}$. (**Lesson 4-4**)

19. Trace the four points at the right. Draw the region of points that are closest to I. (**Lesson 3-9**)

20. Can you draw the figure at the right without lifting your pencil? Explain your answer. (**Lesson 1-3**)

EXPLORATION

21. Suppose the following nonsense statements are true in geometry.

Definition of a *munji* figure: A figure is munji if and only if it is wumji.

Theorem 2: If a figure is munji, then it is not flumsy.

Theorem 3: If a figure is punji, then it is wumji.

Use these theorems to write the following proof using the format of the proofs in this lesson:

Given Figure F is punji.

Prove F is not flumsy.

Lesson
5-4
Proofs Using Transitivity

Vocabulary

interior angles
exterior angles
alternate interior angles
alternate exterior angles
same-side interior angles

▶ **BIG IDEA** One way of proving two things equal or congruent is to show that the things are both equal or congruent to the same thing.

To our knowledge, the first person to write proofs like those used today was the Greek mathematician Thales, in the sixth century BCE. Three hundred years later, Euclid wrote *Elements,* which we mentioned in Lesson 1-5. Euclid's *Elements* consisted of 13 parts, called "books." Within these books, 465 theorems are proved.

The very first proof in *Elements* is related to the following Activity.

Mental Math

Given: "p," "b," "q"

a. What transformation maps the letter "p" to the letter "b"?

b. What transformation maps the letter "p" to the letter "q"?

c. What letter do you get if you rotate the letter "p" 180°?

Activity

MATERIALS DGS or compass and straightedge

Step 1 Construct points A and B.

Step 2 Construct $\odot(A, AB)$.

Step 3 Construct $\odot(B, AB)$.

Step 4 Name $\odot(A, AB) \cap \odot(B, AB) = \{C, D\}$

$\triangle ABC$ is equilateral.

Doing the Activity does not prove that this construction always works. In this lesson, we give a complete proof, similar to what was found in *Elements,* but in modern style and language. The Greek mathematicians wrote paragraph proofs, as do many modern-day mathematicians. The two-column style used here was first used in American geometry books about 110 years ago.

A typical proof of a conditional $p \Rightarrow q$ in geometry has four components. Two components clarify what is being proved:

the **Given** p
and the **Prove** q.

One component is present as an aid to the proof.

a **Drawing**

The fourth component is the **Proof** itself. The proof is a logical argument created with *Conclusions* and *Justifications* by which *q* is shown to follow from *p*. Numbering the steps in the proof connects the conclusions for reference.

Here is the proof that the algorithm in the Activity constructs an equilateral triangle.

Given ⊙*A* with radius *AB*,
⊙*B* with radius *BA*,
⊙*A* ∩ ⊙*B* = {C, D}

Prove △*ABC* is equilateral.

Proof

Drawing

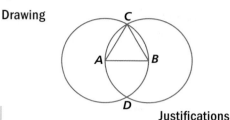

Conclusions	Justifications
1. ⊙*A* with radius *AB*	1. Given
2. *AC* = *AB*	2. All the points on a circle are equidistant from the center of the circle. (definition of circle)
3. ⊙*B* with radius *BA*	3. Given
4. *AB* = *BC*	4. All the points on a circle are equidistant from the center of the circle. (definition of circle)
5. *AC* = *BC*	5. Transitive Property of Equality (with steps 2 and 4)
6. △*ABC* is equilateral.	6. If the measures of all sides of a triangle are equal, then the triangle is equilateral. (definition of equilateral triangle)

Notice how Step 5 uses the conclusions of both Steps 2 and 4 as the antecedent for the conditional used as the justification. This step might be diagrammed like this:

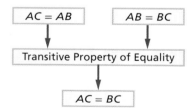

Angles Formed by a Transversal

The Transitive Property of Equality can be used in proving many interesting results, including some theorems about parallel lines and transversals. When two lines are cut by a transversal, the four angles between the lines (∠3, ∠4, ∠5, and ∠6 in the figure at the right) are called **interior angles**. The other angles (∠1, ∠2, ∠7, and ∠8 in the figure at the right) are called **exterior angles**.

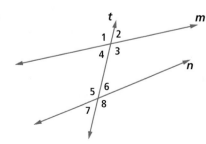

Refer to the figure at the bottom of the previous page. We say that ∠4 and ∠6 are a pair of **alternate interior angles** because they are interior angles on different sides of the transversal and they have different vertices.

 QY1

∠1 and ∠8 are a pair of **alternate exterior angles** in this figure because they are exterior angles on different sides of the transversal and they have different vertices.

 QY2

∠3 and ∠6 are called **same-side interior angles** because they are both interior angles, are on the same side of the transversal, and have different vertices.

 QY3

Read the proof of this theorem carefully. It is a kind of proof you will be asked to learn.

▶ **QY1**

In the figure at the bottom of the previous page, name another pair of alternate interior angles.

▶ **QY2**

Correct this statement: In the figure at the bottom of the previous page, ∠2 and ∠5 are alternate exterior angles.

▶ **QY3**

In the figure at the bottom of the previous page, why are ∠4 and ∠7 not same-side interior angles?

Parallel Lines Theorem

If two parallel lines are cut by a transversal:

1. Alternate interior angles are congruent.
2. Alternate exterior angles are congruent.
3. Same-side interior angles are supplementary.

Given $m \parallel n$ with angles as numbered
Prove $\angle 6 \cong \angle 4$

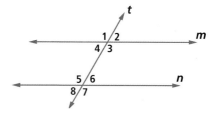

Proof

Conclusions	Justifications
1. $m \parallel n$	1. Given
2. $\angle 6 \cong \angle 2$	2. If lines are parallel, then corresponding angles are congruent. (Corresponding Angles Postulate)
3. $\angle 2 \cong \angle 4$	3. Vertical angles are congruent. (Vertical Angles Postulate)
4. $\angle 6 \cong \angle 4$	4. Transitive Property of Congruence

You are asked to prove Parts 2 and 3 in Questions 8 and 17.

Example 1

In the figure at the right, $n \parallel p$.
If $m\angle 3 = 38$, find $m\angle 1$ and $m\angle 2$.

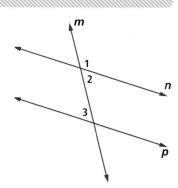

Solution Angles 2 and 3 are alternate interior angles with m as a transversal.
Thus, $m\angle 2 = m\angle 3 = 38$.

Angles 1 and 2 form a linear pair, so
$m\angle 1 = 180 - 38 = 142$.

The converse of each part of the Parallel Lines Theorem is also true. Each converse has a name based on the given information in the theorem.

Alternate Interior Angles Theorem

If two lines are cut by a transversal so that a pair of alternate interior angles are congruent, then the two lines are parallel.

Alternate Exterior Angles Theorem

If two lines are cut by a transversal so that a pair of alternate exterior angles are congruent, then the lines are parallel.

Same-Side Interior Angles Theorem

If two lines are cut by a transversal so that same-side interior angles are supplementary, then the lines are parallel.

GUIDED

Example 2

Prove the Alternate Exterior Angles Theorem using the figure at the right as a drawing.

Given $\angle 5 \cong \angle 3$

Prove $m \parallel n$

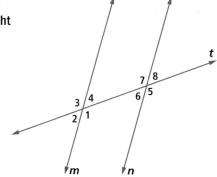

Proof

Conclusions	Justifications
1. $\angle 5 \cong \angle 3$	1. Given
2. $\angle 3 \cong \angle 1$	2. ___?___
3. $\angle 5 \cong \angle 1$	3. ___?___
4. $m \parallel n$	4. If two corresponding angles have the same measure, then the lines are parallel. (Corresponding Angles Postulate)

Here is a summary of postulates and theorems about parallel lines and transversals.

If given *parallel lines*, then	Ways to prove parallel lines
Corresponding angles are congruent. (Corresponding Angles Postulate)	If two lines are cut by a transversal and form congruent corresponding angles (Corresponding Angles Postulate)
Alternate interior angles are congruent. Alternate exterior angles are congruent. Same-side interior angles are supplementary. (Parallel Lines Theorem)	If two lines are cut by a transversal and form congruent alternate interior angles (Alternate Interior Angles Theorem)
	If two lines are cut by a transversal and form congruent alternate exterior angles (Alternate Exterior Angles Theorem)
	If two lines are cut by a transversal and form supplementary same-side interior angles (Same-Side Interior Angles Theorem)

Questions

COVERING THE IDEAS

1. About how long ago did Thales live, and for what is he famous?

2. Euclid's first theorem in *Elements* is about what construction?

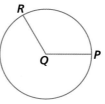

3. In $\odot Q$ at the right, what is the justification for the conclusion $QR = QP$?

4. Name the four components of a proof.

5. In the construction of an equilateral triangle on page 269, suppose $\overline{AD}$ and $\overline{BD}$ are drawn. Give the proof that $\triangle ABD$ is equilateral.

6. **a.** Name the interior angles in the figure at the right.
   b. Name two pairs of alternate interior angles in the figure.
   c. If $m\angle 3 = 140$ and $m\angle 4 = 40$, are m and n parallel?

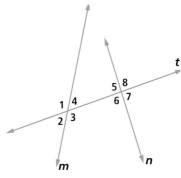

7. **True or False**

a. In the figure at the right, $\overline{AB} \parallel \overline{CD}$.

b. In the figure at the right, $\overline{AD} \parallel \overline{BC}$.

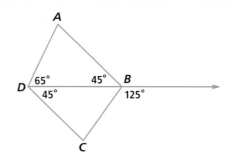

8. Finish this proof of Part 2 of the Parallel Lines Theorem.

Theorem When two lines are parallel, alternate exterior angles are congruent.

Given $m \parallel n$

Prove $\angle 1 \cong \angle 8$

Drawing

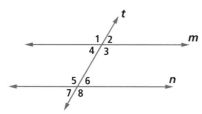

9. The lines in a parking lot are parallel to each other. The width of a handicapped parking space is wider than a regular space. Are the angles in the handicapped space larger than the regular parking spaces? Why or why not?

10. The gate in the fence below is built such that $a \parallel b \parallel c$ and $m\angle 3 = 43$. Find the measures of all of the numbered angles.

11. In this picture of a tennis court, it appears that the sidelines are going to intersect eventually. The sidelines of a tennis court are parallel to each other and perpendicular to the baselines and the net. Trace this picture of a tennis court on your paper. Mark at least two angles in your diagram that must be congruent on the court and will prove that the sidelines on the tennis court are indeed parallel.

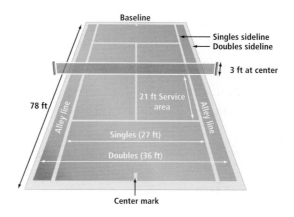

APPLYING THE MATHEMATICS

12. **Fill in the Blanks** If the top of this picnic table is parallel to the seats, fill in the blanks to describe the relationships of ∠1, ∠2, ∠3, and ∠4 and give a reason for each.

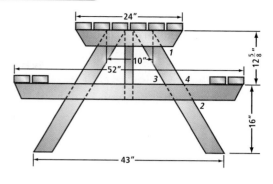

a. ∠1 and ∠2 are ___?___ because ___?___.

b. ∠1 and ∠3 are ___?___ because ___?___.

c. ∠1 and ∠4 are ___?___ because ___?___.

In 13 and 14, a figure is given.

a. Name all pairs of alternate interior angles.

b. Tell which segments will be parallel if these angles are congruent.

13.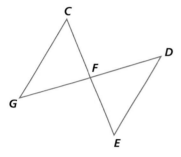

14.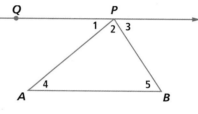

15. Use the diagram at the right and complete this proof by writing the conclusions and justifications.

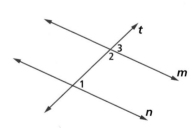

Given B is the midpoint of $\overline{AC}$; C is the midpoint of $\overline{BD}$.

Prove $\overline{AB} \cong \overline{CD}$

16. **Fill in the Blanks** Copy and supply the missing parts in the proof of the following theorem:

Theorem If same-side interior angles are supplementary, then the lines are parallel.

Given ∠1 and ∠2 are supplementary, m∠2 = x.

Prove m ∥ n

Proof It is given that ___?___ and ___?___. Because ∠1 and ∠2 are supplementary, then the measures of ∠1 and ∠2 add to ___?___ and so, m∠1 + x = ___?___. ∠2 and ∠3 are supplementary angles because they are ___?___ and so, x + m∠3 = ___?___. By substitution, m∠1 + x = x + m∠3. So ___?___ by the Addition Property of Equality and ∠1 and ∠3 are ___?___ angles. Therefore, m ∥ n by the ___?___ Postulate, which states that ___?___.

17. Prove Part 3 of the Parallel Lines Theorem. (Hint: Use Question 16 as a guide, but work backward.)

REVIEW

18. The quadrilaterals $ABCD$ and $BCEF$ are congruent. (**Lesson 5-3**)
    a. Draw a picture. (Be careful!)
    b. Make and justify two conclusions about this situation.

19. In the figure below, $ABCDE \cong LMNOP$. Using the CPCF Theorem, what can you say about AD? (**Lesson 5-2**)

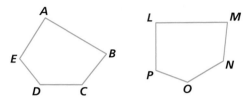

20. **True or False** If $\overline{AB} \cong \overline{EF}$ and $\overline{CD} \cong \overline{GH}$, then $ABCD \cong EFGH$. If true, explain why. If false, draw a counterexample. (**Lesson 5-2**)

21. Let C be the midpoint of $\overline{AB}$, and let T be an isometry. Suppose $T(A) = A$ and $T(B) = B$. What can you say about $T(C)$? Explain. (**Lesson 5-1**)

22. **Multiple Choice** Lines ℓ, m, and n are parallel to the x-axis. Which of the following isometries could be the composite $r_\ell \circ r_m \circ r_n$? (**Lesson 4-7**)
    A reflection over the y-axis
    B reflection over the x-axis
    C translation
    D rotation about the origin

23. Suppose $r_m(P) = Q$ and $r_m(A) = A$. (**Lesson 3-9**)
    a. According to the definition of reflection image, how are m and $\overline{PQ}$ related?
    b. According to the definition of reflection image, what can you conclude about A?
    c. Why does $AP = AQ$?

24. Suppose a network has five odd nodes. (**Lesson 1-3**)
    a. Is the network traversable?
    b. Is it possible to make the network traversable by adding one arc? Why or why not?

EXPLORATION

25. A student argued that the quadrilaterals $ABCD$ and $BCEF$ of Question 18 must be squares. Is this true? If so why? If not, why not? (Remember: A square must have four sides of the same length and four angles of the same measure.)

Proofs Using Reflections

> ▶ **BIG IDEA** When you know that certain points are the reflection images of other points, you can use the definition and preservation properties of a reflection to deduce many things about those points.

Using Preservation Properties of Reflections

In Lesson 3-9 you saw Voronoi diagrams. Recall that these diagrams are constructed by dividing the plane using the perpendicular bisectors of the segments connecting given points. The Voronoi diagram below is defined by points *A, B, C,* and *D.* Suppose these points represent airports on a map. A pilot who needed to make an emergency landing could quickly tell which airport to use by noting the region over which the airplane is flying.

Mental Math

Let *R* be a rotation of 90° about the origin, $X = (1, 0)$, and $Y = (0, 1)$. Find:

a. $R(X)$.

b. $R(Y)$.

c. $R \circ R(X)$.

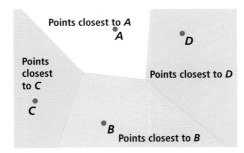

Points closest to *A*

Points closest to *C*

Points closest to *D*

Points closest to *B*

In the diagram, what would happen if the airplane was located on the boundary line between two regions? In this case, the airplane would be the same distance from two of the airports. This is because if a point is on the perpendicular bisector of a segment, then it is equidistant from the endpoints of the segment. Voronoi diagrams are based on this principle. Yet we were not able to prove this principle until now because the proof relies on properties of reflections.

Perpendicular Bisector Theorem

If a point is on the perpendicular bisector of a segment, then it is equidistant from the endpoints of the segment.

Given m is the perpendicular bisector of $\overline{AB}$ and P is on m.

Prove $PA = PB$

Drawing

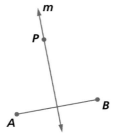

Proof

Conclusions	Justifications
1. m is the perpendicular bisector of $\overline{AB}$, P is on m.	1. Given
2. $r_m(P) = P$	2. definition of reflection
3. $r_m(A) = B$	3. definition of reflection
4. $PA = PB$	4. Reflections preserve distance.

 QY

▸ **QY**

In the figure below, $\overleftrightarrow{MP}$ is the perpendicular bisector of $\overline{TC}$. If $PT = 10$, what is PC?

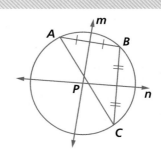

Activity

MATERIALS DGS

Step 1 On a clear DGS screen, construct three noncollinear points A, B, and C.

Step 2 Construct $\overline{AB}$ and $\overline{BC}$.

Step 3 Construct ℓ and m, the perpendicular bisectors of $\overline{AB}$ and $\overline{BC}$.

Step 4 Name $\{P\} = \ell \cap m$.

Step 5 Construct a circle with center at point P and with a radius PA. Make a conjecture regarding the constructed circle and points B and C.

Example 1 exhibits a proof of a possible conjecture from this Activity.

Example 1

Write an argument to complete the proof.

Given $\triangle ABC$, m is the perpendicular bisector of $\overline{AB}$. n is the perpendicular bisector of $\overline{BC}$, $m \cap n = \{P\}$.

Prove A, B, and C lie on $\odot P$ with radius PA.

Proof

Conclusions	Justifications
1. m is the perpendicular bisector of $\overline{AB}$.	1. Given
2. $PA = PB$	2. If a point is on the perpendicular bisector of a segment, then it is equidistant from the endpoints.
3. n is the perpendicular bisector of $\overline{BC}$.	3. Given
4. $PB = PC$	4. If a point is on the perpendicular bisector of a segment, then it is equidistant from the endpoints.
5. $PA = PC$	5. Transitive Property of Equality
6. A, B, and C lie on $\odot P$ with radius PA.	6. A circle is the set of all points the same distance from the center.

Using the Figure Transformation Theorem

You have seen a variety of ways to prove that segments are congruent. You have also seen many ways to prove that angles are congruent. A two-step combination of the Figure Transformation Theorem from Lesson 4-2 and the definition of congruent figures can prove that two figures are congruent whenever the points that determine one figure are the reflection images of the points that determine the other.

Example 2

Prove: If $r_m(A) = P$, $r_m(B) = R$, and $r_m(C) = Q$, then $\triangle ABC \cong \triangle PRQ$.

Solution Remember that in a proof of a conditional $p \Rightarrow q$, p is the "Given" and q is the "Prove."

Given: $r_m(A) = P$, $r_m(B) = R$, and $r_m(C) = Q$

Prove: $\triangle ABC \cong \triangle PRQ$

Drawing:

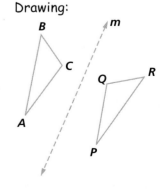

Proof:

Conclusions	Justifications
1. $r_m(A) = P$, $r_m(B) = R$, and $r_m(C) = Q$.	1. Given
2. $r_m(\triangle ABC) = \triangle PRQ$	2. Figure Transformation Theorem
3. $\triangle ABC \cong \triangle PRQ$	3. Definition of congruent figures

Questions

1. Draw a figure in which point G is equidistant from points T and M, but G does not lie on $\overline{TM}$.

2. Suppose t is the perpendicular bisector of $\overline{RB}$ in the figure at the right. Justify each of these conclusions.
   a. K is the reflection image of K over t.
   b. $r_t(R) = B$ c. $KR = KB$

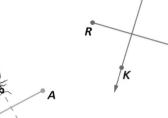

3. Dexter the ant is crawling along the perpendicular bisector of AB. What will always be true about Dexter's distances from A and B?

4. Choose the correct phrase from each pair in parentheses: Any point on the (*bisector, perpendicular bisector*) of a segment is equidistant from the (*endpoints, midpoint*) of the segment.

5. Trace the Voronoi diagram at the beginning of this lesson. Darken the points where a pilot is equidistant from two of the cities A, B, C, or D. Identify all points where a pilot is equidistant from three of these cities.

6. Four cell phone towers (labeled Q, R, S, and T) are shown on the map at the right. Construct a Voronoi diagram to describe the areas served by each of the towers.

7. **Given** $\overline{MN}$ is the perpendicular bisector of $\overline{AE}$.
   Prove $\angle AMN \cong \angle EMN$
   a. Trace the figure at the right and mark the given information on it.
   b. Write the proof.

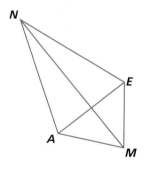

8. In the figure at the right, $\overleftrightarrow{QS}$ is the perpendicular bisector of $\overline{PR}$.
   a. If PS is $13x$, what is SR?
   b. If PR is $12y$, what is PQ?

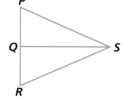

9. Finish the proof of this theorem.

   Theorem The reflection image of a right angle is a right angle.
   Given $\angle ABC$ is a right angle, $r_\ell(\triangle ABC) = \triangle XYZ$.
   Prove $\angle XYZ$ is a right angle.
   Proof It is given that $\angle ABC$ is a right angle and $r_\ell(\triangle ABC) = \triangle XYZ$. The definition of right angle tells us $m\angle ABC = 90$. Reflections preserve angle measure, so …

10. A tree, perpendicular to the ground, stands midway between two stakes. Guy wires are attached from the stakes to a narrow collar around the trunk. Explain why the guy wires must have the same length.

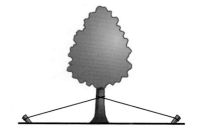

11. **a.** Write the converse of the Perpendicular Bisector Theorem.
    b. Do you think the converse is true?
    c. What would you have to do to show that the converse is true?
    d. What would you do to show that the converse is false?

REVIEW

12. In the portion of a Navajo rug pictured at the right, assume that all the segments that look parallel are parallel. Write an argument to convince a friend that $m\angle 2 = m\angle 3$. (**Lesson 5-4**)

13. Suppose $\triangle EFG$ is an isosceles triangle with $EF = EG$. Explain why G lies on the circle with center E and radius EF. (**Lesson 5-4**)

14. Suppose $r_n(\angle ABC) = \angle DEF$. (**Lesson 5-4**)
    a. Explain why $r_n \circ r_n (\angle ABC) \cong \angle ABC$.
    b. Explain why $m(r_n \circ r_n(\angle ABC)) = m\angle ABC$.
    c. Explain why it is not necessarily true that $AB = DE$.

In 15 and 16, give the justification for the conclusion.

15. Given: m is the perpendicular bisector of $\overline{BC}$, $r_m(A) = D$, and $r_m(B) = C$.
    Conclusion: $PQBA \cong PQCD$ (**Lesson 5-3**)

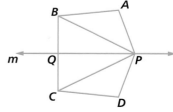

16. Given: $r_t(Y) = W$, $r_t(M) = H$
    Conclusion: $YM = HW$ (**Lesson 5-2**)

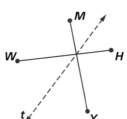

17. Explain why S_2 and $S_{\frac{1}{2}}$ are not isometries, but $S_{\frac{1}{2}} \circ S_2$ is. (**Lessons 4-7, 4-4, 3-7**)

18. Find an equation for the perpendicular bisector of the segment with endpoints $A = (3, 4)$ and $B = (5, 2)$. (**Lessons 3-9, 3-8**)

EXPLORATION

19. In this lesson you saw that the set of all points in the plane equidistant from points A and B is a line.
    a. What is the set of points equidistant from three different collinear points in a plane?
    b. When can there be a point equidistant from four different points?

QY ANSWER

10

Lesson
5-6

Auxiliary Figures and Uniqueness

Vocabulary

uniquely determined

auxiliary figure

non-Euclidean geometries

obtuse triangle

right triangle

acute triangle

▶ **BIG IDEA** Introducing a line or part of a line into a figure can be a helpful strategy in working out a proof.

Some Examples of Uniqueness

The adjective *unique* means "exactly one." When exactly one thing satisfies some given conditions, we say the thing is **uniquely determined**. For instance, your address uniquely determines which building you live in. In algebra, the given condition $4x + 7 = 31$ uniquely determines the value of x.

Mental Math

In the figure below, $\overline{AD} \parallel \overline{BC}$, $\overline{AB} \parallel \overline{DC}$, **and** $m\angle A = 45$. **Find:**

a. $m\angle D$.

b. $m\angle B$.

c. $m\angle C$.

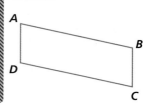

Example 1

Given a segment $\overline{AB}$, which of these things are uniquely determined? If they are not, why not?

a. midpoint of $\overline{AB}$

b. bisector of $\overline{AB}$

c. perpendicular bisector of $\overline{AB}$

Solution

a. Does a segment have exactly one midpoint? **Yes**

b. Does a segment have exactly one bisector? **No. There can be many lines, segments, or rays that pass through the midpoint, and each is a bisector because of the definition of "bisector."**

c. Does a segment have exactly one perpendicular bisector? **Yes**

It is possible to prove that certain figures are unique. For example, in Lesson 5-5, the circle through three noncollinear points A, B, and C is unique because (1) the perpendicular bisector of any segment contains all points equidistant from the endpoints of the segment, (2) the two perpendicular bisectors of $\overline{AB}$ and $\overline{BC}$ intersect at a unique point, and (3) there is a unique circle with a particular center and radius. So we can assert the following theorem.

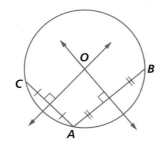

▶ **READING MATH**

The prefix *uni* (from the Latin *unus*), meaning "one," starts many English words. These include *unicycle*, *uniform*, *unilateral*, *union*, and *universal*.

> **Unique Circle Theorem**
>
> There is exactly one circle (a unique circle) through three given noncollinear points.

 QY

What Are Auxiliary Figures?

> ▷ **QY**
>
> Draw two points *A* and *B*. Show that the circle containing *A* and *B* is not uniquely determined.

A segment, line, or other figure that is added to a diagram is called an **auxiliary figure**. The word *auxiliary* means "assisting" or "giving help."

When an auxiliary figure is not uniquely determined, then there are two possibilities: (1) There may be *more than one* figure satisfying the conditions. This is the case with bisectors of segments: (2) There may be *no* figure satisfying the given conditions.

Example 2

In quadrilateral *ABCD*, a student wished to draw as an auxiliary segment, the diagonal $\overline{AC}$ that bisects ∠*A*. Is this always possible? Why or why not?

Solution It is not possible to do this in every quadrilateral. Draw a quadrilateral ABCD. Diagonal $\overline{AC}$ is uniquely determined because points A and C determine a line. However, $\overline{AC}$ does not have to be the bisector of ∠DAB, as the figure shows.

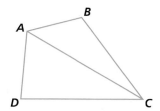

Activity

1. Draw a quadrilateral *ABCD*, different from the one in Example 2, in which the diagonal $\overline{AC}$ *does not* lie on the bisector of ∠*A*.

2. Draw a quadrilateral *EFGH* in which the diagonal $\overline{FH}$ does appear to lie on the bisector of ∠*F*.

How Many Parallels Are There?

As Example 2 demonstrates, uniqueness is not always obvious. The following theorem addresses the question of how many lines can be drawn that are parallel to a given line ℓ and that pass through some given point not on ℓ. In the proof, two auxiliary lines are drawn. We do this to create alternate interior angles that are congruent.

Uniqueness of Parallels Theorem

Through a point not on a line, there is exactly one line parallel to the given line.

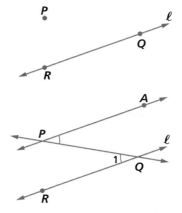

Given Point P not on line ℓ,
points R and Q on line ℓ.

Prove There is exactly one line parallel to ℓ through P.

Proof Draw $\overleftrightarrow{PQ}$. By the Point-Line-Plane Postulate, $\overleftrightarrow{PQ}$ is uniquely determined. We label $\angle PQR$ as $\angle 1$. (See the figure at the right.) Now draw $\overleftrightarrow{PA}$ so that A is on the other side of $\overleftrightarrow{PQ}$ from R and $m\angle APQ = m\angle 1$. $\overleftrightarrow{PA}$ is unique because of the Unique Angle Assumption in the Angle Measure Postulate. $\overleftrightarrow{PA} \parallel \overleftrightarrow{RQ}$ by the Alternate Interior Angles Theorem. So there is at least one line parallel to ℓ through P.

Can there be another parallel? By the Alternate Interior Angles Theorem, $m\angle QPA$ for every line parallel to ℓ through P is the same. Since in a given side of a line there is only one angle with this measure (Angle Measure Postulate), there cannot be more than one parallel. Thus, $\overleftrightarrow{PA}$ is unique and there is exactly one line parallel to ℓ through P.

Playfair's Parallel Postulate

The Uniqueness of Parallels Theorem is important in the history of mathematics. It ultimately changed the entire nature of mathematics. In Euclid's *Elements*, the fifth and final geometric postulate is: *If two lines are cut by a transversal, and the measures of the same-side interior angles sum to less than 180°, then the lines will intersect on that side of the transversal.* This postulate bothered mathematicians, who felt that such a complicated statement should not be assumed true. For 2000 years they tried to prove the fifth postulate from Euclid's other postulates.

After centuries of being unable to prove Euclid's fifth postulate, some mathematicians substituted simpler statements for it. The uniqueness of parallels statement above was first suggested by the Greek mathematician Proclus about 450 CE, but it is known as *Playfair's Parallel Postulate* because it was used by the Scottish mathematician John Playfair in 1795. We were able to prove it as a theorem in this lesson because we assumed the Corresponding Angles Postulate in Lesson 3-6. With that postulate, we proved the first part of the Parallel Lines Theorem and the Alternate Interior Angles Theorem.

Mathematician and geologist John Playfair

By the nineteenth century, other mathematicians had substituted different statements for Playfair's Parallel Postulate. When they assumed *there are no parallels to a line through a point not on it,* they were able to develop a spherical geometry that could apply to the surface of the Earth. When they assumed *there is more than one parallel to a line through a point not on it,* they developed types of geometries for other surfaces. The most notable of these is called hyperbolic geometry. All of these geometries are called **non-Euclidean geometries**. Non-Euclidean geometries are important in physics in the theory of relativity.

These mathematicians greatly influenced *all* later mathematics with their work. For the first time, postulates were viewed as statements assumed true instead of statements definitely true. With this point of view, mathematicians experimented with a variety of algebras and types of geometries formed by modifying or changing postulates. A useful algebra, with some postulates different from those you have studied, is applied in logic and in the operation of computers.

Proving the Triangle-Sum Theorem

In previous courses you learned that the sum of the measures of the three angles in any triangle is 180°. A nice consequence of the Uniqueness of Parallels Theorem is that it enables a short proof of the Triangle-Sum Theorem.

> **Triangle-Sum Theorem**
>
> The sum of the measures of the angles of any triangle is 180°.

Given $\triangle ABC$

Prove $m\angle A + m\angle B + m\angle C = 180$

Proof Draw auxiliary line $\overleftrightarrow{BD}$ with $\overleftrightarrow{BD} \parallel \overleftrightarrow{AC}$. $\overleftrightarrow{BD}$ exists because of the Uniqueness of Parallels Theorem. Pick a point E on $\overleftrightarrow{BD}$ such that B is on $\overline{ED}$. Label angles 1, 2, and 3 as shown.

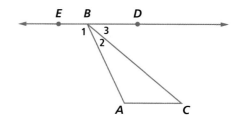

Notice that $\angle A$ and $\angle 1$ are alternate interior angles, as are $\angle C$ and $\angle 3$. Because $\overleftrightarrow{BD} \parallel \overleftrightarrow{AC}$, these alternate interior angles must be congruent by the Parallel Lines Theorem. Thus, $m\angle A = m\angle 1$ and $m\angle C = m\angle 3$. We also know $m\angle EBD = 180$ because it is a straight angle. By the Angle Addition Property, $m\angle 1 + m\angle 2 + m\angle 3 = m\angle EBD$. Now, by substitution, $m\angle A + m\angle B + m\angle C = 180$.

This argument proves the theorem you have used for years.

Because the sum of the measures of the angles of a triangle is fixed at 180, a triangle cannot have more than one angle that is right or obtuse. For this reason we can classify triangles by their largest angle. An **obtuse triangle** is a triangle with an obtuse angle. A **right triangle** is a triangle with a right angle. An **acute triangle** is a triangle with all three angles acute.

Questions

COVERING THE IDEAS

In 1–4, tell whether the figure is or is not uniquely determined in Euclidean geometry and explain why.

1. line parallel to a given line
2. line parallel to two given lines
3. line perpendicular to a given line and through a point not on the given line
4. point equidistant from the endpoints of a given segment
5. Explain why the segment connecting the midpoints of sides $\overline{AB}$ and $\overline{AC}$ of $\triangle ABC$ is uniquely determined.

In 6–8, refer to the Uniqueness of Parallels Theorem.

6. Why is this theorem also called Playfair's Parallel Postulate?
7. What auxiliary figures are used in its proof?
8. Give the justification for each conclusion.
   a. There is a unique line determined by points P and Q.
   b. There is a unique line containing $\overrightarrow{PA}$ so that $m\angle APQ = m\angle 1$ and $\angle APQ$ and $\angle 1$ are alternate interior angles.

9. What postulate in this book substitutes for Playfair's Parallel Postulate and Euclid's fifth postulate?

10. Give an example of a non-Euclidean geometry.

11. Why can't a triangle have one right angle and one obtuse angle?

12. Redraw $\triangle ABC$ from the proof of the Triangle-Sum Theorem. Suppose you were going to redo the proof using an auxiliary line drawn through A instead of an auxiliary line drawn through B.
    a. To what line would the auxiliary line through A be parallel?
    b. Why would the auxiliary line be unique?
    c. Draw the diagram of $\triangle ABC$ and the auxiliary line through A.

APPLYING THE MATHEMATICS

In 13 and 14, tell whether the number described is or is not uniquely determined and why.

13. solution to $x^2 = 25$
14. measure of a right angle

In 15–19, given a quadrilateral *ABCD*, tell whether the auxiliary figure is uniquely determined. If so, make a drawing of this auxiliary figure; if not explain why not. You may find a DGS helpful.

15. line perpendicular to side $\overline{AD}$

16. intersection point of the diagonals $\overline{AC}$ and $\overline{BD}$

17. angle bisector of $\angle ACD$

18. point *M* on side $\overline{AD}$ such that $AM = MD$

19. point of intersection *N* of the perpendicular bisectors of sides $\overline{BC}$ and $\overline{AD}$

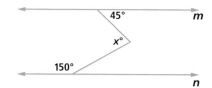

20. Use the figure at the right. Given $m \parallel n$, find *x*.

21. Use the figure for Question 20. Replace 45 by *a*, 150 by *b*, and find a formula for *x* in terms of *a* and *b*. You might find a DGS helpful.

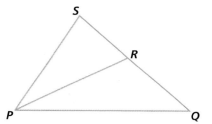

22. Find an equation for the line that is parallel to the line $3x + 4y = 11$, and contains the point $(8, 0)$.

23. Natane was supposed to prove a theorem involving the figure at the right. She decided that she needed an auxiliary line through *R* that was parallel to side $\overline{PQ}$. What justification could she give for this step?

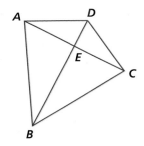

<hr>

REVIEW

24. In the figure at the right $AE = CE$ and $m\angle CED = 90$. Prove that $\triangle BAD \cong \triangle BCD$. (**Lesson 5-5**)

25. Find the center of rotation and magnitude for $r_{x\text{-axis}} \circ r_{y\text{-axis}}$. (**Lesson 4-5**)

26. Consider the lines $y = -4.5$, $y = 13$, $y = 100$, $y = 0$, $y = -28$. The composite of reflections over which two of these lines gives a translation with the greatest magnitude? (**Lesson 4-4**)

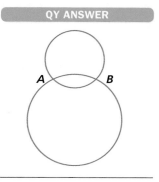

27. Suppose a triangle has sides of length 3, *z*, and $z + 2$, and *z* is an integer. Is *z* uniquely determined? (**Lesson 1-7**)

<hr>

EXPLORATION

28. Use a DGS to complete the following construction:

   Step 1   On a clear DGS screen, construct a triangle $\triangle ABC$.

   Step 2   Construct the line parallel to $\overline{AB}$ through *C*.

   Step 3   Construct the line parallel to $\overline{BC}$ through *A*.

   Step 4   Construct the line parallel to $\overline{AC}$ through *B*.

   Make and dynamically test conjectures about the figure formed when all of these lines are constructed.

QY ANSWER

Sums of Angle Measures in Polygons

Vocabulary

exterior angle of a polygon

▶ **BIG IDEA** The sum of the measures of the interior angles of a convex polygon is determined by the minimum number of triangular regions into which the polygonal region can be divided.

In Lesson 5-6 we proved the sum of the measures of the angles in a triangle is 180. It is natural to wonder if the sum of the measures of the angles in polygons with more sides is constant. It turns out that the Triangle-Sum Theorem allows these questions to be answered for all convex polygons.

Mental Math

The first four terms of a linear sequence are 180, 360, 540, and 720.

a. What is the fifth term?

b. What is the seventh term?

c. 1800 is which term in the sequence?

GUIDED

Example 1

Use the Triangle-Sum Theorem to discover and prove a Quadrilateral-Sum Theorem.

Solution Let S = the sum of the measures of the angles of the convex quadrilateral $QUAD$ at the right.

$$S = m\angle DQU + m\angle U + m\angle UAD + m\angle D$$

Draw auxiliary line segment __?__.

Notice that $\overline{QA}$ has split angles A and Q into two adjacent angles each. Thus, using the Angle Addition Assumption,

$$m\angle UAD = m\angle\underline{\ ?\ } + m\angle\underline{\ ?\ }$$
$$\text{and } m\angle DQU = m\angle\underline{\ ?\ } + m\angle\underline{\ ?\ }.$$

By substituting these expressions for $m\angle UAD$ and $m\angle DQU$,

$$S = (m\angle\underline{\ ?\ } + m\angle\underline{\ ?\ }) + m\angle U +$$
$$(m\angle\underline{\ ?\ } + m\angle\underline{\ ?\ }) + m\angle D.$$

Rearrange and regroup the terms in this equation to get

$$S = (m\angle 3 + m\angle U + m\angle 1) + (m\angle 2 + m\angle D + m\angle 4).$$

We know from the Triangle-Sum Theorem that $m\angle 3 + m\angle U + m\angle 1 = \underline{\ ?\ }$ and $m\angle 2 + m\angle D + m\angle 4 = \underline{\ ?\ }$

So, by substitution, $S = \underline{\ ?\ } + \underline{\ ?\ }$
$$S = \underline{\ ?\ }.$$

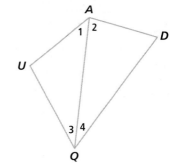

The solution to Example 1 proves the following theorem, which may be familiar to you.

> **Quadrilateral-Sum Theorem**
>
> The sum of the measures of the angles of a convex quadrilateral is 360.

Sums of Angle Measures in Convex Polygons

The sum of the measures of the angles of any convex n-gon can be determined in a manner like that used for quadrilaterals.

Activity 1

Step 1 Draw any convex quadrilateral region.

Step 2 Tear off its corners and place them together so that the angles are adjacent.

Step 3 Describe what happens.

Step 4 Repeat Steps 1–3 for a convex pentagonal region.

Now we turn our attention to the case of all the other convex polygons.

Activity 2

In all of the convex polygons shown below, we have chosen one vertex (A) and drawn all of the possible diagonals from that vertex. Fill in the table to try to find a formula for the sum of the measures of the angles of the n-gon.

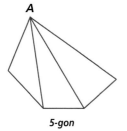
5-gon

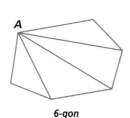
6-gon

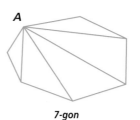
7-gon

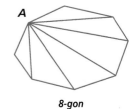
8-gon

Number of sides	5	6	7	8	. . .	n
Number of triangles formed	?	?	?	?	. . .	?
Sum of the measures of the angles in the polygon	?	?	?	?	. . .	?

Activity 2 should have led you to the following theorem.

Polygon-Sum Theorem

The sum of the measures of the angles of a convex n-gon is $(n - 2) \cdot 180$.

STOP **QY**

So far we have focused our attention on the angles within a polygon. These angles are sometimes referred to as *interior angles*. Polygons also have what are known as *exterior angles*.

▶ **QY**

What is the sum of the measures of the angles of any convex nonagon?

Definition of Exterior Angle of a Polygon

An angle is an **exterior angle of a polygon** if and only if it forms a linear pair with one of the angles of a polygon.

In any polygon, two exterior angles can be formed at each vertex by extending the two sides of the angle with that vertex. At the right, angles 1 and 2 are exterior angles of $\triangle XYZ$. The figure at the right below, shows one exterior angle at each vertex of $\triangle ABC$: $\angle DAB$, $\angle EBC$, and $\angle FCA$.

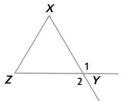

Activity 3

MATERIALS Protractor

Step 1 In the figure there are four angles whose measures are not given. Find the measure of each.

Step 2 Compare the measure of each exterior angle of $\triangle ABC$ with the measure of the two interior angles that are not adjacent to it. For instance, compare m$\angle DAB$ to m$\angle ABC$ and m$\angle BCA$. Do the same for the other two exterior angles. What do you notice?

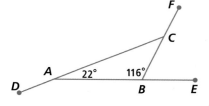

You may have noticed the following result.

Exterior Angle Theorem for Triangles

In a triangle, the measure of an exterior angle is equal to the sum of the measures of the interior angles at the other two vertices of the triangle.

So, for the figure at the right:

m$\angle 1$ = m$\angle 5$ + m$\angle 6$
m$\angle 2$ = m$\angle 4$ + m$\angle 6$
m$\angle 3$ = m$\angle 4$ + m$\angle 5$.

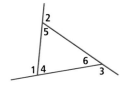

The proof of this theorem is left for you as Question 11.

Because there is a formula for the sum of the measures of the interior angles of a convex n-gon, it is natural to wonder about the exterior angles.

MATERIALS Protractor

The figure at the right is a pentagon with measures of four interior angles given.

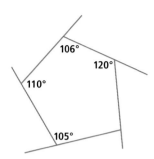

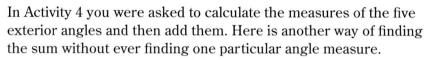

Step 1 Find the measure of the fifth interior angle.

Step 2 Find the measure of each exterior angle shown.

Step 3 Find the sum of the measures of its exterior angles.

In Activity 4 you were asked to calculate the measures of the five exterior angles and then add them. Here is another way of finding the sum without ever finding one particular angle measure.

For a pentagon, the five exterior angles (one at each vertex) form five linear pairs with the five interior angles. The measures of the angles in each linear pair have a sum of 180. This means that the sum of all five pairs is $5(180) = 900$. Because the measures of the interior angles add to $(5 - 2)180$ or 540, what is left is the sum of the exterior angles: $900 - 540 = 360$.

This argument can be extended for any convex n-gon. Consider a single exterior angle at each vertex. Because there are n vertices, these angles are parts of n linear pairs. Linear pairs are supplementary, so each linear pair has a sum of 180. Thus, the sum of all interior and exterior angles (using one exterior angle at each vertex) together is $180n$. The sum of the interior angles in an n-gon is $(n - 2)180$. To get the exterior angles alone, we subtract:

$$180n - (n - 2)180 = 180n - (180n - 360)$$
$$= 180n - 180n + 360$$
$$= 360$$

This proves the following theorem:

Polygon Exterior Angle Theorem

The sum of the measures of the exterior angles of a convex n-gon (one per vertex), is 360.

In other words, the sum of the exterior angles of a convex polygon is a constant! The number of sides does not matter. To help you make sense of this, consider the pentagon shown below at the left. If you view the pentagon from a greater distance (as shown below at the right) you can see the exterior angles begin to resemble the angles around a point, which have a sum of 360.

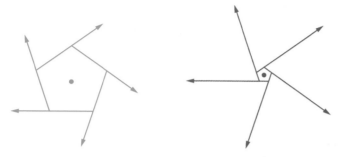

Example 2

Describe how to calculate the measure of one interior angle in a decagon whose angles are all congruent.

Solution 1 Use the Polygon-Sum Theorem. The sum of the measure of the interior angles is (10 - 2)180 = 1440. Dividing this by 10 gives us 144.

Solution 2 Use the Polygon Exterior Angle Theorem. Each exterior angle has measure $\frac{360}{10}$ = 36. Because each interior angle is a supplement of an exterior angle, each interior angle measures 180 − 36 = 144.

Questions

COVERING THE IDEAS

1. Calculate the value of *x* in the figure at the right.

2. **a.** Find the measure of ∠*T* in quadrilateral *TRAP*.

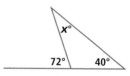

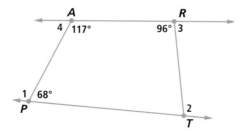

   b. What is m∠1 + m∠2 + m∠3 + m∠4?

3. In the diagram at the right, calculate m∠*F* + m∠*G*.

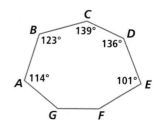

4. Use the figure at the right. Write a proof like that in Guided Example 1 to discover and prove a Pentagon-Sum Theorem.

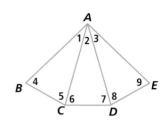

5. If you draw all of the diagonals from one vertex of a convex 30-gon, into how many triangles will these diagonals divide the 30-gon?

6. What is the sum of the measures of the exterior angles (one at each vertex) of any convex hexagon?

7. What is the sum of the measures of the exterior angles (one at each vertex) of any triangle?

8. Find the sum of the measures of the marked angles in the diagram at the right. (In this diagram, the marks do not mean that the angles have equal measure.)

9. A stop sign has the shape of an octagon with 8 congruent angles. What is the measure of each interior angle of a stop sign?

APPLYING THE MATHEMATICS

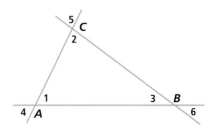

Stop sign in Malay, the official language of Malaysia.

10. In Activity 1 you tore off the corners of convex quadrilateral and convex pentagonal regions and placed the angles of each as adjacent angles to find the sum of their measures. Cut out a larger convex hexagonal region. Tear off its corners and paste each angle on a piece of paper so that each angle you place down is adjacent to the one before it (continuing in the same direction). What happens? Does the result surprise you?

11. Supply the justifications in this proof of the Exterior Angle Theorem for Triangles using the figure at the right.

Given $\triangle ABC$ with exterior angle 4.

Prove $m\angle 4 = m\angle 2 + m\angle 3$.

Proof

Conclusions	Justifications
a. $m\angle 1 + m\angle 2 + m\angle 3 = $ ___?___	a. ___?___
b. $m\angle 1 + m\angle 4 = $ ___?___	b. ___?___
c. $m\angle 1 + m\angle 4 = m\angle 1 + m\angle 2 + m\angle 3$	c. ___?___
d. $m\angle 4 = m\angle 2 + m\angle 3$	d. ___?___

12. If all of the angles of a convex n-gon are congruent, give an expression for the measure of each angle.

13. Use the figure at the right. Find $m\angle 4 + m\angle 5 + m\angle 6$ and justify your answer.

14. The measure of an interior angle of a regular polygon is 5 times the measure of the exterior angle. How many sides does this regular polygon have?

15. An alternative proof of the Polygon-Sum Theorem is as follows: Draw any convex n-gon. Place a point anywhere in the interior of the polygon and call it P. Draw line segments from P to each of the vertices of the polygon.

    a. Given that the polygon has n sides, how many triangles will have P as a vertex?

    b. Write an expression for the sum of the measures of the angles in all of the triangles.

    c. What is the sum of the angles that surround point P? Why?

    d. Use the information from Parts a–c to determine an expression for the sum of the measures of the angles in the n-gon. This expression will probably look a little different from the one stated in the lesson.

    e. Use algebra to show that the expression in the lesson is equivalent to the expression you wrote in Part d.

16. Explain how you could use a protractor and ruler to create a pentagon in which all of the sides are of equal length and all of the angles are of equal measure.

17. Construct a line segment $\overline{AB}$ on your DGS. Construct a regular octagon on which $\overline{AB}$ is a side.

REVIEW

18. **Multiple Choice** Which of the following is not uniquely determined? (**Lesson 5-6**)

    A   a circle on which three given noncollinear points lie

    B   a circle that has a given segment as a diameter

    C   a point that is equidistant from three given noncollinear points

    D   a circle that has a given segment as a radius

19. **Multiple Choice** Which of the following auxiliary figures is always possible to draw? (**Lesson 5-6**)

    A   a diagonal that passes through a given point P in the interior of a polygon

    B   a diameter that passes through a given point Q in the interior of a circle

    C   a line that is parallel to two of the sides of a given parallelogram

    D   a quadrilateral that has four given points in the plane as vertices

20. In the figure at the right, m$\angle A$ = m$\angle CMA$ = x, m$\angle L$ = m$\angle LMC$, and m$\angle L$ = m$\angle LCM$. Find the value of x. (**Lesson 5-6**)

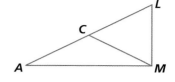

21. Let $P = (4, 0)$ and $Q = (0, 4)$. Find an equation for the line whose points are equidistant from P and Q. (**Lesson 5-4**)

22. Examine the drawing at the right. (**Lesson 5-4**)

    a. Identify all the angles that are alternate interior angles to $\angle 1$.

    b. Are any of the angles you found in Part a congruent to $\angle 1$? Why or why not?

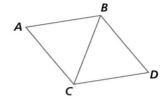

23. In the figure at the right, $\triangle ABC$ and $\triangle DBC$ are equilateral. Prove that $ABDC$ is a rhombus. (A *rhombus* is a quadrilateral whose sides all have equal length.) (**Lesson 5-4**)

24. Emily folded a piece of paper into the shape of a swan. She claimed that the swan and the unfolded paper were congruent because both were made out of the same piece of paper. Is she correct? (**Lesson 5-1**)

25. Suppose k, ℓ, m, and n are four lines, and $k \cap \ell \cap m \cap n = \{P\}$. (**Lesson 4-7**)

    a. What is $r_k \circ r_\ell \circ r_m \circ r_n(P)$?

    b. Explain why $r_k \circ r_\ell \circ r_m \circ r_n$ is not a translation.

    c. What kind of transformation is $r_k \circ r_\ell \circ r_m \circ r_n$?

EXPLORATION

26. Quentin makes the following claim about the Polygon-Sum Theorem to his teacher.

    > A quadrilateral is really just two triangles.
    > A pentagon is really just a quadrilateral and a triangle.
    > A hexagon is really just a pentagon and a triangle.
    > A heptagon is really just a hexagon and a triangle.
    > I can do this forever if you would like.
    > So, you see, all I really need to know is the Triangle-Sum Theorem.

    a. Draw a diagram to explain each of the first four of Quentin's statements.

    b. Is Quentin correct? If so, explain how he would go about calculating the sum of the angles of a 20-gon. Why might he still want to memorize the Polygon-Sum Theorem?

QY ANSWER

$1260°$

Chapter 5 Projects

1 Triangles on Curved Surfaces

A special kind of non-Euclidean geometry is called spherical geometry. In spherical geometry, the points are all on a sphere. To investigate this geometry you will need a sphere that you can write on, a washable marker, 3 or more pieces of string, and a protractor.

a. Choose two points on the sphere. Stretch a piece of string from one point to the other so that the string is as short as possible. This shortest path between two points can be thought of as a "line segment" in spherical geometry. A "line" through two points in spherical geometry is the path on the sphere that contains the shortest path from one point to another. If a line in spherical geometry is drawn, what is the result?

b. Create three points on the sphere. Create a "triangle" and measure its angles to the best of your ability. Do this for several different triangles. Make a conjecture about the sum of the measures of a triangle in spherical geometry.

2 Congruence and Genetics

Human DNA is made up of four basic building blocks. Any two blocks of the same kind should be congruent. Find out what these blocks are. Why do you think it is important that they be congruent? Make a model of a section of human DNA, illustrating the different building blocks.

3 Congruence and Literacy

Johannes Gutenberg is credited with inventing the first printing press—one of the most important inventions ever. The printing press produces many congruent copies of the same image. Find out how the printing press works, and prepare a presentation on this. Include some modern developments of this kind of machine, and how the methods of producing congruent images have changed over time.

4 Proofs as Games

Consider the following game: you are given a certain number (say 15). At each step, you are allowed to add 2 to that number, or to divide it by 3 if it is divisible by 3 (so, if you started at 15, in the second step you could get to 17 or to 5). You are also given a target number (say 10). Your goal is to determine if you can start at the starting number and end at the target number.

a. Is it possible to get from 15 to 9? How many steps would you have to use? Explain your steps.

b. Is it possible to get from 15 to 2? Explain.

c. Invent similar rules for a game, and try to get from a starting number to a target number that you chose.

5 How Long Can Proofs Get?

In this chapter, you encountered proofs that were a few steps long. In mathematics, proofs can get to be extremely long. Use the Internet to find out about a theorem with one of the longest proofs ever. How long was this proof? How long did it take to come up with? How many different people worked on it? Who were these people, and what were their different roles in the proof?

High school student Britney Gallivan (above left) showed that a single piece of paper could be folded in half twelve times. Until her demonstration in 2002, it was thought that seven was the maximum number of times any piece of paper could be folded in half.

6 Star Polygons

Drawn here are two star polygons. (Star polygons are *not* polygons as we have defined "polygon.")

A star polygon S can be formed from any convex n-gon provided n is odd. Draw diagonals from each vertex of P to the two vertices of P that are opposite it. You will wind up with n diagonals that form S. Each pair of consecutive diagonals form one of the n angles of S.

a. Experiment with a DGS to find the sum of the measures of the angles of a star polygon of 5 sides.

b. Experiment with a DGS to find the sum of the measures of the angles of a star polygon of 7 sides.

c. Make a conjecture from your experiments in Parts a and b and try to prove the conjecture.

Chapter 5 Summary and Vocabulary

○ Isometries preserve Angle measure, Betweenness, Collinearity, and Distance (A-B-C-D). As a result, any figure is the same size and shape as its image under an isometry. From this, we define **congruent figures** as any two figures such that there is an isometry that maps one onto the other.

○ Congruence has some properties that are like those of equality: the reflexive, symmetric, and transitive properties. Three other basic properties of congruence are the Segment Congruence Theorem, the Angle Congruence Theorem, and the CPCF Theorem.

○ In a proof of a conditional statement $p \Rightarrow q$, p is the "given," q is the "prove," there is a drawing (when necessary), and a proof to show how q follows from p. Though mathematicians almost always write proofs in paragraphs, in elementary geometry, proofs are commonly written either in two columns or in paragraphs.

○ Most of the proofs in this chapter involve congruence. Common justifications in these proofs are definitions that involve segments of equal length or angles of equal measure, theorems about parallel lines, the congruence theorems, the Transitive Property of Congruence, and properties of reflections.

○ The properties of reflections help to prove that any point on the perpendicular bisector of a segment is equidistant from the endpoints of the segment. This explains why the construction of a circle through three noncollinear points works.

Vocabulary

5-1
*congruent figures
*congruence transformation
*directly congruent
*oppositely congruent

5-2
corresponding parts

5-4
interior angles
exterior angles
alternate interior angles
alternate exterior angles
same-side interior angles

5-6
uniquely determined
auxiliary figure
non-Euclidean geometries
*obtuse triangle
*right triangle
*acute triangle

5-7
exterior angle of a polygon

From the Corresponding Angles Postulate we proved that two lines cut by a transversal are parallel if and only if a pair of **alternate interior angles** are congruent, a pair of **alternate exterior angles** are congruent, or a pair of **same-side interior angles** are supplementary. From this we also can deduce that there is exactly one line parallel to a given line through a point not on the line. This Uniqueness of Parallels Theorem helps deduce the Triangle-Sum Theorem, which is used to prove the Exterior Angle Theorem for Triangles, the Quadrilateral-Sum Theorem, and the formula $S = (n - 2)180$ for the sum, S, of the measures of the interior angles of any convex n-gon. This is then used to deduce the fact that the sum of the measures of one set of exterior angles is 360.

Postulates, Theorems, and Properties

A-B-C-D Theorem (p. 252)
Equivalence Properties of Congruence
  (p. 254)
  Reflexive Property of Congruence
  Symmetric Property of Congruence
  Transitive Property of Congruence
Segment Congruence Theorem
  (p. 258)
Angle Congruence Theorem (p. 258)
Corresponding Parts in Congruent
  Figures (CPCF) Theorem (p. 259)
Parallel Lines Theorem (p. 271)
Alternate Interior Angles Theorem
  (p. 272)
Alternate Exterior Angles Theorem
  (p. 272)

Same-Side Interior Angles Theorem
  (p. 272)
Perpendicular Bisector Theorem
  (p. 277)
Unique Circle Theorem (p. 283)
Uniqueness of Parallels Theorem
  (p. 284)
Triangle-Sum Theorem (p. 285)
Quadrilateral-Sum Theorem (p. 289)
Polygon-Sum Theorem (p. 290)
Exterior Angle Theorem for Triangles
  (p. 290)
Polygon Exterior Angle Theorem
  (p. 291)

Chapter

5 Self-Test

Take this test as you would take a test in class. You will need a calculator. Then use the Selected Answers section in the back of the book to check your work.

In 1 and 2, refer to the figure at the right, in which $RPTMB \cong SFLWB$.

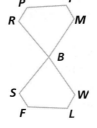

1. $m\angle T = 125$. Which other angle's measure must also equal 125?

2. Which other segment must be congruent to $\overline{SB}$?

3. In the figure below, $m\angle 5 = 37$ and $\ell \parallel m$. What is $m\angle 2$? Justify your answer.

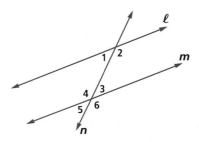

4. Construct an equilateral triangle with perimeter 12 cm.

5. In the figure below, $\overleftrightarrow{CD}$ is the bisector of $\angle ACB$, $m\angle ECA = 2x - 20$, and $m\angle ECB = x + 15$.

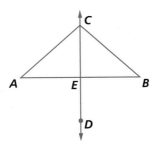

   a. Justify the conclusion $\angle ECA \cong \angle ECB$.

   b. Find the value of x.

6. Write a proof.

   Given $AM = MN$, N is the midpoint of $\overline{MB}$.

   Prove $\overline{AM} \cong \overline{NB}$

7. Two angles of a triangle are congruent. The measure of the third angle is twice the measure of the other two. Find the measure of the third angle.

8. Supply the justification for each conclusion without using the Alternate Interior Angles Theorem.

   Given $\angle 2 \cong \angle 4$, $m \parallel \ell$

   Prove $m \parallel n$

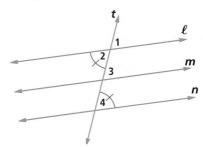

   Proof

Conclusions	Justifications
a. $m \parallel \ell$	a. ___?___
b. $\angle 2 \cong \angle 1$	b. ___?___
c. $\angle 2 \cong \angle 4$	c. ___?___
d. $\angle 1 \cong \angle 4$	d. ___?___
e. $m \parallel n$	e. ___?___

9. All of the interior angles of a decagon are congruent. Find the measure of one of the interior angles.

10. In the figure at the right, $m\angle ABC = m\angle ACB = x$. Find $m\angle EAB$ in terms of x.

11. Write a proof.

    Given $\overleftrightarrow{FT}$ is the perpendicular bisector of $\overline{HE}$ and $\overline{GA}$, $r_{\overline{FT}}(O) = X$.

    Prove $\angle HOG \cong \angle EXA$

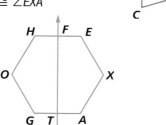

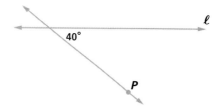

12. Trace $\triangle FST$ and construct a circle through the vertices of the triangle.

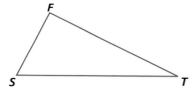

13. Imani was asked to do the following proof.

    Given $\odot O$, $\overline{KI} \cong \overline{IT}$

    Prove $\overline{KE} \cong \overline{ET}$

    She decided that the first step would be to draw the line through I and O and E, so she could show it was a line of symmetry. What is wrong with Imani's reasoning?

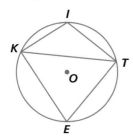

14. In the figure below, Smallville lies 1 km to the north of Littleville.

    a. Trace the figure, and draw the road whose points are equidistant from Smallville and Littleville.

    b. In what directions does this road go?

    • Smallville

    • Littleville

15. Hector has a protractor and wants to make sure the line he draws through P will be parallel to ℓ. Explain what he should do.

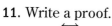

16. In $ABCD$, $\overline{BC} \cong \overline{DA}$, $\angle A \cong \angle B \cong \angle C$. Draw a picture of this situation using tick marks to illustrate the given information.

17. Laura was walking west. She met an obstruction that forced her to turn $30°$ to the right. When she passed the obstruction, she headed west again.

    a. Draw this situation.

    b. How many degrees did she turn the second time? Explain your answer in terms of angles and parallel lines.

Chapter 5 Chapter Review

SKILLS Procedures used to get answers

OBJECTIVE A Identify and determine measures of parts of congruent figures. (Lesson 5-2)

In 1–4, use the figure below, in which *ABCD* ≅ *AEFD*.

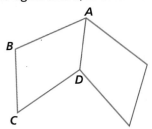

1. Trace the diagram and label points *E* and *F*.

2. Name two pairs of congruent angles.

3. If *FD* = 2″, which other side must have length 2″?

4. Suppose *FD* = 2.6″, *AB* = 3.1″ and *EF* = 2″. Find the perimeter of *ABCDFE*.

In 5–7, refer to the figure below. *O* and *P* are the centers of congruent circles. $\overline{CD}$ and $\overline{LM}$ are diameters.

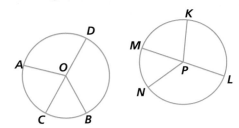

5. If *LM* = 10 cm, find as many other lengths as you can.

6. Suppose ∠*AOB* ≅ ∠*KPN*, *C* corresponds to *M*, m∠*KPN* = 132, and m∠*COA* = 76. Find many other angle measures as you can.

7. Suppose *AO* = 3*x*. What is the length of $\overline{ML}$?

OBJECTIVE B Construct equilateral triangles and construct the circle through three noncollinear points. (Lessons 5-4, 5-5)

8. Trace $\overline{AB}$ below, and construct two equilateral triangles with side length *AB*.

9. Trace the points *P*, *Q*, and *R* below and construct the circle through these three points.

In 10 and 11 refer to the figure below.

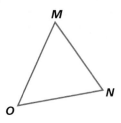

10. a. Trace the triangle, and draw the circle on which its vertices lie.

   b. Draw a point *L* such that *MNOL* is a quadrilateral whose vertices do not all lie on a circle.

11. Construct an equilateral triangle with side length *MO*.

OBJECTIVE C Find lengths and angle measures using properties of the perpendicular bisector and alternate interior angles. (Lessons 5-4, 5-5)

In 12 and 13, refer to the figure below. $\overleftrightarrow{JK}$ is the perpendicular bisector of $\overline{LM}$.

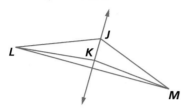

12. If $KL = 7.8$ cm, find MK.

13. Suppose m$\angle MJK = 60$. Find m$\angle MJL$.

In 14–16, refer to the figure at the right, in which $\ell \parallel m$.

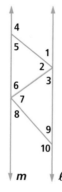

14. If m$\angle 6$ + m$\angle 9 = 94$, find m$\angle 7$.

15. If m$\angle 4 = 130$ and m$\angle 2 = 77$, find m$\angle 7$ + m$\angle 8$.

16. If m$\angle 5$ = m$\angle 6$, what other pairs of angles have equal measure?

OBJECTIVE D Use the Triangle-Sum, Quadrilateral-Sum, and Polygon-Sum Theorems to determine angle measures. (Lessons 5-6, 5-7)

17. The measure of the largest angle of a triangle is twice the measure of the second largest angle and $\frac{10}{3}$ the measure of the smallest angle. Find the measures of the three angles of the triangle.

18. Use the quadrilateral $ABCD$ below. Find the value of y and the measure of each angle.

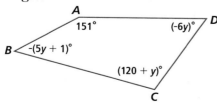

19. a. Find the sum of the measures of the angles of a convex 23-gon.

    b. Find the measure of an interior angle of a regular convex 23-gon.

20. Use the figure below to find the value of x.

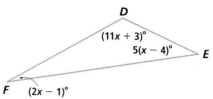

21. Suppose a convex quadrilateral has three angles with measure p and one angle with measure q. A different convex quadrilateral has three angles with measure q and one with measure p. Find the values of p and q.

OBJECTIVE E Use the Exterior Angle Theorem to answer questions about angles of triangles. (Lesson 5-7)

22. Use the figure below to find the values of x and y.

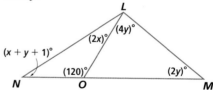

23. In the figure at the right, m$\angle 1 = 110$ and m$\angle 2 = 98$.

    a. Find m$\angle GEH$ + m$\angle EHG$.

    b. Find m$\angle FEH$ + m$\angle EHG$.

    c. Use Parts a and b to find m$\angle FEG$.

    d. Find m$\angle FEG$ using the Triangle-Sum Theorem.

24. In $\triangle ABC$, an exterior angle at $\angle A$ has measure 100, and m$\angle B = 9$m$\angle C$. Find the measures of the angles of the triangle.

PROPERTIES Principles behind the mathematics

OBJECTIVE F Make and justify conclusions about congruent figures.
(Lessons 5-1, 5-2, 5-3)

25. If $QUAR \cong FGON$, list all congruent pairs of segments and angles.

26. Suppose C is on $\overline{AB}$, and $\frac{AC}{CB} = \frac{1}{3}$. Let T be an isometry with $A' = T(A)$, $B' = T(B)$, and $C' = T(C)$. Justify the conclusion $\frac{A'C'}{C'B'} = \frac{1}{3}$.

In 27–28, justify the conclusion.

27. Given: $m\angle PAN = m\angle WFL$
    Conclusion: $\angle PAN \cong \angle WFL$

28. Given: $APEL \cong FRUT$
    Conclusion: The perimeter of $APEL$ is equal to the perimeter of $FRUT$.

29. Suppose $ABCD$ is a quadrilateral, and n is a line so that $r_n(ABCD) = BADC$. Which two segments have n as their perpendicular bisector? Justify your conclusion.

OBJECTIVE G Write proofs using the Transitive Properties of Equality or Congruence. (Lesson 5-4)

30. In the figure at the right, $\angle 1 \cong \angle 2$, and $\angle 2 \cong \angle 3$. Prove that $a \parallel b$.

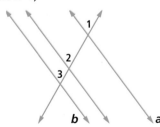

31. In the figure at the right, $m\angle 2 = m\angle 1$. Prove that $\angle 3 \cong \angle 1$.

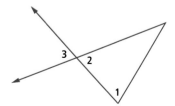

32. In the figure below, $r_{\overline{BC}}(\triangle ABC) = \triangle DBC$, and $r_{\overline{DC}}(\triangle BCD) = \triangle ECD$. Prove that $\triangle ABC \cong \triangle DEC$.

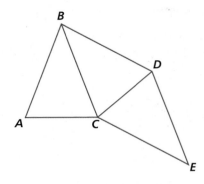

33. Suppose S, T, and R are isometries, and $S \circ T \circ R(\triangle DEF) = \triangle GHI$. Prove that $m\angle D = m\angle G$.

OBJECTIVE H Write proofs using properties of reflections. (Lesson 5-5)

34. In the figure at the right, $r_{\overline{AD}}(\triangle ABC) = \triangle ACB$. Prove that $\overleftrightarrow{AD}$ is the perpendicular bisector of $\overline{BC}$.

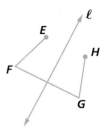

35. In the figure below, $r_m(Q) = X$, $r_m(P) = Y$, $r_m(S) = Z$, and $r_m(R) = W$. Justify the conclusion that $QPSR \cong XYZW$.

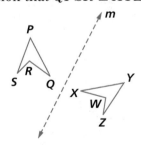

36. In the figure at the right, ℓ is the perpendicular bisector of $\overline{FG}$, and $r_\ell(E) = H$. Prove that $\angle EFG \cong \angle HGF$.

OBJECTIVE I Tell whether auxiliary figures are uniquely determined. (Lesson 5-6)

In 37–41, tell whether the figure described is uniquely determined.

37. Given three points: a triangle on which they all lie

38. Given an angle: a bisector of the angle

39. Given a circle, $\odot O$, and a line, a: a diameter of $\odot O$ that is parallel to a

40. Given three collinear points A, B, and C: a circle where A, B, C all lie on a diameter

41. Given a line, ℓ, and a point, D: a line through D that is perpendicular to ℓ

USES Real-world applications of mathematics

OBJECTIVE J Use the Perpendicular Bisector Theorem and theorems on alternate interior angles in real situations. (Lessons 5-4, 5-5)

42. Car mirrors often have a sign that says: "Objects in the mirror may be closer than they appear." Does the reflection in the car mirror work in the same way as the reflection you studied in this book? Why or why not?

43. The lines in the following figure represent two parallel roads. Suppose a new road is to be built between points M and N. In order to speed up construction, one construction team starts at point M and one at point N. What must be true about the angles $\angle 1$ and $\angle 2$ in order for the road built from M to flow smoothly into the road built from N?

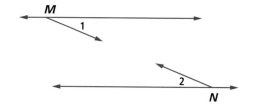

44. A light fixture is suspended from a hook in the ceiling, as shown below. If the lengths of the two cables connecting the endpoints A and B to the hook H are equal, explain why the hook must lie on the perpendicular bisector of $\overline{AB}$.

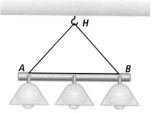

REPRESENTATIONS Pictures, graphs, or objects that illustrate concepts

OBJECTIVE K Draw figures and auxiliary figures to aid proofs. (Lessons 5-6, 5-7)

45. Use tick marks to mark the following information on the figure below.
$RQ = RT$, $OM = OP$, $m\angle TMS = m\angle SMN$, $m\angle TRN = m\angle NRQ$, $m\angle QPR = m\angle OPR$.

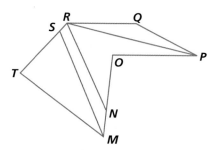

46. Trace the nonagon below and draw diagonals dividing the nonagon into triangles. Explain how to calculate the sum of the interior angles in the nonagon using this construction.

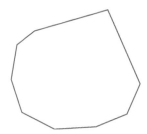

Chapter

6 Polygons and Symmetry

▶ **Contents**

6-1   Reflection Symmetry

6-2   Isosceles Triangles

6-3   Angles Inscribed in Circles

6-4   Types of Quadrilaterals

6-5   Properties of Kites

6-6   Properties of Trapezoids

6-7   Rotation Symmetry

6-8   Regular Polygons

6-9   Frieze Patterns

Nuts, bolts and screws are commonly used to hold things in place. While there are many different sizes of nuts, there are usually only two shapes of nuts that can be purchased at a hardware store, the [regular] hexagonal nut and the square nut, pictured below. The first question that comes to mind is, why are those the only two shapes you can buy?

There are several reasons. First, an adjustable wrench, like the one pictured above, has opposite jaws parallel. The jaws grab the nut and allow a person to turn the nut. That means that to use an adjustable wrench, the opposite sides of a nut must be parallel. It is also reasonable to make the polygon symmetric so its sides can be grabbed from many angles. Squares and regular hexagons have opposite sides that are parallel, while triangles and pentagons do not.

If parallel sides were the only issue, then why not make them all squares or all parallelograms? The answer is that sometimes it is difficult to grab the nut from a specific angle. The hexagon allows six different angles of approach, while the square only allows four. In tight places, it is helpful to have six. But octagons also have opposite sides parallel and they have eight angles of approach, so they would seem to be even better.

The problem with octagons and polygons with more than six sides is that the measure of the angle formed by the sides is too large, so the wrench is likely to slip and might tear off the corners of the nut. That would be a huge problem. This leaves only two shapes workable with a standard wrench.

But there are [regular] pentagonal nuts. You are likely to find them in fire hydrants. A fire department does not want ordinary citizens to be able to open fire hydrants and lower the water pressure. As a result, they often use pentagonal nuts to close the hydrant. Then how do firemen open the hydrant? They have a special wrench that is designed to fit over the nut, not to its side.

Properties of these and other polygons are the subject of this chapter.

Lesson

6-1

Reflection Symmetry

Vocabulary

reflection-symmetric figure

symmetry line

▶ **BIG IDEA** Common basic figures of geometry—segments, angles, and circles—all have reflection symmetry.

In the table below, the letters of the alphabet have been grouped into three categories based on their symmetries. Determine the method used to group them.

Group 1	Group 2	Group 3
F G J L N K P Q Z	**A B C D E M T U V W Y**	**H I O X**

What Is Reflection Symmetry?

When a figure is reflected over a line, there are times when the figure will coincide with its image. Consider the lion face shown at right. When we reflect figure C over line ℓ, the image and the preimage are identical. That is $r_\ell(C) = C$. C is a *reflection-symmetric figure*.

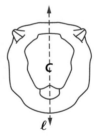

Mental Math

In the diagram below, find

a. $m\angle ICE$.

b. $m\angle ICF$.

c. $m\angle IEG$.

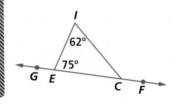

Definition of Reflection-Symmetric Figure, Symmetry Line

A plane figure F is a **reflection-symmetric figure** if and only if there is a line m such that $r_m(F) = F$. The line m is a **symmetry line** for the figure.

Return now to the three groups of letters at the beginning of the lesson. Did you discover how the letters were grouped? Group 1 contains letters that have no reflection symmetry. Group 2 contains letters that have exactly one symmetry line. Group 3 contains letters with two symmetry lines.

 QY

▶ **QY**

Draw all of the symmetry lines for each of the following letters.

E M X H

Applying the Same Reflection Twice

Only if a figure is reflection-symmetric can it be mapped onto itself by a single reflection. But *every* figure can be mapped onto itself by applying the same reflection twice. For example, in the figure at the right, $r_\ell(r_\ell(F)) = F$.

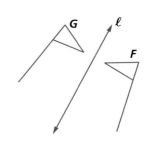

Another way of putting this is to say that whenever $r_\ell(F) = G$, then $r_\ell(G) = F$. We call this property the Flip-Flop Theorem, but you may wish to call it by some other name. Notice that the theorem applies to both points and figures and that after proving the first part of the theorem, we use it as a justification in the proof of the second part.

> **Flip-Flop Theorem**
>
> 1. If F and G are points and $r_\ell(F) = G$, then $r_\ell(G) = F$.
> 2. If F and G are figures and $r_\ell(F) = G$, then $r_\ell(G) = F$.

Given $r_\ell(F) = G$

Prove $r_\ell(G) = F$

Proof of 1 Because F and G are points and it is given that $r_\ell(F) = G$, by the definition of reflection image, ℓ is the perpendicular bisector of $\overline{FG}$. $\overline{FG}$ and $\overline{GF}$ are the same segment, so ℓ is the perpendicular bisector of $\overline{GF}$. Consequently, again using the definition of reflection image, $r_\ell(G) = F$.

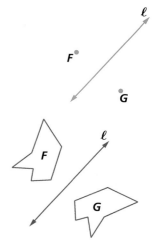

Proof of 2 When F and G are figures, then $r_\ell(F) = G$ means each point on G is the reflection image of a point on F over ℓ, and every point of F has an image on G. Consequently, using Part 1 of this theorem, every point of G has an image on F and every point of G has an image on F. This is what is meant by $r_\ell(G) = F$.

Reflection Symmetry of Segments

How many symmetry lines are there for the segment shown at the right? One obvious symmetry line is the perpendicular bisector of the segment. The not-so-obvious one is the line that contains the segment.

> **Segment Symmetry Theorem**
>
> Every segment has exactly two symmetry lines:
> 1. the line containing the segment, and
> 2. its perpendicular bisector.

Example 1

Write a proof of Part 1 of the Segment Symmetry Theorem.

Solution Let $\overline{AB}$ be any segment. For this proof, we need to show that $\overleftrightarrow{AB}$ is a symmetry line for $\overline{AB}$.

Given: $\overline{AB}$

Prove: $\overleftrightarrow{AB}$ is a symmetry line for $\overline{AB}$.

Proof: A and B are both on $\overline{AB}$, so by the definition of reflection image, $r_{\overleftrightarrow{AB}}(A) = A$ and $r_{\overleftrightarrow{AB}}(B) = B$. Now, by the Figure Transformation Theorem, $r_{\overleftrightarrow{AB}}(\overline{AB}) = \overline{AB}$ because $\overline{AB}$ is determined by its endpoints. Thus, by the definition of a symmetry line, $\overleftrightarrow{AB}$ is a symmetry line for $\overline{AB}$.

The proof of Part 2 is left for you to complete in Question 6.

Reflection Symmetry of Angles

Activity 1

MATERIALS Tracing paper

Trace the angle below and draw its symmetry line.

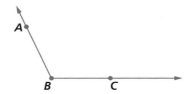

Where did you draw the symmetry line? The correct symmetry line of any angle is the line containing the bisector of the angle. Consider $\angle ABC$ and $\overrightarrow{BD}$ such that $\overrightarrow{BD}$ bisects $\angle ABC$.

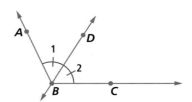

Because $\overrightarrow{BD}$ bisects $\angle ABC$, $m\angle 1 = m\angle 2$. Now reflect $\angle 1$ over $\overleftrightarrow{BD}$. One side of $\angle 1$, $\overleftrightarrow{BD}$, coincides with its image. The other side, $\overrightarrow{BA}$, has $\overrightarrow{BC}$ as its image, because reflections preserve angle measure. Thus, the reflection image of $\angle 1$ is $\angle 2$. This leads to what we call the *Side-Switching Theorem*.

Side-Switching Theorem

If one side of an angle is reflected over the line containing the angle bisector, its image is the other side of the angle.

The Side-Switching Theorem gives every angle its symmetry. Because $r_{\overleftrightarrow{BD}}(\overrightarrow{BA}) = \overrightarrow{BC}$, by the Flip-Flop Theorem, $r_{\overleftrightarrow{BD}}(\overrightarrow{BC}) = \overrightarrow{BA}$. Thus, the sides of the angle reflect onto each other, so the image of $\angle ABC$ is the angle itself. This argument proves the *Angle Symmetry Theorem*.

> ### Angle Symmetry Theorem
>
> The line containing the bisector of an angle is a symmetry line of the angle.

Reflection Symmetry of Circles

Activity 2

Draw a circle with center O and all of its symmetry lines.

It is impossible to draw all of the symmetry lines for a circle because a circle has infinitely many symmetry lines. Any line that passes through the center of the circle is a symmetry line for that circle.

> ### Circle Symmetry Theorem
>
> A circle is reflection-symmetric to any line through its center.

Proof We need to show that the reflection image of any point on the circle also lies on the circle, and that every point on the circle is an image of a point on the circle.

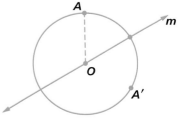

Let A be any point on $\odot O$, let m be any line through the center O, and let $A' = r_m(A)$. O is on m, so $r_m(O) = O$ by the definition of reflection. So $OA = OA'$ because reflections preserve distance. Thus, A' is the same distance from O as A is, so A' is on $\odot O$, by the definition of a circle. This proves that every point on the circle is mapped onto a point on the circle by reflection.

Now we need to show that every point on the circle is the image of a point on the circle. Let B be any point on the circle. Let $r_m(B) = B'$. We know, from the previous paragraph, that B' is a point on the circle. The Flip-Flop Theorem tells us that $r_m(B') = B$. Thus, any point B is the image of a point on the circle. Consequently, $\odot O$ coincides with its image.

Why Are Symmetric Figures Important?

The Reflexive Property of Congruence says that any figure is congruent to itself. Likewise, all of the parts of that figure are congruent to themselves by this reflexive property. These are called *trivial* congruences because they are so obvious.

A reflection-symmetric figure is also congruent to itself. But because a reflection-symmetric figure is its own image under a reflection, it is congruent to itself in a *nontrivial* way. Its parts can correspond to different parts of itself.

Symmetric Figures Theorem

If a figure is symmetric, then any pair of corresponding parts is congruent.

GUIDED

Example 2

The polygon at the right is reflection-symmetric over $\overleftrightarrow{QH}$. This figure is congruent to itself both trivially (where all of the points remain in the same position) and nontrivially (by reflecting the polygon onto itself).

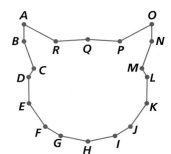

a. Find $r_{\overleftrightarrow{QH}}(ABCDEFGHIJKLMNOPQR)$.
b. Name several congruent angles, sides, and diagonals.

Solution

a. $r_{\overleftrightarrow{QH}}(ABCDEFGHIJKLMNOPQR) = \underline{\quad ? \quad}$

b. Because the figure has symmetry over line $\overleftrightarrow{QH}$, there are 18 pairs of congruent corresponding sides and angles. Some examples are:

$\angle B \cong \angle \underline{\ ?\ }$ $\qquad\qquad$ $\overline{CD} \cong \underline{\ ?\ }$

$\angle J \cong \angle \underline{\ ?\ }$ $\qquad\qquad$ $\overline{MN} \cong \underline{\ ?\ }$

These are true because of the Symmetric Figures Theorem.

There are 67 pairs of diagonals in this figure that are congruent because of the symmetry! Fill in these congruences using the Symmetric Figures Theorem.

$\overline{DJ} \cong \underline{\ ?\ }$ $\qquad\qquad$ $\overline{PM} \cong \underline{\ ?\ }$

Because congruent parts can be found so easily using the Symmetric Figures Theorem, it is very useful to recognize reflection-symmetric figures. For this reason, throughout much of this chapter, polygons are examined for their symmetry.

Questions

COVERING THE IDEAS

1. **Fill in the Blanks** If $r_{\overleftrightarrow{AC}}(ABCD) = ADCB$, then we say that figure $ABCD$ is $\underline{\ ?\ }$ and the line $\overleftrightarrow{AC}$ is the $\underline{\ ?\ }$ for the figure.

2. Draw a quadrilateral $ABCD$ with the property of Question 1.

3. At the beginning of this lesson, letters were categorized by their symmetry. Using the same grouping criteria, do the same for the lowercase letters of the English alphabet.

a b c d e f g h i j k l m
n o p q r s t u v w x y z

4. Below are the 33 letters of the lowercase Russian alphabet.

а б в г д е ё ж з и й к л м н о п
р с т у ф х ц ч ш щ ъ ы ь э ю я

   a. Write the Russian letters that have two symmetry lines.

   b. Write the Russian letters that have exactly one symmetry line.

5. According to the Flip-Flop Theorem, if $r_\ell(\overline{PQ}) = \overline{MN}$, then ___?___.

6. **Fill in the Blanks** Complete this proof of Part 2 of the Segment Symmetry Theorem.

   Given ℓ is the perpendicular bisector of $\overline{AB}$.

   Prove ℓ is a symmetry line for $\overline{AB}$.

   Proof

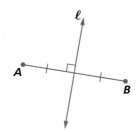

Conclusions	Justifications
a. ℓ is the perpendicular bisector of $\overline{AB}$.	a. Given
b. $r_\ell(A) = $ ___?___ , $r_\ell(B) = $ ___?___	b. Definition of reflection image
c. $r_\ell(\overline{AB}) = $ ___?___ and ___?___	c. ___?___
d. ___?___ is a symmetry line for $\overline{AB}$.	d. Definition of symmetry line

In 7 and 8, describe the symmetry of the figure.

7. $\overrightarrow{AB}$

8. $\overleftrightarrow{CD}$

In 9 and 10, trace the figure and draw all lines that seem to be symmetry lines for the figure. (You might want to check your solutions by folding.)

9.

10.

In **11** and **12**, draw all of the lines that seem to be symmetry lines for the figure.

11.

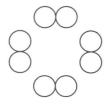

12.

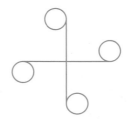

In **13** and **14**, use this fact: If two figures are reflection-symmetric to the same line, then so is their union. Draw all lines of symmetry for the figure.

13. ⊙O ∪ ⊙P

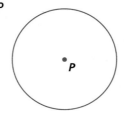

14. ⊙B ∪ ∠ABC

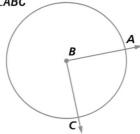

In **15** and **16**, reflect the figure over the line to obtain a hidden word.

15.

16.

17. Many sports are played on fields, courts, or courses that have reflection symmetry.

 a. Name two sports that are played on surfaces that have reflection symmetry. Why do you think symmetry matters in these sports?

 b. Give examples of two sports that occur on surfaces that do not have reflection symmetry. Why do you think symmetry does not matter in these cases?

18. For △ABC, m is a symmetry line.
    a. **Fill in the Blank** $r_m(\triangle ABC) = $ __?__
    b. Name two pairs of angles and two pairs of segments that are congruent by the Symmetric Figures Theorem.

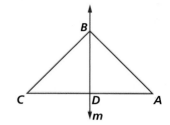

19. In a hexagon, suppose five of the angles have measure x and the sixth has measure 2x + 16. Find x. (**Lesson 5-7**)

20. Let P = (0, 1), Q = (3, 7), M = (5.5, 12), and N = (12.25, 25.5). Is there a point in the plane that is equidistant from P, Q, M, and N? If there is, find it. If not, explain why not. (**Lesson 5-6**)

21. In the figure at the right $\overleftrightarrow{DE} \parallel \overleftrightarrow{AC}$, and m∠B + m∠BDE = 119. Find m∠C. (**Lessons 5-6, 3-6**)

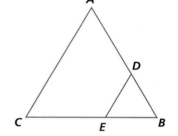

22. If r(ABCD) = EFGH, identify each of the following.
    (**Lessons 5-2, 4-2**)
    a. r(EFGH)
    b. r(C)
    c. r(△ABD)
    d. r(∠GCH)

23. Suppose T is an isometry that preserves orientation, and T ∘ T(P) = P for every point P in the plane. (**Lessons 4-7, 4-6, 4-5**)
    a. Can T be a translation? Why or why not?
    b. Can T be a rotation? If so, give an example of a possible magnitude for T and an impossible magnitude for T. If not, explain why not.

E M H X

24. In this lesson you saw a symmetric face of a lion. Find an image of a human face like Figure I in which the face is looking straight at you. (If you have a digital camera you might be able to use your own face.) Using any graphics software, cut the picture of the face in half like Figure II. Use each half to create a new, perfectly symmetric face by reflecting each half and placing the corresponding images together like Figures III and IV. Do the two new faces look alike? If not, which face did you find most pleasing?

25. The actor Denzel Washington is often viewed as having a very symmetric face. Locate a photograph of him and use the procedure in Question 24 to test this claim.

Lesson

6-2

Isosceles Triangles

Vocabulary

vertex angle of an
isosceles triangle

base of an isosceles triangle

base angles of an
isosceles triangle

median of a triangle

altitude of a triangle

▶ **BIG IDEA** Those triangles that are reflection-symmetric—the isosceles triangles—possess properties that scalene triangles do not possess.

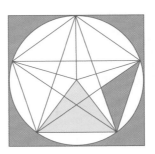

Mental Math

In the figure below,
NICE ≅ *PLAY*.

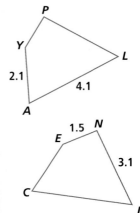

Find:

a. *PL*.

b. *YP*.

c. the perimeter of *NICE*.

Recall that isosceles triangles have at least two sides of equal length. (Triangles that are not isosceles are *scalene*.) Isosceles triangles can be found in rooftops, ice cream cones, and many objects that taper to a point. They are also formed when the endpoints of two radii of a circle are joined.

The angle determined by congruent sides in an isosceles triangle is called the **vertex angle** ($\angle X$ in the figure at the right). The side opposite the vertex angle is called the **base** ($\overline{YZ}$). The other two angles ($\angle Y$ and $\angle Z$) are the **base angles**.

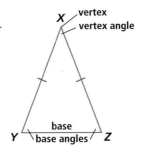

Activity 1

MATERIALS Piece of thin paper

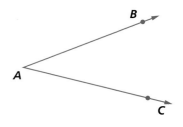

Step 1 With your straightedge, draw an angle and label its vertex A.

Step 2 Identify one side of the angle as $\overrightarrow{AB}$ by marking a point as B.

Step 3 Draw line m, the bisector of angle A, by folding the paper so that the sides of $\angle A$ are on top of one another. Name $C = r_m(B)$.

Step 4 Construct $\overline{BC}$. What type of triangle is $\triangle ABC$? How do you know?

Step 5 How do you know that the fold that goes through point A and the base is a line of symmetry for $\triangle ABC$?

Symmetry of an Isosceles Triangle

The answer to the question in Step 5 of Activity 1 is found in the proof of the following theorem that every isosceles triangle is reflection-symmetric. The proof is long, but the symmetry enables us to deduce rather easily some other properties of isosceles triangles. You should read the proof very slowly and refer to the drawing after every sentence.

Isosceles Triangle Symmetry Theorem

The line containing the bisector of the vertex angle of an isosceles triangle is a symmetry line for the triangle.

Given Isosceles triangle ABC with vertex angle A bisected by line m

Prove m is a symmetry line for $\triangle ABC$.

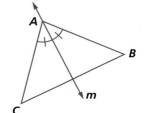

Proof In order to prove that m is a line of symmetry for $\triangle ABC$, we need to show that $r_m(\triangle ABC) = \triangle ACB$. To do this, we need to show that the image of B lies on $\overrightarrow{AC}$ and that its distance from A is AC.

$r_m(A) = A$ by the definition of reflection image. Because m is an angle bisector, when $\overrightarrow{AB}$ is reflected over m, its image is $\overrightarrow{AC}$ by the Side-Switching Theorem. Thus, $r_m(B)$ is on $\overrightarrow{AC}$. Let $B' = r_m(B)$. Because reflections preserve distance, $AB' = AB$. It is given that $\triangle ABC$ is isosceles with vertex angle A, so $AB = AC$. By the Transitive Property of Equality, $AB = AC$. Therefore, B' and C are points on ray $\overrightarrow{AC}$ at the same distance from A, and so $B' = C$. By the Substitution Property, $r_m(B) = C$. By the Flip-Flop Theorem, $r_m(C) = B$. So, by the Figure Reflection Theorem, $r_m(\triangle ABC) = \triangle ACB$, which is the characteristic required for line m to be a symmetry line for the triangle.

Activity 2

MATERIALS DGS

Use a DGS to complete the following construction.

Step 1 Construct a scalene triangle, $\triangle ABC$.

Step 2 Measure AB and AC and keep their measures on the screen.

Step 3 Construct the midpoint of $\overline{BC}$ and label it D.

Step 4 Construct $\overline{AD}$.

(continued on next page)

Step 5 Construct the line ℓ through A and perpendicular to $\overline{BC}$.
Name $E = \ell \cap \overline{BC}$.

Step 6 Construct m, the angle bisector of $\angle BAC$.
Name $F = m \cap BC$.
Your screen should look like the one at the right.

Step 7 Drag point A until $AB = AC$. What happens to $\overline{AD}$, $\overline{AE}$,
and $\overline{AF}$? Then drag A to a different position where
$AB = AC$. Make a conjecture from what you have seen.

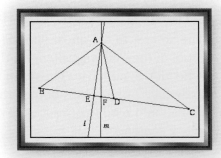

Recall that, in a triangle, the segment from a vertex perpendicular to the line containing the opposite side is called an **altitude of the triangle**. The segment connecting a vertex of the triangle to the midpoint of the opposite side is called a **median of the triangle**. In Activity 2, $\overline{AD}$ is a median and $\overline{AE}$ is an altitude.

Because m is the symmetry line of the isosceles $\triangle ABC$ in the proof of the Isosceles Triangle Symmetry Theorem, it must be the perpendicular bisector of the base $\overline{BC}$. It must also be a median of the triangle because m contains the midpoint of the opposite side. This leads to the following theorem, which is closely related to Activity 1.

Isosceles Triangle Coincidence Theorem

In an isosceles triangle, the bisector of the vertex angle, the perpendicular bisector of the base, and the median to the base determine the same line.

Also, in the previous proof, because m is a symmetry line, $\angle ABC \cong \angle ACB$ by the Symmetric Figures Theorem. This conclusion is an important theorem.

Isosceles Triangle Base Angles Theorem

If two sides of a triangle are congruent, the angles opposite those sides are congruent.

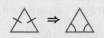

In the language of numbers rather than congruence, the Isosceles Triangle Base Angles Theorem becomes: If two sides of a triangle have equal lengths, then the angles opposite them have equal measures.

The Isosceles Triangle Base Angles Theorem is often useful in proofs in which you know something about segments and need to conclude something related to angles.

Example 1

Given: The figure at the right, with $\overline{AB} \cong \overline{AC}$ and $\angle 1 \cong \angle 2$.

Prove: $\overleftrightarrow{BC} \parallel \overleftrightarrow{ED}$

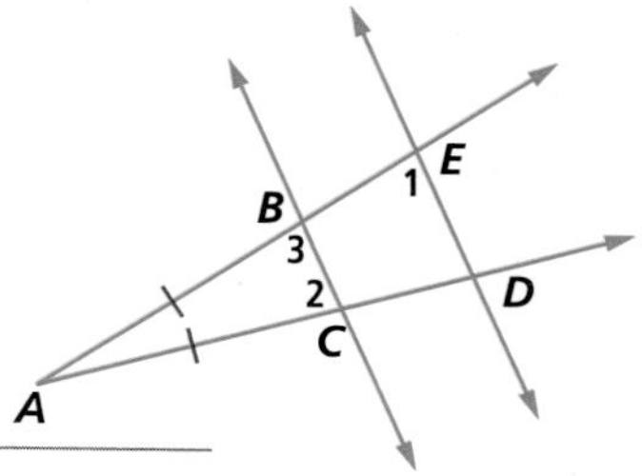

Proof

Conclusions	Justifications
1. $\overline{AB} \cong \overline{AC}$	1. Given
2. $\angle 2 \cong \angle 3$	2. 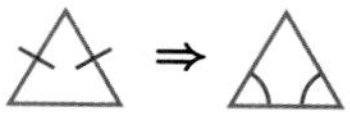 (Isosceles Triangle Base Angles Theorem)
3. $\angle 1 \cong \angle 2$	3. Given
4. $\angle 1 \cong \angle 3$	4. Transitive Property of Congruence
5. $\overleftrightarrow{BC} \parallel \overleftrightarrow{ED}$	5. If corresponding angles are congruent, lines are parallel. (Corresponding Angles Postulate)

The Isosceles Triangle Base Angles Theorem can also be useful in solving problems about angles.

Example 2

Given $\overline{PI} \cong \overline{PT} \cong \overline{PS}$, $\overrightarrow{TP}$ bisects $\angle ITS$, and $m\angle ITS = 148$.

Determine $m\angle IPT$.

Solution $\overrightarrow{TP}$ is an angle bisector, so $m\angle PTI = \underline{\quad?\quad}$.
By the Isosceles Triangle Base Theorem, $m\angle PIT = \underline{\quad?\quad}$.
Because the measures of the angles of a triangle sum to 180,
$m\angle IPT = 180 - \underline{\quad?\quad} = \underline{\quad?\quad}$.

If a triangle is equilateral, then any of its sides can be thought of as the base of an isosceles triangle. You can then prove that all three angles are congruent, and that any of its angle bisectors is a symmetry line for the triangle.

Unequal Sides and Unequal Angles

We have proved that if two sides of a triangle are congruent, then the angles opposite those sides are congruent. A natural question to ask next is, if two sides of a triangle are *not* congruent, can we make a statement about the angles opposite those sides? The following theorem answers that question.

Unequal Sides Theorem

If two sides of a triangle are not congruent, then the angles opposite them are not congruent, and the larger angle is opposite the larger side.

Given $\triangle AXC$ with $AX > AC$

Prove $m\angle ACX > m\angle X$

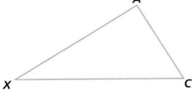

Proof The proof begins by drawing an auxiliary segment to create an isosceles triangle.

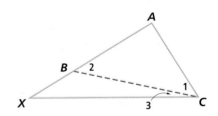

Conclusions	Justifications
1. Identify point B on $\overrightarrow{AX}$ with $AB = AC$, and draw $\overline{BC}$.	1. On a ray, there is exactly one point at a given distance from the endpoint (Number Line Assumption); two points determine a line (Unique Line Assumption).
2. $\triangle ABC$ is isosceles.	2. Definition of isosceles triangle
3. $m\angle 1 + m\angle 3 = m\angle ACX$	3. Angle Addition Postulate
4. $m\angle ACX > m\angle 1$	4. Equation to Inequality Property
5. $m\angle 1 = m\angle 2$	5. Isosceles Triangle Base Angles Theorem

We need to prove that $m\angle ACX$, the angle opposite $\overline{AX}$, is greater than $m\angle X$, the angle opposite $\overline{AC}$. This is done by proving $m\angle ACX > m\angle 1$. Then we note $m\angle 1 = m\angle 2$. Then we prove $m\angle 2 > m\angle X$.

6. $m\angle 2 = m\angle X + m\angle 3$	6. Exterior Angle Theorem for Triangles
7. $m\angle 1 = m\angle X + m\angle 3$	7. Substitution (because $m\angle 1 = m\angle 2$)
8. $m\angle 1 > m\angle X$	8. If $a + b = c$, and a and b are positive, then $c > a$. (Equation to Inequality Property)
9. $m\angle ACX > m\angle X$	9. Transitive Property of Inequality (Steps 4 and 8)

Now suppose that a triangle, $\triangle FST$, has two congruent angles, $\angle F$ and $\angle S$. What can you conclude about its sides? If $ST > FT$, it would follow that $m\angle F > m\angle S$, which is false. If $FT > ST$, it would follow that $m\angle S > m\angle F$. The only choice left is that $FT = FS$. This proves that the converse of the Isosceles Triangle Base Angles Theorem is also true.

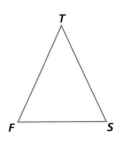

Converse of the Isosceles Triangle Base Angles Theorem

If two angles of a triangle are congruent,
the sides opposite those angles are congruent

A similar argument proves the Unequal Angles Theorem.

Unequal Angles Theorem

If two angles of a triangle are not congruent, the sides opposite
them are not congruent, and the longer side is opposite the
larger angle.

Questions

COVERING THE IDEAS

In 1 and 2, refer to △SOX at the right, in which $\overline{SO} \cong \overline{SX}$.

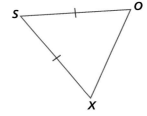

1. In this triangle, identify:
   a. the base.
   b. the vertex angle.
   c. the base angles.
   d. the angles of equal measure.

2. Describe the location of the symmetry line of △SOX.

3. Write in words the theorem represented by these symbols.
   a.
   b.

4. **Multiple Choice** Which are *not* true?
   A Every median of an isosceles triangle is also an angle bisector.
   B If a triangle is isosceles, then the bisector of its vertex angle is
   parallel to the perpendicular bisector of its base.
   C The perpendicular bisector of the base of an isosceles triangle
   contains a median of the triangle.
   D Every angle bisector of an angle of a triangle is also a median
   of the triangle.

5. Use the figure at the right.
   Given $XY = XZ$
   Prove m∠WYQ = m∠RZT

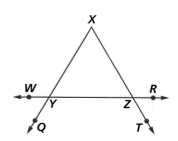

6. Determine whether each statement is *always*, *sometimes but
   not always*, or *never* true.
   a. A triangle that is isosceles is also equilateral.
   b. A triangle that is equilateral is also isosceles.
   c. A triangle that is scalene is also isosceles.

7. How many symmetry lines does each type of triangle have?
   a. equilateral   b. isosceles but not equilateral   c. scalene

8. Why can any side of an equilateral triangle be considered as the base of an isosceles triangle?

9. G. Wilikers, Holly Wood, and Izzie Wright were at three places in their neighborhood as shown at the right. Who is closer to Holly, G. or Izzie?

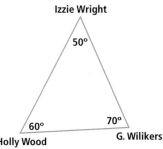

APPLYING THE MATHEMATICS

10. In $\odot O$ at the right, $AB = OB$. Find the measure of $\angle AOB$.

11. In $\triangle SRT$, $SR = ST = 12$ cm. $m\angle S = 60$. Determine RT.

12. Refer to the figure at the right.
    Given $\overline{BE} \parallel \overline{CD}$, $AE = AB$
    Prove $\angle ACD \cong \angle ADC$

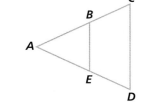

13. In $\triangle ABC$, $m\angle A = 59.7$ and $m\angle B = 60.2$. Order the lengths of the three sides of the triangle from shortest to longest.

14. In nonconvex quadrilateral $RPWT$ at the right, $PW = RW = TW = 18$ in., $m\angle PRW = 40$, and $m\angle WRT = 30$. Determine $m\angle PWT$.

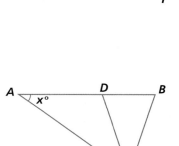

15. Given: $\odot O$; $\overrightarrow{OB}$ and $\overrightarrow{OD}$ bisect $\angle AOC$ and $\angle COE$, respectively; A, O, and E are collinear; and $m\angle BOC = 50$. Find $m\widehat{DE}$.

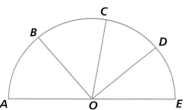

16. In $\triangle BIG$, $BI = 4000$, $IG = 3000$, and $BG = 6000$.
    a. What is the smallest angle of the triangle?
    b. What is the largest angle of the triangle?

17. In the figure at the right, $\triangle ABC$ is isosceles with base $\overline{BC}$ and $\overline{AD} \cong \overline{DC} \cong \overline{BC}$. Find x.

18. The horizontal beam $\overline{RS}$ helps support this roof. To keep the roof from collapsing, the support $\overline{QT}$ is used, where Q is the midpoint of $\overline{RS}$. Parts $\overline{RT}$ and $\overline{ST}$ of the roof are of equal length. Tell whether each statement is true or false.

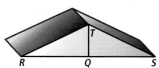

   a. $\overrightarrow{TQ}$ bisects $\angle RTS$.   b. $\overline{TQ} \perp \overline{RS}$.   c. $\overleftrightarrow{QT}$ is vertical.

19. Given: $RS = LS$ and $LP = WP$, as shown at the right. Make and justify some conclusions about angles and lines in the figure.

REVIEW

20. Suppose $\overleftrightarrow{PY}$ is a symmetry line for polygon *POLYGN*, as shown at the right. (**Lesson 6-1**)

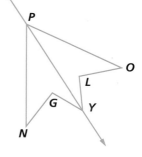

   a. **Fill in the Blank** $r_{\overrightarrow{PY}}(POLYGN) = \underline{\ ?\ }$.
   b. Which sides of *POLYGN* have the same length?
   c. Which angles in the figure have the same measure?

21. **Fill in the Blank** If F and G are figures and $r_m(F) = G$, then $r_m(G) = \underline{\ ?\ }$. (**Lesson 6-1**)

22. The four angles of a quadrilateral have measures x, $2x$, $3x$, and $4x$. If this is possible, determine x and use a protractor to draw such a quadrilateral. If this is not possible, tell why not. (**Lesson 5-7**)

23. Let $A = (-3, 2)$, and $B = (7, 5)$. Let A' and B' be the reflection images of A and B over the y-axis. Find the point of intersection of $\overleftrightarrow{AB}$ and $\overleftrightarrow{A'B'}$. (**Lessons 4-1, 1-5**)

24. Arrange from least to greatest number of sides: rectangle, octagon, pentagon, equilateral triangle, nonagon, heptagon. (**Lesson 2-6**)

EXPLORATION

25. Ima Gardner wants to build a new three-dimensional trellis for her flowers to climb on. She found some directions in an old box. There was only one problem. Some of the directions were torn off. What she had is shown at the right.

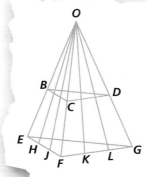

   1. Cut three slats ($\overline{OE}$, $\overline{OF}$, and $\overline{OG}$) the same length.
   2. On each side of the trellis, create isosceles triangles $\triangle OEF$, $\triangle OFG$, and $\triangle OGE$, with base angles of measure 75° and a vertex angle of
   3. Cut and construct supports $\overline{EF}$, $\overline{FG}$, and $\overline{GE}$ to form the base.
   4. Cut and construct supports $\overline{BC}$, $\overline{CD}$, and $\overline{BD}$ that are pa
   5. Divide the angles at O in thirds. Place strings at those angles and attach on base for the flowers to climb.

   a. Finish Step 2 to include the measure of the vertex angle.
   b. What is the measure of the angle between the strings?
   c. Are triangles *OHJ* and *OKL* made by the strings also isosceles? Why or why not?
   d. Using a DGS, find out how long to cut the slats $\overline{EF}$ and $\overline{FG}$ for the bottom of the trellis if the tall slats are 6 feet tall. (Hint: Create a segment 6 cm long for each of the tall supports.)
   e. If m$\angle EFG = 120$, how long should $\overline{EG}$ be?

Lesson

6-3

Angles Inscribed in Circles

Vocabulary

inscribed angle

intercepted arc

▶ **BIG IDEA** The measure of an inscribed angle of a circle is determined completely by the measure of the arc it intercepts.

Two radii in a circle that are not on the same diameter, such as $\overline{PA}$ and $\overline{PC}$, are shown in the figure at the right. Thus $\triangle APC$ is an isosceles triangle. This simple fact, combined with the Isosceles Triangle Base Angles Theorem, helps to prove an important theorem about angles and circles that is not at all obvious.

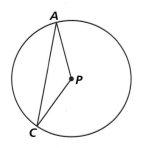

Mental Math

If r_n(*GHOST*) = *ALIVE*, find

a. r_n(*ALIVE*).

b. r_n(*S*).

c. r_n(△*HAV*).

d. r_n(∠*VOE*).

Activity

MATERIALS DGS

Step 1 Recreate the image at the right. Without measuring, compare m∠*A*, m∠*B*, and m∠*C*. Which of these angles do you think has the largest measure? Which has the smallest measure?

Step 2 Replicate this drawing on a DGS. Find the measures of ∠*A*, ∠*B*, and ∠*C*. Does the DGS agree with what you thought in Step 1?

Step 3 Move point *A* around the circle past *B* and *C* all the way to point *D*. Move point *A* onto $\overset{\frown}{ED}$. What happens to m∠*EAD*? Do these results agree with what you found in Step 2?

Step 4 Find m∠*EPD*. How is this angle measure related to what you found in Steps 1–3?

Step 5 Make a conjecture based on your findings in Steps 1–4.

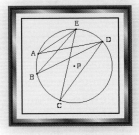

Inscribed Angles in Circles

Angles *A, B,* and *C* in the Activity are said to be *inscribed* in ⊙*P*.

Definition of Inscribed Angle

An angle is an **inscribed angle** in a circle if and only if the angle's vertex is on the circle and each of the angle's sides intersects the circle at a point other than the vertex.

Every inscribed angle intercepts an arc. An angle *intercepts* an arc, forming an **intercepted arc**, if the arc (except for its endpoints) lies in the interior of the angle. The minor arc $\overset{\frown}{AC}$ is an arc intercepted by both the inscribed angle, $\angle ABC$ and the central angle, $\angle APC$. Using what you know about isosceles triangles, it is possible to prove the result you may have conjectured in Step 5 of the Activity.

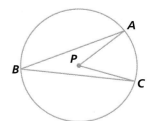

Inscribed Angle Theorem

The measure of an angle inscribed in a circle is half the measure of its intercepted arc.

▶ **READING MATH**

In sports such as football, when a player's pass to a teammate is caught by an opposing player, the opposing player has *intercepted,* or cut off, the pass.

Given $\angle ABC$ inscribed in $\odot O$

Prove $m\angle ABC = \frac{1}{2}m\overset{\frown}{AC}$

Proof The steps of the proof of this theorem depend on the position of the circle's center O relative to the inscribed $\angle ABC$. There are three possibilities. They are referred to as Case I, Case II, and Case III. Each case has its own diagram.

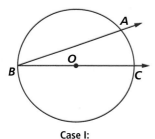

Case I:
O lies on a side of $\angle ABC$.

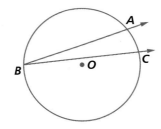

Case II:
O is in the exterior of $\angle ABC$.

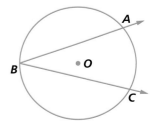

Case III:
O is in the interior of $\angle ABC$.

For all three cases, what is given and what is to be proved are the same.

Case I Draw the auxiliary segment $\overline{OA}$. Because $\triangle AOB$ is isosceles, $m\angle B = m\angle A$ by the Isosceles Triangle Base Angles Theorem. Call this measure x.

By the Exterior Angle Theorem, $m\angle AOC = 2x$.

Because the measure of an arc equals the measure of its central angle, $m\overset{\frown}{AC} = 2x = 2 \cdot m\angle B$.

Solving for $m\angle B$, we see that $m\angle B = \frac{1}{2}m\overset{\frown}{AC}$.

Case I proves that $m\angle B = \frac{1}{2}m\overset{\frown}{AC}$ when one side of $\angle B$ contains the center of the circle. This fact is used in the proofs of the other cases.

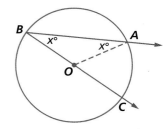

Case II Draw the auxiliary ray $\overrightarrow{BO}$. Label point D as the intersection of $\overrightarrow{BO}$ and the circle.

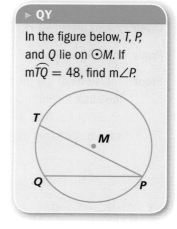

$m\angle ABD = m\angle ABC + m\angle CBD$ Angle Addition Property

$m\angle ABC = m\angle ABD - m\angle CBD$ Addition Property of Equality
(Subtract $m\angle CBD$ from both sides.)

$m\angle ABD = \frac{1}{2}m\widehat{AD} = \frac{1}{2}m\angle AOD$ by Case I.

$m\angle CBD = \frac{1}{2}m\widehat{CD} = \frac{1}{2}m\angle COD$ by Case I.

Substituting,

$m\angle ABC = \frac{1}{2}m\angle AOD - \frac{1}{2}m\angle COD$

$\qquad = \frac{1}{2}(m\angle AOD - m\angle COD)$

$\qquad = \frac{1}{2}m\angle AOC$

$\qquad = \frac{1}{2}m\widehat{AC}$

Case III The proof is like the proof of Case II. You are asked to complete this part of the proof in Question 7.

 QY

▶ **QY**

In the figure below, T, P, and Q lie on $\odot M$. If $m\widehat{TQ} = 48$, find $m\angle P$.

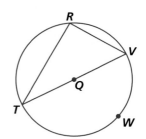

Angles That Intercept Semicircles

Example 1

In the figure at the right, $\overline{TV}$ is a diameter of $\odot O$ and R and W are points on this circle. Explain why the measure of $\angle TRV$ is 90.

Solution $\widehat{TWV}$ is the arc intercepted by $\angle TRV$. $\widehat{TWV}$ is a semicircle, so $m\widehat{TWV} = 180°$. $m\angle TRV = \frac{1}{2}m\widehat{TWV} = \frac{1}{2} \cdot 180 = 90$.

The result of Example 1 was discussed by the Greek mathematician Thales (THAY leez) of Miletas in the 6th century BCE and is often identified with him.

Thales' Theorem

If an inscribed angle intercepts a semicircle, then the angle is a right angle.

This theorem is sometimes stated as "An angle inscribed in a semicircle is a right angle."

The Inscribed Angles Theorem enables you to find measures of many angles in circles. Two examples follow. Others are in Chapter 14.

Example 2

Points B, C, and D lie on $\odot O$. P is a point of $\odot O$ not on $\overgroup{BCD}$ and $m\angle BCD = 140$. Find $m\overgroup{BCD}$.

Solution The arc intercepted by $\angle BCD$ is __?__, so $m\angle BCD = \frac{1}{2}m$__?__.

Then $140° = \frac{1}{2}m$__?__. $m\overgroup{BCD} = 360° - m\overgroup{BPD}$

So $m\overgroup{BCD} = $__?__.

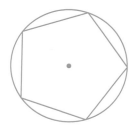

The Inscribed Angle Theorem provides another way to determine the measure of angles in regular polygons.

Example 3

Five points on the circle at the right, equally spaced to form congruent arcs, are connected to form a pentagon. What is the measure of each interior angle of the pentagon?

Solution $360°$ is divided into five equal parts, so the measure of each arc is $\frac{360°}{5} = 72°$. Each angle of the pentagon is an inscribed angle whose intercepted arc is the union of three of the $72°$ degree arcs. So the measure of each angle of the pentagon is $\frac{1}{2}(3 \cdot 72) = 108$.

Check Use the formula for the sum of the measures of the angles of an n-gon: $S = (n - 2)180$. Here $n = 5$, so $S = (5 - 2)180 = 540$. $\frac{540}{5} = 108$. It checks.

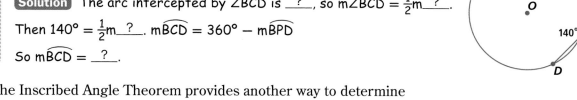

Questions

1. In the figure at the right, $m\angle N = 32$ and $m\angle S = 84$. Find the measure of each arc.
   a. $\overgroup{TN}$ b. $\overgroup{ST}$ c. $\overgroup{NS}$

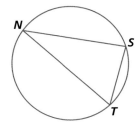

2. In the circle at the right, $m\overgroup{AB} = 52°$ and $m\overgroup{BC} = 84°$.
   a. Find the measures of the three angles of $\triangle ABC$.
   b. Check your answer by using the Triangle-Sum Theorem.

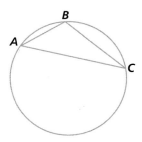

3. In the figure at the right, m∠E = 28, and m∠G = 32.

    a. Find m$\widehat{EF}$, m$\widehat{FG}$, and m$\widehat{EHG}$.

    b. Check your answers by using the fact that the sum of the measures of the arcs of a circle is 360°.

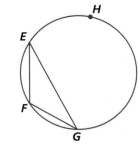

4. In the figure at the right, $\overline{AB}$ contains the center of ⊙O. Explain why ∠ACB ≅ ∠ADB.

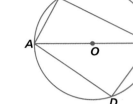

5. Three points, D, E, and F are each 5 cm from point P, m∠DPE = 80, and m∠EPF = 150.

    a. Draw this situation, together with $\overline{DE}$, $\overline{EF}$, $\overline{FD}$, and ⊙(P, PD).

    b. Name two isosceles triangles.

    c. Name one pair of angles that must be congruent by the Isosceles Triangle Base Angles Theorem.

    d. Find m∠EDF.

6. Find all missing angles and arc measures in the figure at the right.

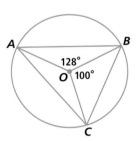

APPLYING THE MATHEMATICS

7. **Fill in the Blanks** Write justifications for the proof of Case III of the Inscribed Angle Theorem.

    Given ∠ABC inscribed in ⊙O.

    Prove m∠ABC = $\frac{1}{2}$m$\widehat{ADC}$

    Proof Draw auxiliary ray $\overrightarrow{BO}$. Name $D = \overrightarrow{BO} \cap ⊙O$.

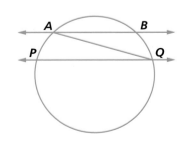

Conclusions		Justifications
a. m∠ABC = m∠ABD + m∠DBC	a.	?
b. = $\frac{1}{2}$m$\widehat{AD}$ + $\frac{1}{2}$m$\widehat{DC}$	b.	?
c. = $\frac{1}{2}$m∠AOD + $\frac{1}{2}$m∠DOC	c.	?
d. = $\frac{1}{2}$(m∠AOD + m∠DOC)	d.	?
e. = $\frac{1}{2}$m∠AOC	e.	?
f. = $\frac{1}{2}$m$\widehat{AC}$	f.	?

8. Prove that the arcs of a circle between two parallel lines have the same measure. That is, given $\overleftrightarrow{AB} \parallel \overleftrightarrow{PQ}$, explain why m$\widehat{AP}$ = m$\widehat{BQ}$. (Hint: Which angles must be congruent as a result of the lines being parallel?)

9. In the figure at the right, $\overline{HW}$ and $\overline{KV}$ are diameters of $\odot Q$ and $m\angle KQW = 78$. Find $m\angle HWV$.

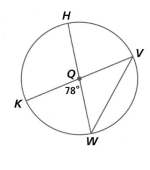

10. Write the justifications in the proof of the following theorem.

Theorem If two inscribed angles in the same circle intercept arcs of the same measure, then the angles are congruent.

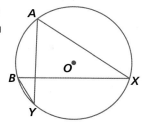

Given $\angle X$ is inscribed in $\odot O$, $\angle X$ intercepts $\widehat{AB}$, $\angle Y$ is inscribed in $\odot O$, and $\angle Y$ intercepts $\widehat{AB}$.

Prove $\angle X \cong \angle Y$

Proof

Conclusions	Justifications
a. $\angle X$ is inscribed in $\odot O$. $\angle X$ intercepts $\widehat{AB}$.	a. ___?___
b. $m\angle X = \frac{1}{2}m\widehat{AB}$	b. ___?___
c. $\angle Y$ is inscribed in $\odot O$. $\angle Y$ intercepts $\widehat{AB}$.	c. ___?___
d. $m\angle Y = \frac{1}{2}m\widehat{AB}$	d. ___?___
e. $m\angle X = m\angle Y$	e. ___?___
f. $\angle X \cong \angle Y$	f. ___?___

11. Identify all angles in the figure at the right that are congruent to the named angle.

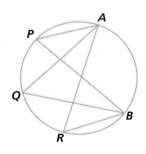

   a. $\angle PAQ$ b. $\angle PBR$ c. $\angle AQB$

12. Ian is practicing kicking soccer balls at a soccer goal. The diagram at the right shows that the scoring angle made by the ball and the two soccer goalposts has a measure of $25°$. Describe all the other locations that Ian could place the soccer ball so that the angle would remain the same and explain why this is so.

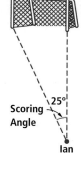

13. Below is a quadrilateral inscribed in a circle. The measures of the arcs are given.

   a. Find $m\angle A$.
   b. Find $m\angle B$.
   c. Find $m\angle C$.
   d. Find $m\angle D$.
   e. How are $\angle B$ and $\angle D$ related?
   f. How are $\angle A$ and $\angle C$ related?

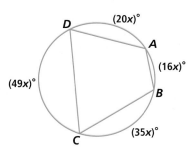

14. a. **Fill in the Blanks** In the figure at the right, quadrilateral
    EFGH is a *cyclic quadrilateral*. This means that all its
    vertices are points on a circle.

    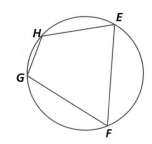

    $m\overset{\frown}{GHE} + m\overset{\frown}{GFE} = \underline{\quad?\quad}$

    $\frac{1}{2}m\overset{\frown}{GHE} + \frac{1}{2}m\overset{\frown}{GFE} = \underline{\quad?\quad}$

    $m\angle F = \frac{1}{2}m\overset{\frown}{GHE}$ and $m\angle H = \frac{1}{2}m\overset{\frown}{GFE}$,

    so $m\angle F + m\angle H = \underline{\quad?\quad}$.

    b. What have you shown about the opposite angles of a cyclic
    quadrilateral?

REVIEW

15. In the figure at the right, $AB = AD$, $BD = DC$, $\overline{AD} \parallel \overline{BC}$,
    and $m\angle C = 70$. Find $m\angle A$. **(Lesson 6-2)**

    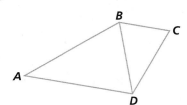

16. Draw an example of a pentagon with no lines of symmetry,
    and one with five lines of symmetry. **(Lesson 6-1)**

17. If the sum of the measures of the angles of a convex
    n-gon is three times the sum of the measures of the angles
    of a hexagon, what is *n*? **(Lesson 5-7)**

18. Explain why the following situation is impossible: $OA = 3$ cm,
    R is a rotation about *O*, $A' = R(A)$, and $AA' = 7$ cm.
    (Lessons 5-1, 1-7)

19. Explain what the Perpendicular to Parallels Theorem states by
    drawing a diagram of the situation. **(Lesson 3-8)**

EXPLORATION

20. Let *A* and *B* be points in the plane. From a point *C*, the field of
    vision occupied by $\overline{AB}$ can be measured by $m\angle ACB$. Call the
    measure of this angle *x*. Suppose $\odot O$ is a circle that contains *A*,
    B, and *C*. Draw a diagram of this situation.

    a. From a point *D* on $\odot O$, what is the angle measure of the field
    of vision taken up by $\overline{AB}$?

    b. For a point *E* in the interior of $\odot O$ and on the same side of
    $\overleftrightarrow{AB}$ as *C*, is the angle measure of the field of vision taken up
    by $\overline{AB}$ greater than, equal to, or less than *x*?

    c. For a point *F* in the exterior of $\odot O$ and on the same side of
    $\overleftrightarrow{AB}$ as *C*, is the angle measure of the field of vision taken up
    by $\overline{AB}$ greater than, equal to, or less than *x*?

    d. What is the set of all points on the same side of $\overleftrightarrow{AB}$ as *C* for
    which the angle measure of the field of vision taken up by
    $\overline{AB}$ is equal to *x*?

Lesson
6-4   Types of Quadrilaterals

► **BIG IDEA**   The most common special types of quadrilaterals are related to each other in a hierarchy that helps in sorting their properties.

The three-sided polygons (triangles) can be classified by the number of congruent sides as *scalene, isosceles,* or *equilateral* or by their largest angle as *acute, right,* or *obtuse.* The four-sided polygons (quadrilaterals) are more diverse, and the classification is more complicated. In this lesson, we give definitions for some special types of quadrilaterals.

The Parallelogram Family

Definition of Parallelogram

A quadrilateral is a **parallelogram** if and only if both pairs of its opposite sides are parallel.

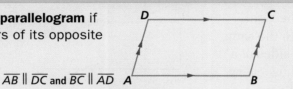

$\overline{AB} \parallel \overline{DC}$ and $\overline{BC} \parallel \overline{AD}$

Definition of Rhombus

A quadrilateral is a **rhombus** if and only if it has four congruent sides.

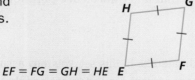

$EF = FG = GH = HE$

Definition of Rectangle

A quadrilateral is a **rectangle** if and only if it has four right angles.

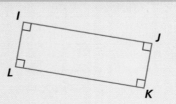

$\angle I$, $\angle J$, $\angle K$, and $\angle L$ are right angles.

Definition of Square

A quadrilateral is a **square** if and only if it has four congruent sides and four right angles.

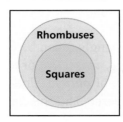

$MN = NO = OP = PM$

$\angle M, \angle N, \angle O,$ and $\angle P$ are right angles.

From their definitions, you can see that a square is a special type of rhombus. (Every square is a rhombus, but not every rhombus is a square.) You can say that the set of squares is a subset of the set of rhombuses. The Venn diagram at the right illustrates the relationship between these sets.

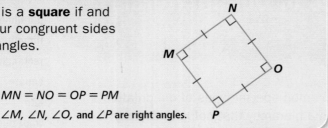

From their definition, it is also the case that every square is a rectangle, as you might expect. This information is summarized in the diagram at the right. This diagram shows part of a hierarchy of quadrilaterals. *If a figure is of any type in a hierarchy, it is also a figure of all types connected above it in the hierarchy.*

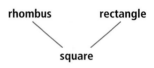

Consider rectangle *IJKL* at the right. Because two perpendiculars to the same line are parallel, the opposite sides of a rectangle must be parallel. Thus, every rectangle is a parallelogram. So we can add *parallelogram* to our hierarchy.

Because all squares are rectangles, and all rectangles are parallelograms, you can conclude that all squares are parallelograms.

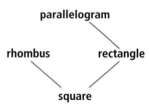

Activity 1

On the next page is a set of quadrilaterals. In this set, you may assume that segments and angles that appear congruent are congruent and that lines appearing to be parallel are parallel.

Step 1 Identify the parallelograms.

Step 2 Identify the rhombuses.

Step 3 Identify the rectangles.

Step 4 Identify the squares.

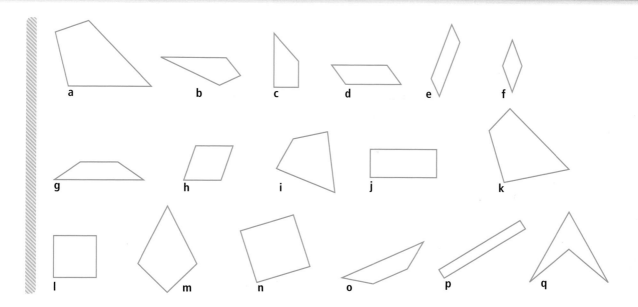

The Kite Family

A fifth type of quadrilateral is formed by the union of two isosceles triangles having the same base, with the base removed. The result is a quadrilateral that resembles a *kite* or arrowhead (or dart).

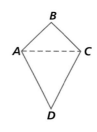

$AB = BC$
$AD = DC$

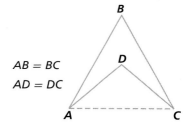

Definition of Kite

A quadrilateral is a **kite** if and only if it has two distinct pairs of consecutive sides of the same length.

Two quadrilaterals are drawn at the right; *EFHG* with exactly three sides congruent, and *IJKL* with four sides congruent (a rhombus). *EFHG* is *not* a kite because the pairs of congruent sides are not distinct, while *IJKL* is a kite because it does have two distinct pairs of consecutive sides the same length.

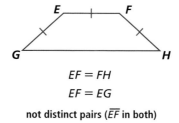

$EF = FH$
$EF = EG$

not distinct pairs ($\overline{EF}$ in both)

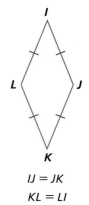

$IJ = JK$
$KL = LI$

two distinct pairs

Thus, every rhombus is a special kite. This information is added to the hierarchy. You also can conclude now that every square is a kite by reading up the hierarchy from square to rhombus to kite.

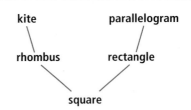

The Trapezoid Family

A figure that is more general than the parallelogram is the trapezoid.

> ### Definition of Trapezoid
>
> A quadrilateral is a **trapezoid** if and only if it has at least one pair of parallel sides.

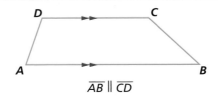

$\overline{AB} \parallel \overline{CD}$

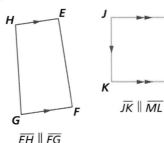

$\overline{JK} \parallel \overline{ML}$

$\overline{EH} \parallel \overline{FG}$

Parallel sides of a trapezoid are called **bases**. At the right, $\overline{AB}$ and $\overline{CD}$ are the bases of trapezoid $ABCD$. $\overline{EH}$ and $\overline{FG}$ are the bases of trapezoid $EFGH$. Either $\overline{JK}$ and $\overline{LM}$ or $\overline{JL}$ and $\overline{KM}$ can be considered as the bases of trapezoid $JKML$. Two consecutive angles whose vertices are endpoints of a single base constitute a pair of **base angles**. For example, $\angle A$ and $\angle B$ are a pair of base angles of trapezoid $ABCD$. Another pair of base angles of trapezoid $ABCD$ is $\angle C$ and $\angle D$. $\angle B$ and $\angle C$ are *not* base angles of trapezoid $ABCD$.

 QY

This terminology enables us to determine a special type of trapezoid, related to isosceles triangles.

> ### Definition of Isosceles Trapezoid
>
> A trapezoid is an **isosceles trapezoid** if and only if it has a pair of base angles equal in measure.

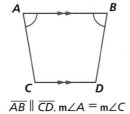

$\overline{AB} \parallel \overline{CD}$, m$\angle A$ = m$\angle C$

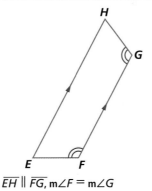

$\overline{EH} \parallel \overline{FG}$, m$\angle F$ = m$\angle G$

> ▶ **QY**
>
> a. Name a pair of base angles of trapezoid *EFGH*.
>
> b. Name a pair of angles of trapezoid *JKML* that are *not* base angles..

> ▶ **READING MATH**
>
> In Britain, the word *trapezium* is used for a trapezoid. But in the United States, *trapezium* means a quadrilateral with *no* parallel sides.

Activity 2

Use the set of quadrilaterals in Activity 1.

Step 1 Identify all the kites.

Step 2 Identify all the trapezoids.

Step 3 Identify all the isosceles trapezoids.

Because a rectangle has opposite sides parallel and all angles congruent, every rectangle is an isosceles trapezoid. Now you can relate all seven types of quadrilaterals we have described in the same hierarchy. We show this with the solid lines below. One other hierarchy relationship will be deduced in the next lesson: Every rhombus is a parallelogram. We show it now with a dashed line. Below is the hierarchy with a drawing of the represented figure surrounding the word.

We call the complete network of connections among the seven types of quadrilaterals the *Quadrilateral Hierarchy Theorem.* This is a way of diagramming the relationships between the types of quadrilaterals that we have discussed. This theorem is very useful because it allows you to relate the properties of specific quadrilaterals.

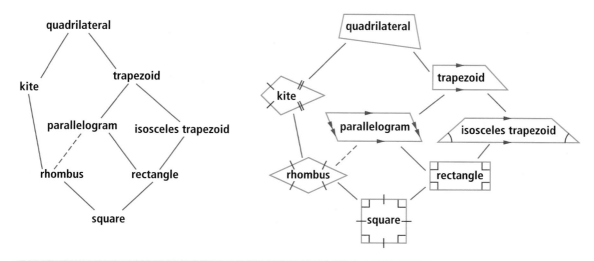

Quadrilateral Hierarchy Theorem

The seven types of quadrilaterals are related as shown in the hierarchy pictured above. Every property true of all figures of one type in the hierarchy is also true of all figures of all types below it to which the first type is connected.

For example, *square* is below *rhombus* in the hierarchy. Thus, any square has all the properties of a rhombus. We say the square *inherits* the properties of a rhombus. Squares and rhombuses are below *kite*. Thus, they inherit all the properties of a kite. In the next few lessons the Quadrilateral Hierarchy Theorem will be used to identify many properties of specific quadrilaterals.

Questions

COVERING THE IDEAS

In 1–5, draw three examples of the quadrilateral named. The first two should be *general* (should not have properties of any figure lower in the hierarchy), while the third example should be *special* (should have a property of a figure lower in the hierarchy.) Using only the definition of the figure, mark parallel segments and mark congruent angles and segments. A sample is given.

Sample: parallelogram

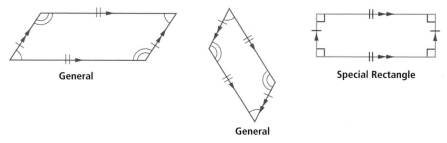

General   Special Rectangle

General

1. rhombus   2. rectangle   3. kite

4. trapezoid   5. isosceles trapezoid

6. Trapezoid $TRAP$ has $\overline{TR} \parallel \overline{AP}$.
   a. Name the bases.
   b. Name two pairs of base angles.

In 7–10, fill in the blanks.

7. The properties of a rectangle found in its definition are that ___?___ and ___?___.

8. The property a rectangle inherits from a parallelogram is that ___?___.

9. The properties of a rhombus found in its definition are that ___?___ and ___?___.

10. The property that a rhombus inherits from a parallelogram is that ___?___.

11. Consider the quadrilaterals at the right. Use only what is marked to fill in the chart, even though other things may look like they are true.

Quadrilateral	Information marked on figure	Name of figure (most specific)
ABCD	?	?
EFGH	?	?
IJKL	?	?
MNOP	?	?

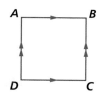

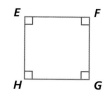

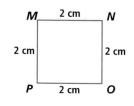

12. What is the most specific quadrilateral name for the outline of each object?

a.

Poster

b.

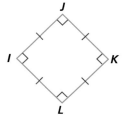

Side of flower bed

c.

Caution sign

APPLYING THE MATHEMATICS

13. In a deck of playing cards, one suit is diamonds. Which type of quadrilateral is shaped like a diamond?

14. **Multiple Choice** A baseball diamond is
    A   a rhombus but not a square.
    B   a square.
    C   both a rhombus and a square.
    D   neither a rhombus nor a square.

15. On a graph, draw a quadrilateral $ABCD$ in which $\overline{AB}$ and $\overline{CD}$ have slope 0, and $\overline{AD}$ and $\overline{BC}$ have slope $\frac{3}{2}$. What kind of quadrilateral have you drawn?

16. Let E be the set of rectangles and H be the set of rhombuses. Describe $E \cap H$.

17. In the figure at the right, $\odot D \cap \odot A = \{B, C\}$.
    a. Write a justification that $AB = AC$.
    b. Write a justification that $CD = BD$.
    c. What kind of quadrilateral is $ABDC$?

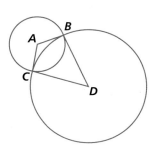

Types of Quadrilaterals   **337**

18. Complete this proof using the figure at the right.

    Given $\overline{AC} \parallel \overline{ED}$, $\overline{BD} \parallel \overline{AE}$, and $\overline{BD} \cong \overline{CD}$.

    Prove *ACDE* is an isosceles trapezoid.

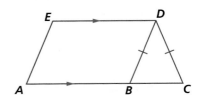

Conclusions	Justifications
a. $\overline{BD} \parallel \overline{AE}$	a. Given
b. $\angle DBC \cong \angle A$	b. ___?___
c. $\overline{BD} \cong \overline{CD}$	c. ___?___
d. $\angle DCB \cong \angle DBC$	d. ___?___
e. $\angle DCB \cong \angle A$	e. ___?___
f. *ACDE* is an isosceles trapezoid.	f. ___?___

19. Which types of quadrilaterals in the hierarchy can be nonconvex? Support your answer with drawings.

REVIEW

20. Prove the following statement: If the vertices of an equilateral triangle lie on a circle, then they divide the circle into three arcs of equal measure. (**Lesson 6-3**)

21. One angle of an isosceles triangle has measure 48. Give two possible pairs of measures for the other two angles. (**Lesson 6-2**)

22. Trace the diagram of the basketball court at the right and draw the lines of symmetry. (**Lesson 6-1**)

23. Suppose all of the angles in a convex octagon are congruent, and an exterior angle is drawn at each vertex. What is the sum of all but one of the exterior angle measures? (**Lesson 5-7**)

24. Which of the four properties mentioned in the A-B-C-D Theorem do size transformations preserve? (**Lessons 5-1, 3-7**)

25. Lisa walked x miles east and y miles north, and ended at a destination that was z miles from her starting point. (**Lesson 1-7**)

    a. Draw a diagram of this situation.

    b. Explain why $x < z$ and $y < z$.

This basketball court is one used by the Federation of International Basketball Associations (FIBA) in international play.

EXPLORATION

26. Biologists classify living things in a hierarchy. Show a hierarchy containing the following terms: *human, dog, animal, mammal, organism, primate, chimpanzee, lion, feline, plant, platypus, whale, duck.*

QY ANSWERS

a. $\angle H$ and $\angle E$; $\angle F$ and $\angle G$

b. $\angle L$ and $\angle K$; $\angle J$ and $\angle M$

Lesson 6-5

Properties of Kites

Vocabulary

ends of a kite

symmetry diagonal

> ▶ **BIG IDEA**   Kites include rhombuses, and rhombuses include squares, so all properties of kites are properties of these other figures as well.

Lockheed Martin's F-117A Nighthawk was the first airplane design whose shape and form contributed to its stealth. The angles between surfaces actually scatter and deflect radar signals making it hard to detect in the air. A view from above reveals many quadrilaterals from the hierarchy. One of its most obvious features is its symmetry. This lesson focuses on kites such as *FAST,* which is outlined on the body of the airplane. The symmetry of kites can be used to deduce some of the characteristics of this quadrilateral.

Mental Math

$\overleftrightarrow{VU}$ **is a symmetry line in the figure below.**

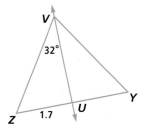

Find:

a. m∠Z.

b. m∠ZUV.

c. ZY.

Constructing a Kite

Because kites must have two distinct pairs of congruent sides, one way to construct a kite precisely is with circles, as in the Activity below.

Activity

MATERIALS DGS

Step 1   Construct ⊙A, point B on ⊙A, and point C not on ⊙A.

Step 2   Construct ⊙C with radius CB.

(continued on next page)

Step 3 Let *D* be the other point of intersection of ⊙*C* and ⊙*A*.

Step 4 Construct polygon *ABCD*.

Step 5 After Step 4, your construction should look like the one at the right. Drag point *C*. For most of the positions of *C*, *ABCD* is a kite. Why?

Step 6 For what positions of *C* is *ABCD* a nonconvex kite?

Step 7 For what positions of *C* is *ABCD* not a kite?

Step 8 Make *ABCD* a convex kite. Construct diagonals $\overline{AC}$ and $\overline{BD}$. Name the point where these diagonals intersect point *E*. Experiment with your sketch to come up with a relationship between the diagonals.

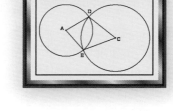

The Activity shows that one way to construct a kite starts with intersecting circles. A kite can also be constructed by taking the union of two isosceles triangles and removing the common side. Points *A* and *C* are then *ends* of the kite. The **ends of a kite** are the common endpoints of its congruent sides.

Reflection Symmetry of a Kite

Because kites can be described using isosceles triangles, many of the theorems from Lesson 6-2 apply to kites.

> ### Kite Symmetry Theorem
>
> The line containing the ends of a kite is a symmetry line for the kite.

Given *ABCD* is a kite with ends *B* and *D*.

Prove $\overleftrightarrow{BD}$ is a symmetry line for *ABCD*.

Drawing

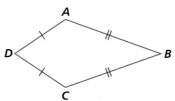 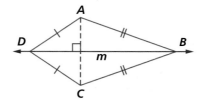

Proof It is given that *ABCD* is a kite with ends *B* and *D*. Construct the auxiliary segment $\overline{AC}$. $\overline{AC}$ has exactly one perpendicular bisector; construct it and call it *m*.

Because m is the perpendicular bisector of $\overline{AC}$, $r_m(A) = C$ and $r_m(C) = A$ by the definition of reflection. B and D are the ends of the kite, so $AB = BC$ and $AD = DC$. So $\triangle ABC$ and $\triangle ADC$ are isosceles triangles. By the Isosceles Triangle Symmetry Theorem, the perpendicular bisector of the base also bisects the vertex angle, which means it contains the vertex.

Thus, the line m contains B and D, so we can name it $\overleftrightarrow{BD}$. By the definition of reflection, $r_{\overleftrightarrow{BD}}(B) = B$ and $r_{\overleftrightarrow{BD}}(D) = D$. By the Figure Transformation Theorem, $r_{\overleftrightarrow{BD}}(ABCD) = CBAD$, which means that $\overleftrightarrow{BD}$ is a symmetry line of $ABCD$.

The diagonal of a kite that is on the symmetry line and contains the ends of the kite is the **symmetry diagonal** of the kite. The above proof has also demonstrated the following theorem.

Kite Diagonal Theorem

The symmetry line of a kite is the perpendicular bisector of the other diagonal and bisects the two angles at the ends of the kite.

GUIDED

Example 1

Given kite *KITE* with ends K and T. If $EM = 8$, $m\angle EKT = 40$, and $m\angle ITK = 23.5$, find as many other lengths and angle measures as you can.

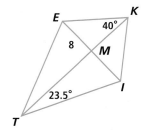

Solution $\overline{KT}$ is a symmetry diagonal for the kite, so $r_{\overleftrightarrow{KT}}(\angle EKT) = \angle IKT$ and $m\angle IKT = \underline{\quad?\quad}$. Also, $r_{\overleftrightarrow{KT}}(\angle ITK) = \angle ETK$, so $m\angle ETK = \underline{\quad?\quad}$.

By the Angle Addition Assumption, $m\angle ETI = 47$ and $m\angle EKI = \underline{\quad?\quad}$. Because KT is the perpendicular bisector of EI, $MI = \underline{\quad?\quad}$ and $EI = \underline{\quad?\quad}$.

The four angles with vertex M each have measure 90. So, from the Triangle-Sum Theorem, $m\angle TIE = \underline{\quad?\quad}$ and $m\angle KEI = \underline{\quad?\quad}$.

Then, because base angles of isosceles triangles are congruent, $m\angle \underline{\quad?\quad} = \underline{\quad?\quad}$ and $m\angle \underline{\quad?\quad} = \underline{\quad?\quad}$. By the Angle Addition Assumption, $m\angle TEK = m\angle TIK = 116.5$.

Symmetry of Rhombuses

The Kite Diagonal Theorem applies to rhombuses and squares, as well as to kites, because they are below kites in the quadrilateral hierarchy. When a kite is a rhombus, any of the opposite vertices are a pair of ends. Thus, a rhombus has two symmetry diagonals. Consequently, the diagonals of a rhombus are perpendicular, and they bisect each other.

> **Rhombus Diagonal Theorem**
>
> Each diagonal of a rhombus is a symmetry line of the rhombus and the perpendicular bisector of the other diagonal.

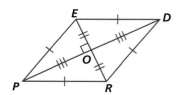

In rhombus $PEDR$ at the right, $\overline{PD}$ is the perpendicular bisector of $\overline{ER}$ and $\overline{ER}$ is the perpendicular bisector of $\overline{PD}$. Because of this and due to the symmetry of any rhombus, many angles in $PEDR$ have the same measure, and many segments have the same length. For instance, O is the midpoint of both $\overline{PD}$ and $\overline{ER}$. Also, because of the two symmetry lines, opposite angles of a rhombus are congruent. That is, $\angle PED \cong \angle PRD$ and $\angle EDR \cong \angle EPR$.

> ### Example 2
> In rhombus $PEDR$, if m$\angle RED = 52$, find each angle measure.
>
> a. m$\angle REP$ b. m$\angle DRE$ c. m$\angle ERP$ d. m$\angle ODE$
>
> **Solution**
>
> a. Each diagonal is an angle bisector of the vertex angles. Thus, m$\angle REP = 52$.
>
> b. $\triangle DER$ is isosceles with base $\overline{ER}$. So, m$\angle DER = $ m$\angle DRE = 52$.
>
> c. Because $\overleftrightarrow{RE}$ bisects $\angle PRD$, m$\angle ERP = 52$.
>
> d. $\overline{PD} \perp \overline{ER}$, so $\angle EOD$ is a right angle. By the Triangle-Sum Theorem, m$\angle ODE = 38$.

Completing the Hierarchy of Quadrilaterals

Notice in Example 2 that there are two pairs of congruent alternate interior angles: $\angle DER$ and $\angle PRE$, and $\angle ERD$ and $\angle REP$. From the first pair, $\overline{ED} \parallel \overline{PR}$, and from the second pair, $\overline{PE} \parallel \overline{RD}$. This allows us to prove the final connection in our hierarchy of quadrilaterals. That is, if a quadrilateral is a rhombus, then it is a parallelogram.

Example 3

Prove that if a quadrilateral is a rhombus, then it is a parallelogram.

Given *RHOM* is a rhombus.

Prove *RHOM* is a parallelogram.

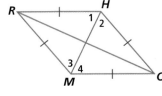

Proof

Conclusions	Justifications
1. *RHOM* is a rhombus.	1. Given
2. $\overline{RO}$ and $\overline{HM}$ are symmetry diagonals of *RHOM*.	2. ___?___
3. $\angle 1 \cong \angle 2$ and $\angle 4 \cong \angle 3$.	3. Symmetric Figures Theorem
4. $\angle 2 \cong \angle 4$	4. ___?___
5. $\angle 1 \cong \angle 4$ and $\angle 2 \cong \angle 3$.	5. ___?___
6. $\overline{RH} \parallel \overline{OM}$ and $\overline{MR} \parallel \overline{OH}$.	6. ___?___
7. *RHOM* is a parallelogram.	7. ___?___

Questions

1. Which quadrilaterals in the hierarchy of quadrilaterals inherit characteristics from kites?

2. **True or False** Every kite is also a rhombus.

3. In kite *BELO* at the right, $BE = BO$ and $EL = OL$.
   a. Name the ends of *BELO*.
   b. Name its symmetry line.
   c. **Fill in the Blank** $r_{\overleftrightarrow{BL}}(\angle BEL) = $ ___?___

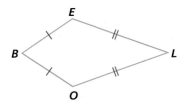

4. Explain why, from their definitions, every square is a kite.

5. a. In the picture of an F-117A at the right, if m$\angle B = 20$, what other angle measure do you know?
   b. What is the symmetry line for the airplane?

6. Draw a nonconvex kite and mark the diagram to show how the Kite Diagonal Theorem applies to nonconvex kites.

7. **Fill in the Blanks** Complete the following proof that the opposite angles of a rhombus *ABCD* are equal in measure.

   The reflection image of $\angle ADC$ over $\overleftrightarrow{AC}$ is ___?___, so m$\angle ADC = $ ___?___ because reflections preserve ___?___. Similarly, the reflection image of $\angle DCB$ over ___?___ is ___?___, so m$\angle DCB = $ ___?___.

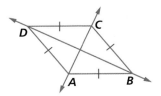

8. Suppose that P and Q are points on the perpendicular bisector of $\overline{XY}$. Prove that $PYQX$ is a kite.

9. In the figure at the right, $\overleftrightarrow{AC}$ bisects both $\angle BCD$ and $\angle BAD$.

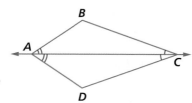

   a. Explain why D must be the reflection image of B over $\overleftrightarrow{AC}$.
   b. Use Part a to prove that $ABCD$ is a kite.

10. In kite $KITE$, $\overline{KT}$ is the symmetry diagonal, T is on the x-axis, $K = (-5, 0)$, and $I = (0, 9)$.

   a. What are the coordinates of point E?
   b. What are the possible coordinates of point T?
   c. For what coordinates of point T is the kite convex?

REVIEW

11. Name all types of quadrilaterals in the quadrilateral hierarchy that are isosceles trapezoids. (**Lesson 6-4**)

12. **True or False** Every square is a trapezoid. (**Lesson 6-4**)

13. In $\odot O$ at the right, $m\angle ABC = 27$ and $m\angle BAD = 63$. Prove that $m\angle ABC = m\angle ABD$. (**Lesson 6-3**)

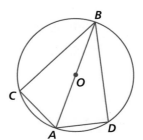

14. Suppose that $\overleftrightarrow{AD}$ is the symmetry line of $\angle BAC$, where D lies in the interior of $\angle BAC$. Let $\overleftrightarrow{AE}$ be the symmetry line of $\angle DAB$, where E lies in the interior of $\angle DAB$. (**Lesson 6-1**)

   a. Draw a picture of this situation.
   b. How is $m\angle EAB$ related to $m\angle BAC$?

15. Is there exactly one circle through any three points? Explain why or why not. (**Lesson 5-6**)

16. The figure at the right shows the hierarchy of quadrilaterals, drawn as a network. Is this network traversable? If not, explain why not. If so, describe a possible path that traverses the network. (**Lesson 1-3**)

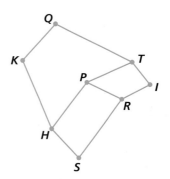

EXPLORATION

17. The name "kite" is given to the quadrilateral studied in this lesson because it resembles a real-life kite. Why do you think that many kites in real life are shaped this way?

Lesson
6-6
Properties of Trapezoids

Vocabulary

corollary

> ▶ **BIG IDEA**   Trapezoids include parallelograms, and parallelograms include rectangles, rhombuses, and squares, so all properties of trapezoids are properties of these other figures as well.

The John Hancock Center in Chicago is one of the tallest buildings in the United States, at 1127 feet. It was the first skyscraper to be built with exterior tube technology. In this design, the exterior shell replaces the old method of building columns up the middle of the structure. This method required 50% less steel than previous methods, and resisted the force of the wind by distributing the force onto all four faces of the building.

Mental Math

△*PRE* is inscribed in ⊙*O*.

Find
a. m$\widehat{EP}$.
b. m$\widehat{ER}$.
c. m$\widehat{RP}$.

The frame of each face consists of six isosceles trapezoids with their diagonals. It uses isosceles trapezoids in much the same way chairs do: the angled sides help distribute the weight over a larger base making it a more stable structure. Six hundred years ago, the Inca used trapezoids in their buildings for the same reason. (See Question 2.) In this lesson, you will learn more about trapezoids.

Angles in Trapezoids

Because trapezoids must have at least one pair of parallel sides, the theorems you studied in Chapters 3 and 5 relating parallel lines and angles can be used to find information about their angles.

GUIDED

Example 1

Consider the figure at the right, in which $\overleftrightarrow{DC} \parallel \overleftrightarrow{AB}$. Fill in the missing justifications.

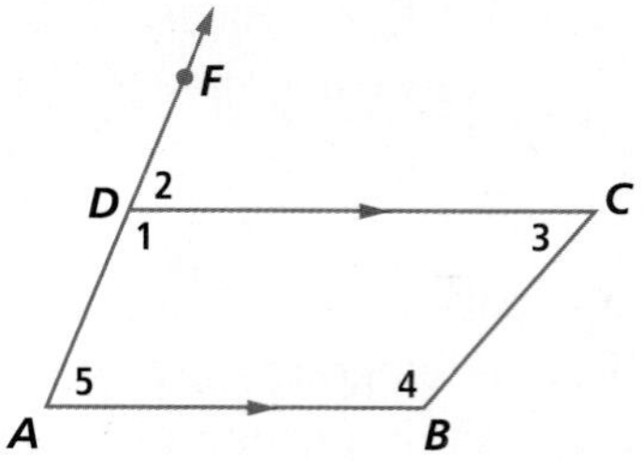

Solution

Conclusions	Justifications
1. $m\angle 1 + m\angle 2 = 180$	1. ___?___
2. $m\angle 2 = m\angle 5$	2. ___?___
3. $m\angle 1 + m\angle 5 = 180$	3. Substitution Property of Equality (Steps 1 and 2)
4. $\angle 1$ and $\angle 5$ are supplementary.	4. definition of supplementary angles

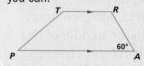

This argument could be repeated with $\angle 3$ and $\angle 4$ and with any trapezoid. The result is the following theorem.

Trapezoid Angle Theorem

In a trapezoid, consecutive angles between a pair of parallel sides are supplementary.

 QY

▶ QY

In trapezoid *TRAP* below, $\overline{TR} \parallel \overline{AP}$. If $m\angle A = 60$, find the measures of as many other angles as you can.

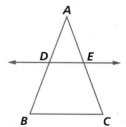

Symmetry of Isosceles Trapezoids

Isosceles trapezoids are related to isosceles triangles in a way that is shown at the right. Let $\triangle ABC$ be isosceles with vertex angle A and base $\overline{BC}$. Draw $\overleftrightarrow{DE} \parallel \overleftrightarrow{BC}$. It is possible to prove that $DECB$ is an isosceles trapezoid. Because an isosceles triangle is reflection-symmetric, you would expect that an isosceles trapezoid is symmetric as well.

Isosceles Trapezoid Symmetry Theorem

The perpendicular bisector of one base of an isosceles trapezoid is the perpendicular bisector of the other base and a symmetry line for the trapezoid.

Given Isosceles trapezoid *ZOID* with $m\angle I = m\angle D$; *m* is the perpendicular bisector of $\overline{DI}$.

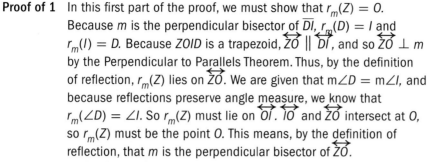

Prove 1. *m* is the perpendicular bisector of $\overline{ZO}$.
2. *m* is a symmetry line for *ZOID*.

Proof of 1 In this first part of the proof, we must show that $r_m(Z) = O$. Because *m* is the perpendicular bisector of $\overline{DI}$, $r_m(D) = I$ and $r_m(I) = D$. Because *ZOID* is a trapezoid, $\overleftrightarrow{ZO} \parallel \overleftrightarrow{DI}$, and so $\overleftrightarrow{ZO} \perp m$ by the Perpendicular to Parallels Theorem. Thus, by the definition of reflection, $r_m(Z)$ lies on $\overleftrightarrow{ZO}$. We are given that $m\angle D = m\angle I$, and because reflections preserve angle measure, we know that $r_m(\angle D) = \angle I$. So $r_m(Z)$ must lie on $\overleftrightarrow{OI}$. $\overleftrightarrow{IO}$ and $\overleftrightarrow{ZO}$ intersect at *O*, so $r_m(Z)$ must be the point *O*. This means, by the definition of reflection, that *m* is the perpendicular bisector of $\overleftrightarrow{ZO}$.

Proof of 2 In this part of the proof, we need to show that $r_m(ZOID) = OZDI$. By the Flip-Flop Theorem, $r_m(O) = Z$. Thus, by the Figure Transformation Theorem, $r_m(ZOID) = OZDI$, and *m* is a symmetry line of *ZOID*.

The Isosceles Trapezoid Symmetry Theorem has a *corollary* (CORE uh lair ee). A **corollary** is an immediate consequence of a theorem. By the Symmetric Figures Theorem, this proves the following theorem.

> ### Isosceles Trapezoid Theorem
>
> In an isosceles trapezoid, nonbase sides are congruent.

Example 2

WXYZ at the right is an isosceles trapezoid with bases $\overline{WZ}$ and $\overline{XY}$.
Find as many lengths and angle measures as you can.

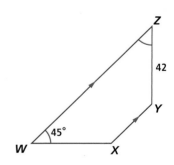

Solution From the Isosceles Trapezoid Theorem, $WX = ZY$, so $WX = 42$. By the Trapezoid Angle Theorem, $m\angle X = 180 - m\angle W = 135 = m\angle Y$. From the definition of isosceles trapezoid, $m\angle W = m\angle Z = 45$.

Rectangles

A rectangle can be considered an isosceles trapezoid in two ways. Either pair of parallel sides can be the bases. This yields another corollary of the Isosceles Trapezoid Symmetry Theorem.

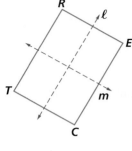

Rectangle Symmetry Theorem

The perpendicular bisectors of the sides of a rectangle are symmetry lines for the rectangle.

Rectangle *RECT* and its two symmetry lines ℓ and *m* are pictured at the right.

Questions

COVERING THE IDEAS

1. To which quadrilaterals on the quadrilateral hierarchy does the Trapezoid Angle Theorem apply?

2. At the right is a photograph of a doorway in the extraordinary complex known as Machu Picchu, built by the Inca in the 15th century. Like many doorways in this complex, it is in the shape of a trapezoid because of the strength of this figure. We have named the trapezoid *INCA*, with $\overline{IN} \parallel \overline{AC}$. If m∠*A* = 84 and m∠*N* = 94, find m∠*I* and m∠*C*.

3. In trapezoid *ABCD* at the right, $\overline{AB} \parallel \overline{DC}$, m∠*A* = 90 and m∠*C* = 30.
   a. Name the bases.
   b. Name a pair of base angles.
   c. Find the measure of as many angles in *ABCD* as you can.

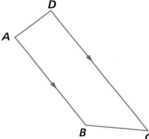

4. Trace the isosceles trapezoid *ROUN* at the right.
   a. Construct the symmetry line for *ROUN*.
   b. Find the center of a circle that contains all four vertices of this trapezoid.
   c. Explain how you found this center.

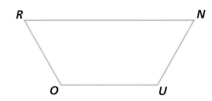

5. In trapezoid *PEZO*, $\overleftrightarrow{OP} \parallel \overleftrightarrow{EZ}$.
   a. Name all pairs of supplementary angles.
   b. Name all pairs of congruent angles.

6. Answer Question 5 if *PEZO* is an isosceles trapezoid.

7. Trace rectangle *RECT* at the right and draw its symmetry lines.

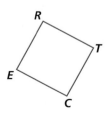

APPLYING THE MATHEMATICS

8. **Given** △*ISO* at the right is isosceles with *IS* = *OS* and $\overline{DZ} \parallel \overline{IO}$.
Supply each justification in this proof that *ZOID* is an isosceles trapezoid.

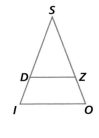

Conclusions	Justifications
a. △*ISO* is isosceles with *IS* = *OS*, $\overline{DZ} \parallel \overline{IO}$.	**a.** ___?___
b. m∠*I* = m∠*O*	**b.** ___?___
c. *ZOID* is a trapezoid.	**c.** ___?___
d. *ZOID* is an isosceles trapezoid.	**d.** ___?___

9. At the right is a street level view of two adjoining sides of the John Hancock Center, described at the beginning of the lesson. *ABCD* is an isosceles trapezoid. If *AD* = 70 m and *DC* = 81 m, what other length(s) do you know?

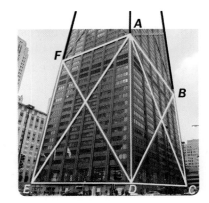

10. Given: *QRST* is an isosceles trapezoid with bases $\overline{QR}$ and $\overline{ST}$.
    a. Prove: *QS* = *RT*. (Hint: Use the reflection symmetry of the figure.)
    b. State the result in words as a theorem.

11. **True or False** Refer to Question 10.
    a. The diagonals of an isosceles trapezoid are equal in length.
    b. The diagonals of a rectangle are equal in length.
    c. The diagonals of a square are equal in length.

12. a. What theorem of this lesson tells you that tennis courts have two symmetry lines?
    b. Why do you think tennis courts are shaped this way?

13. Describe all the symmetry lines of square *FGHI*.

REVIEW

14. In the figure at the right, $\overleftrightarrow{AB}$ is the perpendicular bisector of $\overline{CD}$. Prove that *ACBD* is a kite.
(**Lesson 6-5**)

15. Prove that if both the diagonals in a kite are lines of symmetry for the kite, then the kite is a rhombus. (**Lesson 6-5**)

16. **True or False** If three of the angles in a convex quadrilateral each have measure 90, then that quadrilateral is a parallelogram.
(**Lessons 6-4, 5-7**)

17. In the figure below, O and P are the centers of the circles and $m\widehat{DE} = 140°$. Find $m\angle ABC$ and $m\angle AFC$. (**Lesson 6-3**)

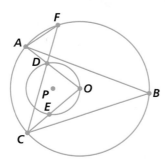

18. Alchemy was a well-respected science in Europe until the 18th century. Alchemists used symbols to describe various chemical elements. Below are three examples. Trace them and draw their lines of symmetry. (**Lesson 6-1**)

   a. silver   **b.** phosphorus   **c.** tin

19. The outer wall of the Pentagon in Washington, D.C. has the shape of a convex pentagon. If a guard walked around the building, following the walls, completing one full circuit, how many degrees did he turn in total? (**Lesson 5-7**)

There are 17.5 miles of corridors inside the Pentagon, yet a person can walk between any two points in the building in just seven minutes.

EXPLORATION

20. June noticed that a quadrilateral can have one, two, or four lines of symmetry, but not three. She was able to create hexagons with one, two, three, and six lines of symmetry, but not with four or five. She made the conjecture that if an n-gon has m lines of symmetry, then m is a factor of n.
   a. Draw a dodecagon with exactly six lines of symmetry.
   b. Draw a dodecagon with exactly four lines of symmetry.
   c. Draw a nonagon with exactly three lines of symmetry.
   d. Can you draw a nonagon with exactly two lines of symmetry?
   e. Do you think June's conjecture is true?

QY ANSWER

$m\angle R = 120$

Lesson
6-7
Rotation Symmetry

Vocabulary

rotation-symmetric figure

center of symmetry

n-fold rotation symmetry

▶ **BIG IDEA** Rotation symmetry is a property of circles, gears, propellers, some polygons, and many designs.

The word "symmetric" is usually applied to butterflies, faces, mirrors, or anything else that has reflection symmetry. However, it is also possible for a figure to have rotation symmetry.

Each of the figures above (the card, the wheel, and the blades of the windmill) could be rotated by a certain number of degrees and it would fit onto itself. This is the idea of *rotation symmetry*.

Mental Math

In the figure below, *STEP* is an isosceles trapezoid with symmetry line *m*.

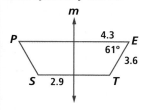

a. Find the measures of angles *P*, *T*, and *S*.

b. Find the lengths of the sides of the trapezoid.

Definition of Rotation-Symmetric, Center of Symmetry

A plane figure *F* is a **rotation-symmetric figure** if and only if there is a rotation *R* with magnitude between 0° and 360° such that $R(F) = F$. The center of *R* is a **center of symmetry** for *F*.

Rotation symmetries are classified by the number of different ways the rotation-symmetric figure could be mapped onto itself using only rotations. Think of the common baby's toy in which shapes are inserted into slots. In the diagram below, the block could be fit into the slot in exactly 5 different ways. The pentagon that is the end of the shape is said to have *5-fold rotation symmetry*.

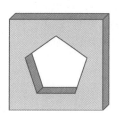

▶ **QY1**

Suppose the block shown at the left will fit directly into the slot. What is the smallest magnitude of rotation that would allow the block to fit into the slot again?

🛑 **QY1**

In general, a figure has **n-fold rotation symmetry** when the smallest magnitude of rotation that will map it onto itself is m, where $m = \frac{360°}{n}$. In the case of the block, $m = \frac{360°}{5} = 72°$.

 QY2

Some figures have both rotation and reflection symmetry. Figures a and d in QY2 have both rotation and reflection symmetry, while Figures b and c have only rotation symmetry.

When Does Reflection Symmetry Imply Rotation Symmetry?

The Two-Reflection Theorem for Rotations states that every rotation is the composite of reflections over intersecting lines. This theorem explains a relationship between the reflection symmetry and the rotation symmetry of a figure.

> **GUIDED**
>
> ### Example
>
> Suppose a figure F has two symmetry lines m and n that intersect at a single point. Does it have to be true that F also has rotation symmetry?
>
> Solution Because m and n are symmetry lines, we know that $r_m(F) = \underline{}$ and $r_n(F) = \underline{}$. By substitution, $r_n(r_m(F)) = F$. Since lines m and n intersect, we also know that $r_n \circ r_m$ is a $\underline{}$ by the Reflection Theorem for Rotations. This rotation maps F onto itself in a nontrivial way; therefore, F has rotation symmetry.

The argument in the example proves the following theorem.

> **Theorem**
>
> If a figure possesses two lines of symmetry intersecting at a point P, then it is rotation-symmetric with a center of symmetry at P.

The result of this theorem is that when a figure has two or more intersecting symmetry lines, the figure also has rotation symmetry. For example, the Rectangle Symmetry Theorem states that the perpendicular bisectors of the sides of a rectangle are symmetry lines.

▶ QY2

Each of the following figures has *n*-fold rotation symmetry. Find both *n* and *m* for each shape.

a.

b.

c.

d.

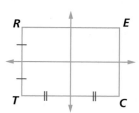

All rectangles have intersecting symmetry lines, so we now know that all rectangles must also have rotation symmetry. Because all squares are rectangles, we know that squares will also have rotation symmetry. The only other quadrilaterals with rotation symmetry are parallelograms. (You will examine the properties of parallelograms in Chapter 7.)

Drawing a Figure with Rotation Symmetry

You have seen figures that possess rotation symmetry. You may wonder how to create these diagrams yourself because drawing them accurately by hand is very difficult. The key to doing this is to recognize two key relationships. First, in each figure with n-fold rotation symmetry, there is a part of the shape that is repeated n times to create the larger figure. Second, the measure of the angle through which that shape is rotated to get to create the shape beside it is $\frac{360}{n}$. In other words, if you were to take a slice of the shape of this magnitude, that slice could be used to create the entire shape. This is shown below.

In QY 2, Figure c has 10-fold rotation symmetry. We have taken a random 36° slice of this figure and cut it out.

 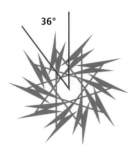 

By rotating that slice 36° nine times, you could reconstruct the entire figure. Notice you only need to rotate the slice nine times because the tenth image would be your starting image.

 QY3

> **QY3**
>
> Is it possible for a figure to have 17° rotation symmetry? Explain your reasoning.

Questions

COVERING THE IDEAS

1. **Fill in the Blank** If a figure F is rotation-symmetric, then there exists a rotation R such that $R(F) =$ __?__.

2. The picture at the right is of a part of a watch mechanism. It has 39 teeth and turns a full revolution in one minute.
   a. What kind of symmetry does the mechanism have?
   b. How many degrees does it turn each second?

3. The figure shown at the right has n-fold rotation symmetry.

   a. What is n?

   b. What is the smallest magnitude of rotation that will map this figure onto itself?

4. A design is created by rotating a figure 45° until a rotation-symmetric figure is formed. Describe the design's rotation symmetry.

5. Create a design like that described in Question 4.

6. Consider the letters at the right.

   a. Which letters have rotation symmetry?

   b. Which letters from Part a also have reflection symmetry?

7. Draw a figure that has 3-fold rotation symmetry, but no reflection symmetry.

8. Draw a figure that has 3-fold rotation symmetry and exactly three symmetry lines.

A B C D E F
G H I J K L M
N O P Q R S T
U V W X Y Z

APPLYING THE MATHEMATICS

9. Is it possible for a figure to have 3 or more symmetry lines, but no rotation symmetry? Explain why or why not.

10. Find an English word that, when typed in capital letters, has 2-fold rotation symmetry.

11. Hexagon $ABCDEF$ at the right has 2-fold rotation symmetry.

    a. If R is the nontrivial rotation that maps $ABCDEF$ onto itself, what is the magnitude of R?

    b. What is $R(ABCDEF)$?

    c. Name all pairs of congruent sides and angles in the figure.

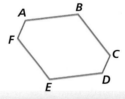

12. Follow the steps to create a rotation-symmetric design using a DGS. Refer to the figure at the right.

    Step 1 Construct points A, B, C, D, E.

    Step 2 Construct quadrilateral $ABCD$ and shade its interior.

    Step 3 Rotate $ABCD$ 45° about a point E not in the interior.

    Step 4 Repeat Step 3 using the image of $ABCD$. Do this until you have a figure that has rotation symmetry.

    Step 5 Drag A, B, C, or D to create a variety of shapes.

    a. Describe the rotation symmetry of your design.

    b. Does the way you have constructed your design ensure that your design will have rotation symmetry?

    c. What would have happened if you had used 40° instead of 45° as the magnitude of rotation?

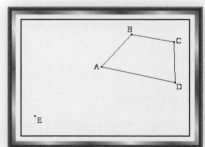

13. Explain why normal playing cards (like the queen pictured at the beginning of this lesson) have 2-fold rotation symmetry.

14. Look around you and describe a shape that you see that has *n*-fold rotation symmetry where:

    a. $n < 3$.   b. $n > 3$.

15. Sometimes it is not so easy to identify the rotation symmetry of a design. The figures below are parts of spirals that spiral out forever. Does the figure have rotation symmetry? If so, describe it.

    a.   b.

REVIEW

16. Let $A = (1, 1)$, $B = (7, 7)$, $C = (5, 6)$ $D = (2, 3)$. $ABCD$ is an isosceles trapezoid. (**Lesson 6-6**)

    a. Which of the sides are the bases?

    b. Find an equation for the line of symmetry of $ABCD$.

17. What is a kite with four ends called? (**Lesson 6-5**)

18. $DEFG$ at the right is a kite with ends D and F. Name as many congruent angles as you can. (**Lesson 6-5**)

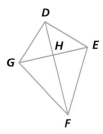

19. In an isosceles triangle, one angle has measure 53. (**Lesson 6-2**)

    a. Give two possibilities for the measures of the other two angles.

    b. If you know that the base is the longest side of the triangle, find the measures of the other two angles. (**Lesson 6-2**)

20. State the Two-Reflection Theorem for Rotations. (**Lesson 4-5**)

EXPLORATION

21. This lesson has focused solely on 2-dimensional rotation symmetry. In 3-dimensional shapes, rotation symmetry refers to a solid shape that, when rotated about a line, produces a shape that is indistinguishable from the original. The line is called the axis of the rotation. How many *axes of rotation* does a cube have?

QY ANSWERS

1. $72°$

2. a. $n = 3, m = 120°$
   b. $n = 2, m = 180°$
   c. $n = 10, m = 36°$
   d. $n = 4, m = 90°$

3. Yes; because it could have $1°$ rotation symmetry, and then 17 of those rotations would give it $17°$ rotation symmetry.

Lesson
6-8
Regular Polygons

Vocabulary

regular polygon

regular *n*-gon

equilateral polygon

equiangular polygon

center of a regular polygon

polygon inscribed in a circle

▶ **BIG IDEA**   Every regular polygon possesses both rotation and reflection symmetry and can be inscribed in a circle.

Since ancient times, domes have been used to cover structures. In the tenth century, Arab and Muslim builders began constructing Star Ribbed Domes. In these domes, a pair of parallel arches is rotated to intersect and produce a star pattern. These star patterns often contain both reflection and rotation symmetry. By rotating the pair of arches, the dome surface is divided into shapes composed of convex polygons with congruent sides and congruent angles. These polygons are *regular polygons*. You can see regular polygons in the photo of the dome in the Great Mosque of Córdoba. Regular polygons exhibit the most reflection and rotation symmetry for their number of sides.

The dome of the Great Mosque of Córdoba in Andalusia, Spain

Mental Math

Jeremy first draws $\overline{MN}$. Then he draws $\overline{TU}$ on the perpendicular bisector of $\overline{MN}$.

a. What kind of figure is *TMUN*?

b. Does *TMUN* have a symmetry line? If so, what is it?

c. Determine whether the following statement is *always, sometimes but not always,* or *never* true: *TMUN* is convex.

Definition of Regular Polygon

A **regular polygon** is a convex polygon whose angles are all congruent and whose sides are all congruent.

The regular polygons with three and four sides have special names. Otherwise, regular polygons are simply called regular pentagons, regular hexagons, and so on. The regular polygon with *n* sides is a **regular *n*-gon.**

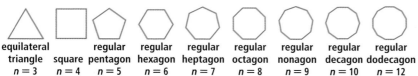

equilateral triangle	square	regular pentagon	regular hexagon	regular heptagon	regular octagon	regular nonagon	regular decagon	regular dodecagon
$n = 3$	$n = 4$	$n = 5$	$n = 6$	$n = 7$	$n = 8$	$n = 9$	$n = 10$	$n = 12$

 QY1

▶ **QY1**

What type of polygon is at the center of the Great Mosque of Córdoba?

Equilateral and Equiangular Polygons

If all sides of a polygon have the same length, the polygon is called **equilateral.** If all angles of a polygon have the same measure, the polygon is called **equiangular.** Rhombuses and squares are equilateral quadrilaterals. Rectangles and squares are equiangular polygons. Regular polygons are both equilateral and equiangular.

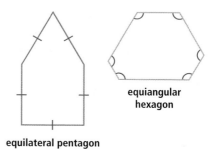

equiangular hexagon

equilateral pentagon

Example

At the right is a photo of a round diamond. At the center of the diamond is a regular octagon. Find the measure of each interior angle of the octagon.

Solution By the Polygon-Sum Theorem, the sum of all the angle measures in an octagon is __?__. Because a regular octagon is equiangular, there are __?__ angles of equal measure. Thus, each angle of a regular octagon has measure __?__.

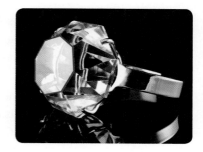

Activity 1

MATERIALS tracing paper

This activity explores the symmetry of regular polygons. Pictured at the right are a regular pentagon and a regular octagon.

Step 1 Trace each polygon and draw all of its symmetry lines.

Step 2 A circle can be constructed containing all the vertices of each of these polygons. How can you determine the center of this circle?

Step 3 Describe the rotation symmetry of each figure.

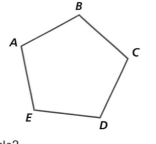

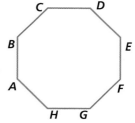

Regular Polygons and Circles

In Activity 1 you should have been able to draw a single circle containing all the vertices of each regular polygon. The center of this circle is the **center of a regular polygon.** To prove this is always possible, look again at the picture of the dome of Córdoba. Find the point in the center. Notice that the lengths of the eight beams coming out from this point to the vertex of the octagon are the same length. This is the key to a proof.

Center of a Regular Polygon Theorem

For any regular polygon, there is a unique point (its center) that is equidistant from its vertices.

Notice that in the drawing accompanying the proof, the entire regular polygon is not drawn. This is because regular polygons look different when they have different numbers of sides.

Given Regular polygon $ABCD$ . . .

Prove There is a point O equidistant from $A, B, C, D, \ldots$.

Proof The consecutive vertices A, B, and C are not collinear (from the definition of polygon), so there is a unique circle containing them by the Unique Circle Theorem. Call its center O. By the definition of circle, $OA = OB = OC$.

Because $AB = BC$ by the definition of regular polygon, quadrilateral $OABC$ is a kite and $\overline{OB}$ is its symmetry diagonal. From the symmetry, and the fact that $\triangle OBC$ is isosceles, $m\angle 1 = m\angle 2 = m\angle 3 = m\angle 4$. All regular polygons are equiangular, so $m\angle ABC = m\angle BCD$. Thus, by Angle Addition and Substitution, $m\angle 2 + m\angle 3 = m\angle 4 + m\angle 5$. Subtracting two of the equal measures, $m\angle 2 = m\angle 5$. So, by transitivity, $m\angle 5 = m\angle 4$. This makes $\overrightarrow{CO}$ the bisector of $\angle BCD$.

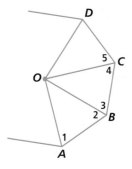

$\triangle BCD$ is isosceles because the polygon is regular and the angle bisector $\overleftrightarrow{CO}$ is its symmetry line. Thus, $r_{\overleftrightarrow{CO}}(B) = D$. Also, because O is on $\overleftrightarrow{CO}$, $r_{\overleftrightarrow{CO}}(O) = O$. Thus, because reflections preserve distance, $OB = OD$. This means that D is on the circle that contains A, B, and C. This circle also contains the next vertex of the regular polygon, and so on. Consequently, one circle contains all the vertices of this regular polygon, and the center of that circle is equidistant from all the vertices of the polygon. This means that the vertices of the polygon all lie on a circle.

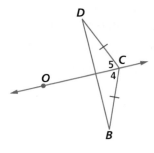

A polygon is **inscribed in a circle** if all of its vertices lie on a circle, so every regular polygon can be inscribed in a circle.

An immediate result of the Center of a Regular Polygon Theorem is that regular n-gons have n-fold rotation symmetry. All we need to show is that there is a rotation that maps every vertex onto an adjacent vertex. In the proof of the Center of a Regular Polygon Theorem, we established that $OABC$ is a kite with symmetry diagonal $\overline{OB}$. By the Kite Diagonal Theorem, $\angle AOB \cong \angle COB$. Let R be the rotation with center O and magnitude equal to $m\angle AOB$. Because these central angles are congruent, $R(A) = B$ and $R(B) = C$. So, we have shown that for any regular n-gon, there exists a rotation that maps any vertex onto an adjacent vertex.

There are n vertices, so it will take n rotations of this magnitude to map any vertex of the regular polygon onto itself. Thus, a regular n-gon has n-fold rotation symmetry. This argument has proved the following theorem.

> ### Regular Polygon Rotation Symmetry Theorem
>
> A regular n-gon has n-fold rotation symmetry.

This theorem enables regular n-gons to be drawn quite easily. Draw the circle first. Choose any point on the circle and rotate it $\left(\frac{360}{n}\right)^\circ$ about the center of the circle. Repeat this process $n - 1$ times and connect the points. The diagram at the right shows how to create a regular hexagon. Notice that in this example, $n = 6$ and the magnitude of the rotation is $\left(\frac{360}{6}\right)^\circ = 60^\circ$.

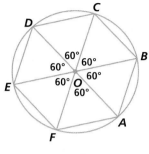

$n = 6$,
a hexagon

Activity 2

Construct a regular 12-gon using the method described above. What is the angle of rotation that you used to create each new point on your circle?

In Activity 1, you should have seen that the two regular polygons have reflection symmetry about a line through their center. This can be proved for any regular polygon using the Regular Polygon Rotation Symmetry Theorem.

Given A regular n-gon F with center O.

Prove F has reflection symmetry about every line containing O and one of its vertices.

Proof Let A be any vertex of F. We want to show that $r_{\overleftrightarrow{OA}}(F) = F$. Let V be any other vertex of F. If V is on $\overleftrightarrow{OA}$, then $r_{\overleftrightarrow{OA}}(V) = V$. If V is not on $\overleftrightarrow{OA}$, then for symmetry, we need to show that $r_{\overleftrightarrow{OA}}(V)$ is another vertex of F. Because all regular polygons have rotation symmetry, there is a rotation that maps V onto A. This same rotation maps A onto another vertex on the polygon. Call this vertex V'. Because rotations preserve distance, $\overline{AV} \cong \overline{AV'}$. By the Center of a Regular Polygon Theorem, $\overline{OV} \cong \overline{OV'}$. This implies that $OVAV'$ is a kite with symmetry diagonal $\overline{OA}$. A symmetry diagonal is a line of symmetry for a kite, so $r_{\overleftrightarrow{OA}}(V) = V'$. Thus, $r_{\overleftrightarrow{OA}}$ maps every vertex of F onto another vertex of F. By the Figure Transformation Theorem, $r_{\overleftrightarrow{OA}}$ maps the entire polygon onto itself. Thus, $\overleftrightarrow{OA}$ is a symmetry line for F.

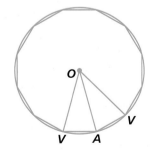

Regular polygons also have reflection symmetry over the perpendicular bisectors of their sides. This fact can be proved using the symmetries we already know.

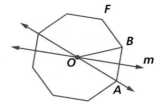

Given Regular polygon F with center O and consecutive vertices A and B.

Prove The perpendicular bisector of side $\overline{AB}$ is a symmetry line for F.

Proof

Conclusions	Justifications
1. $r_{\overleftrightarrow{OA}}(F) = F$	1. Regular polygons have reflection symmetry over a line connecting the center to a vertex.
Let R be the rotation with center O and magnitude m$\angle AOB$.	
2. $R(F) = F$	2. Regular Polygon Rotation Symmetry Theorem
Let m be the perpendicular bisector of $\overline{AB}$.	
3. m bisects $\angle AOB$.	3. Isosceles Triangle Coincidence Theorem
4. $r_{\overleftrightarrow{OA}} \circ r_m = R$	4. Two-Reflection Theorem for Rotations (because $\overleftrightarrow{OA} \cap m = \{O\}$ and the measure of the angle between $\overleftrightarrow{OA}$ and m is $\frac{1}{2}$m$\angle AOB$)
5. $r_{\overleftrightarrow{OA}} \circ r_m (F) = R(F)$	5. Substitution
6. $r_{\overleftrightarrow{OA}} \circ r_m (F) = F$	6. Transitive Property of Equality (Steps 2, 5)
7. $r_{\overleftrightarrow{OA}} \circ r_m (F) = F$, so $r_m (F) = r_{\overleftrightarrow{OA}}(F)$.	7. Flip-Flop Theorem (If $r(A) = B$, then $r(B) = A$.)
8. $r_m (F) = F$	8. Transitive Property of Equality (Steps 7, 1)
9. m is a symmetry line for F.	9. definition of symmetry line

Now we can state the following theorem:

> **Regular Polygon Reflection Symmetry Theorem**
>
> Every regular polygon has reflection symmetry about:
> 1. each line containing its center and a vertex.
> 2. each perpendicular bisector of its sides.

You may have noticed in Activity 1 that when a regular n-gon has an odd number of sides, the perpendicular bisector of one side contains a vertex on the opposite side of the polygon. When n is even, each symmetry line either bisects two opposite angles of the n-gon or is the perpendicular bisector of two opposite sides. Thus, every regular n-gon has n lines of symmetry.

 QY2

> ▶ QY2
>
> How many lines of symmetry are there in a regular 1378-gon?

Questions

COVERING THE IDEAS

1. How are Star Ribbed Domes designed?

2. Draw an equilateral quadrilateral that is not regular.

3. Draw an equiangular quadrilateral that is not regular.

4. Find the measure of each interior angle of a regular decagon.

5. How many symmetry lines does a regular octagon have?

6. How many symmetry lines of a regular pentagon are perpendicular bisectors of sides?

7. Explain how to draw a regular nonagon.

8. What does it mean for a polygon to be inscribed in circle?

9. **Fill in the Blank** A regular heptagon has __?__-fold rotation symmetry.

10. **Fill in the Blank** A regular pentadecagon (15 sides) has __?__ symmetry lines.

11. When is the perpendicular bisector of a side of a regular n-gon also an angle bisector?

12. a. Draw a regular pentagon $ABCDE$.
    b. What is the smallest positive magnitude of rotation that maps $ABCDE$ onto itself?

13. At the right is a picture of the dodecagonal floor of a carousel. If the edge of the floor is a regular polygon, what is the smallest magnitude of rotation that maps the floor onto itself?

APPLYING THE MATHEMATICS

14. The figure at the right is a regular hexagon. Prove that $OENL$ is a rhombus.

In 15 and 16, explain why the shape of the object contains a regular polygon.

15. a nut on a car wheel

16. an open umbrella

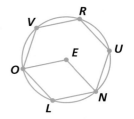

17. There is one value of n for which the n-gon cannot be equilateral without being equiangular and vice versa. What is that value?

18. $DBCFE$ at the right is a regular pentagon. Find m$\angle ADB$ and m$\angle EDA$.

In 19 and 20, give the number of sides in a regular polygon with an interior angle of the given measure.

19. 162

20. 170

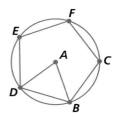

21. Find equations for the four symmetry lines containing the vertices of the regular octagon at the right.

22. The surface of a soccer ball (Figure 1) consists of 32 patches of interlocking regular polygons, 12 of them pentagons, and 20 of them hexagons. The faces are curved rather than flat so that the ball is roughly spherical. These polygons could not tile a floor even if they were flat. (See Figure 2.) What is the measure of the angle between two consecutive hexagons on the net of the ball? (See Figure 3.)

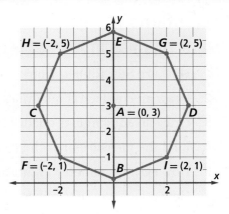

Figure 1

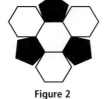

Figure 2

Figure 3

REVIEW

23. Describe the rotation symmetry of the figure at the right. (**Lesson 6 7**)

24. According to the hierarchy of quadrilaterals, what are all the special types of quadrilaterals that must have at least two parallel sides? (**Lesson 6-4**)

25. Suppose the exterior angle to the vertex of an isosceles triangle has measure 108. Which is larger: the length of the base of the triangle, or the lengths of the congruent sides? (**Lesson 6-2**)

26. Figure II at the right is the image of Figure I under an isometry. What type of isometry is it? (**Lesson 4-7**)

27. Suppose $S_{2.5}$ is the size transformation with magnitude 2.5, and that $S_{2.5}(\triangle ABC) = \triangle DEF$. (**Lesson 3-7**)
   a. Name three pairs of congruent angles.
   b. Name three pairs of sides that you know cannot be congruent.

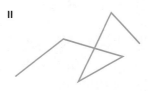

EXPLORATION

28. Construct a regular hexagon with a DGS or with a compass. (Draw a circle, pick a point on the circle, and mark off 6 radii. Connect them to form your hexagon.) Connect four consecutive vertices with segments to form a quadrilateral. What type of quadrilateral did you form? How do you know? Now, try this with other regular polygons with at least five sides. What figure do you get for each? Justify your answer.

Lesson 6-9 Frieze Patterns

Vocabulary

translation symmetry

frieze pattern

▶ BIG IDEA The patterns found on friezes on buildings and in designs that repeat along a line can be classified by their symmetries.

At this point you have studied two basic types of symmetry: reflection symmetry and rotation symmetry. Reflections and rotations are only two of the four isometries. What about translations and glide reflections? Is there such thing as *translation symmetry?* We explore this by translating an image.

A figure F has **translation symmetry** if and only if there is a translation T with nonzero magnitude such that $T(F) = F$. To create a design that has translation symmetry, we need more than a figure and its image, as shown below. This pair of shapes does not have translation symmetry, because every time it is translated, the figure extends in the same direction.

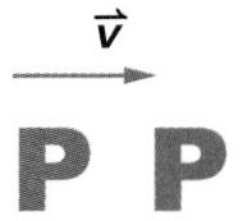

$\vec{v}$
P P

However, if we let this pattern continue forever in both directions and translate the whole infinitely long strip by the vector $\vec{v}$, the image of the strip is indistinguishable from the original. Thus, this infinitely long strip has translation symmetry!

$\vec{v}$
...P P P P P P P P P P P P...

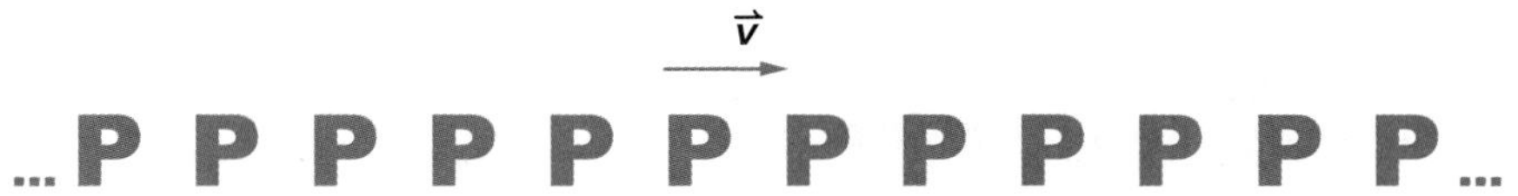

What Is a Frieze Pattern?

The pattern shown above is an example of what is known as a *frieze pattern*. A **frieze pattern** is any pattern in which a single fundamental figure is repeated to form a pattern that has a fixed height but infinite length in two opposite directions. In the pattern above, the fundamental figure is the **P** because it is the smallest portion of the pattern that could be repeated to create the entire strip. (Imagine a cookie cutter that can be flipped, slid, and rotated.)

Mental Math

Multiple Choice E is the midpoint of $\overline{MT}$. Choose the justification that allows you to conclude that $\overline{ME} \cong \overline{ET}$.

A Segment Congruence Theorem

B CPCF Theorem

C definition of congruence

D definition of midpoint

▶ READING MATH

The name "frieze" is an architectural term that comes from the latin *frisium,* which means "embroidered border" or the "decorative strip at the top of the wall."

If you look at the top edge of older buildings, you may see many examples of frieze patterns just by walking down the street. Some other real examples include sidewalks, fences, the carpet in a hallway, a wallpaper border in a room, the border of a web page—any pattern that is in the shape of a strip in which a single pattern is used to create the strip.

frieze from outside of building   frieze in flooring

The Seven Symmetries of Frieze Patterns

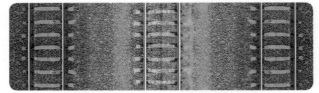

railroad tracks

All frieze patterns possess translation symmetry. Depending upon the frieze pattern, however, there may be other types of symmetries.

Example 1

Describe all of the symmetries of the frieze pattern at the right.

W M W M W M W M W M

Solution This frieze pattern has 180° rotation symmetry about any point halfway up and halfway between an M and a W. (We've

◄—W M W M M·W·M W M W M—►

drawn two such points at the right). It also has reflection symmetry about a vertical line that can be drawn through the middle of the M or the W. Finally, this pattern has glide-reflection symmetry because you can translate the pattern (so that the M is on top of a W) and then reflect the pattern over the horizontal line.

Two frieze patterns are considered to be *equivalent* if they have the same symmetries. For instance, the pattern **... FFFFFFF ...** is equivalent to **... JJJJJJJJ ...** because these two patterns have only translation symmetry. Likewise, **... OOOOO ...** is equivalent to **... XXXXX ...** Remarkably, when we classify friezes by their symmetries, there are only seven different frieze patterns! An example of each type is shown at the top of the next page, along with the symmetries that describe it (we leave translation symmetry out, since all frieze patterns have translation symmetry). To simplify things, we use the terms "vertical" and "horizontal." This assumes that, as you look at it, the pattern extends forever to the left and right.

Why are there only 7 frieze patterns? Why not more? After all, when you look again at the frieze pattern chart you should notice that there are four possible symmetries and each is either a *yes* or a *no*. Because there are 2 options for each, there are $2 \cdot 2 \cdot 2 \cdot 2 = 16$ possible combinations! Yet, there are only 7 patterns. Where did the other 9 go? We consider two of the other 9 in Example 2. You will see others in the questions.

Pattern	180° rotation symmetry	Reflection symmetry over horizontal line	Reflection symmetry over a vertical line	Glide reflection symmetry
1. **FFFFFFFFF** ...	no	no	no	no
2. **NNNNNNNN** ...	yes	no	no	no
3. **DDDDDDDDD** ...	no	yes	no	yes
4. **W W W W W W** ...	no	no	yes	no
5. **HHHHHHHHH** ...	yes	yes	yes	yes
6. **VΛVΛVΛVΛ** ...	yes	no	yes	yes
7. **FᴸFᴸFᴸFᴸFᴸ** ...	no	no	no	yes

STOP QY1

▶ **QY1**

Why is 180° the only magnitude for a rotation symmetry that a frieze pattern can have?

Example 2

Why is there not a frieze pattern on the chart that is marked with no-yes-yes-yes?

Solution Suppose a frieze pattern F had a vertical symmetry line v and a horizontal symmetry line h. In Lesson 5-7, we proved that if a figure has two intersecting symmetry lines, it must also have rotation symmetry. Thus, no-yes-yes-yes is not possible.

STOP QY2

▶ **QY2**

The explanation given in the Solution for Example 2 shows that another combination of symmetries is not possible. Which one?

Questions

COVERING THE IDEAS

1. What is a *frieze pattern?*

2. Why is it impossible to have translation or glide reflection symmetry in a shape that has finite length?

In 3–9, a frieze pattern is given. Refer to the table at the top of this page. Give the number of the frieze pattern that has the same symmetry as this one.

3. ... **QQQQQQQQQ** ... 4. ... **IIIIIIIIIII** ...

5. ... **RᴿRᴿRᴿRᴿ** ... 6. ... **B B B B B B B** ...

7. ... ... 8. ... T ⊥ T ⊥ T ⊥ ...

9. ... ...

APPLYING THE MATHEMATICS

In 10–16, repeat the directions for Questions 3–9.

10.

11.

12.
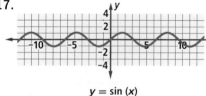

13.

14.

15.

16.

In 17 and 18, refer to the table on page 365. Which frieze pattern best matches the graph of the function, ignoring the axes? (You will study these functions in later mathematics courses.)

17.

$y = \sin(x)$

18.
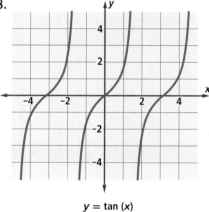

$y = \tan(x)$

19. a. Recall that all frieze patterns have translation symmetry. Use this to explain why if a frieze pattern has a horizontal symmetry line, then it also must have glide reflection symmetry.

   b. What arrangements of yes's and no's (from the chart) does your answer from Part a allow you to eliminate as possible descriptions of frieze patterns?

20. Create one new example of each type of frieze pattern. Number your examples 1–7 to match the patterns in the chart.

In 21–23, recall Lesson 4-8 on transformations in music. Consider the frieze pattern found in the oval note heads only. Assume notes are repeated forever (even though in the music they are not). Give the number of the equivalent frieze pattern.

21. This is from Ludwig van Beethoven's Piano Sonata No. 14 in C-sharp minor, Op. 27, No. 2, the "Moonlight Sonata."

22. This is from Igor Stravinsky's ballet *Petrushka*.

23. This is from Wolfgang Amadeus Mozart's *Requiem Mass*.

REVIEW

24. Suppose ℓ and m are symmetry lines for a regular hexagon. What is the smallest possible value for the measure of the acute angle between ℓ and m? **(Lesson 6-8)**

25. The points $A = (-1, 0)$, $B = (1, 0)$ and $E = (-1, 2\sqrt{3})$ are three of the six vertices of a regular hexagon. **(Lesson 6-8)**
    a. Find the center of the circle on which *A, B,* and *E* lie.
    b. Is this the center of the hexagon? Explain.

26. Draw an example of a figure with 5-fold rotation symmetry that has no lines of symmetry. **(Lesson 6-7)**

27. Determine whether the following statement is *always, sometimes but not always,* or *never* true: A rhombus has exactly two lines of symmetry. **(Lessons 6-4, 6-1)**

28. Explain what the reflexive, symmetric, and transitive properties of congruence say. **(Lesson 5-1)**

EXPLORATION

29. Find one real-life example of at least four of the different types of frieze patterns. For each pattern, provide an image (a sketch or a picture), a description of what pattern it matches, and where you found it. We ask you to find only four because finding examples of all seven is often difficult. Can you find more than four?

Chapter 6 Projects

1 A Special Regular Polygon

In Lesson 6-8, you studied regular polygons. Could you draw one on a piece of paper, using only a compass and a straightedge? The question is much more difficult than it may seem initially. Carl Gauss discovered how to construct a regular 17-gon. He was so proud of his construction that he had it carved into his tombstone. Find a description of this construction and use it to construct a regular 17-gon on a large piece of paper. Leave all of your steps on the paper, and present your work to the class, along with a description of the construction.

2 Upside-Down Names

In his book *Inversions,* Scott Kim writes words and names in such a way that when they are rotated 180°, they spell the same, or a related, word or name. Experiment with writing your own name in such a way. Then try this with several other words. Show your best work.

3 Frieze Patterns

Create an example of each of the seven types of frieze patterns. Make sure to identify which pattern each design represents.

4 Symmetry and Snowflakes

a. Find several pictures of symmetric snowflakes with different shapes and arrange them in a display. What kinds of symmetries do the snowflakes possess?

b. Find out why snowflakes have the symmetries they do, and write a short explanation.

c. Find out how scientists classify the shapes of snowflakes. Contrast two examples of such classifications.

5 Advertising

In many corporate logos, you will find rotation and/or reflection symmetry. Find logos that illustrate the use of each of these and/or combinations of the three, and make a display indicating the symmetry that each has.

6 Using Paper-Folding to Construct a Regular *n*-gon

When you were younger, you may have used paper folding and cutting to construct snowflakes. Find a way to use a similar method to construct regular polygons. Write a set of instructions for constructing a regular *n*-gon using the method you devised. Are there any values of *n* for which your method will not work? Explain any restrictions that you place upon *n*.

Chapter 6

Summary and Vocabulary

In this chapter you studied many polygons and their properties, starting with the symmetry properties of segments, angles, and circles. Segments have two **symmetry lines**, angles have one, and circles have infinitely many. These symmetries are basic in studying symmetries of polygons.

Every isosceles triangle has at least one line of symmetry. This line contains the bisector of the vertex angle, the perpendicular bisector of the base, and the median to the opposite side. Reflection over this line maps one base angle of the triangle onto the other base angle, so they are congruent.

The Isosceles Triangle Base Angles Theorem helps in proving that when two angles are not congruent in a triangle, the opposite sides are not congruent and the larger angle is opposite the larger side. It also helps in deducing its own converse, that congruent angles are opposite congruent sides, and in proving that the measure of an angle inscribed in a circle is half the measure of its intercepted arc.

Hierarchies for types of triangles and quadrilaterals are shown here.

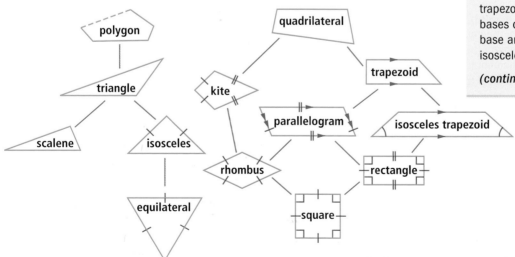

Vocabulary

6-1
reflection-symmetric figure*
symmetry line*

Lesson 6-2
vertex angle of an isosceles triangle
base of an isosceles triangle
base angles of an isosceles triangle
median of a triangle*
altitude of a triangle*

Lesson 6-3
inscribed angle*
intercepted arc

Lesson 6-4
parallelogram*
rhombus*
rectangle*
square*
kite*
trapezoid*
bases of a trapezoid
base angles of a trapezoid
isosceles trapezoid*

(continued on next page)

- The importance of these hierarchies is that any property true for one type is true for figures of every type below it to which it is connected. Thus, the Kite Symmetry Theorem applies to rhombuses and squares. The Trapezoid Angle Theorem can be applied to all special types of quadrilaterals except kites. Similarly, the Isosceles Triangle Symmetry Theorem applies to equilateral triangles as well.

- Many properties of polygons can be deduced from their symmetries. Every isosceles trapezoid has a line of symmetry: the perpendicular bisector of its bases. Also, every kite has a line of symmetry—the line containing the diagonal connecting its ends. From these two symmetries all other line symmetries of quadrilaterals are derived. When a figure has two intersecting lines of symmetry, as rectangles and rhombuses do, then it possesses rotation symmetry.

- The polygons with the most reflection and rotation symmetry for their number of sides are the regular polygons. Every regular n-gon has n symmetry lines that contain all the bisectors of its angles and the perpendicular bisectors of its sides. Every regular n-gon also has **n-fold rotation symmetry.**

- Symmetry can also be observed in frieze patterns. Based on symmetry, there are only seven different kinds of **frieze patterns.**

Vocabulary

Lesson 6-5
ends of a kite
symmetry diagonal

Lesson 6-6
corollary

Lesson 6-7
rotation-symmetric figure*
center of symmetry*
n-fold rotation symmetry

Lesson 6-8
regular polygon
regular n-gon
equilateral polygon
equiangular polygon
center of a regular polygon
polygon inscribed in a circle

Lesson 6-9
translation symmetry
frieze pattern

Postulates, Theorems, and Properties

Flip-Flop Theorem (p. 309)
Segment Symmetry Theorem (p. 309)
Side-Switching Theorem (p. 310)
Angle Symmetry Theorem (p. 311)
Circle Symmetry Theorem (p. 311)
Symmetric Figures Theorem (p. 312)
Isosceles Triangle Symmetry Theorem (p. 317)
Isosceles Triangle Coincidence Theorem (p. 318)
Isosceles Triangle Base Angles Theorem (p. 318)
Unequal Sides Theorem (p. 320)
Converse of the Isosceles Triangle Base
   Angles Theorem (p. 321)
Unequal Angles Theorem (p. 321)
Inscribed Angle Theorem (p. 325)

Thales' Theorem (p. 326)
Quadrilateral Hierarchy Theorem (p. 335)
Kite Symmetry Theorem (p. 340)
Kite Diagonal Theorem (p. 341)
Rhombus Diagonal Theorem (p. 342)
Trapezoid Angle Theorem (p. 346)
Isosceles Trapezoid Symmetry Theorem (p. 346)
Isosceles Trapezoid Theorem (p. 347)
Rectangle Symmetry Theorem (p. 348)
Center of a Regular Polygon Theorem (p. 358)
Regular Polygon Rotation Symmetry Theorem (p. 359)
Regular Polygon Reflection Symmetry Theorem (p. 360)

Chapter 6 · Self-Test

Take this test as you would take a test in class. You will need a calculator and tracing paper. Then use the Selected Answers section in the back of the book to check your work.

1. ℓ and m are symmetry lines for the polygon below.

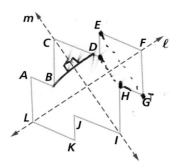

a. Name all the segments congruent to $\overline{ED}$.

b. How is m related to $\angle BCD$? How do you know?

2. Draw all the symmetry lines in the rhombus below.

3. Can there be an equilateral pentagon that is not equiangular? If so, draw one. If not, explain why not.

4. Draw a kite that has rotation symmetry.

5. How many symmetry lines can an isosceles trapezoid have? Draw and label a sketch of each situation.

6. a. Use $\triangle BRT$ below to find x.

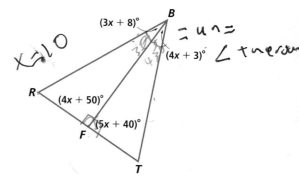

b. Is it true that $BR = BT$? Justify your answer.

7. In $\triangle GJR$, $m\angle G = 40$ and $m\angle J = 80$. Which side of $\triangle GJR$ is the longest?

8. $FPXL$ is an isosceles trapezoid with bases $\overline{FP}$ and $\overline{LX}$. $m\angle P = 110$. Find $m\angle L$.

9. A regular polygon has $100°$ rotation symmetry. What is the least number of sides it can have?

10. Describe the symmetries of the pinwheel below

a. ignoring the colors.

b. considering the colors.

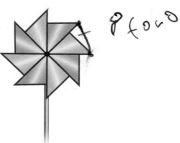

11. If $\overline{MS}$ is a diameter of the circle, what is m∠W?

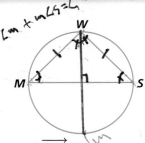

$m∠m + m∠s = ∠w$

12. △RAN is equilateral and $\overrightarrow{NB}$ bisects ∠RNA. Find m$\widehat{RAN}$.

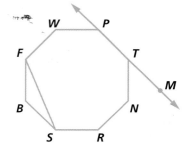

13. Determine whether each statement is *always, sometimes but not always,* or *never* true.

   a. If PQRS is a trapezoid, it is a rectangle.

   b. If a diagonal of STMU is a symmetry line of STMU and the perpendicular bisector of the other diagonal, STMU is a rhombus.

14. PTNRSBFW is a regular octagon.

   a. Find m∠MTN.   b. Find m∠BFS.

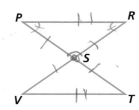

15. Given: $\overline{RP} \parallel \overline{VT}$, ∠R ≅ ∠P
    Prove: $\overline{SV} ≅ \overline{ST}$

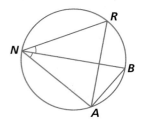

16. FOAM is a rhombus. If FA = 18 cm and m∠O = 60, find the perimeter of FOAM.

17. Draw a hierarchy relating *polygon, equilateral triangle, regular polygon, rhombus, equiangular polygon, square,* and *equilateral polygon.*

18. In regular pentagon MANDC, find m∠CAD and m∠MDC. (Hint: Find some isosceles triangles and their angle measures.)

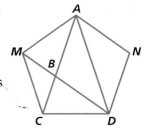

19. Below is part of the Quadrilateral Hierarchy. Every quadrilateral of type G has congruent diagonals. Quadrilateral I is type F, while Quadrilateral II is type H. Using this information, what, if anything, can be concluded about either Quadrilateral I or Quadrilateral II? Explain your answer.

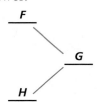

20. Which frieze patterns shown below are equivalent in terms of their symmetries?

Chapter 6 Chapter Review

SKILLS
PROPERTIES
USES
REPRESENTATIONS

SKILLS Procedures used to get answers

OBJECTIVE A Describe the reflection
and rotation symmetry of figures.
(Lessons 6-1, 6-7)

In 1–3, describe the rotation symmetry of the
given figure.

1.

2.

3.

4. The figure at the right
is a regular hexagon.
Trace it and draw all
of its symmetry lines.

In 5 and 6, trace the given figure and draw its
symmetry lines.

5.
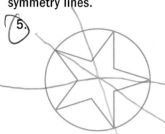

6.

OBJECTIVE B Draw polygons satisfying
various conditions. (Lessons 6-2, 6-4, 6-8)

In 7–12, draw an example of the figure using a ruler,
compass, or protractor.

7. an isosceles right triangle

8. a kite that is not a rhombus

9. a polygon with rotation symmetry that
is not a regular polygon

10. a quadrilateral with exactly one
symmetry line that is not a kite

11. a trapezoid that is not isosceles

12. an equiangular pentagon that is not regular

OBJECTIVE C Apply theorems about isosceles triangles to find angle measures and segment lengths. (Lessons 6-2, 6-8)

13. The exterior angle to the vertex angle of an isosceles triangle has measure 47. Find the measures of the interior angles of the triangle.

In 14 and 15, refer to the triangle at the right in which m∠P = m∠R.

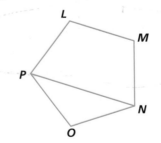

14. If $QR = 5.5''$, and $RP = 3.7''$, find QP.

15. Suppose m∠Q = 40. If M is the midpoint of $\overline{RP}$, find m∠MQR.

16. Suppose △ABC is an isosceles triangle, and m∠A = 56. Draw two possible noncongruent triangles that satisfy these conditions and calculate m∠B and m∠C in both cases.

17. The figure below shows a regular pentagon. Find m∠OPN.

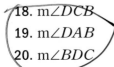

OBJECTIVE D Calculate the measures of inscribed angles from the measures of intercepted arcs, and vice versa. (Lesson 6-3)

In 18–20, use the figure at the right in which O is the center of the circle. Calculate the angle measure.

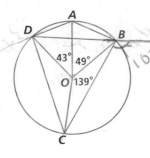

18. m∠DCB

19. m∠DAB

20. m∠BDC

In 21 and 22, refer to the figure below, in which G is the center of the circle, and m∠ROH = 54. Determine the angle measure.

21. m∠HGR

22. m∠ORU

OBJECTIVE E Apply theorems about quadrilaterals and regular polygons to find angle measures and segment lengths. (Lessons 6-4, 6-5, 6-6, 6-8)

23. Find the measure of an interior angle of a regular 13-gon.

In 24 and 25 use the rhombus pictured at the right.

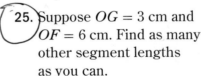

24. Suppose m∠FGO = 64. Find the measures of as many other angles as you can.

25. Suppose $OG = 3$ cm and $OF = 6$ cm. Find as many other segment lengths as you can.

26. $PART$ is an isosceles trapezoid with nonbase sides $\overline{PT}$ and $\overline{AR}$. If m∠T = 63.3, find m∠A.

27. In parallelogram $ABCD$, m∠A = 45.
    a. Draw a picture of $ABCD$.
    b. Find the measures of the other angles of $ABCD$.

PROPERTIES Principles behind the mathematics

OBJECTIVE F Apply properties of symmetry to assert and justify conclusions about symmetric figures. (Lessons 6-1, 6-7)

In 28 and 29, refer to the figure below, in which *m* and *n* are symmetry lines.

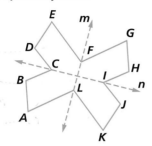

28. Name all the angles that must be congruent to $\angle E$.

29. Name two congruent heptagons in the figure.

In 30 and 31, consider the seven types of quadrilaterals in this chapter.

30. Name two types that possess rotation symmetry.

31. Name all types in which both diagonals are symmetry lines.

32. Determine whether the following statement is *always, sometimes but not always,* or *never* true: A symmetry line in a regular hexagon cuts the hexagon into two congruent quadrilaterals.

33. **True or False** If a pentagon has 5-fold rotation symmetry, then it is regular.

34. In the figure below, *LMNO* is an isosceles trapezoid, *P* is the midpoint of $\overline{LM}$, and *Q* is the midpoint of $\overline{NO}$. Justify the conclusion $\overline{PQ} \perp \overline{LM}$.

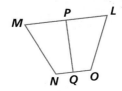

OBJECTIVE G Know the properties of various types of triangles and regular polygons. (Lessons 6-2, 6-8)

35. **True or False** $\triangle ABC$ is isosceles with base $\overline{AB}$ if and only if the perpendicular bisector of $\overline{AB}$ bisects $\angle C$.

36. **True or False** A scalene triangle can have a line of symmetry.

37. State the Regular Polygon Symmetry Theorems.

38. How many of the symmetry lines of a regular heptagon contain diagonals of the figure?

39. a. Explain why the three angle bisectors of the angles of an equilateral triangle intersect at the center of the circle in which the triangle is inscribed.

    b. **True or False** The three perpendicular bisectors of the sides of an equilateral triangle intersect at the center of the circle in which the triangle is inscribed.

OBJECTIVE H From given information, deduce which sides or angles of triangles are smallest or largest. (Lesson 6-2)

40. The exterior angle to the vertex of an isosceles triangle has measure 85°. Which has the greater measure: the base of the triangle, or a nonbase side of the triangle?

41. In a right triangle, the side opposite the 90° angle is called the *hypotenuse*. Explain why the hypotenuse is always the longest side of the triangle.

42. The sides of a triangle have lengths 3 cm, 5 cm, and 6 cm. The angles have measures 30, 56.5, and 93.5. Draw a picture of this situation; showing which angle is opposite which side.

43. **True or False** Suppose that in $\triangle ABC$, $m\angle A = 104$, and in $\triangle DEF$, $m\angle D = 12$. Then we know that $BC > EF$.

OBJECTIVE I Know the properties of the seven special types of quadrilaterals.
(Lessons 6-4, 6-5, 6-6)

In 44–46, determine whether the statement is true or false. If false, draw a counterexample.

44. Every parallelogram is a rhombus.

45. Every equiangular rhombus is a square.

46. Every rectangle is a kite.

In 47–49, identify all the special types of quadrilaterals with the stated property.

47. There is a pair of congruent opposite angles.

48. A symmetry line contains a diagonal.

49. There are two distinct pairs of congruent consecutive angles.

OBJECTIVE J Write proofs using properties of triangles and quadrilaterals.
(Lessons 6-2, 6-4, 6-5, 6-6)

In 50–53, write a proof using the given figure.

50. Given: $TRAP$ is an isosceles trapezoid with bases $\overline{TR}$ and $\overline{PA}$, and m$\angle R$ = m$\angle A$.
Prove: $TRAP$ is a rectangle.

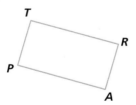

51. Given: $\ell \parallel \overline{MN}$, $\ell \perp \overline{OP}$, m$\angle 1$ = m$\angle 2$
Prove: $MP = NP$

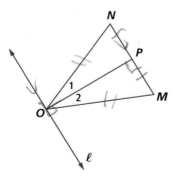

52. Given: Isosceles trapezoid $RSUT$ with bases $\overline{RS}$ and $\overline{TU}$
Prove: $ST = RU$ (Hint: Use symmetry.)

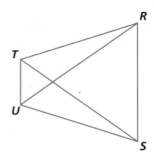

53. Given: $\odot O$ and $\odot P$ intersect at A and B.
Prove: $OAPB$ is a kite.

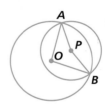

USES Applications of mathematics in real-world situations

OBJECTIVE K Describe the reflection and rotation symmetry in real-world designs.
(Lessons 6-1, 6-7)

54. The image below shows a pattern that was found on an Egyptian tomb in Thebes. Assume that the pattern goes on indefinitely in all directions. Copy the pattern and draw two symmetry lines.

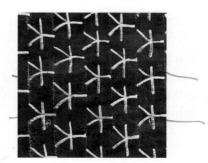

55. The image below shows the flag of the Isle of Man. Describe the rotation symmetry of the figure at the center.

56. Describe the symmetries of the spider web pictured below.

OBJECTIVE L Identify the symmetries of a frieze pattern. (Lesson 6-9)

In 57–60, identify the types of symmetry of the pattern.

57.

58.

59.

60.

REPRESENTATIONS Pictures, graphs, or objects that illustrate concepts

OBJECTIVE M Draw and apply hierarchies of polygons. (Lesson 6-4)

61. Draw a hierarchy relating: *polygon, quadrilateral, parallelogram, kite, rhombus, triangle, isosceles triangle, equilateral triangle.*

62. Draw a hierarchy relating: *polygon, regular polygon, square, quadrilateral, trapezoid.*

63. Draw a hierarchy relating: *quadrilaterals, quadrilaterals with rotation symmetry, quadrilaterals with at least one symmetry line, quadrilaterals with at least two symmetry lines, kite, rhombus, rectangle, square.*

Applications of Congruent Triangles

Contents

7-1   Drawing Triangles

7-2   Triangle Congruence Theorems

7-3   Using Triangle Congruence Theorems

7-4   Overlapping Triangles

7-5   The SSA Condition and HL Congruence

7-6   Tessellations

7-7   Properties of Parallelograms

7-8   Sufficient Conditions for Parallelograms

7-9   Diagonals of Quadrilaterals

7-10 Proving That Constructions Are Valid

Any figure can be reflected, rotated, translated, or glide reflected, and the result is a congruent figure. So why are congruent triangles important? There are several basic reasons.

Triangles are the building blocks of polygons. Recall from Lesson 5-7 that the sum of the measures of the angles of a convex polygon was found by creating triangles inside the polygon. In Chapter 8, the same type of process will be used to determine the areas of polygons. So what you learn about triangles in this chapter will be useful in determining properties of polygons with more sides.

Triangles are strong. Triangles are rigid and are often found in the frameworks supporting bridges, buildings, and other structures. The properties of triangles give strength to quadrilaterals and other polygons.

Triangles are simple to describe. Because triangles are the simplest of polygons, it is possible to determine if two triangles are congruent by knowing only a few facts about them. So you can draw and construct figures, tessellations, and other designs with triangles with only a few measurements. This makes it more efficient to work with triangles than with other figures.

All of these uses rely on the theorems about triangle congruence that you will learn and apply in this chapter.

Lesson 7-1

Drawing Triangles

▶ **BIG IDEA** Some combinations of measures of sides and angles are enough to determine all the other measures of sides and angles of the triangle.

Mental Math

Two angles of a triangle have the given measures. What is the measure of the third angle?

a. 3, 4

b. x, y

c. $75 + a$, $75 - a$

The roofs of many houses are supported by a set of congruent triangular braces known as *trusses*. Supports, like trusses, usually involve triangles because triangles are *rigid*. To examine what this means, try the following activity.

Activity 1

MATERIALS straws, fasteners

Fasten four straws of different lengths together at their ends to create a quadrilateral. Do the same thing with three straws of different lengths. Play with the two figures you have created.

1. What do you think is meant by the statement that triangles are *rigid*?

2. How could you add an additional straw to your quadrilateral to make it rigid?

3. How does your answer to Question 2 relate to your answer from Question 1?

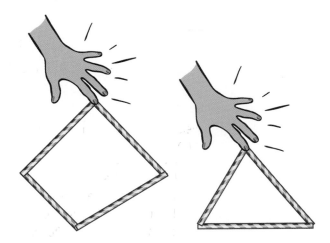

Exploring Sufficient Conditions for Triangles

Suppose you were in charge of setting the trusses for a house and, after setting up all of the trusses you have, you discover that you are one truss short. You now have to call the lumber yard to order one more. How could you ensure that you receive a triangular truss that is congruent to the rest? If you provide the lengths of the three sides and measures of the three angles, must you have them in a particular order? Could you tell less information about the triangle and still ensure that you receive a congruent triangle? Could you just tell the lumberyard the lengths of the three sides? What about the lengths of two sides and the measure of one angle? What about just providing the measures of three angles?

Whenever a set of conditions is enough so that all figures created will be congruent, then we call the set a **sufficient condition** for determining the figure. One way to study whether something is a sufficient condition is to draw several triangles satisfying that condition and see if they are all congruent.

Drawing a Triangle Given Three Sides

In Activity 1 you chose three lengths for the sides of a triangle, and then saw this triangle was rigid. Activity 2 examines whether one set of three lengths is a sufficient condition for determining the triangle.

Activity 2

MATERIALS DGS

Using a DGS, draw triangles with sides 5 cm, 2 cm, and 6 cm.

Step 1 Draw any line and call it m. Choose point A on m.

Step 2 Draw $\odot(A_1, 5 \text{ cm})$. ($\odot A_1$ identifies the first circle with center A.) Name $\odot A_1 \cap m = \{B, C\}$.

Step 3 Draw $\odot(A_2, 6 \text{ cm})$. ($\odot A_2$ identifies the second circle with center A.)

Step 4 a. Draw $\odot(B, 2 \text{ cm})$. Name $\odot A_2 \cap \odot B = \{D, E\}$.

    b. Draw $\overline{AD}$ and $\overline{BD}$. Your drawing should be similar to the one at the right.

    c. To the nearest tenth, what are the lengths of the sides of $\triangle ABD$?

Step 5 a. Measure the angles of $\triangle ABD$ to the nearest degree.

    b. In Steps 2 and 4 you made choices as to which points to call B and D. Would the measures of the angles of $\triangle ABD$ be different if you had named different points? Why or why not?

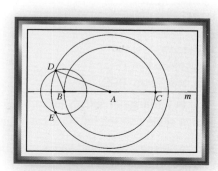

Here is what a DGS screen could look like after completing the steps of Activity 2, hiding the circles, and asking for lengths of sides and measures of angles in △ABD. Do the angle measures in your triangles agree with those given at the right? If so, you might conjecture that the lengths of three segments is a sufficient condition for a triangle. This condition is called the SSS condition because it gives the lengths of the three sides.

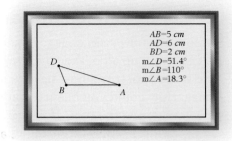

AB=5 cm
AD=6 cm
BD=2 cm
m∠D=51.4°
m∠B=110°
m∠A=18.3°

Drawing a Triangle Given Three Angles

All rectangles have four 90° angles, but not all rectangles are congruent. Is the situation different for triangles? Is knowing the measures of the three angles in a triangle (AAA) a sufficient condition for congruence? This is the subject of Activity 3.

Activity 3

MATERIALS DGS

Step 1 Construct $\overline{AB}$.

Step 2 Construct line ℓ so that A lies on ℓ and $\ell \perp \overline{AB}$.

Step 3 Construct C on ℓ. Draw △ABC and measure its angles.

Step 4 Move C around until m∠C = 50. What must m∠B be in this case?

Compare the triangle you drew to those of some of your classmates. Do they appear to be congruent? Do you think AAA is a sufficient condition?

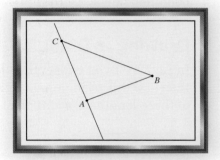

Drawing a Triangle Given Two Sides and an Included Angle

Two sides of a triangle form part of an angle. That angle is said to be *included* by the sides, and is called an **included angle.** In Activity 4 you are asked to draw a triangle given the lengths of two sides and the measure of their included angle. This is called the SAS condition. Notice that the A is between the two Ss to indicate that the angle is included by the sides.

Activity 4

MATERIALS DGS

In a triangular sail △ABC, m∠A = 75°, AB = 4.5 m, and AC = 3 m. Make a scale drawing letting 1 cm in the drawing equal 1 meter of the actual sail.

a. Place △ABC and its measurements on the screen as in Activities 2 and 3.

b. Move point A so $AB = 4.5$ cm (within 0.02 cm).

c. Move point C so that $m\angle A = 75°$.

d. Move point C along $\overline{AC}$ until $AC = 3$ cm, while keeping $m\angle A = 75°$.

e. Record $m\angle B$, $m\angle C$, and BC.

f. Do the measurements in Part e agree (within $2°$ and 0.1 cm) with the measurements on the screen at the right?

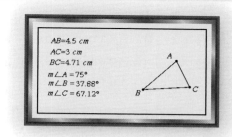

```
AB=4.5 cm
AC=3 cm
BC=4.71 cm
m∠A =75°
m∠B =37.88°
m∠C =67.12°
```

In the Activities in this lesson you were asked to draw a triangle given measures of some of its sides or angles. You then were asked whether SSS, AAA, or SAS was a sufficient condition to determine a unique triangle.

Other combinations of angles and sides are possible. Among these are ASA (two angles and the included side) and AAS (two angles and the nonincluded side). In the Questions you are asked to draw triangles given AA, SA, and SS. These questions ask whether you think everyone else's drawings will be congruent to yours. In other words, are these given conditions sufficient to determine a triangle?

Activity	Given	Name for given condition
2	three sides	SSS
3	three angles	AAA
4	two sides and included angle	SAS

Questions

COVERING THE IDEAS

1. Give three reasons for studying triangles.

2. Explain what is meant by the statement, "triangles are rigid."

In 3–6, a triangle is given. Determine whether the information marked on the triangles is SSS, SAS, ASA, or AAS.

3.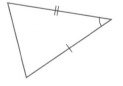

4.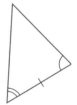

5.

6.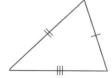

7. Draw a triangle with tick marks where the marked information gives the AAA condition.

8. Name two conditions you think are sufficient conditions for triangle congruence.

APPLYING THE MATHEMATICS

9. $\triangle ABC$ is isosceles with $AB = 5$ cm and m$\angle B = 48$. $\triangle DEF$ is isosceles with $DE = 5$ cm and m$\angle E = 48$. Explain why the triangles do not have to be congruent.

In 10–15, suppose a friend calls you on the phone and asks you for help on a geometry problem that involves a triangle $\triangle ABC$. You don't have the problem in front of you, so she tries to tell you how to draw the triangle.

a. Determine whether the set of instructions would allow you to construct $\triangle ABC$ so that your triangle looks exactly like your friend's triangle.

b. If all triangles you could construct from the given information are congruent, draw that triangle and state which one of the conditions in this lesson is met. If the given information allows for noncongruent triangles to be created, draw at least two noncongruent triangles satisfying the given information.

10. m$\angle A = 90$, m$\angle B = 28$

11. $AB = 3$ cm, m$\angle A = 57$

12. $AB = 10$ cm, $BC = 11$ cm, $AC = 9$ cm

13. $AC = 5$ cm, $BC = 2$ cm

14. $AB = 2.5"$, $BC = 1.5"$, m$\angle B = 30$

15. m$\angle A = 40$, m$\angle B = 60$

16. You are told that a certain desert island is located 50 miles from Abel Island and 40 miles from Bernoulli Island. Abel and Bernoulli Islands are 25 miles apart.

    a. Draw a diagram showing that there are two possible locations for the desert island.

    b. Does your answer contradict the statement that SSS is a sufficient condition for determining a triangle? Why or why not?

In 17–19, use a DGS as in Activities 2 and 3 to explore whether the given condition is sufficient to determine a triangle.

17. AAS   18. ASA   19. SSA

REVIEW

20. In the figure at the right, an isosceles trapezoidal region is cut up into a rectangle and two right triangles. Show how it could be cut up into four right triangles, and how it could be cut up into six right triangles. (**Lesson 6-6**)

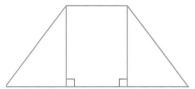

21. Daniel walks one meter in a straight line, then turns 60° to his left. He repeats this five more times, for a total of six turns. At the end, is Daniel back where he started? Draw a picture of the route he took. (**Lesson 5-7**)

22. The numeral 8 can be broken into two congruent pieces in several ways. Draw two of these ways. (**Lesson 5-1**)

23. Suppose ℓ is a line and R is a rotation such that $R(\ell) \parallel \ell$. What can you say about the magnitude of R? (**Lesson 4-5**)

24. When driving, it is common wisdom to put your hands at 10 o'clock and 2 o'clock on the steering wheel. If you kept your hands firmly on the wheel, and rotated the wheel 90° to the left, at what hours would your hands be placed? (**Lesson 3-2**)

EXPLORATION

25. Find the congruent copy of each figure at the right in the picture below.

Masons at Work

book

mug

banana

pushpin

artist's brush

slice of pie

fishhook

musical note

candle

slice of bread

pencil

mitten

Source: Highlights for Children

Lesson

7-2

Triangle Congruence Theorems

Vocabulary

included side

▶ **BIG IDEA** The acronyms SAS, SSS, ASA, and AAS stand for combinations of measures of corresponding sides and angles in two triangles that are sufficient for the congruence of the triangles.

In Lesson 7-1, you saw that all triangles drawn with given lengths of three sides, the SSS condition, seem to be congruent. Here is a proof that the SSS condition is a sufficient condition for congruence. The theorem is widely known as the *SSS Congruence Theorem*.

SSS Congruence Theorem

If, in two triangles, three sides of one are congruent to three sides of the other, then the triangles are congruent.

Given $\overline{AB} \cong \overline{DE}$, $\overline{AC} \cong \overline{DF}$, and $\overline{BC} \cong \overline{EF}$

Prove $\triangle ABC \cong \triangle DEF$

Drawing

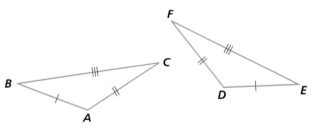

Mental Math

Given a line ℓ and a point P on ℓ:

a. How many different lines through P are there that form a 90° angle with ℓ?

b. How many different lines through P are there that form a 30° angle with ℓ?

c. How many different rays are there with endpoint P that form a 90° angle with ℓ?

Proof It is given that $\overline{AB} \cong \overline{DE}$, $\overline{AC} \cong \overline{DF}$, and $\overline{BC} \cong \overline{EF}$. From the definition of congruence, we know two figures are congruent if and only if one is the image of the other under an isometry. Because $\overline{AB} \cong \overline{DE}$, there is an isometry T with $T(\overline{AB}) = \overline{DE}$ such that $T(A) = D$ and $T(B) = E$. Furthermore, T can be chosen so that $T(C)$ is on the other side of $\overleftrightarrow{DE}$ from F. Label this image of $\triangle ABC$ as $T(\triangle ABC)$ or $\triangle A'B'C'$.

We write the rest of the proof in two-column form. The idea is to prove that $\triangle A'B'C' \cong \triangle DEF$ so that all three triangles are congruent. You should examine the figure after each step.

Conclusions	Justifications
1. $\triangle A'B'C' \cong \triangle ABC$	1. definition of congruence
2. $\overline{BC} \cong \overline{B'C'}$, $\overline{AC} \cong \overline{A'C'}$	2. CPCF Theorem
3. $\overline{B'C'} \cong \overline{EF}$, $\overline{A'C'} \cong \overline{DF}$	3. Transitive Property of Congruence (using Given and Step 2)
4. $CDEF$ is a kite with ends D and E.	4. definition of kite
5. $CDEF$ is reflection-symmetric to $\overleftrightarrow{DE}$.	5. Kite Symmetry Theorem
6. $r_{\overleftrightarrow{DE}}(\triangle A'B'C') \cong \triangle DEF$	6. definition of reflection-symmetric figure
7. $\triangle A'B'C' \cong \triangle DEF$	7. definition of congruence
8. $\triangle ABC \cong \triangle DEF$	8. Transitive Property of Congruence

 QY1

The SAS Condition

A second sufficient condition for congruence is the SAS condition. We could use a proof quite similar to the proof of SSS Congruence Theorem. The proof presented here is slightly different.

SAS Congruence Theorem

If, in two triangles, two sides and the included angle of one are congruent to two sides and the included angle of the other, then the triangles are congruent.

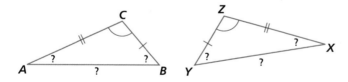

Given $\overline{CB} \cong \overline{ZY}$, $\overline{ZX} \cong \overline{CA}$, and $\angle C \cong \angle Z$

Prove $\triangle ABC \cong \triangle XYZ$

Proof $\angle C \cong \angle Z$, so by the definition of congruence there is an isometry T that maps $\angle C$ onto $\angle Z$, that is, $T(\angle C) = \angle Z$. This also means that $T(C) = Z$. There are now two cases to consider, depending on which side of $\angle C$ the image of $\overrightarrow{CB}$ falls.

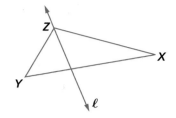

Case I If $T(\overrightarrow{CB}) = \overrightarrow{ZY}$, then $T(\overrightarrow{CA}) = \overrightarrow{ZX}$. $T(C) = Z$ and $CB = ZY$, so it must be true that $T(B) = Y$ because only one point on $\overrightarrow{ZY}$ is at the distance CB from Z. Similarly, $T(A) = X$. This means that $T(\triangle ABC) = \triangle XYZ$, so $\triangle ABC \cong \triangle XYZ$.

Case II If $T(\overrightarrow{CB}) = \overrightarrow{ZX}$, then $T(\overrightarrow{CA}) = \overrightarrow{ZY}$. Draw ℓ, the angle bisector of $\angle Z$. By the Side Switching Theorem, $r_\ell(\overrightarrow{ZX}) = \overrightarrow{ZY}$ and $r_\ell(\overrightarrow{ZY}) = \overrightarrow{ZX}$. Let $G = r_\ell \circ T$. Then $G(\overrightarrow{CB}) = \overrightarrow{ZY}$, and $G(\overrightarrow{CA}) = \overrightarrow{ZX}$. This is exactly what was required in Case I for the transformation T. Repeating the argument from Case I, $G(A) = X$ and $G(B) = Y$. Thus, $G(\triangle ABC) = \triangle XYZ$, so $\triangle ABC \cong \triangle XYZ$.

 QY2

▶ **QY2**

Suppose two triangles both have sides measuring 8 and 9 inches and an angle measuring 70. Can you be sure that the two triangles are congruent? Why or why not?

The ASA Condition

The ASA condition refers to two angles and the **included side**, that is, the side between those two angles. This is the third major triangle congruence theorem.

 QY3

▶ **QY3**

Draw two congruent triangles and mark them with ASA.

> ### ASA Congruence Theorem
>
> If, in two triangles, two angles and the included side of one are congruent to two angles and the included side of the other, then the two triangles are congruent.

Given $\overline{AB} \cong \overline{DE}$, $\angle A \cong \angle FDE$, and $\angle B \cong \angle FED$

Prove $\triangle ABC \cong \triangle DEF$

Drawing

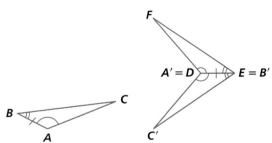

Proof Again consider the image $\angle A'B'C'$ of $\angle ABC$ under an isometry mapping $\overline{AB}$ onto $\overline{DE}$. $\triangle A'B'C'$ and $\triangle DEF$ form a figure much like that in the proof of the SSS Congruence Theorem, but now with two pairs of congruent angles, as shown above.

Again, think of reflecting △A'B'C' over $\overleftrightarrow{DE}$. Because $\overleftrightarrow{DE}$ contains the bisector of ∠C'DF (see Question 15) and of ∠C'EF, we can apply the Side-Switching Theorem. The image of $\overrightarrow{A'C'}$ is $\overrightarrow{DF}$.

Applying the Side-Switching Theorem to ∠C'EF, the image of $\overrightarrow{B'C'}$ is $\overrightarrow{EF}$. This forces the image of C' to be on both $\overrightarrow{DF}$ and $\overrightarrow{EF}$, and so the image of C' is F. Therefore, the image of △A'B'C' is △DEF. This makes △A'B'C' ≅ △DEF and, by the Transitive Property of Congruence, △ABC ≅ △DEF.

Example 1

Prove that △ABC ≅ △TUV.

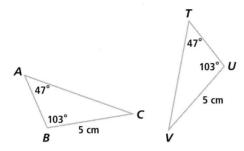

Proof Using the Triangle-Sum Theorem, the third angle in both triangles has a measure of 30. With this third angle, we have the ASA condition: ∠B ≅ ∠U, $\overline{BC} \cong \overline{UV}$, and ∠C ≅ ∠V. Thus, the triangles are congruent by ASA.

Example 1 shows that the AAS condition is a sufficient condition for congruent triangles. It is a direct consequence of the Triangle-Sum Theorem and the ASA Congruence Theorem.

AAS Congruence Theorem

If, in two triangles, two angles and a nonincluded side of one are congruent respectively to two angles and the *corresponding* nonincluded side of the other, then the triangles are congruent.

We leave its proof as Question 16. To make this and other proofs easier, it is useful to have the following theorem.

Third Angle Theorem

If two triangles have two pairs of angles congruent, then their third pair of angles is congruent.

GUIDED

Example 2

Prove the Third Angle Theorem.

Proof

Given: In △ABC and △DEF, ∠A ≅ ∠D and ∠B ≅ ∠E.

Prove: ∠C ≅ ∠F

Drawing:

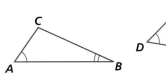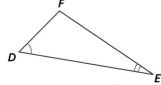

Proof:

Conclusions	Justifications
1. ∠A ≅ ∠D, ∠B ≅ ∠E	1. Given
2. m∠A = m∠D, m∠B = m∠E	2. Angle Congruence Theorem
3. m∠A + m∠B + m∠C = 180 m∠D + m∠E + m∠F = 180	3. _____?_____
4. m∠A + m∠B + m∠C = m∠D + m∠E + m∠F	4. _____?_____ Property of Equality
5. m∠D + m∠E + m∠C = m∠D + m∠E + m∠F	5. Substitution
6. m∠C = m∠F	6. _____?_____ Property of Equality (Add −m∠D + −m∠E to both sides in Step 5.)
7. ∠C ≅ ∠F	7. _____?_____

Questions

COVERING THE IDEAS

1. List four conditions that lead to triangle congruence.

2. Two angles of a triangle have the given measures. What is the measure of the third angle?

   a. 70°, 4°   b. 0.98°, 179°   c. *x, y*

3. **Multiple Choice** The proof of the SSS Congruence Theorem uses the symmetry of which of these figures?

   A   angle

   B   isosceles triangle

   C   kite

In 4–9, write a congruence statement for two triangles with vertices in corresponding order, and state the congruence theorem that justifies the congruence.

4.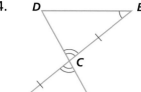

5.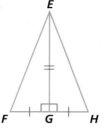

6.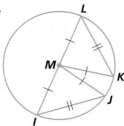

7.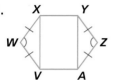

8.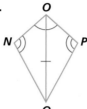

9.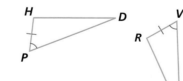

APPLYING THE MATHEMATICS

10. In your own words, describe an isometry that would map △DEF onto △PQR in the figure at the right. (You may make use of the lines on the graph paper if you wish.)

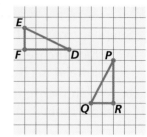

11. In the figure below, $\overline{PH} \cong \overline{VR}$ and $\angle V \cong \angle P$. For each part, a congruence condition is given. Tell what additional information you must know to use this condition.

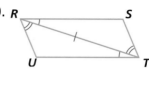

a. AAS Congruence Theorem
b. ASA Congruence Theorem
c. SAS Congruence Theorem
d. SSS Congruence Theorem

12. Sarah made a triangle with a pencil, a pen, and a straw. Why will anyone using objects of the same length get the same triangle?

13. In the hang glider sail pictured at the right, the seam $\overline{FB}$ bisects $\angle SFP$ and $\overline{PF} \cong \overline{SF}$. William only has a pattern for the left side. Explain why this pattern can also be used for the right side.

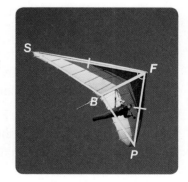

14. In one of the possible proofs of the SSS Congruence Theorem, you come to a place where you know what is marked in the figure at the right.

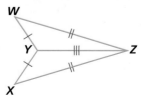

a. What does the Kite Symmetry Theorem tell you about this figure?
b. Explain how this makes △WYZ ≅ △XYZ.

15. Finish this proof.

    Given m∠FDE = m∠C'DE; G, D, and E are collinear.

    Prove $\overleftrightarrow{EG}$ bisects ∠FDC'.

    (This fills in part of the argument in the proofs of the SAS and ASA Congruence Theorems.)

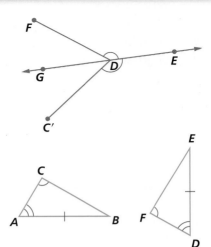

16. Complete this proof of the AAS Congruence Theorem.

    Given ∠A ≅ ∠D; ∠C ≅ ∠F; $\overline{AB}$ ≅ $\overline{DE}$

    Prove △ABC ≅ △DEF

    Proof

Conclusions	Justifications
a. ∠A ≅ ∠D	a. Given
b. ∠C ≅ ∠F	b. Given
c. ∠B ≅ ∠E	c. ___?___
d. $\overline{AB}$ ≅ $\overline{DE}$	d. Given
e. ___?___	e. ___?___

REVIEW

17. Suppose two regular octagons are congruent.
    a. Which theorem explains why these octagons can be inscribed in circles? (**Lesson 6-8**)
    b. Which theorem explains why these circles must have the same radius? (**Lesson 5-2**)

18. **True or False** If △ABC and △ABD are both inscribed in ⊙O, then m∠C = m∠D. (**Lesson 6-3**)

19. Trace the figure at the right, and draw the shortest path from E to F that also touches k. Explain why the path you drew is the shortest path. (**Lesson 4-3**)

20. State the Substitution Property of Equality, and write a short example illustrating it. (**Lesson 3-4**)

EXPLORATION

21. Draw points A, B, C, and D such that each angle in the triangle formed by A, B, and C is congruent to an angle in the triangle formed by A, C, and D, but the triangles are not congruent.

QY ANSWERS

1. Yes; the SSS Theorem guarantees that the two triangles will be congruent.

2. No. We are not sure that the angle is included between the two sides, so we do not have SAS.

3. Answers vary. Sample:

Lesson

7-3

Using Triangle Congruence Theorems

The figure below is a square.

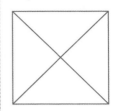

a. How many pairs of alternate interior angles are shown in the figure?

b. How many pairs of vertical angles are there in the figure?

▶ **BIG IDEA** One way to prove the congruence of sides (or angles) is to show that they are corresponding sides (or angles) of congruent triangles.

To use any of the triangle congruence theorems, you need to know only that three particular parts (SSS, SAS, ASA, or AAS) of one triangle are congruent to the corresponding three parts of another triangle. The theorems then enable you to conclude that the triangles are congruent.

In Example 1 below, the SAS Congruence Theorem is used to prove that two triangles are congruent. This requires that two pairs of sides and one pair of included angles must be established as congruent.

Example 1

Given: $\angle B$ and $\angle D$ are right angles, $\overline{BA} \cong \overline{DE}$, $\overline{BC} \cong \overline{DF}$.

Prove: $\triangle EDF \cong \triangle ABC$

Solution It is a good idea to mark a copy of the diagram with the given information. This makes it easier to see how to proceed.

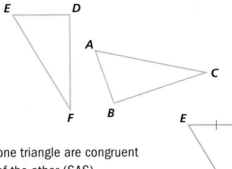

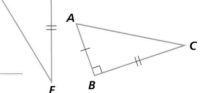

Two sides and the included angle of one triangle are congruent to two sides and the included angle of the other (SAS).

Conclusions	Justifications
1. $\overline{BA} \cong \overline{DE}$	1. Given
2. $\angle D$ and $\angle B$ are right angles.	2. Given
3. $m\angle D = 90, m\angle B = 90$	3. definition of right angle
4. $\angle D \cong \angle B$	4. Angle Congruence Theorem
5. $\overline{DF} = \overline{BC}$	5. Given
6. $\triangle EDF = \triangle ABC$	6. SAS Congruence Theorem (Steps 1, 4, 5)

In the following Activity, no congruent angles or sides are stated as given. But from each piece of given information and the figure, you can deduce that some parts of the triangles are congruent.

Activity

Given: $\overleftrightarrow{AB} \parallel \overleftrightarrow{XY}$; $\overline{AX}$ bisects $\overline{BY}$.

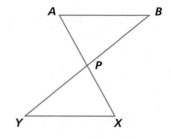

Step 1 Copy the figure and mark the given information onto the drawing.

Step 2 Name the pairs of congruent angles you can get from the given $\overleftrightarrow{AB} \parallel \overleftrightarrow{XY}$.

Step 3 Name the pair of congruent sides you can get from the given $\overline{AX}$ bisects $\overline{BY}$.

Step 4 There is a pair of vertical angles in this figure. Name them.

Step 5 Mark these three pairs of congruent parts on the drawing you made in Step 1. The marks should suggest that the triangles are congruent by either the ASA or AAS Congruence Theorem.

GUIDED

Example 2
Use information from the Activity to prove that $\triangle ABP \cong \triangle XYP$.

Proof

Conclusions	Justifications
1. $\overleftrightarrow{AB} \parallel \overleftrightarrow{XY}$	1. Given
2. $\angle B \cong$ _?_	2. Parallel lines $\Longrightarrow$ alternate interior angles are congruent (Parallel Lines Theorem)
3. $\overline{AX}$ bisects $\overline{BY}$.	3. Given
4. P is the midpoint of $\overline{BY}$.	4. definition of bisector of a segment
5. $\overline{BP} \cong$ _?_	5. definition of midpoint
6. $\angle APB \cong$ _?_	6. Vertical angles are congruent. (Vertical Angles Theorem)
7. $\triangle ABP \cong \triangle XYP$	7. ASA Congruence Theorem (Steps _?_, _?_, _?_)

Parts of Congruent Triangles

When triangles are congruent, every pair of their corresponding parts is congruent by the CPCF Theorem. Thus the SSS, SAS, ASA, and AAS theorems enable you to conclude that the three other pairs of corresponding parts are congruent when you started out knowing about three pairs. That makes these theorems quite powerful. For instance, in Example 2, in an additional Step 8 you could conclude that $\overline{AB} \cong \overline{XY}$ by the CPCF Theorem.

 QY

In Example 3, because there is a circle, there are congruent radii. These are used to obtain congruent triangles. Then the congruent triangles are used to show that two angles are congruent.

> ▸ QY
>
> What else could you conclude in Example 2 using the CPCF Theorem as justification?

Example 3

Given: In the figure at the right, both circles are centered at O and $\overline{BA} \cong \overline{DC}$.

Prove: $\angle AOB \cong \angle COD$

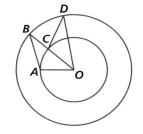

Solution A marked figure is shown below at the right.

We are given that a pair of sides are congruent ($\overline{BA} \cong \overline{DC}$).

The circles give us two more pairs of congruent sides. This is a sufficient condition for congruence by SSS.

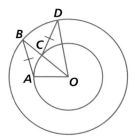

Conclusions	Justifications
1. There are two circles with center O.	1. Given
2. $\overline{OA} \cong \overline{OC}$	2. Radii of a circle are congruent.
3. $\overline{OB} \cong \overline{OD}$	3. Radii of a circle are congruent.
4. $\overline{BA} \cong \overline{DC}$	4. Given
5. $\triangle AOB \cong \triangle COD$	5. SSS Congruence Theorem (Steps 2, 3, 4)
6. $\angle AOB \cong \angle COD$	6. CPCF Theorem

The measures of the angles that were proved to be congruent in Example 3 are the magnitude of the rotation that maps one of the triangles onto the other. Recall that any congruence correspondence means that there is an isometry that maps one figure onto the other. As you learn about congruent triangles, don't forget that sometimes transformations can be used to prove congruence.

GUIDED

Example 4

Given: $\overline{KI} \cong \overline{TI}$; $\overrightarrow{IE}$ bisects $\angle KIT$.

Prove: $\overline{KE} \cong \overline{TE}$

Solution This problem has two quite different possible solutions. There are two triangles that appear to be congruent and the figure is symmetric, so you may be able to use transformation properties or congruent triangles to prove the segments are congruent. We give both proofs for the sake of comparison.

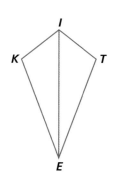

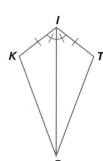

Congruent triangles proof

Conclusions	Justifications
1. $\overline{EI} \cong \overline{EI}$	1. Reflexive Property of Congruence
2. $\overrightarrow{IE}$ bisects $\angle KIT$.	2. Given
3. $\angle KIE \cong \angle TIE$	3. ___?___
4. $\overline{KI} \cong \overline{TI}$	4. Given
5. $\triangle KIE \cong \triangle TIE$	5. ___?___ Congruence Theorem (Steps 1, 3, 4)
6. $\overline{KE} \cong \overline{TE}$	6. ___?___

Transformation proof

Conclusions	Justifications
1. $\overrightarrow{IE}$ bisects $\angle KIT$.	1. Given
2. $\overline{KI} \cong \overline{IT}$	2. Given
3. Draw $\overline{KT}$.	3. Two points determine a line.
4. $\triangle KIT$ is isosceles.	4. ___?___
5. $r_{\overline{IE}}(K) = T$	5. Isosceles Triangle Symmetry Theorem
6. $r_{\overline{IE}}(E) = E$	6. ___?___
7. $r_{\overline{IE}}(\overline{KE}) = \overline{TE}$	7. Figure Transformation Theorem
8. $\overline{KE} \cong \overline{TE}$	8. ___?___

Questions

1. List all the triangle congruence theorems that are used in examples in this lesson.

2. Trace the picture at the right.
   a. Mark your picture to show that the radii of circle O are congruent.
   b. Suppose $\overline{AC} \parallel \overline{OP}$. Mark the angles that must be congruent.
   c. Suppose B is the midpoint of $\overline{AC}$. Mark what you conclude from this.

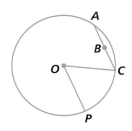

3. Nicky wrote a Step 5 in the proof in Example 1. His conclusion was $\overline{EF} \cong \overline{AC}$. What is the justification?

In 4–6, refer to Example 2.

4. Write the justification for Step 5 as an if-then statement.

5. Chloe said she would have written $\angle A \cong \angle X$ as the conclusion in Step 2 and all the other steps would be the same. This is true, but one justification would change. What is that justification and how would it change?

6. Jamie said he would have written $\angle A \cong \angle X$ and $\angle B \cong \angle Y$ in Step 2. Then he would have skipped Steps 3, 4, and 5 and used the vertical angles in Step 6 to prove the triangles congruent by AAA Triangle Congruence. What is wrong with Jamie's reasoning?

7. Indicate how you could change the proof in Example 3 to prove that $\angle BAO \cong \angle DCO$.

8. Fill in the missing justifications in the following proof.

   Given $\overline{BD} \perp \overline{AC}$; $\overline{BD}$ bisects $\angle ABC$.

   Prove $\triangle ABD \cong \triangle CBD$

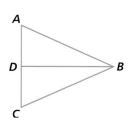

   Proof

Conclusions	Justifications
a. $\overline{BD} \perp \overline{AC}$	a. ___?___
b. $\angle ADB \cong \angle CDB$	b. ___?___
c. $\overline{BD} \cong \overline{BD}$	c. ___?___
d. $\overline{BD}$ bisects $\angle ABC$.	d. ___?___
e. $\angle ABD \cong \angle CBD$	e. ___?___
f. $\triangle ABD \cong \triangle CBD$	f. ___?___

APPLYING THE MATHEMATICS

9. Fill in the missing conclusions in the following proof.

Given M is the midpoint of $\overline{QT}$.
M is the midpoint of $\overline{RS}$.

Prove $\angle R \cong \angle S$

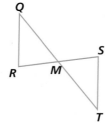

Proof

Conclusions	Justifications
a. M is the midpoint of $\overline{QT}$.	a. Given
b. ___?___	b. definition of midpoint
c. ___?___	c. Vertical angles are congruent.
d. ___?___	d. Given
e. ___?___	e. definition of midpoint
f. ___?___	f. SAS Congruence Theorem (Steps b, c e)
g. ___?___	g. CPCF Theorem

In 10–12, write a complete proof. Include a marked figure.

10. **Given** $\odot O$;
$\overline{AB} \cong \overline{CB}$

Prove $\triangle ABO \cong \triangle CBO$

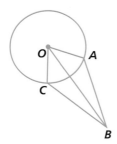

11. **Given** $\overrightarrow{RB}$ bisects $\angle ARC$;
$\angle A \cong \angle B$; $\overline{AP} \cong \overline{BC}$.

Prove $\overline{AR} \cong \overline{BR}$

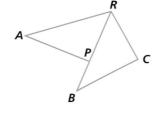

12. **Given** $\odot P$ with diameters
$\overline{SQ}$ and $\overline{RT}$

Prove $\overline{QR} \cong \overline{ST}$

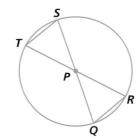

13. In quadrilateral *EFGH*, $\angle E$ is a right angle, $\angle G$ is a right angle, and $\overline{EF} \parallel \overline{HG}$.
    a. Draw an accurate picture of this situation.
    b. Prove that diagonal $\overline{HF}$ splits the quadrilateral into two congruent triangles.

REVIEW

14. a. List all five triangles shown in the figure at the right.
    b. From the marked information, two pairs of triangles can be proved to be congruent. Name them, with vertices in corresponding order. (**Lessons 7-4, 7-2**)

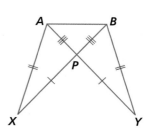

15. Congruent parts are marked in the figure at the right. (**Lesson 7-2**)

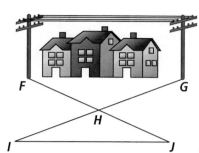

    a. The two triangles can be proved congruent. Which vertices correspond?

    b. Which triangle congruence theorem can be used to prove the triangles congruent?

16. Electric wires run from post to post over the buildings shown at the right. We can't measure the length of wire needed because the buildings are in the way. Place a stake at H. Then mark off J on $\overleftrightarrow{FH}$ so that $FH = HJ$. Now mark off I on $\overleftrightarrow{GH}$ so that $GH = HI$. What triangle congruence theorem indicates that $\triangle FGH \cong \triangle JIH$, so that the distance from I to J is equal to the distance between the posts? (**Lesson 7-2**)

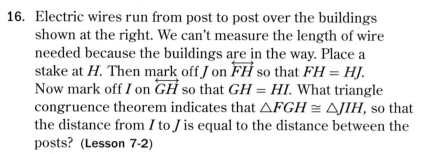

17. **True or False** Tristan and Trisha each made a triangle out of straws with lengths 3 cm, 4 cm, and 6 cm.

    a. The two triangles must be congruent. (**Lesson 7-1**)

    b. The two triangles must have the same orientation. (**Lesson 5-1**)

18. **True or False** If a triangle has two lines of symmetry, then it must have three lines of symmetry. (**Lesson 6-8**)

19. $\triangle ABC$ is inscribed in $\odot P$ and in $\odot Q$. What can you say about P and Q? (**Lesson 6-3**)

20. Suppose T is a translation with magnitude t, R is a translation with magnitude r, and S is a translation with magnitude s. Suppose P is a point so that $R \circ S \circ T(P) = P$. What does the Triangle Inequality Postulate tell you about the magnitudes of these translations? (**Lessons 4-4, 1-7**)

21. Suppose $\overrightarrow{PA}$ is a ray with vertex P and x is a positive number less than 180.

    a. Which postulate of Euclidean geometry says that there are two different rays that intersect $\overrightarrow{PA}$ at P that each form an angle with measure x? (**Lesson 3-1**)

    b. Name the type of isometry that maps one of the rays discussed in Part a onto the other. (**Lesson 4-2**)

EXPLORATION

22. From the two pieces of given information and this figure, make and justify at least three conclusions.

    Given $\overline{AD} \cong \overline{CD}$; $\overrightarrow{DB}$ bisects $\angle ADC$.

QY ANSWER

$\overline{AP} \cong \overline{XP}$, $\angle A \cong \angle X$

Lesson
7-4
Overlapping Triangles

Vocabulary

overlapping figures
nonoverlapping figures

▶ **BIG IDEA** Overlapping triangles in figures can be congruent triangles.

A rectangle is not rigid, which could present a problem if it is used in a bridge support. One of the easiest ways to strengthen a rectangle is to add diagonal supports. These supports create a figure like *ABCDE* in the picture below.

Mental Math

State the SAS Congruence Theorem as an if-then statement.

Brooklyn Bridge in New York City

How many triangles in the marked part of the figure above appear to be congruent? Many people initially only see two pairs of triangles that appear to be congruent: △*AED* and △*BEC,* and △*AEB* and △*DEC.* To prove that the diagonals of *ABCD* are congruent, it is convenient to look at the *overlapping triangles,* such as △*ABC* and △*DCB.* **Overlapping figures** have some part of their interiors in common. (Figures that do not overlap are called **nonoverlapping figures.**)

 QY

When working with overlapping triangles, you might want to:

• Draw the figure twice and shade the triangles.

• Draw the figure and then draw the overlapping triangles separately.

• Ask yourself what transformation was used to form the figure. This can help you find the triangles you should prove congruent or to identify the corresponding parts needed.

▶ **QY**

In figure *ABCDE* above, name another pair of overlapping triangles that have the diagonals as corresponding parts.

Example 1

Given: ∠D ≅ ∠B; $\overline{AM} ≅ \overline{CM}$

Prove: $\overline{BA} ≅ \overline{DC}$

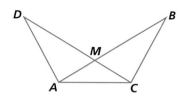

Solution A convenient way to approach proofs is to start by analyzing what you are trying to prove. In this case you need to prove $\overline{BA} ≅ \overline{DC}$. One way to prove that segments in triangles are congruent is the CPCF Theorem.

Look for triangles that have $\overline{BA}$ and $\overline{DC}$ as corresponding parts.

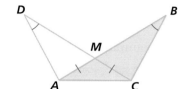

△BCA and △DAC share side $\overline{CA}$ and ∠D ≅ ∠B is given. So, we already know that one pair of angles and one pair of sides of the triangles are congruent. To prove that the triangles are congruent, we need an additional pair of sides or angles that are congruent. Now look at △AMC. Because $\overline{AM} ≅ \overline{CM}$, ∠MCA ≅ ∠MAC by the Isosceles Triangle Base Angles Theorem. This implies that △BAC and △DCA are congruent by the AAS Congruence Theorem. Thus, $\overline{BA} ≅ \overline{DC}$ by the CPCF Theorem.

With overlapping triangles, keeping track of corresponding vertices is important.

GUIDED

Example 2

Given: ∠3 ≅ ∠4; ∠JLN ≅ ∠KLM

Prove: $\overline{JL} ≅ \overline{KL}$

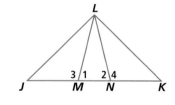

Solution Again a good idea is to use overlapping triangles.

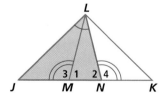 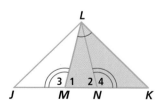

Proof

Conclusions	Justifications
1. ∠1 and ∠3 are a linear pair. ∠2 and ∠4 are a linear pair.	1. Definition of linear pair
2. ∠1 and ∠3 are supplementary. ∠2 and ∠4 are ___?___.	2. ___?___

(continued on next page)

3. $\angle 3 \cong \angle 4$	3. Given
4. ___?___	4. Supplements of congruent angles are congruent.
5. $\overline{LN} \cong \overline{LM}$	5. ___?___
6. $\angle JLN \cong \angle KLM$	6. Given
7. ___?___	7. ASA Congruence Theorem (Steps 4, 5, 6)
8. $\overline{JL} \cong \overline{KL}$	8. ___?___

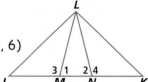

In Example 3, we show the overlapping triangles by separating them.

Example 3

Given: $\overline{DB} \cong \overline{EB}$, $\overline{AD} \cong \overline{CE}$

Prove: $\angle A \cong \angle C$

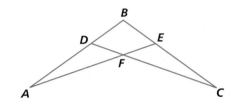

Solution There are two pairs of triangles that have $\angle A$ and $\angle C$ as corresponding parts. You may be inclined to try to prove $\triangle DAF \cong \triangle ECF$ because vertical angles are congruent. Still, there is not enough information to prove this is true. Instead, consider the overlapping triangles $\triangle ABE$ and $\triangle CBD$.

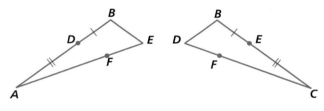

Proof

Conclusions	Justifications
1. $\overline{DB} \cong \overline{EB}$	1. Given
2. $\angle ABE \cong \angle CBD$	2. Reflexive Property of Congruence
3. $\overline{AD} \cong \overline{CE}$	3. Given
4. $AD = CE$, $DB = EB$	4. Segment Congruence Theorem
5. $AB = CE$	5. Additive Property of Equality ($AD + DB = CE + EB$)
6. $\overline{AB} \cong \overline{CB}$	6. Segment Congruence Theorem
7. $\triangle ABE \cong \triangle CBD$	7. SAS Congruence Theorem (Steps 1, 2, 6)
8. $\angle A \cong \angle C$	8. CPCF Theorem

Questions

1. Why is a rectangle not a good figure as a bridge support?

In 2–4, use the figure at the right.

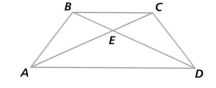

2. Which pair(s) of triangles appear to be congruent and have $\overline{CE}$ and $\overline{BE}$ as corresponding parts?

3. Which pair(s) of triangles appear to be congruent and have $\overline{BD}$ and $\overline{CA}$ as corresponding parts?

4. Which pair(s) of triangles appear to be congruent and have $\overline{BA}$ and $\overline{CD}$ as corresponding parts?

In 5 and 6, draw two overlapping congruent triangles that have:

5. a common angle.

6. a common side.

7. The figure at the right is identical to the figure in Example 1. Provide the justifications for each step.

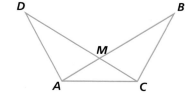

Given $\angle D \cong \angle B$; $\overline{AM} \cong \overline{CM}$

Prove $\angle DAC \cong \angle BCA$

Proof

Conclusions	Justifications
a. $\angle D \cong \angle B$	a. ___?___
b. $\overline{AM} \cong \overline{CM}$	b. ___?___
c. $\angle MAC \cong \angle MCA$	c. ___?___
d. $\overline{AC} \cong \overline{CA}$	d. ___?___
e. $\triangle BCA \cong \triangle DAC$	e. ___?___
f. $\angle DAC \cong \angle BCA$	f. ___?___

8. Use Example 2 as a guide to write a proof.

Given $\angle 2 \cong \angle 3$,
$\angle ACD \cong \angle BCE$

Prove $\angle A \cong \angle B$

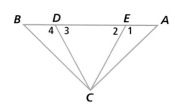

APPLYING THE MATHEMATICS

In 9 and 10, write a proof using the given figure.

9. **Given** $\angle 2 \cong \angle 4$, $\overline{KJ} \cong \overline{KL}$
   Prove $\angle J \cong \angle L$

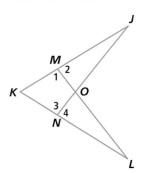

10. **Given** Regular hexagon *ABCDEF*
    Prove $\overline{AC} \cong \overline{BF}$

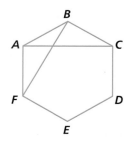

11. Complete the proof.

   Given $\overline{AE} \cong \overline{DE}$, $\overline{EC} \cong \overline{EB}$

   Prove $\angle ABC \cong \angle DCB$

   Proof

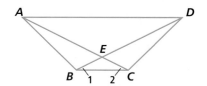

Conclusions	Justifications
a. $\overline{AE} \cong \overline{DE}$, $\overline{EC} \cong \overline{EB}$	a. ___?___
b. $\angle$ _?_ $\cong \angle$ _?_	b. 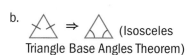 (Isosceles Triangle Base Angles Theorem)
c. $AE = DE$, $EC = EB$	c. ___?___
d. $AC = DB$	d. ___?___
e. $\overline{AC} \cong \overline{DB}$	e. ___?___
f. $\overline{BC} \cong \overline{CB}$	f. ___?___
g. $\triangle$ _?_ $\cong \triangle$ _?_	g. SAS Congruence Theorem (Steps b, e, f)
h. $\angle ABC \cong \angle DCB$	h. ___?___

12. Complete the proof. Parts of the first three steps have been done for you.

Given Points *F, G, H,* and *I* on ⊙*E*; ∠*HGF* ≅ ∠*IFG*

Prove $\overline{HG} \cong \overline{IF}$

Proof

Conclusions	Justifications
a. m∠*GHF* = $\frac{1}{2}$(m$\widehat{GF}$)	a. The measure of an inscribed angle is half the intercepted arc. (Inscribed Angle Theorem)
b. m∠*FIG* = ___?___	b. ___?___
c. ∠*GHF* ≅ ∠*FIG*	c. If two angles have the same measure, then they are congruent.
⋮	⋮

REVIEW

13. In the figure at the right, $AB = CD$ and $\overline{AB} \parallel \overline{CD}$. Prove that $BD = AC$. **(Lesson 7-3)**

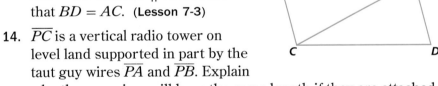

14. $\overline{PC}$ is a vertical radio tower on level land supported in part by the taut guy wires $\overline{PA}$ and $\overline{PB}$. Explain why the guy wires will have the same length if they are attached to the ground at the same distance from *C*. **(Lesson 7-2)**

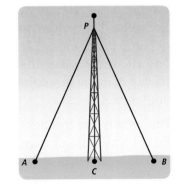

15. **True or False** If a figure has translation symmetry and reflection symmetry, then it must have rotation symmetry. **(Lesson 6-9)**

16. What is the measure of each interior angle of a regular pentagon? **(Lesson 6-8)**

17. Give two definitions of the word *isometry*. **(Lessons 5-1, 4-7)**

18. Trace ⊙*O* at the right, then rotate it 120° and 240° about *C*. **(Lesson 3-2)**

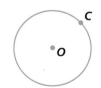

EXPLORATION

19. **Given** ⊙*I*, ⊙*G*, ⊙*H*, with points *L*, *J*, and *K* the other three intersections of each pair of circles

    Prove $\overline{GK} \cong \overline{IJ}$

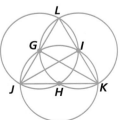

The SSA Condition and HL Congruence

Vocabulary

legs

hypotenuse

concentric circles

▶ **BIG IDEA** The abbreviations SsA and HL stand for two other combinations of measures of corresponding sides and angles in two triangles that are sufficient for the congruence of the triangles.

SSA Condition

Because there are AAS and ASA Congruence Theorems, and there is a congruence theorem with two sides and an included angle (SAS), it is natural to ask what happens if the angle is *not* the included angle. We call this the *SSA condition*.

Examine $\triangle ABC$ and $\triangle XYZ$ at the right. There are two pairs of congruent sides, $\overline{AB} \cong \overline{XY}$ and $\overline{BC} \cong \overline{YZ}$. Also, there is a pair of congruent nonincluded angles, $\angle A \cong \angle X$.

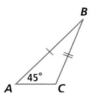

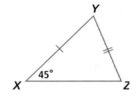

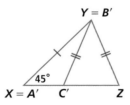

But clearly the triangles are not congruent. In fact, a translation image of $\triangle ABC$ fits snugly into one corner of $\triangle XYZ$. Thus, in general, the SSA condition does *not* guarantee the congruence of triangles.

However, there are times when the SSA condition *does* determine a unique triangle.

For the SSA condition to determine a unique triangle all of the time, the length of the side opposite the given angle must be longer than the other given side. We identify the theorem as the *SsA Congruence Theorem* to emphasize that the longer side (big S) is opposite the given angle. The proof is long, so we number the steps even though this proof is in paragraph form.

SsA Congruence Theorem

If two sides and the angle opposite the longer of the two sides in one triangle are congruent respectively to two sides and the corresponding angle in another triangle, then the triangles are congruent.

Given $\overline{AB} \cong \overline{XY}$, $\overline{AC} \cong \overline{XZ}$, $\angle C \cong \angle Z$, and $AB > AC$.

Prove $\triangle ABC \cong \triangle XYZ$

Drawing

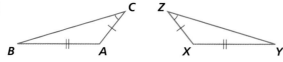

Proof The proof is like the proofs of the SSS and ASA triangle congruence theorems in Lesson 7-2.
In Steps 1–3, we map $\triangle ABC$ onto a conveniently located congruent image $\triangle A'B'C'$.

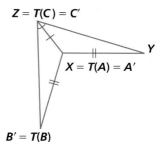

1. Because $\overline{AC} \cong \overline{XZ}$, there is an isometry T with $T(\overline{AC}) = \overline{XZ}$, $T(A) = A' = X$, and $T(C) = C' = Z$. Furthermore, T can be chosen so that $T(B) = B'$ is on the other side of $\overleftrightarrow{XZ}$ from Y.

2. By the definition of congruence, $\triangle ABC \cong \triangle A'B'C'$.

3. It is given that $\angle BCA \cong \angle YZX$. By the CPCF Theorem, $\angle BCA \cong \angle B'C'A'$, and so, by the Transitive Property of Congruence, $\angle B'C'A' \cong \angle YZX$. This makes $\overrightarrow{ZX}$ the bisector of $\angle B'ZY$.

In Steps 4–8, we show that $\triangle A'B'C'$ is congruent to $\triangle XYZ$. This is done by showing that $\triangle XYZ$ is the reflection image of $\triangle A'B'C'$ over $\overleftrightarrow{ZX}$.

4. C' and A' are the points Z and X on the reflection line, so by the definition of reflection, $r(C') = Z$ and $r(A') = X$.

5. To find $r(B')$, consider the auxiliary circle with center X and radius XY. B' is on this circle because $XB' = XY$. $r(B')$ is on the circle because $\overleftrightarrow{XZ}$ contains a diameter, and any line that contains a diameter is a symmetry line for a circle.

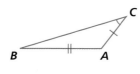

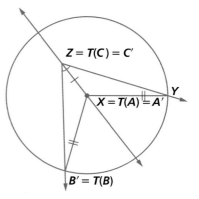

(continued on next page)

6. *Z* is in the interior of this circle because *XY* > *XZ*. Thus $\overrightarrow{ZY}$ intersects the circle in exactly one point. *r*(*B'*) must also be on $\overrightarrow{ZY}$ because of the Side-Switching Theorem. Consequently, *r*(*B'*) = *Y*, the only point of intersection of the circle and $\overrightarrow{ZY}$.

7. By the Figure Reflection Theorem (Steps 4 and 6), *r*(△*A'B'C'*) = △*XYZ*.

8. By the definition of congruence, △*A'B'C'* ≅ △*XYZ*.

9. By the Transitive Property of Congruence (Steps 2 and 8), △*ABC* ≅ △*XYZ*.

Notice that if *XZ* > *XY*, then the given congruent angles are no longer opposite the longer congruent sides. In this case, the point *Z* is outside the circle and the ray intersects the circle in two points. Then there are two choices for *B'* and we cannot conclude that *r*(*B'*) = *Y*. This is why *XY* must be greater than *XZ*.

The proof of the SsA Congruence Theorem is the longest in this book. But there are much longer proofs in mathematics. Some proofs take many pages of writing, and a few proofs are the lengths of books. The SsA Congruence Theorem is not in Euclid's *Elements,* nor is it found in many geometry books.

You now have studied the five triangle congruence theorems: SSS, SAS, ASA, AAS, and SsA. Now we turn to a special case of SsA that is found in most other geometry books, the HL condition.

Actress Danica McKellar co-authored a nine-page proof while a student at UCLA (University of California at Los Angeles).

The HL Condition

Right triangles play an essential role in geometry and mathematics in general. Recall that in a *right triangle,* the **legs** are the sides that include the right angle and the **hypotenuse** is the side opposite the right angle. We know from the Unequal Sides Theorem that in any triangle, the longest side is the side opposite the largest angle. So, if the hypotenuse and one leg of one right triangle are congruent to the hypotenuse and leg of another right triangle, then the triangles satisfy the SsA condition and are congruent. This argument proves the following theorem.

> **HL Congruence Theorem**
>
> If, in two right triangles, the hypotenuse and a leg of one are congruent to the hypotenuse and a leg of the other, then the two triangles are congruent.

▶ **READING MATH**

In this book, the term *legs* is used only for right triangles. However, this word is sometimes used in two other ways in geometry: The legs of a non-equilateral isosceles triangle are the two congruent sides (the sides that are not the base). The legs of a trapezoid that is not a parallelogram are the two nonparallel sides (the sides that are not the bases).

Note that the HL (Hypotenuse–Leg) Congruence Theorem applies only to right triangles. This makes sense because they are the only triangles that have a hypotenuse. To use this theorem in a proof, you must make sure that you are dealing with right triangles.

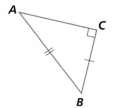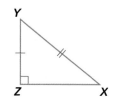

Example 1

$\overline{AB}$ represents a vertical radio tower. Explain why guy wires $\overline{AC}$ and $\overline{AD}$ of the same length will reach the ground at the same distance from B.

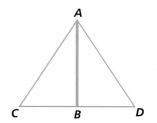

Solution In △ABC and △ABD, $\overline{AB} \cong \overline{AB}$ (Reflexive Property of Congruence), $\overline{AC} \cong \overline{AD}$ (Given), and △ABD and △ABC are right triangles with right angles at B (Given). Thus, △ABC ≅ △ABD by HL Congruence, so BC = BD by the CPCF Theorem.

Once triangles are proved congruent, you can use the CPCF Theorem.

GUIDED

Example 2

Given: ⊙A with diameter $\overline{BC}$, $\overline{CE} \cong \overline{BD}$

Prove: $\overline{CD} \parallel \overline{EB}$

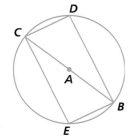

Proof

Conclusions	Justifications
1. ⊙A	1. ___?___
2. ∠D and ∠E are right angles.	2. If an angle is inscribed in a semicircle, then it is a right angle.
3. △CDB and △BEC are right triangles.	3. ___?___
4. $\overline{CB} \cong \overline{BC}$	4. ___?___
5. $\overline{CE} \cong \overline{BD}$	5. ___?___
6. △CDB ≅ △BEC	6. ___?___ Congruence Theorem (Steps 2, 4, and 5)
7. ∠DCB ≅ ∠EBC	7. ___?___
8. $\overline{CD} \parallel \overline{EB}$	8. Alternate Interior Angles Theorem

Questions

COVERING THE IDEAS

In 1 and 2, provide two side lengths and the measure of a nonincluded angle that satisfy the given condition.

1. There are two possible noncongruent triangles

2. There is exactly one possible triangle.

3. Draw a diagram of a right triangle. Label its right angle, legs, and hypotenuse.

4. **Fill in the Blanks** Use either HL or SsA.

   The __?__ condition is a special case of the __?__ condition.

5. When does the SSA condition lead to congruence?

In 6–9, use the information given in the figure. Tell which triangles, if any, are congruent. Explain using the triangle congruence theorems. Be sure to name the vertices in corresponding order.

6.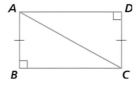

7.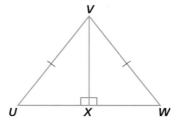

8.

9.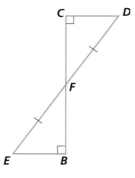

APPLYING THE MATHEMATICS

10. During a fire, two ladders of the same length just reach the second-floor windows. How do you know these ladders are the same distance from the house?

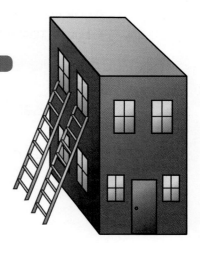

For 11–13, write a proof using the given figure.

11. Given ⊙M with $\overline{PN} \cong \overline{RN}$
Prove PNRT is a kite.

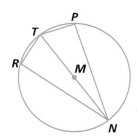

12. Given $\overline{AC} \perp \overline{AB}$,
$\overline{DB} \perp \overline{DC}$,
$\overline{DB} \cong \overline{AC}$
Prove $\angle ECB \cong \angle EBC$

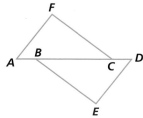

13. Given $\overline{AB} \cong \overline{DC}$, $\overline{AF} \cong \overline{DE}$,
∠F and ∠E are right angles
Prove $\overline{FC} \parallel \overline{BE}$

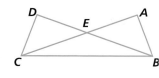

14. If two circles have the same center and different radii, they are called **concentric circles**. The two concentric circles at the right have center F, with $KF < JF$. If $\angle FKJ \cong \angle FHI$, complete the proof that $\triangle FKJ \cong \triangle FHI$.

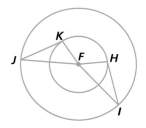

Conclusions	Justifications
a. $KF < JF$	a. ___?___
b. $\overline{JF} \cong \overline{IF}$	b. ___?___
c. $\overline{KF} \cong \overline{HF}$	c. ___?___
d. $\angle FKJ \cong \angle FHI$	d. ___?___
e. $\triangle FKJ \cong \triangle FHI$	e. ___?___

REVIEW

15. In the figure below, FGHI is an isosceles trapezoid. Prove that $GI = FH$. **(Lesson 7-4)**

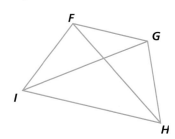

16. In the figure below, △*JKN* is an isosceles triangle with vertex angle *J* and *KL* = *MN*. Prove that m∠*JLM* = m∠*JML*. **(Lessons 7-3, 6-2)**

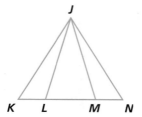

17. Trace the figure below. Use it to create a figure that has 5-fold rotation symmetry about *O*. **(Lesson 6-7)**

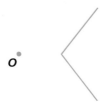

18. From which of the types of quadrilaterals in the quadrilateral hierarchy do parallelograms inherit properties? **(Lesson 6-4)**

19. In △*ABC*, m∠*A* = 97 and *BC* = 3. If you know that *AB* and *AC* are integers, find their lengths. **(Lessons 6-2, 1-7)**

20. **True or False** If the sum of the interior angles in a convex *n*-gon is greater than the sum of the interior angles in a convex *m*-gon, then *n* > *m*. **(Lesson 5-7)**

21. Give an example of a transformation that preserves angles, betweeness, and collinearity, but not distance. **(Lessons 5-1, 3-7)**

EXPLORATION

22. Plot a rectangle *ABCD* on a coordinate plane. Locate *E*, *F*, *G*, and *H* such that they are the midpoints of the consecutive sides of the rectangle.
    a. What special quadrilateral is formed by connecting *E*, *F*, *G*, and *H* in order? Justify your answer.
    b. Find the midpoints of quadrilateral *EFGH* and connect them in order. What special type of quadrilateral is formed? Justify your answer.
    c. Make a conjecture about what would happen if you continue the pattern.

Lesson
7-6 Tessellations

▶ **BIG IDEA** Tessellations are an application of congruence that is both useful and pretty.

The floor of your classroom, unless carpeted or solid-colored, is likely covered with a *tessellation*. When covering a floor, the most obvious concern is that the flooring completely cover the floor without leaving gaps or overlaps. A second concern is cost. With an inexpensive tile, both of these concerns can be addressed with tessellations.

Mental Math

Is 360 evenly divisible by the measure of an interior angle of

a. an equilateral triangle?

b. a regular hexagon?

c. a regular octagon?

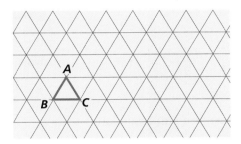

The picture above demonstrates that equilateral triangles can tile the plane without leaving spaces or overlapping. A covering of a plane with nonoverlapping congruent regions is called a **tessellation**. The region is called a **fundamental region** for the tiling. $\triangle ABC$ is a fundamental region for the tessellation shown above. We say that $\triangle ABC$ **tessellates** the plane. A key question is whether a given region can tessellate the plane. One application of this idea is finding the area of a figure. If you know the area of the fundamental region, and how many copies of the region it takes to tessellate a figure, then you know the area of the figure.

Figures That Tessellate

Any triangle can tessellate the plane. Activity 1 uses a DGS to show that this is true for scalene triangles.

Activity 1

MATERIALS DGS

Begin with a clear DGS screen. The screenshot at the right has the steps shown in different parts of the screen to make it clearer. You should build off of the same triangle.

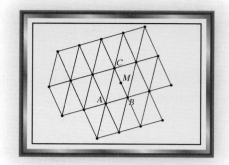

Step 1 Construct scalene $\triangle ABC$.

Step 2 Construct the midpoint of $\overline{BC}$. Call it M.

Step 3 Rotate $\triangle ABC$ 180° about M.

Step 4 Which kind of special quadrilateral is $ABA'C$? How do you know?

Step 5 Repeatedly translate $ABA'C$ by the vectors $\overrightarrow{AC}$ and $\overrightarrow{CA}$.

Step 6 Repeatedly translate the entire figure by the vectors $\overrightarrow{AB}$ and $\overrightarrow{BA}$.

The result covers as much of the plane as you wish.

 QY1

▸ **QY1**

Can any set of congruent triangles tessellate? Why or why not?

Tessellations with Other Polygons

It is not as easy to see, but every quadrilateral region can tessellate! Because the sum of the measures of the angles of a quadrilateral is 360°, it is possible to have one of each angle meeting at a given point. Such a tessellation is shown at the right. (This tessellation was generated using the procedure of Question 6 of this lesson.)

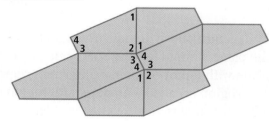

Activity 2

Copy the quadrilateral at the right onto a large sheet of paper. Number the angles. Follow the above pattern or the steps of Question 6 to make enough copies to convince yourself that the quadrilateral tessellates.

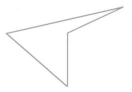

By determining angle measures in figures, you can tell whether certain regular polygons will tessellate.

Example

a. Will a regular pentagon tessellate the plane?

b. Will a regular hexagon tessellate the plane?

Solution To answer these questions, the measure of an individual angle in each figure must be determined.

a. The sum of the measures of the angles in a pentagon is
$(5 - 2) \cdot 180° = 540°$, so each angle in a regular pentagon is $\frac{540°}{5} = 108°$. Because 360 is not evenly divisible by 108, a regular pentagon will not tessellate. At the right is a picture of 3 regular pentagons around a point, but notice that there is a gap that cannot be filled.

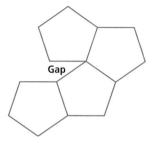
Gap

b. The sum of the measures of the angles in a hexagon is $(6 - 2) \cdot 180° = 720°$, so each angle in a regular hexagon measures $120°$. Because 360 is divisible by 120, a regular hexagon will tessellate. At the right is a sample of such a tessellation.

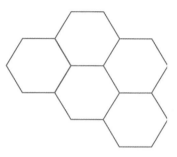

🛑 **QY2**

▶ **QY2**

Name at least one regular polygon with less than 10 sides that does not tessellate.

Tessellations with Fundamental Regions That Are Not Polygons

By modifying the sides of a polygon, you can convert a simple fundamental region into one that has curved sides and could have a complicated shape.

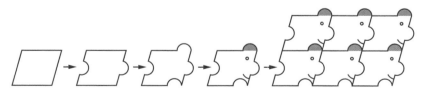

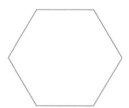

Activity 3

MATERIALS thick paper

Copy the regular hexagon at the right onto a piece of thick paper. Modify it to create a new figure that will tessellate the plane.

The idea of a tessellation is an old one. The word "tessellate" comes from a Latin word meaning "small stone." Small stones, put together into mosaics, covered the floors of many Roman buildings. The Moors, whose religion (Islam) does not allow any pictures of animate objects in their places of worship, used many different tessellations in decorating their mosques.

Tessellations of Other Surfaces

On a globe, you might find the equator (the circle through C, D, and E at the right) and eight lines of longitude such that any two consecutive lines are three hours apart. This tessellates the globe with 16 spherical triangles. Because there are eight in each hemisphere, the angle of each triangle with a vertex at the North Pole N has measure 45°. Not every spherical triangle tessellates the sphere.

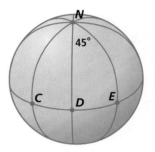

Questions

COVERING THE IDEAS

1. What is a tessellation?

2. Besides floor tiles, what is another application of tessellations?

3. How can you tell if a regular polygon tessellates the plane?

4. What culture inspired tessellations as artwork in buildings?

5. What key question concerning tessellations is discussed in this lesson?

6. a. Using a DGS, tessellate the plane with quadrilateral $ABOC$ at the right, where X, Y, and Z are the midpoints of $\overline{BO}$, $\overline{AB}$, and $\overline{AC}$, respectively.

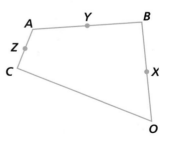

   Step 1 Construct quadrilateral $ABOC$.

   Step 2 Rotate $ABOC$ 180° about X. Call the image $A'OBC'$.

   Step 3 Rotate $A'OBC'$ 180° about Y', where Y' is the image of Y. Call the image $OA'B''C''$.

   Step 4 Rotate $OA'B''C''$ 180° about Z'', where Z'' is the image of the original Z. Call the image $C''A''CO$.

   b. There are now four angles with vertex O. How do you know that these angles fit around O exactly?

   c. What needs to be done to complete the tessellation?

   d. Drag point A so that the original $ABCD$ is nonconvex. Does the quadrilateral still tessellate?

In 7 and 8, trace the figure repeatedly to show part of a tessellation using the given figure as a fundamental region.

7.

8.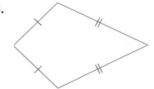

9. Which regular polygon of fewer than 7 sides cannot be a fundamental region for a tessellation?

10. Can a regular octagon tessellate the plane? Explain your answer.

APPLYING THE MATHEMATICS

11. In the first few steps of a tessellation of a quadrilateral at the right, the midpoints *G, E, F,* and *H* seem to be collinear. Explain why this is true.

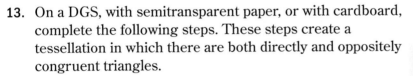

12. Suppose that an architect designed a regular hexagonal floor for a stylish new bathroom. The area of the room is 25 square meters. The area of each tile is 625 square millimeters. Assuming the tiles tessellate the floor, how many will be necessary to cover the floor?

13. On a DGS, with semitransparent paper, or with cardboard, complete the following steps. These steps create a tessellation in which there are both directly and oppositely congruent triangles.

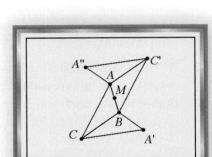

Step 1 Construct △*ABC*.

Step 2 Reflect △*ABC* over $\overleftrightarrow{BC}$. Call the image △*A'BC*.

Step 3 Find the midpoint of $\overline{AB}$; call this point *M*.

Step 4 Rotate *ACA'B* 180° about *M*. Call the image *A"C'BA*.

Step 5 Translate the union of the preimage and image by the vectors $\overrightarrow{CB}$ and $\overrightarrow{BC}$.

14. In the construction of Question 13, prove that *ABA'C* is a kite.

15. In the construction of Question 13, prove that *AC'BC* is a parallelogram.

REVIEW

16. In the figure at the right, *O* is the center of the circle and *LM* = *QP*. Prove that △*LMN* ≅ △*QPR*. (**Lessons 7-5, 6-3**)

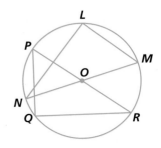

17. **True or False** If two isosceles triangles have congruent vertex angles and congruent bases, then they are congruent. (**Lesson 7-2**)

18. **True or False** No trapezoids have 4-fold rotation symmetry. (**Lessons 6-7, 6-6**)

19. In the figure at the right, m∠V = 130, m∠W = 34, m∠U = 26, and m∠X = 70. Notice that the sum of measures of these angles is not 360. Is this a counterexample to the Quadrilateral-Sum Theorem? Explain why or why not. (**Lesson 5-7**)

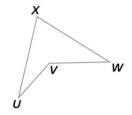

20. In the figure below, ℓ ∥ m and m∠1 = 72. Find the measures of all the other angles. (**Lesson 3-6**)

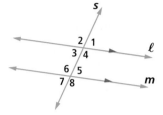

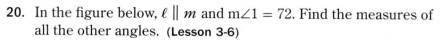
EXPLORATION

21. There are 14 known types of pentagons that will tessellate the plane. The most recent type was discovered in 1985 by Rolf Stein of the University of Dortmund in Germany. At the right is one pentagon of this type. Copy the pentagon. Use the side lengths and angle measure relationships to create a tessellation with this pentagon as the fundamental region.

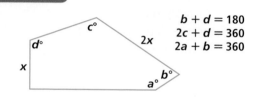

$$b + d = 180$$
$$2c + d = 360$$
$$2a + b = 360$$

22. As you found out in this lesson, not all regular polygons tessellate the plane. **Semiregular tessellations** are tessellations that combine several different regular polygons like in the picture below. If you look at it carefully, you could give a notation based on the polygons at one vertex. This tessellation is often noted as a 6,4,3,4 tessellation because *every* vertex is surrounded by a hexagon (6), square (4), triangle (3), and a square (4). Check another vertex to see the pattern again. Create a different semiregular tessellation and give its notation.

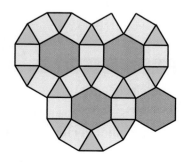

Lesson

7-7

Properties of Parallelograms

▶ **BIG IDEA** The properties of a parallelogram can be deduced either from its rotation symmetry or from congruent triangles formed by its diagonals.

In Chapter 6, you saw some of the properties of trapezoids and kites. In this lesson you will examine parallelograms.

Mental Math

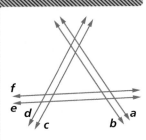

a. Name all pairs of lines in the figure above that appear to be translation images of each other.

b. Name a pair of lines in the figure above that are not translation images of each other.

c. What kind of rotation symmetry does the figure appear to have?

Activity 1

MATERIALS tracing paper

The diagram at the right shows parallelogram *ABCD*. *E, F, G,* and *H* are midpoints of its sides. Use tracing paper to make a copy of the diagram. You will use your tracing paper to explore the symmetry of this parallelogram.

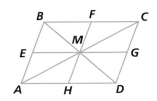

Step 1 Fold your paper to determine whether *ABCD* has reflection symmetry. If so, identify the symmetry lines. If not, how do you know?

Step 2 Turn your paper to determine whether *ABCD* has rotation symmetry. If so, describe the rotation. If not, explain why not.

Step 3 Use your answers to Steps 1 and 2 to make a conjecture about the reflection and rotation symmetry of a parallelogram.

What Properties Do Parallelograms Seem to Possess?

Activity 1 shows that when the diagonals of a parallelogram are drawn, there are many pairs of congruent triangles, segments, and angles. Using the triangle congruence theorems, the theorem at the top of the next page can be proved. We show the proof of Part a and leave the proofs of Parts b, c, and d to you in Questions 6–8.

Properties of a Parallelogram Theorem

In any parallelogram:
    a. opposite sides are congruent.
    b. opposite angles are congruent.
    c. consecutive angles are supplementary.
    d. the diagonals intersect at their midpoints.

Given Parallelogram *ABCD*

Prove $\overline{AB} \cong \overline{CD}$ and $\overline{AD} \cong \overline{CB}$

Proof We wish to prove something about parallelograms, but our previous work has dealt primarily with triangles. So we draw the diagonal $\overline{AC}$ to form $\triangle ACD$ and $\triangle CAB$ with angles numbered as in the drawing.

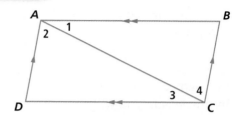

Conclusions	Justifications
1. *ABCD* is a parallelogram.	1. Given
2. $\overline{AD} \parallel \overline{BC}$	2. Opposite sides of a parallelogram are parallel. (definition of a parallelogram)
3. $\angle 2 \cong \angle 4$	3. Parallel lines $\Longrightarrow$ alternate interior angles are congruent. (Parallel Lines Theorem)
4. $\overline{AB} \parallel \overline{CD}$	4. Opposite sides of a parallelogram are parallel. (definition of a parallelogram)
5. $\angle 1 \cong \angle 3$	5. Parallel lines $\Longrightarrow$ alternate interior angles are congruent. (Parallel Lines Theorem)
6. $\overline{AC} \cong \overline{AC}$	6. Reflexive Property of Congruence
7. $\triangle ACD \cong \triangle CAB$	7. ASA Congruence Theorem (Steps 3, 5, and 6)
8. $\overline{AB} \cong \overline{CD}$ and $\overline{AD} \cong \overline{CB}$	8. CPCF Theorem

STOP QY

> **▶ QY**
>
> In the figure of Activity 1, which angles are congruent to
> a. $\angle CMF$?
> b. $\angle CMG$?

GUIDED

Example

In the parallelogram *MATH* at the right, $FT = 5$, $MA = 8$, and $m\angle MHT = 80$. Find the following measurements.

a. *FM*
b. *HT*
c. $m\angle MAT$
d. $m\angle HMA$

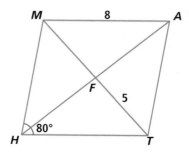

Solution

a. F is the midpoint of $\overline{MT}$, so **FM = __?__**.

b. $\overline{MA}$ and $\overline{HT}$ are opposite sides of a parallelogram, so **MA = HT = __?__**.

c. $\angle MHT$ and $\angle MAT$ are opposite angles of the parallelogram, so **m∠MAT = __?__**.

d. $\angle MHT$ and $\angle HMA$ are consecutive angles of the parallelogram, so, by the Properties of Parallelograms Theorem, $\angle MHT$ and $\angle HMA$ are supplementary. **Thus, m∠HMA = __?__**.

Because of the Quadrilateral Hierarchy Theorem, you can further conclude that the properties of parallelograms apply to all rhombuses, rectangles, and squares. For example, the opposite sides of rhombuses, rectangles, and squares are congruent.

Distance between Parallel Lines

From the properties of a parallelogram comes an important property of distance. The **distance between two parallel lines** is the length of a segment that is perpendicular to the lines with an endpoint on each of the lines.

Distance between Parallel Lines Theorem

The distance between two given parallel lines is constant.

Given $\ell \parallel m$, $\overline{AB} \perp \ell$, and $\overline{KE} \perp \ell$

Prove $AB = KE$

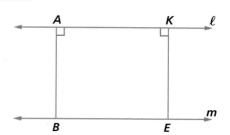

Proof Since it is given that $\ell \parallel m$, $\overline{AB} \perp \ell$, and $\overline{KE} \perp \ell$, by the Perpendicular to Parallels Theorem, $\overline{AB} \perp m$ and $\overline{KE} \perp m$. By the definition above, AB and KE are distances between lines ℓ and m. Because there are four right angles in the figure, $BAKE$ is a rectangle by the definition of a rectangle. By the Quadrilateral Hierarchy Theorem, $BAKE$ is also a parallelogram. By the Properties of a Parallelogram Theorem, opposite sides of $BAKE$ are congruent. So, $AB = KE$ by the Segment Congruence Theorem.

Rotation Symmetry of a Parallelogram

You may already know that parallelograms are rotation-symmetric. The following activity explores this.

Activity 2

MATERIALS DGS

Step 1 Construct a parallelogram *ABCD* by beginning with a segment $\overline{AB}$, finding a translation image $\overline{DC}$, and connecting the appropriate endpoints.

Step 2 Construct diagonals $\overline{AC}$ and $\overline{BD}$ and name $\overline{AC} \cap \overline{BD} = \{E\}$.

Step 3 Rotate *ABCD* 180° about *E*.

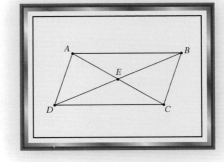

From Activity 2, it appears that, when a parallelogram is rotated 180° about the intersection of its diagonals, the image of the parallelogram coincides with its preimage. This can be shown to be true.

Parallelogram Symmetry Theorem

Every parallelogram has 2-fold rotation symmetry about the intersection of its diagonals.

Given Parallelogram *ABCD*

Prove *ABCD* has 2-fold rotation symmetry with center at $\{E\} = \overline{AC} \cap \overline{BD}$.

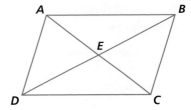

Proof Given parallelogram *ABCD*, let *R* be the rotation 180° about *E*. Because m∠*DEB* = 180 and *E* is the midpoint of $\overline{BD}$ (Properties of a Parallelogram Theorem), *R(B)* = *D* and *R(D)* = *B*. Similarly, because m∠*AEC* = 180 and *E* is the midpoint of $\overline{AC}$, *R(C)* = *A* and *R(A)* = *C*. By the Figure Transformation Theorem, these two facts imply that *R(ABCD)* = *CDAB*, which makes *ABCD* rotation-symmetric. The quotient 360 ÷ 180 equals 2, so the rotation symmetry is 2-fold.

Questions

COVERING THE IDEAS

In 1 and 2, use parallelogram GRIN at the right.

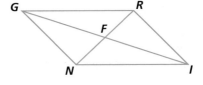

1. If *GR* = 7, *RI* = 5, and *FR* = 3, find
   a. *NI*.   b. *NG*.   c. *FN*.

2. If *GI* = 11 and *RF* = *x*, find
   a. *FI*.   b. *NR*.

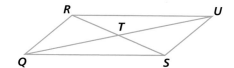

3. In the figure at the right, *QRUS* is a parallelogram with the intersection of the diagonals at point *T.*

    a. Which angles are congruent to ∠*URS*?

    b. Which angles are congruent to ∠*QRS*?

    c. Which angles are supplementary to ∠*RQS*?

    d. Which of the lines are symmetry lines for the parallelogram?

4. Given: $\overline{LM} \parallel \overline{PQ}$, $\overline{MP} \perp \overline{PQ}$, and $\overline{LQ} \perp \overline{PQ}$.

    a. Draw a quadrilateral *LMPQ* that matches this description.

    b. Why does *LQ = PM*?

    c. What is the most specific name for quadrilateral *LMPQ*?

5. In the figure at the right, $\overline{AB} \parallel \overline{DC}$ and $\overline{AD} \parallel \overline{BC}$. Justify each conclusion.

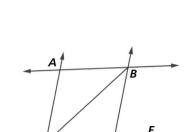

    a. ∠*ADC* ≅ ∠*BCE*   b. ∠*CDB* ≅ ∠*ABD*

    c. $\overline{AB} \cong \overline{CD}$ d. $\overline{DB} \cong \overline{BD}$

    e. △*ADB* ≅ △*CBD*   f. ∠*BAD* ≅ ∠*DCB*

6. Prove Part b of the Properties of a Parallelogram Theorem: In any parallelogram, opposite angles are congruent.

7. Prove Part d of the Properties of a Parallelogram Theorem: In any parallelogram, the diagonals intersect at their midpoints.

8. In the figure at the right, *WXYZ* is a parallelogram. Justify the following conclusions without using the Properties of a Parallelogram Theorem.

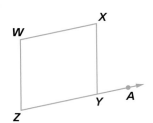

    a. $\overline{WX} \parallel \overline{ZY}$

    b. ∠*AYX* ≅ ∠*YXW*

    c. ∠*AYX* and ∠*XYZ* are supplementary.

    d. ∠*XYZ* and ∠*YXW* are supplementary.

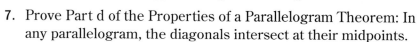

APPLYING THE MATHEMATICS

9. In the figure at the right, *RHOM* is a rhombus with the intersection of the diagonals at point *B*. If m∠1 = 64, find each angle measure.

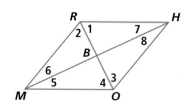

    a. m∠4   b. m∠3   c. m∠*RBH*

    d. m∠*RMO*   e. m∠5   f. m∠7

10. Use the figure at the right. *NOLA* is a parallelogram with *P* the midpoint of diagonal $\overline{OA}$.

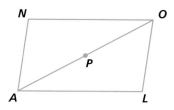

    a. Under a rotation of 180° with center *P*, what is the image of *NOLA*?

    b. If m∠*ANO* = 3*x* + 80 and m∠*OLA* = 68 + 5*x*, find m∠*NAL*.

11. In a 100-meter race, each runner is in a different lane, as shown at the right. What theorem in this lesson justifies that each runner has the same distance to run from start to finish?

12. Given: $NTCR$ is a rectangle and $\overleftrightarrow{EA}$ is the perpendicular bisector of $\overline{RC}$.
Prove: $\overleftrightarrow{EA}$ is the perpendicular bisector of $\overline{NT}$.

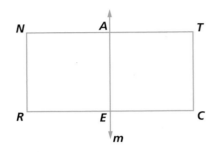

13. **a.** Place the figure names *kite, parallelogram, rectangle, rhombus,* and *square* in this hierarchy.

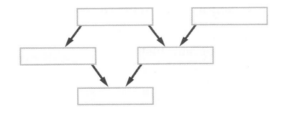

   b. Next to each name, write (rd) if the figure is reflection-symmetric to a diagonal, (rpb) if the figure is reflection-symmetric to its perpendicular bisector of a side, and (R) if the figure is rotation-symmetric.

14. The figure at the right shows a sidewalk constructed by a curb so that $ABDC$, $CDFE$, and $EFHG$ are congruent parallelograms. Explain why $AB = GH$.

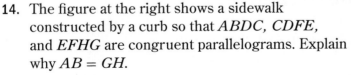

15. Let $ABCD$ be a parallelogram, and let M and N be the midpoints of $\overline{AD}$ and $\overline{BC}$. Draw $\overline{AN}$ and $\overline{CM}$. Prove something about this situation that involves M and N.

16. At the right is a pattern from a Bushongo sewn mat. Thinking of this as part of a tessellation, trace a possible fundamental region. **(Lesson 7-6)**

17. Trace two possible fundamental regions for the tessellation at the right. **(Lesson 7-6)**

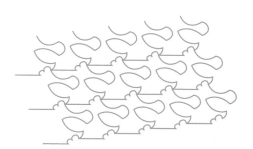

18. **True or False** Suppose $\triangle ABC$ and $\triangle DEF$ are right triangles with congruent hypotenuses. If you know that one of the nonright angles in $\triangle ABC$ is congruent to a nonright angle in $\triangle DEF$, then the triangles are congruent. **(Lesson 7-5)**

19. Given a triangle $\triangle OPQ$, describe how you can find the center of the circle in which $\triangle OPQ$ is inscribed. **(Lesson 6-8)**

20. Determine whether the following statement is *always, sometimes but not always,* or *never* true. Let ℓ be the line with equation $y = 1.5$. If $A = (m, n)$ has integer coordinates, then $A' = r_\ell(A)$ has integer coordinates. **(Lesson 4-1)**

21. **Multiple Choice** A line contains points in n quadrants of the coordinate plane. What values of n are possible? (There may be more than one possible value.) **(Lesson 1-2)**

    A 0   B 1   C 2
    D 3   E 4   F more than 4

22. Draw a parallelogram $ABCD$. On each side of $ABCD$, draw a square. One square has been drawn in the figure below. Connect the centers of symmetry of the four squares. What type of figure is formed?

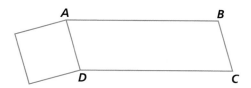

Lesson

7-8

Sufficient Conditions for Parallelograms

> ▶ **BIG IDEA** By using congruent triangles, certain figures can be proved to be parallelograms.

Suppose you want to construct a parallelogram. You know that if you construct a quadrilateral that has two pairs of parallel sides, then by the definition of a parallelogram, that quadrilateral is in fact a parallelogram. Consequently, we say that "two pairs of parallel sides" is a sufficient condition for a quadrilateral to be a parallelogram. Are there other sufficient conditions for constructing parallelograms? We examine this question in this lesson.

Mental Math

a. A quadrilateral is equiangular and equilateral. Is it necessarily a square?

b. A quadrilateral is equiangular. Is it necessarily a rhombus?

c. A rhombus is equiangular. Is it necessarily a square?

Activity 1

MATERIALS DGS

Start with a clear screen.

Step 1 Construct $\overline{AB}$ and $\overrightarrow{EF}$.

Step 2 Translate $\overline{AB}$ by $\overrightarrow{EF}$.

Step 3 Label the image of A as D and the image of B as C.

Step 4 Construct $ABCD$.

It appears that quadrilateral $ABCD$ is a parallelogram.

Step 5 Drag points A and B to various locations on the screen.

Does it appear that $ABCD$ is still a parallelogram?

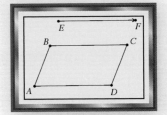

In Activity 1, you constructed a quadrilateral by translating a segment. This created an image segment that was both congruent to the preimage and parallel to it. When the endpoints of these two segments are connected in an appropriate way, a parallelogram is indeed produced. The proof of this statement is left for you to write in Question 9. This result can be stated in the following form: *If a quadrilateral has one pair of sides that are both parallel and congruent, then the quadrilateral is a parallelogram.*

Where Is a Good Place to Find Sufficient Conditions?

You saw in the last lesson that parallelograms have many properties. Some of these are included in the Properties of a Parallelogram Theorem. What about the converses of these statements? Might we look here to find other sufficient conditions for a parallelogram?

In a parallelogram, opposite sides are congruent. Is one pair of congruent opposite sides enough to guarantee that a quadrilateral is a parallelogram? The answer to this is "no." For example, $ABCD$ at the right has a pair of opposite sides that are congruent, but is not a parallelogram.

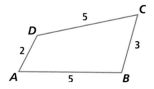

 QY

However, in a quadrilateral, if *both* pairs of opposite sides are congruent, then the quadrilateral is a parallelogram. You are asked to prove this statement in Question 3. Activity 2 shows another example of a sufficient condition for a quadrilateral to be of a special type.

> **QY**
>
> Sketch a figure that shows that having one pair of congruent opposite angles is not a sufficient condition for a quadrilateral to be a parallelogram.

Activity 2

MATERIALS DGS

Step 1 Construct $\overline{AB}$. Name its midpoint P.

Step 2 Construct $\odot P$ with point C on the circle and not on $\overline{AB}$.

Step 3 Construct $\overleftrightarrow{CP}$ with $\odot P \cap \overleftrightarrow{CP} = \{C, D\}$.

Step 4 Hide $\odot P$ and $\overleftrightarrow{CP}$.

Step 5 Construct $\overline{CD}$.
You have created two segments, not necessarily with the same length, that have the same midpoint.

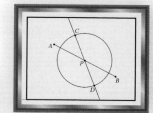

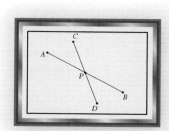

Step 6 Construct quadrilateral $ACBD$.

Step 7 Drag point A to various locations on the screen, maintaining P as the midpoint of $\overline{AB}$.

Step 8 Drag point C, maintaining P as the midpoint of $\overline{CD}$.

Step 9 Make a conjecture as to the shape of quadrilateral $ACBD$.

Activity 2 suggests that if the diagonals of a quadrilateral intersect at their midpoints, then the quadrilateral is a parallelogram. This result and the earlier statements about *both* pairs of opposite sides are two parts of the following theorem.

Sufficient Conditions for a Parallelogram Theorem

If, in a quadrilateral,

    a. one pair of sides is both parallel
       and congruent,

or

    b. both pairs of opposite sides are congruent,

or

    c. the diagonals bisect each other,

or

    d. both pairs of opposite angles are congruent,
then the quadrilateral is a parallelogram.

You are asked to prove Parts a, b, and c in the questions at the end of this lesson. Here is a proof of Part d.

Proof of Part d

Given   Quadrilateral *QUAD* with
       $x = m\angle DQU = m\angle DAU$ and
       $y = m\angle QDA = m\angle QUA$.

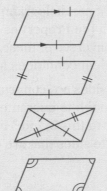

Prove   *QUAD* is a parallelogram.

Proof

Conclusions	Justifications
1. $x + y + x + y = 360$	1. Quadrilateral-Sum Theorem
2. $2x + 2y = 360$	2. Commutative and Distributive Properties
3. $x + y = 180$	3. Multiplication Property of Equality (Multiply both sides by $\frac{1}{2}$.)
4. $y + m\angle QUS = 180$	4. Linear Pair Theorem
5. $y + m\angle QUS = x + y$	5. Transitive Property of Equality (Steps 3 and 4)
6. $m\angle QUS = x$	6. Addition Property of Equality (Add $-y$ to both sides.)
7. $\overline{QD} \parallel \overline{UA}$	7. Alternate Interior Angles Theorem ($\angle Q$ and $\angle QUS$)
8. $\overline{QU} \parallel \overline{DA}$	8. Corresponding Angles Postulate ($\angle QUS$ and $\angle A$)
9. *QUAD* is a parallelogram.	9. definition of parallelogram

Example

Pictured at the right is a figure made out of sticks of two different lengths connected near their ends with small dowel rods. The sticks can pivot around these dowel rods, so the figure is not rigid. If the constructor of this figure used sticks so that the opposite sides were the same length and all of the sticks remain in the same plane, would any variation of this figure always remain a parallelogram?

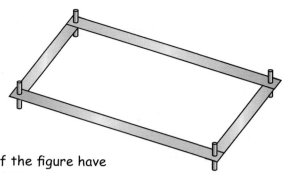

Solution Because both pairs of opposite sides of the figure have the same length, by Part b of the Sufficient Conditions for a Parallelogram Theorem, this figure will always be a parallelogram.

Questions

COVERING THE IDEAS

1. According to the definition of parallelogram, what makes a figure a parallelogram?

2. Give four sufficient conditions for a quadrilateral to be a parallelogram other than its defining characteristics.

3. Prove part b of the Sufficient Conditions for a Parallelogram Theorem: *If both pairs of opposite sides of a quadrilateral are congruent, then the quadrilateral is a parallelogram.*

In 4–8, use the marked diagrams.
a. Is the figure necessarily a parallelogram?
b. If so, give a justification why it is a parallelogram.

4.   5.    6. 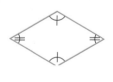   7. 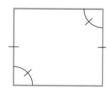   8.

APPLYING THE MATHEMATICS

9. **a.** Use the figure at the right. Write what is given and what needs to be proved to prove Part a of the Sufficient Conditions for a Parallelogram Theorem.

    b. Prove the theorem.

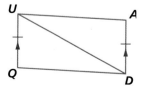

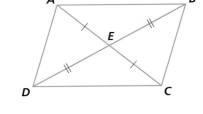

10. **a.** Use the figure at the right. Write what is given and what needs to be proved to prove Part c of the Sufficient Conditions for a Parallelogram Theorem.

   b. Prove the theorem.

11. Two yardsticks and two meter sticks are joined end-to-end to form a quadrilateral. What types of quadrilaterals could they form?

12. Looking at the picture at the right, explain why the keyboard is parallel to the floor.

13. Is it possible for one angle of a parallelogram to have 5 times the measure of another angle of the same parallelogram? If so, draw such a parallelogram. If not, why not?

14. Suppose a figure is known to be a parallelogram.
   a. What one fact about its angles is sufficient for the figure to be a rectangle?
   b. What one fact about its diagonals is sufficient for the figure to be a rectangle?

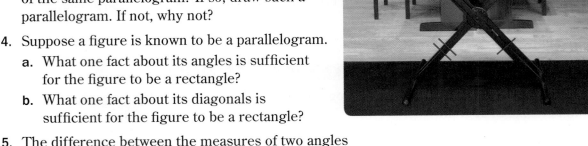

15. The difference between the measures of two angles of a parallelogram is 12°, and all four sides of the parallelogram are the same length. Is this possible? If so, draw such a parallelogram. If not, why not?

16. Suppose $A = (-3, 9)$, $B = (7, 9)$, $C = (14, -2)$ and $D = (4, -2)$.
   a. Determine whether $ABCD$ is a parallelogram.
   b. Justify your answer to Part a.

REVIEW

17. In the figure at the right, $ABCD$ is a parallelogram, $AB = 1$ cm, $CB = 4$ cm, and m$\angle A = 72°$. Find as many other lengths and angle measures as you can. **(Lesson 7-7)**

18. **a.** Draw a parallelogram that has no symmetry lines.
   b. Draw a parallelogram that has two symmetry lines. **(Lesson 7-7)**

19. Draw a nonconvex polygon that can tessellate the plane. **(Lesson 7-6)**

20. Explain why there is no SSSS Congruence Theorem for quadrilaterals by drawing two noncongruent quadrilaterals with four pairs of congruent corresponding sides. **(Lessons 7-1, 6-4)**

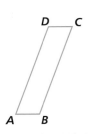

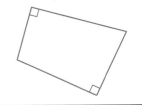
QY ANSWER

Answers vary. Sample:

EXPLORATION

21. Prove or disprove the following: If, in quadrilateral $ABCD$, $AB = CD$ and m$\angle A = $ m$\angle C$, then $ABCD$ is a parallelogram.

Lesson
7-9

Diagonals of Quadrilaterals

> **BIG IDEA** Information about the diagonals of a quadrilateral can be sufficient to determine the type of quadrilateral.

You have seen that the various types of quadrilaterals can be organized into a hierarchy. Now we use that hierarchy to examine properties of diagonals of quadrilaterals. First we summarize what you have learned so far about diagonals.

Properties of Diagonals Already Established

The Kite Diagonal Theorem, which you saw in Lesson 6-5, states that the symmetry diagonal of a kite is the perpendicular bisector of the other diagonal. This means that for a rhombus, *each* diagonal is the perpendicular bisector of the other. From the hierarchy of quadrilaterals, every square must also have this property because every square is a rhombus. To summarize: *Kites, rhombuses, and squares each have a diagonal that is a perpendicular bisector of the other diagonal.*

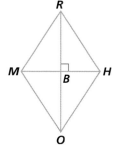

However, it is not the case that every quadrilateral with perpendicular diagonals is a special type.

 QY1

The Properties of a Parallelogram Theorem states that the diagonals of a parallelogram intersect at their midpoints and hence bisect each other. Because the rhombus, the rectangle, and the square are all below and connected to the parallelogram in the quadrilateral hierarchy, this property must be true for these figures as well. To summarize: *The diagonals of parallelograms, rhombuses, rectangles, and squares bisect each other.*

Mental Math

a. Can a convex quadrilateral have two nonintersecting diagonals?

b. Can a nonconvex quadrilateral have two nonintersecting diagonals?

c. Can a convex hexagon have two nonintersecting diagonals?

> ▸ QY1
>
> Draw a quadrilateral with perpendicular diagonals that is not a kite or a trapezoid.

In Lesson 6-6, you learned about symmetries of isosceles trapezoids, and you were asked to prove that the diagonals of an isosceles trapezoid are congruent. Furthermore, the diagonals are split into corresponding congruent parts. Because of the quadrilateral hierarchy, the diagonals of a rectangle and the diagonals of a square are congruent as well. To summarize: *In isosceles trapezoids, rectangles, and squares, corresponding parts of the two diagonals are congruent.*

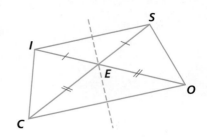

Quadrilaterals other than isosceles trapezoids can have congruent diagonals.

STOP QY2

The above arguments show that the quadrilateral hierarchy provides a way to summarize the properties of diagonals in special types of quadrilaterals.

> ▶ QY2
>
> Draw a quadrilateral with congruent diagonals that is neither a kite nor a trapezoid.

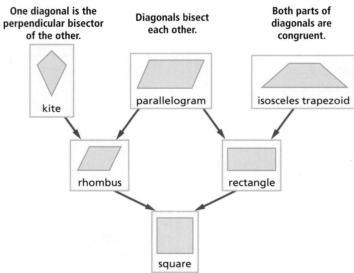

This hierarchy keeps the same general order as the original hierarchy.

Properties of Diagonals as Sufficient Conditions

In Lesson 7-8, the converses of the properties of parallelograms suggested sufficient conditions for parallelograms. This is also true for other special quadrilaterals. The following two activities suggest ways to construct special quadrilaterals. You should think about why these constructions work.

Activity 1

MATERIALS DGS

Step 1 Construct $\overline{AB}$.

Step 2 Construct the perpendicular bisector of $\overline{AB}$.

Step 3 Locate random points C and D on this perpendicular bisector on opposite sides of $\overline{AB}$.

Your figure should look like the one at the right.

Step 4 Connect points to construct quadrilateral $ACBD$.

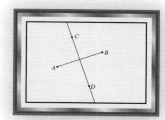

Step 5 Drag A and B to various locations on the screen.

    a. What type of quadrilateral is $ACBD$? What property justifies this?

    b. Suppose that in Step 3 you located only point C on the perpendicular bisector and then reflected C over $\overline{AB}$. Now what shape is quadrilateral $ACBD$? What property justifies this?

What happens if the diagonals of a quadrilateral have the same midpoint and the same length? Activity 2 explores this situation.

Activity 2

MATERIALS DGS

Step 1 Construct $\odot P$.

Step 2 Construct a line through P. Name the points of intersection with the circle A and B.

Step 3 Construct a second line through P. Name the points of intersection with the circle C and D. Your figure should look like the one at the right.

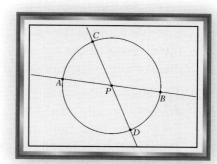

Step 4 Connect points to construct quadrilateral $ACBD$.

Step 5 Drag the two lines you have constructed to determine the type of quadrilateral that $ACBD$ is.

Questions

COVERING THE IDEAS

In 1–3, name all the special types of quadrilaterals that exhibit the property.

1. One diagonal is perpendicular to the other diagonal.
2. The midpoint of one diagonal is also the midpoint of the other diagonal.
3. One diagonal is the same length as the other diagonal.

4. Use the figure from Activity 1. If the diagonals intersect at E, $\overline{AE} \cong \overline{BE}$, and $\overline{CE} \cong \overline{DE}$, prove that $ABCD$ is a rhombus.

5. Describe a method of constructing a rhombus that uses a property of its diagonals.

6. In Activity 2, what must be done in the construction to guarantee that $ADBC$ is a square?

In 7–10, use the hierarchy in the lesson to give the most general name for the quadrilateral.

7.

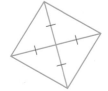

8.

9.

10.

APPLYING THE MATHEMATICS

11. Connie, a building contractor, is checking to be sure that a quadrilateral foundation that was laid is a rectangle. She finds that one angle of the quadrilateral is a right angle and that the diagonals have the same midpoint. Does Connie need to measure any other angles of the quadrilateral? Why or why not?

12. The United States Soccer Federation rules book does not specify an official length or width of the playing field, only that it be rectangular in shape. If the center circle is supposed to be in the exact center of the field, how can this center be determined without any measuring?

13. Woody is building a wooden bed frame that he wants to be sure is rectangular in shape. Before bracing and nailing it in place, he measures the diagonals and confirms that they are equal in length. What else must he confirm about the diagonals before finishing the frame?

14. Refer to the drawing at the right. Prove that if one diagonal of a quadrilateral is the perpendicular bisector of the other diagonal, then the quadrilateral is a kite.

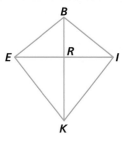

15. An extending gate is shown at the right. Here, $BA = BC = BD = BE$. What type of quadrilateral is $ADCE$ (not drawn)? How do you know? **(Lesson 7-8)**

16. In the drawing below, $ABCD$ is a parallelogram and M is the midpoint of $\overline{AB}$. Prove that $\triangle BMC \cong \triangle AMN$. **(Lessons 7-8, 7-3)**

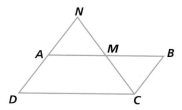

17. Can a rectangle with sides 4 cm and 8 cm be tiled using triangles that have an area of 7 cm²? Explain your answer. **(Lesson 7-6)**

18. Given points A and B, construct two equilateral triangles that have A and B as vertices. **(Lesson 5-4)**

19. A quadrilateral and its diagonals are drawn. Can this drawing be done without crossing a line or lifting the pen? **(Lesson 2-6, 1-3)**

20. State a formula for the area of each of the following figures. **(Previous Course)**

    a. a square with side a b. a circle with radius r
    c. a triangle with height h and base b

21. a. Draw a quadrilateral $ABCD$ and its diagonals $\overline{AC}$ and $\overline{BD}$. Let $E = \overline{AC} \cap \overline{BD}$. Consider the four triangles ABC, BCD, CDA, and DAB created by sides and diagonals and the four triangles AED, AEB, BEC, and CED determined by the sides, the two diagonals, and their intersection point E.

    b. For each of the quadrilaterals in the quadrilateral hierarchy, describe which of the triangles in Part a are congruent if $ABCD$ is that type of quadrilateral and state which isometry maps one triangle onto the other. Also identify which of these triangles are not congruent but have angles of the same measures.

    c. Make a hierarchy like the one in this lesson built on the descriptions given in Part b.

QY ANSWERS

1. Answers vary. Sample:

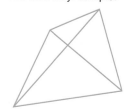

2. Answers vary. Sample:

Lesson

7-10 Proving That Constructions Are Valid

Mental Math

In the figure below, $\overrightarrow{BD}$ bisects $\angle ABC$, and $\overrightarrow{BE}$ bisects $\angle CBD$.

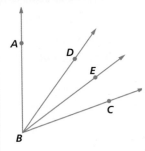

a. If m$\angle CBE = x$, what is m$\angle CBA$?

b. If m$\angle ABD = y$, what is m$\angle EBC$?

c. If m$\angle ABE = z$, what is m$\angle CBD$?

▶ **BIG IDEA** Congruent triangles can be utilized to prove that constructions done by straightedge and compass do in fact construct what they claim to construct.

In previous lessons you have constructed perpendiculars and parallels to lines, and you have constructed triangles and quadrilaterals of various kinds. How can you be certain that these constructions actually produce the figures that were claimed? An essential part of a construction is knowing why it produces exactly what is claimed. That means that a part of each construction is the proof that the steps followed are sufficient to create the shape intended. You now have enough theorems to prove that many constructions work as claimed.

Example

Prove that the perpendicular bisector construction of Lesson 3-9 is valid.

Solution Recall that construction of the perpendicular bisector of $\overline{AB}$ (on page 166) involves the intersection of two circles with centers A and B, each with radius AB, as shown at the right. We need to show that $\overleftrightarrow{CD}$ is the perpendicular bisector of $\overline{AB}$. To do this, draw segments connecting A, C, B, and D to form quadrilateral $ACBD$.

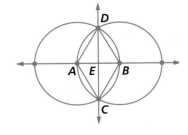

$AD = AB = AC$ because they are radii of $\odot A$. $AB = BC = BD$ because they are radii of circle B. By the Transitive Property of Equality, all of these lengths are equal. Consequently, $ACBD$ is a rhombus. And because each diagonal of a rhombus is the perpendicular bisector of the other, $\overleftrightarrow{CD}$ is the perpendicular bisector of $\overline{AB}$.

Here is a construction with a proof somewhat like that in the Example.

Activity 1

MATERIALS Compass, straightedge

Follow these steps to construct the bisector of an angle AOB.

Step 1 Draw any $\angle AOB$ to start.

Step 2 Construct ⊙(O, OA). (Recall that this is the circle with center O, radius OA, so it contains A.) Name ⊙(O, OA) ∩ $\overrightarrow{OB}$ = {C}.

Step 3 Construct ⊙(A, AO) and ⊙(C, CO). Name ⊙(A, AO) ∩ ⊙(C, CO) = {O, D}

Step 4 Construct $\overrightarrow{OD}$. A figure is shown at the right. It looks like $\overrightarrow{OD}$ bisects ∠AOB. But you cannot be sure without a proof. We leave the proof for you to write in Question 5.

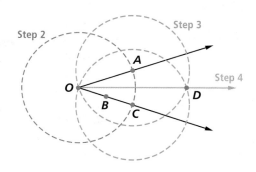

Duplicating an Angle

In Lesson 7-1, you constructed triangles with sides of given lengths. These triangles are congruent because of SSS Congruence. A similar idea enables you to construct an angle congruent to a given angle.

Activity 2

MATERIALS Compass, straightedge

Given ∠AOB and $\overrightarrow{PQ}$, follow these steps to construct an angle congruent to ∠AOB with one side being $\overrightarrow{PQ}$.

Step 1 Draw ∠AOB and $\overrightarrow{PQ}$.

Step 2 Construct ⊙(P, OA). Name ⊙(P, OA) ∩ $\overrightarrow{PQ}$ = A'.

Step 3 Construct ⊙(P, OB).

Step 4 Construct ⊙(A', AB). Name ⊙(P, OB) ∩ ⊙(A', AB) = {B', B''}. Draw $\overrightarrow{PB'}$. Your construction should look something like the one at the right.

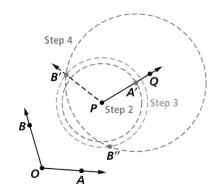

In activity 2, it looks like m∠AOB = m∠A'PB'. A proof shows why.

Given ⊙(P, OA), ⊙(P, OB) and ⊙(A', AB).

Prove m∠A'PB' = m∠AOB.

Proof

Conclusions	Justifications
1. $PA' = OA$	1. definition of ⊙(P, OA)
2. $PB' = OB$	2. definition of ⊙(P, OB)
3. $A'B' = AB$	3. definition of ⊙(A', AB)
4. △OAB ≅ △PA'B'	4. SSS Congruence Theorem
5. m∠AOB = m∠A'PB'	5. CPCF Theorem

Questions

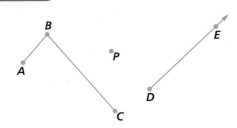

In 1–3, trace only the part of the picture needed for the particular construction.

1. Construct the bisector of $\angle B$.

2. Construct the perpendicular bisector of $\overline{BC}$.

3. Construct an angle congruent to $\angle ABC$ with one side $\overrightarrow{DE}$.

4. In Activity 2, a point B'' is constructed. Write a proof that $\angle A'PB'' \cong \angle AOB$.

5. Complete the proof in Activity 1 that $\overrightarrow{OD}$ is an angle bisector.

   Given $\angle AOB$ and circles O, A, and C as described in the construction.

   Prove $\angle AOD \cong \angle BOD$

In 6 and 7, refer to the picture at the right of the construction of a line through point P perpendicular to line ℓ. P and line ℓ are given.

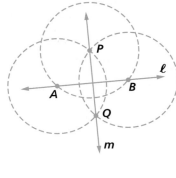

6. Write down the steps of the construction. Follow the steps to construct the perpendicular.

7. Prove that $m \perp \overleftrightarrow{AB}$. (Hint: What kind of quadrilateral is $PAQB$?)

8. Trace the figure at the right and construct a line through C perpendicular to $\overleftrightarrow{GH}$.

9. Draw a large triangle and construct an altitude from one of its vertices.

In 10 and 11, start with a large scalene triangle.

10. Construct the perpendicular bisectors of each of its sides. (If you do this construction perfectly, the lines will be concurrent.)

11. Construct the bisectors of each of the angles. What seems to be true about the intersections of the bisectors?

12. Mr. Gross told Kaitlan to construct a line through P perpendicular to line m in the figure at the right. Kaitlan said that was impossible because the point is too far away from the line. What should Mr. Gross explain to Kaitlan?

REVIEW

13. Draw a quadrilateral whose diagonals do not intersect. What do you think can be said about quadrilaterals with this property? (**Lesson 7-9**)

14. Lola built a rectangular frame out of wood. Unfortunately, when she stood the frame upright, she discovered that she hadn't attached the wooden planks well enough, and the frame bent sideways, like in the figure at the right. Is the top of the frame still parallel to the floor? Explain your answer. (**Lesson 7-8**)

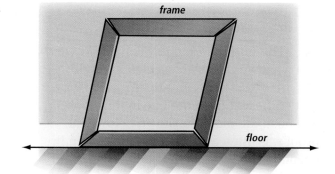

15. **True or False** If a figure can tessellate a square, then it can tessellate the plane. Explain your answer. (**Lesson 7-6**)

16. Consider the parallelogram and the rectangle at the right. Each of them can be cut up into two triangles, such that all four triangles you get by doing this are congruent. Show how to do this. (**Lesson 7-1**)

17. If two segments on the number line overlap, is the length of their union *less than, equal to,* or *greater than* the sum of their lengths? (**Lesson 2-5**)

EXPLORATION

18. In Lesson 3-6, the Corresponding Angles Postulate stated that if two lines are cut by a transversal so that corresponding angles are congruent, then the two lines are parallel. Playfair's Postulate states that given a line and a point not on the line, there is exactly one line through the point parallel to the line. Use a compass, a straightedge, and the Corresponding Angles Postulate to construct a line parallel to a given line through a point not on the line. Try to think of at least two other ways to construct a line parallel to a given line through a point not on the line.

Chapter 7 Projects

1 The "Bridge of Fools"

The fifth theorem in Euclid's *Elements* was what we call the Isosceles Triangle Base Angles Theorem. Euclid's proof was quite different from the one in this book and used the diagram shown below. The proof was often a stumbling block when first encountered by students, and it was called the "Pons Asinorum," or "Bridge of Fools." Here are the given and what is to be proved in terms of the figure.

Given: $\overline{AB} \cong \overline{AC}$, $\overline{AF} \cong \overline{AG}$

Prove: $\angle ABC \cong \angle ACB$

Write a proof. (Hint: You need to find two pairs of congruent overlapping triangles.)

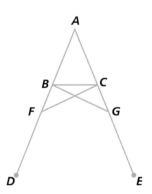

2 Creating Tessellations

In Lesson 7-6, you saw that it is possible to tessellate the plane with irregular shapes. Several methods to make such tessellations start with a simple polygon that can tessellate the plane and modify it. Such tessellations are sometimes called Escher tessellations. Find out how one such method works and use it to create interesting tessellations. Present these to your class.

3 Kaleidocycles

Tessellations can be made into kaleidocycles that are rotating rings made of an even number of tetrahedra (pyramids made with 4 equilateral triangles). The word kaleidocycles comes from the Greek kálos (beautiful) + eîdos (form) + kyklos (ring). Create a kaleidocycle.

4 A Strange Triangle Tessellation

Make a large number of congruent, equilateral triangles (out of paper, fabric, or any other convenient material). Take seven of them, and glue them (or sew them) along their edges so all seven meet at one vertex. (You will need to position the triangles just right in three dimensions in order to do this.) At each vertex of the resulting shape, glue more triangles so that there are always seven to a vertex. If this process is continued many times, the resulting shape will look like a model for the non-Euclidean geometry called hyperbolic geometry.

5 Tiling the Rice Way

Many of the problems examined by mathematicians and scientists throughout history can be understood by people with little or no formal mathematical training. Such people start out solving puzzles and end up proving important discoveries. Marjorie Rice is one such person. Write a paper about her—her background, education, and discoveries.

6 Morley's Theorem

One of the most surprising theorems in all of geometry was discovered in 1904 by geometer Frank Morley. Morley's theorem concerns the trisectors of the angles of any triangle. Consider △*ABC* below. In Figure 1, the trisecting rays of ∠*B* have been drawn. When all six of these rays are drawn, as in Figure 2, points of intersection are found for adjacent rays. △*DEF* is the triangle formed by connecting these points. Morley's theorem is: *In any triangle, the points of intersection of adjacent trisectors of the angles are the vertices of an equilateral triangle.*

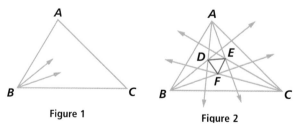

Figure 1   Figure 2

Illustrate this theorem with carefully drawn figures of three noncongruent triangles. Use a protractor and ruler or a DGS. Include measures of all important angles and sides.

7 Use a DGS

a. Construct a parallelogram *ABCD* as in the figure at the top of the next column.

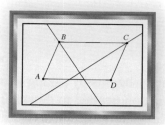

Construct the angle bisectors of ∠*ABC* and ∠*BCD*. Drag a vertex of *ABCD* and change the shape and position of *ABCD*. Make and prove a conjecture about the angle bisectors of consecutive angles of a parallelogram.

b. Construct the other two angle bisectors of the parallelogram as in the figure below.

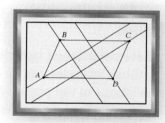

Drag a vertex of *ABCD* and change the shape and position of *ABCD*. Make and prove a conjecture about the angle bisectors of opposite angles of a parallelogram.

c. Drag a vertex of *ABCD* and change the shape and position of *ABCD*. Consider the "inner" quadrilateral defined by the intersection points of the four angle bisectors. Make and prove a conjecture about the shape of this "inner" quadrilateral.

d. Drag *ABCD* so that it becomes a rectangle. Make a conjecture about the "inner" quadrilateral in this case.

e. Drag *ABCD* so that it becomes a rhombus. Make a conjecture about the "inner" quadrilateral in this case. Relate your findings to the properties of a rhombus.

Chapter 7 Summary and Vocabulary

○ The chapter begins by asking what the minimal amount of information needed to determine a unique triangle is. Combinations of three parts were explored to see if they determined a unique triangle. All triangles satisfying any one of four sets of conditions are congruent: SSS, SAS, ASA, and AAS. A fifth condition called SsA holds when the longer of the two given sides is opposite the given angle. A special case of SsA for right triangles is called HL. Using properties of isometries, we deduced that these five conditions are sufficient to show that a pair of triangles are congruent. Once you know that triangles are congruent, you can use the CPCF Theorem (Corresponding Parts of Congruent Figures Theorem) to establish many relationships between parts of figures, such as congruence, perpendicularity, and parallelism.

○ Once you have established sufficient conditions for triangles to be congruent, those triangles can be rearranged in orderly ways to create tessellations of the plane. Quadrilaterals and other figures can also be transformed by isometries to tessellate the plane.

○ The triangle congruence theorems are also used to deduce properties of parallelograms, as well as to find sufficient conditions for quadrilaterals to be parallelograms. Special attention was paid to the relationship between quadrilaterals and their diagonals.

○ The chapter closed by showing how constructions with a compass and straightedge can be proved to be valid. These proofs provide another application of the triangle congruence theorems and the properties of quadrilaterals.

Vocabulary

Lesson 7-1
sufficient condition
included angle

Lesson 7-2
included side

Lesson 7-4
overlapping figures
nonoverlapping figures

Lesson 7-5
*legs
*hypotenuse
*concentric circles

Lesson 7-6
tessellation
fundamental region
tessellates
semiregular tessellation

Lesson 7-7
*distance between
   two parallel lines

Postulates, Theorems, and Properties

SSS Congruence Theorem (p. 386)
SAS Congruence Theorem (p. 387)
ASA Congruence Theorem (p. 388)
AAS Congruence Theorem (p. 389)
Third Angle Theorem (p. 389)
SsA Congruence Theorem (p. 407)
HL Congruence Theorem (p. 408)

Properties of a Parallelogram
   Theorem (p. 420)
Parallelogram Symmetry
   Theorem (p. 422)
Sufficient Conditions for a
   Parallelogram Theorem (p. 428)

Chapter 7 Self-Test

Take this test as you would take a test in class. You will need a ruler, compass, graph paper, and DGS. Then use the Selected Answers section in the back of the book to check your work.

1. SSS is one of six sufficient conditions to prove two triangles are congruent. What are the other five?

In 2–4, state, if possible, the congruence correspondence and the justification. If the figures are not congruent, say so.

2.
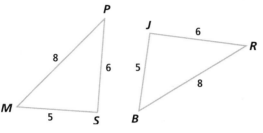

3.
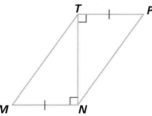

4.
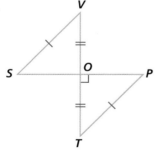

5. Are all triangles with the given measures below congruent? Explain your answer.

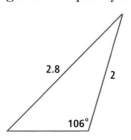

6. Copy trapezoid *TRAP* and tessellate the plane with it.

7. Construct a parallelogram *ABCD* using a DGS. Then construct the diagonals and label the point of intersection *E*.

   a. Describe any symmetries that you are sure exist for *ABCD*.

   b. Drag the vertices so that the sides are all equal lengths. Describe the symmetries that *ABCD* must have now that it might not have had before.

   c. Now drag the vertices again so that all of the angles measure 90 but the sides are not equal lengths. Describe the symmetries that it must have now that it might not have had before.

   d. Finally, drag the vertices to make *ABCD* a square. Does *ABCD* have any symmetries now that were not identified in Parts a–c?

8. Walt bought Kelly a pogo stick for her birthday. The handlebar $\overline{AB}$ is parallel to and congruent to the foot bar $\overline{CD}$. Explain why *ABDC* is a parallelogram.

In 9–11, write a proof.

9. Given: $\overline{AT} \cong \overline{AR}$, $\angle VTA \cong \angle VRA$

   Prove: $\triangle RAS \cong \triangle TAW$

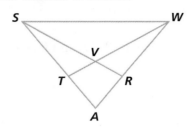

10. Given: $FRWK$ is a parallelogram, $\overline{KT} \cong \overline{RM}$

    Prove: $\overline{SM} \cong \overline{ST}$

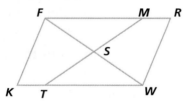

11. Given: $MRPB$ is an isosceles trapezoid with legs $\overline{RP}$ and $\overline{MB}$.

    Prove: $\angle PMB \cong \angle BRP$

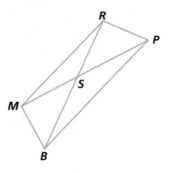

12. Trace the figure below and construct the segment from H to the midpoint of $\overline{ML}$ using a compass and straightedge.

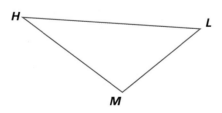

13. Explain why your construction in Question 12 is valid.

14. A carpenter cut two boards, each 8 feet long. He then nailed them to two 12-foot boards. He took a rope and pulled it tight from A to C, then from B to D, being sure he had the same length. Having done that, he was sure he had right angles at A, B, C, and D. Explain how he could be sure.

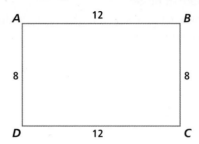

In 15–17, only the diagonals of quadrilateral $ABCD$ are shown. What is the most specific name for the quadrilateral?

15.

16.

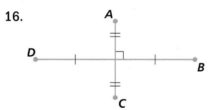

17.

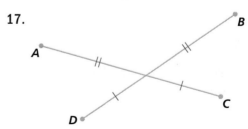

18. Which of the special types of quadrilaterals have perpendicular diagonals?

Chapter 7 Chapter Review

SKILLS Procedures used to get answers

OBJECTIVE A Draw triangles satisfying given conditions and determine whether all such triangles are congruent. (Lessons 7-1, 7-2, 7-5)

In 1–5,
a. Draw a triangle satisfying the given conditions.
b. Will every correct drawing be congruent to yours? Explain your answer.

1. $\triangle ABC$ with $AB = 2$, $BC = 5$, $AC = 7$

2. $\triangle PQR$ with m$\angle P = 90$, $QR = 3$

3. $\triangle HIJ$ with m$\angle H = 120$, $IJ = 7$, $HI = 4$

4. $\triangle OMN$ with m$\angle O = 30$, m$\angle M = 25$, and m$\angle N = 125$

5. $\triangle XYZ$ with $XY = 2$, m$\angle X = $ m$\angle Z = 30$

In 6–8, a triangle is drawn with certain measures indicated. Are all triangles with these measures congruent? Explain your answer.

6.
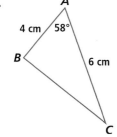

7.
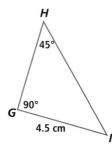

8. *E*
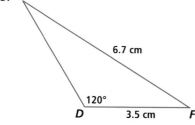

OBJECTIVE B Identify properties of special quadrilaterals. (Lessons 7-7, 7-9)

9. Identify all pairs of segments that are congruent in the figure below.

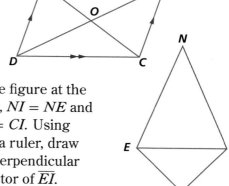

10. In the figure at the right, $NI = NE$ and $CE = CI$. Using only a ruler, draw the perpendicular bisector of $\overline{EI}$. Justify your method.

11. Name the types of quadrilaterals for which the diagonals are congruent.

PROPERTIES Principles behind the mathematics

OBJECTIVE C Determine whether triangles are congruent from given information. (Lessons 7-2, 7-5)

In 12–15, if the two triangles are congruent, justify with a triangle congruence theorem and indicate the corresponding vertices. Otherwise, write *not enough information to tell*.

12.
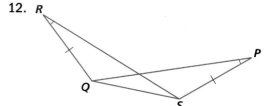

13.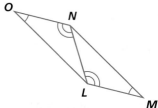

14.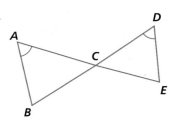

15.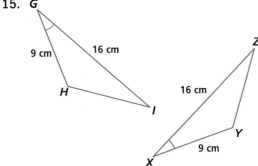

In 16 and 17, two triangles are given with congruent parts marked.

a. Name an additional piece of information that would be enough to guarantee congruence.

b. Name a triangle congruence theorem to justify your answer to Part a.

16.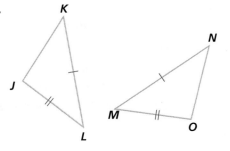

17.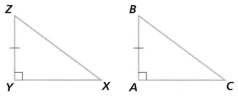

OBJECTIVE D Write proofs that triangles are congruent. (Lessons 7-3, 7-4, 7-5)

In 18–22, write a proof using the given figure.

18. Given: $m\angle PQO = m\angle RQO$, $m\angle POQ = m\angle ROQ$, $PQ = PN$, and $NO = QO$.

    Prove: $\triangle QOR \cong \triangle NOP$

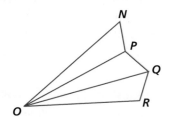

19. Given: $\odot O$ and $\odot P$ intersect at A and B.

    Prove: $\triangle OAP \cong \triangle OBP$

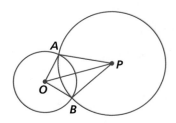

20. Given: $m\angle XWV = m\angle XVW$, $\triangle XYV$ and $\triangle XZW$ are equilateral.

    Prove: $\triangle XYV \cong \triangle XZW$

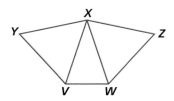

21. Given: $CD = AB$, $\overline{AC}$ is the longest side in $\triangle CDA$ and $\triangle CAB$.

    Prove: $\triangle CAD \cong \triangle ACB$

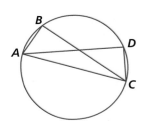

22. Given: $\overline{FD}$ bisects $\angle EDG$, $GD = ED$.

    Prove: $\triangle FED \cong \triangle FGD$

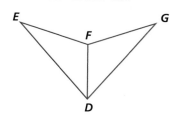

OBJECTIVE E Apply the triangle congruence theorems and the CPCF Theorem to prove that segments or angles are congruent. (Lessons 7-3, 7-4, 7-5)

In 23–28, write a proof using the given figure.

23. Given: $ABCDEFGH$ is a regular octagon.

    Prove: $\overline{CE} \cong \overline{DF}$

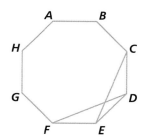

24. Given: $\overline{LM}$ bisects $\angle PLN$, $PL = ML$, and $LN = LO$.

    Prove: $MN = PO$

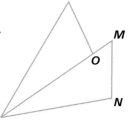

25. Given: $\overleftrightarrow{VW} \parallel \overleftrightarrow{XY}$, $VW = XY$

    Prove: $XZ = ZW$

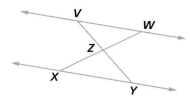

26. Given: $ABCD$ and $XYZW$ are rhombuses, X is the midpoint of $\overline{CD}$, C is the midpoint of $\overline{XY}$, and $\angle B \cong \angle Y$.

    Prove: $\angle YZX \cong \angle BCA$

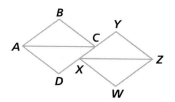

27. Given: $\triangle DFE$ is equilateral, H, I, and G are the midpoints of $\overline{DF}$, $\overline{DE}$, and $\overline{EF}$.

    Prove: $\triangle HIG$ is equilateral.

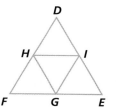

28. Given: $ABCDEF$, $AFGHIJ$, and $AJKLMB$ are regular hexagons.

    Prove: $\triangle FJB$ is equilateral.

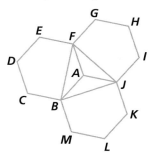

OBJECTIVE F Determine whether conditions are sufficient for parallelograms and other special types of quadrilaterals, and deduce properties of parallelograms. (Lessons 7-7, 7-8, 7-9)

29. **True or False** If a quadrilateral has two sides with length 7 and two sides with length 4, then it is necessarily a parallelogram.

30. In the figure at the right, rotating $ABCD$ 90° about O gives $DABC$. What kind of a special quadrilateral is $ABCD$? Explain how you know.

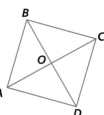

31. In the figure below, $DEFG$ is a parallelogram with $ED = 4$, $EF = 4.5$, $DH = 3.7$, $EH = 2.1$, and m∠$EDG = 60$. Name as many other lengths and angle measures as you can.

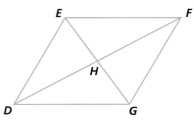

32. Trace the parallelogram at the right and locate its center of symmetry.

33. **True or False** If the two diagonals of a quadrilateral are congruent, then it must be a parallelogram. Explain your answer.

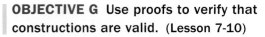

| **OBJECTIVE G** Use proofs to verify that constructions are valid. (Lesson 7-10)

In 34–36, a construction is given. Write a proof that this construction produces the desired result.

34. **Given**   Points A, B, and C.
    Construct Line ℓ containing C such that $\ell \parallel \overleftrightarrow{AB}$
    Step 1 Construct $\overrightarrow{CB}$, $\overrightarrow{AC}$, and $\overrightarrow{AB}$.
    Step 2 Construct line $\overleftrightarrow{CD}$ such that ∠$ACD \cong$ ∠CAB by using the construction to duplicate an angle.
    Step 3 Name $\overleftrightarrow{CD} = \ell$.

35. **Given**   Points A and B.
    Construct Point C such that B is the midpoint of $\overline{AC}$.
    Step 1 Construct $\overleftrightarrow{AB}$.
    Step 2 Construct $\odot(B, AB)$.
    Step 3 Name $\odot(B, AB) \cap \overleftrightarrow{AB} = \{A, C\}$.

36. **Given**   Points A and C.
    Construct Points B and D so that $ABCD$ is a square.
    Step 1 Construct $\overline{AC}$.
    Step 2 Construct m, the perpendicular bisector of $\overline{AC}$.
    Step 3 Name $m \cap \overline{AC} = E$.
    Step 4 Construct $\odot(E, EC)$.
    Step 5 Name $\odot(E, EC) \cap m = \{B, D\}$.

USES Applications of mathematics in real-world situations

| **OBJECTIVE H** Use theorems about triangles and parallelograms to explain real-life situations. (Lessons 7-2, 7-5, 7-7, 7-8, 7-9)

37. $\overline{RA}$ and $\overline{AB}$ are beams of the same length at the end of a roof. $\overline{AP}$ is perpendicular to $\overline{RB}$. Explain why $\overline{RP}$ and $\overline{PB}$ have the same length.

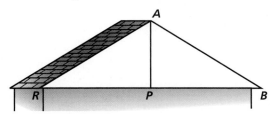

38. Many athletic contests are played on a court or a field that is in the shape of a quadrilateral. Which quadrilateral is the most common? Why do you think there is so little variety in the shape of playing fields and courts?

39. What special type of quadrilateral has four lines of reflection symmetry?

40. A park ranger used the following method to estimate the width of a river. Standing at point P directly across from the tree, the ranger walked 10 paces to X, placed a stick at X, then walked 10 paces to Q. The ranger then turned 90° and walked to point R, where the ranger, stick, and tree were lined up.

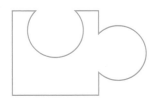

a. Which segments have length equal to the width of the river?

b. Why does this method work?

OBJECTIVE I Draw tessellations of real objects. (Lesson 7-6)

41. A factory cuts out machine parts from a sheet of metal. Explain why this process is less wasteful if it is possible to tessellate the plane with this machine part.

42. The puzzle piece in the figure at the right can be used to tessellate the plane. Trace it, and show how this can be done.

43. Draw a possible fundamental region of the tessellation shown in the pattern below.

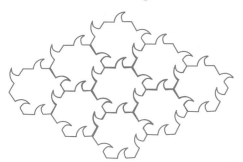

44. A section of a honeycomb looks like a tessellation of the plane by regular hexagons. Draw a section of a honeycomb.

REPRESENTATIONS Pictures, graphs, or objects that illustrate concepts

OBJECTIVE J Represent relationships between quadrilaterals using the hierarchy based on diagonals. (Lesson 7-9)

45. True or False Every quadrilateral in which the diagonals are congruent is an isosceles trapezoid.

46. Draw a hierarchy for the following quadrilaterals, based on the properties of their diagonals: kite, isosceles trapezoid, parallelogram, square.

47. Name all the special types of quadrilaterals whose diagonals bisect each other.

OBJECTIVE K Construct basic shapes with a compass and a straightedge. (Lesson 7-10)

48. Trace the figure at the right. Construct a point that is equidistant from A, B, and C.

49. Trace the figure below. Using only a ruler and compass, find whether the bisectors of the angles of $\triangle DEF$ intersect at a point.

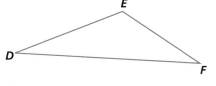

50. Trace the figure at the right. Construct a point Z on $\overrightarrow{XP}$ such that $\triangle XYZ$ is an isosceles triangle.

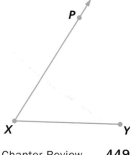

Chapter

8 Lengths and Areas

Contents

8-1   Perimeter

8-2   Fundamental Properties of Area

8-3   Areas of Irregular Figures

8-4   Areas of Triangles

8-5   Areas of Quadrilaterals

8-6   The Pythagorean Theorem

8-7   Special Right Triangles

8-8   Arc Length and Circumference

8-9   The Area of a Circle

Basketball courts vary somewhat depending on the level of play. Pictured to the right is the way that a college basketball court must be laid out according to National Collegiate Athletic Association Regulations, as seen from the top. The smallest circles are the baskets. The large circle contains the center jump region. Semicircles are parts of the boundaries of four regions. The entire court is a 50-foot by 90-foot rectangle. At the bottom is a scorer's table, another rectangle. It is recommended that there be a 10-foot wide region between the court and the seats.

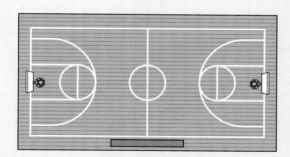

The details of assembling a court involve a considerable amount of geometry. You have seen formulas for perimeter and area in previous years. In this chapter you will see how these formulas can be deduced from basic properties.

Segments and arcs painted on the court involve the boundaries of many regions. The regions they bound must be the same from court to court. The amount of material needed for the court is an area problem. Some regions are painted, so areas of parts of the court may need to be calculated in order to know how much paint is needed. All the distances are specified and accurate measurements for the court layout must be computed precisely in order to comply with the rules of basketball.

Lesson

8-1 Perimeter

▶ **BIG IDEA** A figure's perimeter is a measure of its boundary.

Boundaries of regions come in all shapes and sizes. Fences, walls, rivers, and roads are some common boundaries. Measuring and finding the lengths of these boundaries are important in many applications.

Mental Math

The network below is not traversable. The lengths of the paths are marked. If you were to trace the network without taking your pencil off the page, what is the least total distance you would have to cover?

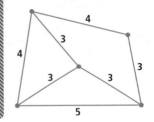

Activity

Major Martinez lives on Manhattan in New York City. Every morning, he goes for a jog. On Manhattan, the streets are $\frac{1}{20}$ of a mile apart and the avenues are $\frac{1}{5}$ of a mile apart. The major believes he jogs at about a 10-minute-per-mile pace. One morning, the major started his jog at 8th Ave. and 14th St. Here is a record of the route he took:

- Started at 8th Ave and 14th St.
- Continued on 14th St to 6th Ave.
- Followed 6th Ave to 23rd St.
- Followed 23rd St to 2nd Ave.
- Followed 2nd Ave to 34th St.
- Followed 34th St to 8th Ave.
- Followed 8th Ave back to 14th St.

1. What kind of shape did the major's route create?

2. How far did he jog this morning?

3. If it took him 40 minutes, did he run faster or slower than a 10-minute-per-mile pace?

The route given in the Activity is a polygon. The distance you found in Question 2 of the Activity is the *perimeter* of the polygon.

Definition of Perimeter

The **perimeter** of a polygon is the sum of the lengths of its sides.

If a polygon has n sides, you can always find its perimeter by adding the n lengths. But when some sides are known to have the same length, you can replace repeated additions with multiplication.

GUIDED

Example 1

Write a formula for the perimeter of each of the following shapes.

a. a rhombus with side length s

b. a parallelogram with adjacent side lengths a and b

c. the perimeter of a regular hexagon with side length s

d. the perimeter of a regular n-gon with side length s

Solution Let p be the perimeter of the polygon.

a. A rhombus has 4 equal sides of length s, so $p = $ _____?_____.

b. The opposite sides of a parallelogram are congruent. Thus, we can say that there are two sides of length a and two of length b, so $p = a + a + b + b$. This can be simplified to $p = $ _____?_____.

c. There are 6 sides of length s, so $p = $ _____?_____.

d. There are n sides of length s, so $p = $ _____?_____.

It is not important that you memorize the formulas in Example 1. Rather, it is important to know how to express the relationship between the sides and the perimeter in the most efficient way.

 QY

▶ **QY**

The perimeter of a regular heptagon is 42 cm. What is the length of one side?

Example 2

Suppose a kite is drawn in which the ratio of the lengths of two of the sides is 3:2. If the perimeter of the kite is 40 inches, what are the lengths of the sides?

Solution Begin by sketching a kite and marking the lengths. Because the sides are in the ratio 3:2, you can let one side have length $3x$ and the other $2x$. Then, because the figure is a kite, two sides have each length.

$$p = 2x + 3x + 2x + 3x$$
$$40 = 10x$$
$$4 = x$$

So, the side lengths are $2 \cdot 4 = 8$ cm and $3 \cdot 4 = 12$ cm.

Check Because $12:8 = 3:2$, and the perimeter is 40 cm, the solution checks.

Example 3

Suppose n plots of land, each a w-by-h rectangle, are next to each other, as shown at the right. A fence is to be built to surround and separate the plots. How long a fence is needed?

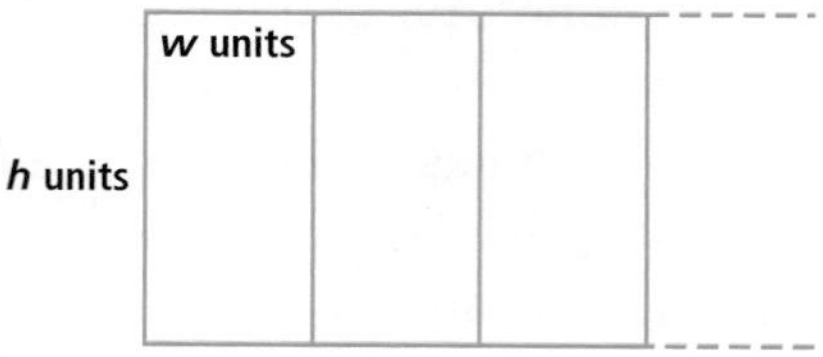

w units

h units

Solution Look for a pattern.

Number of plots	Amount of fence needed
1	$2h + 2w$
2	$3h + 4w$
3	$4h + 6w$
⋮	⋮
n	$(n + 1)h + 2nw$

Fence is needed for 1 more height than there are plots and 2 times as many widths than there are plots.

Check Substitute numbers for n, h, and w. For instance, let $h = 70$, $w = 60$, and $n = 3$. The expression $(n + 1)h + 2nw$ should yield the right amount of fence.

Questions

COVERING THE IDEAS

In 1–3, refer to the Activity.

1. What was the length of the first straight path in the major's jogging route?

2. Is the polygon created by the major's jogging path convex or nonconvex?

3. What is the shortest driving distance from 6th Ave. ∩ 9th St. to 6th Ave. ∩ 38th St.?

In 4–7, find the perimeter.

4. a square with side length x units

5. a kite with side lengths m inches and n inches

6. a regular 12-gon in which s is the length of each side

7. a regular pentagon in which the sides measure 15 yards each

APPLYING THE MATHEMATICS

8. Finish the check of Example 3.

9. If the ratio of the length to the width of a rectangle is 8:5 and the perimeter of the rectangle is 520, what are the dimensions of the rectangle?

10. **a.** Imagine that Example 3 pictures 4 tennis courts, each 60 feet by 120 feet. If a 10-foot high fence is placed around each court, what total length of fence is needed?

   b. Generalize Part a if there are n tennis courts.

11. Refer to the Activity. Suppose the major wants to jog along the rectangular route created by 14th St., 10th Ave., 42nd St., and 2nd Ave. He claims that he is running in "a big square."

   a. Assuming the streets meet at right angles, does this route form a square?

   b. What is the length of the route described above?

   c. If he jogs at his 10-minute-per-mile pace, how long will it take him to complete this route?

12. **a.** The perimeter of a parallelogram is 42 cm. One side is 13 cm long. What are the lengths of the other three sides?

   b. The perimeter of a parallelogram is 42 cm. One side is x cm long. What are the lengths of the other three sides?

13. **Fill in the Blanks** Essence found another way to answer the question of Example 3. Complete her solution.

   The union of the n plots of land is one big rectangular region with dimensions h by __?__. The perimeter of this big region is __?__. Then __?__ pieces of fence, each of length __?__, are needed to split the n plots of land. Multipying the distances found in the third and fourth blanks, and adding that to the distance in the second blank, we get __?__, which algebra shows to be equivalent to $hn + h + 2nw$.

14. If x is the length and y is the width of a rectangle, find and graph all integer pairs (x, y) that create a rectangle with a perimeter of 12 cm.

15. **a.** **True or False** If the perimeter of one square is equal to the perimeter of another, the two squares are congruent.

   b. Name another type of polygon for which equal perimeters determine congruent figures.

16. What happens to the perimeter of a rectangle if it is changed in the following ways?

   a. each side length is multiplied by 3

   b. 3 units are added to each side length

17. Find the perimeter of the polygon created when you connect the points listed below in the order that they are given.
   (4, 3), (4, 6), (–3, 6), (–3, 8), (–5, 8), (–5, 0) (4, 0), (4, 3)

18. Equilateral triangles are placed next to each other in a frieze pattern. Eight triangles are shown here.

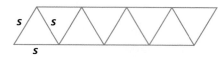

   a. If there are 100 triangles, what is the total length of all the sides in the figure?

   b. If there are n triangles, what is the total length of all the sides in the figure?

REVIEW

19. Draw a kite whose diagonals do not intersect. **(Lesson 7-8)**

20. Draw a counterexample to show that there is no SSSSS Congruence Theorem for pentagons. **(Lesson 7-1)**

21. Suppose that all of the angles in a polygon are right angles. **(Lesson 5-7)**

   a. Is the polygon necessarily a rectangle?

   b. Would your answer to Part a change if you knew that the polygon was convex?

22. **True or False** When doing a construction, you are allowed to use a protractor to draw a 58° angle. **(Lesson 3-9)**

23. A triangle has sides of lengths x, y and 17. Write inequalities describing the possible values for x and y. **(Lesson 1-7)**

24. One inch is exactly 2.54 centimeters. About what percent of an inch is 2 centimeters? **(Previous Course)**

EXPLORATION

25. Matilda Iculous is having a dinner party. She has two tables, that each seat 4 people per side and 2 on each end. As a result, Mat knows she can seat a maximum of 24 people if she uses the tables separately. She can also place the tables next to each other in any way to form one larger table. One possible arrangement is shown at the right. Because Mat wants all of her guests to feel welcome, she wants to make sure that every guest has a seat and that every seat is filled so that every person is seated next to two others. What are the possible numbers of guests that Mat can invite if she wishes all of the seats to be filled?

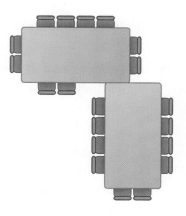

26. Design a route for the major from the Activity so that he can jog 3 miles without ever jogging on the same section of street more than once.

Lesson

8-2

Fundamental Properties of Area

> ▶ **BIG IDEA** Formulas for the areas of figures in the plane can be deduced from four fundamental properties of area.

What Is Area?

Area is a measure of the space covered by a two-dimensional region. The region may be small, like a microchip in a computer, or it may be large, like a country. One way to determine the area of a region is to try to tessellate it with a fundamental region whose area you already know. If this is possible, then it is possible to calculate the area of the region just by counting how many times the fundamental region was used in the tessellation.

Usually, the fundamental region is a square, and so we say that area is measured in **square units**. For instance, the rectangle at the right has dimensions 5 units and 4 units. Its area is 20 square units because 20 unit squares cover the region.

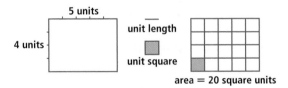

Fundamental Properties of Area

Whenever the two dimensions of a rectangle are integers, it can be tessellated by unit squares and the area of the rectangle can be expressed as an integer. You can either count all of the squares directly or multiply the two dimensions. You can see above that multiplying 4 by 5 is a way of counting the squares. (There are 4 rows with 5 squares in each row.) What happens, however, if the dimensions are not both integers?

Suppose a room is 2.75 meters long and 2.25 meters wide, as shown at the right. We put this on a 0.25 meter grid to help demonstrate the results.

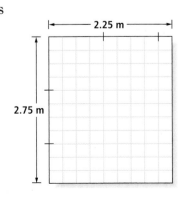

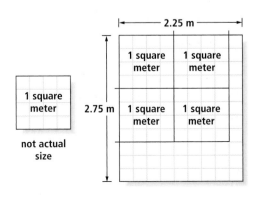

Notice two things. First, we can only put 4 complete squares in the room. Also, one square meter is made up of 16 smaller squares. Outside of the 4 complete squares, there are 35 smaller squares, each of which is $\frac{1}{4} \cdot \frac{1}{4} = \frac{1}{16}$ of a square unit. The total area, in square meters, is $4 + \frac{35}{16} = 4 + 2\frac{3}{16} = 6\frac{3}{16} = 6.1875$ square meters. This is exactly what you get when you multiply 2.25 meters by 2.75 meters. As a result, it should make sense that you can find the area of a rectangle by multiplying the length and the width, even if those lengths are not integers.

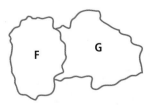

This one situation has applied the four fundamental properties of area that we assume.

Area Postulate

a. *Uniqueness Property*
Given a unit region, every polygonal region has a unique area.

b. *Congruence Property*
Congruent figures have the same area.

c. *Additive Property*
The area of the union of two nonoverlapping regions is the sum of the areas of the regions.

d. *Rectangle Formula*
The area of a rectangle with dimensions ℓ and w is ℓw. $(A = \ell w)$

The Uniqueness Property means that for a given region, there exists exactly one measure of its area. This area can be expressed using different units (square feet, square inches, etc.), but these numbers must be equivalent expressions representing equal amounts of space on the plane.

The Congruence Property means that when two figures F and G are congruent, they have equal areas. In other words, if $F \cong G$, then Area(F) = Area(G). This property may seem obvious to you because congruent figures have congruent corresponding parts.

The Additive Property says that when two regions do not overlap one another, the total area of their union is the sum of their areas. Consider the regions at the right. Because they do not overlap, the Additive Property states that the area of the region $F \cup G$ is Area(F) + Area(G).

 QY1

When a polygonal region is the union of rectangular regions, you can apply the Area Postulate to find its area.

▶ **QY1**

Is it true that the perimeter of $F \cup G$ is the sum of the perimeters of F and G? Why or why not?

Activity

Step 1 Break this polygonal region into smaller regions and calculate its area.

Step 2 Find another way to determine its area.

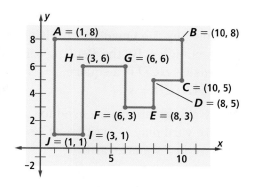

Example 1

Painters have been hired to paint one wall of an auditorium. They need to calculate how many gallons of paint they need to purchase for the job. The specifications are as follows:

> The wall is 60 ft long and 18 ft high.
>
> There are 5 windows along the wall that measure 3 ft by 7 ft each.
>
> A gallon of paint will cover approximately 350 square feet.
>
> The wall will need 2 coats of paint.
>
> Paint costs $14.95 per gallon.
>
> What will the paint cost for this job?

Solution Draw a rough picture. The area of the wall including windows is $60 \text{ ft} \cdot 18 \text{ ft} = 1080 \text{ ft}^2$.

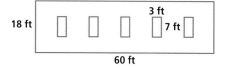

Each window has an area of $3 \text{ ft} \cdot 7 \text{ ft} = 21 \text{ ft}^2$.

There are 5 windows, so the total area taken up by windows is $5 \cdot 21 \text{ ft}^2 = 105 \text{ ft}^2$.

Thus, the area needing paint is $1080 \text{ ft}^2 - 105 \text{ ft}^2 = 975 \text{ ft}^2$.

To determine how many gallons of paint are needed, divide $\dfrac{975 \text{ ft}^2}{350 \frac{\text{ft}^2}{\text{gal}}} \approx 2.79 \text{ gal}$. However, this is only enough for one coat.

Two coats would require twice this much, or about 5.58 gallons. Because paint is typically bought in gallons, the painters will need to buy 6 gallons of paint to complete this job. Thus, the total paint cost will be $14.95 \cdot 6 = 89.70.

Changing Units of Area

From the Rectangle Formula, other area formulas can be deduced. For instance, a special case of the Rectangle Formula is that a formula for the area of any square with side s is $A = s^2$.

Suppose $ABCD$ is a square with side s. Then $ABCD$ is a rectangle because, by the Quadrilateral Hierarchy Theorem, all squares are rectangles. By the Rectangle Formula, Area($ABCD$) $= \ell w$. Because $ABCD$ is a square, $\ell = s$ and $w = s$. By substitution, Area($ABCD$) $= s^2$.

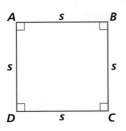

Many abbreviations are used for units of area. For example, one square yard may be written as 1 sq yd or 1 yd^2. How many square feet are in a square yard?

$$1 \text{ yd} \cdot 1 \text{ yd} = 1 \text{ yd}^2$$

Because 1 yd $= 3$ ft, you can use substitution.

$$3 \text{ ft} \cdot 3 \text{ ft} = 1 \text{ yd}^2$$
$$9 \text{ ft}^2 = 1 \text{ yd}^2$$

That is, 9 square feet $= 1$ square yard, as pictured at the right. In general, the area of a square unit is the unit length squared. Thus, because 1 meter is equal to 100 centimeters, 1 square meter is equal to 10,000 square centimeters. This next example demonstrates this idea.

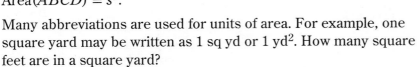

1 yd = 3 ft
1 yd² = 9 ft²

 QY2

▸ QY2

How many square millimeters are in a square meter? (Remember: 1 meter $=$ 1000 millimeters.)

Example 2

Suppose a farm is $\frac{1}{2}$ mile long by $1\frac{1}{2}$ miles wide. How many square feet of land is this?

Solution 1 First convert each dimension to feet.
(Remember: 1 mi $= 5280$ ft.)
So, $\frac{1}{2}$ mi $= \frac{1}{2} \cdot 5280$ ft $= 2640$ ft
$1\frac{1}{2}$ mi $= 1\frac{1}{2} \cdot 5280$ ft $= 7920$ ft
$A = 2640$ ft $\cdot 7920$ ft $= 20{,}908{,}800$ ft^2

Solution 2 First find the area in square miles.
$A = \frac{1}{2}$ mi $\cdot 1\frac{1}{2}$ mi $= \frac{1}{2} \cdot \frac{3}{2}$ mi$^2 = \frac{3}{4}$ mi^2
1 mi$^2 = (5280$ ft$)^2 = 27{,}878{,}400$ ft^2
So, substituting, $A = \frac{3}{4} \cdot 27{,}878{,}400$ ft$^2 = 20{,}908{,}800$ ft^2.

Questions

COVERING THE IDEAS

1. Suppose a figure F is reflected over a line m, and then its image is rotated 90° about a point P. Will the final image have the same area as F? Why or why not?

2. Draw a $1\frac{3}{4}''$ by $2''$ rectangle. Mark off unit squares to show on your drawing why the area of this rectangle is $3\frac{1}{2}$ square inches.

3. Quadrilateral $ABCD$ at the right is a rectangle. Measure its length and width with a ruler.

   a. Give its area in square centimeters.

   b. Give its area in square millimeters.

4. Find the area of the polygon with vertices $O = (0, 0)$, $G = (4, 0)$, $A = (4, 5)$, $X = (6, 5)$, $E = (6, -2)$, and $H = (0, -2)$.

5. In the figure at the right, all of the angles are right angles. Use this diagram to find the area of each polygon.

   a. $ABGH$ b. $CDEG$ c. $BIFEDC$ d. $ABCDEH$

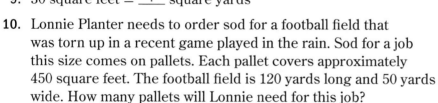

In 6–9, fill in the blank.

6. $1 \text{ ft}^2 = \underline{\quad?\quad} \text{ in}^2$

7. $6.3 \text{ km}^2 = \underline{\quad?\quad} \text{ m}^2$

8. 1 square centimeter $= \underline{\quad?\quad}$ square millimeters

9. 30 square feet $= \underline{\quad?\quad}$ square yards

10. Lonnie Planter needs to order sod for a football field that was torn up in a recent game played in the rain. Sod for a job this size comes on pallets. Each pallet covers approximately 450 square feet. The football field is 120 yards long and 50 yards wide. How many pallets will Lonnie need for this job?

11. If Leticia's farm is a square piece of land $\frac{1}{2}$ mile on a side, what is its area?

APPLYING THE MATHEMATICS

12. The illustration at the right shows an aerial view of the Millennium Biltmore Hotel in Los Angeles. The dimensions of some of the sides are given. If the building architect needs to find the area of the roof, determine this number in at least two different ways. (Assume that dimensions that appear to be equal actually are and that all angles are right angles.)

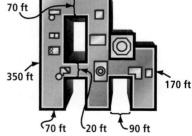

13. In some states, regulations for child care facilities require a minimum of 35 square feet per child. However, some research has shown that 50 square feet of area is optimal. Suppose a child care facility contains two rectangular rooms, one 30 feet by 42 feet and the other 26 feet by 34 feet.

    a. If a state followed the 35 square feet per child rule, how many children would the state allow this facility to handle?

    b. If this facility wanted to follow the 50 square feet per child rule, how many children could they accept?

In **14** and **15**, use the following information. An *acre* is a unit of land area in which 640 acres = 1 square mile.

14. **a.** How many square feet are in an acre?
    b. If a house is built on a 0.25-acre lot, how many square feet are in the lot?

15. In 2006, a wildfire was reported to have devastated 40,200 acres north of Palm Springs, California. How many square miles is this?

16. A rectangular pool is 4 meters wide and 6 meters long. The owner of the pool wants to create a rectangular walkway around the pool so that the walkway is 2 meters wide. What is the area of the walkway, in square meters?

17. A rectangular pen needs to be constructed out of 100 feet of fencing. Let ℓ be the length of the pen.
    a. Find an expression for the width w of the pen in terms of ℓ.
    b. Use your answer from Part a to give an expression for the area of the rectangle in terms of ℓ.

18. Suppose $P = (-2, 4)$, $Q = (-2, -3)$, $PQRS$ is a rectangle, and the area of $PQRS$ is 42. Find the two possible locations of R and S.

REVIEW

19. If a regular 14-gon has perimeter 7, what is the length of each side? (**Lesson 8-1**)

20. In the figure at the right, $\overline{AB}$ was rotated about O to get $\overline{CD}$. Explain why $ABDC$ is not a parallelogram. (**Lesson 7-8**)

21. **a. True or False** If two isosceles triangles have two pairs of congruent sides, then they are congruent.
    b. Would your answer in Part a be different if the triangles were equilateral? Explain your answer. (**Lesson 7-2**)

22. Explain how the Isosceles Trapezoid Symmetry Theorem can be used to show that the diagonals in an isosceles trapezoid are congruent. (**Lesson 6-6**)

EXPLORATION

23. **a.** Suppose a rectangle has an area of 12 square units. Find a relationship between the length ℓ and the width w of such a rectangle. Graph all of the possible pairs in the coordinate plane.
    b. Using a DGS, create a fixed area rectangle as in Part a. The rectangle you create should allow the user to drag one side to make it smaller or larger while the other sides change in order to maintain the given area.

QY ANSWERS

1. Answers vary. Sample: No, because part of the borders overlap.

2. 1,000,000

Lesson
8-3

Areas of Irregular Figures

> ▶ **BIG IDEA** Areas of irregular figures can be estimated by covering the figures with smaller and smaller square grids.

Here is a aerial photo of Christmas Island, also known as Kiritimati, part of the Republic of Kiribati in the Pacific Ocean. It is the largest coral atoll in the world. The island was uninhabited when Captain James Cook landed there on December 24, 1777, but had 5115 residents as of the 2005 census.

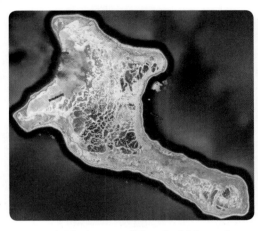

Mental Math

a. How many whole segments of length 3 ft could you place side by side on a segment of length 29.23 ft?

b. In Part a, what is the length of the part of the large segment that is not covered by the smaller segments?

c. Twenty-five segments of length 3 ft are placed side by side on a large segment. At the end, only 1.2 ft of the large segment remains uncovered by smaller segments. What is the length of the large segment?

It is natural to want to find the area of Christmas Island. Although the exact area cannot be found, you can create a sequence of approximations that gets closer and closer to the area.

Activity

Follow these steps to estimate the area of Christmas Island.

Step 1 We have superimposed a tessellation of congruent squares on the map, as shown at the right. Each square has a side of 8 miles, so its area is 64 square miles.

a. Count the number of squares that are entirely in the island. Call this number I (for inside). Here there is one square entirely in the island, so $I = 1$.

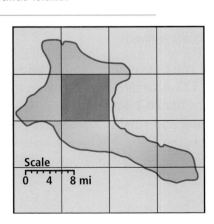

Scale

0   4   8 mi

(continued on next page)

b. Count the number of squares that are partially island and partially water. Call this number B (for boundary). We find $B = 11$. For these squares, instead of trying to estimate how much of each square is inside and adding the parts, it is easier to assume that, on average, half of each boundary square is inside. So add $\frac{1}{2}B$ to the number in Part a.

c. Let U be the area of each square. So multiply $I + \frac{1}{2}B$ by U to estimate the total area. Here $U = 64$ square miles, and we find $(I + \frac{1}{2}B) \cdot U = (1 + \frac{1}{2} \cdot 11) \cdot 64 = 416$ square miles. This is the first estimate.

Step 2 At the right, each square has side 4 miles, so its area is 16 square miles. Repeat Parts a–c of Step 1 for this new grid to obtain a second estimate. You may wish to compare your results with someone else.

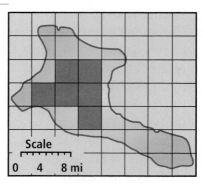

Step 3 Below, each square has side 2 miles, so its area is 4 square miles. Repeat Parts a–c of Step 1 for this new grid to obtain a third estimate.

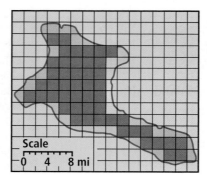

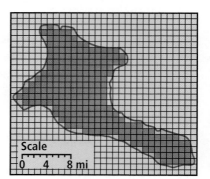

Step 4 At the right, each square has side 1 mile, so its area is 1 square mile. Repeat Parts a–c of Step 1 for this new grid to obtain a fourth estimate.

The above procedure can be continued using grids with smaller and smaller squares. The estimates can be made to differ from the actual area by no more than 0.1 square mile, 0.01 square mile, or even less. When smaller and smaller squares are used, we say that the area of the island is the *limit* that the estimates approach. The actual area of Christmas Island is 248 square miles.

Questions

COVERING THE IDEAS

1. Give three reasons people might have for estimating the area of an island.

2. Four tessellations of squares are used in this lesson to estimate the area of Christmas Island. What is the area of an individual square

   a. in the second tessellation?

   b. in the third tessellation?

   c. in the final tessellation?

3. a. Was your final estimate in the Activity greater than or less than the first estimate of 800 square miles?

   b. Was it greater than or less than the official area of 248 square miles?

4. Suppose you find E squares entirely inside a region and P squares partially inside the region. If each square has area U square units, what formula gives an estimate, in square units, for the region's area?

5. The two paw prints below are congruent. The grid squares on the left are $\frac{1}{4}$ inch on a side, and the grid squares on the right are $\frac{1}{8}$ inch on a side. Estimate the area of the outlined paw print using the given grid.

   a.
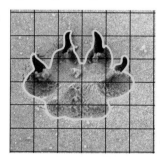
   b.
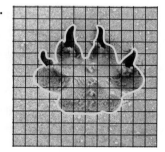

6. **Fill in the Blank** The area of a region is the ___?___ of the estimates made using finer and finer grids.

APPLYING THE MATHEMATICS

In 7 and 8, use this outline of Lake Okeechobee in Florida and the scale indicating lengths in miles. Lake Okeechobee is the fourth largest natural lake entirely within the continental United States.

7. Trace the lake's outline onto a sheet of paper. Make a grid of squares 10 miles on a side and put the grid on top of the lake. Use the grid and the method of this section to estimate the area of Lake Okeechobee.

8. Repeat Question 7, but with a grid of squares 5 miles on a side.

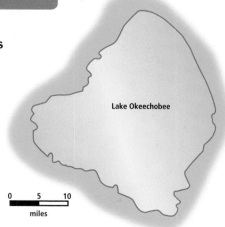

Lake Okeechobee

0   5   10
miles

9. Estimate the area bounded by the *x*-axis, the graph of $y = \sqrt{x}$, and the line $x = 9$. (You can graph $y = \sqrt{x}$ on a calculator and turn on a grid.) Do this by counting squares.

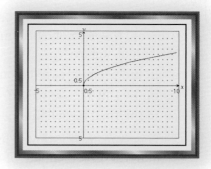

10. A different way of estimating the area of an irregular figure is to count all the squares that are more than half-filled by the figure. Then use this count as an estimate of the area of the figure.

   a. Use this method on the drawing of Christmas Island with the squares that have area 4 square miles.

   b. Use this method on the drawing of Christmas Island with the squares that have area 1 square mile.

   c. Which method do you think is better, this method or the one in the Activity?

REVIEW

11. a. Suppose that the area of a rectangle is *a* square units. What is the area of the image of the rectangle under a size transformation with magnitude 10?

   b. Suppose that a polygon has area *b* square units, and that it can be broken up into three rectangles. What is the area of the image of the polygon under a size transformation with magnitude 10? (**Lessons 8-2, 3-7**)

12. In the figure at the right, the outer square has side length 7 cm, and the inner square has side length 5. What is the area of the shaded region? Explain your answer. (**Lesson 8-2**)

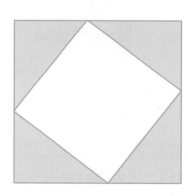

13. Trace the figure below, and construct a rectangle that has *A, B,* and *C* as vertices. (**Lesson 7-10**)

14. Devon wants to check if △DEF is congruent to △GHI. First he measures m∠D and m∠G. Then he measures $\overline{DE}$ and $\overline{GH}$. Finally, he measures $\overline{HI}$ and $\overline{EF}$. After all of these measurements, he concludes that the triangles are congruent. Which Triangle Congruence Theorem did he use, and what can you say about m∠D? (**Lesson 7-5**)

15. Explain why it is impossible for a quadrilateral to have sides with lengths 1, 1, 1, and 4. (**Lesson 1-7**)

16. Let $A = (2, 7)$ and $B = (7, -5)$. Find the distance between A and B. (**Previous Course**)

17. Religions practiced on Christmas Island are Buddhism (36%), Islam (25%), Christianity (18%), Taoism (15%), and other (6%). Make a circle graph of this information. (**Previous Course**)

EXPLORATION

18. **Multiple Choice** A *lattice point* is a point whose coordinates are integers. When the vertices of a polygon are lattice points, there is a formula for its area. The formula is known as Pick's Theorem. Use the polygon at the right and test with other polygons to answer this question. Let P be the number of lattice points *on* the polygon. Let I be the number of lattice points *inside* the polygon. Which is the polygon's area (in square units)?

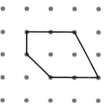

A $\frac{1}{2}P + I - 1$

B $\frac{1}{2}P + I$

C $\frac{1}{2}P + I + 1$

D $\frac{1}{2}(P + I)$

19. From the Internet, an atlas, or other source, trace the outline of a lake, island, small country, or other region. Construct a grid of squares to place over your outline. Use the grid to estimate the area of the region. Compare the area you find with the area as indicated in the reference you used.

Areas of Triangles

Vocabulary

height of a triangle

altitude of a triangle

▶ **BIG IDEA** From the Rectangle Area Formula, area formulas for right triangles and any triangles can be deduced.

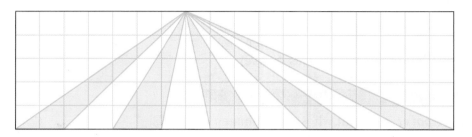

Mental Math

Simplify.

a. $\sqrt{8} \cdot \sqrt{8}$

b. $\sqrt{2} \cdot \sqrt{8}$

c. $\sqrt{50}$

The diagram above is constructed on a grid of congruent unit squares. The shaded triangles are clearly not congruent, but they have the same area. The reason why is found in this lesson.

An Area Formula for a Right Triangle

Triangles and some polygonal shapes are so common that formulas have been developed to give their areas. All of these formulas can be derived using the Area Postulate.

It is easy to deduce a formula for the area of any right triangle. Consider $\triangle ABC$, where $\angle B$ is the right angle, as shown at the right. Rotate $\triangle ABC$ about M, the midpoint of the hypotenuse. This is the same transformation that you used previously to make a tessellation. The image is $\triangle CDA$. Because $\angle BAC$ and $\angle ACB$ are complementary, all four angles of quadrilateral $ABCD$ are right angles, so $ABCD$ is a rectangle. By the Congruence and Additive Properties of the Area Postulate, the area of each triangle is half the area of the rectangle. This argument proves the Right Triangle Area Formula.

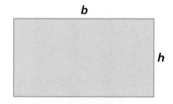

Area = bh

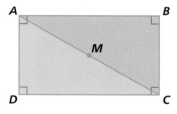

Right Triangle Area Formula

The area of a right triangle is half the product of the lengths of its legs.

$$A = \tfrac{1}{2}hb$$

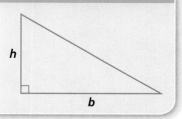

Example 1

Find the area of △DOG at the right.

Solution △DOG is a right triangle with legs 9 cm and 12 cm in length, so

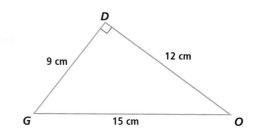

$$\text{Area}(\triangle DOG) = \frac{1}{2} \cdot 9 \text{ cm} \cdot 12 \text{ cm}$$
$$= 54 \text{ cm}^2.$$

An Area Formula for Any Triangle

From the area of a right triangle, it is possible to derive a formula for the area of any triangle. Remember that an **altitude of a triangle** is the perpendicular segment from a vertex to the line containing the opposite side. An altitude may lie inside, outside, or on a side of a triangle.

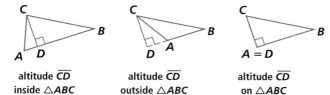

altitude $\overline{CD}$ | altitude $\overline{CD}$ | altitude $\overline{CD}$
inside △ABC | outside △ABC | on △ABC

In each drawing above, $\overline{CD}$ is the altitude to side $\overline{AB}$ of △ABC. The length of an altitude is called the **height of the triangle** with that base. In all cases, the same simple formula for the area of the triangle can be deduced.

Triangle Area Formula

The area of a triangle is half the product of the length of a side (the base) and the altitude (height) to that side.

$$A = \frac{1}{2}hb$$

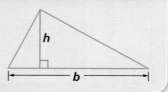

Proof

Case I Altitude *inside* △ABC

The altitude splits △ABC into two right triangles.
Let $BD = x$ and $DC = y$. Then $b = x + y$.

$\text{Area}(\triangle ABC) = \text{Area}(\triangle ABD) + \text{Area}(\triangle ADC)$	Additive Property of Area
$= \frac{1}{2}hx + \frac{1}{2}hy$	Right Triangle Area Formula
$= \frac{1}{2}h(x + y)$	Distributive Property
$= \frac{1}{2}hb$	Substitution Property

Case II Altitude *outside* △ABC

The area of △ABC can be found by subtracting the areas of the two right triangles.

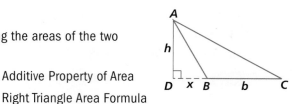

Area(△ABC) = Area(△ADC) − Area(△ADB)   Additive Property of Area

$= h(x + b) - \frac{1}{2}hx$ Right Triangle Area Formula

$= \frac{1}{2}hx + \frac{1}{2}hb - \frac{1}{2}hx$ Distributive Property

$= \frac{1}{2}hb$ Subtraction (Subtract $\frac{1}{2}hx$ from $\frac{1}{2}hx$.)

Case III Altitude on △ABC

In this case, the triangle is a right triangle, so the formula is already known to apply.

Activity

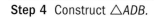

MATERIALS DGS

Step 1 Start with a clear *DGS* screen. Construct $\overline{AB}$ and point C not on $\overline{AB}$.

Step 2 Construct a line parallel to $\overline{AB}$ through point C.

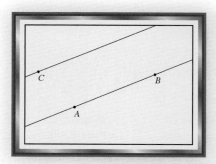

Step 3 Hide point C and construct another point D on the parallel line.

Step 4 Construct △ADB.

Step 5 Drag point D to various locations on the screen. Are the various triangles formed all congruent?

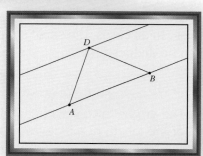

Step 6 Measure the area of △ADB and drag point D again.

Make a conjecture about your findings. Justify your conjecture.

GUIDED

Example 2

Find the area of each triangle in the figure at the right.

a. △ABD   b. △ACD   c. △ADE   d. △ABE

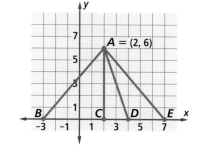

Solution Notice that $\overline{AC}$ is an altitude of each named triangle and that $AC = 6$.

a. Area(△ABD) $= \frac{1}{2} \cdot AC \cdot BD$

$= \frac{1}{2} \cdot \underline{\ ?\ } \cdot \underline{\ ?\ }$

$= \underline{\ ?\ }$

b. $Area(\triangle ACD) = \frac{1}{2} \cdot AC \cdot CD$

$= \frac{1}{2} \cdot \underline{\quad?\quad} \cdot \underline{\quad?\quad} = \underline{\quad?\quad}$

c. $Area(\triangle ADE) = \frac{1}{2} \cdot AC \cdot \underline{\quad?\quad}$

$= \frac{1}{2} \cdot \underline{\quad?\quad} \cdot \underline{\quad?\quad} = \underline{\quad?\quad}$

d. $Area(\triangle ABE) = \frac{1}{2} \cdot \underline{\quad?\quad} \cdot \underline{\quad?\quad}$

$= \frac{1}{2} \cdot \underline{\quad?\quad} \cdot \underline{\quad?\quad} = \underline{\quad?\quad}$

STOP QY

▶ QY

Refer back to the shaded triangles at the beginning of the lesson. What is the area of each triangle?

Questions

COVERING THE IDEAS

In 1 and 2, give a formula for the area of the figure.

1. right triangle
2. any triangle

In 3–5, sketch the altitude of the triangle from vertex A.

3.

4.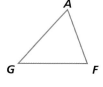

5.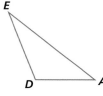

In 6–8, find the area of the triangle.

6.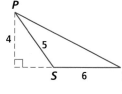

7.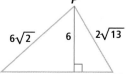

8.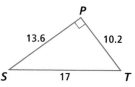

9. Refer to the figure at the right. Find the area of each triangle.
   a. $\triangle KJM$
   b. $\triangle KLM$
   c. $\triangle KJL$

10. Explain why the 5 shaded triangles in the figure at the beginning of the lesson have the same area.

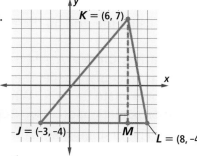

APPLYING THE MATHEMATICS

In 11–13, trace the triangle and draw its three altitudes.

11.

12.

13.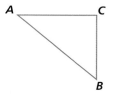

14. By measuring, estimate the area of $\triangle LMN$ at the right to the nearest square centimeter.

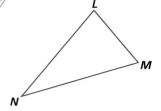

15. The picture at the right shows the triangular region in San Francisco formed by Market Street, 16th Street, and Dolores Street, with the approximate dimensions. Find the area of this triangular region.

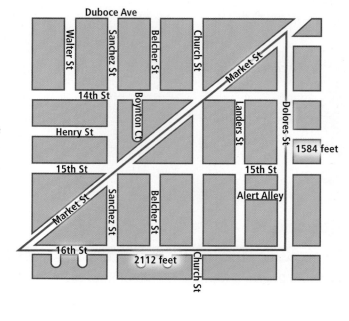

16. Find the length of the side of a triangle that has an area of 238 cm² and whose altitude to that side has a length of 17 cm.

17. A triangle has an area that is equal to that of a rectangle with dimensions of 7 cm by 13 cm. Find the length of a side of this triangle if the altitude to that side has a length of 10 cm.

18. A triangle undergoes a size transformation of magnitude 3. What happens to the given quantity?
    a. the perimeter of the triangle
    b. the area of the triangle

19. Refer to the figure at the right, with $EF = 7$, $GF = 8$, and $EK = 4$. Find the given quantity.
    a. Area($\triangle EGF$)   b. GH

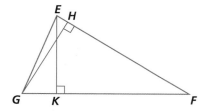

REVIEW

20. At the right, each grid square at the left is 120 miles on a side, and each grid square at the right is 60 miles on a side. Estimate the area of Texas: **(Lesson 8-3)**
    a. using the grid on the left.
    b. using the grid on the right.

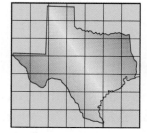
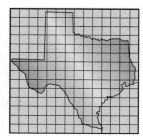

21. **Multiple Choice** A grid is used to approximate the area of an irregular shape, then a second, finer grid, is used to approximate the area. What can be said about these two approximations? **(Lesson 8-3)**
    A The first approximation is always larger.
    B The second approximation is always larger.
    C The two approximations are always the same.
    D The first approximation may be larger, smaller, or equal to the second approximation.

22. Suppose that *F* and *G* are figures that do not overlap. How is the area of their union related to the sum of their areas? (**Lesson 8-2**)

23. Sun was asked to describe a construction where she is given a segment with length ℓ, and has to produce a segment with length 2ℓ. Sun answered: "That's easy! I'll just draw a line, and use my ruler to measure a segment with length 2ℓ." Why is Sun's answer not a valid construction? (**Lessons 7-10, 3-9**)

24. Refer to the figure at the right.
    Given: $\odot O$, and $m\angle DBC = m\angle ACB$.
    Prove: $EB = EC$. (**Lesson 7-4**)

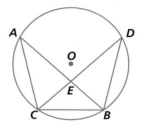

25. Let $A = (3, 2)$, $B = (-4, 5)$, $C = (2, 6)$. Is $\triangle ABC$ isosceles? Explain your answer. (**Previous Course**)

EXPLORATION

26. At the right, the named points are equally spaced along the rectangle *ADEH*. Each point *A*, *B*, *C*, and *D* is connected to each point *E*, *F*, *G*, and *H*. How many triangles in the drawing have the same area as $\triangle CHF$? Name them.

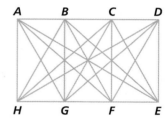

27. A different formula from the one developed in this lesson for the area of a triangle is known as *Hero's* or *Heron's formula*, after the Greek mathematician and physicist Hero (or Heron) of Alexandria, who lived about 50 CE. The formula uses the *semiperimeter s* of a triangle, which is defined to be half the perimeter of the triangle. If a triangle $\triangle ABC$ has sides of lengths *a*, *b*, and *c*, Hero's formula is Area$(\triangle ABC) = \sqrt{s(s-a)(s-b)(s-c)}$.

    a. Use Hero's formula to find the area of a right triangle with legs of lengths 10 cm and 24 cm, and a hypotenuse of length 26 cm. Check your answer by computing this area using the method of this lesson.

    b. Find the exact area of a triangle whose sides have lengths 14 units, 17 units, and 19 units.

    c. Use a DGS to construct a triangle. Measure the lengths of the sides and determine the area using Hero's formula. Then use the DGS Area tool to determine the area. Do these numbers always appear to be the same?

8-5

Areas of Quadrilaterals

Vocabulary

altitude of a trapezoid

height of a trapezoid

> ▶ **BIG IDEA** From the Triangle Area Formula, area formulas for trapezoids, parallelograms, and quadrilaterals with perpendicular diagonals can be deduced.

With the ability to find the area of a triangle, you can find the area of any polygon by dividing it into triangular regions. But for some types of quadrilaterals, you do not have to go through all that work. There are formulas for their areas.

Mental Math

Suppose $x > 0$. Simplify.

a. $\sqrt{x} \cdot \sqrt{x}$

b. $\sqrt{6y} \cdot \sqrt{y}$

c. $\sqrt{2z} \cdot \sqrt{18z}$

Finding the Area of a Trapezoid

The **altitude of a trapezoid** refers to any segment from one base perpendicular to the other. The **height of a trapezoid** is the length of that segment. This length is constant because the bases are parallel. In trapezoid $TRAP$ at the right, the lengths of the bases (8 and 13) are given, along with the height (6). This is enough to find the area.

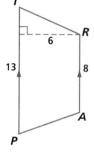

First split the trapezoid into triangles $\triangle TRP$ and $\triangle RAP$. Notice that the height of the trapezoid is the height of each triangle and the bases of the trapezoid are the corresponding bases of the triangles, so the area of each triangle can be found. Then apply the Additive Property of Area to find the area of the trapezoid.

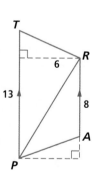

$$\begin{aligned} \text{Area}(TRAP) &= \text{Area}(\triangle TRP) + \text{Area}(\triangle RAP) \\ &= \tfrac{1}{2} \cdot 6 \cdot 13 + \tfrac{1}{2} \cdot 6 \cdot 8 \\ &= 39 + 24 \\ &= 63 \end{aligned}$$

A Formula for the Area of Any Trapezoid

The ideas used in the specific instance above can be applied to deduce a formula for the area of any trapezoid. A bonus is that a trapezoid area formula applies to all of the special kinds of quadrilaterals that are below the trapezoid in the hierarchy of quadrilaterals.

Trapezoid Area Formula

The area of a trapezoid equals half the product of its altitude and the sum of the lengths of its bases.

$$A = \tfrac{1}{2}h(b_1 + b_2)$$

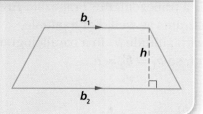

Given *ZOID* is a trapezoid with altitude h and bases b_1 and b_2.

Prove Area(*ZOID*) $= \tfrac{1}{2}h(b_1 + b_2)$

Drawing

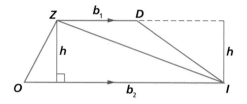

Proof

Conclusions	Justifications
1. *ZOID* is a trapezoid with altitude h and bases b_1 and b_2.	1. Given
2. Area(*ZOID*) = Area($\triangle ZDI$) + Area($\triangle ZOI$)	2. Additive Property of Area
3. Area(*ZOID*) $= \tfrac{1}{2} \cdot h \cdot b_1 + \tfrac{1}{2} \cdot h \cdot b_2$	3. Triangle Area Formula
4. Area(*ZOID*) $= \tfrac{1}{2}h(b_1 + b_2)$	4. Distributive Property

Recall that $\tfrac{1}{2}(b_1 + b_2)$ is the *arithmetic mean* or *average* of b_1 and b_2. Rearranging the Trapezoid Area Formula to $A = \dfrac{b_1 + b_2}{2}h$ allows you to see that the area of a trapezoid can be stated as the average of the lengths of its bases times the height.

GUIDED

Example 1

Compute the area of *QRST* at the right.

Solution By the Two Perpendiculars Theorem, $\overline{QR} \parallel \overline{TS}$, so $\overline{QR}$ and $\overline{TS}$ are the bases of a trapezoid. Apply the Trapezoid Area Formula with $b_1 = \underline{\ ?\ }$, $b_2 = \underline{\ ?\ }$, and $h = \underline{\ ?\ }$.

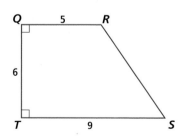

$$\text{Area (QRST)} = \tfrac{1}{2}h(b_1 + b_2)$$
$$= \tfrac{1}{2} \cdot \underline{\ ?\ } (\underline{\ ?\ } + \underline{\ ?\ })$$
$$= \underline{\ ?\ }$$

Check your answer by drawing the altitude from R to $\overline{TS}$, creating a rectangle and a right triangle. Find their areas and add.

Areas of Parallelograms

Since every parallelogram is a trapezoid, the Trapezoid Area
Formula applies to parallelograms as well. For example,
GRAM is a parallelogram with altitude h. In a parallelogram,
opposite sides are congruent, so $b_1 = b_2 = b$.

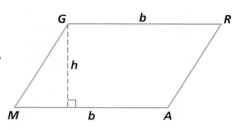

$$\text{Area}(GRAM) = \tfrac{1}{2}h(b_1 + b_2)$$
$$= \tfrac{1}{2}h(b + b)$$
$$= \tfrac{1}{2}h(2b)$$
$$= hb$$

Parallelogram Area Formula

The area of a parallelogram is the
product of the length of one of its
bases and the altitude to that base.

$$A = hb$$

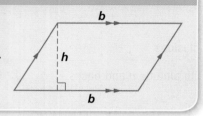

 QY1

▶ **QY1**

Find the area of
parallelogram *TRIG*.

Areas of Quadrilaterals with Perpendicular Diagonals

What happens when you know that a quadrilateral has perpendicular
diagonals? Would knowing the lengths of the diagonals give you a
shortcut to finding its area?

 QY2

▶ **QY2**

Which special
quadrilaterals always have
perpendicular diagonals?

GUIDED

Example 2

Given: Quadrilateral *ABCD*; $\overline{AC} \perp \overline{BD}$.

Prove: $\text{Area}(ABCD) = \tfrac{1}{2}(BD)(AC)$.

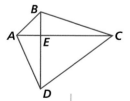

Proof

Conclusions	Justifications
1. $\overline{AC} \perp \overline{BD}$	1. ___?___
2. $\text{Area}(\triangle ABC) = \tfrac{1}{2}(BE)(AC)$	2. Triangle Area Formula
3. $\text{Area}(\triangle ADC) =$ ___?___	3. ___?___
4. $\text{Area}(ABCD) = \text{Area}(\triangle ABC) + \text{Area}(\triangle ADC)$	4. Area Postulate (Additive Property)
5. $\text{Area}(ABCD) =$ ___?___	5. ___?___
6. $\text{Area}(ABCD) = \tfrac{1}{2}(BD)(AC)$	6. ___?___

Perpendicular Diagonals Quadrilateral Area Formula

The area of a quadrilateral with perpendicular diagonals equals half the product of its diagonals. When d_1 and d_2 are the lengths of its diagonals,

$$A = \tfrac{1}{2}d_1d_2.$$

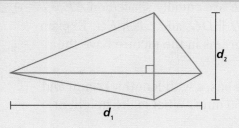

Questions

COVERING THE IDEAS

1. Describe a process for finding the area of any polygon.

2. Define *altitude of a trapezoid*.

3. In finding the area of a trapezoid, you use the length of an altitude of the trapezoid. Why is this length uniquely determined?

4. Refer to the figure at the right.
    a. Name the bases and altitude of trapezoid *BIOR*.
    b. Find Area(*BIOR*).

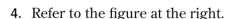

5. A park is in the shape of a trapezoid, as shown at the right. Find its area in square miles.

6. A rhombus has diagonals with lengths 6 cm and 15 cm. What is its area?

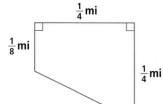

In 7 and 8, refer to the figure at the right.

7. Find the area of trapezoid *ABEF*.

8. Find the area of trapezoid *ABCF* in two different ways.

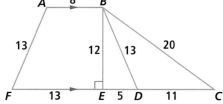
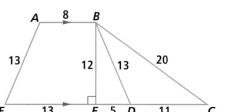

In 9 and 10, find the area of the given quadrilateral.

9.

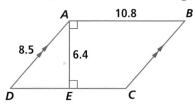

10.

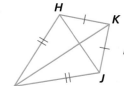

$HJ = 7$, $KG = 11$

APPLYING THE MATHEMATICS

11. Points A, B, C, D, and E are equally spaced on $\overleftrightarrow{AE}$, $\overleftrightarrow{AE} \parallel \overleftrightarrow{FG}$, $\overleftrightarrow{AF} \parallel \overleftrightarrow{CG}$, $\overleftrightarrow{FB} \parallel \overleftrightarrow{DG}$, and $\overleftrightarrow{CF} \parallel \overleftrightarrow{EG}$. Explain why the three parallelograms in the picture have the same area.

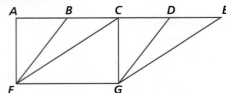

12. A trapezoid has a pair of bases with lengths 12 mm and 18 mm and an area of 150 mm². What is the height of the trapezoid?

13. Find the area of $ZOID$ at the right.

14. The perimeter of a trapezoid is 45 with the nonparallel sides having lengths of 9 and 11. If the altitude of this trapezoid is 8, what is its area?

15. Consider the trapezoid $MILK$ at the right. Show that Area($\triangle MRK$) = Area($\triangle IRL$).

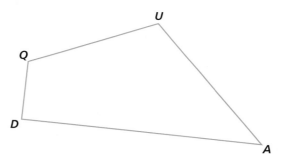

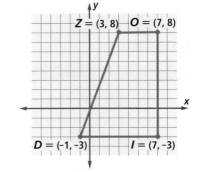

16. Use measuring tools to find the area of $QUAD$ in square millimeters.

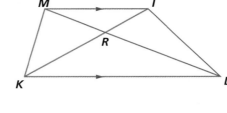

17. Franklin Benjamin is building a kite like the one shown at the right. He purchased sticks with lengths 15" and 10". He needs to purchase kite material to cover the surface of the kite. What is the area of the surface covering of the kite (without accounting for enough to fold over the sticks)?

18. Draw two quadrilaterals, one convex (but not a rectangle) and one nonconvex, each with an area of exactly 24 square units.

REVIEW

19. Express the areas of the following triangles in terms of the variables that are lengths in the figure. (Lesson 8-4)
    a. $\triangle ABC$
    b. $\triangle ABD$
    c. $\triangle ADE$

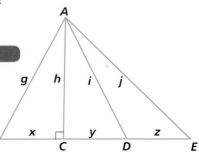

20. A tailor wants to use the pattern below to make a jacket, and he needs to know how many square yards of cloth to buy. Explain how he could do this. (**Lesson 8-3**)

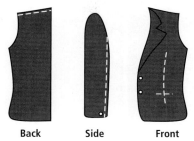

Back   Side   Front

21. The image at the right shows a tessellation of figure *F*. (**Lessons 8-2, 7-6, 1-3**)

    a. Suppose that the area of *F* is 3.5. What is the area of the large figure?

    b. Is it possible to draw the entire figure without retracing lines and without lifting your pen? Why or why not?

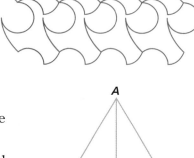

22. Suppose a square is cut up into two triangles. Does the total perimeter increase, decrease, or stay the same? (**Lesson 8-1**)

23. Suppose A, C, and O are points such that $AO = CO$. A is rotated $120°$ about O to get point B. What are the two possible values for m$\angle ACB$? (**Lesson 6-3**)

24. In the figure at the right, $\triangle ABC$ is an equilateral triangle, and $\overline{AD}$ bisects $\angle BAC$. Prove that $2BD = AB$. (**Lesson 6-2**)

25. Simplify $\sqrt{125}$. (**Previous Course**)

EXPLORATION

26. In the two quadrilaterals below, d is the length of a diagonal in the interior of the quadrilateral, and h_1 and h_2 are lengths of the altitudes from the vertices to the diagonal. Compute the areas of the two quadrilaterals using the formula for the area of a triangle. Use your findings to determine a general formula for the area of a quadrilateral.

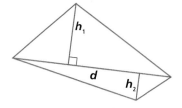

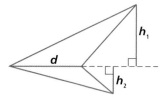

Lesson 8-6

The Pythagorean Theorem

Vocabulary

Pythagorean triple

▶ **BIG IDEA** The Pythagorean Theorem, one of the most useful theorems in all of mathematics, can be proved using the Area Postulate.

If the area of a square is A and its sides have length x, then $x \cdot x = A$ and x is called the positive square root of A, written $\sqrt{A}$. It is perhaps surprising that, by using areas of squares, lengths of sides of triangles can be found.

 QY1

Mental Math

In the figure below, $r_\ell(\triangle ABC) = \triangle A'B'C'$. How many pairs of congruent triangles are there in the figure?

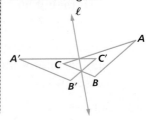

Activity 1

MATERIALS protractor or compass

Step 1 Use a protractor or compass to draw a right triangle HFG with $FG = 2$ cm and $FH = 3$ cm. Make your measurements as accurate as possible.

Step 2 Trace three other copies of the triangle to form the quadrilateral $FBCD$ like the one at the right.

Step 3 **a.** What justifies the conclusion that $FBCD$ is a square?
b. What is the area of $FBCD$?

Step 4 **a.** Measure the sides and angles of $GHLN$.
b. What type of figure is it?

Step 5 Use the following relationship to find the area of $GHLN$:
Area($GHLN$) = Area($FBCD$) − Area($\triangle HFG$) − Area($\triangle LDH$) − Area($\triangle NCL$) − Area($\triangle GBN$).

Step 6 Use the area of $HLNG$ to find the length of $\overline{GH}$.

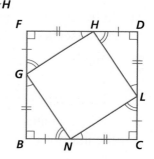

▶ **QY1**

If the area of a square is 60 cm², what is the length of a side of the square?

History of the Pythagorean Theorem

In Activity 1, you used squares to find the length of $\overline{GH}$ from the lengths of $\overline{FH}$ and $\overline{FG}$. Generalizing the process, we can prove the Pythagorean Theorem, which is regarded by many people as the most famous theorem in the world. The Pythagorean Theorem relates the lengths of the three sides of any right triangle.

> ### Pythagorean Theorem
>
> In any right triangle with legs of lengths a and b and hypotenuse of length c, $a^2 + b^2 = c^2$.

The Pythagorean Theorem is named after Pythagoras, the Greek mathematician who proved it in the 6th century BCE. His was the earliest proof known to the western world until recent times. We are now aware of proofs from cultures throughout the world, and it is not clear where or when the first proof originated. What we call the "Pythagorean Theorem" was known to the Babylonians before 1650 BCE and possibly in India before 800 BCE.

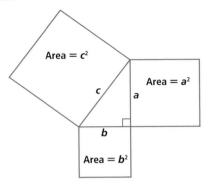

The Pythagorean Theorem can also be interpreted as a theorem about areas. It is called "The Theorem of Three Squares" in Japan.

> ### Pythagorean Theorem (alternate statement)
>
> In any right triangle, the sum of the areas of the squares on its legs equals the area of the square on its hypotenuse.

A collection of 370 different proofs of the Pythagorean Theorem was compiled by Elisha Loomis in 1940. It includes proofs by the 12th century Indian mathematician Bhaskara; by the 15th century Italian painter, sculptor, architect, and engineer Leonardo Da Vinci; by the 19th century president of the United States, James A. Garfield; and by many others. These and most early proofs of the Pythagorean Theorem deduce the theorem from the areas of triangles and rectangles. The proof we give in this lesson is of this type. It dates at least back to the year 1733, but it is similar to early Chinese proofs. It may have been the proof Pythagoras actually used! Our proof uses the process of Activity 1.

A Proof of the Pythagorean Theorem

Given A right triangle with legs a and b and hypotenuse c

Prove $a^2 + b^2 = c^2$

Proof The right triangle *FHG* and its congruent copies (by SAS congruence) are shown at the right. The outer quadrilateral is a square because each side has length $(a + b)$ and it has four right angles. The inside quadrilateral also is a square because each side has length c and each angle is a right angle.

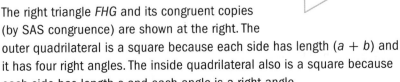

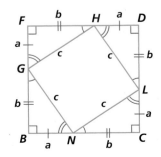

(continued on next page)

QY2

Now we find the area of the inner quadrilateral as you did in Activity 1.

> Area($GHLN$) = Area($FDCB$) − Area(4 right $\triangle$s)

Since each side of $FDCB$ is $a + b$, its area is $(a + b)^2$. Each of the four right triangles has area $\frac{1}{2}ab$, so the area of $GHLN$ is $(a + b)^2 - 4 \cdot \frac{1}{2}ab = (a^2 + 2ab + b^2) - 2ab = a^2 + b^2$. Because the inner square has sides of length c, its area can also be given by c^2. The areas are equal, so $c^2 = a^2 + b^2$.

STOP QY3

The Pythagorean Theorem is useful in many kinds of problems.

Example 1

The length of a soccer field must be between 100 and 130 yards. The width must be between 50 and 100 yards. What is the range for the length of the diagonal of the field?

Solution Let d stand for the length of the diagonal.

The smallest field possible is 100 yards by 50 yards.
The diagonal for that field can be found using $100^2 + 50^2 = d^2$.
Then $d^2 = 12{,}500$ and $d = \sqrt{12{,}500} = 111.8$ yd.

The largest field possible is 130 yards by 100 yards.
The diagonal for that field can be found using $130^2 + 100^2 = d^2$.
Then $d^2 = 26{,}900$ and $d = \sqrt{26{,}900}$ yd. So the length of the diagonal ranges from about 111.8 yd to 164.0 yd.

Converse of the Pythagorean Theorem

Activity 2 asks you to explore whether the converse of the Pythagorean Theorem might be true.

Activity 2

MATERIALS DGS

Step 1 Construct a scalene $\triangle ABC$.

Step 2 Construct a square with side $\overline{BC}$ that is in the exterior of $\triangle ABC$. We will refer to this as "the square on $\overline{BC}$."

Step 3 Construct squares on the sides $\overline{BA}$ and $\overline{AC}$.

Your figure should look similar to the one at the right.

Step 4 Measure $\angle BCA$.

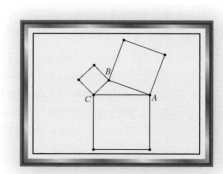

▶ **QY2**

What is the relationship between $\angle FGH$ and $\angle FHG$? How does that guarantee that $\angle HGN$ is a right angle?

▶ **QY3**

If the longest side of a right triangle has length 85 cm and the shortest side has length 13 cm, what is the length of the third side?

Step 5 Measure the areas of the three squares.

Step 6 Calculate the sum of the areas of the squares on sides $\overline{BC}$ and $\overline{AC}$.

Step 7 Drag point B (which will change the measure of $\angle BCA$), and compare the area of the square on side $\overline{BA}$ with the sum calculated in Step 6.

Step 8 $\angle BCA$ is the angle across from the square on side $\overline{BA}$. When this area is less than the sum of the areas of the squares on the sides $\overline{BC}$ and $\overline{CA}$, what type of angle is $\angle BCA$?

Step 9 When this area is greater than the sum of the areas of the squares on the sides $\overline{BC}$ and $\overline{CA}$, what type of angle is $\angle BCA$?

Step 10 Is there a third possibility?

You know that the converse of a theorem is not necessarily true. The above activity should lead you to the conclusion that in the case of the Pythagorean Theorem, the converse is true.

Pythagorean Converse Theorem

Suppose a triangle has sides of lengths a, b, and c.
If $a^2 + b^2 = c^2$, then the triangle is a right triangle.

Given Triangle ABC with sides of lengths a, b, and c, and with $c^2 = a^2 + b^2$

Prove $\triangle ABC$ is a right triangle.

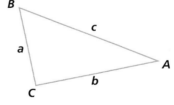

Proof Consider a second right triangle XYZ with legs of lengths a and b and hypotenuse of length z, as drawn at the right.

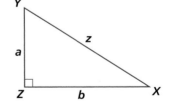

By the Pythagorean Theorem, in $\triangle XYZ$, $a^2 + b^2 = z^2$.
But it is given that $a^2 + b^2 = c^2$.
So, by the Transitive Property of Equality $z^2 = c^2$.
Taking positive square roots of each side, $z = c$.

Thus, the three sides of $\triangle ABC$ are congruent to the three sides of $\triangle XYZ$. So, by SSS Congruence, $\triangle ABC \cong \triangle XYZ$. Because $\angle Z$ is a right angle, by the CPCF Theorem, $\angle C$ is a right angle. Thus, $\triangle ABC$ is a right triangle.

It is also true that if $a^2 + b^2 \neq c^2$, then a, b, and c cannot be the lengths of the sides of a right triangle. Furthermore, you should have seen in Activity 2 that if c^2 is greater that $a^2 + b^2$, then the angle opposite the side of length c is obtuse. If c^2 is less than $a^2 + b^2$, then the angle is acute.

 QY4

▶ **QY4**

A triangle has sides of length 7 cm, 8.2 cm, and 10.8 cm. Is it a right triangle?

To make a right triangle, the ancient Egyptians took a rope with 12 equally spaced knots in it, and then bent it in two places to form a triangle with sides of lengths 3, 4, 5. The angle between the sides of length 3 and 4 was used as a right angle. Does $\angle C$ measure exactly 90°? Yes, because $3^2 + 4^2 = 5^2$, the angle at C must measure exactly 90°.

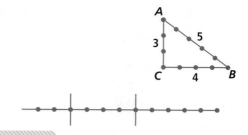

Example 2

A set of three whole numbers that can be the lengths of the sides of a right triangle is called a **Pythagorean triple**. Decide if {7, 22, 28} is a Pythagorean triple. If not, decide whether a triangle with these side lengths is acute or obtuse.

Solution Take the two smallest numbers, 7 and 22, and see if the sum of their squares equals the square of the larger number, 28.

Does $7^2 + 22^2 = 28^2$? No, because $7^2 + 22^2 = 533$ and $28^2 = 784$. Thus {7, 22, 28} is not a Pythagorean Triple, and because 784 > 533, the largest angle is obtuse and this is an obtuse triangle.

Questions

COVERING THE IDEAS

1. If a square has area of 625 m², what is the length of one of its sides?

2. a. What is the English translation of the Japanese name for the Pythagorean Theorem?
   b. What does the Pythagorean Theorem have to do with area?

3. From the reading, which cultures contributed a proof of the Pythagorean Theorem?

4. Name two famous people who created proofs of the Pythagorean Theorem.

5. In an episode of *The Simpsons,* Homer quotes the Scarecrow from *The Wizard of Oz* (1939): "The sum of the square roots of any two sides of an isosceles triangle is equal to the square root of the remaining side." This statement is meant to be the Pythagorean Theorem, but it is not correct. Indicate what is wrong with the Scarecrow's quote of the Pythagorean Theorem and correct it.

6. Use the figure at the right.
   a. What is the area of each triangle?
   b. What is the area of the large square?
   c. What is the area of the tilted square in terms of x and y?
   d. What is z in terms of x and y?

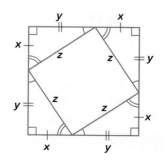

In 7–9, find the exact length of the missing side.

7.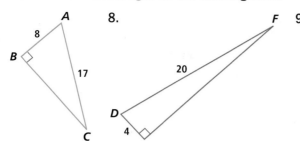

8.

9.

10. A steel rod is 14 feet long. Can it be laid down flat on the floor of a rectangular room that measures 9' by 12'?

In 11–14, determine if a triangle with given side lengths is a right, acute, or obtuse triangle.

11. 25, 22, 12   12. 9, 40, 41   13. 11, 17, 24   14. 5, 13, 12

APPLYING THE MATHEMATICS

15. Find the perimeter of a rhombus with diagonals 12 and 16.

16. Find the area of $\triangle STP$ if $\overline{TP}$ is the diameter of the circle at the right.

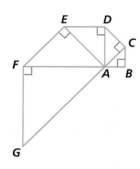

17. In the figure at the right, find RA if $\overline{TR} \parallel \overline{AP}$.

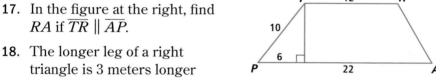

18. The longer leg of a right triangle is 3 meters longer than the shorter leg. The hypotenuse is 6 meters longer than the shorter leg. Find the lengths of the sides of the triangle.

19. Each triangle at the right is an isosceles right triangle. If $AB = 1$, find AG.

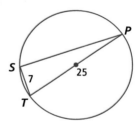

20. Below is a diagram containing the dimensions of a lacrosse field. Suppose Rebecca runs the length of the diagonal $\overline{AC}$ of the field to reach a ball. At the same moment, Haley is in the center E of the defensive area. She runs to the corner F of the sideline and the defensive area and then runs along the sideline to C. Each girl runs 2.8 yards per second. Who will reach the loose ball first? By how many seconds?

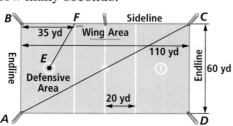

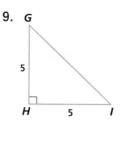

The Pythagorean Theorem   **485**

21. Often in construction, to assure that an angle is a right angle, construction workers mark off 6 ft on one side of the angle and 8 ft on the other side, and then measure the hypotenuse connecting the 6 ft and 8 ft legs. If the hypotenuse is not 10 ft, then they have to adjust the angle and re-measure. Why does the 10-ft hypotenuse assure that the angle is right?

22. In the figure at the right, $\overline{PR} \perp \overline{MT}$, $\overline{MP} \perp \overline{PT}$, $RT = 9$, $RP = 12$, and $RM = 16$. Find the perimeter of $\triangle PMT$.

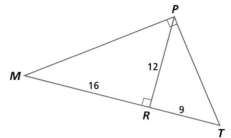

REVIEW

23. Surveyors were hired to find the area of the empty lot $\triangle DNF$. First, they marked off an east-west line $\overleftrightarrow{EW}$ through A. Then they measured the distances from F, D, and N to $\overleftrightarrow{EW}$ and recorded them. What is the area of the lot? (**Lesson 8-5**)

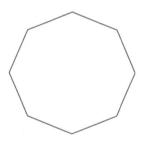

Segment	$\overline{DC}$	$\overline{NB}$	$\overline{FG}$	$\overline{CB}$	$\overline{BG}$
Distance (in feet)	80	200	160	66	250

24. Draw a counterexample to the following conjecture: If a triangle has a side with length b cm, then the area of the triangle is at least b cm^2. (**Lesson 8-4**)

25. Trace the regular octagon at the right.
    (**Lesson 6-8**)
    a. How many different points are equidistant from all of its vertices?
    b. Find one such point.

26. In a circle, what is the difference between a minor arc and a major arc? (**Lesson 3-1**)

EXPLORATION

27. You can create Pythagorean triples using the Fibonacci sequence 1, 1, 2, 3, 5, 8, 13, 21, . . . , in which each term after the second term equals the sum of the two preceding terms. It can be proved that any four consecutive terms of this sequence can be used to create a Pythagorean triple. Choose four consecutive terms.

    Let a = the product of the first and the last terms.

    Let b = twice the product of the middle terms.

    a. Use the Pythagorean Theorem to find c. (c should be an integer.)
    b. Use this algorithm to find five sets of Pythagorean triples.

QY ANSWERS

1. $\sqrt{60}$ cm (or $2\sqrt{15}$ cm)

2. They are complementary. Because $\angle FHG \cong \angle NGB$ and $m\angle FHG + m\angle FGH = 90$, $m\angle NGB + m\angle FGH = 90$. Then, because $m\angle FGH + m\angle HGN + m\angle NGB = 180$, $m\angle HGN = 90$.

3. 84 cm

4. No, because $10.8^2 = 116.64$, while $7^2 + 8.2^2 = 116.24$.

8-7

Special Right Triangles

Vocabulary

45-45-90 triangle

30-60-90 triangle

apothem

▶ **BIG IDEA**   The lengths of the sides of right triangles with 30° or 45° angles are related in simple ways.

The Moseleys are preparing for an outdoor party at their new home. There are a number of things that need to be done before the party. For instance, they would like to tile the floor of their regular octagonal gazebo. The cost of the tile is $2.50 per square foot.

To ensure they do not overpay, they would like to find the exact area of the octagonal base. The length of a side of the octagon is 4 ft.

The Moseleys know they can find the area of the gazebo floor by dividing the region into triangles. To find the exact area of those triangles, they can use properties of some special right triangles. These you will see in this lesson.

Mental Math

a. A right triangle has both legs of length 6. What is the length of its hypotenuse?

b. A triangle has sides of length 2, 5, and 6. Is the triangle a right triangle?

c. One right triangle has a leg of length 7. A second right triangle has a hypotenuse of length 7. Which triangle has the longer hypotenuse?

Lengths in Isosceles Right Triangles

Certain right triangles have well-known relationships among their sides and angles. These special triangles occur in many situations that involve other polygons. For instance, by drawing a diagonal of a square, two isosceles right triangles are formed. The Isosceles Triangle Base Angles Theorem tells us that the base angles of an isosceles triangle are congruent. Because one angle is right, the Triangle-Sum Theorem tells us that the measure of each base angle of an isosceles right triangle must be 45°. For this reason, an isosceles right triangle is called a **45-45-90 triangle.** In a 45-45-90 triangle, if you know the length of one side, you can find the lengths of the others.

Suppose the congruent sides of one of these triangles have length x and the hypotenuse has length c, as in the triangle at the right.

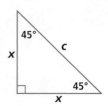

By the Pythagorean Theorem, $c^2 = x^2 + x^2$.
So $c^2 = 2x^2$.
Taking the positive square root, $c = x \cdot \sqrt{2}$.

The result is a relationship among the sides of any isosceles right triangle.

> ### Isosceles Right Triangle Theorem
> In an isosceles right triangle, if a leg has length x, then the hypotenuse has length $x\sqrt{2}$.

GUIDED

Example 1

Find the missing side lengths in each isosceles right triangle.

a.
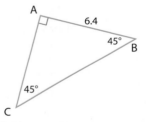

b.
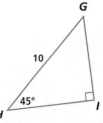

Solution For each triangle, label each side with x or $x\sqrt{2}$. Remember, the hypotenuse is the longest side of a right triangle. Because $x\sqrt{2} > x$, the $x\sqrt{2}$ must always be the length of the hypotenuse.

a. If $x = 6.4$, then $AC = $ __?__ and $BC = $ __?__ .

b. If $x\sqrt{2} = 10$, then $x = \dfrac{10}{\sqrt{2}} = \dfrac{10 \cdot \sqrt{2}}{\sqrt{2} \cdot \sqrt{2}} = \dfrac{10\sqrt{2}}{2} = 5\sqrt{2}$.

$x = $ __?__ , $HI = $ __?__ , $GI = $ __?__

STOP QY1

45-45-90 triangles are useful in finding area because they allow you to find the height of the triangle from just one side length.

▶ QY1

Check your answers to Guided Example 1 using the Pythagorean Theorem.

Example 2

The bottom step of the Moseleys' pool needs to be replaced. The cost of the repair is based on the size of the step. A side view of the step is shown at the right. $FABD$ is a trapezoid with $\overline{AB} \parallel \overline{FD}$ and $\overline{FD} \perp \overline{DB}$. The Moseleys know that $AB = BD$, $AD = 12$ in., and $FD = 18$ in. Round the area of the side of the step to the nearest square inch.

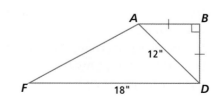

Solution You can find the area of this region by finding the area of each triangular region and using the Additive Property of Area. $\angle FDB$ is a right angle because it is a same-side interior angle with $\angle ABD$ and $\overline{AB} \parallel \overline{FD}$. Thus, BD is the height of both triangles AFD and ABD. Because $AB = DB$, $\triangle ADB$ is an isosceles right triangle. **Let $x = AB$**. Then, by the Isosceles Right Triangle Theorem,

$$x\sqrt{2} = 12$$
$$x = \frac{12}{\sqrt{2}}$$
$$x = 6\sqrt{2}.$$

So $AB = BD = 6\sqrt{2}$.

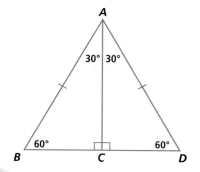

$$\text{Area}(\triangle AFD) = \tfrac{1}{2}(18)(6\sqrt{2})$$
$$= 54\sqrt{2} \text{ in}^2$$
$$\text{Area}(\triangle ABD) = \tfrac{1}{2}(6\sqrt{2})(6\sqrt{2})$$
$$= 36 \text{ in}^2$$
$$\text{Area}(FABD) = 54\sqrt{2} + 36$$
$$\approx 112 \text{ in}^2$$

Lengths in 30-60-90 Triangles

When an altitude is drawn in an equilateral triangle, two triangles with angles 30°, 60°, and 90° are formed. This special right triangle is called a **30-60-90 triangle.** Again the lengths of all sides can be determined if one side length is known.

30-60-90 Triangle Theorem

In a 30-60-90 triangle, if the length of the shorter leg is x, then the length of the longer leg is $x\sqrt{3}$ and the length of the hypotenuse is $2x$.

Given $\triangle ABC$ with $m\angle A = 30$, $m\angle B = 60$, $m\angle C = 90$. The shorter leg is opposite the smaller acute angle, so let $BC = x$.

Prove 1. $AB = 2x$
      2. $AC = x\sqrt{3}$

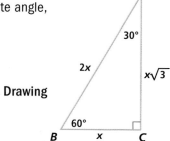

Drawing

Proof The idea is to think of $\triangle ABC$ as half an equilateral triangle and use the Pythagorean Theorem.

Proof of 1 Reflect $\triangle ABC$ over $\overleftrightarrow{AC}$. Let $D = r_{\overleftrightarrow{AC}}(B)$. Reflections preserve distance, so $CD = x$. Because reflections preserve angle measure, the image $\triangle ADC$ is a 30-60-90 right triangle with $m\angle ACD = 90$, $m\angle ADC = 60$, and $m\angle CAD = 30$. Thus, B, C, and D are collinear, and the big triangle ABD has three $60°$ angles, making it equilateral with $AB = BD = 2x$.

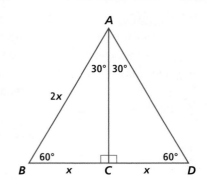

Proof of 2 Now, using AB, you can apply the Pythagorean Theorem to $\triangle ABC$ and find AC.

$$(AC)^2 + (BC)^2 = (AB)^2$$
$$(AC)^2 + x^2 = (2x)^2$$
$$(AC)^2 + x^2 = 4x^2$$
$$(AC)^2 = 3x^2$$

Taking the positive square roots of each side,

$$AC = x\sqrt{3}.$$

STOP QY2

> ▶ **QY2**
>
> If the shortest side of a 30-60-90 triangle has length 70 cm, what are the lengths of the other two sides?

Finding the Areas of Regular Polygons

All regular polygons can be triangulated from their center. Suppose you begin with a regular n-gon with side length s. If you know the length a of a segment connecting the center of the polygon to the midpoint of a side, then you have enough information to create a formula for the area of the regular polygon.

Activity 1

Refer to the three regular polygons drawn at the right. (One drawing is incomplete.)

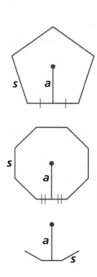

Step 1 If you draw segments joining the center of the regular polygon to its vertices, you create triangles. Explain how you know that these triangles are isosceles.

Step 2 Explain how you know that a is the height of each of these isosceles triangles.

Step 3 In terms of s and a, what is the area of each isosceles triangle?

Step 4 Write a formula for the area of any regular n-gon in terms of s, a, and n.

Step 5 Write a formula for the perimeter p of any regular n-gon with side length s.

Step 6 Use your answers to Steps 3 and 4 to create a formula for the area of any regular polygon in terms of a and p.

In the Activity, the length *a* is known as the *apothem*. The **apothem** of a regular polygon is the segment joining the center of the polygon to the midpoint of any side. In Step 6, you were deducing the following formula for the area of a regular polygon.

> ### Regular Polygon Area Formula
>
> The area of a regular polygon is half the product of the length of its apothem *a* and its perimeter *p*.
>
> $$A = \tfrac{1}{2}ap$$

▶ **READING MATH**

The word *apothem* is pronounced with emphasis on the first syllable and with a short "a" as in "apple":

A-poth-em

Sometimes special right triangles are formed by the apothem.

Example 3

Find the area of a regular hexagon with side length 10.

Solution Draw regular hexagon *HIJKLM* with apothem $\overline{OP}$.

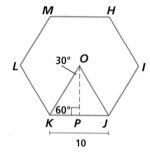

The perimeter of the hexagon is $6(10) = 60$. To find the apothem, notice that $KP = 5$ because it is half of the side length. Also notice that $m\angle KOJ = \frac{360}{6} = 60$, so $m\angle POK = 30$. Thus, $\triangle KPO$ is a 30-60-90 triangle in which 5 is the length of the shorter leg. So, $OP = 5\sqrt{3}$.

$$
\begin{aligned}
\text{Area}(HIJKLM) &= \tfrac{1}{2}ap \\
&= \tfrac{1}{2}(5\sqrt{3})(60) \\
&= 150\sqrt{3}
\end{aligned}
$$

Activity 2

Find the area of the floor of the Moseleys' gazebo. Recall that the gazebo floor is a regular octagon with side length 4 ft. Before you begin, it is important to realize that the method used in Example 3 does not work in this case. This is because the measure of each of the vertex angles of the isosceles triangles formed by joining the center to the vertices is $\frac{360}{8} = 45$. Thus, when the apothem is drawn and it bisects that angle, two angles with measure 22.5 are created. This will not result in special right triangles. We provide a hint in the diagram at the right to get you started.

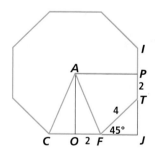

Questions

1. What are the measures of the three angles in an isosceles right triangle?

2. **Fill in the Blank** In an isosceles right triangle, the hypotenuse is __?__ times the length of a leg.

3. Describe the relationship between the lengths of the sides of a 30-60-90 triangle.

4. If an isosceles triangle has a $45°$ angle, must it be a right triangle?

In 5–8, find the exact length of the missing sides.

5.

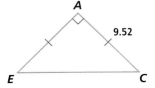

6.

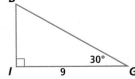

7. a 45-45-90 triangle with hypotenuse 16

8. a 30-60-90 triangle with hypotenuse 26

9. State a formula for the area of a regular polygon.

10. Find the exact area of a regular hexagon with side length 12 cm.

11. Find the area of a regular heptagon with side length 15 m and apothem with length approximately 10.32 m, to the nearest tenth of a square meter.

12. Recalculate the area of the floor of the Moseleys' gazebo, but this time do it by dividing the octagon into two trapezoids and a rectangle as shown at the right.

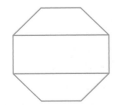

13. In the parallelogram at the right, $\overline{WE} \cong \overline{EO}$. If $WF = 10$, find the exact area of the parallelogram.

14. $ABDC$ is a trapezoid with $\overline{AB} \parallel \overline{CD}$. Using the information in the figure below, find the exact perimeter of $ABCD$.

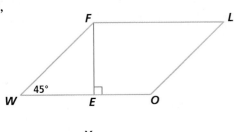

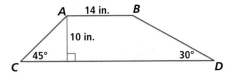

15. Find the area of $\triangle XYZ$ at the right in terms of h.

16. Find the area of an equilateral triangle with side s.

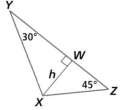

REVIEW

17. A square is inscribed in a circle with diameter 14. What is the area of the square? (**Lesson 8-6**)

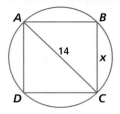

18. *HLAF* is a convex kite. $HA = 30$, $HL = 25$, $FA = 17$. Find the area of the kite. (**Lessons 8-6, 8-5, 6-5**)

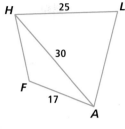

19. In the figure at the right $\overline{AC} \parallel \overline{DE}$, $DE = 30$, $AC = 50$, the area of $\triangle BDE$ is 450, and the area of $\triangle ABC$ is 1250. Find the height of $ADEC$. (**Lessons 8-5, 8-4**)

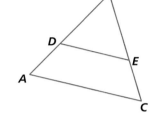

20. Is the quadrilateral below a parallelogram? How do you know? (**Lesson 7-8**)

21. A periscope is a device that uses mirrors to allow a person in a submarine to see objects on the surface. Use the figure at the right to illustrate how this is possible. (**Lesson 4-4**)

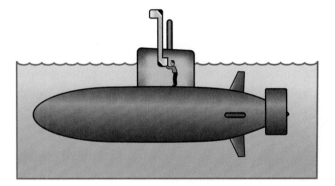

EXPLORATION

22. There are many triangles that could be considered special. One such triangle is called the *heptagonal triangle*. It is formed by the side length a, a shorter diagonal length b, and a longer diagonal length c of a regular heptagon.

    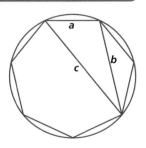

    a. Use a DGS to create a regular heptagon.

    b. Measure c, a, and b. Calculate $\left(\frac{c}{a}\right)^2 + \left(\frac{a}{b}\right)^2 + \left(\frac{b}{c}\right)^2$. Your answer should be very close to a whole number. Which whole number?

    c. Measure the three angles of this triangle. How do they seem to be related?

Lesson
8-8
Arc Length and Circumference

▶ **BIG IDEA** A circle's perimeter is called its circumference and is π times the circle's diameter.

Mia and Rosa are on a bike ride. Mia's bike has 29″-diameter wheels and Rosa's bike has 26″-diameter wheels. If Mia bikes at a rate of 270 revolutions per minute and Rosa bikes at a rate of 300 revolutions per minute, who will ride farther in 1 hour and by how much?

In order to answer this question, you need to know how far each bike travels with one revolution of the wheel. This distance is the *circumference* of the circular wheel. Just as the word *perimeter* describes the distance around a polygon, the word **circumference** describes the distance around a curve such as a circle or an ellipse.

As you have learned in earlier years, the ratio $\frac{\text{circumference}}{\text{diameter}}$ for any circle is constant. This ratio is the famous number pi, denoted by the Greek letter π.

Mental Math

a. A regular octagon has perimeter 100 cm. What is the length of one of its sides?

b. A regular hexagon has perimeter 72 cm. The total length of k sides of the hexagon is 36 cm. What is the value of k?

c. A regular 20-gon has perimeter 30 cm. How many of its sides make up 30% of its perimeter?

Definition of π

$\pi = \frac{C}{d}$, where C is the circumference and d the diameter of a circle.

The number π is irrational; π cannot be written either as a finite decimal or as a simple fraction. The decimal for π is infinite. Your calculator has a key for π that approximates π to 8 or more decimal places beginning 3.14159265.... Many people use 3.14 or 3.14159 or $\frac{22}{7}$ as estimates, but you should use the π key on your calculator to minimize errors due to rounding. Solving the defining equation $\pi = \frac{C}{d}$ for C gives a formula for the circumference of any circle.

Circle Circumference Formula

If a circle has circumference C, diameter d, and radius r, then $C = \pi d$ or $C = 2\pi r$.

 QY

▶ QY

The diameter of the Moon at its equator is approximately 2160 miles. Calculate the circumference of the equator of the Moon using the estimate 3.14 and the π key. What is the difference between the two values?

Example 1

Recall Rosa and Mia's bike ride from the beginning of the lesson. Find the circumference of the wheels of each bike to the nearest tenth of an inch.

Solution Rosa's bike: __?__ · $\pi \approx$ 81.7 in.
Mia's bike: __?__ · $\pi \approx$ 91.1 in.

Exact and Approximate Answers

When you are asked to give an *exact* circumference, you should leave π in the answer; for example, in the circle at the right, $C = 9\pi$ meters. If you are not asked for an exact answer, then you multiply by the approximation given by your calculator. You may be told how many places to which to round your answer. Example 2 answers the question posed at the beginning of the lesson.

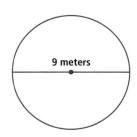

9 meters

Example 2

Who rode farther in one hour, Rosa or Mia? By how much? Round to the nearest foot.

Solution One revolution moves the bike the length of the circumference. We rounded this distance in the last example, but it is useful to keep all calculations exact until the last step to avoid rounding errors.

Circumference of Rosa's bike wheel = 26π in.
Circumference of Mia's bike wheel = 29π in.

At 300 revolutions per minute, Rosa travels
$$300 \, \tfrac{\text{rev}}{\text{min}} \cdot 26\pi \, \tfrac{\text{in.}}{\text{rev}} = 7800\pi \, \tfrac{\text{in.}}{\text{min}}.$$

At 270 revolutions per minute, Mia travels
$$270 \, \tfrac{\text{rev}}{\text{min}} \cdot 29\pi \, \tfrac{\text{in.}}{\text{rev}} = 7830\pi \, \tfrac{\text{in.}}{\text{min}}.$$

There are 60 minutes in 1 hour, so in one hour:

Rosa's distance = $7800\pi \, \tfrac{\text{in.}}{\text{min}} \cdot 60$ min = $468{,}000\pi$ inches.

Mia's distance = $7830\pi \, \tfrac{\text{in.}}{\text{min}} \cdot 60$ min = $469{,}800\pi$ inches.

Thus Mia rode farther by
$$469{,}800\pi - 468{,}000\pi = 1800\pi.$$
$$1800\pi \text{ inches} \cdot \tfrac{1 \text{ ft}}{12 \text{ in.}} = 150\pi \text{ ft.}$$

Now, substitute for π and round to the nearest foot: 150π ft $\approx$ 471 ft.

Mia rode about 471 feet farther than Rosa.

Due to the difference in circumferences, the two wheels of this bicycle will go through different numbers of revolutions per minute even though they will travel the same distance.

$C = \pi d$ or $C = 2\pi r$
$R = \frac{1}{2} d$

Computing Arc Length

If you know the circumference of a circle and the degree measure of an arc, you can also calculate the length of that arc. This is because the degree measure of an arc can be used to calculate the fraction of the circumference occupied by the arc. For instance, a 90° arc has a length equal to $\frac{90}{360} = \frac{1}{4}$ of the circumference. Likewise, a 72° arc has a length equal to $\frac{72}{360} = \frac{1}{5}$ of the circumference. In general, the length of an arc on a circle with circumference C can be calculated in the following way:

$$\text{length of arc} = \frac{\text{degree measure of arc}}{360°} \cdot C = \frac{\text{degree measure of arc}}{360°} \cdot 2\pi r$$

GUIDED

Example 3

In $\odot D$, $DF = 10$ cm and $m\angle FDE = 60$. Find the exact length of $\widehat{FE}$.

Solution $m\angle FDE = 60$, so $m\widehat{FE} = \underline{\ ?\ }$. Thus, $m\widehat{FE}$ is $\frac{?}{360}$ of the circumference of $\odot D$. So,

$$m\widehat{FE} = \frac{?}{360} \cdot (2\pi r)$$

$$= \frac{?}{6} \cdot 2 \cdot \pi \cdot \underline{\ ?\ }$$

$$= \underline{\ ?\ } \text{ cm.}$$

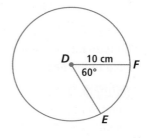

Example 4

Mikayla is cutting a freshly made pie. The radius of the pie is 5 in. She begins cutting at the center of the pie. The first piece has an arc length of approximately 3 in. At about what angle should she cut the next piece if she wants it to be the same size as the first?

Solution Let $x =$ the measure of the arc. Because the measure of an arc is equal to the measure of its central angle, you need to determine the measure of the 3-inch arc. The length of the arc is a fraction of the arc's circumference. This fraction is defined in terms of the measure $x°$ of the arc because the measure of an entire circle is always 360°. Thus,

$$3 = \frac{x}{360} \cdot 10\pi$$

$$\frac{3}{10\pi} = \frac{x}{360}$$

$$10\pi x = 1080$$

$$x \approx 34$$

Mikayla should cut the second piece at an angle of about 34°.

Questions

1. **Fill in the Blank** Polygon is to perimeter as circle is to ___?___.

2. Define π.

3. In calculations, why is it important to use the π key on your calculator instead of 3.14?

4. **Multiple Choice** The circumference of a circle is 15π inches. What is its radius?

   A 15 in.   **B** 15π in.   **C** 7.5 in.   **D** 7.5π in.

5. Consider $\odot A$ at the right.
   a. Find its exact circumference.
   b. Round its circumference to the nearest thousandth.
   c. Find the exact length of $\overset{\frown}{CB}$.
   d. Round the length of $\overset{\frown}{CB}$ to the nearest thousandth.

6. In Example 2 of this lesson, would Mia have biked a longer distance than Rosa if Rosa biked at 310 revolutions per minute?

7. Explain the difference between arc length and arc measure.

8. What is the degree measure of a 4π-cm arc in a circle of radius 8 cm?

9. $\odot E$ has radius 6 inches. F is rotated about E to F'. What magnitude of rotation (to the nearest 0.001°) would allow the length of $\overset{\frown}{FF'}$ to be 8 inches?

10. Each circular layer of Lakota's two-layer birthday cake is trimmed with a strip of pink icing. The top layer has radius 8 in. and the bottom layer has radius 10 in. Her grandfather cuts the top layer into 8 pieces and the bottom layer into 10 pieces. How much longer is the strip of pink icing on a slice of the bottom layer than the strip on a slice of the top layer?

11. In $\odot A$, the length of $\overset{\frown}{CB}$ is 8π. Find CB.

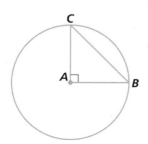

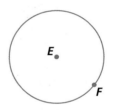

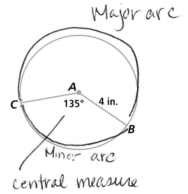

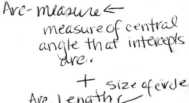

12. The 100-foot clock tower in Rome, Georgia is more than 100 feet tall. The minute hand is 4 feet, 3 inches long, and the hour hand is 3 feet, 6 inches long. How many inches farther does the tip of the minute hand travel than the tip of the hour hand in one day?

13. A standard running track consists of two semicircles connected by 100 m straightaways. The figure below displays the inside lane of the track. What is the radius of each semicircle if the inside circumference of the track is 400 m in length?

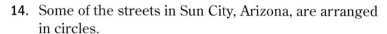

100 m

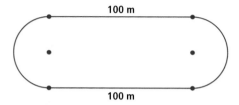

100 m

14. Some of the streets in Sun City, Arizona, are arranged in circles.

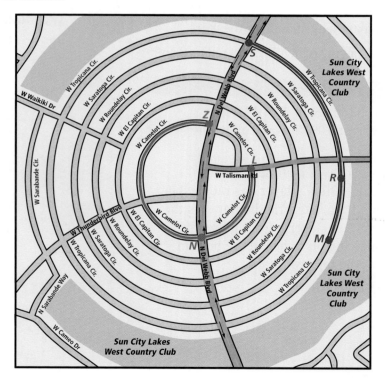

Lorenzo walks the path indicated by $\overarc{LNZ}$. $m\overarc{LNZ} \approx 297°$ and the radius of the circle containing it is about 203 m. Marisol walks the path indicated by $\overarc{MRS}$. $m\overarc{MRS} \approx 97°$ and the radius of the circle containing the arc is about 534 m. Who walks farther, and by about how much?

15. Suppose 10 meters of thread fits on a spool in 100 turns. What is the diameter of the spool?

16. Suppose it takes you 150 seconds to walk around a circular garden. At this rate, about how long would it take you to walk straight through the garden along a diameter?

REVIEW

17. A regular nonagon has apothem 3 cm and area 27 cm^2. Find the length of its side. **(Lesson 8-7)**

18. A fast pitch softball field has bases that are vertices of a square 60 feet apart. How far is a throw from third base to first base, to the nearest foot? (The throw is along a diagonal of the square.) **(Lesson 8-6)**

In 19–22, give a formula for the indicated quantity. (Lessons 8-5, 8-1)

19. area of a trapezoid
20. area of a parallelogram
21. perimeter of a square
22. perimeter of a kite

23. Four points are in a plane. No three are on the same line. How many lines do the four points determine? **(Lesson 1-5)**

24. In Example 2, how fast is Mia traveling in $\frac{\text{kilometers}}{\text{hour}}$? **(Previous Course)**

EXPLORATION

25. If you are standing on the equator, you are traveling in a circle whose center is the center of Earth. Use the library or Internet to find the diameter of this circle. Use this information to calculate how fast you are moving in a circle whose center is due to the rotation of Earth.

QY ANSWER

Using 3.14, the circumference is 6,782.4 mi. Using the π key, the answer is about 6,785.84 mi. The difference is about 3.44 mi.

Lesson

8-9

The Area of a Circle

Vocabulary

sector of a circle

▶ **BIG IDEA** From the Parallelogram Area Formula and the definition of π, the famous formula $A = \pi r^2$ for the area of a circle can be deduced.

Around 250 BCE, the great inventor and geometer Archimedes used the perimeter of regular polygons to find an approximation for the circumference of a circle, and with that, a value for pi that is between $3\frac{10}{70}$ and $3\frac{10}{71}$. In the Activity, you will use a similar method to show the development of the formula for the area of a circle.

Mental Math

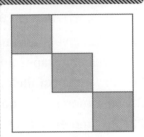

The figure above shows a square with smaller squares inside it. Each small square has sides of length s. Find the area of:

a. the large square.

b. the shaded region.

c. the unshaded region.

Activity

MATERIALS DGS, or protractor, ruler, tracing paper

Step 1 Either with a DGS or with a protractor and ruler, carefully draw an isosceles triangle with a vertex angle of 36°. Call this △ABC, as at the right.

Step 2 Repeatedly rotate △ABC about point A until you have created a regular decagon shown below on the left.

Step 3 **a.** Draw a circle through all the base angle vertices of these isosceles triangles.

b. How do you know that the same circle will go through all of these vertices?

Step 4 Copy the isosceles triangles (tracing paper may be helpful here) and circular arcs and rearrange them horizontally into a frieze pattern, as started at the right.

Step 5 **a.** Trace a quadrilateral that approximates the boundary of the figure you formed in Step 4.

b. What type of quadrilateral is it?

Step 6 **a.** Estimate the area of the quadrilateral you drew in Step 5.

b. How did you arrive at your estimate?

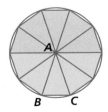

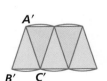

A Proof That $A = \pi r^2$

By generalizing the idea in the Activity, you can see where the formula $A = \pi r^2$ comes from. A circle is not a polygon. But it can be approximated as closely as you want by a polygon. We get its area by using *sectors*. A **sector of a circle** is a region bounded by two radii and an arc of a circle. Dividing a circle into sectors is similar to the method you used to divide regular polygons into triangles.

The circle at the right has radius r. It is split into congruent sectors. Each sector resembles an isosceles triangle with altitude r and a curved base. On the right, the sectors are rearranged as in the Activity. Together they form a figure that is like a parallelogram with height r. Because the union of all the arcs is the circle, each base of the "parallelogram" has length half the circumference of the circle, or $\frac{1}{2}C$. Since $C = 2\pi r$, $\frac{1}{2}C = \pi r$.

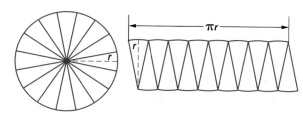

As the number of sectors increases, the figure they form more closely resembles a parallelogram with base πr and height r. That parallelogram has area $A = bh = \pi r \cdot r = \pi r^2$. The areas of the "parallelograms" become better and better approximations to the area of the circle. *We say that the limit of the area of the parallelogram is the area of the circle.* This argument proves a famous formula.

Circle Area Formula

The area A of a circle with radius r is πr^2.

$$A = \pi r^2$$

Example 1

The figure at the right shows an optical computer disk. Data are recorded only on the region between the outside circle and the circle with diameter 4.6 cm. Find the recordable area to the nearest thousandth of a square centimeter.

Solution Find the area of the middle and subtract it from the area of the entire disc. The outside diameter is 12 cm and the inside diameter is 4.6 cm, so the outside radius is $r_1 = 6$ cm and the inside radius is $r_2 = 2.3$ cm.

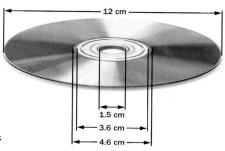

Area(outside circle) − Area(inside circle) $= \pi r_1{}^2 - \pi r_2{}^2 =$
$\pi(r_1{}^2 - r_2{}^2) = \pi(6^2 - 2.3^2) = 30.71\pi \approx 96.478$ cm²

Areas and Probability

Areas can be used to calculate probability. When a point is chosen at random out of a region, the probability that it will be in a selected smaller region is the ratio:

$$\text{probability} = \frac{\text{area of selected region}}{\text{area of entire region}}.$$

Example 2

The Rippit Goode tool company manufactures circular saw blades for a furniture manufacturing company. After several saw blades broke, the quality control department decided to sample the blades to determine the probability that the break would extend into particular regions. They divided a blade into 8 regions. What is the probability that a break will extend into region A?

Solution The probability is $\frac{\text{area of region } A}{\text{area of saw blade}}$.

Area of saw blade $= \pi(5 \text{ in.})^2 = 25\pi \text{ in}^2$

Area of smaller circle: area $(\odot(O, 2.5 \text{ in.})) = \pi(2.5 \text{ in.})^2 = 6.25\pi \text{ in}^2$

Area of region $A = \frac{25\pi - 6.25\pi}{4} \text{ in}^2 = \frac{18.75}{4}\pi \text{ in}^2$

Probability $= \frac{\text{area of region } A}{\text{area of saw blade}} = \frac{\frac{18.75}{4}\pi \text{ in}^2}{25\pi \text{ in}^2} = 0.1875$

If the place where the break occurs is completely random, then the probability that the blade will break in region A is 0.1875, or about 19%.

Areas of Sectors

Just as an arc length is a fraction of the circumference of the circle, the area of a sector is a fraction of the area of the circle.

Example 3

In $\odot O$ at the right, the radius is 22. Find the area of the shaded sector.

Solution $m\angle AOB = 24$, so the shaded sector is $\frac{24}{360}$ of $\odot O$.

Area of the shaded sector $= \frac{24}{360}$ (Area of $\odot O$)

$$= \frac{1}{15}\pi(22)^2$$

$$= \frac{484}{15}\pi$$

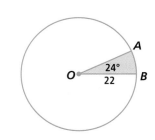

Questions

1. What is the area of a circle with a radius of 10 cm?

2. **Fill in the Blanks** A circle is the ___?___ of a sequence of regular polygons as the number of sides gets ___?___.

3. When a circle is split into congruent sectors and the sectors are arranged to form a figure that is like a parallelogram,
   a. what is the height of the parallelogram?
   b. what is the base of the parallelogram?
   c. what is the area of the parallelogram?

4. Suppose an FM radio station's transmitter allows the station to be heard clearly within 30 miles of the station. What is the listening area, to the nearest square mile?

5. If the diameter of the bull's-eye in the target at the right is 3 in. and the target has a diameter of 20 in., what is the probability that a dart that randomly hits the target will be a bull's-eye?

6. a. Find the area of a 150° sector in a circle with radius 12 cm.
   b. Find the perimeter of that sector.

7. $\odot B$ at the right contains point C. $\overline{BC}$ is a diameter of $\odot A$. What is the ratio of the area of circle A to the area of circle B?

8. Suppose a circle is drawn through all the vertices of a regular hexagon.
   a. If the radius of the circle is 100 mm, what is the area of the hexagon?
   b. What percent of the area of the circle is outside the hexagon?

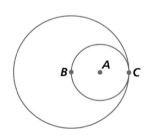

9. Suppose the owners of the radio station in Question 4 wanted to double the number of square miles that they could broadcast their signal. What distance would the new signal need to be broadcasted to accomplish this?

10. Jenny Phan, an engineer for the Cool Blades Fan
    Company is designing a new window fan. She wants to
    use a regular polygon with a side of 10.0 inches to house
    the motor.

    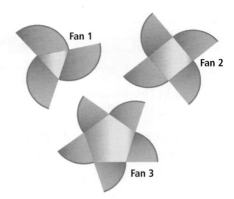
    Fan 1
    Fan 2
    Fan 3

    a. Find the area of a blade for each fan in terms of π.
       Which fan uses the most material to build only its
       fan blades?
    b. Find the total area of the polygonal face of the motor
       housing and the blades for each fan. (Use 6.9 inches
       for the apothem of the regular pentagon)
    c. Jenny plans to trim the blades of the fan as shown
       (brown lines on the picture) with a strip of metal edging.
       Find how much she will need to trim each fan. Which fan
       uses the most material?

**In 11 and 12, a dog is leashed to the corner of a rectangular
40'-by-25' building, as shown. The dog cannot enter the interior
of the building.**

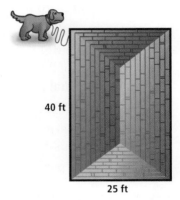
40 ft
25 ft

11. What is the area of the region in which the dog can roam if it is
    on a 20-foot leash?

12. What is the area of the region in which the dog can roam if it is
    on a 40-foot leash?

13. Anne Chovini is starting up a pizza restaurant. She charges
    $2.40 for a 45° sector of a pizza with radius 12 inches. If she
    charges the same price per square inch, how much should she
    charge for a 30° slice of a pizza with diameter 16 inches?

14. Which takes up a greater fraction of
    area: a square inscribed in a circle or
    a circle inscribed in a square?

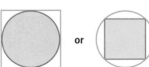
or

15. The shape below is created using
    two arcs with a radius of 5 cm inside of a square whose sides
    measure 5 cm. Find the area of the shaded region.

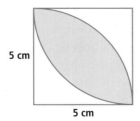
5 cm
5 cm

REVIEW

16. The London Eye is a large Ferris wheel in London. It is 135 meters in diameter, and takes half an hour to complete a full turn. If a person rides it for an hour, how far will the person travel? **(Lesson 8-8)**

17. Two legs of a right triangle have lengths x and $2x + 6$. The length of the hypotenuse is 39. Find the lengths of the legs of the triangle. **(Lesson 8-6)**

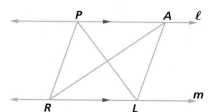

The London Eye was completed in 1999.

In **18** and **19**, use the figure at the right.

18. **Multiple Choice** Which of the following is true? **(Lesson 8-4)**

    A Area $(\triangle RLP) >$ Area $(\triangle RLA)$

    B Area $(\triangle RLP) =$ Area $(\triangle RLA)$

    C Area $(\triangle RLP) <$ Area $(\triangle RLA)$

19. **Multiple Choice** If q is the perimeter of $\triangle RLP$ and r is the perimeter of $\triangle RLA$, then **(Lesson 8-1)**

    A $q > r$.

    B $q = r$.

    C $q < r$.

    D the relationship between q and r cannot be determined.

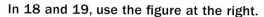

20. Region F has area 3, region G has area 7, and region H has area 10. Suppose that you know that region F is contained in region G and that regions G and H overlap. What can you say about the area of the union of F, G, and H? **(Lesson 8-2)**

21. In the regular hexagon at the right, the line cutting the hexagon forms a triangle and a heptagon. What other types of polygons can be created by cutting the hexagon in the figure with a line? **(Lesson 2-6)**

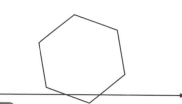

EXPLORATION

22 **a.** Recall that a *lattice point* is a point with integer coordinates. On a sheet of graph paper draw a circle whose center is a lattice point, and whose radius is the length of 1 square. Repeat this 9 times, increasing the radius by one square length each time, until you have 10 circles. For each circle, count the number of lattice points contained inside or on the circle. For each circle, find its area.

    b. How does the area of the circles compare to the number of lattice points contained in the circle? Why do you think this might be true?

Chapter 8 Projects

1 Quadrature of the Lune

A *lune* is a figure bounded by two arcs. In 460 BCE, the Greek geometer Hippocrates of Chios tried to "square a lune." That is, he tried to construct a square equal in area to the lune. He was able to prove that the area of the lune bounded by $\overset{\frown}{PQR}$ and $\overset{\frown}{PR}$ has the same area as the right triangle *OPR*.

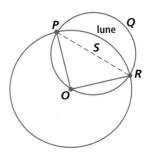

a. Construct a lune with circle *O* and with m∠*POR* = 90. The center of semicircle $\overset{\frown}{PQR}$ is the midpoint *S* of $\overline{PR}$.

b. Find the area of the triangle and the lune if the radius of circle *O* is 1 unit.

c. Generalize your findings for a circle of any radius.

d. Is the theorem of Hippocrates true if m∠*POR* ≠ 90? Why or why not?

e. Find a proof of Hippocrates' theorem. Write out the proof in a clear way, and explain it to your classmates.

2 Broadcast Areas

Radio stations have maps of the areas in which their signals are received. Contact four radio stations with various tower locations, broadcast frequencies, and transmission powers to find out their signal areas. Use the formulas and techniques of this chapter to calculate their broadcast areas. Do research to discover what affects the size and shape of these areas. Which factors are permanent? Which factors change as conditions change? Which other factors affect the size of a region in which a station can be received?

3 U-Turns and Area

Suppose you hold a piece of chalk against a flat surface, and wish to move the chalk along the surface so that in the end it is facing a direction opposite of what it was originally facing. When you turn the chalk, it leaves its marks on the surface. What is the smallest possible area for this region? This problem is called the *Kakeya problem,* named for Soichi Kakeya, the Japanese mathematician who proposed it in 1917. It was the key to the solution of a very famous problem in advanced mathematics in the 20th century.

a. Using chalk and a blackboard, try to make the area as small as you can get it (Hint: Notice that the name of this project involves U-turns.) If you find a way of doing this with a particularly small area, trace it.

b. Look up the Kakeya problem, and find out what the smallest possible area for turning the chalk is.

c. The solution of the problem involves a construction that uses triangles, transformations, and area. Look up this solution, and write a description of the basic construction of these triangles.

4 Hopscotch

Hopscotch is a game with many names and variations. Hopscotch began in ancient Britain as a training exercise for Roman foot soldiers (dressed in full armor) to improve their footwork, similar to training drills for football players. Some of these training fields were over 100 feet long. Roman children drew their own smaller courts in imitation of the soldiers, added a scoring system and "Hopscotch" spread throughout Europe. The game is called "Rayuela" in Argentina, "Marelles" in France, "Tempelhuepfen" in Germany, "Ekaria Dukaria" in India, "Hinkelbaan" in the Netherlands, and "Pico" in Vietnam.

Each player has a marker, usually a common stone. The first player tosses his marker into the first square. The marker must land completely within the designated square without touching a line or bouncing out.

a. Draw the diagrams of "hopscotch" courts from three different countries and label each with dimensions, the name of the game and the country of origin. (If measurements are not given, the courts are often created such that the parallel lines are equally spaced.)

b. Find the areas of each of the different regions on the court.

c. In each region on the court, write the probability for a stone that is randomly thrown to land in that spot, assuming that it lands on the court.

5 Famous Proofs of the Pythagorean Theorem

In Lesson 8-6 you saw a proof of the Pythagorean Theorem. Of all of the theorems in mathematics, the Pythagorean Theorem likely has the most proofs.

a. At least how many different proofs of the Pythagorean Theorem are known?

b. Find a list of proofs of the theorem, and the drawings that come with the proofs. Copy a large collection of these drawings to make an artistic display.

c. Find three proofs other than the ones you saw in the lesson, and present them.

6 Maps Distort Area

Look at a map of the world. Find out the area of Greenland and the area of the United States. Do these areas seem to agree with what you see on the map? Find out what a Mercator projection is, how maps of Earth are made to fit on flat pieces of paper, and how this distorts area.

Due to Greenland's proximity to the Geomagnetic North Pole, the Aurora Borealis is visible from much of the country from late August through late April.

Chapter 8 Summary and Vocabulary

- This chapter is devoted to deriving and applying formulas for area, perimeter, and circumference. **Perimeter** and **circumference** measure the boundary of a figure. An equilateral n-gon with sides of length s has perimeter ns. A circle with diameter d has circumference πd. The length of an arc of a circle is a fraction of that circumference.

- In contrast to perimeter, **area** measures the region enclosed by a figure. This region can be estimated by using congruent squares. With finer and finer grids, even the areas of irregular shapes can be estimated.

- A rectangle with dimensions h and b has area hb. Splitting it with a diagonal, two congruent right triangles are formed. Each has area $\frac{1}{2}hb$. By splitting any triangle into two right triangles, its area can be shown to be $\frac{1}{2}hb$. Using this, the area of any regular polygon with perimeter p and apothem a can be shown to be $\frac{1}{2}ap$. Putting two triangles together, the area of any trapezoid is $\frac{1}{2}hb(b_1 + b_2)$. A special case of a trapezoid is a parallelogram, whose area is hb.

- This chapter contains some of the most important formulas in geometry. Areas of right triangles and squares help to develop the **Pythagorean Theorem**: In a right triangle with legs a and b and hypotenuse c, $c^2 = a^2 + b^2$. The area of a parallelogram can be used to derive the formula $A = \pi r^2$ for the area of a circle.

- The Pythagorean Theorem leads to results in special right triangles. In an isosceles right triangle (also known as a **45-45-90 triangle**) with leg of length x, the hypotenuse has length $x\sqrt{2}$. In a **30-60-90 triangle** with length of the shorter leg x, the length of the longer leg is $x\sqrt{3}$ and the length of the hypotenuse is $2x$.

Vocabulary

8-1
perimeter

8-2
area
square units

8-4
height of a triangle
altitude of a triangle

8-5
altitude of a trapezoid
height of a trapezoid

8-6
Pythagorean triple

8-7
45-45-90 triangle
30-60-90 triangle
apothem

8-8
circumference
π (pi)

8-9
sector of a circle

Postulates, Theorems, and Properties

Area Postulate (p. 458)
  Uniqueness Property
  Congruence Property
  Additive Property
  Rectangle Formula
Triangle Area Formula (p. 469)
Trapezoid Area Formula (p. 475)
Parallelogram Area Formula (p. 476)

Pythagorean Theorem (p. 481)
Pythagorean Converse Theorem
  (p. 483)
Isosceles Right Triangle Theorem
  (p. 488)
30-60-90 Triangle Theorem (p. 489)
Circle Circumference Formula (p. 494)
Circle Area Formula (p. 501)

Chapter 8 Self-Test

Take this test as you would take a test in class. You will need a calculator. Then use the **Selected Answers** section in the back of the book to check your work.

1. All angles in the diagram are right angles. $PS = 12$, $ST = 20$, $PR = 8$, and $WT = 8$.

   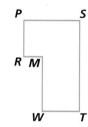

   a. Find the perimeter of *PSTWMR*.

   b. Find the area of *PSTWMR*.

2. Assume you know formulas for the areas of a rectangle and right triangle. Use one or both formulas to explain why the area of $\triangle MNP$ below is $\frac{1}{2}bh$.

   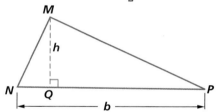

3. a. Find the perimeter of the parallelogram below.

   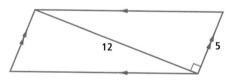

   b. Find the area of this parallelogram.

4. A square has area 64. Find the length of a diagonal of the square.

5. An isosceles trapezoid with perimeter 48 has bases 20 and 8. Find the area of the trapezoid.

6. Each stair goes up 8" and back 12". Find the area of the 12-gon that is on the side of the stairs. Show your work.

   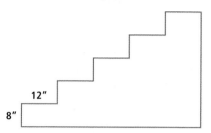

7. On a regulation NCAA basketball court there is a region consisting of a semicircle and a rectangle. Find the area of the region.

8. Two runners start in the same spot and run in opposite directions for 60 yards. They both turn left and run 80 yards. How far apart are they?

9. In the map and grid shown below, each small square of the grid has side 9.5 miles. Use the method of Lesson 8-3 to estimate the area of the island of Hawaii.

   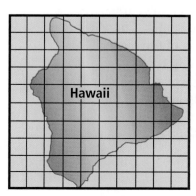

10. A regular hexagon has perimeter 72 millimeters. Find the area of the hexagon.

11. A lawn sprinkler sprays water a distance of 20 ft and moves through an angle of 80°. What is the perimeter of the sector of lawn that is sprayed?

12. Suppose a *squircle* is the boundary of a region formed by drawing two semicircles whose diameters are the opposite sides of a square, as shown below with square $ABCD$. Find a formula for the area of a squircle if $AB = x$.

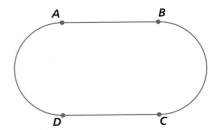

13. In the figure below, m$\angle N = 60$, m$\angle M = 30$, $\overline{PS} \parallel \overline{NM}$, $SM = 12$, and $PS = 8\sqrt{3}$. Find the area of $PSMN$.

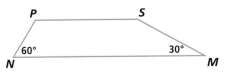

14. On a coordinate plane, $A = (-4, 2)$, $B = (0, 8)$, $C = (6, 8)$, and $D = (5, 2)$. Determine the area of $ABCD$.

15. The two circles pictured have the same center. The radius of the larger circle is 10; the radius of the smaller circle is 8. Find the area of the region between the two circles.

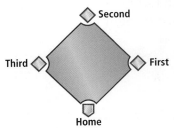

16. At a popular pizza parlor in 2006, a 9″ diameter pizza cost $12.30. What was the cost of one square inch of pizza?

17. Janelle places a 2-inch wide frame around a 4-by-6 inch wide rectangular photograph. What is the outside perimeter of the frame?

18. A standard baseball diamond is a square with the bases at vertices of the square. The distance between the bases in a Major League diamond is 90 feet. In a Little League diamond it is 60 feet. A player standing on first base throws the ball to third base. How much farther is the throw in the Major Leagues than in Little League?

19. Suppose $\triangle XYZ$ has sides of length x, y, and z. What kind of angle is $\angle Z$ if

a. $x^2 + y^2 = z^2$?   b. $x^2 + y^2 < z^2$?

c. $x^2 + y^2 > z^2$?

20. Fenway Park in Boston looks roughly like the diagram below from an overhead view. How far is it from Home (H) to point A, at the far end of left field? The units are feet. Assume the left field wall is perpendicular to the left field foul line.

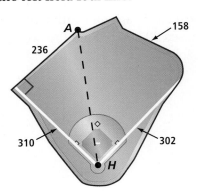

Chapter 8 Chapter Review

SKILLS
PROPERTIES
USES
REPRESENTATIONS

SKILLS Procedures used to get answers

OBJECTIVE A Calculate the perimeters of parallelograms, kites, and equilateral polygons from appropriate lengths, and vice versa. (Lesson 8-1)

In 1–4, give the perimeter of the figure.

1. a regular 17-gon with side length 4.5 cm

2. a parallelogram with two adjacent sides with lengths 7.25 cm and 4.75 cm

3. a square whose area is 144 square inches

4. a kite in which two noncongruent sides have lengths x and y

5. A rectangle has perimeter 37 feet. One of the sides of the rectangle is 3 feet long. What are the lengths of the other sides?

6. The perimeter of a regular nonagon is 2 cm. What is the length of a side of the nonagon?

OBJECTIVE B Describe or apply a method for determining the area of an irregularly shaped region. (Lesson 8-3)

7. The large square in the grid below has sides of length 8 cm. Use the grid to estimate the area of the given polygonal region.

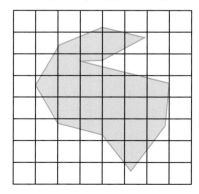

In 8 and 9, use the given figure to estimate the area of the island.

8.

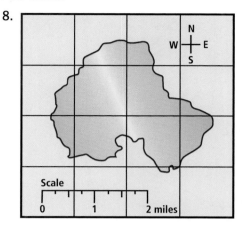

9.

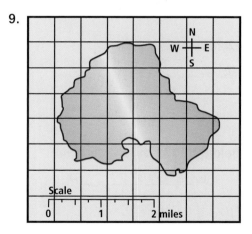

10. Explain how you could estimate the area of the region below.

OBJECTIVE C Calculate areas of squares, rectangles, parallelograms, trapezoids, triangles, and regular polygons from relevant lengths. (Lessons 8-2, 8-4, 8-5, 8-7)

In 11–14, calculate the area of the figure.

11. rhombus *PHIL* with $PI = 2y$ and $HL = 13x$

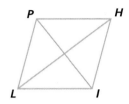

12. $\triangle JKL$

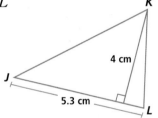

13. trapezoid *NICE*

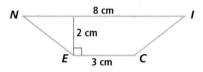

14. parallelogram *OMWH*

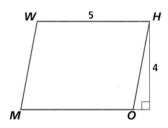

15. The area of a trapezoid is 85 cm². The height of the trapezoid is 5 cm, and one of its bases has length 15 cm. What is the length of the other base?

16. Give dimensions for one rectangle with perimeter 40 feet, and with area less than 100 square feet.

OBJECTIVE D Calculate lengths and measures of arcs, the circumference, and the area of a circle from the measures of relevant lengths and angles, and vice versa. (Lessons 8-8, 8-9)

17. A circle has circumference 7 cm. What is its radius?

18. Calculate the exact area of a circle with diameter 3 cm.

19. Calculate the length of an arc with measure 30° in a circle with radius 9 in. to the nearest hundredth.

20. In $\odot T$ below, $OT = 5$ cm and $m\angle OTP = 40$. Find the area of the shaded region.

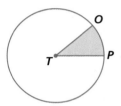

21. *A* and *B* are points on $\odot O$ with radius 4 cm. If the area of the sector determined by *A* and *B* is 4π cm², find $m\angle AOB$.

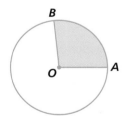

OBJECTIVE E Apply the Pythagorean Theorem to calculate lengths of segments and areas in right triangles and other figures. (Lesson 8-6)

22. The length of the hypotenuse in a right triangle is 1.7 m. The length of one of the legs is 1.5 m. Find the area of the triangle.

23. The area of a rectangle is 420 square feet. One of the sides of the rectangle has length 12 feet. Find the length of a diagonal of the rectangle.

In 24 and 25 find the length of the missing side.

24.

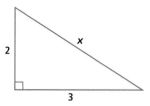

25.

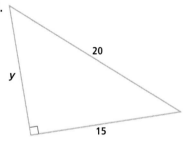

26. Figure *RNDF* below is a parallelogram. Use the given lengths to find the value of *x*.

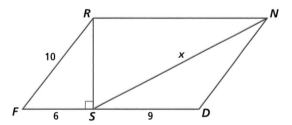

OBJECTIVE F Apply the Pythagorean Converse Theorem. (Lesson 8-6)

In 27–30, determine whether a triangle whose sides have the given lengths is a right triangle.

27. 39, 80, 89

28. $\sqrt{2}, \sqrt{3}, \sqrt{5}$

29. 5, 8, 9

30. $\sqrt{11}, 4, \sqrt{5}$

31. Use the given lengths to determine if △*ABC* is a right triangle. Explain your answer.

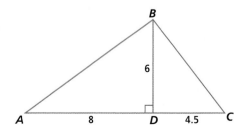

OBJECTIVE G Apply the special right triangle relationships to find lengths and areas. (Lesson 8-7)

32. One angle in a right triangle has measure 30°. The length of the hypotenuse is 30. Find the area of the triangle.

33. The figure at the right shows a regular hexagon in which the apothem is 2.6 cm and *HI* = 3. Find the area of the hexagon.

34. The perimeter of an isosceles right triangle is $10 + 5\sqrt{2}$. Find the lengths of the sides of the triangle.

35. In the figure at the right, *OAPS* and *ILDM* are squares, and the points *I, L, D,* and *M* are the midpoints of the segments of *OAPS*. If the area of *ILDM* is 25 cm², find the area of *OAPS*.

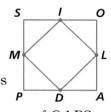

PROPERTIES Principles behind the mathematics

OBJECTIVE H Tell how to derive formulas for area. (Lessons 8-2, 8-3, 8-4, 8-9)

36. Several rectangles are cut out of a square with sides of length *s,* as shown in the figure below. Find the area of the shaded region in terms of *s* and *x*.

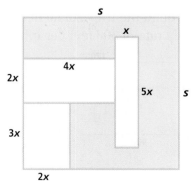

37. Explain how the formula for the area of a parallelogram is derived from the formula for the area of a trapezoid.

38. Explain how the formula for the area of a general triangle is derived from the formula for the area of a right triangle.

39. Suppose n copies of figure F completely tile a square with sides of length T. What is the area of figure F? Explain your answer.

40. A circle of radius 2 cm is cut out from a circular disc of radius 5 cm. What is the area of the resulting figure?

USES Applications of mathematics in real-world situations.

OBJECTIVE I Apply perimeter formulas for parallelograms, kites, and regular polygons to real situations. (Lesson 8-1)

41. The four bases of a baseball field form a square in which adjacent vertices are 90 ft apart. If Jerome can run 16 miles per hour, how long will it take him to run once around the bases?

42. An architect is commissioned to design a building whose roof is in the shape of a regular heptagon, with sides of length 40 meters. If she wishes to construct a rail along the outer edge of the roof, how long will the rail be?

43. How long would it take for a person to walk around the trapezoidal field pictured here at a rate of 300 feet per minute?

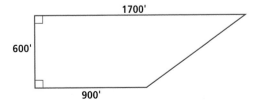

44. Daniella wishes to build a kite with a frame made of thin wood rods. She decides that two adjacent sides should be 1 foot and 2.5 feet long. What is the total length of wood rods she will need to purchase?

OBJECTIVE J Apply the Pythagorean Theorem to real situations. (Lesson 8-6)

45. If a 24-foot ladder reaches 22 feet high on a wall, how far away from the wall is the bottom of the ladder? Round your answer to the nearest tenth of a foot.

46. In order to get to her friend's house, Simone has to walk half a mile east and 400 feet south. How far would the trip have been if she could walk to her friend's house in a straight line?

47. A reporter in a helicopter is hovering above a building at a height of 200 meters. Two hundred meters away from the foot of the building is a fire that the reporter is photographing. How far away is the reporter from the fire?

48. Greg bought a new pole lamp, and is worried it might not form a 90° angle with the floor. Describe how he could check if this is indeed the case, using only a tape measure.

OBJECTIVE K Apply formulas for areas of squares, rectangles, parallelograms, trapezoids, and triangles to real situations. (Lessons 8-2, 8-4, 8-5)

49. A triangular piece of fabric is needed for a sail. If the sail is to be 16' high and 18' long at the base, about how much fabric will be used?

50. A company is hired to create a square park in a city. They produce a model of the park, whose sides have length 1.5 meters. If the scale of this model is 1:300, what will the area of the park be?

51. A child painted the figure below in perspective.

 a. If the road in the picture is supposed to be a rectangle that is 5' wide and 400' long, what is its area?

 b. When drawn in perspective, the road looks like a trapezoid. If this trapezoid is 1" wide at the top, 8.5" wide at the bottom, and 8.4" high, what is its area?

52. Carina wishes to tile her bathroom floor using the parallelogram tile below. If the area of the floor is 9 m², at least how many of the tiles will be necessary?

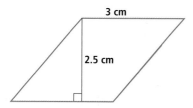

OBJECTIVE L Apply formulas for the area and circumference of a circle to real situations. (Lessons 8-8, 8-9)

53. A meteorologist's weather balloon malfunctions and falls to the ground. The meteorologist calculates that it would have to hit the ground within 3 miles of the point at which it malfunctioned.

 a. What is the area of the region that the meteorologist needs to search?

 b. If the balloon landed at random, what is the probability that it landed within one mile of the point in which it malfunctioned?

54. Earth is about 93 million miles from the Sun. Assume that Earth's orbit is a circle, and that it completes one revolution every 365 days. How fast does Earth travel, in miles per hour?

55. Sky is doing a project on carrots. He slices one, and draws the picture below. What is the area of the slice that is not included in the inner circle?

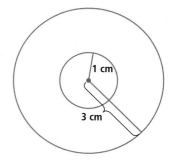

56. Alexus bought 200 meters of fencing, to build a play area for her dog. If she uses all of the fencing, which play area will have the larger area—a square or a circle?

REPRESENTATIONS Pictures, graphs, or objects that illustrate concepts

OBJECTIVE M Determine the area of a polygon in the coordinate plane. (Lessons 8-2, 8-4, 8-5)

57. A trapezoid has vertices $(7, 3)$, $(7, -10)$, $(-3, 2)$, and $(-3, -5)$. Find the area of the trapezoid.

58. A triangle has vertices $(-0.2, 1.3)$, $(1.87, 1.3)$, and $(2.4, 1.9)$. Find the area of the triangle.

59. A quadrilateral has vertices $(0, 0)$, (x, y), $(3, 0)$, $(x + 3, y)$. Find its area.

60. A rectangle has vertices $(1, 1)$, $(1, x)$, $(x, 1)$, and (x, x). Find all possible values of x, if you know that the area of the rectangle is 8.

Three-Dimensional Figures

Contents

9-1   Points, Lines, and Planes in Space

9-2   Prisms and Cylinders

9-3   Pyramids and Cones

9-4   Drawing in Perspective

9-5   Views of Solids and Surfaces

9-6   Spheres and Sections

9-7   Reflections in Space

9-8   Making Polyhedra and
        Other Surfaces

9-9   Surface Areas of Prisms
        and Cylinders

9-10 Surface Areas of Pyramids
        and Cones

cylinder

cone

pyramid

prism

Virtually everything that you see or touch is the surface of a 3-dimensional figure. These figures come in all sorts of shapes and sizes, and their surfaces may be smooth, rough, or pointed. Because 3-dimensional figures may have great complexity, we study them by breaking them down into simpler figures. For example, the sandcastle at the left is made up of a number of different types of 3-dimensional figures.

A key to understanding these figures is having the vocabulary to talk about them, their parts, and their properties. By the end of this chapter, you should be able to identify, picture, and describe, by hand or with a computer, the properties of prisms, cylinders, pyramids, cones, and spheres.

These geometric solids are everywhere you look. They can be found in and on the buildings that you walk into and pass everyday. A new pencil is likely a prism. A grapefruit is a sphere.

Blocks you played with in your younger years are boxes. And sandcastles can be made by clustering 3-dimensional solids in interesting groups.

The activities in this chapter will give you tools to deduce formulas for calculating surface area and prepare you to find the volume of these and other 3-dimensional solids.

Two cubes, two spheres and three cylinders...

Lesson 9-1
Points, Lines, and Planes in Space

Vocabulary

angle formed by a line and a plane

line perpendicular to a plane

foot of a segment

parallel planes

distance between parallel planes

distance to a plane from a point

skew lines

dihedral angle

edge of a dihedral angle

perpendicular planes

▶ **BIG IDEA** Properties of planes are deduced from the properties of points, lines, and planes in the Point-Line-Plane Postulate.

Most of the figures you have seen in the previous chapters lie in a single plane. Their points are *coplanar.* You can think of a plane as being flat and having no thickness, like a tabletop that goes on forever in all directions. In fact, you can draw a plane on a page in the same way as you would draw a tabletop.

The figure below at the left shows plane figures on a page as seen from the top. The figure at the right is drawn to represent a 3-dimensional view of a plane in space, and looks like a tabletop.

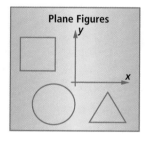

Mental Math

a. One plane separates space into how many regions?

b. Two intersecting planes separate space into how many regions?

b. Three intersecting planes separate space into how many regions?

Assumed Properties of Points, Lines, and Planes

Recall from Chapter 1 that the three terms *point, line,* and *plane* are undefined because, in different geometries, they have different meanings. In Lesson 1-5, you saw three parts of the Point-Line-Plane Postulate in Euclidean geometry. Now we introduce two more parts of that postulate. Each of these corresponds to a property of points and lines in two dimensions that you have seen before.

Point-Line-Plane Postulate (Expanded)

a. *Unique Line Assumption*
Through any two points there is exactly one line. If the two points are in a plane, the line containing them is in the plane.

b. *Number Line Assumption*
Every line is a set of points that can be put into a one-to-one correspondence with the real numbers, with any point on it corresponding to 0 and any other point corresponding to 1.

c. *Dimension Assumption*
(1) There are at least two points in space.
(2) Given a line in a plane, there is at least one point in the plane that is not on the line.
(3) Given a plane in space, there is at least one point in space that is not in the plane.

d. *Unique Plane Assumption*
Through three noncollinear points, there is exactly one plane.

e. *Intersection Assumption*
Two different planes either do not intersect or intersect in exactly one line.

STOP **QY1**

Let the tips of your fingers represent three points. If you hold them so they are not collinear, exactly one plane contains all three. If your fingertips are collinear, many planes contain them. All those planes intersect in the line of your fingertips. The picture below illustrates this.

> **QY1**
>
> Which other part of the Point-Line-Plane Postulate earlier in this book is the 2-dimensional counterpart to Part d of the Expanded Point-Line-Plane Postulate?

Fingertips noncollinear
Exactly one plane

Fingertips collinear
Many planes

This is why we say that three noncollinear points *determine* a plane.

Because two points determine a unique line, lines are named by two of the points on them, $\overleftrightarrow{PQ}$ for example. Similarly, because three noncollinear points determine a unique plane, we name planes by three such points, plane *ABC,* for example. When appropriate, lines are named by a single letter. Likewise, planes are sometimes named by single capital letters, for example plane *X* or plane *Y.*

STOP **QY2**

> **QY2**
>
> Name the plane pictured below in two different ways.
>
>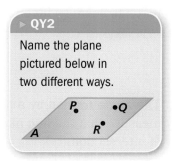

It is also the case that there is exactly one plane containing a line and a point not on the line. Here is why: Let the line be ℓ and the point be P. Because of the Number Line Assumption of the Point-Line-Plane Postulate, ℓ contains two points A and B. Now A, B, and P cannot be collinear, because then P would be on ℓ. So A, B, and P are noncollinear points and there is exactly one plane through them. All the points of ℓ are in this plane because of the Unique Line Assumption, that if two points are in a plane, the line containing them is in the plane. This argument proves the following theorem.

Unique Plane Theorem

There is exactly one plane through a line and a point not on that line.

Think of two parts of planes: a door and the wall it is in. As you open the door, the intersection of the door and the wall is the line containing the hinges of the door. Now fix a finger in the way of the door. There is exactly one position of the door that touches your finger. The door represents the unique plane that contains the hinges and your finger.

You can see many other examples of intersecting planes around you: two adjacent walls; a wall and the ceiling; a wall and the floor. These intersecting planes are usually *perpendicular*. To understand when planes are perpendicular, we need to examine angles formed by a line and a plane.

The Angle Formed by a Line and a Plane

The photograph at the right shows the Iwo Jima Memorial statue at Arlington Cemetery in Washington, D.C. Below is a model of the flagpole and the ground. You can see that the flagpole forms angles with any line in the plane of the ground that contains A, the point of intersection. Here it looks as if $m\angle BAC \approx 59$, $m\angle BAD \approx 90$, and $m\angle BAE \approx 121$. We need to decide which angle should be viewed as *the* angle the flagpole makes with the ground. Fortunately, that has been decided for us.

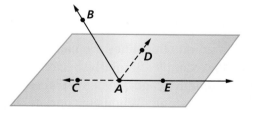

The **angle formed by a line and a plane** is the angle of smallest measure that the line makes with rays in the plane from the point of intersection.

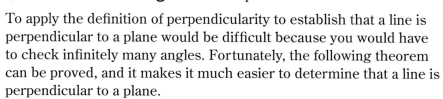

For the Iwo Jima statue, the angle is in the position of ∠*BAC*, and so the measure of the angle between the flagpole and the plane is 59°.

If all the angles a flagpole makes with lines on the ground are right angles, then the flagpole is perpendicular to the plane. A line (or segment or ray) is **perpendicular to a plane** if and only if the line and plane intersect and the line is perpendicular to every line in the plane passing through the point of intersection. If one endpoint of the segment is in the plane, then that point of intersection is often called the **foot of the segment** in that plane.

To apply the definition of perpendicularity to establish that a line is perpendicular to a plane would be difficult because you would have to check infinitely many angles. Fortunately, the following theorem can be proved, and it makes it much easier to determine that a line is perpendicular to a plane.

Line-Plane Perpendicular Theorem

If a line is perpendicular to two different lines at their point of intersection, then it is perpendicular to the plane that contains those lines.

The proof of this theorem is quite complicated, and we omit it here.

Parallel Planes

The ideas of parallel and perpendicular lines in two dimensions have counterparts in three dimensions. A plane can be parallel to a plane. Two different planes are **parallel planes** if and only if they are identical or have no point in common. The **distance between parallel planes** is the length of a segment perpendicular to the planes with an endpoint in each plane. The **distance to a plane from a point** not on it is measured along the perpendicular segment to the plane from the point. A line is parallel to a plane if they do not intersect in a single point.

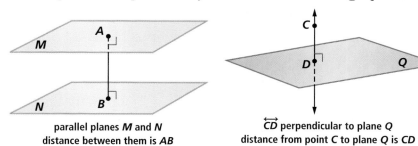

parallel planes *M* and *N*
distance between them is *AB*

$\overleftrightarrow{CD}$ perpendicular to plane *Q*
distance from point *C* to plane *Q* is *CD*

Two lines in space can intersect in exactly one point, be parallel, or *skew*. **Skew lines** are lines that do not intersect and are not in the same plane.

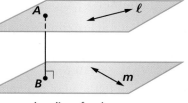

skew lines ℓ and *m*

Activity

Use the cube at the right to match the item in the first column to its description in the second column.

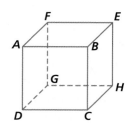

a. plane *AFB* and plane *CHG*

b. $\overleftrightarrow{GH}$ and plane *ADF*

c. $\overleftrightarrow{EH}$ and plane *FBC*

d. $\overleftrightarrow{FE}$ and $\overleftrightarrow{BC}$

e. plane *EBC* and plane *DGH*

f. $\overleftrightarrow{FE}$ and $\overleftrightarrow{AB}$

i. skew lines

ii. parallel lines

iii. a line perpendicular to a plane

iv. a line parallel to a plane

v. two intersecting planes

vi. two parallel planes

Dihedral Angles and Perpendicular Planes

Recall from previous lessons that when two lines intersect at a point, four angles are formed. The point of intersection is the common vertex for all the angles. In much the same way, when two planes intersect in a line, four **dihedral angles** are formed by the adjacent half planes and the line of intersection. The line of intersection of these planes is the **edge of the dihedral angle.**

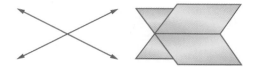

Dihedral angles are measured in the following way. Suppose the planes *X* and *Y* intersect in line ℓ. Pick a point *P* on ℓ. Draw $\overrightarrow{PD} \perp \ell$ in *X* and $\overrightarrow{PC} \perp \ell$ in *Y*. The measures of the dihedral angles are the measures of the four angles formed by $\overleftrightarrow{PC}$ and $\overleftrightarrow{PD}$. If $\overleftrightarrow{PC} \perp \overleftrightarrow{PD}$, then the planes are *perpendicular*. Two planes are **perpendicular planes** if and only if they form a 90° dihedral angle.

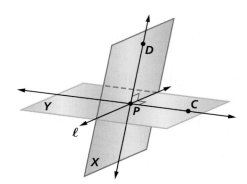

Questions

COVERING THE IDEAS

In 1 and 2, a property of a line is given. What is the corresponding property of a plane?

1. Two points determine a line.

2. Two lines are either identical, intersect in exactly one point, are non-intersecting and parallel, or are skew.

3. What part of the Expanded Point-Line-Plane Postulate guarantees that a three-legged footstool does not wobble?

4. What part of the Expanded Point-Line-Plane Postulate is related to the flatness of a plane?

5. **True or False** Suppose line m intersects plane X. If m is perpendicular to any two lines n and p in X, then m is perpendicular to X.

In 6–9, draw a picture of the situation.

6. two intersecting planes that are not perpendicular

7. two perpendicular planes

8. two skew lines

9. two parallel planes

10. In this picture of a needle intersecting a plane, which angle seems to be the angle between the needle and the plane?

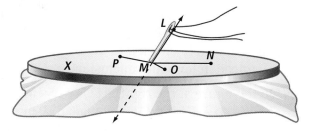

APPLYING THE MATHEMATICS

In 11–14, draw a picture.

11. two planes perpendicular to a third plane

12. two intersecting lines that are both parallel to a plane

13. two parallel lines that are both parallel to the same plane

14. two skew lines that are both parallel to the same plane

15. Points A, B, and C are vertices of a triangle.
    a. How do you know there is a unique plane that contains A, B, and C?
    b. Let X be the name of the unique plane from Part a. How you know that $\overleftrightarrow{AB}$, $\overleftrightarrow{BC}$, and $\overleftrightarrow{AC}$ must entirely lie in plane X? (This proves that a triangle determines a plane.)

16. Explain why there is exactly one plane through two intersecting lines.

17. Consider the photograph at the right of a sculpture by Dustin Shuler in Berwyn, Illinois, called "The Spindle." Explain how it illustrates that the following statement is false: If two lines in space are each perpendicular to the same line, then the two lines are parallel.

18. Is it possible for four lines to all be skew to each other? If so, draw such a situation. If not, why not?

19. In the figure at the right, C, D, A, and P are coplanar, as are E, B, A, and P. Angles have measures as follows:

    m$\angle DAP = 120$ m$\angle EAC = 110$ m$\angle CAP = 90$
    m$\angle BAC = 80$ m$\angle BAP = 90$ m$\angle BAD = 100$
    m$\angle EAP = 140$ m$\angle EAD = 175$

    Find the measure of the acute dihedral angle formed by plane AEB and plane ADC. The intersection of the two planes is $\overleftrightarrow{AP}$.

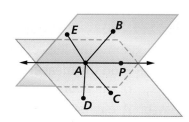

20. Refer to the drawing at the right. Some chairs wobble.

   a. What would have to be true for the chair not to wobble?

   b. What property explains why the chair wobbles?

   c. Why is it better to mount a camera on a tripod rather than the chair?

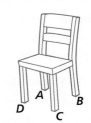

21. You are setting a fence post in the ground and you want it to be perpendicular to the ground. From at least how many locations do you need to check it to make sure it is perpendicular? Give a justification for your conclusion.

22. In the figure at the right, $\overleftrightarrow{AD}$ is perpendicular to plane E. Points A, P, B and C are coplanar. A and B do not lie in plane E, while P and C do. Determine the measure of the angle $\overrightarrow{PB}$ makes with plane E if m$\angle BPA = 63$.

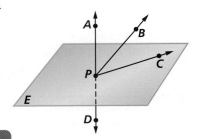

REVIEW

23. The dartboard at the right was on Donte's wall. One day, he took the dartboard down, and counted the holes in it. There were a total of 350 holes that seemed evenly spaced. Of these, five were within circle 9. The radius of the board is 8 inches. Find a good estimate for the radius of circle 9. **(Lesson 8-9)**

24. A circle, a square, and an equilateral triangle all have perimeter 1 cm. Which has the largest area? **(Lessons 8-9, 8-8, 8-7, 8-1)**

25. Alejandra has 12 matches of the same length, and wishes to arrange them in the shape of a right triangle. Find all of the possible lengths for the sides of the triangle. **(Lesson 8-6)**

26. In the figures at the right, a square was cut out of the rectangular region and placed along its top side.

   a. Does one shape have a greater perimeter than the other? If so, which one? **(Lesson 8-1)**

   b. Does one shape have a greater area than the other? If so, which one? **(Lesson 8-2)**

27. The figure at the right is a regular pentagon. This figure can not tessellate the plane. Trace the figure, and show how you could cut it up into two shapes, each of which can tessellate the plane. **(Lesson 7-6)**

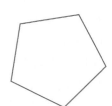

EXPLORATION

28. Examine the legs of a chair at home or in school.

   a. Find the measure of the angle each leg makes with the plane of the floor.

   b. Are most of the chair legs you see perpendicular to the floor?

QY ANSWERS

1. the first part of the Unique Line Assumption

2. A, PQR

Lesson 9-2

Prisms and Cylinders

Lesson
9-2

▶ **BIG IDEA**   Prisms and cylinders are two special types of cylindrical surfaces.

Recall that polygons and polygonal regions are different. A polygon is the boundary of a polygonal region. The region is the union of the boundary and its interior.

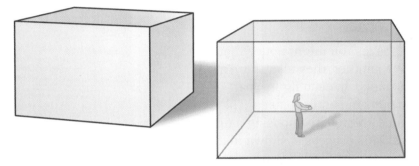

A similar distinction is made in three dimensions, as shown by the above drawings. The person in the space at the right is in the interior of a *surface* consisting of six rectangular regions in six different planes. Roughly, a **surface** is a 3-dimensional figure that separates space into two regions. If one of the regions is *bounded,* all distances between points in the region are less than some fixed number, then that region is the **interior of the surface** and the other region is the **exterior of the surface.** A **solid** is the union of a surface and its interior. For example, a balloon is a surface, while a bowling ball is a solid. The figure at the left above pictures a solid. When drawing, you can distinguish a solid from a surface by shading and showing no hidden lines.

Not all surfaces have interiors and exteriors. Some surfaces (like planes) separate space into two unbounded regions. But the surfaces you will study in this chapter are bounded.

Vocabulary

surface
interior of a surface
exterior of a surface
solid
face of a surface
edge of a surface
vertices of a surface
cylindrical solid
cylindrical surface
bases of a cylindrical solid
lateral surface
lateral face of a prism
height, altitude of a solid
right solid
oblique solid
cylinder
prism
lateral edge of a prism
regular prism
cube

Mental Math

a. *F* and *G* are two non-overlapping regions, each with an area of 19. What is the area of their union?

b. *F* and *G* are two regions, each with an area of 19. The area of their union is 30. Do they overlap?

c. In Part b, what is the area of the overlap between the two regions?

Cylindrical Solids and Surfaces

The words *cube, cylinder,* and *prism* refer to types of 3-dimensional figures called *cylindrical solids.*

When a plane intersects a surface in a polygonal region, that region is called a **face of the surface** or face of the solid the surface determines. The sides of the polygonal region are called **edges of the surface,** and the vertices of the polygonal region are called **vertices of the surface.** For example, in the diagram at the right, M is a vertex, $\overline{MN}$ is an edge, and rectangle $MNPQ$ is a face.

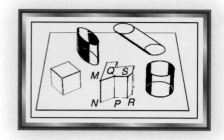

 QY1

▶ QY1

Name another vertex, edge, and face that are shown in the diagram above.

Cylindrical Solids and Surfaces

Examine the figures in the diagram above. You might notice that, while they are very different from one another, they also share a few important characteristics. For instance, in each figure, part of the surface consists of congruent regions in parallel planes. This suggests that all of these shapes can be created by choosing a figure and then translating it through space. This process is shown in the diagram below, where the regular pentagon at the left is translated by $\overrightarrow{PQ}$ to create the 3-dimensional figure at the right.

When a solid 3-dimensional shape is created using the method described above, the shape is called a *cylindrical solid.*

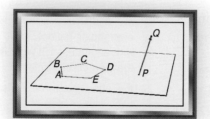 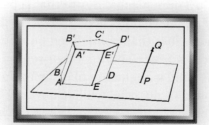

> ### Definitions of Cylindrical Solid, Cylindrical Surface
>
> A **cylindrical solid** is the set of points between a region and its translation image in space, including the region and its image. The boundary of a cylindrical solid is called a **cylindrical surface.**

In a cylindrical solid, the region that was translated and its image are its **bases.** In the figure above at the right, the bases are pentagons $ABCDE$ and $A'B'C'D'E'$. Bases are always congruent and are always in parallel planes.

The rest of the surface of the solid is called the **lateral surface.** When the bases are polygons, the lateral surface is made up of **lateral faces,** such as parallelogram $A'E'EA$ in the figure on the previous page. The intersection of two lateral faces is called a **lateral edge.** The **height** or **altitude of the solid** is the distance between the planes of its bases. In the figure, the height is $D'F$. When the translation vector is perpendicular to the planes containing the bases, we say that the solid is a **right solid;** otherwise it is an **oblique solid.**

Example

The surface at the right is made up of six rectangles and two regular hexagons. In the solid shown at the right, identify a base, a lateral face, and a lateral edge. Finally, tell whether the solid is right or oblique.

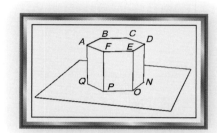

Solution Hexagon ABCDEF is a base. Rectangle AFPQ is a lateral face. $\overline{DN}$ is a lateral edge. The figure is right because the lateral edges are perpendicular to the base.

Two types of cylindrical surfaces are particularly common and important.

Definitions of Cylinder, Prism

A **cylinder** is a cylindrical surface whose base is a circle.

A **prism** is a cylindrical surface whose base is a polygon.

While the base of a cylinder must be a circle, the base of a prism can be any polygonal region. We name a prism by the name of its base. For instance, if the base is a triangle, the prism is called a *triangular prism*. The faces of the lateral surface of a prism are always parallelograms. Many prisms that you will encounter will be right prisms whose bases are regular polygons. These are called *regular prisms*. A **regular prism** is a right prism whose bases are regular polygons. Thus, the shape in the Example is a regular hexagonal prism. If we wanted to indicate that we are including the interior of this figure, we would add the word *solid* to our description, so it would be a solid regular hexagonal prism. The prism above the definition of cylindrical solid on the previous page is a solid oblique pentagonal prism.

 QY2

One kind of prism that you have seen before is a *cube*. A **cube** is a right square prism in which all 6 faces are congruent squares.

▷ QY2

If a figure is a regular polygonal prism, describe the shapes that make up its lateral surface.

Hierarchy of Cylindrical Surfaces

From their definitions, the various types of cylindrical surfaces fit nicely into a hierarchy.

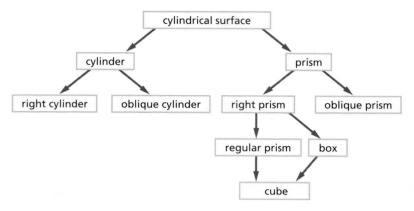

Activity

MATERIALS DGS-3D

Step 1 Use the following instructions to construct a pentagonal prism using a DGS-3D.

a. Construct a point *P* on the plane.

b. Construct a line through *P* that is perpendicular to the plane.

c. Construct vector $\vec{v}$ on the perpendicular line.

d. Construct a regular pentagon on the plane. Your screen should look like the one at the right.

e. Construct a pentagonal prism using the pentagon and the vector.

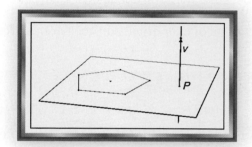

Step 2 Copy the table below. Rotate the pentagonal prism in space and carefully count each element to complete the first row of the table.

Name	Faces	Vertices	Edges
Pentagonal Prism	?	?	?
Hexagonal Prism	?	?	?
Octagonal Prism	?	?	?
⋮	⋮	⋮	⋮
n-gonal Prism	?	?	?

Step 3 Complete the table. If you need to use a DGS to construct the hexagonal and octagonal prisms, use the instructions from Step 1 but use a different shape in Part d. The last row of the table asks you to generalize your findings for any prism in which the base has *n* sides.

Step 4 How many vertices, edges, and faces does a 100-gonal prism have?

QY3

Questions

COVERING THE IDEAS

1. What is the difference between a surface and a solid?

2. In the figure below, $\triangle ABF$ is a translation image in space of $\triangle DCE$.

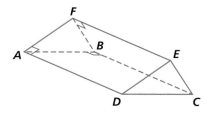

a. What is the general name for the figure?
b. Name the bases.
c. What is the specific name for the figure?
d. Name the lateral faces.
e. List the edges.

In 3 and 4, refer to the regular hexagonal prism in the example.

3. Name the four pairs of opposite faces.
4. Name all edges that are parallel to $\overline{EF}$.

5. **True or False** The height of a right prism equals the length of a lateral edge.

6. **Multiple Choice** Which is *not* true?
   A Every prism is a cylindrical surface.
   B Every cylinder is a cylindrical surface.
   C Every prism is a cylinder.

7. Explain why a cube is a regular prism.

8. The solid at the right was created by translating a regular decagonal region through space. Name this shape as specifically as possible.

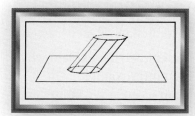

APPLYING THE MATHEMATICS

In 9–11, use the fact that the lateral edges of an oblique cylinder or prism are not perpendicular to the planes of the bases. Such cylindrical solids seem to lean. The amount of lean can be measured from the perpendicular. For example, the cylinder at the right has a 49° lean.

9. Sketch a cylinder that has a 20° lean.

10. Sketch a rectangular prism with a 70° lean.

11. How does the amount of lean compare with the measure of the angle a lateral edge of the figure makes with the plane of a base?

12. The right cylinder pictured at the right has height 10 cm and base area 25π cm². Find PQ.

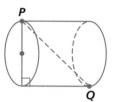

13. The base of the regular triangular prism below has perimeter 12 units and the height of the prism is 16 units.
    a. Find the area of one of its lateral faces.
    b. Find the area of one of its bases.

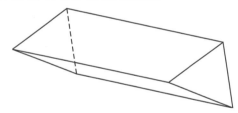

14. a. How many vertices does a prism have if its base is an octagon?
    b. What if its base is an 18-gon?

15. A restaurant uses drinking glasses that are right cylinders 7 inches high and have a diameter of 3 inches. They want their straws to always stick at least 2 inches out of the glass. What is the shortest straw they should buy?

16. Consider an oblique square prism.
    a. How many vertices does the prism have?
    b. How many faces are squares?
    c. What kind of shapes are the lateral faces?

17. **True or False** If the statement is true, explain why. If it is false, explain why or provide a counterexample.
    a. All of the lateral edges of a prism are congruent.
    b. The lateral edges of a prism are perpendicular to the bases.
    c. The lateral faces of a prism are always congruent.
    d. A cube is the only prism in which all of the faces (including bases and lateral faces) are congruent.

18. In the right triangular prism at the right, m∠NMP = 30, $\overline{MN} \perp \overline{NP}$, MP = 12, and NPOS is a square. Find MW.

19. A cube is outlined by sticks, so it is hollow and you can stick your hand through it. If one of the sticks has length s, what is the total length of the sticks needed to make the cube?

REVIEW

20. When lines ℓ and p are both perpendicular to line m, each of the following is possible. Draw a convincing picture of each situation. **(Lesson 9-1)**
    a. ℓ ∥ p
    b. ℓ ⊥ p
    c. ℓ and p are skew.

21. A basketball player dribbles a ball and it returns back up to his hand. If the path the ball travels is a straight line, what is true about this line and the plane described by the floor? **(Lessons 9-1, 4-3)**

22. In the figure at the right, O is the center of both circles. The inner circle has radius 2, the outer circle has radius 3, $m\widehat{AC} = m\widehat{BD} = 90°$, and CD = 1. Find the length of the longest path in the figure from A to B that never retraces itself. **(Lesson 8-8)**

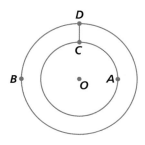

23. Consider the rectangular region at the right. **(Lesson 8-1)**
    a. If you had a pair of scissors, and made exactly one straight cut taking part of the figure away, could either figure have a greater perimeter than the rectangle?
    b. Trace the rectangle, and draw where you could make two straight cuts (possibly not through the rectangle) so that the resulting figure would have a greater perimeter than the original rectangle.

24. Let A = (0, 0) and B = (1, 0). Let T be a translation with translation vector $\vec{v}$, T(A) = C, and T(B) = D. Suppose that the segments $\overline{AB}$ and $\overline{CD}$ intersect. **(Lesson 4-6)**
    a. What can you say about the direction of $\vec{v}$?
    b. What can you say about the magnitude of $\vec{v}$?

EXPLORATION

25. Some solid prisms have special properties relative to light. What are these properties?

26. Find two nonmathematical meanings of the word *lateral* in a dictionary. How is the mathematical meaning of this word related to these nonmathematical meanings?

QY ANSWERS

1. Answers vary. Sample: P, $\overline{PR}$, QPRS

2. congruent rectangles

3. 15

Lesson 9-3

Pyramids and Cones

Vocabulary

conic surface

base, apex of a conic solid

pyramid

conic solid

cone

lateral edges, base edges of a pyramid

faces, lateral faces of a pyramid

right pyramid

oblique pyramid

regular pyramid

axis of a cone

right cone

oblique cone

lateral surface, lateral edge of a cone

height, altitude of a pyramid or cone

slant height of a pyramid

slant height of a cone

▶ **BIG IDEA** Pyramids and cones are two special types of conic surfaces.

Pyramids

Beginning around 2000 years ago, Greek and Roman writers identified Seven Wonders of the World (now sometimes called the Seven Wonders of the Ancient World). Of course, their world was limited to what they knew, so these wonders were all in Europe, North Africa, or the Middle East. Today, only one of these wonders survives: the three pyramids at Giza near Cairo in Egypt, shown on page 517. These pyramids date from 2600 to 2800 BCE, making them almost 5000 years old. They were built as funeral chambers and monuments to pharaohs.

Pyramids are still used as shapes in building important structures. Two 20th century buildings are pictured here. The Transamerica Pyramid in San Francisco, on the right, was built in 1972. The pyramid entrance to the Louvre art museum in Paris, France, below, was completed in 1989.

Pyramids are examples of *conic solids*. The boundary of a conic solid is a **conic surface**.

Mental Math

The figure below shows a square with its diagonals that is divided into four smaller squares. How many right triangles are there in the picture?

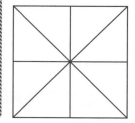

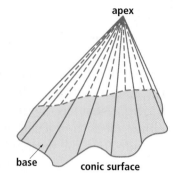

apex

base   conic surface

Definitions of Base, Apex, Conic Solid, Pyramid, Cone

Given a region in a plane (the **base**) and a point (the **apex**) not in the plane of the base, a **conic solid** is the set of points between the base and the apex, including the base and the apex.

A **pyramid** is the surface of a conic solid whose base is a convex polygonal region.

A **cone** is the surface of a conic solid whose base is a circular region.

For example, a solid pyramid like one of the pyramids in Egypt is the set of points on segments joining a square region (its base) and a point (its apex) above the middle of the base.

Parts of a pyramid are named in the figure at the right. The segments connecting the apex of the pyramid to the vertices of the base are **lateral edges.** The other edges are called **base edges.** The polygonal regions formed by the edges are the **faces** of the pyramid. All faces other than the base are triangular regions. These triangular regions are the **lateral faces** of the pyramid.

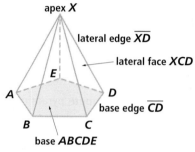

apex X

lateral edge $\overline{XD}$

lateral face XCD

base edge $\overline{CD}$

base ABCDE

As with prisms, pyramids are classified by their bases. There are triangular pyramids, square pyramids, pentagonal pyramids, and so on. If the base of a pyramid is rotation-symmetric and the segment connecting the apex to the center of symmetry of the polygonal base is perpendicular to the plane of the base, then the pyramid is called a **right pyramid.** If a pyramid is not right, it is called an **oblique pyramid.** A **regular pyramid** is a right pyramid whose base is a regular polygon. The Egyptian pyramids and the pyramid entrance to the Louvre are regular square pyramids.

 QY1

▶ QY1

Draw an oblique square pyramid with lateral edges $\overline{RS}$ and $\overline{TS}$.

Cones

From its definition, a cone is like a pyramid in that it has one base and an apex. The difference is that the base of a cone is a circle.

The line through the apex and the center of the circle is the **axis** of the cone. Other terms used for cones are like those used for pyramids. When the axis is perpendicular to the plane of the circle, the cone is called a **right cone.** Otherwise the cone is an **oblique cone.** The surface of a cone other than the base is the **lateral surface** of the cone. A **lateral edge** of a cone is any segment whose endpoints are the apex and a point on the circle.

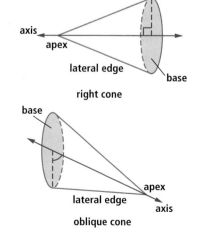

axis

apex

lateral edge

base

right cone

base

lateral edge

apex

axis

oblique cone

Heights of Pyramids and Cones

The **altitude** of a pyramid or cone is the line segment from the apex to the base that is perpendicular to the plane containing the base. The **height** of a pyramid or cone is the length of its altitude. This length can also be called the *altitude,* just like the term *radius* can be used to refer to either a length or a segment. In a right cone or pyramid, the altitude (*h* in the pyramid and cone at the right) joins the apex to the center of symmetry of the base.

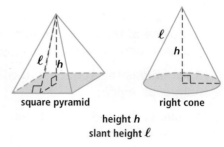

square pyramid   right cone

height *h*
slant height *ℓ*

Each lateral face of a regular pyramid is an isosceles triangle congruent to all the other lateral faces. The length of the altitude of any of these triangles (*ℓ* in the pyramid above) is called the **slant height** of the pyramid. A right cone also has a slant height; it is the length of a lateral edge.

 QY2

▶ **QY2**

Which is longer, the slant height or the height of a pyramid?

Example

The regular square pyramid shown at the right has height $AB = 10$.
If $BC = 7$, find its slant height AC.

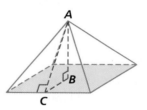

Solution Use the Pythagorean Theorem.

$$AC^2 = AB^2 + BC^2$$
$$= 10^2 + 7^2$$
$$= 149$$
$$AC \approx \sqrt{149}$$
$$\approx 12.2$$

Hierarchy of Conic Surfaces

At the right is a hierarchy of conic surfaces. You should compare this hierarchy with the hierarchy of cylindrical surfaces. They are very much alike.

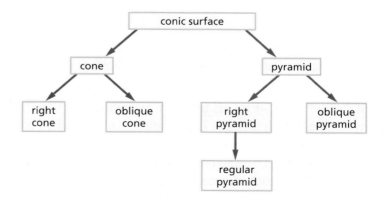

Notice the similarities between cones and pyramids, and between cylinders and prisms. Later in this chapter and in Chapter 10, you will see similarities in the formulas for the volumes and surface areas of these figures.

Activity

MATERIALS DGS-3D

Step 1 Use the following instructions to construct a regular pentagonal pyramid using a DGS-3D.

a. Construct a regular pentagon on the plane.

b. Construct a line that passes through the center of the pentagon and is perpendicular to the plane. Construct a point P on this line. Your screen should look like the one at the right.

c. Construct a pentagonal pyramid using the point P as the apex and the pentagon as the base.

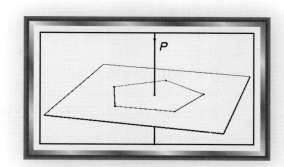

Step 2 Copy the table below. Rotate the pentagonal pyramid in space and carefully count each element to complete the first row of the table.

Name	Faces	Vertices	Edges
Pentagonal Pyramid	?	?	?
Hexagonal Pyramid	?	?	?
Octagonal Pyramid	?	?	?
⋮	⋮	⋮	⋮
n-gonal Pyramid	?	?	?

Step 3 Complete the table. If you need to use a DGS to construct the hexagonal and octagonal pyramids, use the instructions from Step 1, but use a different polygon. The last row of the table asks you to generalize your findings for any pyramid in which the base has n sides.

Step 4 How many vertices, edges, and faces does a 100-gonal pyramid have?

 QY3

▶ **QY3**

If a pyramid has 30 vertices, how many sides does the base have?

Questions

COVERING THE IDEAS

1. A sketch of a regular square pyramid is shown at the right.
   a. Identify its base.
   b. Identify its apex.
   c. How many lateral edges are there? Identify them.
   d. How many lateral faces are there? Identify them.
   e. How many base edges are there? Identify them.

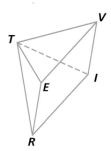

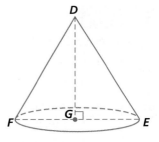

2. A sketch of a right cone is shown at the right, with G the center of the circular base and $\overline{DG}$ perpendicular to the plane of the circle.

   a. Identify the axis of the cone.   **b.** Identify a lateral edge.

   c. Identify the apex.

   d. What is the height of the cone?

   e. What is the slant height of the cone?

   f. What does $\overline{FE}$ represent?   **g.** What does $\overline{GE}$ represent?

In 3–6, identify the type of figure drawn.

3.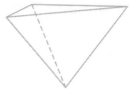

4.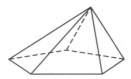

5.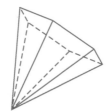

6.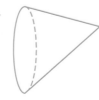

7. The figure at the right is a regular pentagonal pyramid with $CA = 12$ and $AP = 4$.

   a. Find its height.   **b.** Find its slant height.

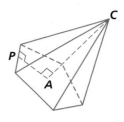

8. **a.** State the generalization that you found in the Activity.

   b. Would your generalization still work if your pyramids were not regular?

APPLYING THE MATHEMATICS

9. Repeat the Activity, but this time start with a circle in the plane. Explain why your generalization is true or not true.

10. Given a regular square pyramid, arrange each part in order from smallest to largest: the length of a lateral edge, the height of the pyramid, the slant height of the pyramid.

11. **Fill in the Blank** In a regular pyramid, the slant height of the pyramid is the ___?___ of the isosceles triangle that is a face.

12. Consider a right cone with altitude 10 feet and slant height of 12 feet. Find the diameter of the circle that is the base.

13. A *truncated pyramid* or *truncated cone* is the part of the pyramid or cone on or between the plane of its base and a parallel plane. Draw a truncated hexagonal pyramid.

14. If the altitude of a regular square pyramid has length 8 in. and the square has sides of length 4 in., find:

    a. the slant height.   **b.** the length of a lateral edge.

15. The Great Pyramid of Khufu is a regular square pyramid with a base edge of length 241 meters and a slant height of 195 meters. Find the height of the pyramid.

16. You can make a cone by cutting a part of a piece of paper. Suppose a cone is made from a quarter disc with the dimensions as shown at the right. The paper is then rolled so that $\overline{AB}$ and $\overline{AC}$ coincide and a cone is formed.

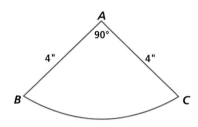

    a. What is the circumference of the base of this cone?

    b. What is the radius of the base of the cone (to the nearest tenth of an inch)?

    c. What is the height of the cone?

    d. What is the slant height of the cone?

REVIEW

17. The diagram at the right shows a staircase. If the staircase is a solid, give the best name for the solid. **(Lesson 9-2)**

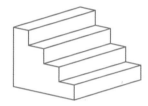

18. Sergio is playing golf and is about to hit with his driver. He knows that 90% of the time, his ball will land between 280 and 310 yards away on a line that is within 6° of where he is aiming. What is the area of the region in which Sergio's ball will land 90% of the time? **(Lesson 8-9)**

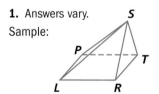
19. When she got her driver's license, Kelly's dad gave her his old car. The car's mileage was 72,123 miles. Given that the diameter of Earth is about 7926 miles, if this car had been traveling along the equator, how many times would it have circled the globe with its mileage? **(Lesson 8-8)**

20. In 450 BCE, the Greek historian Herodotus claimed that the area of a face of each great pyramid of Egypt equaled the area of its base. Using the information from Question 15, was Herodotus correct? **(Lesson 8-3)**

21. Refer to $\triangle ABC$ at the right. Which point on $\overline{AC}$ is closest to point B? How do you know? **(Lesson 6-2)**

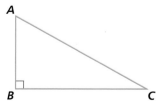

EXPLORATION

22. In Question 15, you are given the length of an edge of a base of the Great Pyramid of Khufu, in meters. What is the area of the base, to the nearest *acre*? (Use the fact that 640 acres = 1 square mile.)

Lesson
9-4   Drawing in Perspective

Vocabulary

vanishing point

> ▸ **BIG IDEA**   Realistic drawings of figures can be achieved by using perspective.

Below at the left, a cube is drawn as it would look in many if not most mathematics books and in many technical drawings. Notice that its back face *EFGH* is the same size as its front face *ABCD*. The advantage of this kind of drawing is that you can easily see that certain sides, angles, and faces are congruent.

Mental Math

a. A right triangle has legs that are 2 meters long. What is its area?

b. A right triangle has legs that are between 100 and 200 meters long. Give upper and lower bounds for its area.

c. A square has an area of 100 square kilometers. What is the length of its sides?

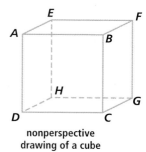

nonperspective
drawing of a cube

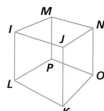

perspective drawing of a cube

However, artists trying to achieve realism would not draw a cube in this way. This is because, in the real world, the back of a cube is farther away from the viewer than the front. The back looks smaller, as in the figure at the right above, where the back face *MNOP* looks smaller than the front face *IJKL*.

To draw a three-dimensional figure on a plane surface, a technique called *perspective drawing* can be used. Perspective drawing was developed by the Italian architects Leon Battista Alberti (1404–1472) and Filippo Brunelleschi (1377–1446) in order to make their drawings more realistic. The main idea in perspective is taken from a familiar situation: a drawing of railroad tracks.

We know that railroad tracks are parallel and thus do not intersect, but in life and in many drawings, they look as if they will meet if extended. The point where they seem to meet is called a **vanishing point.**

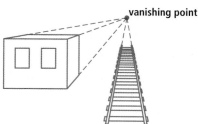

vanishing point

In this lesson, you will see and draw examples of one-point and two-point perspective drawings. These have one and two vanishing points, respectively, and are the simplest to draw. There are also three-point, four-point, five-point, and six-point perspective drawings. These are more difficult and studied in classes devoted to drawing.

One-Point Perspective

The drawing of the railroad tracks at the bottom of the previous page exhibits one-point perspective. Notice how the parallel tracks and all lines parallel to them in the drawing seem to meet at one point. That point is the vanishing point for this drawing. In a one-point perspective drawing, the single vanishing point is at eye level.

In a one-point perspective drawing, there is one vanishing point. Horizontal lines on the object remain horizontal and parallel. Vertical lines on the object remain vertical and parallel. Parallel lines that go from front to back in the picture intersect at the vanishing point.

In the painting below, Thomas Talbot Bury has used perspective drawing to create a feeling of depth. You can see that the sides of the walkway meet at the vanishing point. Likewise, the windows along the left wall are drawn to the same vanishing point. Artists create a sense of three-dimensional space by their choice of a horizon line and vanishing point(s).

Activity 1

MATERIALS Pencil and straightedge or DGS

Follow these steps to draw a cube in one-point perspective using your DGS or a pencil and a straightedge.

Step 1 Draw a square *ABCD* with horizontal and vertical sides. Then draw a point *P* that is outside the square, above and to the left of the square.

Step 2 Draw very light segments connecting *A*, *B*, *C*, and *D* to *P*. Pick a point *E* on $\overline{AP}$ so that *E* looks like it could be one of the back vertices of the cube. (*EA* must be less than *AB*.)

Step 3 Now think of *E* as if *E* is the image of *A* under a size change with center *P*. Draw the size change image of the rest of square *ABCD*. This image will be the back of the perspective drawing of the cube.

Step 4 Darken the segments connecting the back vertices of the cube to the corresponding front vertices.

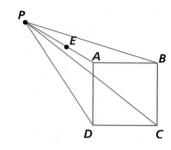

Notice how your finished drawing shows the cube as seen from a location above and to the left of the cube.

 QY1

> ▶ **QY1**
>
> In Activity 1, suppose you wanted to show a cube from a location above and to the right of the cube. Where would you put *P*?

Two-Point Perspective

The photograph at the right shows two-point perspective. In two-point perspective, there are two vanishing points, both on the horizon. Vertical lines on the object remain vertical and parallel. However, any pictured horizontal lines (white in the picture) meet at one of the two vanishing points. The picture looks as if you are viewing it from on the same level as the horizon, that is, ground level. In creating a two-point perspective drawing, it is best if the two vanishing points are on the same horizontal line and far apart, one on each side of the sheet of paper, as shown here. It is often the case that an edge of the figure is closest to the viewer, as shown by the red line drawn over the picture above.

left vanishing point

right vanishing point

Activity 2

MATERIALS Pencil and straightedge or DGS

Follow these steps to draw a cube in two-point perspective using a pencil and a straightedge or your DGS.

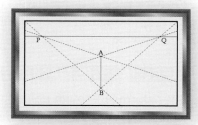

Step 1 Draw a horizontal line (the horizon line) above where you want the cube to be and place two vanishing points P and Q far apart on the line.

Step 2 Draw the front vertical edge $\overline{AB}$ of the cube where you want it. Then draw the light lines $\overleftrightarrow{AP}$, $\overleftrightarrow{AQ}$, $\overleftrightarrow{BP}$, and $\overleftrightarrow{BQ}$ as shown at the right. These lines will contain the four nonvertical edges of two faces of the cube.

Step 3 Let C be a point on $\overrightarrow{BP}$ so that C is a little closer to B than A is to B. Draw a vertical segment from C intersecting $\overrightarrow{AP}$ at D. ($ABCD$ will be a face of the cube.) Draw the light lines $\overleftrightarrow{CQ}$ and $\overleftrightarrow{DQ}$.

Step 4 Let E be the point on $\overrightarrow{BQ}$ so that E is a little closer to B than A is to B. Draw a vertical segment from E intersecting $\overrightarrow{AQ}$ at F. ($ABEF$ will be another face of the cube.) Draw the light lines $\overleftrightarrow{EP}$ and $\overleftrightarrow{FP}$.

Step 5 You will now have all but one of the edges of your cube drawn. Draw the last edge. Then darken the segments to show the visible edges of the cube.

Notice how your finished drawing shows the cube as seen from a location above the cube and in front of $\overline{AB}$.

 QY2

▶ **QY2**

In a perspective drawing of a solid cube, how many edges of the cube would be visible and how many would be hidden?

Questions

COVERING THE IDEAS

1. **a.** In a perspective drawing, do railroad tracks meet in the distance?
   b. In the real world, do railroad tracks meet in the distance?

2. What is a vanishing point?

3. In realistic drawings, do artists use perspective?

4. Consider these four drawings.
   a. Which are drawn in perspective?
   b. Trace each perspective drawing and find its vanishing point.

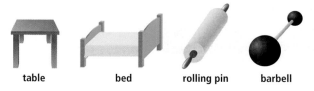

table   bed   rolling pin   barbell

5. Create a one-point perspective drawing of a cube as seen from below and to the right of the cube.

6. Create a two-point perspective drawing of a cube as seen from below the cube.

7. Look back at the perspective drawing of the room at the beginning of Lesson 9-2. How many vanishing points are in that drawing?

APPLYING THE MATHEMATICS

In 8–10, trace enough points on each perspective drawing to find its vanishing point(s). Indicate what type of perspective is illustrated.

8.

9.

10.

11. Make a perspective drawing with one vanishing point of a box that is not a cube.

12. Make a perspective drawing with one vanishing point of a square pyramid. (Hint: Draw the base in perspective as if it were the bottom of a cube. Then place the vertex of the pyramid in an appropriate place.)

13. Consider the drawings of the two buildings at the right.

    a. Consider the longest vertical segments on the drawing of each building. Which of these segments appears to be longer, the one on the front building or the one on the back building?

    b. Measure to determine which segment is longer.

    c. Why might the segment on the back building seem to be longer?

14. Draw your first and last names using a two-point perspective drawing.

REVIEW

15. A famous conic solid appears on the one-dollar bill. What is it, and what kind of conic solid is it? (**Lesson 9-3**)

16. Desiree's baby sister has a block with ○ ● ■ ★ ▲ **A** on the six faces. Three views of the block are shown here. (**Lesson 9-2**)

Which pictures are on opposite sides of the block?

In 17–19, describe the following figures. (**Lesson 5-5**)

17. given a point A, the set of all points B with $AB = 1$

18. all of the points that are equidistant from two given points A and B

19. all of the points that are equidistant from three given noncollinear points A, B, and C

20. The figure at the right shows two points, A and B, and a river. A person wishes to leave point A, travel to the river, and then reach point B. Another person wishes to leave point B, travel to the river, and reach point A. Suppose both of them want to take the shortest path possible, leave at the same time, and walk at the same speed. Trace the picture, and draw the point C at which the two people will meet. (**Lesson 4-3**)

21. Define the reflection image of point P over line ℓ. (**Lesson 4-1**)

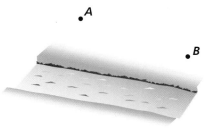

EXPLORATION

22. In a newspaper, magazine, or book, find a picture that does not appear to be done in perspective. Make your own drawing of it using one-point or two-point perspective.

QY ANSWERS

1. outside the square, above and to the right of the cube

2. 9 visible, 3 hidden

Lesson 9-5
Views of Solids and Surfaces

Vocabulary

view of a 3-dimensional figure

isometric drawing

▶ **BIG IDEA** Views of 3-dimensional objects from different directions are useful in describing them in 2-dimensions.

One of the world's most recognized jetliners is shown in the photograph at the right. The drawings below show three pictures from three different positions relative to the jetliner. These diagrams are called **views** because they are two-dimensional representations of a 3-dimensional object that is a nonperspective drawing. Views are often (but not always) drawn from one of six directions: front, back, top, bottom, left, or right.

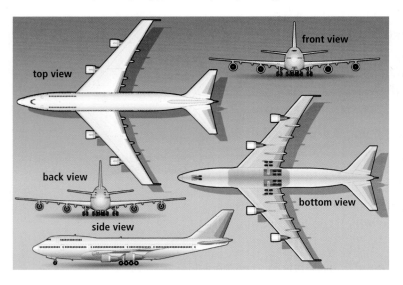

These drawings are used by manufacturers of the airplanes and their parts, the people who build hangars for the planes, the architects and engineers who plan the runways for the planes, and aircraft controllers who have to help the pilots navigate the planes. So it is important that all the people working on a project with this plane understand the views.

Example 1

The drawings below show three views of a solid.

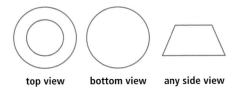

top view   bottom view   any side view

What kind of solid is being shown?

Solution By analyzing the views, you can find some clues. The top view has two concentric circles, indicating that the final shape has a top and bottom that both have circular 2-dimensional cross sections, but they are not the same size. One possibility is that this solid is a truncated cone.

 QY1

When making blueprints or technical drawings for a building, an architect or engineer will create views of the building from the different sides. These views are called *elevations*. Elevations provide a scale model of each side of the building, and help determine what the final building will look like.

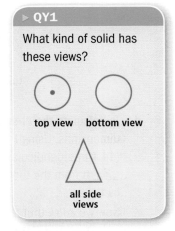

▶ **QY1**

What kind of solid has these views?

top view   bottom view

all side
views

GUIDED

Example 2

Below are two different elevations of a shed that is in the shape of a prism.

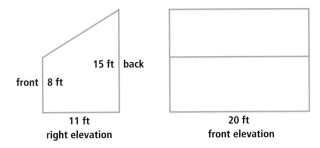

right elevation   front elevation

a. What type of prism is it?
b. How high is the peak (highest point) of the shed?
c. Draw an elevation from the back of the shed.
d. Find the area of the roof of the shed.

(continued on next page)

Solution

a. The right and front elevations indicate that the shed looks like the figure at the right. This shape is a ___?___.

b. The elevations show that **the peak of the shed is __?__ ft high.**

c. The right and front elevations indicate that **from the back, the building is a __?__-by-__?__ rectangle.**

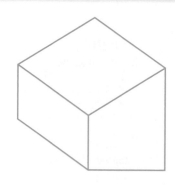

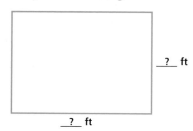

__?__ ft

__?__ ft

d. To calculate the area of the roof, you must first find its dimensions. Using the right elevation, draw the auxiliary line (11 ft) perpendicular to the back wall. This divides the wall __?__ ft from the top. Then one side of the roof can be found by the ___?___.

Now we know that the rectangular roof is about 20 feet by __?__ feet. Thus, **the area of the roof is about __?__ square feet.**

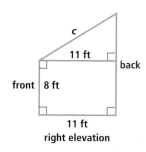

right elevation

Another way of representing views is by using drawings on isometric graph paper, like those in Figures 1 and 2 at the right. *Isometric graph paper* consists of dots located at the vertices of a tessellation of equivalent triangles. Drawings using these dots are called **isometric drawings.**

Figure 1   Figure 2

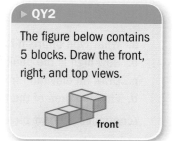

Activity

Use a DGS-3D or a set of cubes to build these figures and paper and pencil to draw the views.

Step 1 Figure 1 at the right is made up of cubes. Build this figure.

Step 2 Look down on the figure that you have built. Draw a top view.

Step 3 Look at the figure from the front (from point A). Draw a front view.

Step 4 Look at the figure from the right side (from point B). Draw a right view.

Step 5 Repeat Steps 1–4 with Figure 2, which has 12 cubes. To identify a right view for figure 2, look at your figure from point C.

STOP QY2

▶ QY2

The figure below contains 5 blocks. Draw the front, right, and top views.

front

Questions

COVERING THE IDEAS

1. What three different terms are found in this lesson for a 2-dimensional representation of a 3-dimensional figure?

2. Look back at the picture of the airplane at the beginning of the lesson. If you were to draw the back view of the plane, what would be different from that of the front view?

3. **Multiple Choice** Which of the following could *not* be a view of a right cone?

 A   B   C

4. Views of a solid you have studied are shown below. Name the solid.

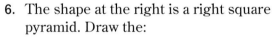

front and back view   left and right view

5. The figure at the right is created using 5 congruent blocks. Each block is 10" long and has square ends that are 4" on an edge. Draw each of the following views, including the dimensions.
   a. left view   b. top view   c. back view

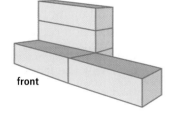

front

6. The shape at the right is a right square pyramid. Draw the:
   a. bottom view.
   b. top view.
   c. front view.

 front

APPLYING THE MATHEMATICS

7. Draw two noncongruent surfaces that could have the figure at the right as a view.

8. Suppose the base of a right cone is resting on the ground. The diameter of the right cone is 6 meters and the apex of the cone is 4 meters above the ground.
   a. Sketch a side view of this cone and label it with the information given.
   b. Use your side view to calculate the distance from the apex to a point on the edge of the circular base (its slant height).

9. Some farm buildings are in the shape of half-cylinders. In these buildings, the front and back ends are semicircular regions. One such building is shown at the right. Suppose the building is 40 ft long and 26 ft wide.

a. Draw a top view of the building. Include all measurements.

b. Draw a side view of the building. Include all measurements.

c. How much floor space (the area of the floor) does the building have? What view is most helpful when calculating this?

In 10 and 11, draw the front, back, left, and right views of the figure.

10. oblique cylinder

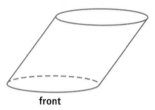

front

11. oblique cone

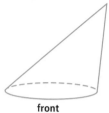

front

12. The diagram at the right shows the top view of a structure created using blocks like those in Question 5. The number within each rectangle denotes how many blocks are stacked in that space.

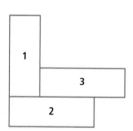

a. Draw the front view.

b. Draw the right view.

c. Using the dimensions given in Question 5, mark the dimensions of each sketch.

In 13 and 14, Shanna drew the figure with 9 cubes. Draw the top, right, and front views of each figure.

13.

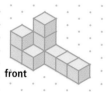

front

14.

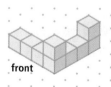

front

15. Suppose you have a cube whose edges are 2 cm long. If you look at the cube from a side and rotate it about an axis that is perpendicular to your line of sight and to the floor, the side view you see is a rectangle whose height is constant, but whose length changes. Draw and give the dimensions of the smallest and largest rectangular views that you would see as you turn the cube.

REVIEW

16. In the regular square pyramid at the right, AX is the height, and B is the midpoint of $\overline{CD}$. If $AX = 18$ and $CD = 20$, find:

    a. XB. (Lesson 9-3)

    b. Area($\triangle ABC$). (Lesson 8-4)

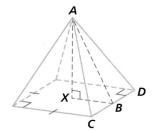

17. **True or False** Consider the Sun, Earth, and Pluto as points. Then, at any given moment the Sun, Earth, and Pluto are all contained in a single plane. (Lesson 9-1)

18. In the figure at the right, $\triangle ABC$ is a right triangle, m$\angle C = 30$, and $AD = DC$. Prove that $\triangle ADB$ is equilateral. (Lesson 8-7)

19. Explain why it makes sense for a honeycomb to be made out of hexagons, but not out of pentagons. (Lesson 7-6)

20. a. How many more diagonals does a hexagon have than a pentagon?

    b. How many more diagonals does a decagon have than a nonagon?

    c. Generalize Parts a and b and explain why your generalization is true. (Lesson 2-6, Previous Course)

EXPLORATION

21. Choose an object in your home or classroom and draw it from four viewpoints: top, front, right, and left. For what kind of object will all the views differ?

22. a. Using a globe, draw Earth, including all visible land masses, as seen from below the South Pole.

    b. Is this view the same as a bottom view?

QY ANSWERS

1. right circular cone

2. front

right view

top view

Lesson
9-6
Spheres and Sections

Vocabulary

sphere

radius of a sphere

center of a sphere

diameter of a sphere

great circle of a sphere

hemisphere

small circle of a sphere

plane section

2-dimensional cross section

conic sections

▸ **BIG IDEA**   The intersection of a plane with a sphere is either a single point or a circle.

Ancient Greek astronomers deduced that Earth was shaped like a sphere based on the fact that Earth casts a consistent round shadow on the Moon during lunar eclipses. Their theory was strengthened by the fact that, at sea, people saw only the tops of ships that were far away. Nevertheless, many Europeans in the Middle Ages and Renaissance believed that the world was flat. That belief persisted until 1522, when the voyage begun by Ferdinand Magellan, a Portuguese navigator, completed the first circumnavigation of Earth.

Mental Math

a. What is the measure of the arc intercepted by the minute and hour hands of a clock at 4 o'clock?

b. What is the measure of an inscribed angle that intercepts the arc described in Part a?

c. At what time is the measure of the arc between the hour and the minute hands 60°?

The Sphere

A *sphere* is a 3-dimensional counterpart of the circle. Notice that the terminology of circles extends to spheres.

Definitions of Sphere, Radius, Center

A **sphere** is the set of points in space at a certain distance (its **radius**) from a point (its **center**).

A sphere is a surface like a table tennis ball or a bubble, not a solid figure. A **radius of a sphere** is any segment connecting the center of the sphere to a point of the sphere. A **diameter of a sphere** is any segment connecting two points of the sphere that contains the center of the sphere.

To draw a sphere, draw a circle. Then, to give the illusion of depth, draw an oval through the middle using dashed lines for the top half of the oval. You may need to add the center, the radius, the diameter, or arcs of the circle, as in the sphere with center *A* shown at the right.

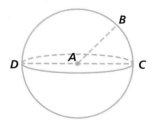

As with a circle, the center and all points inside the sphere are not points of the sphere. If you wish to include the interior of the sphere, then you need to call it a *solid sphere* (just as you did for prisms and pyramids). To picture a solid sphere, you may wish to shade the drawing, as shown at the right.

STOP QY1

▶ QY1

Name the center, a diameter, and two radii in the sphere on page 550.

Sections of a Sphere

In 1884, Edwin A. Abbott wrote *Flatland,* a novel about a 2-dimensional world. The characters in the novel are squares, triangles, other polygons, circles, and line segments. The story hinges on a sphere from a 3-dimensional world passing through Flatland and befriending a square. As the sphere passes through the plane of Flatland, the sphere appears to the square as a point, then a circle getting bigger and bigger, for a while. Then the circle gets smaller and smaller until it becomes a point and disappears. The square is seeing a part of the *cross section* of the sphere as it intersects the plane.

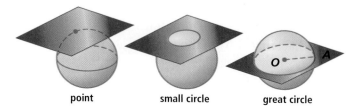

A scene from *Flatland: The Movie,* 2007.

Two types of intersections can occur with the intersection of a sphere and a plane. The intersection is a single point if the plane just touches the sphere; it is a circle otherwise. (In Question 16, you are asked to prove that the intersection is a circle.) If the plane contains the center of the sphere, the intersection is called a **great circle** of the sphere. A great circle (shown below at the right) splits the sphere into two **hemispheres.** Otherwise, the intersection is called a **small circle** (below in the center).

point   small circle   great circle

Both great circles and small circles are examples of *plane sections* of a sphere.

> ### Definition of Plane Section
>
> A **plane section** of a 3-dimensional figure is the intersection of that figure with a plane.

Plane sections are also known as **2-dimensional cross sections.** Some plane sections in this book are shaded for clarity, even when the section is the boundary of the shaded region.

Activity 1

MATERIALS DGS-3D

The goal is to create a model of the sphere passing through the plane in *Flatland*.

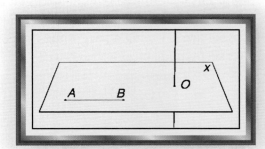

Step 1	Construct point *O* on a plane *X*. Construct a line perpendicular to plane *X* through *O*. Construct a point *P* on that line as shown at the right.
Step 2	Construct $\overline{AB}$ as shown.
Step 3	Hide point *O* and $\overleftrightarrow{OP}$ but don't hide *P*.
Step 4	Construct a sphere using *P* as the center and *AB* as its radius.
Step 5	Your sphere can be moved up and down by dragging point *P*. Drag point *P* so that the sphere intersects the plane. Construct the intersection curve of the sphere and the plane as shown in the diagram at the right.
Step 7	Hide the sphere.
Step 8	Recreate the scene from *Flatland* by dragging point *P* up and down. As *P* approaches the plane, you should see a circle of increasing size. What is the largest radius this circle will have and when will this occur?

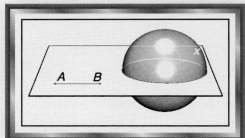

Earth as a Sphere

Earth is almost a solid sphere. One of its great circles is the equator. Points on the equator are about 6378 km (or 3963 miles) from the center of Earth. However, Earth has been slightly flattened by its rotation. The North and South Poles are about 6357 km (or 3950 miles) from the center of Earth.

Below are two sketches of Earth. The sketch at the left is of Earth as seen from slightly north of the equator, so the equator is tilted. Notice how the oval representing the circle of the equator is widened to give the illusion of looking at it from above. Then the South Pole cannot be seen. The sketch at the right is as seen from the plane of the equator.

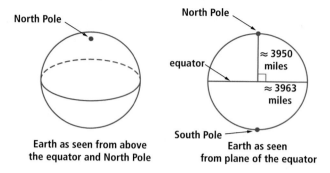

Earth as seen from above
the equator and North Pole

Earth as seen
from plane of the equator

 QY2

In about 230 BCE, the Greek mathematician Eratosthenes used the following information to estimate the circumference of Earth. He noticed that at noon on the summer solstice (on June 21st), the Sun was directly overhead at Syene. He knew that Alexandria was about 5000 stades due north of Syene. (The *stade* was a Greek unit of length equal to about 517 feet.) In another year, he calculated that at noon on the summer solstice in Alexandria, the Sun was 7.2° away from overhead.

Because Alexandria is due north of Syene, Eratosthenes concluded that the distance between the cities was a fraction of Earth's circumference. He assumed that the Sun was far enough away that rays from anywhere on Earth to it would be parallel, as indicated in the diagram at the right.

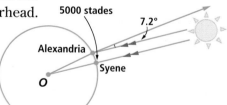

> **QY2**
>
> Find the difference in the circumferences of Earth around the equator and around the poles.

GUIDED

Example

Use the above figure to show how Eratosthenes used this information to find the circumference of Earth.

Solution Let C be the circumference of Earth. Due to the Corresponding Angles Postulate, $m\angle O = \underline{\quad?\quad}$

Thus the distance from Alexandria to Syene is $\frac{?}{360} \cdot C$.

So, $\frac{?}{360}C = 5000$ stades.

Solving this equation, $C = \underline{\quad?\quad}$ stades

$\approx \underline{\quad?\quad}$ feet

$\approx \underline{\quad?\quad}$ miles.

 QY3

> **QY3**
>
> Today, with accurate readings from space, we know that the circumference of the section created by intersecting a plane through the North and South Poles is about 24,860 miles. How does this compare with Erastothenes' measurement?

Plane Sections of Prisms and Cylinders

In *Flatland,* the square can only infer the sphere's shape by looking at its cross sections. While *Flatland* is fiction, the same idea is very commonly used today with all sorts of shapes. Biologists use plane sections of tissue to study a tissue's cell structure. These sections are thin enough so that light from a microscope will shine through them. Doctors use a CAT (*c*omputerized *a*xial *t*omography) scanner to take x-rays of plane sections of the human body for diagnostic purposes. Engineers and architects use cross sections to describe the shapes of things they are designing or examining.

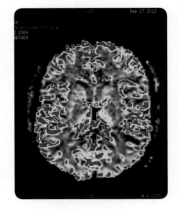

A plane section of the brain of a sleeping person.

Activity 2

MATERIALS DGS-3D or model of a cube (optional)

The figure at the right shows a cube in which the vertices and the midpoints of the edges are labeled. Identify the plane that cuts the cube to produce the plane sections described below. Recall that planes can be identified using three noncollinear points. If you find that you are having a hard time visualizing this, you can create a DGS 3-D cube and try out the planes that you think might work. A physical model of a cube might also be helpful.

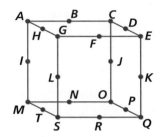

Description of Plane Section	Plane
Example: square	HDP
1. scalene triangle	?
2. isosceles triangle	?
3. equilateral triangle	?
4. rectangle that is not a square	?
5. isosceles trapezoid that is not a rectangle	?
6. pentagon	?
7. hexagon	?

Activity 2 shows that a variety of shapes are possible for the plane sections of a cube. A variety of plane sections are possible also for a prism or a cylinder, depending on whether the intersecting plane

1. is parallel to the bases,

2. is not parallel to and does not intersect a base or bases, or

3. intersects one or both bases.

Activity 3

Consider the figures below.

plane sections parallel to bases

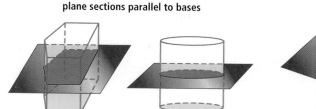

plane sections not parallel to bases

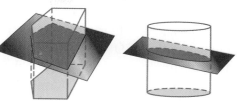

Step 1 If the intersecting plane is parallel to the bases, make a conjecture about the relationship between the base and the section.

Step 2 For a prism, if the intersecting plane is neither parallel to nor intersecting the bases, make a conjecture about the relationship between the number of sides of the base and the number of sides of the section.

Step 3 For a cylinder, if the intersecting plane is neither parallel to nor intersecting the bases, make a conjecture about the shape of the section.

When a plane intersects a base as well as lateral faces, more sections are possible. Figure I at the right shows a plane intersecting a solid pentagonal region. The section is the triangular region ABC where $\overline{AB}$ is on the back face, $\overline{AC}$ is on the side face, and $\overline{BC}$ is on the bottom face of the prism. Figure II shows a plane intersecting a cylinder. The section is a rectangle.

plane sections intersecting one or both bases

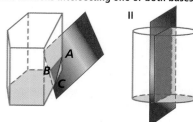

Activity 4

Make drawings like the two plane sections of prisms I and II shown at the right above, but with a prism that has a quadrilateral base.

Plane Sections of Pyramids and Cones

For pyramids and cones, sections parallel to the base have shapes similar to the base, but they are smaller. You can sketch them by drawing segments or arcs parallel to the base.

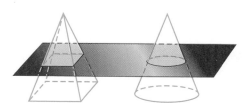

Conic Sections

Below at the far left are two right conical surfaces with the same axis, formed by rotating a line intersecting the axis about that axis. The plane sections formed are called the **conic sections.** The conic sections describe orbits of planets and paths of balls and rockets. Their focusing properties are used for long-range navigation (LORAN), telescopes, headlights, satellite dishes, flashlights, and whispering chambers. You are likely to have graphed parabolas in your study of algebra and you may have graphed hyperbolas, too. You are likely to study all the conic sections in later mathematics courses.

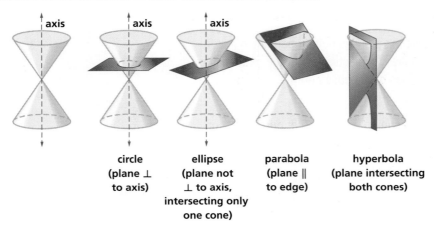

circle	ellipse	parabola	hyperbola
(plane ⊥ to axis)	(plane not ⊥ to axis, intersecting only one cone)	(plane ∥ to edge)	(plane intersecting both cones)

Questions

COVERING THE IDEAS

1. Draw a sphere with center P and a diameter $\overline{AK}$ that is not horizontal.

2. **Fill in the Blank** Complete the analogy: *Circle is to 2 dimensions as sphere is to ___?___.*

3. Define: *plane section.*

4. Name all possible types of plane sections of the object.

   a. a sphere   **b.** an orange   **c.** a straw

5. **True or False** Every cross section of a prism with an n-gon base is an n-gon.

6. **True or False** The diameter of the equator is the same as the diameter of the plane section of Earth containing the poles.

7. **a.** Name the four types of conic sections.

   b. Name three uses of conic sections.

In 8–11, draw a box with a plane section that satisfies the condition.

8. parallel to a face
9. not parallel to a face
10. intersecting four faces and not containing any vertex of the cube
11. intersecting all six faces and not containing any vertex of the cube

12. If Eratosthenes were alive today, he could repeat his calculation of the circumference of Earth using places in the United States. Louisville, Kentucky is 408 miles due north of Montgomery, Alabama. At a given moment, the Sun is 6° lower in the sky in Louisville than it is in Montgomery. From this information, what estimate would Eratosthenes obtain for the circumference of Earth?

APPLYING THE MATHEMATICS

In 13–15, copy the figure shown.

a. Sketch a plane section parallel to the bases.

b. Sketch a plane section not parallel to and not intersecting the base(s).

c. Name the shape of each section.

13.
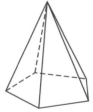
right pentagonal pyramid

14.

right cone

15.
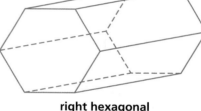
right hexagonal prism

16. Here is a proof that the intersection of a sphere and a plane not through its center is a circle. Given is sphere O and plane M, intersecting in the curve containing A and X as shown below. Fill in the justifications.

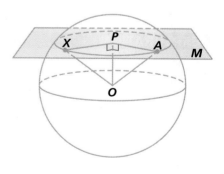

Proof Let P be the foot of the perpendicular from point O to plane M. Let A be a fixed point and X be any other point on the intersection.

a. $\overline{OP} \perp \overline{PA}$ and $\overline{OP} \perp \overline{PX}$ because ___?___.

b. $\overline{OP} \cong \overline{OP}$ because of the ___?___.

c. $\overline{OA} \cong \overline{OX}$ because ___?___.

d. $\triangle OPX \cong \triangle OPA$ by the ___?___.

e. $\overline{PX} \cong \overline{PA}$ by the ___?___.

Thus, any point X on the intersection lies at the same distance from P as A does. So, by the definition of circle, the intersection of sphere O and plane M is the circle with center P and radius PA.

17. Consider the figure at the right, in which the radius PC of the cross section is equal to the perpendicular distance QP from the center of the cross section to the sphere. Suppose the area of the cross section is 9π square meters. If the radius of the sphere is 5 meters, find the distance from the center of the sphere to the cross section.

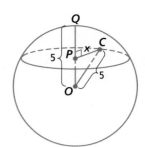

REVIEW

18. Here are three views of a building. **(Lesson 9-5)**
    a. How tall in stories is the building?
    b. How many blocks long is the building from front to back?
    c. Where is the tallest part of the building?
    d. Create an isometric drawing of the building.

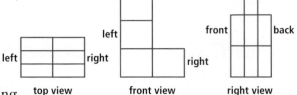

19. Mayra cut a paper party hat that was in the shape of a cone along a straight line, and spread it out on the table to get the figure at the right. Trace the figure, and draw a segment whose length is equal to the slant height of the hat (before Mayra cut it). **(Lesson 9-3)**

20. Suppose $\triangle ABC$ is a triangle with no lines of symmetry. Explain why all of the angles in the triangle have different measures. **(Lesson 6-2)**

21. Let A and B be two points in the plane. Describe the set of all points that are rotation images of B about A. **(Lesson 3-2)**

22. Look at the networks at the right. For each one, let V equal the number of nodes, and E equal the number of arcs. **(Lessons 2-7, 1-3)**
    a. Calculate $E - V + 1$ for each of the networks.
    b. Draw a network where $E - V + 1 = 4$.

EXPLORATION

23. Earth is often described as an *oblate spheroid*. The *oblateness* of a planet measures how much a planet bulges at the equator.
    a. Look in a reference for a formula for oblateness.
    b. What is the oblateness of Earth?
    c. Which major planet has the greatest oblateness?
    d. Which major planet is most like a sphere?

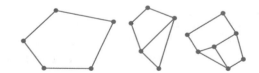

QY ANSWERS

1. The center is A, $\overline{DC}$ is a diameter, $\overline{AB}$ and $\overline{AC}$ are two radii.

2. $2\pi \cdot 3963 - 2\pi \cdot 3950$ $= 2 \cdot \pi \cdot 13 = 26\pi$ miles.

3. approximately 1.5%

Lesson
9-7
Reflections in Space

Vocabulary

perpendicular bisector of a
  segment (in space)

reflection image of a point
  over a plane

reflecting plane, plane of
  reflection

congruent figures (in space)

reflection-symmetric space
  figure

symmetry plane

▶ **BIG IDEA**   Reflections over planes in space and 3-dimensional reflection-symmetry are the 3-dimensional counterparts of reflections over lines in a plane and 2-dimensional reflection-symmetry.

Many of the properties of reflections in two dimensions carry over to three dimensions. For example, when you look in a mirror, the mirror appears to lie halfway between you and your image. It contains the midpoint of the segment connecting a point to its image. That is, the mirror *bisects* the segment. Also, an imaginary line from the tip of your nose to its image will always be perpendicular to the mirror.

In general, a plane M is the **perpendicular bisector** of a segment $\overline{AB}$ if and only if $M \perp \overline{AB}$ and M contains the midpoint of $\overline{AB}$. This enables 3-dimensional reflections (over planes) to have the same defining condition as their 2-dimensional counterparts (over lines).

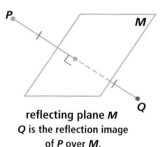

reflecting plane *M*
Q is the reflection image
of *P* over *M*.

Mental Math

Consider this 5-pointed star.

a. What kind of rotation symmetry does this figure have?

b. How many lines of symmetry does the figure have?

c. Is the figure translation-symmetric?

Definition of Reflection Image of a Point over a Plane

For a point P which is not on a plane M, the **reflection image of P over M** is the point Q if and only if M is the perpendicular bisector of $\overline{PQ}$.

For a point P on a plane M, the reflection image of P over M is P itself.

In the definition, plane M is the **reflecting plane, plane of reflection,** or mirror. Reflections over planes preserve the same properties as their 2-dimensional relatives. That is, reflections in space preserve angle measure, betweenness, collinearity, and distance. So the definition of congruence in the plane can be extended to three dimensions.

Definition of Congruent Figures (in Space)

Two figures F and G in space are **congruent figures** if and only if G is the image of F under a reflection or composite of reflections.

As in two dimensions, 3-dimensional reflections reverse orientation. You are oppositely congruent to your mirror image.

Symmetry Planes

A 2-dimensional figure is reflection-symmetric if and only if it coincides with its image under some reflection. There is a corresponding definition for three dimensions.

> **Definitions of Reflection-Symmetric Figure, Symmetry Plane**
>
> A space figure F is a **reflection-symmetric figure** if and only if there is a plane M (the **symmetry plane**) such that $r_M(F) = F$.

The existence of symmetry planes is used to help rock collectors, chemists, and geologists to identify types of crystals. Biologists refer to reflection symmetry as *bilateral symmetry*. Many animals have bilateral symmetry.

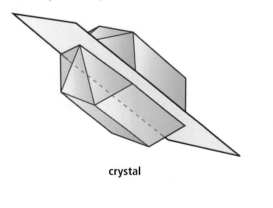

crystal

beetle

The body surfaces of most people have approximate bilateral symmetry. The right side and left side are reflection images of each other over the plane that goes through the middle of the body from head to toe. Although your right hand and your left hand are nearly congruent, you cannot slide or turn a right hand onto a left hand, because they have different orientation. Your right hand is oppositely congruent to your left hand.

Space figures may have any number of symmetry planes, from zero to infinitely many. In the right cylinder at the top left of the next page, any plane containing $\overline{PQ}$, the segment connecting the centers of its bases, is a vertical symmetry plane. Therefore, it has infinitely many vertical symmetry planes and one horizontal symmetry plane. The regular triangular pyramid next to the cylinder has exactly 3 symmetry planes, one of which is drawn.

right cylinder

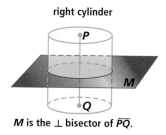

M is the ⊥ bisector of $\overline{PQ}$.

right triangular pyramid

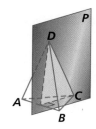

P is the ⊥ bisector of $\overline{AB}$.

Any right prism will have at least as many symmetry planes as the base has symmetry lines, plus an additional symmetry plane parallel to its bases. A right pyramid that is not a triangular pyramid has exactly as many symmetry planes as the base has symmetry lines.

Example

Determine the number of symmetry planes for a regular hexagonal prism.

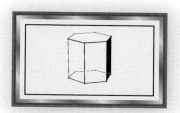

Solution One symmetry plane is the plane passing through the midpoints of the lateral edges.

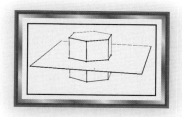

Because the base has 6 symmetry lines, there are 6 symmetry planes perpendicular to the bases and containing the symmetry lines of the bases. One of these is shown at the right. **There are 7 symmetry planes.**

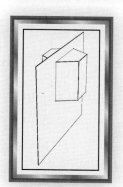

 QY

> QY

How many symmetry lines would the figure in the Example have if it were a regular hexagonal pyramid instead of a regular hexagonal prism?

Questions

COVERING THE IDEAS

In 1–5, a figure is given.

a. Indicate the number of symmetry planes.

b. Describe the locations of these planes relative to the figure.

1.

right circular
cylinder

2.

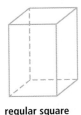

regular square
prism

3.

top

4.

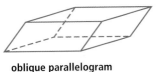

oblique parallelogram
prism

5.

right cone

6. How many symmetry planes does a regular pentagonal prism have?

7. **Multiple Choice** The symmetry plane of a human standing upright is

   A   parallel to the ground about waist high.

   B   perpendicular to the ground, halfway between the front and back.

   C   perpendicular to the ground, halfway between right and left sides.

8. Name a property of a figure not preserved by reflections in space.

APPLYING THE MATHEMATICS

9. a. How many symmetry planes does a regular square pyramid, the shape of the Pyramids at Giza, have?

   b. Draw such a pyramid with one of its symmetry planes.

10. Draw the symmetry plane of the oblique cylinder at the right.

11. Is it possible for a prism to have only one plane of symmetry? Explain why or why not, and sketch one if it is possible.

12. **True or False** The cross section formed by any symmetry plane of a sphere is a great circle of the sphere. Explain your answer.

13. How many symmetry planes does a cube have?

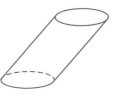

REVIEW

14. Tia and Eva cut an orange, whose shape is approximately a sphere with radius 6 cm, into two halves. Both of them agree that the halves have the same size. What is the approximate area of the cross section that was made by cutting the orange? **(Lessons 9-6, 8-9)**

15. The figure at the right depicts a front view of 27 unit cubes arranged to form a $3 \times 3 \times 3$ cube. Suppose that the unit cube marked with an X is removed. **(Lesson 9-5)**

    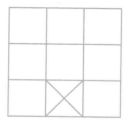

    a. From which views is it possible to tell for sure that this arrangement is missing a unit cube: front, back, sides, top, or bottom?

    b. Describe a view from which it is possible to tell that a part of the cube was taken out.

16. Draw a 3-dimensional solid that has the front, right, and bottom views as shown below. **(Lesson 9-5)**

    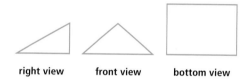

    right view   front view   bottom view

17. When drawn in perspective, the front and back sides of an object appear to be the same size. Which of the sides is bigger in reality? **(Lesson 9-4)**

18. A pyramid has a base that is a regular n-gon. If two such pyramids are glued together at their bases, how many faces, edges, and vertices will the resulting shape have? **(Lesson 9-3)**

19. Points A, B, and C are noncollinear. Points D and E are equidistant from A, B, and C. **(Lesson 9-1)**

    a. Can A, B, C, D, and E all lie in the same plane? Why or why not?

    b. Draw a possible picture of the above situation.

EXPLORATION

20. a. What is radial symmetry?

    b. Find a picture of some living or inanimate objects that possess radial symmetry.

    c. Must an object that has radial symmetry also have reflection symmetry?

Lesson
9-8

Making Polyhedra and Other Surfaces

Vocabulary

polyhedron, polyhedrons, polyhedra

faces of a polyhedron

edges of a polyhedron

vertices of a polyhedron

tetrahedron

hexahedron

convex polyhedron

regular polyhedron

net

frustum of a pyramid

▶ **BIG IDEA** Polyhedra, cylinders, and cones can be created from 2-dimensional nets.

Prisms and pyramids are special kinds of *polyhedra*. Polyhedra are the 3-dimensional analogues of polygons. Roughly speaking, a **polyhedron** is a 3-dimensional surface that is a union of polygonal regions (its **faces**) and has no holes. The sides of these faces are called **edges.** The endpoints of edges are its **vertices.** Every edge is contained in exactly two faces. The intersection of any two faces is either an edge of both faces, or a vertex of both faces. The plural of *polyhedron* is either **polyhedrons** or **polyhedra.**

Polyhedra can be classified by the number of faces. Below are pictured a solid **tetrahedron** (4 faces) and a surface that is a **hexahedron** (6 faces). The tetrahedron has 4 vertices and 6 edges.

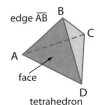

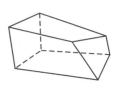

tetrahedron   hexahedron

Both a box and a cube are special types of hexahedra. In fact, all prisms and pyramids are polyhedra because all of their faces are polygons. However, there are many polyhedra that are neither prisms nor pyramids.

STOP QY1

Technically, neither a polygon nor a polyhedron can be a convex set, but a polygonal region and a solid can be. Still, we call polygons and polyhedra convex if they are boundaries of a convex set. A polyhedron is a **convex polyhedron** if and only if the segment connecting any two points of it is contained within the polyhedron and its interior.

Regular polyhedra are the 3-dimensional counterparts to regular polygons. A **regular polyhedron** is a convex polyhedron in which all faces are congruent regular polygons and the same number of edges intersect at each of its vertices. There are only five regular polyhedra; they are pictured at the top of the next page.

Mental Math

a. How many of the faces of a right hexagonal prism are hexagons?

b. How many of the faces of a right hexagonal prism are rectangles?

c. How many faces does a right 25-gonal prism have?

▶ **QY1**

How many vertices and edges does a hexahedron have?

| regular tetrahedron (4 faces) | cube (6 faces) | regular octahedron (8 faces) | regular dodecahedron (12 faces) | regular isosahedron (20 faces) |

Nets for Polyhedra

A **net** is a 2-dimensional figure that has been created by unfolding the faces of a 3-dimensional surface along its edges. For instance, when cuts are made along some of the edges of a cube, then the cube can be flattened out.

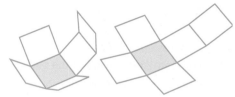

If you choose different edges to cut a cube, you may obtain two nets that are not congruent to each other. There are 11 different noncongruent nets of a cube, but only two for a tetrahedron. For a dodecahedron and an icosahedron, there are 43,380 different noncongruent nets!

Nets for Pyramids

What would a net for a pyramid look like? What shapes can the base be? What shapes are the sides? Activity 1 explores the net of a hexagonal pyramid.

Activity 1

MATERIALS heavy paper, scissors, tape, compass, straightedge

Follow these steps to construct a regular hexagonal pyramid.

Step 1 Construct a regular hexagon on a piece of heavy paper.

Step 2 On each side of the hexagon, construct congruent isosceles triangles with the base being a side of the hexagon. What must be true of the heights of the triangles so that your net will form a pyramid?

Step 3 Cut out your net. Fold it on the sides of the hexagon.

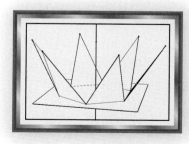
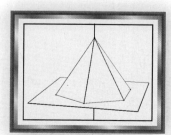

(continued on next page)

Step 4 Tape the sides together to form the pyramid.

Step 5 What would happen to the height of the pyramid if the altitudes of your triangles get longer or shorter?

 QY2

▶ **QY2**

Sketch a net for a regular octagonal prism.

GUIDED

Example

At the right is a net for a nonregular 3-dimensional solid made of rectangles and pentagons. Describe the surface, justifying your conclusions.

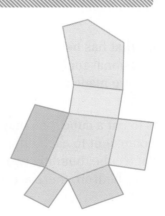

Solution There are __?__ faces. __?__ of the faces are rectangles, while the other two are congruent ___?___. The congruent ___?___ are the two ___?___ of the surface, indicating the surface might be a ___?___. Because the ___?___ have the same height, they are the lateral faces of the surface. So the surface is a ___?___.

Nets for Cylinders and Cones

The lateral surface of a cylinder can be rolled out into a rectangle. Let d be the diameter of the base circle. Then the rectangle must have one side of length πd, to match the circumference of the bases. The other side of the rectangle is the height h of the cylinder.

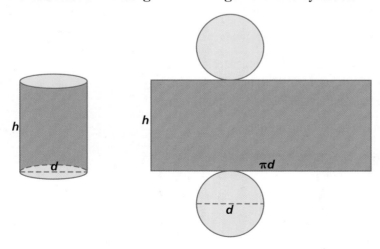

Activity 2

MATERIALS heavy paper, compass, scissors, tape

Step 1 On a piece of heavy paper, construct a circle with a radius of 10 cm and cut it out.

Step 2 Draw any two radii of that circle that form an obtuse angle and cut out the sector determined by the radii.

Step 3 Using each piece of the circle, tape along the radii of the parts of the circle to form two cones.

Step 4 After forming the cones, what is the name for the point at the top of the cone and what was that in the original circle?

Step 5 Each cone was made with the same length of the radius, but your cones are not the same height. Why is this the case?

 QY3

Questions

> ▶ **QY3**
>
> Given one circle, how could you form two cones that have the same height?

COVERING THE IDEAS

1. Fill in the hierarchy of 3-dimensional surfaces at the right using the following terms: regular tetrahedron, polyhedron, prism, tetrahedron, pyramid.

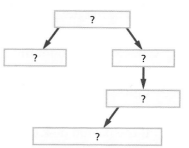

2. **Matching** In the left column are words associated with polygons. In the right column are words associated with polyhedra. Match the corresponding words.

   a. side
   b. regular polygon
   c. vertex
   d. quadrilateral
   e. 12-gon

   i. dodecahedron
   ii. face or edge
   iii. regular polyhedron
   iv. tetrahedron
   v. vertex

In 3 and 4, a statement is given about polygons. Change the italicized words to obtain a corresponding statement about polyhedra.

3. In a *polygon,* each *vertex* is the intersection of two *sides.*

4. A *20-gon* has 20 *sides* that are congruent *segments.*

5. Why is a cylinder not considered to be a polyhedron?

6. Draw 3 noncongruent nets for a cube.

7. **a.** Draw a net for a regular pentagonal pyramid.
   b. Cut out your net and fold it to make the pyramid.

8. At the right is a diagram of a cardboard box that will be printed and used for a cereal box. The dotted lines represent the cuts that will be made so that the flaps will fold together.

   a. Is this a net for a polyhedron? Explain.

   b. Once assembled, the dimensions of the box will be 12" × 4" × 8". How much area is available to print on the front of the box (face *D*) ?

   c. What is the total area of faces *A, B, C,* and *D*?

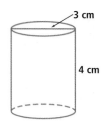

9. Suppose you wanted to make the cylinder pictured at the right out of cardboard. Draw a net you could use. Make the net actual size and find the area for each region of the net. Leave your answers in terms of π.

10. Draw a net for a regular triangular prism with each edge with a length of 2". Make the net actual size and put the areas for each section on the net.

APPLYING THE MATHEMATICS

11. Examine the 5 regular polyhedra pictured in the beginning of this lesson. Count the number of faces (*F*), vertices (*V*) and edges (*E*) of each polyhedron. A famous formula called Euler's Formula states that $F + V - E = 2$. Does it hold true for these 5 regular polyhedra?

12. If the partial disk at the right is cut out of cardboard and $\overline{AB}$ is moved to coincide with $\overline{AC}$, a cone will be formed. Let $AB = 4$ in. and m∠*BAC* = 100. Cut out the shape and make the surface.

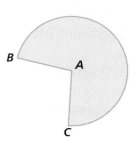

13. **Multiple Choice** Which is *not* a net for a cube?

   A   B   C   D

14. Draw a nonconvex polyhedron.

15. Pictured at the right are side, front, and top views of a solid. Draw a net for this solid.

left view

front view

top view

16. A **frustum of a pyramid** is the polyhedron that results from slicing a pyramid with a plane parallel to its base and using the section of the plane as a face.

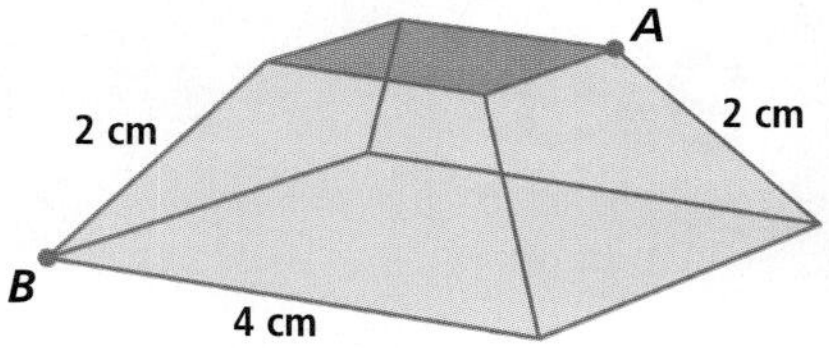

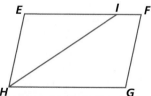
2 cm   A   2 cm

B

4 cm

   a. Draw a net for a frustum of a regular square pyramid like the one pictured at the right.
   b. Draw the top, bottom, and side views of this frustum.
   c. Find the areas of each face of this solid. (Hint: Use the net.)

17. Make up a definition for a *frustum of a cone* that is similar to the definition of a frustrum of a pyramid in Question 16.

18. Describe the cross section of the frustum of a cone by a plane that contains the axis of the cone.

REVIEW

19. Use perspective to draw a cube and one of its symmetry planes. **(Lessons 9-7, 9-4)**

20. Suppose a saw cuts a piece of wood that is in the shape of a rectangular solid. Explain with a drawing how the cross section could be a triangle. **(Lesson 9-6)**

21. In one episode of the cartoon *Inuyasha,* the characters confront a creature who seems fierce when viewed from the front, but cannot be seen when viewed from the side. How is such a thing possible? **(Lesson 9-5)**

22. A circle has an area of 12π square inches. What is its diameter? **(Lesson 8-9)**

23. In the figure at the right, $HEFG$ is a parallelogram and $\overrightarrow{HI}$ bisects $\angle EHG$. Prove that $EH = EI$. **(Lessons 6-4, 6-2, 5-4)**

E   I   F

H   G

EXPLORATION

24. Use a DGS-3D to explore the frustum of a square pyramid.
   Step 1 Construct a regular square pyramid based on Steps 1–3 of Activity 1, starting with a square.
   Step 2 Construct a plane parallel to the base of the pyramid that cuts the pyramid between its vertex and its base.
   Step 3 Construct the plane section that is the intersection of the parallel plane and the pyramid.
   Step 4 Hide the parallel plane.
      a. What are the shapes of the base and the "top"?
      b. What is the relation between these?
      c. What are the shapes of the lateral faces of the frustum?

QY ANSWERS

1. 8 vertices and 12 edges.

2. Answers vary. Sample:

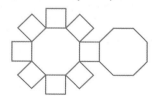

3. Cut the circle in half along a diameter. Tape the halves together and the cones will be the same height.

Lesson

9-9

Surface Areas of Prisms and Cylinders

Vocabulary

surface area, S.A.

lateral area, L.A.

▶ **BIG IDEA** A formula for the surface area of a right prism or cylinder can be found by examining its net.

The cost of any container, from a suitcase to a new house, from a paper bag to a hot air balloon, depends on the amount of material used to make it. The sum of the areas of all the faces or surfaces that enclose a solid is called its **surface area**, which we abbreviate as **S.A.**

Consider a brown paper lunch bag. It is approximately a rectangular prism with the top base missing. Because each face of a rectangular prism is a rectangular region, the surface area of the bag is the sum of the areas of five rectangles. These areas are easily found by examining the net for the bag.

Mental Math

Order the following numbers from smallest to largest: the area of a square with sides of length 2, the area of a circle with radius 1, the area of a circle with radius 2, the area of a square with sides of length 1.

Example 1

A brown paper lunch bag has dimensions 5 in. × 4 in. × 10 in. What is its surface area?

Solution Draw the bag and a net to represent its surfaces. Because the bag has no top, the net only shows the five rectangles that make up the surface of the bag.

Notice that the net is the union of a 4-by-5 rectangle (the bottom of the bag) and an 18-by-10 rectangle (the lateral surfaces of the bag put together). Thus, the surface area is found by finding the sum of these two areas.

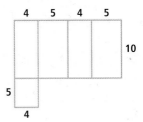

$$\text{Surface area} = \text{area of base} + \text{area of lateral surface}$$
$$= 5 \cdot 4 + 18 \cdot 10$$
$$= 20 + 180$$
$$= 200 \text{ in}^2$$

Recall from Lesson 9-2 that the lateral surface of a prism is the part of the surface that is not a base. In Example 1, the sides of the paper bag make up the lateral surface. The area of the lateral surface is called the **lateral area.** To calculate the lateral area, we unwrapped the lateral surface to make a rectangle. The length of the rectangle was the perimeter of the base of the bag (18 in.) and the height of the rectangle was the height of the bag (10 in.). It turns out that the lateral surface of any right cylindrical solid can be unwrapped to form a rectangle whose length is the perimeter of the base and whose height is the height of the solid. This leads to the following formula.

Lateral Area Formula for Right Cylindrical Solids

The lateral surface area, L.A., of any right cylindrical surface is the product of its height h and the distance p around its base.
$$\text{L.A.} = ph$$

The formula above applies to the lateral area of any right cylindrical solid. If the right cylindrical solid is a prism, then p represents the perimeter of the base. If the cylindrical solid is a cylinder, then p represents the circumference of the base. If the cylindrical solid is like the one in Example 2 below, p just represents the distance around the base. In all cases, however, the same formula applies because the lateral surface can be unwrapped to form a $p \times h$ rectangle.

Example 2

The figure at the right is half of a solid cylinder. Find its exact lateral area.

Solution The base of the figure is the union of a semicircle and a segment.

$$p = \text{semicircle length} + \text{segment length}$$
$$= \tfrac{1}{2}\pi\,(6) + 6$$
$$= 3\pi + 6$$

Now we apply the formula for the lateral area L.A.

$$\text{L.A.} = ph$$
$$= (3\pi + 6)8$$
$$= 24\pi + 48$$

Surface Area

Every cylindrical solid has two congruent bases. The bases have the same area. So, by definition of surface area, the next theorem holds for every cylindrical solid.

Surface Area Formula for Cylindrical Solids

The surface area S.A. of any cylindrical solid is the sum of its lateral area L.A. and twice the area B of a base.

$$S.A. = L.A. + 2B$$

Although the above formula applies to any cylindrical solid, if the solid is not a right cylindrical solid, then the lateral surface area may not be easy to find. However, the total area is still just the sum of the lateral area and the two bases, so the above formula still applies.

 QY1

▶ **QY1**

What is the surface area of the solid from Example 2?

GUIDED

Example 3

Find an expression for the surface area of a cylinder with radius r and height h.

Solution The bases are circles.

The area of each base is ____?____.
The circumference of each base is ____?____.
So the lateral area is ____?____.

Now substitute into the formula for the surface area.

$$
\begin{aligned}
S.A. &= L.A. + 2B \\
&= \underline{\quad?\quad} + 2(\underline{\quad?\quad}) \\
&= \underline{\quad?\quad}
\end{aligned}
$$

 QY2

▶ **QY2**

A regular prism has bases with perimeter p and apothem a. The height of the prism is h. Find an expression for the surface area of this prism.

Example 4

Some park entrances have concrete cylinders at the beginning of pedestrian walkways to prevent cars from entering. If each cylinder is 1.3 meters high and the diameter of each base is 0.5 meter, how many concrete cylinders can be painted with a can of paint that covers 30 square meters?

Solution To determine how many cylinders can be painted with one can of paint, first find the surface area of the cylinder. The bottom of the concrete cylinder does not need to be painted because it is fixed to the ground. Because the diameter of the base is 0.5 meter, the radius of the base is 0.25 meter.

$$
\begin{aligned}
S.A. &= L.A. + B \\
&= ph + \pi r^2 \\
&= 2\pi rh + \pi r^2 \\
&= 2\pi(0.25)(1.3) + \pi(0.25)^2 \\
&= 0.65\pi + 0.0625\pi \\
&= 0.7125\pi \text{ m}^2
\end{aligned}
$$

Because 1 can of paint covers 30 m² and each concrete cylinder's surface area is 0.7125πm², 1 can of paint will cover $\frac{30}{0.7125\pi} \approx 13.40$ concrete cylinders. So one can will cover 13 full concrete cylinders.

 QY3

To find the surface area of complex figures such as the square nut in the next example, you will need to sum the areas of each surface.

▶ **QY3**

In Example 4, if there are 20 cylinders and each concrete cylinder is to get three coats of paint, how many cans of paint are needed?

GUIDED

Example 5

A manufacturer is going to apply an anticorrosive coating to millions of square nuts. In order to calculate the amount of coating to order, the manufacturer needs to know the surface area of each nut. The square nut at the right has a square side of length 1.5". The cylinder hole is 1" in diameter. The thickness of the nut is 0.875". Find the surface area of the nut to the nearest thousandth of a square inch.

Solution The surface of the figure is composed of the figures at the right.

top and bottom:
2 squares,
circles removed

L.A. of box:
4 rectangles

L.A. of cylinder:
rectangle

The top and bottom face each have area __?__² − π(__?__)² = __?__.

Each rectangle of the L.A. of the box has area __?__.

The L.A. of the cylinder is __?__.

$d = 1.0$ so $r = 0.5$.

$$
\begin{aligned}
S.A. &= \text{Top and Bottom} + \text{L.A. of box} + \text{L.A. of cylinder} \\
&= 2(\underline{\ ?\ }) + 4(\underline{\ ?\ }) + \underline{\ ?\ } \\
&= \underline{\ ?\ } \text{ exactly}
\end{aligned}
$$

S.A. ≈ 10.928 square inches, to the nearest thousandth of a square inch.

Questions

COVERING THE IDEAS

1. **Fill in the Blanks** The surface area of any prism or cylinder is the ___?___ of its lateral area and ___?___ the area of its base.

2. a. What is meant by the *lateral area* of a surface?
   b. What is a formula for the lateral area of a right prism?
   c. What is a formula for the lateral area of a right cylinder?

3. Tell whether or not the unit can be used to describe surface area.
   a. meters   b. ft^3 c. in^2
   d. square kilometers   e. cubic yards

4. A grocery bag has dimensions as pictured at the right.
   a. Draw a net for the bag.   b. What is its surface area?

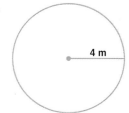
17"
7"   12"

In 5 and 6, a net for the surface of a solid is given.
a. Draw the solid.
b. Calculate its lateral area.
c. Calculate its surface area.

5.

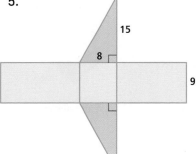

15

8

9

6.

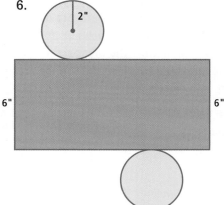

2"

6"

6"

In 7 and 8, each figure is translated 10 m in the direction
perpendicular to the plane in which the figure lies.
a. Find the lateral surface area of the resulting solid.
b. Find its surface area.

7.

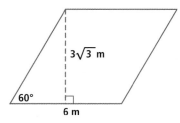

$3\sqrt{3}$ m

60°

6 m

8.

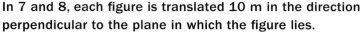

4 m

APPLYING THE MATHEMATICS

9. Consider a box with dimensions x, y, and z.
   a. Find its surface area in terms of x, y, and z.
   b. Triple each dimension and find the new surface area.
   c. Divide your answer in Part b by the answer in Part a.
   d. Make a conjecture based on your results in Parts a–c.

10. How many square inches of sheet metal are needed to make a soup can that is $2\frac{2}{3}$" in diameter and 4" high?

11. A manufacturing company constructed boxes by cutting square pieces out of an 8' × 10' rectangle. The edges are then folded up to make an open box. Find an expression for the surface area of the open box in terms of the length x of the side of the square.

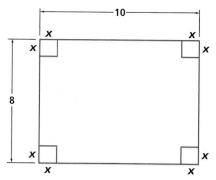

12. Right trapezoidal prism fixtures are often used to display rings in jewelry store windows. They are often covered with a velvet-like material. How many square inches of the velvet-like material are needed to cover the fixture at the right? (The bottom does not need to be covered.)

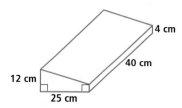

13. A pineapple was cut to make a right cylinder and then cored as shown in the figure at the right. The pineapple has a height of 6 in. and a diameter of 5 in. The diameter of the core is 1 in. Find the surface area of the cored pineapple.

14. The Giant's Causeway in Northern Ireland is a mass of basalt prism-shaped columns packed tightly together. The Causeway stones were formed as a result of rock crystallization under conditions of accelerated cooling. Most of the prisms are hexagonal. The average height of a column is about 14 m. The average diameter of the columns is 45 cm. Assuming that the average column is a regular prism, find the surface area of an average column of the Giant's Causeway.

15. Some toilet paper rolls contain 300 sheets, each 4.5 in. by 4 in. (The perforated side measures 4 in.) Describe the 2-dimensional figure that you would create by unrolling an entire roll. What is the area, in square feet, of that figure?

16. A dairy tank in the shape of a right cylinder is 102" in diameter and 255" high. How many cans of paint are needed to cover the lateral area and top with two coats of paint? Assume that 1 can of paint covers 350 square feet.

17. In a box, the length of the rectangular base is double its width. The height of the box is triple its width. If the surface area of the box is 352 square inches, find the width of the box.

18. The rectangular prism pictured at the right has a square base. Find its surface area.

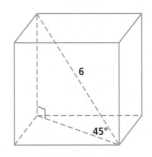

19. Suppose the surface area of a regular triangular prism is four times the area of one base. If the height of the prism is $\frac{1}{2}$ ft, find the length of each side of the triangle.

REVIEW

20. Draw a net for a regular hexagonal prism. (**Lesson 9-8**)

21. When examining the cross section of a tree, each ring represents a year of life for the tree. Draw an example of such a section for a nine-year-old tree. (**Lesson 9-6**)

22. A tennis ball can holds three tennis balls. The balls touch each other and the top, bottom, and sides of the cylindrical can. The diameter of a tennis ball is 2.5 inches. What is the ratio of the height of the can to the circumference of the can? (**Lesson 9-6**)

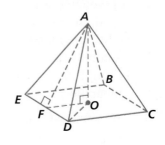

23. $ABCDE$ at the right is a regular pyramid with height AO. $DE = 18$ and $AO = 15$. Find each number to the nearest tenth. (**Lesson 9-3**)
    a. slant height
    b. BD
    c. length of a lateral edge
    d. area of $\triangle AOD$

24. In the figure at the right, $ABCD$ is a rectangle. Prove that Area($\triangle DEC$) = Area($\triangle ADE$) + Area($\triangle EBC$). (**Lesson 8-4**)

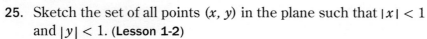

25. Sketch the set of all points (x, y) in the plane such that $|x| < 1$ and $|y| < 1$. (**Lesson 1-2**)

EXPLORATION

26. The *Menger sponge* is a shape achieved by starting with a cube, and removing smaller and smaller cubes from it.
    a. Look up the Menger sponge, and explain how this shape is created.
    b. If the cube you start with has sides of length 1, calculate the initial surface area and the surface area after 1 step.
    c. Do some research to find an approximation for the surface area after 2 steps.
    d. Make a conjecture about the surface area and volume of the resulting shape after n steps.

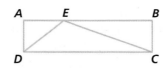

QY ANSWERS

1. $33\pi + 48$
2. $ph + ap$
3. 5

Lesson
9-10
Surface Areas of Pyramids and Cones

> ▶ **BIG IDEA** A formula for the surface area of a regular pyramid or right cone can be found by examining its net.

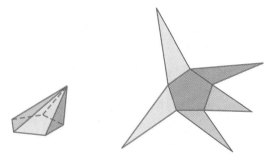

Mental Math

Let $P = (0, 7)$.

a. Suppose P is translated by the vector $\langle 3, 2 \rangle$. Find the coordinates of its image P' under this transformation.

b. Suppose P is translated by the composite of the translation with vector $\langle 3, 2 \rangle$ and the translation with vector $\langle -7, 3 \rangle$. Find the coordinates of P'', its image under this transformation.

c. The image of P''' under the translation T is the point P. Find the vector for this translation.

Above are an oblique pentagonal pyramid with a regular base and its net. Notice that the triangles forming the lateral surface are not congruent. As a result, to calculate the lateral area you must find the area of each of the five triangles and add them.

Regardless of how you calculate the lateral area, however, the surface area of any pyramid or cone is always found by adding its lateral area and the area of its one base.

Surface Area Formula for Pyramids and Cones

The surface area, S.A., of any pyramid or cone is the sum of the lateral area L.A. and the area B of its base.
$$\text{S.A.} = \text{L.A.} + B$$

Lateral Areas of Regular Pyramids

There is no simple formula for the lateral area of every pyramid in terms of lengths of its sides. However, when the pyramid is regular, there is a simple formula for the lateral area. This is a result of the fact that, in a regular pyramid, the lateral faces are all congruent triangles.

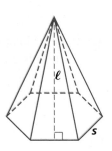

GUIDED

Example 1

Suppose a regular pyramid has n lateral faces and base edge s. Find an expression for the lateral area of a regular pyramid in terms of the perimeter p of its base and its slant height ℓ.

Solution Each triangle on the lateral surface has area ____?____.
Because there are n congruent triangles, multiply this area by n to get the lateral area.

$$\text{L.A.} = \left(\tfrac{1}{2}s\ell\right)n$$
$$= \tfrac{1}{2}s\ell n$$

Now $p =$ ____?____, so you substitute p into the formula to get

$$\text{L.A.} = \tfrac{1}{2}\ell p.$$

The formula $\text{L.A.} = \tfrac{1}{2}\ell p$ provides a quick way to calculate the lateral area of a regular pyramid. A visual way of remembering this formula is to think of removing the triangular faces from the lateral surface and rearranging them as shown in the Activity below.

Activity

Consider a regular pentagonal pyramid such as that drawn below at the left, with each edge of the base having length s and the slant height of the pyramid having length ℓ. The 5 lateral faces can be rearranged to form a trapezoid, as shown.

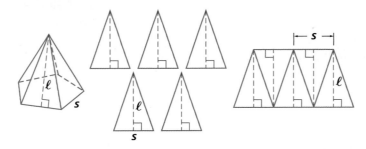

Step 1 Write a formula for the area of this trapezoid in terms of s and ℓ.

Step 2 In the formula you wrote above, for what expression can you substitute p, the perimeter of the base?

Step 3 After making the substitution in Step 2, show that your expression for the area of the trapezoid can be rewritten as $\tfrac{1}{2}\ell p$.

STOP QY

▶ QY

In the Activity, the pyramid has an odd number of faces.

a. What would change in the Activity if the pyramid had an even number of faces?

b. Would the formula for the lateral area be the same?

Example 2

Each side of the base of the Great Pyramid of Khufu at Giza has length 230 meters. The height of this regular square pyramid is 139 meters. Estimate its surface area.

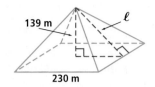

139 m

ℓ

230 m

Solution You must first calculate the slant height ℓ. The slant height is the hypotenuse of a right triangle whose legs are the height of the pyramid and half the side of the base.

$$\ell^2 = 115^2 + 139^2$$
$$\ell^2 = 32{,}546$$
$$\ell = \sqrt{32{,}546} \approx 180$$

Because the perimeter of the base is 920 m, you have what you need to calculate the lateral area.

$$\text{L.A.} = \tfrac{1}{2}\ell p = \tfrac{1}{2}(920)\sqrt{32{,}546} \approx 82{,}986.3 \text{ m}^2$$

The square base has an area of 52,900 m².

So the surface area $\approx 82{,}986.3 + 52{,}900 \approx 136{,}000$ m².

If you forget the formula for the lateral area of a pyramid, you can calculate the area without it by using triangles. However, the formula L.A. $= \frac{1}{2}\ell p$ is still very useful because it applies to cones as well. To understand why this is true, think of a series of regular pyramids in which you add more and more sides to the base, as shown below.

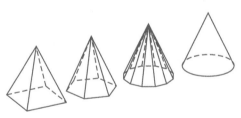

The more sides you add, the more closely the base resembles a circle. Suppose you continue adding sides forever. The solid you get will be indistinguishable from a cone. So, for a cone, the only difference in calculating the lateral area is in the calculation of the perimeter of its base.

Lateral Area Formula for Regular Pyramids and Right Cones

The lateral area L.A. of a regular pyramid or right cone is half the product of its slant height ℓ and the perimeter (circumference) p of the base.

$$\text{L.A.} = \tfrac{1}{2}\ell p$$

Example 3

Find the lateral area of the cone shown at the right.

Solution The slant height is given to be 10 inches. Because the right triangle is a 30-60-90 triangle, the radius of the base is 5 inches. Consequently, the circumference C of the base is found by: $C = 2\pi r = 2\pi \cdot 5 = 10\pi$ inches. Thus L.A. $= \frac{1}{2}\ell p = \frac{1}{2} \cdot 10 \cdot 10\pi = 50\pi \approx 157.07$ square inches.

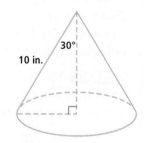

In the beginning of this lesson we showed an oblique pyramid and you could see that, given the necessary measurements, calculating the lateral area is a manageable task because the lateral faces are simple polygons. For a cone, however, this is not the case. No simple formula exists for calculating the lateral area of an oblique cone.

Questions

COVERING THE IDEAS

1. **Matching** Match the relationship among the area B of the base, the lateral area L.A., and the surface area S.A. with the figure.

   a. pyramid   i   S.A. $=$ L.A. $+ B$
   b. prism   ii S.A. $=$ L.A. $+ 2B$
   c. cone
   d. cylinder

2. Find the lateral area of the pyramid entrance to the Louvre, shown on page 532. Its base is a square with side 35 meters and its slant height is 27 meters.

In 3 and 4, a surface is described. Give exact answers.
a. Draw a picture.   b. Find its lateral area.   c. Find its surface area.

3. a regular hexagonal pyramid with $2\sqrt{2}$-inch lateral edges and 2-inch base edges

4. a cone with height 16 cm and radius 3 cm

APPLYING THE MATHEMATICS

In 5 and 6, find the exact surface area of each figure and the surface area rounded to the nearest tenth.

5. right cone

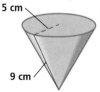

6. regular hexagonal prism

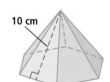

7. Some grain bins are in the shape of a cylinder with a right conical roof, as shown below. Suppose the radius of the bin is 12 ft, the height of the wall is 15 ft, and the height of the bin is 20 ft. What is the total surface area of the bin?

8. Some houses, like the one shown at the right, have roofs that are in the shape of right pyramids with rectangular bases. The base of the front part of the house shown in the photo is rectangular, 20 ft by 30 ft. The walls of the house are 9 ft high and the peak of the roof is 14 ft high.

   a. Why does the formula L.A. $= \frac{1}{2} \ell p$ not help you to find the area of the roof?

   b. Find the surface area of the roof.

9. A sector of a circle is a net for the lateral surface of a cone.

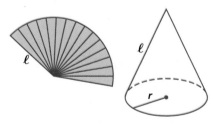

   If you divide the sector into smaller sectors as shown above, the sectors can be placed together to form a trapezoidal shape (as is done in this lesson with pyramids) that approaches a rectangle as the number of slices increases.

   a. Draw a diagram of the rectangular region that results from the activity described above.

   b. What are the dimensions of the rectangle in terms of ℓ and r (the radius of the original cone)?

10. Suppose the net shown at the right is turned into a cone.

   a. Find the slant height of the cone.

   b. Calculate the circumference of the base.

   c. Calculate the altitude of the cone.

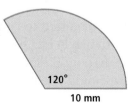

120°

10 mm

REVIEW

11. The Azrieli towers in Tel-Aviv, Israel, are three skyscrapers shaped like cylindrical solids. One has a square base, one has a circular base, and one has an equilateral triangle as a base. The perimeter of the equilateral triangle is 171 meters, and the height of that tower is 169 meters. Calculate its lateral area. **(Lessons 9-9, 8-7)**

12. Suppose you have three squares and two equilateral triangles, all with sides of the same length. Describe a polyhedron you could make by gluing together these shapes. **(Lesson 9-8)**

13. A designer used a computer to make a 3-dimensional image of one of the letters of the alphabet. The images below show three of the plane sections of the letter. The planes with which the sections were taken are all parallel, and the sections go from higher in the model to lower. Which letter did the designer make? **(Lesson 9-6)**

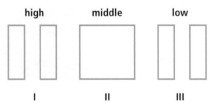

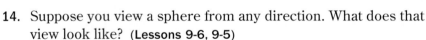

14. Suppose you view a sphere from any direction. What does that view look like? **(Lessons 9-6, 9-5)**

15. **True or False** The lateral faces of every prism are parallelograms. **(Lesson 9-2)**

16. A platform is being suspended from the ceiling using cables. In order to ensure that the platform is level (parallel to the ground), Sierra measures the distance to the ground from two separate points on the platform and finds those distances to be equal. Assuming that Sierra measured correctly, is this enough information to determine that the platform is level? Explain why or why not. **(Lesson 9-1)**

17. What are the four assumed properties of area? **(Lesson 8-2)**

EXPLORATION

18. Distance runners will often take an ice bath in extremely cold water following a race in order to speed their recovery. Kourtney is planning to run a marathon, so she wants to buy some ice to use in her post-race ice bath. Her choices at the store are a 10-pound block of ice or 10 pounds of small ice cubes. Which should she buy if she wants to make her bath water as cold as possible as quickly as possible, and why?

QY ANSWERS

a. The shape would be a parallelogram.

b. yes

Chapter 9 Projects

1 Regular Polyhedra

Use cardboard and tape to construct models of the regular polyhedra from the nets provided. The patterns shown below should be enlarged. Cut along solid lines, fold along dotted lines.

tetrahedron

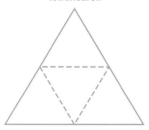

cube

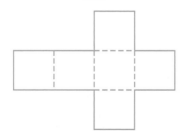

dodecahedron

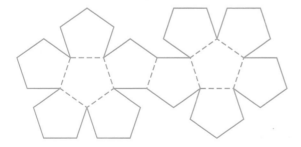

2 Star Polyhedra

Two famous families of polyhedra are the platonic solids mentioned in Lesson 9-8 and the Kepler-Poinsot polyhedra. Find pictures of all of the Kepler-Poinsot polyhedra, construct nets for them, and make models of two or more of them.

octahedron

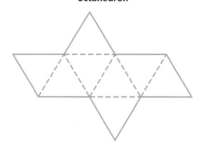

icosahedron

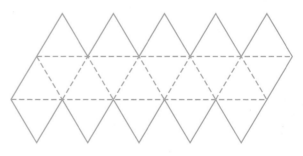

3 Plane Sections of a Cube

Make a series of large drawings of all the possible plane sections of a cube. How many are there? What is the minimum and maximum number of possible sides of a plane section of a cube? Make a conjecture about the minimum and maximum number of sides of the plane section of any regular prism with bases that are n-gons.

4 Three-Dimensional Paradoxes

The artist M. C. Escher is famous for drawing pictures that use tricks of perspective to draw things that could not exist in the real world. Find three such drawings, and present them to your class. Find out what methods Escher used to make these pictures. Make a sketch of a picture with a similar idea.

5 Making Maps

Mapmakers have the challenge of drawing a spherical object (Earth) on a flat piece of paper. Find out at least three different methods mapmakers use. Find a map illustrating each method, and write out each method's advantages and disadvantages.

6 Structures

Copy pictures or take photographs of buildings, dwellings, or other structures with a shape approximating each of the types of surfaces studied in this chapter; cylinders, prisms, cones, pyramids, other polyhedra, and spheres. (Do not use examples shown in this chapter.) Identify the type of perspective illustrated in the photo. Arrange this information nicely on a poster or in a small booklet.

7 Folding Planes

You have probably folded a piece of paper into an airplane in the past.

a. Discuss the similarities and differences between folding planes and making nets for surfaces.

b. Find several patterns for planes and make them out of paper. Keep track of which planes fly better than others.

c. What features of a paper airplane seem to make it fly the farthest, stay in the air the longest, and be more accurate?

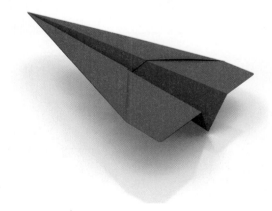

Chapter 9 Summary and Vocabulary

○ The purpose of this chapter is to familiarize you with the common 3-dimensional figures, the figures of solid geometry. To accomplish this, you should know their definitions and how they are related, be able to sketch them, identify plane sections, draw views from different positions, and be able to make some of them from 2-dimensional nets. Below is a hierarchy relating many of these surfaces.

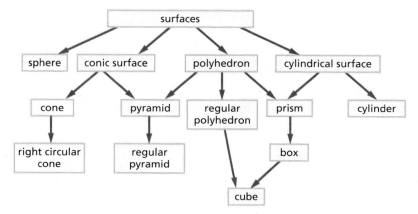

○ While there are many different kinds of 3-dimensional shapes, we focus on the ones in the hierarchy. This is because they tend to be less complicated than irregular shapes and therefore more accessible to study. It is also useful to note that many irregular figures often are built from regular ones, or from figures that can be approximated by symmetric shapes. Earth can be approximated by a sphere, your neck can be approximated by a cylinder and a volcano can be approximated by a cone.

○ Many ideas from two dimensions extend to three. The basic properties of planes are given in the Extended Point-Line-Plane Postulate. Like lines, planes may be perpendicular or parallel to each other, and planes can be perpendicular or parallel to lines. Circles and spheres have the same defining property, except spheres are in three dimensions. Reflections and reflection-symmetry are defined the same way in three dimensions as in two except that the reflecting line is replaced by a reflecting plane.

Vocabulary

Lesson 9-1
angle formed by a line and a plane
*line perpendicular to a plane
foot of a segment
parallel planes
distance between parallel planes
distance to a plane from a point
*skew lines
dihedral angle
edge of a dihedral angle
perpendicular planes

Lesson 9-2
surface, solid
interior, exterior of a surface
face, edge, vertices of a surface
*cylindrical solid
bases of a cylindrical solid
lateral surface
lateral face of a prism
cylindrical surface
height, altitude of a solid
right solid, oblique solid
*cylinder, *prism
regular prism, cube
lateral edge of a prism

Lesson 9-3
*conic surface, *conic solid
base, apex of a conic solid
*pyramid, *cone
lateral edges, base edges of a pyramid
faces, lateral faces of a pyramid
right pyramid, oblique pyramid, regular pyramid
axis of a cone
right cone, oblique cone

(continued on next page)

○ Architects use views to describe 3-dimensional figures in two dimensions. The difficulty of representing a 3-dimensional object on a 2-dimensional piece of paper is approached by drawing different views of the object and by drawing in perspective. It is also useful to examine the 2-dimensional net that can be folded to create the solid.

○ The surface area of a solid is the area of its surface. If the solid is a right cylindrical solid, the lateral area is determined by unwrapping the lateral surface to form a rectangle whose length is the distance around the base and whose height is the height of the solid. This gives the formula L.A. $= ph$.

○ If the solid is a regular pyramid, the lateral surface can be divided into congruent triangles. The sum of the areas of these triangles leads to the formula L.A. $= \frac{1}{2}\ell p$. The lateral area of a cone is determined by comparing it to a regular pyramid with a large number of faces. The formula is the same as for a regular pyramid, but in this case p represents the circumference of the base of the cone.

Postulates, Theorems, and Properties

Point-Line-Plane Postulate (Expanded)
    Unique Line Assumption (p. 518)
    Number Line Assumption (p. 519)
    Dimension Assumption (p. 519)
    Unique Plane Assumption (p. 519)
    Intersection Assumption (p. 519)
Unique Plane Theorem (p. 520)
Line-Plane Perpendicular Theorem
    (p. 521)

Lateral Area Formula for Right
    Cylindrical Solids (p. 571)
Surface Area Formula for Cylindrical
    Solids (p. 572)
Surface Area Formula for Pyramids
    and Cones (p. 577)
Lateral Area Formula for Regular
    Pyramids and Right Cones
    (p. 579)

Vocabulary

Lesson 9-3 (cont.)
lateral surface, lateral edge
  of a cone
height, altitude of a
  pyramid or cone
slant height of a pyramid
slant height of a cone

Lesson 9-4
vanishing point

Lesson 9-5
view of a 3-dimensional
  figure
isometric drawing

Lesson 9-6
*sphere
*radius, center; diameter of
  a sphere
*great circle of a sphere
hemisphere
*small circle of a sphere
*plane section
2-dimensional cross
  section
conic sections

Lesson 9-7
perpendicular bisector of a
  segment (in space)
*reflection image of a point
  over a plane
reflecting plane, plane of
  reflection
*congruent figures
  (in space)
*reflection-symmetric
  space figure
symmetry plane

Lesson 9-8
*polyhedron, polyhedrons,
  polyhedra
faces, edges, vertices of
  a polyhedron
tetrahedron, hexahedron
convex polyhedron
regular polyhedron
net
frustum of a pyramid

Lesson 9-9
surface area, S.A.
lateral area, L.A.

Chapter

9 Self-Test

Take this test as you would take a test in class. You will need a calculator. Then use the Selected Answers section in the back of the book to check your work.

1. In how many different planes can one circle be? Explain.

2. Draw two perpendicular planes.

3. A certain right triangular prism has exactly one symmetry plane. Draw the prism and its symmetry plane.

4. Draw a top, side and front view of a 5-step open stepladder.

5. Given these three views of the object pictured below, what might it be?

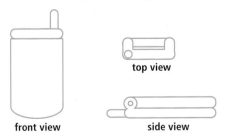

front view   side view

top view

6. If a cone has slant height 20 and height 16, what is the diameter of the base?

7. Given the net of the right pyramid below, find the surface area of the pyramid formed when the net is folded up.

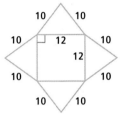

8. When the net below is folded into a cube, which face is opposite 3?

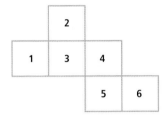

9. Draw a net for the cylinder below. Give lengths of all segments and any radii on your net.

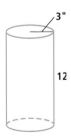

10. A small circle of a globe is 6″ from the center of the globe. If the circle has area 36π square inches, what is the length of the diameter of the globe?

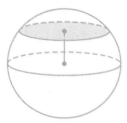

11. The Pyramid of Cestius is one of the best-preserved ancient buildings in Rome. The base of this regular square pyramid is a square with sides 100 feet. The height is 125 feet. Find the lateral area of the pyramid.

12. Can a lateral face of a prism be a triangle? Why or why not?

13. In the right triangular prism pictured below, $BC = 5$, $AB = 4$, and $AE = 12$. Find the surface area of the prism.

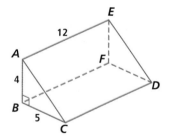

14. Using the prism in Question 13, name
    a. its two bases.
    b. a lateral face.
    c. a lateral edge

15. This hollow air conditioning duct is an open-ended cylinder with diameter 28 inches. If its length is 10 feet, what is the area of the material needed to build it?

16. Draw views of the regular hexagonal prism below as seen from the front, side, and top.

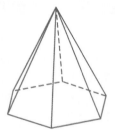

17. a. How many planes of symmetry does a regular heptagonal prism have?
    b. How many of these planes are perpendicular to its base?

18. Draw an example of one possible cross section of the Liberty Bell below with:
    a. a plane parallel to the ground.
    b. a plane perpendicular to the ground.

In **19** and **20**, redraw the figure below
19. in one-point perspective.
20. in two-point perspective.

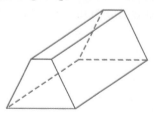

Chapter 9 Chapter Review

SKILLS Procedures used to get answers

OBJECTIVE A Draw common 3-dimensional figures. (Lessons 9-1, 9-2, 9-3, 9-6)

In 1–7, draw each figure.

1. two parallel planes intersected by a line at a 30° angle
2. a line that is perpendicular to a plane
3. a right prism with a nonconvex base
4. an oblique cone
5. two perpendicular planes
6. a sphere with a plane section through its center
7. a cube and a right pyramid whose base is one of the sides of the cube

OBJECTIVE B Give views of a figure from the top, sides, or bottom. (Lesson 9-5)

8. Give each view of the regular square pyramid below.

   a. top   b. front   c. right

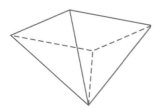

9. Give each view of the oblique cylinder below.

   a. top   b. front   c. right

front

10. Give each view of the house below.

    a. top   b. front   c. right

OBJECTIVE C Calculate surface areas and lengths in prisms, cylinders, pyramids, and cones. (Lessons 9-2, 9-3, 9-9, 9-10)

11. The figure below shows a cube with sides of length 3.

    a. Calculate HB.

    b. Calculate HD.

    c. Calculate the lateral area of the cube.

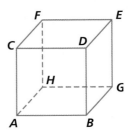

12. In a regular pentagonal pyramid, the length of each side of the base is 1.5 cm, and the slant height is 3.5 cm.

    a. Find the lateral area of the pyramid.

    b. Find the area of a lateral face.

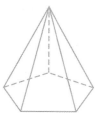

In 13 and 14, refer to the regular hexagonal prism below.

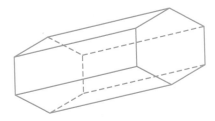

13. If the height of the prism is 4, and the length of the side of a base is 1, find the lateral area.

14. If the height of the prism is 5, and the lateral area is 32, find the surface area of the prism.

15. A right cone has a slant height of 30 cm, and its lateral area is 400 cm². Find the radius of its base.

16. Find the surface area of the oblique square prism below.

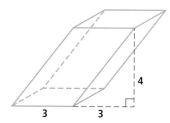

OBJECTIVE D Make and analyze perspective drawings. (Lesson 9-4)

In 17 and 18, draw the figure in perspective.

17.

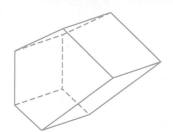

18.

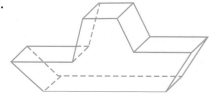

In 19 and 20, an image is given in perspective. Trace its outline, and show a vanishing point.

19.

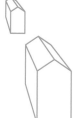

20.

PROPERTIES Principles behind the mathematics

OBJECTIVE E Make conclusions based on the Point-Line-Plane Postulate. (Lesson 9-1)

21. Three points lie on exactly one plane. What can you say about those points?

22. Line ℓ lies in plane X.
    a. How many different planes are there through ℓ?
    b. How many of these are perpendicular to X?

23. The intersection of $\overleftrightarrow{PQ}$ and plane A contains at least two points. What can you say about the relationship between the points and the plane?

24. Can two planes intersect in exactly two points? Why or why not?

OBJECTIVE F Identify parts of common 3-dimensional figures. (Lessons 9-2, 9-3)

25. Identify the requested part of the figure below.
    a. the altitude
    b. the base
    c. a lateral face

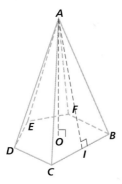

26. In a cube, how many different pairs of faces could be bases of the cube?

27. How many different pairs of faces could be bases for a regular hexagonal prism?

28. The base of a pyramid is an n-gon. How many faces does the pyramid have?

29. Draw a truncated cone.

OBJECTIVE G Distinguish 3-dimensional figures by their defining properties. (Lessons 9-2, 9-3, 9-6)

30. **True or False** No pyramid can have a pair of parallel faces.

31. **True or False** Three small circles on the same sphere can have the same radius.

32. In one polyhedron, all of the sides are parallelograms. In the other, all of them are triangles. If one is a pyramid and the other is a prism, which one is which?

33. **True or False** Any polyhedron whose faces are all rectangles must be a box.

34. A 3-dimensional figure you saw in this chapter is defined as the set of all points in space equidistant from a given point. What is this figure?

OBJECTIVE H Determine symmetry planes in 3-dimensional figures. (Lesson 9-7)

35. At least how many planes of symmetry does a box have?

36. When does a pyramid have a plane of symmetry?

37. How many planes of symmetry does a regular prism with a 12-gon base have?

In 38 and 39, tell how many symmetry planes the figure has.

38.

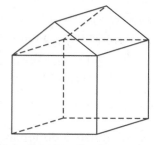

triangular prism placed on a cube

39.

a right cone placed on a right cylinde

USES Applications of mathematics in real-world situations.

OBJECTIVE I Draw plane sections of real life 3-dimensional objects. (Lesson 9-6)

In 40 and 41, refer to the picture at the right of a tuning fork. Describe where to place a plane so that its intersection with the tuning fork will be the given shape.

40. two disjoint rectangles

41. a circle

In 42 and 43, suppose someone is standing on level ground wearing the baseball cap shown below. Draw an example of one possible section of the hat with the given plane.

42. a plane perpendicular to the ground

43. a plane parallel to the ground

44. The picture below shows some sections of oranges. For one of these oranges, draw what a section would have looked like if the orange were cut in a plane that is perpendicular to the shown section.

OBJECTIVE J Apply formulas for lateral and surface area to real situations. (Lessons 9-9, 9-10)

45. A can of vegetables is a right cylinder. The radius of its base is 5 cm and its surface area is 160π cm^2. How tall is the can?

46. A paper holder for an ice cream cone is in the shape of a right cone. If its slant height is 8 cm and its radius is 2.5 cm, what is its lateral area?

47. The sides of the base of an ancient square pyramid are 80 cubits (an ancient unit). The pyramid is 40 cubits high. Find its lateral area.

48. Shayna and Regina each have a box to decorate. The dimensions of Shayna's box are twice the dimensions of Regina's. Shayna claims that the surface area of her box is twice the surface area of Regina's box. Regina disagrees. Who is right? Explain your answer.

REPRESENTATIONS Pictures, graphs, or objects that illustrate concepts

OBJECTIVE K From a net, make a surface, and vice versa. (Lesson 9-8)

49. Tell whether each figure below could be a net for a cube.

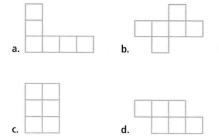

a.   b.

c.   d.

50. Draw a net for a hexagonal right pyramid.

51. Draw a net for a regular octagonal prism with height 4 and base with edge 5.

52. Name the surface that could be made from the net below.

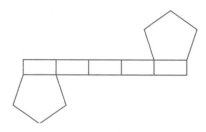

53. Draw a net for a cylinder with height 7 and base with radius 4.

OBJECTIVE L From 2-dimensional views of a figure, determine the 3-dimensional figure. (Lesson 9-5)

In 54 and 55, use the given views of the buildings.

a. How many stories tall is the building?

b. How many sections long is the building from front to back?

c. Where is the tallest part of the building located?

54.

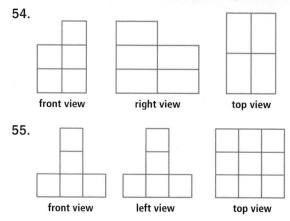

front view   right view   top view

55.

front view   left view   top view

56. Kendall saw the front and right side views of a cake, shown below. Draw two different possible top views of this cake.

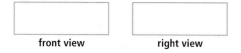

front view   right view

57. a. What common object is pictured below?

   b. Draw the side view for this object.

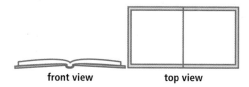

front view   top view

Chapter

10 Formulas for Volume

Contents

Formulas for Volume

10-1 Fundamental Properties of Volume

10-2 Multiplication, Area, and Volume

10-3 Volumes of Prisms and Cylinders

10-4 Volumes of Pyramids and Cones

10-5 Organizing and Remembering Formulas

10-6 The Volume of a Sphere

10-7 The Surface Area of a Sphere

The packaging of a product needs to account for the appearance, cost, strength, size and shape of the container. Both surface area and volume are important. Surface area determines how much material is necessary to create the container. Volume measures the capacity of the container, that is, how much can be placed inside the container. Sometimes the shape and size of the container together influence the quality of the final product. This is especially true in the packaging of high-quality ice cream.

One of the critical considerations for the commercial production of ice cream is keeping all of the ice cream at a uniform temperature of between –5° and 0°F.

If the ice cream at the center of the container is far from the surface, there is a greater chance it will be warmer than the ice cream close to the surface. For this reason, many companies sell their ice cream in containers that are in the shape of truncated cones, called *frustums*.

You may have noticed that frustums tend to be small, with pint or quart capacity, but not gallon or half gallon capacity. This keeps as much of the ice cream as possible close to the surface of the container, allowing the temperature to be kept uniform.

Another advantage of the frustum is that the shape does not tessellate, so there is always air surrounding the container. In addition, the bottom circle of the container is smaller than the top, so the containers can be stacked. The bottom is also indented so there is room for cold air between the containers.

These considerations ensure that the quality of the final product will be consistently high. Understanding the container's surface area, shape, and volume enables the manufacturers to provide a better product. In this chapter, you will examine ideas and problems that will help you to better understand the volumes of figures.

Lesson
10-1
Fundamental Properties of Volume

Vocabulary

volume

cubic units

unit cube

box

cube root

▶ **BIG IDEA** Formulas for the volumes of figures can be deduced from fundamental properties of volume.

Volume is a measure of how much a 3-dimensional surface will hold, its *capacity*. Volume is also the measure of the amount of material in a 3-dimensional solid. An empty swimming pool, for instance, can be thought of as a 3-dimensional surface. Even though the pool has no top and even when it is not filled with water, it still has volume, which is the amount of water it would hold if filled. That same pool, when filled, would represent a 3-dimensional solid and has the same volume as the capacity of the empty pool.

Volume is usually measured in **cubic units.** The cube at the right has edges of length 1 unit, so the area of each face is 1 square unit. Because there are six faces, the surface area of the cube is 6 square units. Its volume is quite different from its surface area. The volume of the cube is 1 cubic unit or 1 unit³. For this reason it is called the **unit cube.**

1 cubic unit

A sugar cube has a volume of about 1 cubic centimeter, or 1 cm³. An average refrigerator has between 18 and 26 cubic feet of usable storage space.

A **box** is the everyday name for a right rectangular basic prism. We begin our study of volume with this shape.

▶ **READING MATH**

In medicine, a cubic centimeter is often called a *cc* (see see).

Example 1

What is the volume of a shoe box with base $12\frac{1}{2}$" by 6" and height 4"?

Solution 1 Count. The volume is the number of cubes with edge 1" that will fill the box. Because the height is 4", there will be 4 layers of cubes. In each layer, there will be 12 · 6 whole cubes and $\frac{1}{2}$ · 6 half cubes. So each layer will have 72 + 3 or 75 cubic inches of volume. The volume is 4 · 75, or 300 cubic inches.

Solution 2 Multiply. The number of cubes of the base is $12\frac{1}{2}$ · 6, or 75. Multiply this by the number of layers, 4. The volume is 4 · $12\frac{1}{2}$ · 6, or 300. So the volume is 300 in³.

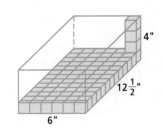

QY1

▶ QY1

What is the surface
area of the shoe box of
Example 1?

Volume Postulate

By counting, you can tell that the volume of the box in Example 1 is
the product of its three dimensions. It is obvious that this is true for
simple numbers, but will it also be true for numbers like $\sqrt{2}$ or π? That
the volume of a box is the product of its dimensions for any positive
real number dimensions is one of the four fundamental properties
of volume we assume. We call this group of assumptions the Volume
Postulate. Notice how these compare to the fundamental properties
of area assumed in the Area Postulate in Lesson 8-2.

Volume Postulate

a. *Uniqueness Property*
Given a unit cube, every polyhedral region
has a unique volume.

b. *Congruence Property*
Congruent figures have the same volume.

c. *Additive Property*
The volume of the union of two nonoverlapping solids is the
sum of the volumes of the solids.

d. *Box Formula* The volume of a box with dimensions ℓ, w, and h
is ℓwh. ($V = \ell wh$)

In symbols, if Volume(S) refers to the volume of a solid S, and A and
B are nonoverlapping solids, then the Additive Property states:

$$\text{Volume}(A \cup B) = \text{Volume}(A) + \text{Volume}(B).$$

Using this property, volumes of various figures can be calculated.

GUIDED

Example 2

Solids I and II are each the union of 6 unit cubes. Give the volume and
surface area of each solid.

Solution The volume of each solid is the same because it is the
sum of the volumes of the six congruent cubes.

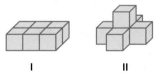

I   II

$$\text{Volume(I)} = \text{Volume(II)} = \underline{\ ?\ } \text{ cubic units}$$

Solid I can be viewed as a prism in which each base is a __?__-by-__?__
rectangle. Its lateral surface consists of __?__ unit squares, and each base
has area __?__ square units.

(continued on next page)

$$\text{S.A.} = \text{L.A.} + 2 \cdot B$$
$$= \underline{\ ?\ } + 2 \cdot \underline{\ ?\ }$$
$$= \underline{\ ?\ }.$$

The surface area of solid I is __?__ square units.

Solid II is not a prism. Find the surface area by counting the squares that cover the solid. Putting them in groups helps keep track of them. There are __?__ bottom faces, __?__ top faces, and __?__ faces each for right, left, front, and back. The surface area of solid II is __?__ square units.

Contrasting Volume and Surface Area

Guided Example 2 illustrates that solids may have equal volumes but unequal surface areas. However, many people judge the volume of a container by its surface area. They think, "If it looks bigger, then it holds more."

Pictured at the right are a salad dressing bottle and a mayonnaise container. The two containers hold the same amount; they have the same volume (16 fl oz). The salad dressing bottle has greater surface area, which may give the impression that it has greater volume and holds more.

Liquid Volume

In the first paragraph of the lesson, we indicated that volume is *almost always* in cubic units. In both the metric system and the U.S. customary system, special units are used to measure the volumes of liquids. These units are the *liter* and the *gallon*. These liquid units can be converted to cubic inches or cubic centimeters using the conversion equations in the first row of the table below. Packaged liquids you might buy in a store, such as juices and yogurt, are often measured in smaller units, such as the milliliter, quart, or pint. Some conversion facts between liquid units are in the second row of the table.

Table of Liquid Measures	
U.S. Customary System	**Metric System**
1 liquid gallon (gal) = 231 cubic inches	1 liter (L) = 1000 cubic centimeters
1 liquid quart (qt) = $\frac{1}{4}$ liquid gallon = 32 fluid ounces (fl oz)	1 milliliter (mL) = $\frac{1}{1000}$ liter
1 liquid pint (pt) = 16 fluid ounces (fl oz)	

STOP QY2

▸ QY2

How many pints are in a gallon?

These units can be converted to each other using methods you have learned in previous courses.

Example 3

The salad dressing container pictured earlier contains 16 fl oz of dressing. How many cubic inches is this?

Solution $16 \text{ fl oz} = 16 \text{ fl oz} \cdot \dfrac{1 \text{ gal}}{128 \text{ fl oz}} \cdot \dfrac{231 \text{ in}^3}{1 \text{ gal}}$

$= 28.875 \text{ in}^3$

16 fl oz of salad dressing occupies 28.875 cubic inches of volume.

The Volume of a Cube

The formula for the volume of a cube is a special case of the formula for the volume of a box, in which each dimension is the same.

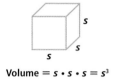

Volume $= s \cdot s \cdot s = s^3$

Cube Volume Formula

The volume V of a cube with edge s is s^3: $V = s^3$.

Because of the cube formula, we call s^3 the "cube of s" or "s cubed." If a cube has volume 8, its edge will satisfy $s^3 = 8$, so $s = 2$. We say that 2 is the **cube root** of 8, and write $2 = \sqrt[3]{8}$. In general, $\sqrt[3]{y}$ is the edge of a cube whose volume is y. It is also the case that, when $y \geq 0$, $\sqrt[3]{y} = y^{\frac{1}{3}}$.

Definition of Cube Root

x is the **cube root** of y, written $y = \sqrt[3]{y}$ or $x = y^{\frac{1}{3}}$, if and only if $x^3 = y$.

Calculators differ in the keys they employ to calculate a cube root. You should learn the sequence of keys on your calculator that will display the cube root of a number.

Example 4

The previous example showed that the volume of the salad dressing container was 28.875 cubic inches. If this quantity of dressing were sold in a cubic container, what would be the length of an edge of that cube?

(continued on next page)

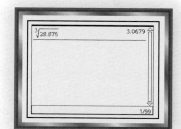

Solution Let s be the length of an edge. Because $s^3 = 28.875$, s is exactly $\sqrt[3]{28.875}$. A calculator shows $\sqrt[3]{28.875} \approx 3.068$. The length of an edge is approximately 3.07 inches.

Check Is $(3.07)^3 \approx 28.875$? Yes, $(3.07)^3 \approx 28.934443$, which is reasonably close.

Questions

COVERING THE IDEAS

1. **a.** If two solids have the same volume, must they have the same surface area?

   b. If two solids have the same surface area, must they have the same volume?

2. **a.** What does volume measure for a 3-dimensional solid?

   b. What does volume measure for a 3-dimensional surface?

3. Explain the difference between surface area and volume.

4. Find the volume of a box with dimensions 6", 8", and 1".

   a. in cubic inches   **b.** in cubic feet

5. An open shopping bag's base is 12" by 7" and its height is 17".

   a. Find the volume of the bag.

   b. Find the surface area of the bag.

6. A flat top is put on the shopping bag in Question 5.

   a. By how much does the top increase the surface area?

   b. By how much does the top increase the volume?

7. Explain in your own words what the Additive Property in the Volume Postulate says.

8. Name four commonly used units of volume.

9. What is the volume of a pint of milk in cubic inches?

10. Give a formula for the volume of a cube.

11. **Multiple Choice** Suppose $p = \sqrt[3]{q}$. Then which one of the following is true?

    A If p is the length of an edge of a cube, then q is the volume of the cube.

    B If p is the volume of the cube, then q is the length of an edge of the cube.

    C If p is the surface area of the cube, then q is its volume.

    D If p is the volume of the cube, then q is its surface area.

12. Use your calculator to complete this table of cube roots at the right. Round answers to the nearest hundredth.

n	$\sqrt[3]{n}$
1	?
3	?
5	?
7	?
9	?
11	?

13. A cube has volume 60 cm³. What is the length of an edge
   a. exactly? b. to the nearest hundredth?

APPLYING THE MATHEMATICS

14. Which has a greater volume, a 5-gallon bucket or a cubic foot of water, and by how much?

15. Three noncongruent cardboard boxes each have volume 72 cubic inches. What might be the dimensions of the boxes?

16. Celeste needs to buy an air conditioner for her apartment. The floor plan for her apartment is shown at the right. In the plan, the first dimension of the room is the width measured left to right and the second dimension is the length. The table at the right below shows the recommended air conditioner size by volume of the space to be cooled.

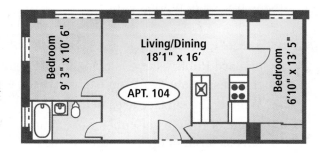

   a. If you consider the floor of the apartment to be one big rectangle, what are the dimensions of that rectangle?
   b. Assume the ceiling is 9 feet high throughout. What is the volume of the apartment?
   c. How large of an air conditioner should Celeste buy?

Maximum Apartment Size (ft³)	Recommended Air Conditioner Size (Btu)
3,600	10,000
4,600	12,000
6,000	15,000
8,000	18,000
11,000	24,000

17. Some people use the formula $V = Bh$ for the volume of a box. (You may wish to look at the picture in Example 1 when you answer this question.)
   a. What is B in this formula?   b. Why does this formula work?

18. How many cubic centimeters are there in 3.2 liters?

19. 1 in. = 2.54 cm exactly.
   a. What is the volume, in cubic centimeters, of a cube with edge length 1 in.?
   b. **Fill in the Blank** To the nearest hundredth, 1 cubic inch ≈ __?__ cubic centimeters.

20. People who give blood give about a pint at a time. How many cc's of blood is this?

21. Drinking water is vital for human existence. Some studies recommend that people should consume about 2 liters of water per day. If you followed these guidelines, how long would it take you to drink all of the water in a 5000-gallon tanker truck? (1 gallon ≈ 3.79 liters)

22. a. 1 yard = __?__ feet   b. 1 square yard = __?__ square feet
    c. 1 cubic yard = __?__ cubic feet

23. A farmer wants to build a bin to hold chicken feed. He wants the bin to hold 400 bushels of feed (1 bushel ≈ 1.244 cubic feet). If he builds the bin in the shape of a cube, how long should he make one edge of the cube?

REVIEW

24. A standard Rubik's cube is made up of 27 smaller cubes whose sides measure about 1.9 cm each. (**Lesson 9-9**)
    a. What is the surface area of a standard Rubik's cube?
    b. If one of the corner cubes were removed, what would be the surface area of the resulting shape?

25. Nolan had a log in the shape of a right cylinder. He leans the log diagonally against a wall, and claims that its shape is now an oblique cylinder. Is Nolan correct? Explain your answer. (**Lesson 9-2**)

26. **True or False** If plane A is perpendicular to planes B and C, then planes B and C are parallel. (**Lesson 9-1**)

27. In a plane, suppose regions A and B are disjoint. Let C be the union of A and B. Suppose that the area of C is 42, and that the area of B is 12 more than the area of A. Find the areas of A and B. (**Lesson 8-2**)

28. Expand the following products. (**Previous Course**)
    a. $(a + 2b)(3a - b)$ b. $(x^2 + 2)(x + 3)$ c. $(t^2 + p^2)(t^2 - p^2)$

EXPLORATION

29. You have three boxes. One is 2 in. by 6 in. by 8 in. Another is a cube with edge 4 in. The third box is 1 in. by 18 in. by 8 in. What single kind of object might best fit into each box?

30. Two polyhedra made up of six cubes each are shown in Example 2 of this lesson. Describe two noncongruent polyhedra that are each the union of eight cubes so that
    a. one of the polyhedra has the least surface area possible.
    b. the other polyhedron has the greatest surface area possible.
    Find the surface area of each polyhedron.

Feliks Zemdegs won the 3-by-3-by-3 event at the Rubik's Cube world championship in Las Vegas in July 2013.

QY ANSWERS

1. 298 in²

2. 8

Lesson
10-2 Multiplication, Area, and Volume

▶ **BIG IDEA** Multiplication of three polynomials can be represented by volumes of boxes.

To obtain the area of a rectangle, you merely have to multiply its length by its width. Any positive numbers can be the length and the width. Thus, the area of a rectangle is a *model* for the multiplication of two positive numbers. This enables multiplication to be pictured. For instance, the diagram below shows that
$3.5 \cdot 2.3 = (3 + 0.5)(2 + 0.3) = 6 + 0.9 + 1 + 0.15 =$
$3 \cdot 2 + 3 \cdot 0.3 + 0.5 \cdot 2 + 0.5 \cdot 0.3.$

Mental Math

a. Two squares have the same perimeter. Are they necessarily congruent?

b. Two squares have the same area. Are they necessarily congruent?

c. Two rectangles have the same area. Are they necessarily congruent?

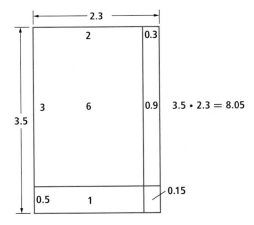

In your study of algebra, you may have seen that the multiplication of two polynomials can also be pictured by area. For example, in algebra you may have seen this picture of $(a + b)(c + d)$.

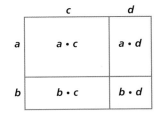

Area of largest rectangle
$= (a + b)(c + d)$
$= ac + ad + bc + bd$

Example 1

a. What binomial multiplication is pictured at the right?

b. Multiply the binomials.

c. Check your result by letting $x = 3$.

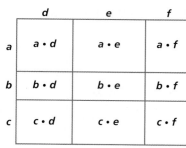

	$3x$	4
$2x$	$6x^2$	$8x$
7	$21x$	28

Solution

a. The height of the largest rectangle is $2x + 7$. Its width is $3x + 4$. The area of the rectangle is the product $(2x + 7)(3x + 4)$.

b. Add the areas of the 4 smaller rectangles.
$6x^2 + 8x + 21x + 28 = 6x^2 + 29x + 28$
Thus, $(2x + 7)(3x + 4) = 6x^2 + 29x + 28$.

c. Let $x = 3$. Then $(2x + 7)(3x + 4) = (13)(13) = 169$.
$6x^2 + 29x + 28 = 6(9) + 29(3) + 28 = 54 + 87 + 28 = 169$
Because both answers match, the answer is correct.

The multiplication of two polynomials with more terms can also be pictured with an area diagram. Here is a picture of the multiplication of two trinomials $(a + b + c)(d + e + f)$.

	d	e	f
a	$a \cdot d$	$a \cdot e$	$a \cdot f$
b	$b \cdot d$	$b \cdot e$	$b \cdot f$
c	$c \cdot d$	$c \cdot e$	$c \cdot f$

$$A = (a + b + c)(d + e + f)$$
$$= ad + ae + af + bd + be + bf + cd + ce + cf$$

Picturing Multiplication with Volume

The box formula $V = \ell wh$ involves multiplication of *three* numbers. So the volume of a box can model the multiplication of three polynomials. The biggest box below has dimensions $a + b$, $c + d$, and $e + f$, so its volume is the product of these three binomials. But its volume also is the sum of the volumes of the eight smaller boxes.

Notice that the product of the three binomials consists of all possible products in which one factor is taken from the first binomial, one from the second, and one from the third.

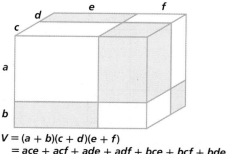

$$V = (a + b)(c + d)(e + f)$$
$$= ace + acf + ade + adf + bce + bcf + bde + bdf$$

Activity

Draw each of the 8 smaller boxes that make up the above figure, and label the length, width, and height of each.

Changing Dimensions of a Box

If you multiply each side of a box by a number, the change in volume is easy to find.

Example 2

A box has dimensions ℓ, w, and h. If the length is multiplied by 2, the width is multiplied by 3, and the height is multiplied by 4, how is the volume of the box changed?

Solution Draw a picture of the situation.

The new volume is $(2\ell)(3w)(4h)$, which is $24\ell wh$, or 24 times the original volume.

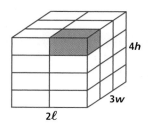

> ▶ **QY**
>
> In Example 2, if you multiplied the length by 3 and left the other dimensions alone, how would the volume of the box change?

🛑 **QY**

Example 3 demonstrates how complicated the change in volume is if you *add* to the dimensions of a figure.

Example 3

The length of a box is increased by 3, the width is increased by 5, and the height is increased by 6. How is the volume of the box changed?

Solution Let the original dimensions of the box be x, y, and z. So the original volume is xyz. The new dimensions are x + 3, y + 5, and z + 6.

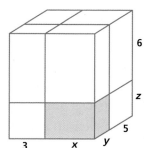

The new volume is $(x + 3)(y + 5)(z + 6)$.

First multiply two of the binomials, and then multiply this product by the third binomial. Use the Distributive Property.

$$(x + 3)(y + 5)(z + 6) = (x + 3)(yz + 5z + 6y + 30)$$
$$= x(yz + 5z + 6y + 30) + 3(yz + 5z + 6y + 30)$$
$$= xyz + 5xz + 6xy + 30x + 3yz + 15z + 18y + 90$$

To find how much the volume increased, subtract the original volume (xyz) from the new volume.

The volume increased by $5xz + 6xy + 30x + 3yz + 15z + 18y + 90$.

(continued on next page)

Check Your answer should be valid for any x, y and z. Try it letting $x = 2$, $y = 4$ and $z = 10$. With these values:

> Original Volume $= (2)(4)(10) = 80$
> New Volume $= (2 + 3)(4 + 5)(10 + 6) = 720$
> Increase in Volume $= 720 - 80 = 640$

Now substitute $x = 2$, $y = 4$, and $z = 10$ into your answer, hoping to get 640.

> $5xz + 6xy + 30x + 3yz + 15z + 18y + 90$
> $= 5(2)(10) + 6(2)(4) + 30(2) + 3(4)(10) + 15(10) + 18(4) + 90$
> $= 640$

It checks.

Questions

COVERING THE IDEAS

1. Use the area model for multiplication to write $4.7 \cdot 5.2$ as the sum of four products.

2. **a.** What multiplication is pictured at the right?
   b. Multiply the binomials.
   c. Check your result by letting $x = 5$.

3. **a.** What multiplication is pictured at the right?
   b. Multiply these trinomials.
   c. Check your answer using $a = 2$, $b = 3$, $c = 4$, and $d = 10$.

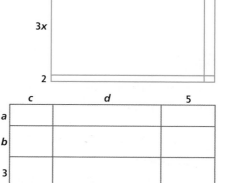

4. **a.** What multiplication is pictured at the right?
   b. Multiply the three binomials.
   c. Check your answer by letting $x = 7$, $y = 11$, and $z = 5$.

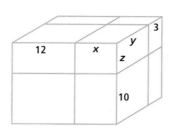

In 5–7, a box has dimensions ℓ, w, and h. For each problem, draw a picture and describe how the volume is changed if the box's dimensions are changed in the indicated way.

5. Its length is multiplied by 3 and all other dimensions stay the same.

6. Its length, width, and height are all multiplied by 3.

7. 3 is added to the length, and all other dimensions stay the same.

APPLYING THE MATHEMATICS

In 8 and 9, use this information. Many bricks used for building houses are 8 inches long, 4 inches wide, and $2\frac{1}{4}$ inches deep.

8. An architect wants to use a different brick that is 1.5 times as long and 1.2 times as wide, but leave the depth alone. What is the volume of the new brick?

9. The architect wants to make the original brick x times as long and y times as wide, leaving the depth alone.
   a. Explain how this changes the volume of the original brick.
   b. How might this affect the cost of the brick?
   c. How might this affect the cost of the house the architect is building?

10. The area of a rectangle is $x^2 + 4x + 3$. The length of the rectangle is $x + 1$.
    a. Find the width of the rectangle.
    b. Draw a picture of this.

11. Consider the rectangle at the right.
    a. Write an equation with x, y, and z expressing that the area of the large rectangle is equal to the sum of the areas of the two smaller rectangles.
    b. What property of real numbers does this illustrate?

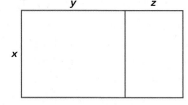

12. Expand $(x + 5)^3$:
    a. by multiplying $(x + 5)(x + 5)(x + 5)$.
    b. by finding the volume of an appropriate cube and by finding the volume of the eight boxes that make up the cube.

13. Mr. Moon said, "Here is a box with dimensions x, y, and z. If I replace it with a box with dimensions $\sqrt{x}$, $\sqrt{y}$, and $\sqrt{z}$, what will the new volume be?" Luna replied, "the square root of the volume of the original box." Is Luna right?

14. Start with a cube with a 1 cm long edge. Increase its edge length by 0.2 cm. Do this again and again to see how each addition of 0.2 cm affects the volume of the cube.
    a. Complete the table at the right.
    b. Graph the six pairs (x, y).

x = length of edge in centimeters	y = volume in cubic centimeters
1.0	?
1.2	?
1.4	?
1.6	?
1.8	?
2.0	?

15. A box company increased the length of a box by 3 inches and kept the other dimensions the same. This change increased the volume by 20%. What was the original length?

16. One dimension of a box is increased by 7 units. Another dimension is decreased by 4 units. The third dimension remains the same. Is it possible for the resulting box to have the same volume as the original box? Why or why not?

REVIEW

17. The figure at the right shows the front view of a 3-dimensional solid. Each of the squares is 20 cm high by 20 cm wide. The solid is 30 cm deep. Find its volume. **(Lessons 10-1, 9-5)**

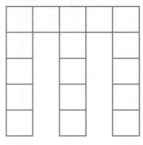

18. **a.** Draw a net for the lateral surface of a regular octagonal prism with height 7 and a base with sides of length 4. **(Lesson 9-8)**

    b. What is the lateral area of the figure in Part a? **(Lesson 9-9)**

19. The top of the carousel at the right is a cylinder 3'7" high. The diameter of the cylinder is 24'. To the nearest square foot, what is its lateral area? **(Lesson 9-9)**

20. **True or False** When you look in a mirror, the image you see is oppositely congruent to you. **(Lesson 9-7)**

21. State the formula for the area of a parallelogram and describe how this formula can be derived from the area of a trapezoid. **(Lesson 8-5)**

22. Give an example of a transformation that preserves angle measure, betweenness, collinearity, but not distance. **(Lesson 5-1)**

EXPLORATION

23. A box can be studied by placing it on a 3-dimensional coordinate system.

    a. Coordinates of three vertices are missing on the picture at the right. Find the coordinates.

    b. Calculate the volume of this box.

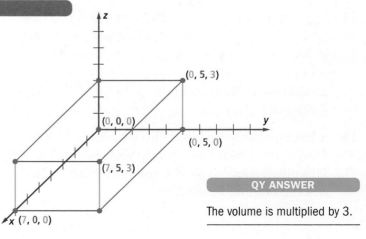

QY ANSWER

The volume is multiplied by 3.

Lesson
10-3 Volumes of Prisms and Cylinders

> ▶ **BIG IDEA** From the Box Volume Formula and Cavalieri's Principle, volume formulas for any cylindrical solids can be deduced.

The standard measure of crude oil in the United States is the barrel, which has a capacity equal to 42 gallons. In 2006, total U.S. consumption averaged about 20.7 million barrels every day. To give a dramatic presentation of this fact to the public, many news agencies translate this to how much oil covers a football field. If 20.7 million barrels of oil were poured into a cylindrical tank whose base is the size of a football field, how high would the tank reach? To answer this question, we look at the way that volume is found for any cylindrical surface. The diagram at the right shows such a tank placed next to the Washington Monument, which is 555 feet high.

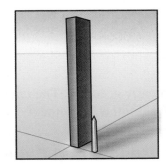

Volumes of Right Prisms and Cylinders

Consider a cylindrical surface whose base is a prism or cylinder having an area of B. Then we think of B unit squares covering the region. We can think this way even if B is not an integer.

area = **B** square units

If a prism with this base has height 1 unit, the prism contains B unit cubes, and so the volume of the prism is B cubic units. This is pictured in the middle figure at the right. The bottom figure at the right is a prism with this base and height h. That prism has h times the volume of the middle prism, and so its volume is Bh. This argument shows that, if a right prism or cylinder has height h and a base with area B, then its volume is Bh.

volume = **B** cubic units

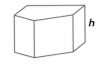
volume = **Bh** cubic units

Now you can determine how tall a prism would have to be to contain the amount of oil consumed every day in the United States, if the base of the prism is the size of a football field.

Example 1

An American football field is 120 yards long (with the end zones) and 50 yards wide. If a rectangular prism were built on a football field that would contain the amount of oil consumed in the United States in a day in 2006, how high would the tank be? Make a guess before you go on.

Solution It helps to draw a picture, like the one at the right. The volume V of the tank is found using $V = Bh$, where

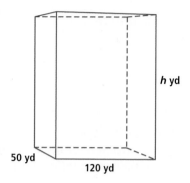

$$B = 120 \text{ yd} \cdot 50 \text{ yd}$$
$$= 360 \text{ ft} \cdot 150 \text{ ft}$$
$$= 54{,}000 \text{ ft}^2.$$
So $V = 54{,}000h \text{ ft}^3.$

Now convert the usage in the United States from barrels to cubic feet.

$$20.7 \cdot 10^6 \text{ barrels} \cdot \frac{42 \text{ gal}}{1 \text{ barrel}} \cdot \frac{1 \text{ ft}^3}{7.48 \text{ gal}} \approx 116{,}229{,}947 \text{ ft}^3$$

Equate the two expressions for V and solve for h.

$$116{,}229{,}947 = 54{,}000h$$
$$h \approx 2152 \text{ ft}$$

The picture on page 609 is nearly accurate. The tank would be almost 4 times the height of the Washington Monument.

Volumes of Oblique Prisms and Cylinders

Now suppose you have an *oblique* prism or cylinder. Recall that in these figures, the lateral edges are not perpendicular to the planes of the bases. Pictured at the right are a right prism and an oblique prism with congruent bases and equal heights.

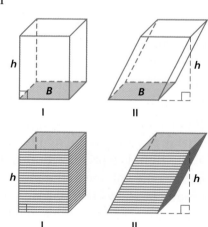

Imagine Prism I to be made up of a stack of thin slices like congruent sheets of paper. Shift the slices of the first stack until it takes the form of Prism II.

Notice that the height, area of the base, and number of slices are the same in Prism I and Prism II. Thus, it makes sense that

$$\text{Volume (Prism II)} = \text{Volume (Prism I)}.$$

Or, because they have equal heights and bases,

$$\text{Volume (Prism II)} = Bh.$$

Cavalieri's Principle

The key ideas of this argument are: (1) the prisms have their bases in the same planes; (2) each slice is parallel to the bases; and (3) the slices in each prism have the same area. The conclusion is that these solids have the same volume. The first individuals to use these ideas to obtain volumes seem to have been the Chinese mathematician Zu Chongzhi (429–500) and his son Zu Geng. However, in the West, Bonaventura Cavalieri (1598–1647), an Italian mathematician, first realized the importance of this principle, and in the West it is named after him. *Cavalieri's Principle* is the fifth and last part of the assumed statements about volume.

Volume Postulate

e. *Cavalieri's Principle*
Let I and II be two solids included between parallel planes. If every plane P parallel to the given planes intersects I and II in sections with the same area, then Volume(I) = Volume(II).

At the right, plane P is parallel to the planes Q and R containing the bases, and all three solids have bases with area B. Because plane sections X, Y, and Z are translation images of the bases (this is how prisms and cylinders are defined), they also have area B. Thus, the conditions for Cavalieri's Principle are satisfied. These solids have the same volume. But we know the volume of figure II, the box.

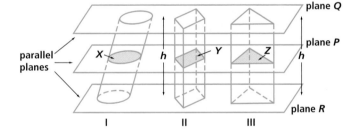

$$\text{Volume(II)} = \ell \cdot w \cdot h$$
$$= B \cdot h.$$

Thus, using Cavalieri's Principle,

$$\text{Volume(I)} = B \cdot h \text{ and}$$
$$\text{Volume(III)} = B \cdot h.$$

This proves the following theorem for *all* cylinders and prisms.

Prism-Cylinder Volume Formula

The volume V of any prism or cylinder is the product of its height h and the area B of its base.

$$V = Bh$$

Example 2

A cylindrical duct is used to expel hot air from a basement clothes dryer. The length of the duct is 12 ft. The duct goes down at an angle so that the bottom ring of the duct in the basement is 10 feet lower than the top. The duct diameter is 14 inches. Find the volume of the air that is in the duct.

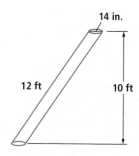

14 in.

12 ft

10 ft

Solution The important idea is that the length of the duct makes no difference in finding the volume of air. Because $V = Bh$, all you need is the area of the base and the height. Change the height to inches.

$$V = Bh$$
$$B = \pi \cdot r^2 = \pi \cdot \underline{\ ?\ }^2 = \underline{\ ?\ } \text{ in}^2$$
$$\text{Thus } V = \underline{\ ?\ } \cdot 120 \approx 18{,}473 \text{ in}^3.$$

There are about 18,473 cubic inches or about 10.7 cubic feet of air in the duct.

Example 3

The pentagonal prism shown at the right was constructed using pentagon *ABCDE* and vector $\overrightarrow{EF}$, where *F* is in plane *Y*. Suppose you move *FGHIJ* in that plane. What will happen to the volume of the prism? Why?

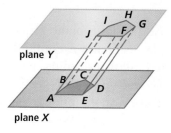

plane Y

plane X

Solution The height of the prism is the distance between the two planes. So, no matter where the point F is placed on plane Y, the height of the prism is always the same. The base area is unaffected by moving FGHIJ. Thus, because the height and the area of the base remain constant, the volume of the prism remains constant as well.

Questions

1. A cubic foot of liquid is how many gallons?

2. The United States Strategic Petroleum Reserve was 687.9 million barrels as of July 2006.

   a. If you filled up a football field with this many barrels of oil, how high would the prism be?

   b. Using the average daily consumption of the United States in 2006, calculate how many days worth of oil was in the Reserve.

3. **Multiple Choice** In this lesson, a stack of paper is used to illustrate all but which one of the following?

   A Cavalieri's Principle

   B that an oblique prism and a right prism can have the same volume

   C that the volume of an oblique prism is *Bh*

   D that a cylinder and a prism have the same volume formula

In 4–9, find the volume of each solid.

4.

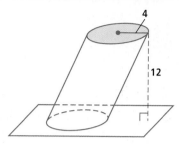

5.

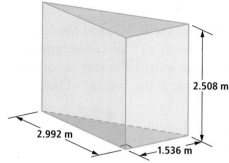

6. a regular hexagonal prism whose base has edge 5 meters, and whose height is 20 meters

7. the oblique prism with rectangular bases drawn at the right

8. a right rectangular prism whose base is 3 feet by 7 feet, and whose height is 10 feet

9. a sewer pipe 100 feet long with a radius of 24 inches.

10. Cavalieri's Principle was discovered by mathematicians of what two nationalities?

11. State Cavalieri's Principle.

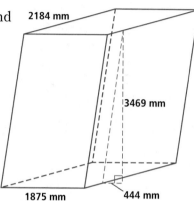

APPLYING THE MATHEMATICS

12. Suppose the Roman arch below is made of solid concrete. The bases of the columns are squares and the arch is a semicircle. How much concrete was used?

13. In the Georgia Aquarium, the Ocean Voyager display contains 6 million gallons of water. The viewing window is made from acrylic and is 61 ft wide, 24 ft high and 2 ft thick. A recent price for acrylic is about 8 cents per cubic inch. How much would that sheet of acrylic cost?

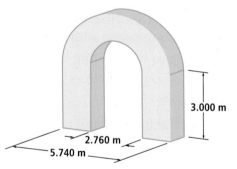

14. If a cylinder has a height h and base with radius r, find a formula for its volume in terms of h and r.

15. The volume of an oblique prism is 42 cubic meters. Its height is 7 meters. Find the area of its base.

16. A fish tank for jellyfish is called a *Kreisel tank,* from the German word for a spinning top. The spinning of the cylindrical tank forces the jellyfish to stay suspended and not stick to the sides. What is the volume of the Kreisel tank at the right?

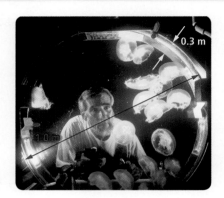

17. Suppose you double the height and radius of a cylinder. What is the relationship between the volumes of the smaller and bigger cylinders?

18. A milliliter of water has a mass of 1 gram and occupies 1 cm^3 of space. What mass of water (to the nearest gram) will fill a cylindrical can that is 15 cm high and has radius 3 cm?

REVIEW

19. Model $(a + 4)(b + c)$ with the area of a rectangle and compute the product. (**Lesson 10-2**)

20. A box with dimensions 1 meter, x meters, and y meters has volume 21 cubic meters and surface area 62 square meters. Find x and y. (**Lessons 10-1, 9-9**)

21. The figure at the right shows two intersecting planes X and Y. Describe where a third plane Z could be placed so its intersection with this figure will look like (**Lesson 9-1**)
    a. two intersecting lines.
    b. two parallel lines.
    c. one line.

22. In the figure at the right, $\overline{AC}$ is a diameter in the circle, $AB = 9$, $AC = 12$. Find the area of the part of the circle that is outside of $\triangle ABC$. (**Lessons 8-9, 8-6, 6-3**)

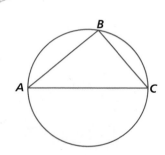

23. Each of the following is the area formula for a certain type of quadrilateral. Name the quadrilateral that goes with each formula. (**Lesson 8-5**)
    a. $A = s^2$
    b. $A = hb$
    c. $A = \frac{1}{2}h(b_1 + b_2)$
    d. $A = \ell w$

24. Solve $A = \pi r^2 h$ for r where $r > 0$. (**Previous Course**)

EXPLORATION

25. Find and describe one other mathematical contribution made by Zu Chongzhi, and one other mathematical contribution made by Bonaventura Cavalieri.

Lesson 10-4

Volumes of Pyramids and Cones

▶ **BIG IDEA** By splitting a prism into three pyramids and using Cavalieri's Principle, volume formulas for any conic solids can be deduced.

Recall that the area formula for a right triangle was found by splitting a rectangle into two parts. It may come as a surprise to you that the volume formula for a pyramid can be found by splitting a prism into three parts. Each part will be a triangular pyramid. Such a pyramid has four vertices, and any one of them could be its apex. So, if the vertices are W, X, Y, and Z, and we want Y to be the apex, we will identify the pyramid as Y-XWZ. The apex comes first. The other three vertices can be in any order.

Mental Math

a. A circle has a radius of 5 meters. Give an exact expression for its area.

b. A circle has a circumference of 10π meters. Give an exact expression for its area.

c. A circle has an area of 100π square meters. Give an exact expression for its radius.

Activity 1

MATERIALS DGS-3D

Step 1 Construct and label △ADC in the base plane, as shown below.

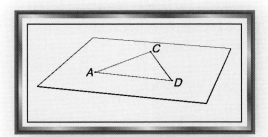

Figure 1

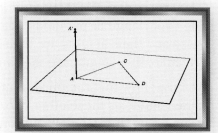

Figure 2

Step 2 Construct a point A' above the base plane and construct the vector $\overrightarrow{AA'}$.

Step 3 Translate △ADC using the vector $\overrightarrow{AA'}$. Label the top of the prism △$A'D'C'$.

The region between △ADC and △$A'D'C'$ is a triangular prism. The task is to show that you can fill this entire triangular region with 3 pyramids.

(continued on next page)

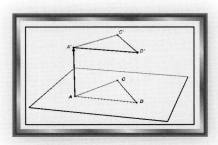

Figure 3

Step 4 Construct the triangular pyramid C'-ADC and color this surface green.

Step 5 Construct the triangular pyramid A-$A'D'C'$ and color this surface blue.

Step 6 Construct $\triangle C'DD'$ and then triangular pyramid A-$C'DD'$. Color this red.

Step 7 Hide $\triangle ADC$, $\triangle A'D'C''$, and $\triangle C'DD'$ so that your diagram looks like the one at the right.

Rotate your final figure so that you can see the three triangular pyramids that make up the original triangular prism. Look at it from various viewing angles to try to determine the relation between these three pyramids and the original prism.

Step 8 Save your final figure as it will be needed again later in the lesson.

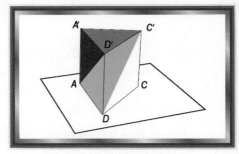

Figure 4

Creating a Prism from Pyramids

Consider the triangular pyramid C'-ADC from Figure 4 in Activity 1. Let $B = \text{Area}(\triangle ADC)$ and let $h = $ height of the pyramid. We want to compute the volume of pyramid C'-ADC.

From Activity 1, you can see that three triangular pyramids can be put together to form a triangular prism. The two bases of this prism ($\triangle ADC$ and $\triangle A'D'C'$) are congruent because they were formed by translating the one triangular base to the other. The height of this prism is the same as the height h of the triangular pyramid C'-ADC. The two triangular pyramids C'-ADC and A-$A'D'C'$ have congruent bases and their heights are the same, namely h. By Cavalieri's Principle, they must have the same volume.

We now show that the third triangular pyramid A-$DC'D'$ has the same volume as these other two. To do so, look at the face $D'C'CD$ of the prism. It is a parallelogram with diagonal $\overline{C'D}$. This means $\triangle D'C'D \cong \triangle CDC'$.

The pyramids A-$DC'D'$ and A-$C'DC$ have the same volume because the height of each is the perpendicular distance from point A to the plane of parallelogram $D'C'CD$, and the bases are congruent triangles. But pyramid A-$C'DC$ is the same pyramid as C'-ADC that was created in the Activity (only the order of the vertices is changed). So by the Transitive Property of Equality, the three pyramids have the same volume. Therefore, each pyramid has a volume that is one-third the volume of the prism.

Consequently, the volume of pyramid C'-$ADC = \frac{1}{3}Bh$.

Activity 2

MATERIALS DGS-3D

Confirm that the volume of pyramid C'-$ADC = \frac{1}{3}Bh$.

Step 1 Start with Figure 4 from Activity 1.

Step 2 Have the DGS compute the volume of each of the three triangular pyramids and the volume of the triangular prism.

Your results in Activity 2 should numerically verify that the volume of any triangular pyramid is one-third the volume of the triangular prism with a congruent base and the same height.

Volumes of Any Pyramids

Cavalieri's Principle can be used to show that the volume of any pyramid or cone is equal to the volume of a particular pyramid of the same height with a triangular base of equal area. To do this, it is necessary to prove that all the cross sections have the same area. For instance, if a cone has base area 8π units^2 and height h units, a triangular pyramid with the same volume can be constructed if its base has area 8π units^2 and its height is h units. Furthermore, these need not be regular pyramids or right cones.

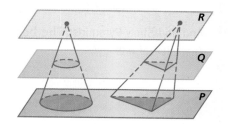

This argument proves the following theorem.

Pyramid-Cone Volume Formula

The volume V of any pyramid or cone equals one-third the product of its height h and its base area B.

$$V = \frac{1}{3}Bh$$

Thus, a pyramid or a cone that has the same base area and same height as a prism or cylinder has $\frac{1}{3}$ the volume of that prism or cylinder.

Often the volume of a pyramid is easier to calculate than its surface area.

 QY

> ▶ **QY**
>
> Find the volume of a cone with height 10 cm and a base of radius 3 cm.

GUIDED

Example 1

The four glass regular square pyramids of the Muttart Conservatory in Edmonton, Alberta, Canada come in two sizes. The Temperate and Tropical Pyramids are 24 m high and the area of each base is 660 m². The Arid and Show Pyramids are 18 m tall and the lengths of the sides of their bases are 19.5 m. Find the volume of the pyramids.

Solution Use the Pyramid-Cone Volume Formula.

$$V = \frac{1}{3}Bh$$

Temperate and Tropical Pyramids: $B = \underline{\;?\;}$ m² and $h = \underline{\;?\;}$ m.

$$V = \frac{1}{3} \cdot (\underline{\;?\;}) \cdot (\underline{\;?\;})$$
$$= \underline{\;?\;} \text{ m}^3$$

Arid and Show Pyramids: The base is a square, so $B = \underline{\;?\;}$ m².
The value of h is given.

$$V = \frac{1}{3} \cdot (\underline{\;?\;}) \cdot (\underline{\;?\;}) = \underline{\;?\;} \text{ m}^3$$

Example 2

If an ice cream cone has a height of 10 cm and a volume of 90 cm³, what is the radius of its base?

Solution Use the Pyramid-Cone Volume Formula.

$$V = \frac{1}{3}Bh$$

In this case, $V = 90$ cm³ and $h = 10$ cm.

10 cm

Substituting, $\quad 90 = \frac{1}{3} \cdot B \cdot 10$

$$90 = \frac{10}{3}B$$

$$B = 27 \text{ cm}^2$$

The base is a circle, so we know that $B = \pi r^2$.

Consequently, $\quad \pi r^2 = 27$

So, $\quad r^2 = \frac{27}{\pi}$

$$r = \sqrt{\frac{27}{\pi}} \approx 2.93 \text{ cm}.$$

With environmental concerns in mind, manufacturers are beginning to make edible containers for outdoor events and amusements parks.

Example 3

The waffle bowl at the right is a frustum of a cone. A sketch of the full cone is shown below. What is the volume of the waffle bowl?

Solution Use the Pyramid-Cone Volume Formula $V = \frac{1}{3}Bh$. First find the volume of the whole cone. Because the frustum does not include the bottom cone, subtract the volume of the bottom cone (cut off that volume) from the volume of the whole cone to find the volume of the frustum.

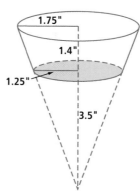

Volume (whole cone) $= \frac{1}{3}(1.75)^2 \pi\ (4.9)$

Volume (bottom cone) $= \frac{1}{3}(1.25)^2 \pi\ (3.5)$

Volume (frustum) = Volume (whole cone) −

Volume (bottom cone)

$= \frac{1}{3}(1.75)^2 \pi\ (4.9) - \frac{1}{3}(1.25)^2 \pi\ (3.5)$

$\approx 9.988\ \text{in}^3 \approx 10\ \text{in}^3$

▶ **READING MATH**

The word *frustum* comes from a Latin word meaning "a piece broken off."

Questions

COVERING THE IDEAS

In 1 and 2, refer to the splitting of a prism into pyramids in Activity 1.

1. Name the triangle congruence theorem that justifies $\triangle D'C'D \cong \triangle CDC'$.

2. Why do the pyramids in each pair have equal volume?
   a. $C'\text{-}ADC$ and $A\text{-}A'D'C'$
   b. $A\text{-}C'DC$ and $C'\text{-}ADC$
   c. $A\text{-}DC'D'$ and $A\text{-}C'DC$

3. In the box at the right, T is the center of the rectangle $UVWX$. How does the volume of the pyramid $T\text{-}PQRS$ compare to the volume of the box?

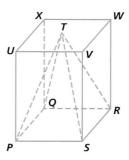

In 4–6, find the volume of the solid.

4. the pyramid with right triangular base ABC and height DB

5. the trapezoidal pyramid with height 10

6. a right cone with height 24 and base diameter 16

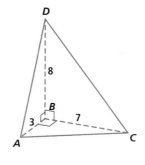

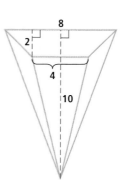

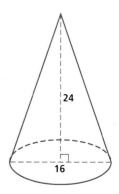

7. In the figure below, the cone sits atop a cylinder. If the height of the cone and that of the cylinder are the same, how do their volumes compare?

8. The Carysfort Reef Lighthouse is six miles into the Atlantic Ocean off of Key Largo, Florida. Its main house is a frustum of a cone approximately 39 feet in diameter at the base and 32 feet in diameter at the top. The height of the main house is 20 feet, and the vertex of the cone is approximately 94 feet above that. Find the volume of the main house to the nearest hundred cubic feet.

Carysfort Reef Lighthouse

APPLYING THE MATHEMATICS

9. Find the volume of a regular hexagonal pyramid with base edge of length 6 and a height of 10.

10. Miamisburg Mound in Ohio, pictured at the right, is in the shape of a cone and is believed to have been constructed by the Adenas at some time between 800 BCE and 100 CE. Its height was originally 70 feet and its base has a circumference of 877 feet. What was the original volume of this mound?

11. What happens to the volume of a cone if its height is kept the same but the radius of the base is multiplied by 5?

12. Consider the water cooler cup pictured at the right.
    a. How many cubic centimeters of liquid will it hold?
    b. How many times will it need to be used in order to fill a liter jug?
    c. How much paper is needed to make the cup?

13. The figure at the right was created by rotating a line in space about a vertical axis and cutting it off with two parallel planes, both at a vertical distance of 10 from the vertex. If the circumference of each congruent base is 20π, find the total volume of the two parts.

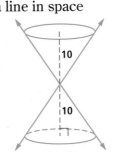

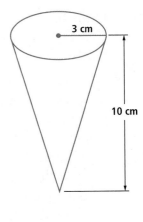

3 cm

10 cm

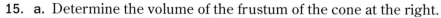

14. A 3-foot tall cone of shelled corn just touches the wall of a circular silo with diameter 15 feet. How many bushels of shelled corn are in the silo? (There are 1.24 cubic feet in 1 bushel.)

15. a. Determine the volume of the frustum of the cone at the right.

    b. Now consider the frustum that is 2 cm inside of this one in every direction. Its radii are 3 cm and 2 cm, the distance between the circles is 6 cm, and the total height of the cone is 18 cm. Determine the volume of this frustum.

    c. What is the ratio of your answers to Parts b and a?

    d. Refer back to page 595 and explain what this calculation has to do with the shape of ice cream containers.

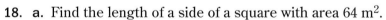

REVIEW

16. Helen lives in an apartment that has a floor plan with an area of 850 square feet. The ceiling in her apartment is 8 feet high. **(Lesson 10-3)**

    a. What is the volume of her apartment?

    b. Andrew lives in an apartment with the same floor plan, but the ceiling is 9 feet high. What is the volume of Andrew's apartment?

17. Consider the side view of a staircase pictured at the right. Each stair is 7" high, 12" long and 22" wide. **(Lesson 10-3)**

    a. Calculate the volume of the staircase by calculating the volume of each stair and adding up the volumes.

    b. Explain how you could use Cavalieri's Principle to calculate the volume.

18. a. Find the length of a side of a square with area 64 m^2.

    b. Find the length of an edge of a cube with volume 64 m^3.

    c. Your answers to Parts a and b should be integer lengths. Find a number different from 64 that would also yield integer lengths for the answers to these parts. **(Lessons 10-1, 8-2)**

19. In the detective show *Columbo,* Columbo solves one case using the following fact: Aquariums are required to have one gallon of water for each 1-inch length of fish in them. If an aquarium has volume 75 m^3, how many 2-inch fish could be put in it? **(Lesson 10-1)**

20. The sphere pictured at the right has radius 3, and center O, its intersection with plane A is a great circle. Plane B is h units above plane A. The intersection of plane B with the sphere has area 25. Find the value of h. **(Lesson 9-6)**

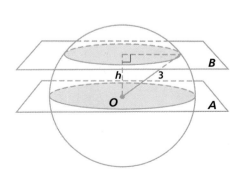

21. Gabrielle was investigating an isometry T using a DGS. She started with $\triangle ABC$ and found $T(\triangle ABC)$, $T \circ T(\triangle ABC)$, and $T \circ T \circ T(\triangle ABC)$. She continued doing this for a long time. In each case she got a triangle with the same orientation as $\triangle ABC$. All the triangles fit in the same screen. What kind of isometry is T? How do you know? (**Lesson 4-7**)

EXPLORATION

22. The volume of a cone is one-third the volume of a cylinder with the same base and same height. Find an example to show that this statement is not true if "volume" is replaced by "surface area."

23. Use a DGS-3D.

    Step 1 Construct a triangle in a plane.

    Step 2 Construct a vector in the plane and translate the original triangle by this vector, thus constructing a second triangle congruent to the first. (Figure 1)

    Step 3 Construct a point above the plane and then construct a plane parallel to the first plane through this point. (Figure 2)

    Step 4 Construct 2 other points in this parallel plane.

    Step 5 Construct triangular pyramids using the points from Step 4 as the vertices of the pyramids and the triangles from Step 2 as the bases. (Figure 3)

    Step 6 Measure the volume of each of the triangular pyramids.

Drag either of the vertices of the pyramids. What is true about the volumes of the two pyramids? Explain why this is true.

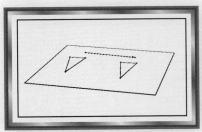

Figure 1

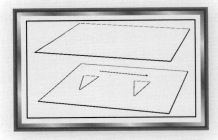

Figure 2

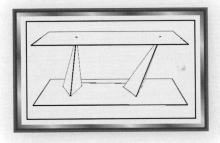

Figure 3

QY ANSWER

30π cm^3

Lesson

10-5 Organizing and Remembering Formulas

> **BIG IDEA** Strategies for deducing some formulas from others make it necessary to memorize only a small number of surface area and volume formulas.

Some facts need to be memorized because they are not related to each other, such as the first few places in the decimal for π. But you cannot always trust your memory. So, even if you are one of those people who like to memorize things, it helps to have ways of organizing information that can help you remember things.

Formulas That Can Be Deduced

Suppose an equilateral triangle has sides of length 73.42. You know that its perimeter is $3 \cdot 73.42$. You do not memorize this information. You do not even need to memorize that the perimeter p of an equilateral triangle with side s can be found using the formula $p = 3s$, because you can easily deduce the formula from the definitions of equilateral triangle and perimeter.

Similarly, you know that the perimeter p of a square with side s is found using the formula $p = 4s$. Again, this is a formula you do not have to memorize.

 QY1

If you remember definitions, you do not even have to memorize a formula for the perimeter p of a regular n-gon with side s. There are n sides, each of length s, so $p = \underbrace{s + s + \ldots + s}_{n \text{ addends}} = ns$. *If you can quickly deduce a formula, then you do not need to memorize it.*

Remembering Surface Area and Volume Formulas

In Chapter 9, you encountered formulas for surface area and lateral area for 3-dimensional figures of two basic types: right cylindrical surfaces (prisms and cylinders) and right conic surfaces (pyramids and cones). In this chapter, you have seen volume formulas for cylindrical and conic surfaces. Which should you remember? The answer is simple: *Remember the formulas that apply to the most figures.*

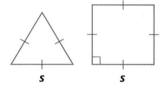

> **QY1**
>
> What is a formula for the perimeter p of a regular octagon with side s?

There are eight—only eight—basic formulas for surface areas and volumes that you will encounter in this book. So far you have seen six of the eight. The table below also includes the other two formulas for the surface area and volume of a sphere, which you will see in the next two lessons. The table is organized into columns based on the number of bases a figure has. Prisms and cylinders have two parallel bases. Cones and pyramids each have one base. Spheres have no bases.

	Cylindrical surfaces (Prisms/Cylinders) (two parallel bases)	Conic Surfaces (Pyramids/Cones) (one base)	Spheres (no bases)
Lateral Area	L.A. = ph (right cylinders only)	L.A. = $\frac{1}{2}\ell p$ (right surfaces only)	
Surface Area	S.A. = L.A. + $2B$	S.A = L.A. + B	S.A. = $4\pi r^2$
Volume	$V = Bh$	$V = \frac{1}{3}Bh$	$V = \frac{4}{3}\pi r^3$

If you organize the formulas in this way, you will find them easier to remember because the links between them are clearer. Of course, one thing is obvious from the table: *To use a formula, you must know what each variable in the formula represents.* To make it easier to memorize that, the first letters are always the first letters of the quantity the variables represent.

Deriving Formulas

The formulas $V = Bh$ and $V = \frac{1}{3}Bh$ are not very specific. They do not distinguish between a figure with a circular base and one with a polygonal base. This is done on purpose to avoid the creation and memorization of lots of new formulas. You can substitute formulas for perimeter and area that you know into formulas for volume to avoid the memorization of extra formulas.

For instance, the formula L.A. = $2\pi rh$ gives the lateral area of a right cylinder. This formula is not in the table because a few general ideas help reduce the load of formulas to remember. *To find the lateral area and surface area for right cylinders and cones, substitute the area and circumference of a circle, πr^2 for B and $2\pi r$ for p, in the corresponding formulas for cylindrical and conic surfaces.*

GUIDED

Example 1

Write a formula for the volume of a cone in terms of its radius r and height h.

Solution Because a cone is a conic surface, $V = $ ____?____. Because the base is a circle, $B = $ ____?____. Thus, by substitution, $V = $ ____?____.

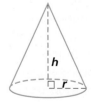

If you you can make the substitutions quickly, you can avoid learning a special formula for the lateral area of a right cylinder or a right cone.

 QY2

In Lesson 10-3, you saw a formula for the volume of a box: $V = \ell w h$. From this formula and the postulates about volume, all the other volume formulas can be deduced. The process we used—the process of proof—is the most powerful idea of all for remembering formulas. *If you cannot remember a formula, try to derive it from some simpler formulas you know to be true.* The difficulty with this advice is that it usually takes some time to do a proof, but often there is no time. So, if you do not want to spend your time proving them, you must either learn some formulas by heart or have access to a list of the formulas.

▶ **QY2**

What is a formula for the lateral area L.A. of a right cone whose base has radius r and whose height is h?

Using General Formulas

Another way to avoid learning lots of formulas needlessly is to *use general formulas to get formulas for special types of figures.*

Example 2

A bread holder is in the shape of a quarter-cylinder as shown at the right.

a. Express the volume V of the bread holder in terms of r and h.

b. Find the volume of the bread holder if $h = 16"$ and $r = 8"$.

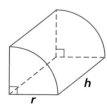

Solution

a. Notice that the base of this solid is a quarter of a circular region, so the base area is $B = \frac{1}{4}\pi r^2$. Substituting $\frac{1}{4}\pi r^2$ in for B in the formula $V = Bh$ yields the formula $V = \frac{1}{4}\pi r^2 h$.

b. Substitute for r and h in the formula you found in Part a.

$$V = \frac{1}{4}\pi r^2 h$$
$$= \pi \cdot 8^2 \cdot 16$$
$$= 256\pi \approx 804 \text{ in}^2$$

Questions

COVERING THE IDEAS

1. What are the best geometry formulas to remember?

In 2 and 3, choose from the following: boxes, cones, cylinders, prisms, pyramids, spheres.

2. In which figures does S.A. = L.A. + 2B?

3. In which figures does S.A. = L.A. + B?

4. Consider the formula L.A. = ph.
   a. What does each variable represent?
   b. To what figures does this formula apply?

5. Deduce a formula for the surface area of a cube with side length s from the formula for the surface area of a right prism.

In 6 and 7, a right cone has slant height ℓ and its base has radius r.

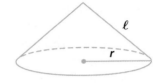

6. Find a formula for its L.A. in terms of ℓ and r.
7. Find a formula for its S.A. in terms of ℓ and r.

8. Find a formula for the volume of a cylinder in terms of its height h and the radius r of its base.

9. Consider the bread holder in Example 2.
   a. Express the area of the curved surface of in terms of r and h.
   b. Find the area of the curved surface if $h = 16"$ and $r = 8"$.

10. Describe the process for obtaining special formulas for cones and cylinders.

APPLYING THE MATHEMATICS

11. Suppose a cone has a height that is twice the radius r of its base.
    a. Find a formula for the volume of this cone in terms of r.
    b. If the cone has a volume of 24π, what is its radius?

12. Many cylindrical shapes in the real world are hollow (like piping, tin cans, etc.), but they all have thickness. Let R be the outer radius of the shape and let r be the inner radius.

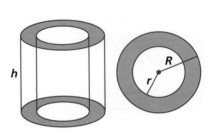

   a. Find an expression for the thickness of the walls of this cylinder.
   b. Write an expression for the volume of the material used to create this solid in terms of R, r, and h.
   c. Suppose a piece of PVC pipe 50 cm long has an inner radius of 50 mm and a thickness of 4 mm. Calculate the amount of plastic contained in this section of pipe.

13. When grain bins are full, farmers sometimes dump extra corn onto the ground in temporary piles. When dumped from a grain elevator, the corn creates a conical pile whose height is always about $\frac{2}{10}$ of its diameter, d.
    a. Draw a diagram of this cone using d as the only variable.
    b. Write a formula for the volume of a conical pile in terms of d.
    c. When a pile of corn is on the ground, one of the only accessible measurements is its circumference. Suppose a pile of corn has a circumference of 100 ft. About how many bushels of corn are contained in the pile? (1 bushel $\approx$ 1.24 ft^3)

14. **a.** Find a formula for the volume of a prism whose bases are regular hexagons and whose height is equal to the length e of the sides of its base. Leave your formula in simplified radical form.

    b. If the edge of the prism is 10 cm, find the exact volume of the prism.

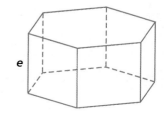

15. **a.** A regular tetrahedron is a pyramid whose faces are all equilateral triangles. Find a formula for the volume of a regular tetrahedron in terms of its edge length, s. The height of the pyramid is $\sqrt{\frac{2}{3}}s$.

    b. If the tetrahedron has edges that measure 10 cm in length, find its volume.

REVIEW

16. Concrete is typically measured in cubic yards. Suppose you buy one cubic yard of concrete to create a sidewalk. The sidewalk is 3 feet wide and 3 inches deep. How long a sidewalk can you create? **(Lesson 10-3)**

17. Perform the multiplication $(x - 3)(y + 3)(3z)$. **(Lesson 10-2)**

18. When all of the dimensions of a cube are multiplied by 7, the resulting cube has a volume of 1000. What are the lengths of the edges of the original cube? **(Lesson 10-1)**

19. Suppose the number of cubic centimeters in the volume of a cube is the same as the number of square centimeters in this cube's surface area. Find the dimensions of the cube. **(Lessons 10-1, 9-9)**

20. **Multiple Choice** Suppose n is a very large number, and x is the measure of an interior angle in a regular n-gon. Which of the following numbers is x closest to? **(Lesson 6-8)**

    A $120°$ **B** $180°$ **C** $90°$ **D** $360°$

21. Consider the top pictured at the right. If you wished to draw this picture on a piece of paper (not including the dashed lines), could you do this without lifting your pencil? **(Lesson 1-3)**

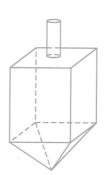

EXPLORATION

22. **a.** Give a possible set of dimensions for a cylinder whose surface area is 400π.

    b. Give dimensions for a second cylinder, not congruent to the first, whose surface area is 400π.

QY ANSWERS

1. $p = 8s$

2. L.A. $= \pi r \sqrt{r^2 + h^2}$

Lesson
10-6 The Volume of a Sphere

▶ **BIG IDEA** By thinking of a sphere as a union of many "almost pyramids," a formula for the volume of a sphere can be deduced.

Review of Volume Formulas

Here is how the volume formulas of this chapter were developed. We began with the Volume Postulate in Lesson 10-1.

$$V = \ell w h \quad \text{(volume of a box)}$$

Cavalieri's Principle was then applied and the following formula was deduced in Lesson 10-3.

$$V = Bh \quad \text{(volume of a prism or cylinder)}$$

A prism can be split into three pyramids with the same height and base. Using Cavalieri's Principle again led to a formula for the volume of a conic solid in Lesson 10-4.

$$V = \frac{1}{3}Bh \quad \text{(volume of a pyramid or cone)}$$

In this lesson, still another application of Cavalieri's Principle results in a formula for the volume of a sphere. Along the way, figures that look quite different are shown to have the same volume.

Comparing a Sphere with Another Surface

The sphere below has a diameter $2r$, and the cylinder and double cone each have a height of $2r$. Each cone inside the cylinder has height r. An amazing result, discovered by Archimedes (287–212 BCE), is that the volume of the sphere at the left equals the volume *between the cylinder and the two cones*. The argument that follows uses plane sections and Cavalieri's Principle to prove this result.

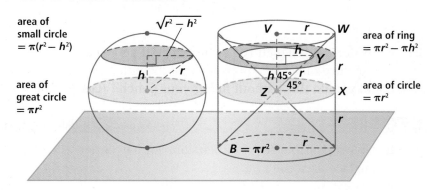

Consider the area of a plane section h units from the center of a figure created by removing a double cone from a cylinder. This section is the shaded ring in the figure on the right at the bottom of page 628. The area of the ring is equal to the difference between the areas of the circles that are its boundary. The outside circle of the ring has radius r. The inside circle has radius h.

In the double cone, $\triangle VWZ$ is an isosceles right triangle because $VW = VZ = r$. Thus, $m\angle WZV = 45$. So m $\angle YZX = 45$ and $\triangle XYZ$ is also an isosceles right triangle. Consequently, $XY = XZ = h$.

$$\text{Area}_{\text{ring}} = \pi r^2 - \pi h^2 = \pi(r^2 - h^2)$$

Now, consider the area of the shaded section of the sphere. The shaded section is a small circle of the sphere. It is h units from the center. Because the radius of the sphere is r, by the Pythagorean Theorem, the radius of the shaded circle is $\sqrt{r^2 - h^2}$. The area of the small circle is found using the formula for the area of a circle.

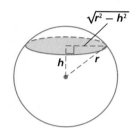

$$A_{\text{smallcircle}} = \pi\left(\sqrt{r^2 - h^2}\right)^2 = \pi(r^2 - h^2)$$

Thus, the shaded sections have equal area.

Finding the Formula

Because the area of the small circle equals the area of the ring no matter what height h is chosen, Cavalieri's Principle can be applied. The volume of the sphere is thus the difference between the volume of the cylinder $(B \cdot 2r)$ and the volume of the two cones (each with volume $\frac{1}{3}B \cdot r$).

$$\begin{aligned}
\text{Volume of sphere} &= (B \cdot 2r) - 2 \cdot (\tfrac{1}{3}B \cdot r) \\
&= 2Br - \tfrac{2}{3}Br \\
&= \tfrac{4}{3}Br
\end{aligned}$$

But here the base of the cone and cylinder is a circle with radius r. So $B = \pi r^2$. Substituting,

$$\begin{aligned}
\text{Volume of sphere} &= \tfrac{4}{3} \cdot \pi r^2 \cdot r \\
&= \tfrac{4}{3}\pi r^3.
\end{aligned}$$

Sphere Volume Formula

The volume V of any sphere is $\frac{4}{3}\pi$ times the cube of its radius r.
$$V = \tfrac{4}{3}\pi r^3$$

STOP QY

The Sphere Volume Formula seems first to have been discovered by Archimedes. In fact, Cavalieri studied Archimedes' way of calculating the volume before he stated his principle.

▶ QY

Express the volume of a sphere in terms of its diameter d.

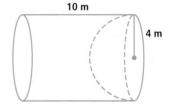

Example 1

The figure at the right represents a solid right cylinder of height 10 m with a hemisphere of radius 4 m removed. Find the exact volume of the figure.

Solution The volume of the figure is equal to the volume of the hemisphere subtracted from the volume of the cylinder. The radius of the cylinder and the radius of the sphere are the same, 4 m.

$$V_{figure} = V_{cylinder} - V_{hemisphere}$$
$$= \pi r^2 h - \frac{1}{2}(\frac{4}{3}\pi r^3)$$
$$= \pi \cdot 4^2 \cdot 10 - \frac{1}{2}(\frac{4}{3}\pi \cdot 4^3)$$
$$= 160\pi - \frac{128}{3}\pi \text{ m}^3$$
$$= 117\frac{1}{3}\pi \text{ m}^3$$

GUIDED

Example 2

The diagram at the right shows the layers of Earth. Assuming Earth and all its layers are spheres, use the information in the diagram to approximate the volume of the outer core to the nearest 100,000,000 cubic kilometers.

Solution The radius of the inner core is ___?___. The radius of the sphere containing the inner and outer core is ___?___. So, the volume of the outer core is equal to the volume of a sphere of radius ___?___ minus the volume of a sphere of radius ___?___.

$$V_{outer\ core} = \frac{4}{3}\pi(\underline{\ ?\ })^3 - \frac{4}{3}\pi(\underline{\ ?\ })^3$$
$$= \frac{4}{3}\pi(\underline{\ ?\ }^3 - \underline{\ ?\ }^3)$$
$$= \frac{4}{3}\pi(\underline{\ ?\ }) \approx 168{,}900{,}000{,}000 \text{ km}^3$$

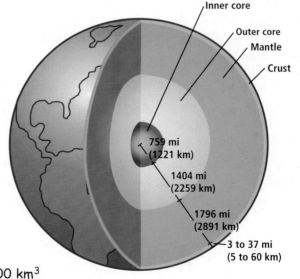

Questions

COVERING THE IDEAS

1. What was Archimedes' discovery relating to the volume of a sphere?

2. If the areas of plane sections of a cylinder minus a double cone are equal to the areas of plane sections of a sphere, what statement justifies that their volumes must also be equal?

In 3 and 4, use the drawings at the right of a sphere and a cylinder with a double cone inside it.

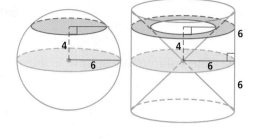

3. **a.** Give the area of the smaller shaded cross section of the sphere.

   b. Give the area of the shaded ring between the cylinder and the cone.

4. Give the volume of

   a. the cylinder.   **b.** the sphere.

   c. the two cones together.

   d. the solid region between the cylinder and the two cones.

5. The sphere at the right has radius 10. The small circle shown is 6 units from the center of the circle. Find the radius of the small circle.

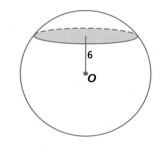

6. Standard billiard balls are 2.25″ in diameter. Find the volume of a standard billiard ball to the nearest tenth of a cubic inch.

7. The figure at the right is composed of a hemisphere on top of a box with square bases and height 12 m. A side of the square is 4 m. Find the volume of the figure.

APPLYING THE MATHEMATICS

8. The Hagia Sophia in Istanbul was built in 532 CE. Its great dome is a hemisphere with diameter 107 feet. Find the volume of the dome to the nearest thousand cubic feet.

9. An ice cream cone is 14 cm high and has base diameter 5 cm. Is the volume of the cone greater or less than a spherical scoop of ice cream 6 cm in diameter?

10. The gasoline industry uses barges to move fuel along coastal and inland waterways. A double-hull barge can carry about 4.8 million liters of gasoline. The gasoline is then transported by tank trucks that bring it to gas stations. Many gas stations store their gas in tanks that are right cylinders with a hemisphere at each end. Approximately how many tanks of gas with the dimensions seen in the figure below can be filled with the fuel from one barge? (1 liter ≈ 0.0353147 cubic foot).

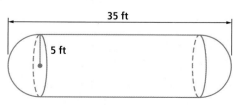

11. In the figure at the right, a section of a sphere was removed, leaving two perpendicular semi-circular faces. Express the volume of the figure in terms of the radius r of the sphere.

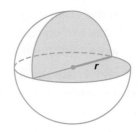

12. Suppose the radius of one sphere is equal to the diameter of another sphere. What is the ratio of the volume of the small sphere to the volume of the larger sphere?

13. The volume of a sphere is 468 cubic meters. Find its diameter to the nearest meter.

REVIEW

14. A right cone has height 7 cm and volume $147{,}000\pi$ cm³. Find the radius of its base. **(Lesson 10-4)**

15. **a.** Write Cavalieri's Principle as a conditional. **(Lessons 10-3, 2-2)**
    b. Write the converse of the conditional you wrote in Part a. Is this converse true? **(Lesson 2-3)**

16. A box has volume 24. A cube is drawn that has the same volume as the box. A second cube is drawn whose edges have half the length of the edges of the first cube. The second cube is cut into two boxes, each having half of its volume. Finally, a third cube is drawn that has the same volume as one of these boxes. What are the lengths of the edges of the third cube? **(Lesson 10-1)**

17. A right square pyramid and a right cone have the same lateral area, and the same slant height. The base of the pyramid has sides of length 0.8. Find the radius of the base of the cone. **(Lesson 9-10)**

18. Given: $ABCD$ is an isosceles trapezoid, with bases $\overline{AB}$ and $\overline{CD}$.
    Prove: $DE = EC$ **(Lessons 7-3, 6-6)**

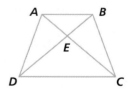

19. In the figure below, $\overline{FG} \parallel \overline{HI}$, and $HI = 7FG$. Give an example of a transformation P such that $P(\overline{HI}) = \overline{FG}$. **(Lesson 3-7)**

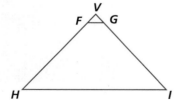

EXPLORATION

20. Archimedes is considered to be one of the greatest mathematicians of all time. He wanted to have one specific diagram on his tombstone. Research Archimedes to find out what the diagram was and what it represented.

Lesson
10-7
The Surface Area of a Sphere

> ▶ **BIG IDEA** From the formula for the volume of a sphere, a
> formula for the surface area of a sphere can be deduced.

You might be wondering why a lesson on surface area is at the end of
a chapter on volume. The answer is simple. We derive the formula for
the surface area of a sphere from the volume formula.

To calculate surface area, we created a net of the surface and found
the area of each part of the net. This assumes that the surface can be
unfolded into parts that are flat. But no part of a sphere's surface can
be made to be flat. Another strategy is needed.

The Biosphere of Montreal, Quebec, Canada

Breaking the Sphere into Pyramids

Even though there is no simple net for a sphere, we can still derive
the formula for the surface area. The idea is to consider a solid
sphere as being made up of "almost pyramids." Think of each
pyramid as having its apex at the center and its base being a triangle
covering a region as in the geodesic dome pictured above. One such
"pyramid," with height h and base area B, is drawn in the sphere at
the right. The solid is not exactly a pyramid because its base is not
exactly a polygon. Even so, when the "almost pyramid" is small, its
volume is close to that of a pyramid, $\frac{1}{3}Bh$. Because h equals r, the
radius of the sphere, each "almost pyramid" has a volume of $\frac{1}{3}Br$.

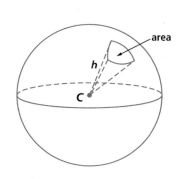

Now break up the entire sphere into "almost pyramids" with base areas B_1, B_2, B_3, B_4, and so on. The sum of all the Bs is the surface area of the sphere. The volume V of the sphere is the sum of the volumes of all the "almost pyramids" with base area B_1, B_2, B_3, B_4, . . .

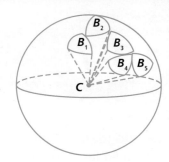

$$V = \tfrac{1}{3}B_1 r + \tfrac{1}{3}B_2 r + \tfrac{1}{3}B_3 r + \tfrac{1}{3}B_4 r + \dots$$
$$V = \tfrac{1}{3}r(B_1 + B_2 + B_3 + B_4 + \dots)$$
$$V = \tfrac{1}{3}r \cdot \text{S.A.}$$

Finding the Formula

Now substitute $\tfrac{4}{3}\pi r^3$ for the volume V.

$$\tfrac{4}{3}\pi r^3 = \tfrac{1}{3}r \cdot \text{S.A.}$$

To solve for the surface area, multiply both sides by 3.

$$4\pi r^3 = r \cdot \text{S.A.}$$

Now divide by r.

$$4\pi r^2 = \text{S.A.}$$

Thus, the surface area of any sphere with radius r is $4\pi r^2$.

> ### Sphere Surface Area Formula
>
> The surface area of a sphere with radius r is $4\pi r^2$.
> $$\text{S.A.} = 4\pi r^2$$

This formula indicates that the surface area of a sphere is equal to 4 times the area of a great circle of the sphere.

Example 1

Earth is almost a sphere with a diameter of approximately 7900 miles. Find the surface area of Earth.

Solution Because the diameter is 7900 miles, the radius is 3950 miles. Thus, S.A. $= 4\pi(3950)^2 \approx 196{,}000{,}000$ square miles.

 QY1

▶ **QY1**

The area of the United States is about 3,700,000 mi². About what percent of Earth's surface is the United States?

Example 2

Find the volume of a sphere with a surface area of 64π cm².

Solution First find the radius.

$$4\pi r^2 = 64\pi$$

$$\frac{4\pi r^2}{4\pi} = \frac{64\pi}{4\pi}$$

$$r^2 = 16$$

$$r = 4 \text{ cm}$$

Now substitute this radius into the volume formula.

$$V = \frac{4}{3}\pi r^3$$

$$= \frac{4}{3}\pi (4)^3$$

$$= \frac{4}{3}\pi (64)$$

$$= \frac{256}{3}\pi \text{ cm}^3$$

Both the surface area and volume of a hemisphere are half that of a sphere.

GUIDED

Example 3

The top of the fuel tank shown at the right is used on aerospace flights. It is a hemisphere with a 42-inch diameter. Find the lateral area and volume of this hemisphere.

Solution Start with the formulas for the surface area and volume of a sphere. S.A. = __?__ and V = __?__. Because a hemisphere is half of a sphere, divide the expressions by __?__ to get formulas for the lateral area and volume of a hemisphere. L.A. = __?__ and V = __?__. Now substitute __?__, the radius of the fuel tank hemisphere. L.A. = __?__ ≈ __?__ in² and V = __?__ ≈ __?__ in³.

 QY2

Questions

COVERING THE IDEAS

1. **Fill in the Blank** The surface area of a sphere is __?__ times the area of one of its great circles.

2. Calculate the surface area of a sphere with radius 5".

3. A sphere has a diameter of 12 cm.
   a. Calculate the exact surface area of this sphere.
   b. Approximate this surface area to the nearest tenth of a square centimeter.

> **QY2**
>
> Write a formula for the surface area of a hemisphere (including the circular base) in terms of its radius.

4. Find the exact volume of a sphere with a surface area of 400π in^2.

5. A sphere is created using 750 cubic centimeters of clay. What is the surface area of this sphere?

6. The round part of one soup ladle is a hemisphere and holds 8 ounces, or approximately 237 cubic centimeters of liquid. What is the surface area of the ladle, to the nearest square centimeter? Will the ladle fit into a container with a diameter of 9 cm?

APPLYING THE MATHEMATICS

7. Find the formula for the surface area of a sphere in terms of its diameter d.

8. The equatorial diameter of Jupiter is approximately 11 times as great as the Earth's diameter.
   a. How do the surface areas of the planets compare?
   b. How do the volumes of the planets compare?
   c. How do your answers in Parts a and b relate to the number 11?

9. A sphere of radius r fits exactly into a cylinder, touching the cylinder at the top, bottom, and sides as shown in the figure at the right. How does the total surface area of the sphere compare to the lateral surface area of the cylinder?

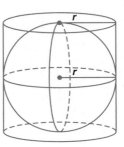

10. Softballs and baseballs are sized by the circumference of a great circle of the balls. A baseball is 9 inches in circumference and softballs come in three sizes: 11 inches, 12 inches, and 16 inches.
    a. Find the amount of material it would take to cover each of the balls (surface area).
    b. Is there a relationship between the ratios of the radii of any two balls and their respective surface areas?

11. The figure at the right was created using a right cone, a right cylinder, and a hemisphere.
    a. Find the total volume of the solid.
    b. Find the total surface area of the shape.

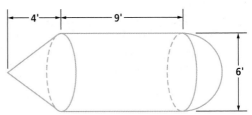

12. The radius of Earth is approximately 3950 miles, and $\frac{2}{3}$ of the surface is water. The radius of the Moon is approximately 1740 miles, and none is water. Which object has more land area, the Moon or Earth, and how do you know?

13. Write a formula for the surface area of a sphere in terms of C, the circumference of a great circle on the sphere.

REVIEW

14. The mean diameter of the Sun is 109 times the mean diameter of Earth. Then the volume of the Sun is how many times the volume of Earth? **(Lesson 10-6)**

15. A tent is in the shape of a right square pyramid. It is 5 feet high, and its base has sides of length 4 feet. What is the tent's volume? **(Lesson 10-4)**

16. You can make the lateral surface of two different cylinders by rolling a 210-by-297 mm piece of paper along one or another of its sides.
    a. What is the lateral area of each cylinder? **(Lesson 9-9)**
    b. Which has more volume? **(Lesson 10-3)**

17. Explain why the two stacks of coins in the figure at the right have the same volume. **(Lesson 10-3)**

18. Consider two cylindrical jars of jam. One jar of jam is twice as tall as the other, but only half as wide. Which jar holds more jam? **(Lesson 10-3)**

19. Suppose solid A has volume 78 cm^3 and solid B has volume 63 cm^3. **(Lesson 10-1)**
    a. Find the greatest possible volume of the union of the two solids.
    b. Find the least possible volume of the union of the two solids.

EXPLORATION

20. Another way to derive the formula for the surface area of a sphere from the volume formula is to imagine a sphere made of thin plastic. Let the plastic sphere be t units thick.

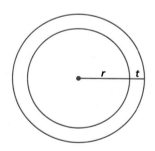

    a. By subtracting the volume of the small sphere from the large sphere, find an expression for the volume of the plastic sphere in terms of r and t.
    b. Imagine that this plastic was laid out to form a cylindrical solid whose base area is the surface area of the sphere. The height of this cylindrical solid is t. Thus, the volume of this prism is $V = $ S.A. $\cdot t$. This expression and the expression in Part a are both expressions for the volume of the plastic, so they can be set equal to one another. Use this to find an expression for S.A. in terms of r and t.
    c. Now we let the thickness of the plastic become smaller and smaller until it is infinitely small. What happens to the expression as t approaches 0?

Chapter 10 Projects

1 What Is Teddy's Surface Area and Volume?

Find as many different methods as you can to measure the volume and the surface area of a stuffed animal. Measure these quantities using several of these methods. For each method you come up with, discuss what factors contribute to its accuracy and its inaccuracy.

2 Lengths and Areas on the Surface of Earth

a. Suppose you drew an extremely large circle on the face of Earth, measured its diameter with a long tape measure, and measured its circumference by walking around it. If you did this, the ratio of circumference to diameter would not be π. Would it be greater than or less than π? Explain how you arrived at your answer.

b. If you followed the instructions from Part a and the circle you created was a great circle for Earth, what would the ratio of the circumference to the "diameter" of this circle be? Write an explanation with a diagram explaining your answer.

3 Making Cones

In Chapter 9, you learned that a sector of a circle is a net for a cone. Draw a number of circles of equal diameter and cut out different sized sectors to form cones. Calculate the volume of these cones and attempt to answer the question: What is the measure of the central angle of the sector that yields the cone with the most volume?

4 Packing Spheres

Suppose you had a large pile of oranges of the same radius, and wished to pack them in a large box. The problem of finding the most efficient way to do this is called Kepler's problem. It was a famous unsolved problem of mathematics until a proof of the best way was given in 1997.

Find the most efficient way to pack spheres. What is the efficiency of that method?

5 What Is the Volume of a Spherical Cap?

Suppose an empty plastic sphere is filled less than halfway with water. The depth of the water, at its deepest point, is h and the radius of the sphere is r.

a. Refer to Archimedes's development of the formula for the volume of a sphere. Draw a diagram to demonstrate that the portion of the double-cylinder/cone shape will have the same volume as the partially-filled sphere.

b. If the radius of the sphere is 10 cm and the depth of the water is 4, find the volume of the water by using Archimedes' diagram. (Hint: Use isosceles right triangles.)

c. Create a formula for the volume of the water in terms of h and r. This shape is formally called a *spherical cap*.

Chapter 10 Summary and Vocabulary

○ The lateral and surface areas of a 3-dimensional figure measure its boundary, which is 2-dimensional. So these areas, like the areas you studied in Chapter 8, are measured in square units. Volume measures the space enclosed by a 3-dimensional surface, or the amount of material inside a 3-dimensional solid. **Volume** is measured in **cubic units**. Beginning with the formula for the volume of a box, $V = \ell wh$, and using the principles in the Volume Postulate, formulas for the volumes of other figures were developed in this chapter. Pyramids and cones were shown to have one-third the volume of the prism or cylinder with congruent bases and the same height. The volume of a sphere was shown to be the same as the volume of a cylinder of the same height minus two cones.

○ There are only eight basic formulas for volume and surface area of standard shapes. Two of the surface area formulas were developed in the last chapter but are included here for completeness. These are key formulas to remember. In all cases r represents radius, h represents height, ℓ represents slant height, p represents perimeter, and B represents area of a base.

○ Knowing the eight formulas below allows you to derive other formulas. The formulas help show that if you multiply any one dimension of a 3-dimensional figure by a certain number, the volume of the figure is multiplied by that number.

Vocabulary

Lesson 10-1
volume
cubic units
unit cube
box
cube root

	Cylindrical surfaces (Prisms/Cylinders) (two parallel bases)	Conic Surfaces (Pyramids/Cones) (one base)	Spheres (no bases)
Lateral Area	L.A. = ph (right cylinders only)	L.A. = $\frac{1}{2}\ell p$ (right surfaces only)	
Surface Area	S.A. = L.A. + 2B	S.A = L.A. + B	S.A. = $4\pi r^2$
Volume	$V = Bh$	$V = \frac{1}{3}Bh$	$V = \frac{4}{3}\pi r^3$

Postulates, Properties, and Theorems

Volume Postulate
  Uniqueness Property (p. 597)   Box Volume Formula (p. 597)
  Congruence Property (p. 597)   Cavalieri's Principle (p. 611)
  Additive Property (p. 597)

Take this test as you would take a test in class. You will need a calculator. Then use the Selected Answers section in the back of the book to check your work.

1. A right cylinder has a diameter of 3 inches and a height of 6 inches.

   a. Find the exact volume of the cylinder.

   b. Find the number of fluid ounces the cylinder will hold to the nearest hundredth of an ounce. (One cubic inch is approximately 0.554 fluid ounce.)

2. A sugar cube has an edge of 0.8 cm. What is the volume of the cube?

3. The base of a rectangular prism has an area of 40 square inches. How tall must the prism be to have a volume of 300 cubic inches?

4. Find the length of a side of a cube that has the same surface area as the right triangular prism below.

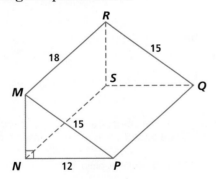

5. Suppose a 10-by-50-by-2 box and a cube have the same volume. Which has a greater surface area?

6. The circumference of the great circle of a baseball cannot be less than 9″ and cannot be greater than $9\frac{1}{4}$″.

   a. Determine the least possible volume for a legal baseball, to the nearest cubic inch.

   b. Determine the greatest possible surface area for a legal baseball, to the nearest square inch.

7. The right cone pictured below has diameter 10″ and a slant height of 13″. Determine the exact volume of the cone.

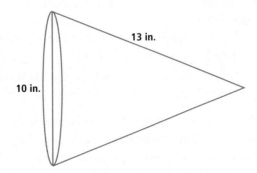

13 in.

10 in.

8. A cube has edges of length x. The length of the cube is multiplied by 3, the width is halved, and the height is not changed.

   a. Write a formula for the volume of the new box.

   b. Is the volume of the new box greater than, less than, or equal to the original cube?

9. A cylinder has height and diameter w. Hemispheres are placed at both ends of the cylinder. Find a formula for the volume of the new shape in terms of w.

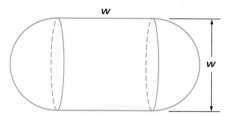

10. Give two expressions for the volume of the box pictured below.

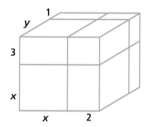

11. The medium and large cups at most fast food restaurants are made so that the top bases are equal in area, and the bottom bases are equal in area, as shown in the figure below. This allows restaurants to save money because they only have to buy one size of lid for both cups. If the volume of the medium-sized cup is V, can you use Cavalieri's Principle to determine the volume of the large-sized cup? Explain your answer.

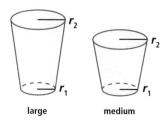

large   medium

12. Determine the volume of the right trapezoidal prism pictured below.

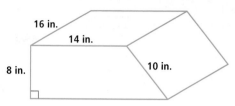

13. The bases of the right cylindrical solid pictured below are 60° sectors of a circle with radius 6. The height of the solid is 12.

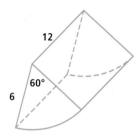

a. Determine the exact volume of the solid.

b. Determine the exact surface area of the solid.

14. A sphere with radius 15 inches sits inside another sphere with the same center and a radius $\frac{3}{4}$-inch longer. Find the volume of the space in between the two spheres.

15. Find the volume of a pyramid to the nearest cubic centimeter if its base is a regular hexagon with sides 60 cm and each lateral edge has length 100 cm.

16. Give a formula from which you can deduce formulas for the volume of a cube, a right rectangular prism (box), a cylinder, and a triangular prism.

Chapter 10 Chapter Review

SKILLS Procedures used to get answers

OBJECTIVE A Calculate the volumes of cylindrical solids from appropriate lengths, and vice versa. (Lesson 10-3)

1. Find the exact volume of a cylinder with height 5 and a base with radius 4.5.

2. Find the volume in cubic feet of a box that is 4" by 8" by 9".

3. The base of the prism shown below is a right triangle with legs of lengths 5 and 12. The distance between the bases of the prism is 24. Find the volume of the prism.

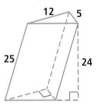

4. A cylinder has volume 150 cm². Its height is twice the radius of its base. Calculate the radius of its base, rounded to the nearest tenth.

5. The surface area of a cube is 834 cm². Find its volume.

6. If a cylinder is to have a volume of 30π cubic units and a base with radius 3, what must its height be?

7. Find the area of the base of a prism whose volume is 20 cm³ and whose height is 15 cm.

OBJECTIVE B Calculate the volumes of conic solids from appropriate lengths, and vice versa. (Lesson 10-4)

8. Find the volume of the cone shown below.

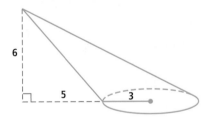

9. Find the volume of a square pyramid whose base has a diagonal of length $\sqrt{2}$ and whose height is 17.

10. The slant height of a regular right hexagonal pyramid is 12. The perimeter of the base is also 12.

    a. What is the height of the pyramid?

    b. What is the volume of the pyramid?

11. The great pyramid in Giza, Egypt originally had a square base with sides of length 440 cubits, and a height of 280 cubits. (A cubit is a measure of length.) What was its original volume?

12. A cone has volume 480 m³. Its height is four times the radius of its base. Find the height of the cone.

13. A cube has sides of length 2. Each side of the cube is the base of a right pyramid whose vertex is outside the cube. The heights of the pyramids are 1, 2, 3, 4, 5, and 6. Find the volume of this solid.

OBJECTIVE C Calculate the surface area and volume of a sphere from appropriate lengths. (Lessons 10-6, 10-7)

14. Give the exact surface area and volume of a sphere with radius 85.

15. A sphere has surface area 148π cm². What is its exact volume?

16. A sphere has volume 482 cm³. What is its surface area, rounded to the nearest hundredth?

17. The cross section of a sphere is this great circle with center O. Find the volume of the sphere, rounded to the nearest tenth.

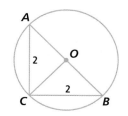

18. This figure shows a cross section of a sphere inscribed in a cube. The circle is a great circle. The volume of the cube is 343 ft³. Find the exact surface area of the sphere.

PROPERTIES Principles behind the mathematics

OBJECTIVE D Develop formulas for specific figures from more general formulas. (Lessons 10-1, 10-5)

19. A right cone has base radius r and height $r + t$. Find a formula for its volume.

20. A cone with base of radius r has a lateral surface area of $2\pi r^2$. Find a formula for its volume.

21. A cone has slant height ℓ and its base has radius r. Find a formula for its volume.

22. A sphere has the same surface area as a cube with sides of length s. Find a formula for the volume of the sphere in terms of s.

23. The height of a tetrahedron whose sides have length m is $\frac{\sqrt{6}}{3}m$. Find a formula for the volume of the tetrahedron. (Recall that the base of a tetrahedron is an equilateral triangle.)

OBJECTIVE E Compare the surface areas and volumes of related figures. (Lesson 10-5)

24. A sphere and a cube have the same surface area. Which has the greater volume?

25. A right cone and a cylinder have congruent bases and the same height. Which has the greater volume?

26. The dimensions of a right rectangular prism are x, y, and z. A right rectangular pyramid has a base with sides $2x$ and $3y$, and the same volume as the prism. What is the height of the pyramid?

27. A sphere has radius r. Find the lengths of the edges of a cube that has the same surface area as the sphere.

28. If a pyramid and prism have the same base and equal heights, how do their volumes compare?

29. If a cylinder has a base with twice the area of a cone's base, but has half the height of the cone, how do their volumes compare?

OBJECTIVE F Determine what happens to the surface area and volume of a figure when dimensions are multiplied by some number. (Lesson 10-2)

30. All of the dimensions of a box are halved.
    a. What happens to its surface area?
    b. What happens to its volume?

31. One dimension of a cube is multiplied by 8, one dimension is multiplied by 3, and the third dimension is multiplied by 0.2.
    a. What happens to the surface area?
    b. What happens to the volume?

32. By how much would all three dimensions of a box have to be multiplied in order for the new volume to be 27 times the volume of the original box?

33. The diameter of a pizza is doubled. If its thickness remains the same, how do the volumes of the two pizzas compare?

34. The diameter of one sphere is 7.5 times the diameter of another sphere. How do the volumes of the two spheres compare?

35. One of the dimensions of a box is multiplied by 5.

    a. What happens to the volume of the box?

    b. Is there enough information to say exactly what happens to the surface area?

OBJECTIVE G Know the conditions under which Cavalieri's Principle can be applied. **(Lessons 10-3, 10-4)**

36. A cylinder and a hemisphere have bases of the same area and have equal heights. Can Cavalieri's Principle be applied in this situation? Why or why not?

37. The figure below shows two right prisms with the same height. Can Cavalieri's Principle be used to deduce that the prisms have the same volume? Why or why not?

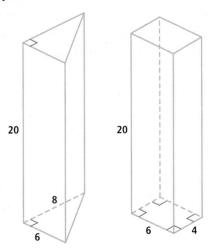

38. In deriving the formula for the volume of a sphere, a plane section of a sphere was shown to be equal in area to what other figure?

39. Suppose two cylindrical solids have congruent bases and equal heights. One of the solids is oblique and one is right.

    a. **True or False** The solids have the same volume.

    b. **True or False** The solids have the same surface area.

USES Applications of mathematics in real-world situations.

OBJECTIVE H Use volume in relation to liquid capacity. **(Lesson 10-1)**

40. A bottle contains $1\frac{1}{2}$ liters of juice. How many cubic centimeters of juice is this?

41. A pool is 13 meters by 3 meters by 2.5 meters. Find the volume in liters of the water needed to fill the pool.

42. Which is more: a quart of milk, or the amount of milk that fills a box measuring 5" by 7" by 7"?

43. Which is more: 0.4 pint or 5.4 liters?

44. From a reservoir containing 10,000 cubic meters of water, 7000 liters are removed. How much water will be left, in cubic millimeters?

OBJECTIVE I Apply the formula for surface area of a sphere to real situations. **(Lesson 10-7)**

45. The diameter of Mars at the equator is approximately 6805 kilometers. Suppose a satellite takes pictures of the surface of Mars. Each picture covers a 20-kilometer by 20-kilometer square. At least how many pictures would be needed to cover the whole surface?

46. The area of Antarctica is approximately 14,000,000 km². The radius of Earth at the equator is about 6378 km. What percentage of the surface of Earth does Antarctica cover?

47. To the nearest square foot, how much rubber is needed to make a kickball 8.5″ in diameter?

48. The WNBA uses a spherical basketball that is 28.5 inches in circumference, an inch smaller than the NBA's ball. What percent of the surface area of an NBA ball is a WNBA ball?

OBJECTIVE J Apply formulas for volumes to real situations. (Lessons 10-1, 10-3, 10-4, 10-6)

49. A cylindrical can of rubber balls is 15″ tall and 5″ wide.

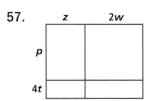

   a. If two rubber balls are fit snugly inside the can, what percent of the can do the balls fill?
   b. If three balls are fit snugly inside the can, what percent of the can do the balls fill?

50. A cylindrical paper cup has a base with radius 7 cm and height 15 cm. How many cc's of liquid can this paper cup hold?

51. A mineshaft is in the shape of a cylinder, 10 feet in diameter and half a mile deep. What is the volume of the dirt that was displaced to create the shaft?

52. The great pyramid in Giza, Egypt is a regular pyramid with a square base. The base of the pyramid has sides of length 230.4 meters. The height of the pyramid is 138.8 meters. The pyramid is constructed of limestone. Estimate the weight of the pyramid, assuming it is entirely full of limestone and that one cubic meter of limestone weighs 1500 kg.

53. The radius of Earth at the equator is about 6378 km. If a box with the volume of Earth had a base 100 km by 100 km, what would its height be?

REPRESENTATIONS Pictures, graphs, or objects that illustrate concepts

OBJECTIVE K Represent products of two (or three) numbers or expressions as areas of rectangles (or volumes of boxes and vice versa). (Lesson 10-2)

In 54 and 55 perform the multiplication.
54. $(3q + 7)(4p - 5)$
55. $(2x + y)(y - 3z)$

In 56 and 57, give two expressions for the area of the rectangle.

56.
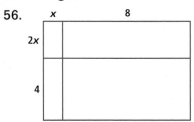

57.
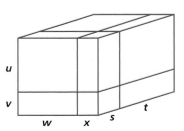

58. Give two expressions for the volume of the box below.

Chapter

11

Indirect Proofs and Coordinate Proofs

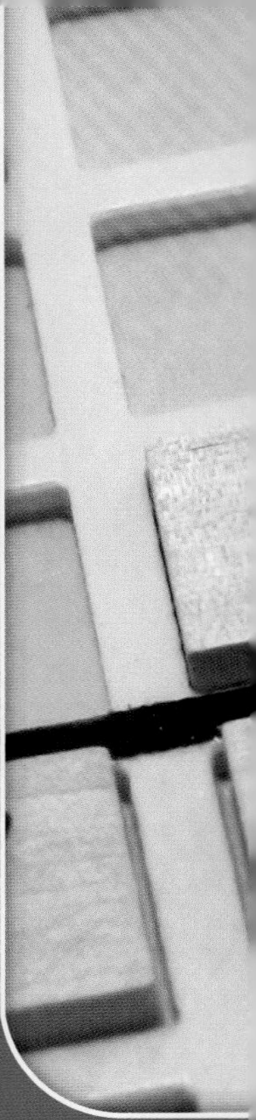

Contents

11-1 Ruling Out Possibilities

11-2 The Logic of Making Conclusions

11-3 Indirect Proof

11-4 Proofs with Coordinates

11-5 The Pythagorean Distance Formula

11-6 Equations of Circles

11-7 Means and Midpoints

11-8 Theorems Involving Midpoints

11-9 Three-Dimensional Coordinates

Almost all of the proofs that you have done in this course have been *direct proofs.* They proceed directly from given information to a conclusion. In this chapter, you will see two special kinds of proof. *Indirect proofs* start from a conclusion different from the one you want. *Coordinate proofs* are proofs involving figures that are defined by ordered pairs of real numbers on the coordinate plane. Along the way, you will work with the logic of proofs.

One bit of the logic of indirect proofs is to rule out possibilities. Sudoku number puzzles (Japanese: 数独) involve indirect reasoning. A typical grid for a Sudoku puzzle is shown below. The object is to put a digit in each empty square so that each row and each column contains all the digits 1 through 9. Such an array is called a *Latin square.*

For the sake of discussion, number the nine 3 × 3 sub-squares like the keys on a telephone, where subsquare 1 is the top left.

In subquares 4 to 6, there are already two 2s: one in subsuqare 4 and the other in subsquare 6. Therefore, we can try to place a 2 in subsquare 5. Suppose it is in the first row of that subsquare. This cannot be true because there is already a 2 in the first row of subsquare 4. Suppose it is in the third row. This also violates the rules, so it must be in the middle row of the subsquare. Of the three cells, two are in columns that already contain a 2. Therefore, the next 2 must be in the third column of the middle row of subsquare 5.

This type of reasoning, where you make hypothetical conjectures and find a contradiction is called *indirect reasoning*.

Indirect reasoning makes it possible to solve problems we could not do otherwise. Likewise, a greater variety of problems could be solved when René Descartes connected algebra and geometry using a coordinate system. You have already used coordinates to solve problems involving lines and transformations. In this chapter, coordinates are used to prove theorems about polygons and to describe circles.

Lesson
11-1
Ruling Out Possibilities

▸ **BIG IDEA** When the solution to a problem is one of a finite number of possibilities, an effective strategy may be to eliminate possibilities until only one is left.

The Law of Ruling Out Possibilities

In the short story "A Scandal in Bohemia," by Sir Arthur Conan Doyle, the famous fictional detective Sherlock Holmes remarks, "Once you eliminate the impossible, whatever remains, no matter how improbable, must be the truth." For instance, if Sherlock Holmes knows that either the maid or the butler committed a crime, and he determined it was not the maid, then the butler must be guilty. We call this principle of reasoning the *Law of Ruling Out Possibilities*.

Law of Ruling Out Possibilities

When statement *p* or statement *q* is true, and *q* is not true, then *p* is true.

This can be extended to "If you know that one of a set of possibilities must be true, and you can eliminate all but one, that one must be true."

The Law of Ruling Out Possibilities is used even by very young children. If a child knows that a toy is either in a parent's right hand or in the parent's left hand, and the right hand is opened and found empty, the child will know to look in the left hand for the toy.

In mathematics, ruling out possibilities is often easy. For example, you know every angle in a triangle is either acute, right, or obtuse. If $\angle A$ in $\triangle ABC$ is not acute or right, then, using the Law of Ruling Out Possibilities, $\angle A$ is obtuse. There is no other possibility.

In real life, if words are not carefully defined, you may not be able to rule out possibilities so easily. For example, if a person is not young, that does not necessarily mean the person is old.

Sudoku puzzles can be solved by repeated application of the Law of Ruling Out Possibilities.

Mental Math

Fill in the Blanks

a. If a triangle is not scalene, it must be ___?___.

b. If an integer is neither negative nor positive, it must be ___?___.

c. If a positive integer is prime and not odd, then the integer is ___?___.

Activity 1

Copy and solve this Sudoku puzzle. Remember that each of the numbers 1 through 9 must be in each row, each column, and each square of nine numbers only once.

9	4	7	2			3	8	5
5		6			4		1	9
2	8	1	5		3			
7	5			2		1		4
		2	3	7	1	9		
8		3		4			6	2
			1		2	8	7	6
1	2		4			5		3
6	7	5			9	4	2	1

Solving Logic Puzzles

In the questions for this lesson, you are asked to solve some logic puzzles. These puzzles use the idea of ruling out possibilities again and again. Notice how little information is given and how much you can deduce. The same happens in geometry and other branches of mathematics.

Here are some hints for doing these puzzles:

1. Logic puzzles take a lot of time and analysis, so do not hurry.

2. Construct a grid and place Xs in squares whenever something *cannot* occur. Place an O in the square when the situation *must* occur.

Example

Aaliyah, Marissa, Jordan, Hassan, and Ian each have a different hobby. Their hobbies are chess, fencing, sailing, hockey, and knitting. From the clues below, determine which hobby each student has.

(1) Aaliyah likes either sailing, hockey, or fencing.
(2) Jordan likes either hockey or knitting.
(3) Marissa does not like fencing, hockey, or sailing.
(4) Ian does not like sailing, knitting, fencing, or hockey.
(5) Hassan does not like fencing.

In fencing, points are scored when the tip of the blade touches a valid target area on the opponent.

Solution To solve this puzzle, the grid at the right can be used. The first clue tells you that Aaliyah does not like chess and does not like knitting. Two X_1s in the third row of the grid show this. (We call it X_1 so you can tell it comes from Clue 1.) The fourth clue tells you the hobbies Ian does not like. The four X_4s in the fifth row show this. Of course, this means that Ian likes chess. We show this with an O. Now we know that no one else likes chess, so we place Xs in the rest of the cells in the column labeled "Chess."

To continue, reread the clues and see what you can eliminate based on the clues and the grid on which you are recording your information. Look carefully at the grid. You now should know Aaliyah's hobby and the hobbies Aaliyah doesn't have. Put an O in the box showing that Aaliyah has fencing as a hobby and an X to show the things that are now ruled out as her hobbies. You are asked to complete this puzzle in Question 7.

Sometimes a single piece of given information may yield many conclusions, as in the following Activity.

Activity 2

When they were kids, Tina, Ernest, Charles, Lisa, and Jennifer were best friends. Each of them had a dream job. The jobs were actor, astronaut, writer, firefighter, and engineer. When they grew older they all got their dream jobs. These days they are called Ms. Apple, Mr. Leonards, Mr. Jameson, Ms. Nathan, and Ms. Willows, though not necessarily in that order. From the clues below, match each person's first name, last name, and profession.

(1) Tina and Lisa are not actors, and neither is Ms. Willows.
(2) Jennifer is not the firefighter.
(3) The astronaut is either Tina or Ms. Apple.
(4) Neither Ernest nor Ms. Nathan is the firefighter or actor.
(5) Mr. Leonards, Ernest, and the writer have all recently traveled to Europe.

Step 1 Again, we use a grid. The grid should have a space for each *last name-first name* pair, and for each *first name-job* pair, and also each *job-last name* pair.

Step 2 Look at Clue 1. Write five conclusions that you can make, looking for possibilities to rule out.

Step 3 Mark the grid with Xs or Os for the conclusions made in Step 2.

Step 4 Read Clue 2, make your conclusion(s), and mark them on the grid.

Step 5 Repeat Step 4 for the remaining clues and then start over again with Clue 1 until you have found each person's last name and occupation.

Questions

COVERING THE IDEAS

1. You examine a coin. The face that shows up is "heads." You conclude that the face on the other side is "tails." What principle of reasoning have you used?

2. **Fill in the Blank** If statement m or statement n is true, and m is not true, then __?__ must be true.

3. If A, B, and C are three different points, then either A, B, and C are collinear or there is a triangle ABC.

   a. If you know that A, B, and C are noncollinear, what can you conclude?

   b. What logical principle have you used in part a?

In 4–6, refer to Activity 2.

4. From Clue 3 alone, who is known not to be the astronaut?

5. Write at least two conclusions that follow from Clue 1.

6. Write at least two conclusions that follow from Clue 5.

7. Finish the Example of this lesson.

APPLYING THE MATHEMATICS

In 8 and 9, make a conclusion from the given information.

8. Lines m and n in space are neither skew nor parallel.

9. Point A is neither on nor inside the circle with center C and radius 5.

10. The **Trichotomy Law** for real numbers says: "Of two real numbers a and b, either $a < b$, $a = b$, or $a > b$, and no two of these can be true at the same time." You know that $\pi \neq 3.14159$. What can you conclude by using the Trichotomy Law?

11. James Fitzgerald, sculptor and logician, was complimented on his ability to carve a lifelike tiger from a piece of wood. He replied, "It's easy. You start with a block of wood and cut away all parts that do not look like a tiger." What Law of Logic was he using?

12. Jasmine was given a parallelogram that is not a rhombus, a triangle, a kite that is not a rhombus, and a square, and told to paint each with yellow, blue, red, or green paint. Each is labeled as Polygon 1, Polygon 2, Polygon 3, or Polygon 4. Make a grid and tell Jasmine what color to paint and what number to mark on each shape using the clues below.
    (1) The parallelograms and Polygon 4 are not green.
    (2) Polygon 1 and Polygon 3 have diagonals that are perpendicular.
    (3) The regular quadrilateral is not yellow or red.
    (4) The yellow polygon has two pairs of opposite sides congruent.
    (5) Polygon 1 does not have four congruent sides.

13. In the puzzle at the right, each square is to contain a digit from 1 to 9. No digit can appear twice in the same row or in the same column. The sums of each row and column are given. Find one of the two solutions.

			10
			19
			21
19	24	7	

14. Three students (Chance, Paco, and Tallulah) were placed in a group to do a geometry project. Each student had a favorite geometric shape (hyperbola, sphere, or triangle) and each student had a favorite mathematician from history (Euclid, Nikolai Ivanovich Lobachevski (1792–1856), and Emmy Noether (1882–1935)). Further, each mathematician had a favorite geometric shape. Assume these five statements:
   (1) Chance is not the student that liked the hyperbola.
   (2) Noether favored the sphere.
   (3) Euclid liked the triangle.
   (4) Paco doesn't like complicated figures such as the sphere or the hyperbola.
   (5) If a student favored a mathematician, and the mathematician favored a shape, then the student favored that shape.
   Write each triplet (student, shape, mathematician) of items that belong together. (You may wish to make one or more tables like the one below to record matches and mismatches.)

	Hyberbola	Triangle	Sphere
Chance			
Paco			
Tallulah			

REVIEW

15. A cylinder has height h and bases with radius r. Two hemispheres are glued onto the two bases of the cylinder. A front view of this figure is shown at the right.
   a. Give a formula for the volume of the figure. (Lessons 10-6, 10-3)
   b. Give a formula for the surface area of the figure. (Lessons 10-7, 9-9)

16. A container in the shape of a box is x feet wide, x feet long, and y feet high. The container is filled to the brim with water. How high would a container containing the same volume of water be, if it were y feet wide and y feet long? (Lesson 10-2)

17. **True or False** Any two nets for the same polyhedron will have the same area. (Lesson 9-8)

18. Name the special kinds of quadrilaterals in which the diagonals intersect each other and are perpendicular to each other. (Lesson 7-9)

19. Suppose $r_{y\text{-axis}}(x^2, 1) = (x^2 - 4, 1)$. Find x. (Lesson 4-2)

20. Simplify. **(Lesson 1-1)**

    a. $\left|5\frac{1}{3}\right| + |-12|$ **b.** $\left|5\frac{1}{3} - 12\right|$ **c.** $\sqrt{\left(5\frac{1}{3} - 12\right)^2}$

EXPLORATION

21. Below are the 15 clues to one version of a famous and difficult logic puzzle called "Who Owns the Zebra?" People have reported that Lewis Carroll or Albert Einstein wrote it, but the earliest appearance seems to be in 1962, after these men had died.

    (1) There are five houses in a row, each of a different color and inhabited by people of different nationalities, with different pets, drinks, and flowers.

    (2) The English person lives in the red house.

    (3) The Spaniard owns the dog.

    (4) Coffee is drunk in the green house.

    (5) The Ukrainian drinks tea.

    (6) The green house is immediately to the right (your right) of the ivory house.

    (7) The geranium grower owns snails.

    (8) Roses are in front of the yellow house.

    (9) Milk is drunk in the middle house.

    (10) The Norwegian lives in the first house on the left.

    (11) The person who grows marigolds lives in the house next to the person with the fox.

    (12) Roses are grown at the house next to the house where the horse is kept.

    (13) The person who grows lilies drinks orange juice.

    (14) The Japanese person grows gardenias.

    (15) The Norwegian lives next to the blue house.

    Who drinks water? And who owns the zebra?

22. Try your hand at writing your own puzzle problem. Have a friend or family member try and solve your puzzle.

23. Logic puzzles like those in this lesson can be found in many puzzle books. Find an example of a logic puzzle different from the ones in this lesson, and solve it.

Lesson
11-2
The Logic of Making Conclusions

▶ **BIG IDEA** Deductive proof is based on a small number of laws of logic.

Sudoku puzzles have become popular in spite of the fact that when you begin, it may look like there is no solution and you may not know where to start. However, by carefully using logic and ruling out possibilities, you can find a solution. The logic used in Sudoku puzzles is the same logic that is used in mathematics. Instead of reasoning from clues, in mathematics we reason from postulates, definitions, and previously proved theorems. Instead of finding out what number goes in what space, we try to determine what is true about figures, numbers, or other objects.

Deduction depends on some general assumed patterns of reasoning. These general patterns are the laws of logic.

Mental Math

On average, how fast is Malcolm driving, in miles per hour, if he drives 20 miles in

a. 20 minutes.

b. 15 minutes.

c. 30 minutes.

d. 40 minutes.

The Law of Detachment

To begin the study of logic, examine the proof below. Two conclusions and one justification have been highlighted.

Given $CA = CB$, $\angle MCA \cong \angle MCB$

Prove $\angle A \cong \angle B$

Proof

Conclusions	Justifications
1. $CA = CB$, $\angle MCA \cong \angle MCB$	1. Given
2. $\overline{CM} \cong \overline{CM}$	2. Reflexive Property of Congruence
3. $\triangle AMC \cong \triangle BMC$	3. SAS Congruence Theorem
4. $\angle A \cong \angle B$	4. CPCF Theorem (If two figures are congruent, then corresponding parts of those figures are congruent.)

If you let p stand for the statement "$\triangle AMC \cong \triangle BMC$" and q stand for the statement "$\angle A \cong \angle B$," the pattern of reasoning in Steps 3 and 4 in green above looks like what is shown at the right.

Conclusions	Justifications
p	
q	$p \Rightarrow q$

If p is accepted as true and $p \Rightarrow q$ is accepted as true, then the conclusion q must be accepted as true. This is a fundamental principle of logic. We call it the *Law of Detachment* because q is detached from $p \Rightarrow q$.

> ### Law of Detachment
>
> From a true conditional $p \Rightarrow q$ and a true statement p, you may conclude q.

 QY1

Be careful using the Law of Detachment. Consider Statements 1 and 2 below to be true.

> Statement 1: *If it rains, then the baseball game will be cancelled.*
>
> Statement 2: *It didn't rain.*

No conclusion is possible about the game from Statement 1 and Statement 2. Statement 1 tells what will happen if it rains. You can only make a conclusion using $p \Rightarrow q$ when the antecedent p is true. It could be that the baseball game might be cancelled even if it didn't rain. The game might be cancelled for the lack of umpires, for example.

The Law of Transitivity

Consider the following statements:

> Statement 3: *If an angle has measure 90°, then it is a right angle.*
>
> Statement 4: *If an angle is a right angle, then its sides are perpendicular.*

From these you can conclude:

> Statement 5: *If an angle has measure 90°, then its sides are perpendicular.*

The general form of this logic can be described with variables.

$$\begin{array}{ll} \text{If} & p \Rightarrow q \\ \text{and} & q \Rightarrow r, \\ \text{then} & p \Rightarrow r. \end{array}$$

This is called the *Law of Transitivity of Implication,* or the *Law of Transitivity* for short.

> ### Law of Transitivity
>
> If $p \Rightarrow q$ and $q \Rightarrow r$ are true, then $p \Rightarrow r$ is true.

 QY2

> **QY1**
>
> Give another example of the use of the Law of Detachment in the proof on the previous page.

> **QY2**
>
> Using just the Law of Transitivity, what can be concluded from Statements 6 and 7 below?
>
> Statement 6: *If you miss the bus, you will get home late.*
>
> Statement 7: *If you get out of school late, you will miss the bus.*

Example 1

What conclusion can be made using all three of the statements below?

(1) Theorem: The diagonals of any rectangle are congruent.

(2) Definition: If quadrilateral *ABCD* has 4 right angles, then *ABCD* is a rectangle.

(3) Given: *ABCD* has 4 right angles.

Solution Rewrite Statement 1 as an if-then statement.

If a figure is a ___?___, then ___?___.

Use the Law of Transitivity on Statement 2 and the rewritten Statement 1 to make a conclusion.

If quadrilateral ABCD ___?___, then ___?___.

Let p = "*ABCD* has 4 right angles" and use the Law of Detachment with your result from Step 2 to make a conclusion about *ABCD*.

___?___

The Negation of a Statement

The **negation** of a statement p, written **not-p**, is a statement that is false whenever p is true, and true whenever p is false. For instance, let p be the statement *Line a is perpendicular to line b.* Then the statement *Line a is* not *perpendicular to line b* is the negation of p, not-p.

Sometimes there are many ways of writing a negation. Let p = *It is raining.* Here are two ways of writing the negation of p:

not-p = *It is not raining.*

not-p = *It is raining* is not true.

Caution: The statement q = *It is snowing* is not the negation of *It is raining* because it is possible for both p and q to be false at the same time.

Sometimes there are special terms that are appropriate for negations. For instance, if r = $\triangle ABC$ *is isosceles*, here are two ways of writing the negation of r:

not-r = $\triangle ABC$ *is not isosceles.*

not-r = $\triangle ABC$ *is scalene.*

The verb "$\neq$" usually signifies the negation of "=". The negation of "$x = 2$" is "$x \neq 2$," and vice versa.

In English, it is often not considered good writing to use a double negative. However, in mathematics, double negatives can be useful. If you say △ABC *is not scalene,* then you mean *It is not true that* △ABC *is not isosceles.* This means △ABC *is isosceles.*

> **Double Negative Property**
>
> The statement not-(not-*p*) is true exactly when *p* is true, and false exactly when *p* is false.

Inverses and Contrapositives

Recall that the converse of the conditional statement $p \Rightarrow q$ is $q \Rightarrow p$. Let p = *the dimensions of a rectangle are 3 and 5* and q = *its area is 15.*

Original	$p \Rightarrow q$:	If the dimensions of a rectangle are 3 and 5, then its area is 15.
Converse	$q \Rightarrow p$:	If the area of a rectangle is 15, then its dimensions are 3 and 5.

Here the original conditional is true, and its converse is false.

Negating *both* the antecedent and the consequent of the original conditional gives a new conditional of the form **not-*p* $\Rightarrow$ not-*q*,** called the **inverse** of the original.

Inverse	not-$p \Rightarrow$ not-q:	If the dimensions of a rectangle are not 3 and 5, then its area is not 15.

This inverse is false. A rectangle with dimensions 7.5 and 2 does not have dimensions 3 and 5, but it has area 15.

However, if both parts of an original true conditional are negated and its antecedent and consequent are switched, a second *true* statement appears. This statement, of the form **not-*q* $\Rightarrow$ not-*p*,** is called the **contrapositive** of the original.

Contrapositive	not-$q \Rightarrow$ not-p:	If the area of a rectangle is not 15, then its dimensions are not 3 and 5.

Notice that the contrapositive is true, just as $p \Rightarrow q$ is true.

> ▶ **READING MATH**
>
> The English prefix *contra-*, which means "against," "opposite," and "contrasting," comes from the Latin *contra*, meaning "against." In addition to *contrapositive*, some words starting with *contra* are "contraband," "contradict," and "contrary."

GUIDED

Example 2

Given the conditional *If n ≠ 5, then n = 3,* write its converse, inverse, and contrapositive. Indicate which of the four statements are true and which are false.

(continued on next page)

Solution Let p = "n ≠ 5." Let q = "n = 3."

Converse (q ⇒ p): If ___?___, then ___?___.

Inverse (not-p ⇒ not-q): If ___?___, then ___?___.

Contrapositive (not-q ⇒ not-p): If ___?___, then ___?___.

The ___?___ and ___?___ are true. The ___?___ and ___?___ are false.

Both Guided Example 2 and the rectangle example preceding it verify the following law of logic.

Law of the Contrapositive

A conditional (*p* ⇒ *q*) and its contrapositive (not-*q* ⇒ not-*p*) are either both true or both false.

 QY3

▶ **QY3**

Fill in the Blank Finish this statement so that both it and its contrapositive are false: If a figure is a rectangle, then it is a ___?___.

You might notice that using the Double Negative Property, the contrapositive of the inverse of a conditional statement is the converse of the original conditional statement. This means that both the converse of a conditional statement and the inverse of a conditional statement have the same truth value, that is, they are either both true or both false.

Two statements are **logically equivalent** if and only if they are both true or both false at the same time. The statements above show that any conditional statement and its contrapositive are logically equivalent, as are the converse and the inverse of the original conditional.

The map at the right of the province of Alberta in Canada, illustrates this further. The statement *If you are in Medicine Hat, then you are in Alberta* can be considered as the original conditional statement. It is a true statement. The contrapositive is: *If you are not in Alberta, then you are not in Medicine Hat.* This is also a true statement. The converse, *If you are in Alberta, then you are in Medicine Hat* and the inverse, *If you are not in Medicine Hat, then you are not in Alberta* are both false statements.

Example 3

Prove *If a quadrilateral is not a kite, then it is not a square.*

Solution This statement is true because it is the contrapositive of If a quadrilateral is a square, then it is a kite. If a statement is true, by the Law of the Contrapositive, its contrapositive is true.

Using Laws of Logic to Solve Puzzles

Lewis Carroll (1832–1898), the author of *Alice's Adventures in Wonderland* and *Through the Looking Glass,* was a logician. That is, he studied the process of reasoning. He was a professor at Oxford University in England and wrote some books on logic. Here is a puzzle from one of his books.

Lewis Carroll is the pen name of Charles L. Dodgson.

Example 4

Given these three statements, what can you conclude?

(1) *Babies are illogical.*
(2) *Any person who can manage an alligator is not despised.*
(3) *Illogical persons are despised.*

Solution To avoid lots of writing, use variables to name the parts of the statements. Let:

B = A person is a baby.
D = A person is despised.
M = A person can manage an alligator.
I = A person is illogical.

Now

(1) becomes B ⇒ I.
(2) becomes M ⇒ not-D.
(3) becomes I ⇒ D.

From (1) and (3) you can use the Law of Transitivity to conclude B ⇒ D. The contrapositive of (2) is D ⇒ not-M, which you can conclude because of the Law of the Contrapositive. Using the Law of Transitivity again, B ⇒ not-M. This is what you can conclude: If a person is a baby, then that person cannot manage an alligator. Or, written in a shorter way: A baby cannot manage an alligator.

Questions

COVERING THE IDEAS

1. What logical principle is used in the proof on the first page of this lesson?

2. **Matching** Match each bit of reasoning with one of the choices below.

     i. Law of Detachment   **ii.** Law of Transitivity

     iii. Law of the Contrapositive   **iv.** poor reasoning

   a. *If you study, then you will pass.*
     This means if you didn't pass, then you didn't study.

   b. *If you study, then you will pass.*
     Belle studied. She must have passed.

   c. *If you study, then you will pass.*
     So if you pass, then you must have studied.

   d. *If you study, then you will pass.*
     Herman didn't study. So he will not pass.

3. *If a triangle is a right triangle, then the midpoint of the longest side is equidistant from the three vertices of the triangle.*

   a. Write the converse of the given statement.

   b. Write the inverse of the given statement.

   c. Write the contrapositive of the given statement.

   d. Which of statements a–c and the given statement are true?

4. Write the negation of this statement: $\angle A$ is acute.

5. Write the negation of this statement without using the word "not": $\triangle ABC$ *is scalene.*

6. The number $\frac{1}{3}$ is not irrational.

   a. Is this a double negative?

   b. How else might this statement be written?

In 7–9, for each of the following, write *true*, *false*, or *cannot tell*.

7. If a statement is true, then its contrapositive is ___?___.

8. If a statement is true, then its converse is ___?___.

9. If not-p is true, then p is ___?___.

10. Assume these three statements are true.

   (1) *If you wear a tie, then you are a snazzy dresser.*

   (2) *If you do not have shiny shoes, then you did not use shoe polish.*

   (3) *All snazzy dressers use shoe polish.*

   What can be concluded using all of the statements?

APPLYING THE MATHEMATICS

11. A large store has the slogan *If we don't have it, then you don't need it.*

    a. What is the contrapositive of this statement?

    b. Do you think this contrapositive is the intended message?

In 12–14, if possible, write a statement about prisms in which:

12. The statement is true and the converse is also true.

13. The statement is true but the converse is false.

14. The statement is true but the contrapositive is false.

15. Here is a solution to an equation, with statements and justifications.

    $p: 2x^2 + x - 3 = 0$ Given equation

    $q: x = \dfrac{-1 \pm \sqrt{1^2 - 4 \cdot 2 \cdot (-3)}}{2 \cdot 2}$ Quadratic Formula

    $r: x = -1.5$ or $x = 1$ Evaluate the expression.

    Describe how the Laws of Detachment and Transitivity are used in this solution.

16. A police officer said, "It cannot be said that Taylor did not fail to stop for the stop sign." This is a triple negative that could be written in the form not-(not-(not-p)).

    a. Identify the statement p.

    b. If p is true, not-(not-(not-p)) is ___?___.

    c. In simple language, what do you think the officer wanted to say?

17. A political candidate said, "If George Washington were alive today, we wouldn't have income taxes." Comment on this candidate's logic.

REVIEW

18. A sphere with radius r fits snugly into a cube. **(Lessons 10-7, 9-9)**

    a. What is the length of one of the edges of the cube?

    b. What is the ratio of the surface area of the sphere to the surface area of the cube?

19. Deja has toy blocks that are in the shape of cubes with edges of length 2 cm. She wants to construct a sphere of radius 20 cm, or at least as close as possible to this sphere as she can. **(Lessons 10-6, 10-3)**

    a. What is the volume of a sphere with radius 20 cm?

    b. How close to this volume is it possible to get using the building blocks?

20. The ancient Greek mathematicians were concerned with the following problem: Given a cube, is it possible to construct a square whose area is equal numerically to the volume of the cube? Suppose a cube has edges of length a. What is the length of a side of a square whose area is the same as the volume of the cube? **(Lessons 10-1, 8-2)**

21. Consider the photo of an ambulance at the right. **(Lesson 9-7)**

    a. What would the lettering on the hood of the ambulance look like when viewed through a rear view mirror?

    b. If there were a sign on the back of the ambulance that read "AMBULANCE," how would it make sense to write it? Why?

22. Find the length of the hypotenuse of the right triangle with vertices $(0, 0)$, $(7, 0)$, and $(7, -2)$. **(Lesson 8-6)**

23. Let ℓ be the line with equation $y = 3x + 7$. Let m be the line passing through $(1, 8)$ that is perpendicular to ℓ. Write an equation for m in standard form. **(Lesson 3-8)**

24. Using a number line, find the distance between points with coordinates of -7.3 and 8.6. **(Lesson 1-1)**

EXPLORATION

25. There is an old tale about a land that was inhabited by two tribes of people. One tribe was called the "Truth Tellers" because they always told the truth. The other tribe was called the "Liars" because they never told the truth. There was no way to tell a Truth Teller from a Liar other than to ask a question and determine from the answer if the person was telling the truth or not. Also, there was a law in the land that an inhabitant was only allowed to answer one question from a tourist. A tourist was visiting the land and came to a fork in a road. The tourist knew that one road led to the town, while the other road led to a forest. The tourist did not want to go to the forest, but did want to go to town. There was one inhabitant standing at the fork who knew the land well. The tourist asked the inhabitant exactly one question. With the answer, the tourist proceeded directly to town. What was the question the tourist asked?

Lesson 11-3

Indirect Proof

Vocabulary

direct reasoning

direct proof

indirect reasoning

indirect proof

contradictory statements

▶ **BIG IDEA**   Indirect proof is based on the idea of reasoning from a supposition until a contradiction is obtained, and then, as a result, asserting that the supposition must be false.

A lawyer, summing up a case for the jury, said "The prosecutors claim that my client is guilty. Let us suppose, for a moment, that he is guilty. If my client is guilty, then he must have been at the scene of the crime when the crime was committed. But as you remember, we brought in witnesses and telephone records that demonstrate that my client was 50 miles away at the time. So he wasn't at the scene of the crime when it was committed. He could not have been both at the scene of the crime and also 50 miles away. Our assumption has led to a contradiction; therefore the supposition must be false and my client is not guilty."

In this summation, the lawyer has used *indirect reasoning*.

In **direct reasoning**, a person begins with given information known to be true. The Laws of Detachment and Transitivity are used to reason from that information to a conclusion. The proofs you have written so far in this book have been **direct proofs**.

In **indirect reasoning**, a person examines and tries to rule out all the possibilities other than the one thought to be true. This is exactly what you did in solving the logic puzzles of Lesson 11-1. You marked Xs in boxes to show which possibilities could not be true. When you had enough Xs, you knew that the only possibility left must be correct.

The client must have cringed when he heard his lawyer say, "Let us suppose, for a moment, that he is guilty." But this is an effective argument using indirect reasoning. If not-p leads to a contradiction, not-p must be false. This makes p true. This is the fifth and last law of logic discussed in this book.

Law of Indirect Reasoning

If valid reasoning from a statement p leads to a false statement, then p is false (so not-p is true).

A proof that uses indirect reasoning is called an **indirect proof**.

It is helpful to think about an indirect proof as having three parts.

Supposition: Begin by supposing the negation of what you want to prove. (If you want to prove p, suppose not-p.)

Deduction to Contradiction: Use valid reasoning to show that the supposition leads to a false statement.

Final Conclusion: Use the Law of Indirect Reasoning to conclude what you want to prove.

 QY

▶ **QY**

Identify these three parts of the lawyer's argument.

Notice that the first part, the *supposition,* is quite different from the first step in a direct proof. In a direct proof, the first step usually is one of the given statements that we know to be true. A supposition, on the other hand, is just something that is supposed without knowing whether it is true or false, in the hope that it turns out to be false.

Contradictory Statements

In the deduction part of an indirect proof, you need to show that a statement is false. But how can you tell if it is false? One way to tell that a statement is false is if you know its negation is true. For instance, suppose you know $y = 5$ is true. Then $y \neq 5$ must be false. Suppose you know $\triangle ABC$ is isosceles. Then the statement *$\triangle ABC$ is scalene* must be false.

You also know a statement is false if it contradicts another statement known to be true. Two statements p and q are **contradictory** if and only if they cannot both be true at the same time. For instance, if Jayla is a freshman, then she cannot be a junior. If you know x is positive, then x cannot be negative. If you know a triangle is scalene, then it cannot be isosceles. A statement and its negation are always contradictory.

Indirect Proofs You Have Seen

One of the first proofs in this book, for the Line Intersection Theorem on page 34, was indirect. Here it is reproduced with the three parts identified.

Theorem Two different lines intersect in at most one point.

Proof *Supposition:* Suppose that two lines intersect in two different points P and Q.

Deduction to Contradiction: Then through P and Q there are two lines. But the Unique Line Assumption of the Point-Line-Plane Postulate indicates that there is exactly one line through two points.

Final Conclusion: This contradiction indicates that the supposition is false. So, applying the Law of Indirect Proof, two different lines cannot intersect in two different points.

When you have encountered an equation or system with no solution, you may have been doing an indirect proof without knowing it.

Example 1

Show that the system $\begin{cases} 5x + 20y = 24 \\ x + 4y = 3 \end{cases}$ has no solution.

Solution *Supposition:* Suppose there is a solution.

Deduction to Contradiction: Then, to find the solution, multiply both sides of the second equation by 5.

$$\begin{cases} 5x + 20y = 24 \\ 5x + 20y = 15 \end{cases}$$

Subtracting the second equation from the first,

$$0 = 9.$$

This contradicts the fact that $0 \neq 9$.

Final Conclusion: So the supposition must be false. Consequently, applying the Law of Indirect Reasoning, there is no solution.

A Famous Indirect Proof from Euclid's Elements

Euclid's *Elements* cover more than geometry. Included among the topics are theorems about prime numbers. Recall that the *prime numbers* are those positive integers that are divisible only by themselves and 1. In order, the prime numbers are 2, 3, 5, 7, 11, 13, 17, 19, 23, . . . Recall that a positive integer greater than 1 that is not prime is called a *composite number.*

For the following proof, you need to know two things Euclid had already proved: First, if a number n is divisible by a particular prime number, then $n + 1$ cannot also be divisible by that prime. Second, every integer has a unique factorization into primes.

Infinitude of Primes Theorem

There are infinitely many prime numbers.

Proof The proof is indirect.

Supposition: Suppose that there are only finitely many primes. Then there is a largest prime. Call that number P.

Deduction to Contradiction: Let n be the number that is 1 greater than the product of all the prime numbers: $n = 2 \cdot 3 \cdot 5 \cdot 7 \cdot \ldots \cdot P + 1$. Now either n is prime or n is composite. If n is prime, then because it is obviously larger than P, we have a contradiction of the supposition. If n is composite, then it is factorable into primes. But it cannot have any of the primes from 2 to P as a factor because it is 1 greater than a multiple of them. So it would have to have a prime greater than P as a factor. This also contradicts the supposition.

Final Conclusion: Because in both cases the supposition leads to a contradiction, the supposition is false. Applying the Law of Indirect Proof, there are not finitely many prime numbers. There must be infinitely many primes.

Still Another Indirect Proof

Here is an example using content from this book.

Example 2

Prove: In a scalene triangle, no median is an altitude.

Solution First set up the given, what is to be proved, and draw a picture as you would do with a direct proof.

Given: $\triangle ABC$ is scalene.
       $\overline{CM}$ is a median of $\triangle ABC$.
Prove: $\overline{CM}$ is not perpendicular to $\overline{AB}$.
Supposition: Suppose $\overline{CM} \perp \overline{AB}$.
Deduction to Contradiction:

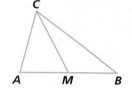

Conclusions	Justifications
1. $\angle AMC \cong \angle CMB$	1. $\perp$ lines $\Rightarrow \cong$ right $\angle s$
2. $\overline{CM}$ is a median.	2. Given
3. $\overline{AM} \cong \overline{MB}$	3. definition of median
4. $\overline{CM} \cong \overline{CM}$	4. Reflexive Property of Congruence
5. $\triangle AMC \cong \triangle BMC$	5. SAS (Steps 1, 3, 4)
6. $\overline{AC} \cong \overline{BC}$	6. CPCF Theorem
7. $\triangle ABC$ is isosceles.	7. definition of isosceles triangle
8. $\triangle ABC$ is scalene.	8. Given

Final Conclusion: Since the supposition $\overline{CM} \perp \overline{AB}$ leads to contradictory statements (Steps 7 and 8), it must be false. So, by the Law of Indirect Reasoning, $\overline{CM}$ is not perpendicular to $\overline{AB}$.

The form of Example 2 can serve as a model for other indirect proofs.

GUIDED

Example 3

Prove that no convex quadrilateral has four acute angles.

Solution

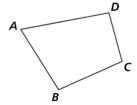

Given: Quadrilateral ABCD

Prove: ___?___

Proof: Supposition: Suppose ABCD has ___?___.

Deduction to Contradiction: Then, by the definition of ___?___, $m\angle A < 90$, $m\angle B < 90$, ___?___ < 90, and ___?___.

Adding the sides of these four inequalities,
$$m\angle A + m\angle B + m\angle C + m\angle D < 90 + 90 + 90 + 90$$
$$m\angle A + m\angle B + m\angle C + m\angle D < 360.$$

But, by the Quadrilateral-Sum Theorem, ___?___ $= 360$.

Final Conclusion: Because the supposition leads to a contradiction, it must be false. Therefore, by ___?___, ___?___.

Questions

COVERING THE IDEAS

1. What are the three parts of an indirect proof?

2. a. What is a contradiction?
   b. What are contradictory statements?

3. State the Law of Indirect Reasoning.

4. Give an example of a contradiction different from any in this lesson.

5. In an indirect argument to prove that two coplanar lines are parallel, what supposition might you start with?

In 6 and 7, statements p and q are given.
a. Are p and q contradictory?
b. Explain your answer.

6. p: $JKLM$ is a kite.
   q: $JKLM$ is a rectangle.

7. p: $\overline{AB} \perp \overline{BC}$
   q: $m\angle ABC = 88$

8. Pretend you are about to prove the following theorem with an indirect proof: *If P is a point not on line m, then there is not more than one line through P parallel to line m.* What would you have to show in the deduction part of the proof?

9. Use an indirect argument, as in the Examples, to show that the lines with equations $2x - y = 5$ and $3y = 6x + 5$ are parallel without using the idea of slope.

10. Refer to the proof that there are infinitely many prime numbers.

   a. Explain how you can tell that no prime number less than 13 is a factor of $2 \cdot 3 \cdot 5 \cdot 7 \cdot 11 + 1$.

   b. $2 \cdot 3 \cdot 5 \cdot 7 \cdot 11 \cdot 13 + 1 = 59 \cdot 509$, and both 59 and 509 are prime. What does this fact have to do with the proof?

11. Use an indirect proof to prove: *In a scalene triangle, no altitude is a median.*

12. a. Prove that no convex quadrilateral has four obtuse angles.

   b. Can a convex quadrilateral have three obtuse angles? If so, draw such a quadrilateral. If not, why not?

13. Use an indirect proof.

   Given $\triangle ABC$.

   Prove $\triangle ABC$ cannot have two right angles.

APPLYING THE MATHEMATICS

14. Bianca just turned 16, the legal age for driving in her state. Bianca's father said, "A kid of your age should not be driving." Bianca replied, "You may be right." (She now had his attention.) Bianca continued, "Look what happens if I don't drive. I will not be able to run errands for Mom. I will have to be driven to school, and the school functions you want me to attend, or not be able to attend them. This will disrupt both of your schedules. I think this may contradict what you want to happen." Bianca stopped here and let her father come to a conclusion.

   a. What conclusion could Bianca's father make?

   b. Identify the three parts of an indirect proof in Bianca's logic.

15. As a decimal, $\sqrt{2} = 1.41421\ldots$. As a decimal, $\frac{1393}{985} = 1.41421\ldots$. Show, by indirect reasoning, that $\sqrt{2} \neq \frac{1393}{985}$. (Hint: Begin by assuming the two numbers are equal. Then square both sides of the equation.)

16. Use an indirect proof to show that the equation $5(3x - 12) = 3(5x - 18)$ has no solution.

17. Write an indirect proof that "There is no integer that is greater than all other integers." (Hint: If you consider a largest integer *N*, can you create a larger one?)

18. Prove that if a rectangle is not a square, its diagonals cannot be perpendicular.

REVIEW

19. Consider the statement: *If Jimmy shows up on time, I will eat my hat!* Write its converse, inverse, and contrapositive. **(Lesson 11-2)**

20. Make a conclusion using all three of the following true statements. **(Lesson 11-2)**
    (1) Every square is a rhombus.
    (2) The diagonals of a kite are perpendicular.
    (3) If a figure is a rhombus, then it is a kite.

21. Trace the perspective view of the room shown at the right. **(Lesson 9-4)**
    a. Draw a door on the back wall in perspective.
    b. On the right wall, draw a window in perspective.
    c. Draw a rectangular rug on the floor in perspective.

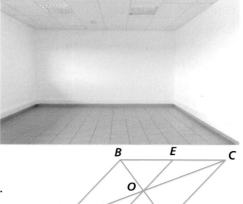

22. Given: *ABCD* is a parallelogram.
    Prove: $\triangle AOF \cong \triangle COE$ **(Lesson 7-7)**

23. a. Find the slope of the line with equation $9x + 8y = 36$.
    b. Give an equation of a line that is perpendicular to the line from Part a. **(Lessons 3-8, 1-2)**

24. Give a good definition and a bad definition of circle. **(Lesson 2-4)**

25. Are the points $(2, 7)$, $(-3, 14)$, and $(101, 102)$ collinear? How do you know? **(Lesson 1-2)**

26. Explain why $(a - b)^2 = (b - a)^2$ for all values of *a* and *b*. **(Previous Course)**

EXPLORATION

27. At the right, points *A* through *G* are vertices of a cube. Use either a direct proof or an indirect proof to show that $\overline{AC}$ and $\overline{CE}$ are not perpendicular.

28. Write a story similar to the one in Question 14. Then identify the parts of an indirect proof in the story you wrote.

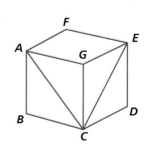

Supposition (Step 1): The lawyer supposes his client was guilty. Deduction to Contradiction (Step 2): The lawyer shows that his client could not have been present at the scene of the crime at the time of the crime. Final Conclusion (Step 3): The lawyer concludes that his client is not guilty.

Proofs with Coordinates

Vocabulary

convenient location
for a figure

▶ **BIG IDEA** Coordinate proofs are like other proofs, but start with a figure graphed on a coordinate plane.

So far, almost every proof in this book has been in *synthetic geometry*, where points are locations. Because figures in coordinate geometry have the same properties as those in synthetic geometry, it is not surprising that there are proofs in coordinate geometry. They have conclusions and justifications just like other proofs. However, because the proofs use coordinates, the justifications are often from algebra or arithmetic.

Recall that the slope m of a line determined by two points (x_1, y_1) and (x_2, y_2) is defined as $m = \frac{y_2 - y_1}{x_2 - x_1}$. Recall also the Parallel Lines and Slopes Theorem: Two nonvertical lines are parallel if and only if they have the same slope. These ideas are utilized in Example 1.

Mental Math

A sphere's radius is greater than 4 inches. What can you conclude about

a. the area of a great circle of the sphere?

b. the sphere's surface area?

c. the sphere's volume ?

Example 1

Consider quadrilateral *ABCD* with vertices $A = (3, 3)$, $B = (5, 10)$, $C = (15, 10)$, and $D = (13, 3)$. Prove that *ABCD* is a parallelogram.

Solution Draw and label a picture, as done at the right. In the drawing, it appears that *ABCD* is a parallelogram.

ABCD is a parallelogram if both pairs of opposite sides are parallel. Recall that two lines are parallel if they have the same slope. So calculate the slopes of the sides of *ABCD*.

To stress the justification of each step, this proof is written in two-column form.

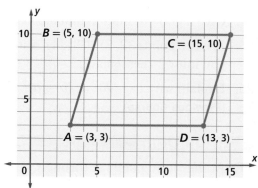

Conclusions	Justifications
1. slope of $\overline{AD} = \frac{3-3}{13-3} = \frac{0}{10} = 0$	1. definition of slope
slope of $\overline{BC} = \frac{10-10}{15-5} = \frac{0}{10} = 0$	
slope of $\overline{DC} = \frac{10-3}{15-13} = \frac{7}{2}$	
slope of $\overline{AB} = \frac{10-3}{5-3} = \frac{7}{2}$	
2. $\overline{AD} \parallel \overline{BC}$, $\overline{DC} \parallel \overline{AB}$.	2. Parallel Lines and Slopes Theorem
3. $ABCD$ is a parallelogram.	3. definition of parallelogram

The proof in Example 1 is about a single parallelogram. To prove a theorem that applies to all parallelograms, the coordinates of the vertices must be variables.

Example 2

Prove that the diagonals of a square $SQRE$ with $S = (-a, -a)$, $Q = (-a, a)$, $R = (a, a)$, and $E = (a, -a)$ are perpendicular.

Proof The Perpendicular Lines and Slopes Theorem tells us that the diagonals are perpendicular if the product of their slopes is -1. Calculate the slopes of the diagonals:

Slope of $\overline{SR} = \frac{a-(-a)}{a-(-a)} = \frac{a+a}{a+a} = \frac{2a}{2a} = 1$

Slope of $\overline{EQ} = \frac{a-(-a)}{-a-a} = \frac{a+a}{-2a} = \frac{2a}{-2a} = -1$

The product of the slopes is $1 \cdot -1$, or -1.

The diagonals of $SQRE$ are perpendicular by the Perpendicular Lines and Slopes Theorem.

The last statement of the proof in Example 2 is imperative. Without it, your proof is like an incomplete sentence. You must justify why your calculations prove the statement that is to be proved.

Convenient Locations

The coordinates of the vertices of the square in Example 2 were carefully chosen to be representative of a general square. By using variables as coordinates, the length of the side of the square is arbitrary. But the coordinates of the square were chosen such that the center is the origin. Still, the above proof holds for any square; not just those centered at the origin. Why? Because no matter where a square is located, a coordinate plane can be established with the origin at the center of the square and the x-axis and y-axis each parallel to two sides of the square.

The coordinates given in Example 2 are said to be a *convenient location* for the square. A **convenient location for a figure** is one in which its *key points* are described with the least number of different variables. In Example 2, the vertices of the square are described with just one variable, *a*. This was possible because all sides of a square share one relationship: The lengths are the same. Additional variables are necessary when there are multiple relationships among sides of the polygon. For example, in a general rectangle, two pairs of opposite sides are congruent, but they are not all congruent to each other. Thus, two variables are necessary.

 QY1

The location in Example 2 turns out to be convenient because the square is rotation-symmetric with center of rotation (0, 0). When a polygon is reflection-symmetric, a convenient location can usually be found by locating the polygon so that it is symmetric to the *x*-axis or the *y*-axis. Otherwise, a convenient location can be found by placing the polygon with one vertex at (0, 0) and another vertex on one of the axes.

 QY2

To remember convenient locations, recall how to find certain reflection and rotation images on a coordinate plane.

The reflection image of (*a*, *b*) over the *x*-axis is (*a*, –*b*).
The reflection image of (*a*, *b*) over the *y*-axis is (–*a*, *b*).
The image of (*a*, *b*) under a rotation of 180° about the origin is (–*a*, –*b*).

The point (*a*, *b*) and the three images (*a*, –*b*), (–*a*, *b*) and (–*a*, –*b*) are the vertices of a rectangle. This is another convenient location for a rectangle. Here are some convenient locations for some of the figures you have studied.

▶ QY1

In the rectangle below, each horizontal side has length *a* and each vertical side has length *b*. Complete the convenient locations of its vertices.

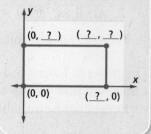

▶ QY2

Name three other types of quadrilaterals that are rotation-symmetric.

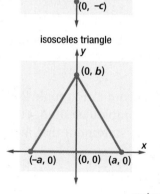

kite

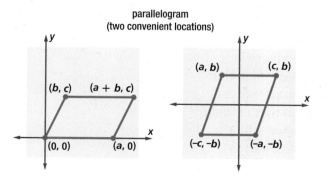

isosceles triangle

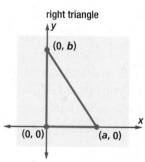

right triangle

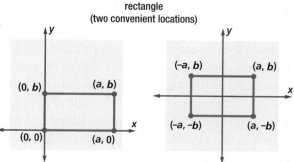

parallelogram (two convenient locations)

rectangle (two convenient locations)

Notice how the convenient locations take advantage of the symmetry of the polygons. If a polygon is rotation-symmetric, like the parallelogram, then the origin can be at the center of the rotation. If a polygon is reflection-symmetric, like the kite, then the line of symmetry can be placed along either the *x*-axis or the *y*-axis.

STOP QY3

Once you have a figure in a convenient location, coordinate proofs of parallelism and perpendicularity can be rather straightforward. Just calculate and compare slopes and remember to justify how the slopes help to prove or disprove your statement.

> **QY3**
>
> Use the convenient location for the parallelogram, with one vertex at (0, 0), to show that the opposite sides of a parallelogram are parallel.

Questions

COVERING THE IDEAS

1. **Fill in the Blank** To prove a theorem that applies to all figures of a given type, the nonzero coordinates of the vertices must be ___?___.

2. Create a convenient location for a square that has one vertex at the origin, one side on the *x*-axis, and a side of length *s*.

3. In the lesson, a convenient location is shown for a kite whose symmetry diagonal lies on the *y*-axis. Create a convenient location for a kite whose symmetry diagonal is on the *x*-axis.

4. If $C = (9, 7)$, $N = (3, 6)$, $J = (-3, 2)$ and $O = (4, 3)$, prove that *CNJO* is *not* a trapezoid.

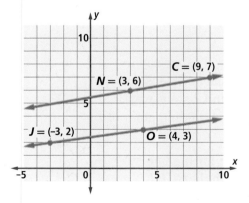

In 5 and 6, draw a figure of the indicated type in a convenient location on a coordinate plane.

5. right triangle with legs of length 14 and 17

6. isosceles triangle that is symmetric to the *x*-axis

7. *MARK* is a rhombus with vertices $M = (-3, 0)$, $A = (-2, 5)$, $R = (3, 6)$ and $K = (2, 1)$. Prove that the diagonals of *MARK* are perpendicular.

8. A rhombus in a convenient position has vertices $(a, 0)$, $(0, b)$, $(a, 0)$, and $(0, -b)$.

   a. Draw such a rhombus.

   b. Explain why the diagonals of the rhombus are perpendicular.

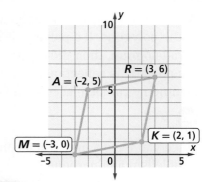

APPLYING THE MATHEMATICS

9. Give the vertices of a general isosceles trapezoid in a convenient location that is symmetric to the *y*-axis.

10. Given: Quadrilateral *KAYL*, with $K = (-c, -d)$, $A = (-a, b)$, $Y = (c, d)$, and $L = (a, -b)$, where $a + c \neq 0$.

    Prove: *KAYL* is a parallelogram.

11. Fill in the missing coordinate for the convenient location for the equilateral triangle at the right.

12. Prove that the diagonals of a kite are perpendicular.

In 13 and 14, consider quadrilateral *SANG* with $S = (0.3, -0.3)$, $A = (0.4, -0.1)$, $N = (-0.2, 0.2)$, and $G = (-0.3, 0)$.

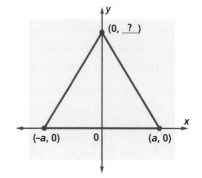

13. a. Find the image of *SANG* under a size change of magnitude 10 and center at the origin. Call the image *S'A'N'G'*.

    b. Prove that *S'A'N'G'* is a parallelogram.

    c. From this, explain why *SANG* must be a parallelogram.

14. Prove: *SANG* is a rectangle.

15. Tiana and Angel are on a treasure hunt. Angel's map says, "Go 3 steps north and 2 steps east." Tiana starts 4 steps south of Angel. Her map says, "Go 3 steps south and 2 steps west." Describe the figure formed by connecting the ending point of each girl to each starting point.

16. Consider the figure below. Find an equation for the line containing point *B* that would create a rectangle *BCDE*.

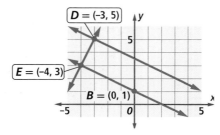

REVIEW

17. The famous baseball player Yogi Berra was also famous for many memorable quotations. He once said: "If you can't imitate him, don't copy him." Write down the converse, the inverse, and the contrapositive of this statement. (**Lesson 11-2**)

18. In Lewis Carroll's book *Alice's Adventures in Wonderland,* the following line appears: "Contrariwise," continued Tweedledee, "if it was so, it might be; and if it were so, it would be; but as it isn't, it ain't. That's logic." What principle of logic is Tweedledee using in his argument? (**Lesson 11-1**)

19. One box contains a sphere of radius 2, and another box contains two spheres each of radius 1. If both boxes are as small as possible, which has the greater volume? (**Lesson 10-7**)

Yogi Berra played for the New York Yankees from 1946 to 1963, usually as a catcher.

20. Suppose Ty, Rachelle, and Makayla all live on the same side of the street. Ty lives between Rachelle and Makayla. He lives 350 yards from Makayla, and Makayla and Rachelle live 975 yards apart. (**Lesson 1-6**)
    a. Picture the situation on a number line and label the given distances.
    b. How far does Ty live from Rachelle?

21. Full-grown zebras can range from 46 to 55 inches high at the shoulder and their weights can range from 550 to 650 pounds. Let h be these possible heights and w be these possible weights. (**Previous Course**)
    a. Graph all possible ordered pairs (h, w).
    b. Describe the graph.

22. Simplify the expression $\dfrac{\sqrt{x^4}}{\sqrt{x^2}}$. (**Previous Course**)

23. Simplify the expression $(-(x - y))^2 - (x + y)^2$. (**Previous Course**)

EXPLORATION

24. Three vertices of a parallelogram are $(-2, 0)$, $(2, 3)$, and $(3, -1)$.
    a. Find at least two possible locations of the fourth vertex.
    b. Are there other possible locations? Explain.

25. The equations of three lines are given $\overleftrightarrow{BC} = -\frac{1}{2}x + 2$, $\overleftrightarrow{CD} = 2x$, and $\overleftrightarrow{DA} = -\frac{1}{2}x - 2$. Find an equation for $\overleftrightarrow{AB}$ so that $ABCD$ is a rectangle.

QY ANSWERS

1. first quadrant rectangle with vertices $(0, 0)$, $(a, 0)$, (a, b), and $(0, b)$

2. parallelogram, rectangle, rhombus

3. The slopes are $\frac{c - 0}{b - 0} = \frac{c}{b}$, $\frac{c - 0}{(a + b) - a} = \frac{c}{b}$, $\frac{0 - 0}{a - 0} = 0$, and $\frac{c - c}{(a + b) - b} = 0$. Opposite sides have the same slope, so they are parallel.

Lesson
11-5

The Pythagorean Distance Formula

Vocabulary

taxicab distance

▶ **BIG IDEA** From the Pythagorean Theorem, a formula can be obtained for the distance between any two points in the plane.

In Lesson 11-4, you saw that by putting figures on the coordinate plane, you could prove that certain lines are parallel or perpendicular. This lesson shows how to determine whether segments on the coordinate plane are congruent.

Distances on the Coordinate Plane

In some rural areas of the United States, there are roads 1 mile apart going north, south, east, and west. Juan's house is 3 miles east and 5 miles north of Cierra's house, as shown on the grid at the right with Cierra's house at point (0, 0) and Juan's house at (3, 5).

When Cierra goes from her house to Juan's house along the roads, she travels 8 miles. This distance, along horizontal and vertical lines, is sometimes called the **taxicab distance** from one place to another.

A bird flying from Cierra's house to Juan's house would fly directly along $\overline{CJ}$, taking the shortest distance. This distance is the length of the hypotenuse of a right triangle with legs 3 miles and 5 miles. It can be found using the Pythagorean Theorem. It is the distance "as the crow flies" and is the usual distance between points in geometry. You can also think of this distance as the magnitude of a vector starting at (0, 0) and ending at (3, 5).

🛑 **QY1**

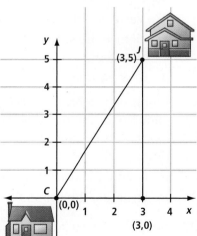

Any time that you need to find the distance between two points on the coordinate plane on an oblique line, you can draw a right triangle whose hypotenuse is the distance between the two points.

GUIDED

Example 1
Find the distance d between $(-26, 7)$ and $(28, 25)$ to the nearest tenth.

> **Mental Math**
>
> State the fewest number of coins (pennies, nickels, dimes, quarters) you could have if you had
>
> **a.** 27¢.
>
> **b.** $1.45.
>
> **c.** 99¢.

> ▶ **QY1**
>
> Use the Pythagorean Theorem to find the distance between Cierra's and Juan's houses, to the nearest tenth of a mile.

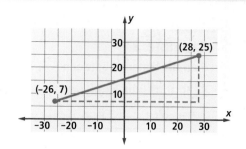

Solution Plot the points. Extend a vertical line from one point and a horizontal line from the other and find the coordinates of the third vertex of a right triangle. The lengths of the legs are __?__ and __?__.

Now, use the Pythagorean Theorem.

$$d^2 = \underline{}^2 + \underline{}^2$$
$$= \underline{}$$
$$d = \sqrt{\underline{}}$$
$$d \approx \underline{}$$

The above process can always be used to calculate distances on an oblique line, but it is very useful to have a general formula.

Let $P = (x_1, y_1)$ be the first point and $R = (x_2, y_2)$ be the second point. First find a point Q so that $\overline{PR}$ is the hypotenuse of a right triangle PQR. Such a point is $Q = (x_2, y_1)$. Then $\overline{PQ}$ is a horizontal segment and $\overline{QR}$ is a vertical segment. You can view these segments as on a horizontal or vertical number line and thus their lengths are $|x_2 - x_1|$ and $|y_2 - y_1|$ respectively. So,

$$PR^2 = PQ^2 + QR^2$$
$$= |x_2 - x_1|^2 + |y_2 - y_1|^2$$
$$= (x_2 - x_1)^2 + (y_2 - y_1)^2.$$

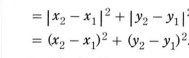

Taking the square roots of both sides,

$$PR = \sqrt{(x_2 - x_1)^2 + (y_2 - y_1)^2}.$$

This is a formula you should learn. We call it the *Pythagorean Distance Formula* to distinguish it from taxicab distance. That name also reminds you of its origin. But most people just call it the "Distance Formula." When we do not mention a type of distance between points A and B, and when we write AB, we mean this Pythagorean distance.

> **Theorem (Pythagorean Distance Formula on the Coordinate Plane)**
>
> The distance d between two points (x_1, y_1) and (x_2, y_2) on the coordinate plane is
> $$d = \sqrt{(x_2 - x_1)^2 + (y_2 - y_1)^2}.$$

With the Pythagorean Distance Formula, the distance of Guided Example 1 can be calculated without drawing a figure.

 QY2

▸ **QY2**

Use the Distance Formula to find the distance in Example 1.

Example 2

Find the distance between $(-5, 7)$ and $(-15, 24)$, to the nearest tenth.

Solution Substitute $(-5, 7)$ and $(-15, 24)$ into the distance formula.

$$d = \sqrt{(x_2 - x_1)^2 + (y_2 - y_1)^2}$$
$$= \sqrt{(\underline{\ ?\ } - \underline{\ ?\ })^2 + (\underline{\ ?\ } - \underline{\ ?\ })^2} \quad \text{Substitute for } x_1, x_2, y_1, \text{ and } y_2.$$
$$= \sqrt{(\underline{\ ?\ })^2 + (\underline{\ ?\ })^2} \quad \text{Work first inside parentheses.}$$
$$= \sqrt{100 + 289} \quad \text{Simplify.}$$
$$= \sqrt{389}$$
$$\approx 19.7$$

 QY3

> ▶ **QY3**
>
> Use the Distance Formula to find the magnitude of the vector from Cierra's house to Juan's house at the beginning of this lesson.

Using the Distance Formula in Proofs

With the Distance Formula, you can prove that segments on the coordinate plane are congruent by showing their lengths are equal. Example 3 contains a coordinate proof of a theorem you have known for some time.

Example 3

Using coordinates, prove that the diagonals of an isosceles trapezoid are congruent.

Solution First state the given and what you need to prove in terms of a figure.

Given: TRAP is an isosceles trapezoid.

Prove: $AT = PR$

One of the convenient locations for any isosceles trapezoid *TRAP* is with the y-axis as its symmetry line.

Let $T = (a, 0)$, $P = (-a, 0)$, $A = (-b, c)$, and $R = (b, c)$.

Now draw a figure.

Using the Pythagorean Distance Formula,

$$AT = \sqrt{(-b - a)^2 + (c - 0)^2}$$
$$= \sqrt{(-b - a)(-b - a) + c^2}$$
$$= \sqrt{b^2 + 2ab + a^2 + c^2}$$

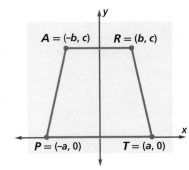

$$PR = \sqrt{(b - -a)^2 + (c - 0)^2}$$
$$= \sqrt{(b + a)(b + a) + (c - 0)^2}$$
$$= \sqrt{b^2 + 2ab + a^2 + c^2}$$

Thus, $AT = PR$.

Notice that the coordinate proof is straightforward, but it does require knowledge of algebra and square roots.

Questions

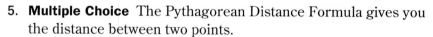

In 1–3, use the drawing at the right.

1. a. What are the coordinates of Z?
   b. Find the taxicab distance between S and N.
   c. Find SN using the lengths ZN and ZS.
2. a. What are the coordinates of L?
   b. Find the taxicab distance between L and A.
   c. Find LA using the Pythagorean Distance Formula.
   d. Find KA.
   e. Find $LK + KA$.
3. Find SN using the Pythagorean Distance Formula.

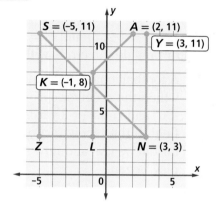

4. Give the distance between (a, b) and (c, d) in terms of a, b, c, and d.

5. **Multiple Choice** The Pythagorean Distance Formula gives you the distance between two points.

   A along horizontal and vertical segments.

   B as the crow flies.

   C as a taxicab would go.

6. $\triangle BIG$ has $B = (-600, 100)$, $I = (1000, -100)$, and $G = (300, 800)$. Is $\triangle BIG$ scalene, isosceles but not equilateral, or equilateral?

7. To get to a hospital from the middle of a nearby town, you can drive 7 miles east, turn right and go 12 miles south, and then turn right again and go 2 miles west. By helicopter, how far is it from the middle of the town to the hospital? (Ignore the altitude of the helicopter.)

8. Use a coordinate proof to show that the diagonals of a rectangle are congruent.

9. On a map, it can be seen that Brandon lives 1 mile east and 0.2 mile south of school, while Mollie lives 0.4 mile west and 0.8 mile south of school.

    a. Draw a graph with appropriate coordinates for the school, Brandon's residence, and Mollie's residence.

    b. Using taxicab distance, who lives closer to the school, Brandon or Mollie?

    c. Using Pythagorean distance, who lives closer to the school?

10. An ant starts at a foot of a staircase and crawls to the base of the fourth step. If each step is 8 inches tall and 10 inches wide, how far is the ant from where it began?

11. Let $A = (-2, 3)$ and $B = (2, 4)$. Is the point $C = (-1, 7)$ on, inside, or outside the circle with center A that contains B?

12. A football coach explains a "shoot route" passing play in which the receiver sprints up the field 4 yards, then turns $-90°$ and runs for 2 yards, then turns $90°$ and runs up the field. If the receiver runs 7 yards after his second turn, how far did he run and how far is he from where he was when the play began?

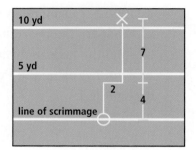

13. In the video game below, the mouse runs 2 units east, 3 units north, 2 units east, 3 units north, 2 units east and then 5 units south. How far is the mouse from where it began?

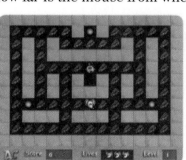

14. Let $A = (x_1, y_1)$ and $B = (x_2, y_2)$. Let S be a size transformation of magnitude 5. Show that the distance between $S(A)$ and $S(B)$ is 5 times the distance between A and B.

15. C and D are points on the coordinate plane. The x-coordinate of D is 4 less than the x-coordinate of C. The y-coordinate of D is 2 more than the y-coordinate of C. Find CD.

16. a. Find the exact magnitude of a vector from $(-4, 5)$ to $(7, -11)$.

    b. Find the magnitude of a vector from (d, e) to (f, g).

REVIEW

17. Let $A = (3, 4)$, $B = (1.2, -6)$, $C = (-11, -3.2)$, and $D = (-7, 9)$. Prove that $ABCD$ is not a trapezoid. (**Lesson 11-4**)

18. Suppose that $PENTA$ is a pentagon in the coordinate plane, and that none of its vertices are on the x- and y-axes. Use the Law of Ruling Out Possibilities to show that at least two of the vertices of $PENTA$ lie in the same quadrant. (**Lessons 11-2, 11-1**)

19. **a.** What does it mean for two segments to bisect each other? (**Lesson 2-4**)
    b. Name all of the special types of quadrilaterals whose diagonals bisect each other. (**Lesson 7-9**)

20. **a.** Describe how you could divide the interior of a regular n-gon into n congruent regions.
    b. Describe how you could divide the interior of a regular 12-gon into 4 congruent regions.
    c. Suppose $n = k\ell$, where k and ℓ are integers. Describe how you could divide the interior of a regular n-gon into k congruent regions. (**Lesson 6-8**)

21. Let A be the set of all points (x, y) in the coordinate plane such that x and y are both integers. What is the image of A under translation by the vector $(-3, 2)$? (**Lesson 4-6**)

22. What is the mean of x and y? (**Previous Course**)

EXPLORATION

23. Let $C = (10, 15)$ and $A = (7, -2)$. Suppose E, I, O, and U are four other points with $AC = EC = IC = OC = UC$. Give possible coordinates of E, I, O, and U.

Lesson
11-6
Equations of Circles

Vocabulary

line tangent to a circle

unit circle

▶ **BIG IDEA** As they do for lines, there are equations that describe circles.

Recall that any line in the coordinate plane has an equation of the form $Ax + By = C$, where A and B are not both zero. The converse is also true. That is, if the set of (x, y) pairs that satisfy an equation of this form are graphed, those points constitute a line. In this lesson, we find an equation for any circle in the coordinate plane. It is a wonderful application of the Pythagorean Distance Formula.

An Equation for a Circle

Call the radius of a circle r and let its center be (h, k). Let (x, y) be any point on this circle. The goal is to find an algebraic relationship between x, y, h, k, and r.

Because the point (x, y) is on a circle of radius r, we know that the distance between (x, y) and (h, k) is always r. This we can describe using the Pythagorean Distance Formula:

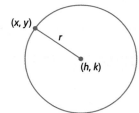

$$\sqrt{(x - h)^2 + (y - k)^2} = r.$$

This is an equation for the circle, but most people prefer equations without square roots. So we square both sides of the equation to get rid of the radical.

$$(x - h)^2 + (y - k)^2 = r^2$$

Theorem (Equation for a Circle)

The circle with center (h, k) and radius r is the set of points (x, y) satisfying the equation

$$(x - h)^2 + (y - k)^2 = r^2.$$

For example, the circle with center $(8, 15)$ and radius $\sqrt{3}$ has equation $(x - 8)^2 + (y - 15)^2 = 3$.

Mental Math

Zahara's phone company charges long-distance rates on calls more than 25 miles from her house.

a. What is the area of Zahara's local calling region?

b. Zahara's parents live 16 miles south and 12 miles east of her. Is a call to her parents billed as long distance?

c. Zahara's brother lives 10 miles east and 24 miles north of her. Is a call to him long distance?

Example 1

Write an equation for the circle graphed at the right.

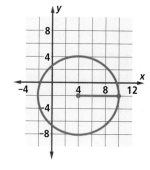

Solution Every circle has an equation of the form

$$(x - h)^2 + (y - k)^2 = r^2.$$

The center of this circle is (4, –2) and its radius is 6. Substitute $h = 4$, $k = -2$, and $r = 6$ into the equation

$$(x - 4)^2 + (y - -2)^2 = 6^2.$$

Simplify the equation a little.

$$(x - 4)^2 + (y + 2)^2 = 36$$

Check From the graph, notice that the point (10, –2) is on this circle. Test to see that its coordinates satisfy the equation.
$(10 - 4)^2 + (-2 + 2)^2 = 6^2 + 0^2 = 36$. Yes, it checks.

STOP QY1

By substituting for one coordinate of a point on a circle, you can use the equation for the circle to find the other coordinate.

> **QY1**
>
> Write an equation of the circle centered at the origin whose radius is 3.

Example 2

What are the coordinates of the points of intersection of the circle with equation $(x - 4)^2 + (y + 2)^2 = 36$ (from Example 1) and the line $x = 7$?

Solution Estimate first. Trace the graph of Example 1. Add the line $x = 7$. There seem to be two points of intersection, one near (7, 3) and the other near (7, –7).

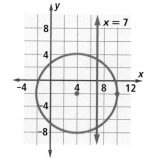

To find the exact coordinates, use the equation of the circle.

$$(x - 4)^2 + (y + 2)^2 = 36$$
$$(7 - 4)^2 + (y + 2)^2 = 36$$
$$9 + (y + 2)^2 = 36$$
$$(y + 2)^2 = 27$$
$$y + 2 = \pm \sqrt{27}$$
$$y = -2 \pm \sqrt{27}$$

The exact coordinates are $(7, -2 + \sqrt{27})$ and $(7, -2 - \sqrt{27})$. Because $\sqrt{27} \approx 5.2$, the points of intersection are near (7, –2 + 5.2) and (7, –2 – 5.2), or near (7, 3.2) and (7, –7.2).

This fits the estimate made by sight.

GUIDED

Example 3

Give the center and radius of the circle whose equation is
$x^2 + (y + 5)^2 = 100$.

Solution This problem comes down to determining the values of h, k, and r in the general equation of a circle. Compare the given equation to that general equation.

$$(x - h)^2 + (y - k)^2 = r^2$$
$$x^2 + (y + 5)^2 = 100$$

So $h = \underline{\ ?\ }$, $k = \underline{\ ?\ }$, and $r = \underline{\ ?\ }$.
The center $(h, k) = \underline{\ ?\ }$ and the radius $r = \underline{\ ?\ }$.

 QY2

> ▶ **QY2**
>
> Find the coordinates of four points that lie on the circle in Guided Example 3. Hint: The easiest points to find are those directly above, below, to the left, and to the right of the center.

Example 4

Is the point (–8, 2) on, inside, or outside the circle from Guided Example 3?

Solution This point is a point on the circle if it satisfies the equation

$$x^2 + (y + 5)^2 = 100.$$

Substituting $x = -8$ and $y = 2$ into this equation, the left side becomes

$$(-8)^2 + (2 + 5)^2 = 64 + 49 = 113.$$

Because $113 \neq 100$, this point is not on the circle. The distance from (–8, 2) to the center is $\sqrt{113}$, which is greater than the radius, so the point (–8, 2) is outside the circle.

Activity

MATERIALS DGS

Step 1 Use a DGS to construct the circle with equation
$(x + 4)^2 + (y - 3)^2 = 25$. This equation is of the form
$(x - h)^2 + (y - k)^2 = r^2$, where $h = -4$,
$k = 3$, and $r = 5$.

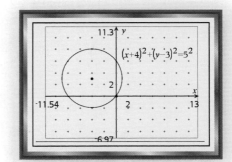

Step 2 Translate the circle to the right and to the left. Which of h, k, and r change? Which stay the same? Describe how the equation changes as the circle moves to the right and to the left.

Step 3 Describe how to transform the circle so that r remains constant but h and k change.

Step 4 Describe how to move the circle so that h and k remain the same but r changes.

Step 5 Describe how to move the circle so that k remains constant but r and h change.

Example 5

The line $y = 11$ intersects the circle $(x + 6)^2 + (y - 8)^2 = 25$ at two points. Find those two points of intersection.

Solution To find these two points, substitute 11 for y in the equation of the circle. Then solve the resulting equation for x.

$$(x + 6)^2 + (y - 8)^2 = 25$$
$$(x + 6)^2 + (11 - 8)^2 = 25$$
$$(x + 6)^2 + (3)^2 = 25$$
$$(x + 6)^2 + 9 = 25$$
$$(x + 6)^2 = 16$$
$$x + 6 = \pm\sqrt{16}$$
$$x = -6 \pm 4$$

So, $x = -10$ or $x = -2$.

When $x = -10$, $y = 11$, and when $x = -2$, $y = 11$. Thus, the two points of intersection are $(-10, 11)$ and $(-2, 11)$.

 QY3

▷ **QY3**

What happens when you let $x = 0$ in the equation of the circle of Example 5 to try to find where the circle intersects the line $x = 0$? Use the graph to explain geometrically what the algebra tells you.

Questions

COVERING THE IDEAS

1. Write an equation for the circle with radius r and center (h, k).

2. **Fill in the Blank** The equation for a circle is derived from the _____?_____ Formula.

In 3–6, an equation for a circle is given. Find the center and exact radius of the circle.

3. $(x - 5)^2 + (y + 7)^2 = 49$

4. $x^2 + (y - 8)^2 = 13$

5. $2 = x^2 + y^2$

6. $\sqrt{(x + 2)^2 + (y - 4)^2} = 0.36$

7. **a.** Graph the circle whose equation is $(x + 5)^2 + y^2 = 16$.
   b. Give four points that lie on the circle in Part a.

8. The point $(3, 6)$ is on the circle whose center is $(1, 8)$. Find an equation for this circle.

9. Find the coordinates of all points of intersection of the circle $(x - 5)^2 + (y - 2)^2 = 36$ and the line with the given equation.
   a. $x = 1$ **b.** $y = -3$ **c.** $y = 9$

10. **Multiple Choice** Where is the point (10, -5) in relation to the circle with equation $(x + 10)^2 + (y - 5)^2 = 100$?

    A   It is the center of the circle.

    B   It is inside the circle but is not the center.

    C   It is on the circle.

    D   It is outside the circle.

APPLYING THE MATHEMATICS

11. A circle has point $A = (-7, 4)$ as its center and contains point $B = (-10, 1)$.

    a. Find an equation for this circle.

    b. Find the circle's exact circumference.

    c. Find the exact area of the circle.

12. A circle whose area is 100π is drawn with center $(3, 0)$. What is an equation for this circle?

13. A **line is tangent to a circle** if they have exactly one point in common. Write the equations of the two horizontal lines that are tangent to the circle $(x + 2)^2 + (y - 4)^2 = 81$.

14. The **unit circle** is the circle with center $(0, 0)$ and radius 1.

    a. Find an equation for this circle.

    b. Identify the coordinates of eight points on this circle. (Hint: Four points are on the axes.)

15. In the town of Cartesia, the post office is the center of the coordinate system and the local jail has coordinates $(-4, 2)$. (The jail is 4 miles west and 2 miles north of the post office.) A criminal has escaped from the jail and is traveling on foot. It is estimated that the criminal can travel at most 8 miles per hour.

    a. Write an equation for the circle made up of farthest possible locations of the criminal 1 hour after the escape.

    b. Write an inequality that represents the possible location of the criminal t hours after the escape.

REVIEW

16. Suppose that the distance between $(x, 3)$ and $(2x, 5)$ equals 3. **(Lesson 11-5)**

    a. Write an equation you could solve in order to find the possible values of x. (Hint: You can write this equation without using a square root.)

    b. Find the possible values for x.

    c. Use the Distance Formula to check that the points you found are indeed 3 units apart.

17. An eccentric millionaire threw a party for seven of his friends to show a rare diamond that he acquired. At one point in the evening, the diamond disappeared. The host assembled all of the suspects together in a room, and then said: "Before the party started, I told five of you that the diamond presented here was actually just a copy. Therefore, you wouldn't have stolen it. I spent the whole evening talking to Nadia. If she had stolen the diamond, I would have seen her. Therefore, I suspect, the thief is Glenn!" Explain the two laws of logic that the millionaire used to make his deduction. (**Lessons 11-2, 11-1**)

18. For a high school production of *Hamlet,* the director wants a backdrop picturing a river with a far-off castle in the background. Should the backdrop be painted using perspective? Why or why not? (**Lesson 9-4**)

19. Suppose that the sides of a triangle have lengths x, x, and $x\sqrt{2}$.
    a. Explain why this must be a right triangle. (**Lesson 8-6**)
    b. Explain why this must be a 45-45-90 triangle. (**Lesson 8-7**)

20. In $\triangle ABC$, $AB = AC$, and D and E are the midpoints of $\overline{AB}$ and $\overline{AC}$, respectively. Prove that m$\angle AED$ = m$\angle ADE$. (**Lesson 6-2**)

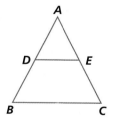

21. Suppose set A contains every point in the coordinate plane whose y-coordinate is greater than its x-coordinate, whose y-coordinate is less than 6, and which lies to the left of the y-axis. (**Lesson 1-2**)
    a. Write the inequalities that define the region A.
    b. Draw the region A in the plane.

22. Suppose $\overleftrightarrow{AB}$ and $\overleftrightarrow{BC}$ have the same slope. What can you conclude about A, B, and C? (**Lesson 1-2**)

QY ANSWERS

1. $x^2 + y^2 = 9$

2. Answers vary. Sample: $(0, 5)$, $(0, -15)$, $(10, -5)$, $(-10, -5)$.

3. Substitute $x = 0$ to get $36 + (y - 8)^2 = 25$. Thus, $(y - 8)^2 = -11$. This equation has no solutions, because the circle does not intersect the y-axis.

EXPLORATION

23. A *lattice point* is a point with integer coordinates. Write an equation for a circle on which there are no lattice points and whose interior contains no lattice points.

24. Find all the points (x, y) that satisfy the equations $x^2 + y^2 = 4$ and $y = 2x$.

Lesson

11-7 Means and Midpoints

▶ **BIG IDEA**   The coordinates of the midpoint of a segment are the averages of the coordinates of the endpoints of the segment.

A Geometric Interpretation of the Mean

Suppose you score 92 and 77 on two tests. Your average or mean score is the sum of these numbers divided by 2. It is

$$\frac{92 + 77}{2} = 84.5.$$

The mean has a physical interpretation. Think of a ruler with hooks for attaching weights. (Assume the ruler itself is weightless.) If equal weights are hung from 92 and 77, the ruler will be horizontal if it is hung from a string at 84.5, or if it is placed on a sharp object at 84.5.

Mental Math

Multiple Choice Which is the best buy?

A A 48-oz bottle of a liquid for $4.50

B A half-gallon container of the same liquid for $5.00

C A pint container of the same liquid for $2.00

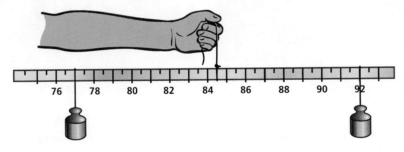

Notice that 84.5 is the *midpoint* of the segment connecting 92 and 77. This is easy to see, using the distance between points on a number line:

$$|84.5 - 92| = |\text{-}7.5| = 7.5$$

and

$$|84.5 - 77| = |7.5| = 7.5.$$

Using algebra, the midpoint can be shown to be the mean of the endpoints for any segment on a number line.

Theorem (Midpoint Formula on a Number Line)

On a number line, the midpoint of the segment with endpoints x_1 and x_2 has coordinate $\frac{x_1 + x_2}{2}$.

Proof Draw a picture. Let A and B be the endpoints of the segment, with coordinates x_1 and x_2, respectively. Let M be the point with coordinate $\frac{x_1 + x_2}{2}$. To show that M is the midpoint of $\overline{AB}$, we need to show that $AM = MB$. Calculate the distance using the Distance Postulate.

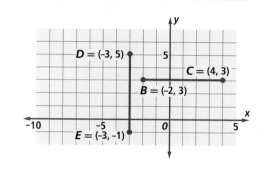

$$AM = \left|\frac{x_1 + x_2}{2} - x_1\right| = \left|\frac{x_1 + x_2 - 2x_1}{2}\right| = \left|\frac{x_2 - x_1}{2}\right|$$

$$MB = \left|x_2 - \frac{x_1 + x_2}{2}\right| = \left|\frac{2x_2 - (x_1 + x_2)}{2}\right| = \left|\frac{x_2 - x_1}{2}\right|$$

So $AM = MB$. Thus, M is the midpoint of $\overline{AB}$.

A Formula for the Midpoint in Two Dimensions

If a segment is horizontal or vertical, the one coordinate of both its endpoints will be the same. The midpoint will also have that coordinate. Then take the average of the unequal coordinates of the endpoints to be the other coordinate of the midpoint of the segment.

GUIDED

Example 1

a. Find the coordinates of the midpoint of $\overline{DE}$.
b. Find the coordinates of the midpoint of $\overline{BC}$.

Solution

a. The coordinates of the midpoint of $\overline{DE}$ are $\left(-3, \frac{? + ?}{2}\right) = (-3, 2)$
b. The coordinates of the midpoint of $\overline{BC}$ are ($\underline{\ ?\ }$, $\underline{\ ?\ }$).

If a segment is neither horizontal nor vertical, then the coordinates of the midpoint can be calculated by finding the average of the x-values and the average of the y-values.

Theorem (Midpoint Formula on the Plane)

In the coordinate plane, the midpoint of the segment with endpoints (x_1, y_1) and (x_2, y_2) is $\left(\frac{x_1 + x_2}{2}, \frac{y_1 + y_2}{2}\right)$.

Given $\overline{AB}$ with $A = (x_1, y_1)$, $B = (x_2, y_2)$, and

$$M = \left(\frac{x_1 + x_2}{2}, \frac{y_1 + y_2}{2}\right)$$

Prove M is the midpoint of $\overline{AB}$.

Proof As in the proof of the Midpoint Formula on a Number Line, we show that $AM = MB$. Use the Distance Formulas.

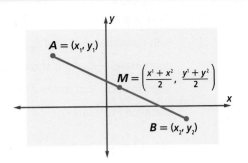

$$AM = \sqrt{\left(\frac{x_1 + x_2}{2} - x_1\right)^2 + \left(\frac{y_1 + y_2}{2} - y_1\right)^2}$$

$$= \sqrt{\left(\frac{x_1 + x_2 - 2x_1}{2}\right)^2 + \left(\frac{y_1 + y_2 - 2y_1}{2}\right)^2}$$

$$= \sqrt{\left(\frac{x_2 - x_1}{2}\right)^2 + \left(\frac{y_2 - y_1}{2}\right)^2}$$

$$MB = \sqrt{\left(x_2 - \frac{x_1 + x_2}{2}\right)^2 + \left(y_2 - \frac{y_1 + y_2}{2}\right)^2}$$

$$= \sqrt{\left(\frac{2x_2 - (x_1 + x_2)}{2}\right)^2 + \left(\frac{2y_2 - (y_1 + y_2)}{2}\right)^2}$$

$$= \sqrt{\left(\frac{x_2 - x_1}{2}\right)^2 + \left(\frac{y_2 - y_1}{2}\right)^2}$$

Thus, $AM = MB$. Now we need to show that M is on $\overline{AB}$. For this, we calculate the slopes of $\overleftrightarrow{AM}$ and $\overleftrightarrow{MB}$.

$$\text{slope of } \overleftrightarrow{AM} = \frac{\frac{y_1 + y_2}{2} - y_1}{\frac{x_1 + x_2}{2} - x_1} = \frac{y_1 + y_2 - 2y_1}{x_1 + x_2 - 2x_1} = \frac{y_2 - y_1}{x_2 - x_1}$$

$$\text{slope of } \overleftrightarrow{MB} = \frac{y_2 - \frac{y_1 + y_2}{2}}{x_2 - \frac{x_1 + x_2}{2}} = \frac{2y_2 - (y_1 + y_2)}{2x_2 - (x_1 + x_2)} = \frac{y_2 - y_1}{x_2 - x_1}$$

The slopes are equal, so $\overleftrightarrow{AM} \parallel \overleftrightarrow{MB}$. Both lines contain point M, so $\overleftrightarrow{AM} = \overleftrightarrow{MB}$ and M is on $\overleftrightarrow{AB}$. So M is the midpoint of $\overline{AB}$.

Fortunately, applying the Midpoint Formula is much easier than proving it.

 QY1

> ▶ **QY1**
>
> Find the midpoint of segment $\overline{PQ}$ when $P = (-12, 9)$ and $Q = (1, 5)$.

Example 2

If $A = (2, -1)$ and $B = (6, 7)$, write an equation of the perpendicular bisector of $\overline{AB}$.

Solution The perpendicular bisector of $\overline{AB}$ passes through the midpoint of $\overline{AB}$ and has a slope that is the opposite reciprocal of the slope of $\overline{AB}$.

Let M be the midpoint of $\overline{AB}$. From the Midpoint Formula on the Plane, $M = \left(\frac{2 + 6}{2}, \frac{-1 + 7}{2}\right) = (4, 3)$. The slope of $\overline{AB}$ is $\frac{7 - (-1)}{6 - 2} = 2$.

Thus, the slope of the perpendicular bisector is $-\frac{1}{2}$.

This is enough information to write an equation for the line.

$$y = -\frac{1}{2}x + b$$

Now substitute $x = 4$ and $y = 3$ to solve for b.

$$3 = -\frac{1}{2}(4) + b$$
$$3 = -2 + b$$
$$5 = b$$

So, an equation of the perpendicular bisector of $\overline{AB}$ is
$y = -\frac{1}{2}x + 5$.

If you know the endpoint of a segment and its midpoint, you can find the other endpoint by working backward.

Example 3

M is the midpoint of $\overline{AB}$. If $A = (-6, 3)$ and $M = (4, 7)$, find the coordinates of B.

Solution Let $B = (x, y)$. To find x and y, use the Midpoint Formula on the Plane.

$$M = \left(\frac{x_1 + x_2}{2}, \frac{y_1 + y_2}{2}\right)$$

So
$$(4, 7) = \left(\frac{-6 + x}{2}, \frac{3 + y}{2}\right).$$

Consequently,
$$4 = \frac{-6 + x}{2} \quad \text{and} \quad 7 = \frac{3 + y}{2}$$
$$8 = -6 + x \qquad\qquad 14 = 3 + y$$
$$x = 14 \qquad\qquad\qquad y = 11.$$

The coordinates of B are $(14, 11)$.

 QY2

Questions

COVERING THE IDEAS

1. The U.S. Weather Bureau calculates the mean temperature for a particular place on a particular day by averaging the high and low temperatures. Find the mean temperature in Chicago on February 7, 2007, when the high was 10°F and the low was –2°F, a record low for that date.

2. Give a number-line interpretation of the mean in Question 1.

▶ **QY2**

Check Example 3 by finding the midpoint of $\overline{AB}$ when $A = (-6, 3)$ and $B = (14, 11)$.

3. **a.** You scored 82 on your first test. What would you need to score on your next test to have a mean score of 90 for the two tests?

   b. You scored 82 on your first test. What would you need to score on your next test to have a mean score of m for the two tests?

4. **a.** Give the coordinates of the midpoint of the segment connecting (u, v) to (u, w).

   b. **Multiple Choice** The segment in Part a:

   A is oblique.　　**B** is vertical.　　**C** is horizontal.

   D can be either vertical or oblique, depending on the values of the variables.

   E can be either horizontal or oblique, depending on the values of the variables.

5. Find the midpoint of the segment with endpoints $(-0.09, 12)$ and $(0.3, -4)$.

6. Find an equation for the perpendicular bisector of the segment connecting $(0, 0)$ to $(4, 10)$.

7. Give an equation for the set of all points that are equidistant from $(3, -8)$ and $(11, 12)$.

8. The center of a circle is $(4, 5)$. One endpoint of a diameter of the circle is $(-2, -3)$. Find the coordinates of the other endpoint.

APPLYING THE MATHEMATICS

9. The graph at the right shows the total energy consumption in the United States (in quadrillion Btu) in 1990 and 2006.

   a. What are the coordinates of the midpoint of $\overline{PQ}$?

   b. What does each coordinate of the midpoint represent?

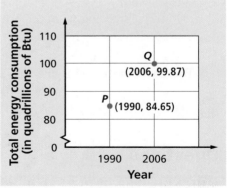

10. According to the College Board, the average published tuition and fees for a 4-year private college in the school year ending in 1996 was \$15,489. By the end of 2006, the cost had risen to \$21,235.

    a. Assuming the increase in tuition and fees continues at a constant rate, find the cost of tuition and fees in 2016.

    b. Why is the task of Part a appropriate to this lesson?

In **11** and **12**, determine whether the given quadrilateral is a parallelogram by using the theorem that states that a quadrilateral is a parallelogram if and only if its diagonals bisect each other.

11. *ABCD,* where $A = (-3, 2)$, $B = (0, 2)$, $C = (1, -2)$, and $D = (-3, -1)$

12. *EFGH,* where $E = (5, 2)$, $F = (9, 0)$, $G = (8, -3)$, and $H = (4, -1)$

13. A rhombus *RHMB* in standard position has $R = (c, 0)$, $H = (0, d)$, $M = (-c, 0)$, and $B = (0, -d)$.
    a. Give the coordinates of the midpoints of each of its sides.
    b. Explain why the midpoints of the sides of *RHMB* are vertices of a rectangle.
    c. Give equations for the lines of symmetry of the rectangle.
    d. Give equations for the lines of symmetry of the rhombus.

14. Suppose the endpoints of a diameter of a circle are (-2, 6) and (4, -18).
    a. Locate the center of the circle.
    b. Determine the exact radius of the circle.
    c. Find an equation for the circle.

REVIEW

15. Sydney draws a circle centered at the origin with an area of 12.25π.
    a. What is the radius of the circle? **(Lesson 8-9)**
    b. Give the coordinates of the points where the circle intersects the *x*-axis. **(Lesson 11-6)**
    c. Give the coordinates of the points where the circle intersects the *y*-axis. **(Lesson 11-6)**
    d. Write an equation for the circle. **(Lesson 11-6)**

16. Without expanding, explain why
$$\sqrt{(x-5)^2 + (y-7)^2} = \sqrt{(5-x)^2 + (7-y)^2}.$$ **(Lesson 11-5)**

17. State the Isosceles Trapezoid Symmetry Theorem. **(Lesson 6-6)**

18. Consider the drawing of a miniature golf hole at the right. Draw a possible path a ball could take from *A* to *B*, hitting two walls on the way. **(Lesson 4-3)**

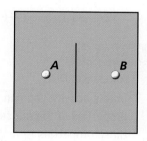

EXPLORATION

19. Choose any three points on the coordinate plane that you would like to be the vertices of a triangle *ABC*.
    a. Find the midpoints of the three sides and use these to find equations for the three medians of the triangle. (Recall that a *median* of a triangle is a segment connecting a vertex to the midpoint of the side opposite the vertex.)
    b. Solve a system to find the point of intersection of two of the medians. Then show that the third median contains that point of intersection. This will prove that the three medians of your triangle are concurrent.

QY ANSWERS

1. (-5.5, 7)

2. $\left(\dfrac{-6+14}{2}, \dfrac{3+11}{2}\right) = (4, 7)$

Lesson
11-8
Theorems Involving Midpoints

Vocabulary

midsegment of a trapezoid

midsegment of a triangle

medial triangle

▶ **BIG IDEA** There are surprising relationships among segments connecting midpoints of sides of polygons.

To talk about segments that connect midpoints, it helps to have language specifically describing them. The segment joining the midpoints of the nonparallel sides of a trapezoid is called the **midsegment of the trapezoid** for those bases. A **midsegment of a triangle** connects the midpoints of any two sides, the third side of the triangle being the base for that midsegment. By moving D in trapezoid $ABCD$ along $\overline{CD}$ to C, you can see that the midsegment of the trapezoid becomes the midsegment of $\triangle ABC$.

Mental Math

What is the magnitude of the rotation about the origin under which the image of $(-a, b)$ is

a. $(-b, -a)$?

b. (b, a)?

c. $(a, -b)$?

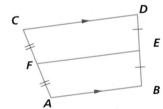

$\overline{EF}$ is a midsegment of trapezoid $ABCD$.

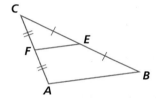

$\overline{EF}$ is a midsegment of $\triangle ABC$.

Activity

MATERIALS DGS

In this activity, you can use a DGS to help you discover some properties of the midsegment of a trapezoid. Then you can apply these properties to realize some properties of the special case of the triangle.

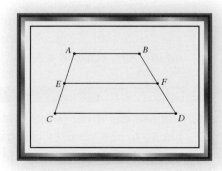

Step 1 Construct trapezoid $ABDC$ with bases $\overline{AB}$ and $\overline{CD}$.

Step 2 Construct the midpoint of $\overline{AC}$; name it E.

Step 3 Construct the midpoint of $\overline{DB}$; name it F.

Step 4 Construct $\overline{EF}$ (the midsegment of trapezoid $ABDC$).

Step 5 Find the lengths of $\overline{EF}$, $\overline{AB}$, and $\overline{CD}$.

Step 6 Drag various parts of the trapezoid and observe the relationship between the lengths. Make a conjecture about the relationship between the length of the midsegment of a trapezoid and the length of the bases.

Step 7 Find the slopes of $\overline{EF}$, $\overline{AB}$, and $\overline{CD}$. Drag various parts of the trapezoid and observe the relationship between the slopes.

Step 8 Drag point D towards C until CD is as close to zero as you can make it. Now you have a triangle.

Step 9 Drag points A and B, observing the relationship between the slopes and the lengths of $\overline{AB}$ and the midsegment of the triangle $\overline{EF}$.

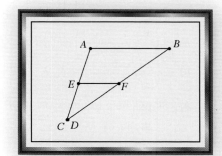

Midsegment Theorems

In the Activity, you were asked to make conjectures. It is likely that those conjectures were quite similar to the statements in the following theorems.

> ### Midsegment of a Trapezoid Theorem
>
> The midsegment of a trapezoid is parallel to the bases and its length is equal to the average of the lengths of the two bases.

You are asked to prove this theorem in Question 4.

> ### Midsegment of a Triangle Theorem
>
> A midsegment of a triangle is parallel to and half the length of the base of the triangle.

GUIDED

Example 1

Prove the Midsegment of a Triangle Theorem.

Solution In the proof of this theorem, we use $2a$, $2b$, and $2c$ for the coordinates of vertices in convenient locations for a triangle. This avoids fractions when coordinates of midpoints are calculated.

Given $\triangle PQR$ with $Q = (0, 0)$, $R = (2a, 0)$, $P = (2b, 2c)$; M the midpoint of $\overline{PQ}$; and N the midpoint of $\overline{PR}$. Here is a drawing.

Prove You need to prove two things:
    a. that the midsegment is parallel to the base, and
    b. that the midsegment's length is half that of the base.

In terms of the diagram, that means to prove:

a. _____?_____

b. _____?_____

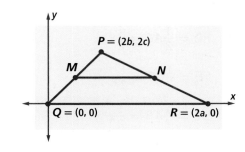

(continued on next page)

First, find the coordinates of M and N.

$M = \left(\dfrac{?\,+\,?}{?}, \dfrac{?\,+\,?}{?}\right) = (\underline{\,?\,}, \underline{\,?\,})$

$N = \left(\dfrac{?\,+\,?}{?}, \dfrac{?\,+\,?}{?}\right) = (\underline{\,?\,}, \underline{\,?\,})$

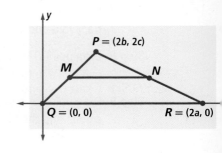

$P = (2b, 2c)$

M

N

$Q = (0, 0)$

$R = (2a, 0)$

a. Slopes help to prove $\overline{MN} \parallel \overline{QR}$.

    slope of $\overline{MN} = \underline{\,?\,}$

    slope of $\overline{QR} = \underline{\,?\,}$

  $\overline{MN} \parallel \overline{QR}$ because $\underline{\,?\,}$.

b. To prove $MN = \dfrac{1}{2} QR$, use the Distance Formula.

$$MN = \sqrt{((b + a) - b)^2 + (\underline{\,?\,} - \underline{\,?\,})^2}$$

$$= \sqrt{a^2 + \underline{\,?\,}}$$

$$= |a|$$

$$QR = \sqrt{(2a - 0)^2 + \underline{\,?\,} - \underline{\,?\,})^2}$$

$$= \sqrt{4a^2}$$

$$= 2\,|a|$$

The absolute value is necessary in the preceding two steps because a is a variable that could have a negative value, but MN and QR are lengths, which cannot be negative. Likewise, the square root symbol always refers to the nonnegative square root.

Thus, $2MN = QR$, so $MN = \dfrac{1}{2}QR$.

STOP **QY**

▶ **QY**

$ADCB$ is a trapezoid. E, F, and G are the midpoints of $\overline{CD}$, $\overline{AB}$, and $\overline{AC}$, respectively. If $BC = 8$ and $AD = 14$, find FE and FG.

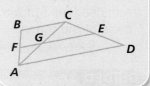

C

B

E

G

F

D

A

Example 2

$\triangle FDE$ is formed by connecting the midpoints of $\triangle ABC$. What is the ratio of the perimeter of $\triangle FDE$ to the perimeter of $\triangle ABC$?

Solution Let $x = AC$, $y = BC$, and $z = AB$, so the perimeter of $\triangle ABC = x + y + z$.

Use the Midsegment of a Triangle Theorem.

$FD = \dfrac{1}{2}AC = \dfrac{1}{2}x$

$FE = \dfrac{1}{2}BC = \dfrac{1}{2}y$

$DE = \dfrac{1}{2}AB = \dfrac{1}{2}z$

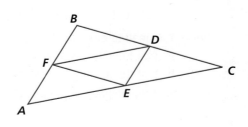

B

D

F

C

E

A

The perimeter of $\triangle FDE$ is $\frac{1}{2}x + \frac{1}{2}y + \frac{1}{2}z = \frac{1}{2}(x + y + z)$.

The ratio $\dfrac{\triangle FDE}{\triangle ABC} = \dfrac{\frac{1}{2}(x + y + z)}{(x + y + z)} = \dfrac{1}{2}$.

$\triangle FDE$ in Example 5 is called the **medial triangle** of $\triangle ABC$.
Example 2 proves that the perimeter of a triangle is double that of its
medial triangle.

Questions

COVERING THE IDEAS

In 1 and 2, use the figure at the right. Tick marks indicate
congruent segments.

1. If $OP = 8$, find KL and MN.
2. If $m\angle KOP = 52$ and $m\angle KRN = 83$,
   find the measures of all other angles in the figure.
3. In $\triangle ABC$, $\overline{AD}$ is a median.

   a. Find the coordinates of D.
   b. Is $\overline{AD} \perp \overline{BC}$?

4. Prove the Midsegment of a
   Trapezoid Theorem using the
   convenient location given.

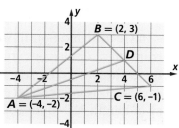

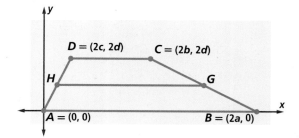

APPLYING THE MATHEMATICS

5. Use the formula $A = \frac{1}{2}h(b_1 + b_2)$ to find another formula for the
   area of a trapezoid that involves the length of its midsegment.

6. In the figure at the right, $\overline{CF}$ is an altitude and $\overline{ED}$ is a
   midsegment.

   a. Why is $\angle CEJ \cong \angle CAF$?   b. Why is $\angle CFA \cong \angle CJE$?

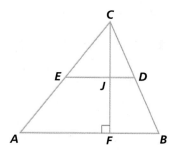

7. Jacqui is building a racetrack for her toy cars. She builds a piece of ramp with vertical beams 8 and 17.5 inches high, and wants to add a vertical support halfway between the two beams. Jacqui has marked the halfway point between the existing beams. How tall must the support be, and how will Jacqui know the support is vertical?

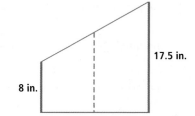
17.5 in.

8 in.

8. In the figure at the right, $HIJK$ is a rectangle, $IK = 30$, and $D, E, F,$ and G are midpoints. Prove that the perimeter of $DEFG$ is 60.

9. Prove that a rhombus is formed by connecting midpoints of consecutive sides of a rectangle. (You may find the figure for Question 8 handy.)

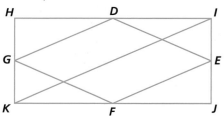

10. $\overline{AD}$, the midsegment of trapezoid $LMNO$ at the right, intersects diagonals $\overline{NL}$ and $\overline{MO}$ at points B and C. Find an expression for the length of $\overline{BC}$ in terms of the lengths of the bases $\overline{LO}$ and $\overline{MN}$. (Hint: $BC = AD - AB - CD$.)

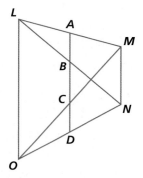

11. Some Native American temples in Central and South America were built in the shape of a frustum of a regular pyramid. Recall that a frustum of a pyramid is the solid formed by removing a top part of the pyramid. Suppose a solid regular square pyramid originally had a base with edges of length 120 meters and a height of 100 meters. If the pyramid is cut off halfway up, what is the surface area of the frustum to the nearest hundred square meters? (Do not include the base.)

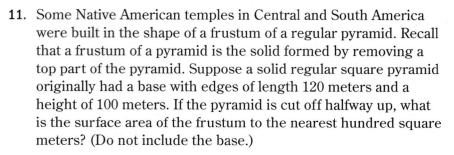

REVIEW

12. On a road trip, Tate wants to stop halfway between Cheyenne, Wyoming, and Salt Lake City, Utah. He is 55 miles away from Cheyenne, and Salt Lake City is 440 miles farther along the same road. How far away is Tate from where he wants to stop?
(**Lesson 11-7**)

13. Suppose the endpoints of a diameter of a circle are (6, 2) and (–4, –8).
    a. Locate the center of the circle.
    b. Determine the exact radius of the circle.
    c. Find an equation for the circle. **(Lessons 11-7, 11-6)**

14. Suppose $A = (a, b)$ and $B = (x, y)$.
    a. What is the distance between A and B according to the Distance Formula? **(Lesson 11-5)**
    b. Let T be the translation by the vector (3, 6). Use the Distance Formula to find the distance between $T(A)$ and $T(B)$. How is this distance related to your answer from Part a? **(Lessons 11-5, 4-6)**

15. A rectangle in the coordinate plane has vertices at (6, –4), (–6, –4), (–6, 4), and (6, 4). Write a set of inequalities that describes the interior of the rectangle. **(Lesson 1-6)**

16. State the contrapositive of the Isosceles Triangle Base Angles Theorem. **(Lessons 11-2, 6-2)**

17. A large sphere has three times the surface area of a smaller one. How do the volumes of these two spheres compare? **(Lessons 10-7, 10-6)**

EXPLORATION

18. Prove the following theorem, known as Varignon's Theorem: If the midpoints of the sides of any quadrilateral are connected in order, then the quadrilateral formed is a parallelogram. Prove that this is always a parallelogram. (Hint: Draw the diagonals of the original quadrilateral and prove that the quadrilateral formed by the midpoints has sides parallel to these diagonals.)

19. In Question 18, you proved that the quadrilateral formed by connecting the midpoints of any quadrilateral is a parallelogram (called the Varignon parallelogram). Use a DGS to construct quadrilateral $ABCD$ and Varignon parallelogram $EFGH$ (where E, F, G, and H are the midpoints of the sides of $ABCD$). What would need to be true of $ABCD$ in order to guarantee that $EFGH$ is a
    a. rhombus? Explain.
    b. rectangle? Explain.
    c. square? Explain.

QY ANSWER

$FE = 11, FG = 4$

Lesson 11-9

Three-Dimensional Coordinates

Vocabulary

2-dimensional coordinate system

3-dimensional coordinate system

z-coordinate

axes

ordered triple

x-, y-, z- axis

3-space

right tetrahedron

▶ **BIG IDEA** In 3-dimensional space, where a point is represented by an ordered triple (x, y, z), formulas for distance and midpoints are similar to the corresponding formulas for 2-dimensional space.

Coordinatizing Space

In the coordinate plane, each point can be located by an ordered pair of numbers. For this reason, we call the use of ordered pairs a **2-dimensional coordinate system.** Points in space can be located using a **3-dimensional coordinate system.** For instance, you can locate points in a room by letting the origin be a corner of the room where two walls and the floor intersect. With two coordinates, x and y, you can describe the location of any point on the floor. To describe an object in the room which is not on the floor, you can use a third number to indicate the height from the floor. This third number is the **z-coordinate.** The three lines where the walls and floor meet are the **axes** of this 3-dimensional coordinate system.

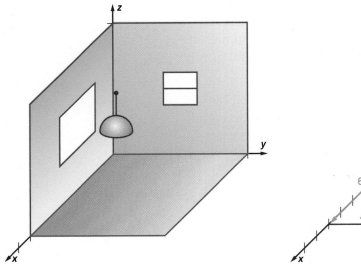

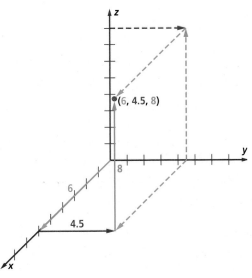

For example, consider a lamp that is hanging from a ceiling in a room. The lamp is attached to the ceiling at a point that is 8 feet off the floor, 6 feet from a corner in the *x*-direction, and 4.5 feet from that corner in the *y*-direction, as shown on the previous page. This lamp can be identified by the **ordered triple** (6, 4.5, 8). The *x*-coordinate is 6, the *y*-coordinate is 4.5, and the *z*-coordinate is 8.

Now imagine extending each axis in its negative direction, as shown in the figure at the right. The three axes are called the **x-axis,** the **y-axis,** and the **z-axis.** The ordered triple (*x, y, z*) represents a point in 3-dimensional space, called **3-space** for short. We have previously referred to this simply as *space.* It is common to also call it 3-space to distinguish it from a 2-dimensional situation. The position of a point is given by its three distances and directions from the origin (0, 0, 0).

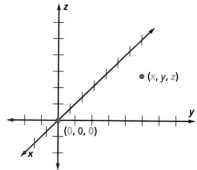

Caution: Normally, when you draw a 2-dimensional plane, the *x*-axis is horizontal and the *y*-axis is vertical. When 3-dimensional coordinates are used, the *x-y* plane is usually drawn to look horizontal and the *z*-axis is a vertical line.

In the drawings of 3-dimensional coordinate systems above, it appears that the *x*-axis is coming out of the plane determined by the paper. This text will always have the *x*-axis coming out of the paper. If you use other materials to look at 3-D graphs, make sure to check the labeling of the axes because other sources may label the axes differently.

Example 1

Plot the point *P* = (−5, 3, 4) on a 3-dimensional coordinate axis.

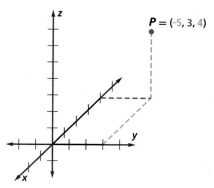

Step 1 Draw a 3-dimensional coordinate system.

Step 2 Graph (−5, 3) in the *x-y* plane.

Step 3 Construct dashed segments from (−5, 3) perpendicular to the axes.

Step 4 From (−5, 3), construct a dashed segment 4 units up, parallel to the *z*-axis.

Sometimes, to give the impression of depth with a point such as (−5, 3, 4), the edges of a rectangular box with vertices (0, 0, 0), (−5, 0, 0), (0, 3, 0), and (0, 0, 4) are sometimes included with the hidden edges represented by dashed segments, as shown on the next page.

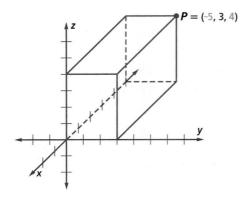

$P = (-5, 3, 4)$

STOP QY

► QY

Plot the point
$Q = (-2, -4, 5)$ on a
3-dimensional coordinate
system using a box with
appropriately dashed
edges to give the
impression of depth.

The Distance Formula in Space

Many of the formulas for 2-dimensional coordinates have
counterparts in three dimensions. For instance, the Pythagorean
Distance Formula states that the distance between (x_1, y_1) and

(x_2, y_2) is $\sqrt{(x_2 - x_1)^2 + (y_2 - y_1)^2}$. Its counterpart is the

Three-Dimension Distance Formula.

> **Theorem (Three-Dimension Distance Formula)**
>
> The distance d between two points (x_1, y_1, z_1) and (x_2, y_2, z_2) is
> given by
> $$d = \sqrt{(x_2 - x_1)^2 + (y_2 - y_1)^2 + (z_2 - z_1)^2}.$$

Given $P = (x_1, y_1, z_1)$ and $Q = (x_2, y_2, z_2)$

Prove $PQ = \sqrt{(x_2 - x_1)^2 + (y_2 - y_1)^2 + (z_2 - z_1)^2}$

Proof We apply the Pythagorean Theorem twice.
In right triangle PRT, $PT^2 = RT^2 + PR^2$.
In right triangle PTQ, $PQ^2 = PT^2 + QT^2$.
Substituting $RT^2 + PR^2$ for PT^2 in the second equation
gives $PQ^2 = RT^2 + PR^2 + QT^2$. Consequently,
$PQ^2 = (x_2 - x_1)^2 + (y_2 - y_1)^2 + (z_2 - z_1)^2$.
So, taking the nonnegative square root of each side,
$$PQ = \sqrt{(x_2 - x_1)^2 + (y_2 - y_1)^2 + (z_2 - z_1)^2}$$

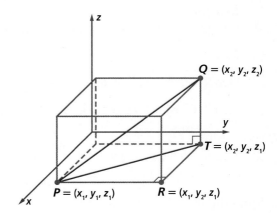

GUIDED

Example 2

When $R = (-5, -3, 4)$ and $T = (3, 7, -2)$, calculate RT exactly and to the nearest tenth.

Solution Let R = $(-5, -3, 4) = (x_1, y_1, z_1)$ and
T $= (3, 7, -2) = (x_2, y_2, z_2)$.

$$RT = \sqrt{x_1 - x_2)^2 + (y_1 - y_2)^2 + (z_1 - z_2)^2}$$
$$= \sqrt{(\underline{?})^2 + (\underline{?})^2 + (\underline{?})^2}$$
$$= \sqrt{(\underline{?}) + (\underline{?}) + (\underline{?})}$$
$$= \sqrt{(\underline{?})} \text{ exactly}$$
$$\approx \underline{?}, \text{ to the nearest tenth}$$

The Three-Dimension Distance Formula helps answer this question: What is the length of the diagonal of a box? A box with dimensions ℓ, w, and h is shown at the right, conveniently located with one endpoint of the diagonal at the origin and the other at (ℓ, w, h).

The diagonal has length QP.

$$QP = \sqrt{(\ell - 0)^2 + (w - 0)^2 + (h - 0)^2}$$
$$= \sqrt{\ell^2 + w^2 + h^2}$$

This argument proves the following formula.

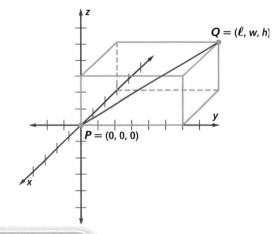

Theorem (Box Diagonal Formula)

The length d of the longest diagonal in a box with dimensions ℓ, w, and h is given by
$$d = \sqrt{\ell^2 + w^2 + h^2}.$$

Notice the similarity between this formula and the Pythagorean Theorem. If both sides of the equation are squared, then $d^2 = \ell^2 + w^2 + h^2$.

Spheres

Just as the Pythagorean Distance Formula leads directly to an equation for any circle in the coordinate plane, the Three-Dimension Distance Formula enables us to find an equation for any sphere. In Lesson 11-6, an equation for the circle with center (h, k) and radius r was shown to be $(x - h)^2 + (y - k)^2 = r^2$. Every sphere has an analogous equation in 3-space.

Theorem (Equation for a Sphere)

The sphere with center (h, k, j) and radius r is the set of points (x, y, z) satisfying the equation

$$(x - h)^2 + (y - k)^2 + (z - j)^2 = r^2.$$

Given The sphere with center (h, k, j) and radius r

Prove The sphere is the set of points (x, y, z) that satisfy the equation
$(x - h)^2 + (y - k)^2 + (z - j)^2 = r^2.$

Proof Because r is the radius, by the definition of sphere, the distance from (h, k, j) to (x, y, z) is r. That distance is given by the Three-Dimension Distance Formula:

$$\sqrt{(x - h)^2 + (y - k)^2 + (z - j)^2} = r.$$

Squaring both sides of this equation,

$$(x - h)^2 + (y - k)^2 + (z - j)^2 = r^2.$$

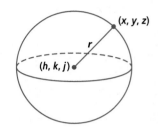

GUIDED

Example 3

Find an equation for the sphere with center at $(-3, 2, 1)$ and radius 7.

Solution Here $(h, k, j) = (\underline{\ ?\ }, \underline{\ ?\ }, \underline{\ ?\ })$ and $r = \underline{\ ?\ }$ in the equation for a sphere.

Substitute these values into the equation for a sphere.

$$(x - \underline{\ ?\ })^2 + (y - \underline{\ ?\ })^2 + (z - \underline{\ ?\ })^2 = \underline{\ ?\ }$$

Simplify by changing any subtraction of a negative number to the addition of its opposite. An equation is

$$(\underline{\ ?\ })^2 + (\underline{\ ?\ })^2 + (\underline{\ ?\ })^2 = \underline{\ ?\ }.$$

Check The point $(4, 2, 1)$ is on this sphere. Check that its coordinates satisfy the equation you have written.

The Midpoint Formula in Space

The Midpoint Formula from Lesson 11-7 also extends easily to three dimensions.

Theorem (Three-Dimension Midpoint Formula)

The midpoint of the segment with endpoints (x_1, y_1, z_1) and (x_2, y_2, z_2) is

$$\left(\frac{x_1 + x_2}{2}, \frac{y_1 + y_2}{2}, \frac{z_1 + z_2}{2} \right).$$

You can verify the Three-Dimension Midpoint Formula by calculating the distances between the endpoints and the midpoint, just as in two dimensions.

 QY2

▸ **QY2**

Find the midpoint of $\overline{AB}$, where $A = (5, 11, 6)$ and $B = (-1, 4, 9)$.

Questions

COVERING THE IDEAS

1. **Fill in the Blank** Any point in three dimensions can be located with an ordered ___?___.

2. Draw a 3-dimensional coordinate system and plot the points $A = (-3, -4, 7)$ and $B = (0, 0, -4)$.

3. One place in a coal mine is 800 feet east, 620 feet south, and 290 feet below the entrance to the mine. On a 3-dimensional coordinate system in which north, east, and up are the positive directions, what are the coordinates of this point?

4. After Example 1, a box is shown with one vertex at $P = (-5, 3, 4)$. Give the coordinates of the other seven vertices of the box.

5. Find the distance between $D = (2, -3, 6)$ and $F = (5, 4, -2)$.

6. In the box at the right, calculate NP and NK.

7. Find an equation for the sphere with radius 5 and center $(-2, 4, -6)$.

8. Find an equation for the sphere with radius $\frac{1}{2}$ and center $(0, 0, 0)$.

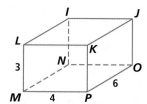

9. A particular sphere has the equation $(x - 4)^2 + y^2 + (z + 7)^2 = 32$.
   a. What is the center of the sphere?
   b. What is the exact radius of the sphere?

10. Find the midpoint of $\overline{RK}$ if $R = (6, -3, 5)$ and $K = (-4, 7, -11)$.

APPLYING THE MATHEMATICS

11. A regular square pyramid $R\text{-}UBAD$ has base vertices $U = (2, 1, 7)$, $B = (8, 1, 7)$, and $A = (8, -5, 7)$.
    a. Find the coordinates of D, the fourth base vertex.
    b. If the height of the pyramid is 12, find two possible locations for the apex R.
    c. Find the exact slant height of the pyramid.
    d. Find the exact length of a lateral edge of the pyramid.
    e. Find the exact lateral area of the pyramid.
    f. Find the volume of the pyramid.

12. A 3-dimensional figure has its base in the *x-y* plane. The base is a circle in the *x-y* plane with equation $(x - 3)^2 + (y + 4)^2 = 36$. The apex of this figure is at $(3, -4, 8)$.
   a. What specific shape is this figure?
   b. What is the volume of this figure?
   c. What is the lateral area of this figure?

13. A sphere has a diameter with endpoints $(1, -6, 5)$ and $(5, 4, -3)$.
   a. Find an equation for this sphere.
   b. Find the volume of this sphere.
   c. Find the surface area of this sphere.

14. An air traffic controller at Cherry Capital Airport in Traverse City, Michigan notices an airplane that is 3 miles east of the airport, 8 miles north of the airport, flying at an altitude of 2 miles. He notices another plane that is 15 miles west of the airport, 13 miles south of the airport, flying at an altitude of 1 mile. At that point, how far apart are the airplanes, rounded to the nearest tenth of a mile?

15. $\triangle ABC$ has vertices $A = (2, 5, 7)$, $B = (0, 7, 3)$ and $C = (0, -5, 9)$.
   a. Find the coordinates of the midpoint of side $\overline{AB}$.
   b. Find the distance from the midpoint of $\overline{AB}$ to vertex C.

16. A trunk is the shape of a box with dimensions 2.5 feet by 4.5 feet by 1.5 feet. To the nearest inch, what is the length of the longest dowel rod that can fit in this trunk?

REVIEW

17. In the figure at the right, *D*, *E*, and *F* are the midpoints of the three sides of $\triangle ABC$. Suppose $AB = 2x$, $AC = 2y$, and $BC = 2z$.
   a. Express the lengths *ED*, *EF*, *DF*, *AD*, *DC*, *CF*, *FB*, *BE*, and *EA* in terms of *x*, *y*, and *z*. **(Lesson 11-8)**
   b. Explain why the four small triangles that make up $\triangle ABC$ are all congruent. **(Lesson 7-2)**

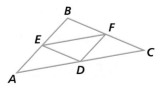

18. **Multiple Choice** Which information does *not* determine an equation of a circle? **(Lesson 11-6)**
   A   the coordinates of the center, and the length of the radius
   B   the coordinates of the two endpoints of a diameter
   C   the coordinates of the center, and one of the points on the circle
   D   the coordinates of two points on the circle

19. Given: not-*p* → *q*, *p* → *r*, and not-*q*.
   Draw a conclusion. **(Lesson 11-2)**

20. **Skill Sequence** Solve. **(Previous Course)**

    a. $\dfrac{14}{x} = \dfrac{84}{11}$ b. $\dfrac{y+2}{3} = \dfrac{4}{5}$ c. $\dfrac{4}{z} = \dfrac{z}{10}$

21. If a person can walk a mile in 13 minutes, how far can the person walk at that rate in 30 minutes? **(Previous Course)**

EXPLORATION

22. A **right tetrahedron** is a tetrahedron in which one vertex is the intersection of three mutually perpendicular planes. You can think of a right tetrahedron as formed by a plane obliquely slicing a box near one of its corners. Every right tetrahedron can be placed on a 3-dimensional coordinate system with the convenient coordinates $(0, 0, 0)$, $(a, 0, 0)$, $(0, b, 0)$, and $(0, 0, c)$. As you know, the Pythagorean Theorem asserts that the sum of the areas of the squares on legs of a right triangle is equal to the area of the square on the hypotenuse of that triangle. Explore the following generalization of that theorem to right tetrahedrons.

    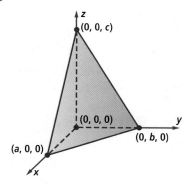

    a. Pick some numerical values for a, b, and c.
    b. Find the areas of the four faces of the tetrahedron. (Hint: Finding the areas of three faces is not difficult. Use Heron's Formula to find the area of the fourth face.)
    c. Find a relationship among the four areas.
    d. Repeat Parts a–c with another set of values for a, b, and c. Does the same relationship hold? Use a DGS-3D to verify the relationships you have found.

23. Find convenient locations of each of the following figures on a 3-dimensional coordinate system.
    a. cube
    b. rectangular solid
    c. regular square pyramid

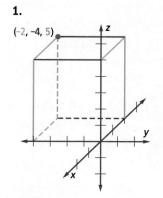

Chapter 11 Projects

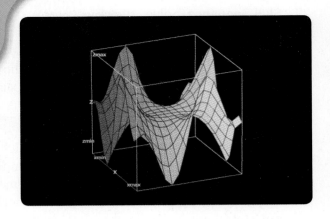

1 Truth Tables

Examine a book on symbolic logic to find out what is meant by a *truth table*. Then show how each of the five rules of logic in this chapter can be shown to be valid using truth tables.

2 Contour Maps

Contour maps can give a visual representation of the elevations of Earth's surface, depth of an ocean, amount of rainfall, temperatures, wildfire activity, and much more. Research contour maps and explain how they are made and read.

3 Photogrammetry

Photogrammetry is measuring objects (2-dimensional or 3-dimensional) from photographs (photo-grammes) and producing 3-dimensional maps like you find on the Internet. Its most important feature is the fact that the objects are measured without being touched. Therefore, the term *remote sensing* is used by some people to refer to photogrammetry. Find some examples of photogrammetry and explain its uses.

4 Graphing Equations

In this chapter you found equations describing a circle or a sphere. Find a program that can graph solutions to equations in three dimensions. Experiment with several equations to see what kind of shapes you can make. Then look up equations for famous surfaces that have pretty shapes. (One example is a saddle curve.). Display what you think is the prettiest of the shapes that you created or found.

5 Equations for Planes

A line in a 2-dimensional rectangular coordinate system is the set of points (x, y) that satisfy an equation of the form $ax + by = c$. Similarly, in a 3-dimensional rectangular coordinate system like the one in Lesson 11-9, a plane is the set of points (x, y, z) that satisfy an equation of the form $ax + by + cz = d$. Find a graphing utility that graphs planes. Use it to graph planes that intersect all three axes, exactly two axes, and exactly one axis. Show these graphs.

Chapter 11

Summary and Vocabulary

● Every valid mathematical argument follows specific rules of logic. In this chapter, five rules are stated.

Law of Detachment From a statement or given information p and a conditional $p \Rightarrow q$, you may conclude q.

Law of Transitivity If $p \Rightarrow q$ and $q \Rightarrow r$, then $p \Rightarrow r$.

Law of the Contrapositive A conditional $(p \Rightarrow q)$ and its contrapositive (not-$q \Rightarrow$ not-p) are either both true or both false.

Law of Ruling Out Possibilities When p or q is true and q is not true, then p is true.

Law of Indirect Reasoning If reasoning from a statement p leads to a false conclusion, then p is false.

● The last three of these laws comprise the basic logic used in indirect proofs. If you can prove that the contrapositive of a statement is true, then the statement is true. If you can rule out all possibilities but one, then the possibility left is true. If you reason from the negation of what you want to prove and arrive at a contradiction, then the negation is false, so what you want to prove must be true.

● In coordinate geometry, figures are described by equations or by giving coordinates of key points. To deduce a general property of a polygon using coordinates, it is efficient to place the polygon in a **convenient location** on the coordinate plane. Either the figure is placed with one vertex at the origin and one or more sides on the axes, or the figure is placed so that it is symmetric to the x-axis or y-axis.

● Three formulas are involved in the coordinate proofs of this chapter. Let (x_1, y_1) and (x_2, y_2) be two points. The slope of the line through them, $\frac{y_2 - y_1}{x_2 - x_1}$, gives a way of telling whether this line is perpendicular or parallel to another. The distance between these points, deduced using the Pythagorean Theorem, is $\sqrt{(x_2 - x_1)^2 + (y_2 - y_1)^2}$. This gives a way to tell whether segments are congruent. The midpoint of the segment joining the points is $\left(\frac{x_1 + x_2}{2}, \frac{y_1 + y_2}{2}\right)$. Many theorems that involve midpoints can be proved using the Midpoint Formula on the Plane.

Vocabulary

11-2
negation (not-p)
inverse (not-$p \Rightarrow$ not-q)
contrapositive
   (not-$q \Rightarrow$ not-p)
logically equivalent

11-3
direct reasoning,
   direct proof
indirect reasoning,
   indirect proofs
contradictory statements

11-4
convenient location for
   a figure

11-5
taxicab distance

11-6
line tangent to a circle
unit circle

11-8
midsegment of a trapezoid
midsegment of a triangle
medial triangle

11-9
2-dimensional coordinate
   system
3-dimensional coordinate
   system
z-coordinate
axes
ordered triple
x-, y-, z- axis
3-space
right tetrahedron

○ Just as there are equations for lines, there are equations for circles. An equation for the circle with center (h, k) and radius r is $(x - h)^2 + (y - k)^2 = r^2$.

○ A coordinate geometry of three dimensions can be built by extending the 2-dimensional coordinate system. An ordered pair becomes an **ordered triple.** The equation for a circle has an analogous equation for a sphere. The Distance Formula and the Midpoint Formula have their 3-dimensional counterparts. Many properties of 3-dimensional figures can be deduced using 3-dimensional coordinates. One example given in this chapter was the formula for the length of a diagonal of a box.

Postulates, Properties, and Theorems

Law of Ruling Out Possibilities
   (p. 648)
Trichotomy Law (p. 651)
Law of Detachment (p. 655)
Law of Transitivity (p. 655)
Double Negative Property (p. 657)
Law of the Contrapositive (p. 658)
Law of Indirect Reasoning (p. 663)
Infinitude of Primes Theorem
   (p. 665)
Theorem (Pythagorean Distance
   Formula on the Coordinate Plane)
   (p. 677)
Theorem (Equation for a Circle)
   (p. 682)
Theorem (Midpoint Formula on a
   Number Line) (p. 688)

Theorem (Midpoint Formula on the
   Plane) (p. 689)
Midsegment of a Trapezoid Theorem
   (p. 695)
Midsegment of a Triangle Theorem
   (p. 695)
Theorem (Three-Dimension Distance
   Formula) (p. 702)
Theorem (Box Diagonal Formula)
   (p. 703)
Theorem (Equation for a Sphere)
   (p. 704)
Theorem (Three-Dimension Midpoint
   Formula) (p. 704)

Chapter 11 Self-Test

Take this test as you would take a test in class. You will need a calculator. Then use the Selected Answers section in the back of the book to check your work.

1. Consider the statement *If a figure is a hexagon, then it is a polygon.*

   a. Is the statement true?

   b. Write the inverse of this statement.

   c. Show that the inverse is false by drawing a counterexample.

2. Consider the statement *If two angles are adjacent, then they form a linear pair.*

   a. Write the contrapositive of the statement.

   b. Is the contrapositive true?

3. $\angle A$ in $\triangle ABC$ is neither acute nor obtuse.

   a. What can you conclude?

   b. What law of logic have you used?

4. Write an argument to show why no triangle can have three angles all with measures less than 50°.

5. Four antique playing cards, the Jack, Queen, King, and Ace, are from four different centuries, the 17th, 18th, 19th, and 20th. From the clues below, match each card to its century. (Note: The 17th century covers the years from 1601–1700, the 18th, 1701–1800; and so on.)

   (1) The Queen is older than the King.

   (2) The Jack is exactly 100 years older than the Queen.

   (3) The Ace is older than the Jack.

6. What (if anything) can you conclude using all of the following statements?

   (1) All babies are happy.

   (2) If someone is teething, then that person is a baby.

   (3) Dante is sad.

In 7 and 8, *D*, *E*, and *F* are midpoints of the sides of $\triangle ABC$ at the right.

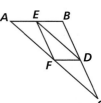

7. Prove that $BDFE$ is a parallelogram.

8. If $AB = 11$ and $BC = 22.3$, find as many other lengths as you can.

9. Let $R = (3, 4)$, $S = (8, 4)$, and $T = (11, 8)$. Find the perimeter of $\triangle RST$.

10. For the circle with the equation $(x + 1)^2 + (y - 9)^2 = 25$, determine

    a. the center.

    b. the radius.

    c. one point on the circle.

11. **Multiple Choice** Quadrilateral $ABCD$ has coordinates $A = (3, 6)$, $B = (7, 9)$, $C = (13, 1)$, and $D = (9, -2)$. Most specifically, $ABCD$ is a

    A rectangle.　　　　B rhombus.

    C kite.　　　　　　D parallelogram.

12. Find the coordinates of the midpoint of the segment with endpoints $(0, -4, 8)$ and $(14, 82, -16)$.

In 13 and 14, a rhombus *RHOB* is located on the coordinate system below. *E, I, M,* and *U* are midpoints of the sides.

13. Give the coordinates of *E, I, M,* and *U*.

14. Prove that *EIMU* is a rectangle.

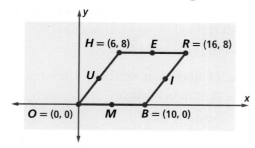

15. Give convenient coordinates for the vertices of a kite.

16. The endpoints of a diameter of a sphere are (3, 2, –1) and (4, 9, 7).

   a. What are the coordinates of the center of the sphere?

   b. What is the radius of the sphere?

17. Marco is 200 yards east and 150 yards south of a tall building. His friend Nina is in the building 123 yards above ground. How far apart are they, to the nearest yard?

18. Prove that the diagonals of the quadrilateral below have the same midpoint.

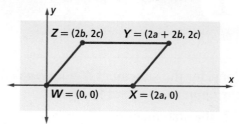

19. Given: $\odot O$; $\overline{OB}$ is not an altitude of $\triangle AOC$.

   Prove: $\overrightarrow{OB}$ does not bisect $\angle COA$.

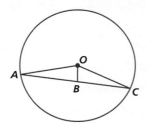

20. In the state of Kansas, Goodland is about 60 miles north and 12 miles west of Selkirk, and Garden City is about 36 miles south and 39 miles east of Selkirk. What is the flying distance from Goodland to Garden City?

Chapter 11 Chapter Review

SKILLS
PROPERTIES
USES
REPRESENTATIONS

SKILLS Procedures used to get answers

OBJECTIVE A Determine the distance between two points in the coordinate plane. (Lesson 11-5)

In 1–4, find the distance between the given points.

1. $(0, 0)$ and $(-1, 3)$ 2. $(3, 3)$ and $(3, 3)$
3. $(-2, 2)$ and $(1.5, -1.5)$
4. $(101, 127)$ and $(-20, 35)$
5. A triangle has vertices at $(1, 2)$, $(10, 2)$, and $(1, 14)$.
   a. Find its perimeter.
   b. Is it scalene, isosceles but not equilateral, or equilateral?
6. A quadrilateral $QUAD$ has vertices $Q = (-2, 3)$, $U = (4, 1)$, $A = (-2, -6)$, and $D = (7, 0)$. Which is its longest side?
7. Write an expression for the distance between (x, y) and $(2x, 2y)$.

OBJECTIVE B Determine the coordinates of the midpoint of a segment in the coordinate plane. (Lesson 11-7)

In 8 and 9, determine the midpoint of the segment with given endpoints.

8. $(8, 3)$ and $(7, 2)$
9. $(-123, 145)$ and $(289, -342)$
10. Write an expression for the midpoint of the segment with ends $(-x, x)$ and $(2, 7x)$.
11. $(1, 2)$ is the midpoint of $(3, 8)$ and (x, y). Find $x + y$.
12. $(3, 2)$ is the midpoint of a segment with one endpoint at $\left(\frac{1}{2}, \frac{7}{2}\right)$. Find the coordinates of the other endpoint.

OBJECTIVE C Apply the Midsegment of a Trapezoid and Midsegment of a Triangle Theorems. (Lesson 11-8)

In 13–16, refer to the figure below, in which $\triangle DEF$ is the medial triangle of $\triangle ABC$.

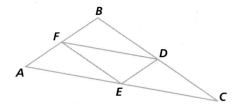

13. Identify three trapezoids in the figure.
14. If $FD = 12$ and $BD = 10$, identify as many other lengths as you can.
15. If the perimeter of $\triangle FDE$ is 17, what is the perimeter of $\triangle ABC$?
16. Prove that $m\angle DFE = m\angle FEA$.

OBJECTIVE D Find distances and coordinates of midpoints in 3-dimensional space. (Lesson 11-9)

In 17 and 18, find the distance between the given points.

17. $(-17, 3, 5)$ and $(-89, 4, -33)$
18. $(x, y, 0)$ and $(0, x, y)$

In 19 and 20, find the coordinates of the midpoint of the segment with given endpoints.

19. $(7, 2, -7)$ and $(0, 9, 0)$
20. $\left(\frac{2}{3}, \frac{4}{9}, -2\right)$ and $\left(5, \frac{1}{9}, \frac{4}{5}\right)$
21. $(1, 3, 7)$ is one endpoint of a diameter of a sphere. $(0, 6, 8)$ is the center of the sphere.
   a. Find the coordinates of the other endpoint of the diameter.
   b. Find the length of the diameter.

PROPERTIES Principles behind the mathematics

OBJECTIVE E Write the converse, inverse, or contrapositive of a conditional. (Lesson 11-2)

In 22–26, write a. the converse, b. the inverse, and c. the contrapositive of the statement. d. Tell which (if any) of these are true.

22. If a quadrilateral has four right angles, then it is a square.

23. If $x = 4$, then $x^2 = 16$.

24. If a figure is a cylindrical solid, then it is a cube.

25. If $AB = BC$, then B is the midpoint of $\overline{AC}$.

26. If three points are coplanar, then they are collinear.

OBJECTIVE F Follow the basic laws of logic to make conclusions. (Lessons 11-1, 11-2, 11-3)

In 27–29, use all of the statements.

a. What can you conclude?

b. What laws of reasoning did you use?

27. (1) $\ell \parallel m$ or $j \parallel k$
    (2) Lines ℓ and m intersect.

28. (1) Deanna and Diego each prefer either even numbers or odd numbers.
    (2) Deanna and Diego's preferences are different.
    (3) Deanna prefers even numbers.

29. (1) A lies on $\odot O$.
    (2) $OA \neq OB$

30. Ivy wished to solve the equation $x^2 = x^2 + 1$. After simplifying the equation, she got the equation $0 = 1$.
    a. What should Ivy conclude?
    b. What law of logic should she use to make this conclusion?

31. Suppose $p \Rightarrow$ not-q and you know that q is true. What can you conclude about p?

OBJECTIVE G Write indirect proofs. (Lesson 11-3)

In 32–35, give indirect proofs.

32. Prove that this system has no solution.
$$\begin{cases} 3x - y = 5 \\ -9x + 3y = -20 \end{cases}$$

33. Suppose $ABCD$ is a quadrilateral in which no two of the sides have the same length. Prove that the diagonals of $ABCD$ do not bisect each other.

34. Suppose that $\triangle ABC$ is an equilateral triangle inscribed in a circle of diameter d. Prove that none of the sides of $\triangle ABC$ have length d.

35. Use an indirect proof to show that there are infinitely many integers whose units digit is 0.

OBJECTIVE H Use coordinate geometry to deduce properties of figures and prove theorems. (Lessons 11-4, 11-5, 11-6, 11-7)

36. Prove that the square with vertices at (a, a), $(a, -a)$, $(-a, -a)$, and $(-a, a)$ can be inscribed in a circle with center $(0, 0)$ and radius $a\sqrt{2}$.

37. Use coordinates to prove that the diagonals in a rectangle bisect each other.

38. Prove that the triangle with vertices $(3t, 0)$, $(0, 3t)$, and $(6t, 3t)$ is a right triangle.

39. Let $ABCD$ be the quadrilateral with vertices $A = (-12, 6)$, $B = (6, 24)$, $C = (22, 12)$, $D = (14, -2)$. Show that the quadrilateral formed by connecting its midpoints is a parallelogram.

40. The triangle with vertices $X = (3, 0)$, $Y = (-3, 0)$, and $Z = (0, 3)$ is inscribed in a circle.
    a. Show that $\angle Z$ is a right angle.
    b. Find an equation for the circle.

USES Real-world applications of mathematics

OBJECTIVE I Apply laws of logic in real situations. (Lessons 11-1, 11-2, 11-3)

In 41–43, given the statements, what can you conclude by using the laws of logic?

41. (1) On Friday afternoons, Sophia either fences or goes ballroom dancing.

(2) One Friday afternoon, Sophia did not show up at the dance hall.

42. (1) Prior to the year 2006, Pluto was considered to be a planet.

(2) Malik has a diagram of the solar system in which Pluto is listed as a planet.

43. (1) If you live in Paris, then you live in France.

(2) If you live in France, then you cannot see the Northern Lights from your house.

(3) Jacqueline lives in Paris.

44. Keiko's teacher said that if you study hard for the test, then you will pass. Keiko did not study hard. Is it correct to deduce that she will not do well? Why, or why not?

OBJECTIVE J Apply the Distance and Box Diagonal Formulas in real situations. (Lessons 11-5, 11-9)

45. A storage unit is in the shape of a box whose dimensions are 7 feet by 7 feet by 5 feet. To the nearest foot, what is the length of the longest thin metal rod that can be fit into the storage unit?

46. Elena is 13 blocks east and 3 blocks north of the center of town. Denzel is 5 blocks east and 9 blocks north of the center of town. Each city block is an eighth of a mile long. What is the distance between the two of them, in a straight line?

47. Darnell and Serena are in two hot air balloons. Darnell's balloon is 3 miles west, 2 miles north, and 300 feet under Serena's balloon. How far apart are the two balloons, in a straight line?

48. A car drove 7 miles north and 8 miles east at a speed of 35 miles per hour. A swallow flew in a straight line from the starting point of the car to its destination at a speed of 24 miles per hour. Which of the two got to the destination first?

REPRESENTATIONS Pictures, graphs, or objects that illustrate concepts

OBJECTIVE K Graph and write an equation for a circle or a sphere given its center and radius, and vice versa. (Lessons 11-6, 11-9)

49. What is an equation for the circle with center (102, –89) and radius 7.5?

50. What is an equation for the sphere with center (2, 8, 5) and radius 4?

In 51 and 52, graph the equation.

51. $(x - 3)^2 + (y - 2)^2 = 49$

52. $x^2 + (y + 6)^2 = 25$

53. Find the center and radius of the sphere with equation $(x + 3)^2 + (y + 12)^2 + (z - 17)^2 = 6.25$.

OBJECTIVE L Give convenient locations for triangles and quadrilaterals in the coordinate plane. (Lesson 11-4)

In 54–56, give convenient locations for the figure.

54. an isosceles triangle whose vertex lies on the x-axis

55. an isosceles trapezoid that is symmetric with respect to the y-axis

56. a rectangle symmetric to the x- and y-axes whose length is twice its height

Chapter

12 Similarity

▶ **Contents**

12-1 Size Transformations Revisited

12-2 Review of Ratios and Proportions

12-3 Similar Figures

12-4 The Fundamental Theorem of Similarity

12-5 Can There Be Giants?

12-6 The SSS Similarity Theorem

12-7 The AA and SAS Triangle Similarity Theorems

Figures that are size change images of congruent figures are called *similar figures.* Similar figures may or may not have the same size. Similar figures are found both in fun and serious pursuits. The dolls shown above are Russian nesting dolls called "matryoshka dolls" and are approximately similar. They fit inside one another and are like the kachina dolls found in the carvings of Native Americans and other cultures.

Model planes, model cars, model trains, model ships, dolls, and doll houses can all be considered as objects that are similar to real figures, played with for enjoyment. Clothes designers, inventors, architects, and city planners use designs that are similar to a real object to see how the object would look without having to make it actual size. Scientists magnify small things like insects or the atom, or make models of large objects like Earth or our solar system, in order to study them.

The concept of similarity is as important in analyzing figures as the concept of congruence. Certain conditions on triangles force them to be similar, just as there are conditions that force congruence. In this and the next chapter, you will study the basic properties of similar figures and the transformations relating them, and see a few of their many applications.

Lesson

12-1 Size Transformations Revisited

Vocabulary

size change, size
  transformation with any
  center

magnitude, size-change factor

dilation, dilatation

expansion

contraction

identity transformation

▶ **BIG IDEA** Size transformations can be accomplished with any point as center and do not require coordinate geometry.

The outline of a sailboat shown below is a picture that you saw in Lesson 3-7. In that lesson, you enlarged or reduced the size of a figure in the coordinate plane by multiplying all of its coordinates by a size-change factor k. The transformation that maps (x, y) onto (kx, ky) was called the *size transformation* or *size change with center (0, 0) and magnitude k*, or S_k for short.

To accomplish S_k, you need to know the coordinates of the main points of a figure and you are restricted to size changes centered at the origin. To study similar figures, we need to obtain size, change images without using coordinates and with any point as the center.

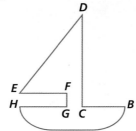

Mental Math

Find the exact length of a side of

a. a rhombus with perimeter 27 cm.

b. a cube with surface area 42 in².

c. a cube with volume 512 yd³.

Activity 1

MATERIALS tracing paper
The goal of this activity is to apply a size change of magnitude 1.5 to the sailboat pictured above.

Step 1 Trace the picture of the sailboat onto a piece of paper. Place a point *O* on the paper somewhere close to the boat.

Step 2 To create the enlarged image of the sailboat, you need to find the image of each point on the sailboat and then connect the image points. Here is how to find the image of point *D*.

a. Draw $\overrightarrow{OD}$.

b. Measure *OD*.

c. Locate the point on $\overrightarrow{OD}$ that is 1.5 times as far from *O* as *D* is. Label that point *D'*.

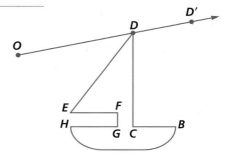

Step 3 Repeat the process for each named point on the sailboat. Connect these points to create an enlarged version of the boat.

Activity 1 shows how to enlarge an image by a factor of 1.5. If you wanted to reduce the size of the image, you could follow the same procedure but multiply by a number between 0 and 1. For example, if in Step 2c you found the point that is 0.1 times as far from O as D is, and then repeated that in Step 3, you would create an image that was one-tenth the size of the original.

Size Changes with Any Center

Because each preimage point has a unique image, the procedure described in Activity 1 defines a transformation. Had point O been $(0, 0)$ on a coordinate system, this would have been the size transformation $S_{1.5}$. By not using coordinates, we are able to consider size changes with any point as the center.

Definitions of Size Change (Size Transformation) with Any Center, Magnitude, and Size-Change Factor

Let O be a point and k be any nonzero real number. For any point P, let $S(P) = P'$ be the point on $\overleftrightarrow{OP}$ with $OP' = k \cdot OP$ in the direction of $\overrightarrow{OP}$ if k is positive and in the direction opposite $\overrightarrow{OP}$ if k is negative. Then S is the **size change** or **size transformation** with center O and **magnitude** or **size-change factor** k.

> **READING MATH**
>
> If you have an eye exam, the eye doctor may put drops in your eyes to dilate (enlarge) the pupils of your eyes. This allows the doctor to see the back of your eyes and check for eye diseases.

Some textbooks and software refer to a size change as a **dilation** or **dilatation.** When $|k| > 1$, S is called an **expansion.** When $0 < |k| < 1$, S is a **contraction.** When $k = 1$, S is the **identity transformation** because each point coincides with its image. The size change on the previous page is an expansion with magnitude 1.5 and center O.

Example 1

A size transformation has mapped $\triangle ABC$ to $\triangle A'B'C'$. Trace the triangles and find the center and magnitude of the size transformation.

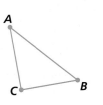

Solution The triangles are traced below. The line through a point and its image contains the center of the size transformation. So, if you draw two such lines, their intersection is the center of the size transformation. We draw $\overleftrightarrow{AA'}$ and $\overleftrightarrow{BB'}$, intersecting at O. To find the magnitude of the size transformation, measure the lengths.

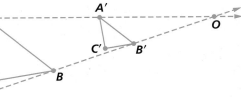

$OB' = 2.25$ cm
$OB = 4.5$ cm

The ratio $\frac{OB'}{OB} = \frac{2.25}{4.5} = \frac{1}{2}$ gives the magnitude of the size transformation with center O.

In Lesson 3-7, we proved that the transformation S_k with $S_k(x, y) = (kx, ky)$ preserves angle measure, betweenness, and collinearity. If we can show that S_k is the same transformation as the size change S with magnitude k defined on the preceding page, then the properties proved for S_k must also be true for the size changes as we have defined them in this lesson. To do this, we show that a point and its image under S_k satisfy the defining properties of a size change with center O and magnitude k.

Theorem

The transformation S_k that maps (x, y) onto (kx, ky), with $k \neq 0$, is a size transformation S with center $(0, 0)$ and magnitude k.

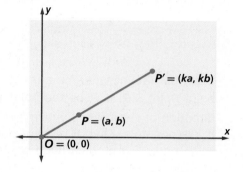

Given   Points $O = (0, 0)$, $P = (a, b)$, and $P' = (ka, kb)$

Prove   1. P' is on $\overleftrightarrow{OP}$
              2. $OP' = |k| \cdot OP$

Proof of 1   The slope of $\overleftrightarrow{OP}$ is $\frac{b - 0}{a - 0}$, or $\frac{b}{a}$. The slope of $\overleftrightarrow{OP'}$ is $\frac{kb - 0}{ka - 0}$, or $\frac{kb}{ka}$, which equals $\frac{b}{a}$ because $k \neq 0$. Because the slope of $\overleftrightarrow{OP}$ is equal to the slope of $\overleftrightarrow{OP'}$, the lines $\overleftrightarrow{OP}$ and $\overleftrightarrow{OP'}$ are parallel. And because there is only one line parallel to $\overleftrightarrow{PP'}$ through O, the lines $\overleftrightarrow{OP}$ and $\overleftrightarrow{OP'}$ are identical and the points O, P, and P' must be collinear.

Proof of 2   Use the Pythagorean Distance Formula.
$$OP = \sqrt{(a - 0)^2 + (b - 0)^2} = \sqrt{a^2 + b^2}$$
$$OP' = \sqrt{(ka - 0)^2 + (kb - 0)^2}$$
$$= \sqrt{(ka)^2 + (kb)^2}$$
$$= \sqrt{k^2(a^2 + b^2)}$$
$$= |k| \sqrt{a^2 + b^2}$$
But $\sqrt{a^2 + b^2} = OP$, so by substitution $OP' = |k| \cdot OP$.

Since the size transformations defined in Lesson 3-7 give the same images as the size transformations defined in this lesson, all size changes have the following preservation properties, which were proved in Lesson 3-7.

Size-Change Preservation Properties Theorem

Every size transformation preserves:
1. angle measure.
2. betweenness.
3. collinearity.

We turn now to the most important property of size transformations, which has to do with the sizes of figures and their size transformation images. By now it is probably already obvious to you, that a size transformation of magnitude $k > 0$ creates an image where lengths are k times as long as on the preimages.

Size-Change Distance Theorem

Under any size change with magnitude $k \neq 0$, the distance between any two image points is $|k|$ times the distance between their preimages.

Given P, Q, and a size transformation S_k;
$S_k(P) = P'$ and $S_k(Q) = Q'$

Prove $P'Q' = |k| \cdot PQ$

Proof   Let $P = (a, b)$ and $Q = (c, d)$.

From the Pythagorean Distance Formula,

$PQ = \sqrt{(c - a)^2 + (d - b)^2}$.

By definition of S_k, $P' = (ka, kb)$ and $Q' = (kc, kd)$.

Also from the Pythagorean Distance Formula,

$$P'Q' = \sqrt{(kc - ka)^2 + (kd - kb)^2}$$
$$= \sqrt{(k(c - a))^2 + (k(d - b))^2}$$
$$= \sqrt{k^2(c - a)^2 + k^2(d - b)^2}$$
$$= \sqrt{k^2\left((c - a)^2 + (d - b)^2\right)}$$
$$= \sqrt{k^2}\,\sqrt{(c - a)^2 + (d - b)^2}$$
$$= |k|\,\sqrt{(c - a)^2 + (d - b)^2}$$
$$= |k| \cdot PQ.$$

Thus, $P'Q' = |k| \cdot PQ$.

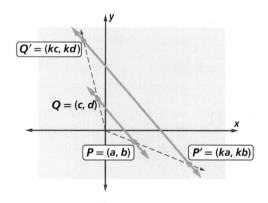

Put another way, the *ratio* of the length of an image to the length of its preimage is $|k|$. That is, $P'Q' = |k| \cdot PQ$ implies that $\dfrac{P'Q'}{PQ} = |k|$.

GUIDED

Example 2

Suppose S is a size transformation of magnitude 0.3. If $\triangle ABC$ has a perimeter of 70 cm, find the perimeter of $S(\triangle ABC)$.

Solution   Let $\triangle A'B'C' = S(\triangle ABC)$, let $p = $ the perimeter of $\triangle ABC$, and let $p' = $ the perimeter of $\triangle A'B'C'$.

By the definition of perimeter, $p = AB + BC + AC = \underline{\quad ? \quad}$ cm.

(continued on next page)

For the same reason, $p' = \underline{\ ?\ } + \underline{\ ?\ } + \underline{\ ?\ }.$

$$= 0.3 \cdot AB + 0.3 \cdot BC + 0.3 \cdot AC$$
$$= 0.3\,(\underline{\ ?\ } + \underline{\ ?\ } + \underline{\ ?\ })$$
$$= 0.3p$$
$$= 0.3 \cdot \underline{\ ?\ }$$
$$= \underline{\ ?\ }\ cm$$

Activity 2 looks at the effect of changing the center or magnitude of a size transformation.

Activity 2

MATERIALS DGS

Step 1 Create quadrilateral $ABCD$ and point O.

Step 2 Let S be a size transformation with center O and magnitude $k > 1$. Construct $S(ABCD)$ and name it $A'B'C'D'$. We choose $k = 1.8$ as shown at the right.

Step 3 Describe what happens when O is in the interior of $ABCD$ and k is positive. (Be careful, there are two cases.)

Step 4 If $k > 0$ and O is placed on vertex A, what happens to the position of A'? Why?

Step 5 Let $k = -1$ and drag $ABCD$. What isometry is equivalent to a size transformation of magnitude -1?

Step 6 The size transformation of magnitude k and center O maps $ABCD$ onto $A'B'C'D'$. Describe the size transformation that maps $A'B'C'D'$ back onto $ABCD$.

Step 7 Create another point P and find the image of $A'B'C'D'$ under a size transformation of magnitude m with center P. Name the new quadrilateral $A''B''C''D''$.

This quadrilateral is the image of $ABCD$ under the composite of two size transformations. Is the composite a size transformation? If so, what is its magnitude?

Steps 1 and 2

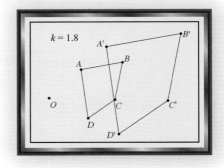

Steps 5–7

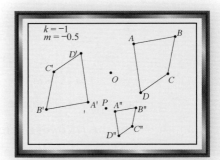

Summary of Basic Properties of Size Changes

With the Size-Change Distance Theorem, you have seen the most basic properties relating figures and their size change images. If a figure G is the image of a figure F under a size change with magnitude $k \neq 0$, then:

1. Size changes preserve betweenness and collinearity: The image of a line is a line.

2. Corresponding lines, rays, or line segments on F and G are parallel, implying that size changes preserve angle measure: Angles on G have the same measure as corresponding angles on F.

3. Size changes multiply distance by the absolute value of the size-change magnitude. The distance between two points on G is $|k|$ times the corresponding distance between two points on F.

The preservation of angle measure, betweenness, and collinearity implies that the images of figures are determined by the images of the key points. Thus, if you want to find the image of a triangle, you simply find the image of the three vertices (instead of finding the image of infinitely many points on the segments). If you want to find the image of a circle, simply find the image of the center and one point on the circle! This idea is the same as found in the Figure Reflection Theorem in Lesson 4-2.

> ### Figure Size-Change Theorem
>
> If a figure is determined by certain points, then its size-change image is the corresponding figure determined by the size-change images of those points.

You applied this theorem when you constructed the image of the sailboat at the beginning of this lesson.

Questions

COVERING THE IDEAS

1. Trace the figure at the right. Let S be the size transformation with center O and magnitude $\frac{2}{3}$. Construct $S(\triangle ABC)$.

2. In Question 1, where do the lines $\overleftrightarrow{AA'}$ and $\overleftrightarrow{BB'}$ intersect?

3. If the size transformation from Question 1 were changed to have a magnitude of $-\frac{2}{3}$, construct $S(\triangle ABC)$.

4. Let S be a size transformation with magnitude 4. If $PQ = 12$ cm, how long is $S(\overline{PQ})$?

5. Use the figure at the right. Find the center and magnitude of the size transformation that maps $ABCD$ to $A'B'C'D'$.

6. Suppose S is a size transformation with center Q and magnitude $\frac{4}{3}$. Let $S(\triangle PIG) = \triangle COW$.
   a. If $PI = 9$ cm, what is CO?
   b. If $m\angle G = 55$ and $m\angle P = 85$, what is the measure of angle O?

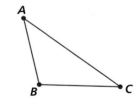

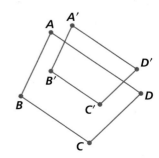

7. **Fill in the Blanks** Every size transformation preserves
   ___?___, ___?___, and ___?___.

8. Use the Figure Size-Change Theorem to find the image of circle A shown at the right under the size change S with center O and magnitude 3.

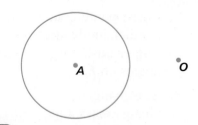

APPLYING THE MATHEMATICS

9. The figure at the right shows the mighty fencer Arim. Suppose you wanted to create his opponent Mira, who is exactly 0.9 times his size, facing him, and standing on the same flat surface.
   a. Describe how to accomplish this task with a composite of a size transformation and an isometry.
   b. Could this task be accomplished by a single size transformation with negative magnitude? Explain.

10. In the circle at the right, C is the center of the larger circle and $\overline{CE}$ is the diameter of circle D, the image of circle C under a size change.
    a. Find the center and magnitude of the size change that maps circle C onto circle D.
    b. Find the center and magnitude of the size change that maps circle D onto circle C.

11. In the diagram at the right, a size transformation has mapped $\overline{FS}$ onto $\overline{F'S'}$. If $SO = 8$ cm, $OF = 6$ cm, $OF' = 9$ cm, and $FS = 4$ cm, find SS' and $F'S'$.

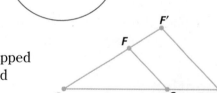

12. In the diagram at the right, a size transformation mapped FOX to $F'OX'$.
    a. What is the center of this size transformation?
    b. Find the magnitude of this size transformation.
    c. Calculate XO.

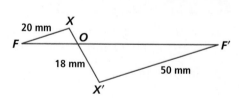

13. Prove: If $S_k(\triangle ABC) = \triangle A'B'C'$, then $\dfrac{AB}{BC} = \dfrac{A'B'}{B'C'}$.

14. Let S_k and $S_{\frac{1}{k}}$ be size transformations with the same center. Prove that the composite of these size transformations is the identity transformation. (Recall that the size change of magnitude 1 is the identity transformation.)

15. Recall that an equation of a circle with center (h, k) and radius r is $(x - h)^2 + (y - k)^2 = r^2$.

    a. Write an equation for the circle with center $(4, -2)$ and radius 5.

    b. Find an equation for the image of the circle from Part a under the size transformation with center $(0, 0)$ and magnitude 3.

    c. Find an equation of the image of the circle from Part a under the size change with center $(4, -2)$ and magnitude 3.

16. In 3-dimensional space, the size transformation with center $(0, 0, 0)$ and magnitude k maps (x, y, z) onto (kx, ky, kz).

    a. If $M = (-1, 3, 6)$, find the coordinates of $S_4(M)$. Call this point M'.

    b. If $N = (2, 7, -6)$, find the coordinates of $S_4(N)$. Call this point N'.

    c. Using the Three-Dimension Distance Formula, verify that $M'N' = 4 \cdot MN$.

REVIEW

17. In 3-dimensional space, what is the set of all points whose distance from the origin is 3, and whose x-coordinate is 0? (**Lesson 11-9**)

18. Refer to the figure at the right.
    Given $\angle B$ and $\angle C$ are right angles, $AB = AC$.
    Prove $ABDC$ is a kite. (**Lesson 7-5**)

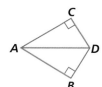

19. During a 24-hour day, what is the total magnitude of rotation through which the minute hand on a clock goes? (**Lesson 3-2**)

20. Simplify the expression $\dfrac{\frac{x^2}{6}}{\frac{x}{3}}$. (**Previous Course**)

21. Cedric was born 30 cm tall. When he was 20 years old, he was 190 cm tall. (**Previous Course**)

    a. If Cedric grew at a constant rate, how many centimeters per year did he grow?

    b. If the growth was at a constant rate, how tall was Cedric when he was 12 years old?

    c. Do you think the rate of his growth from birth to age 20 was constant? Why, or why not?

EXPLORATION

22. Generalize the results of Question 15.

Lesson

12-2

Review of Ratios and Proportions

Vocabulary

ratio

proportion

terms of a proportion

extremes

means

▶ **BIG IDEA** Size transformations and their images lead to figures in which corresponding lengths are in equal ratios.

You have often seen ratios and proportions. Here is a review of some of the language of ratios and proportions.

Ratios and Proportions

A **ratio** is a quotient of two numbers, $\frac{m}{n}$ or m/n. Sometimes the ratio m/n is written $m{:}n$. In a ratio $\frac{m}{n}$, m and n must be quantities of the same kind, such as lengths, populations, or areas. (If the quantities are of different kinds, $\frac{m}{n}$ is called a *rate*.)

A statement that two ratios are equal is called a **proportion.** Each equation below is a proportion.

$$\frac{CB}{C'B'} = \frac{5}{3} \qquad \frac{2}{7} = \frac{x}{9} \qquad \frac{y+3}{5} = \frac{7}{y} \qquad \frac{A'B'}{AB} = \frac{B'C'}{BC}$$

Proportions arise when a figure is transformed by a size change. For instance, when $\triangle ABC$ is a size transformation image of $\triangle XYZ$, you can say, "The sides of the triangles are proportional." This means the three ratios of corresponding sides are equal.

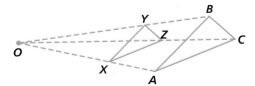

That is, $\frac{AB}{XY} = \frac{AC}{XZ} = \frac{BC}{YZ}$. Any of the three ratios is the magnitude of the size change. You can pick any two of these to form a proportion.

$$\frac{AB}{XY} = \frac{AC}{XZ} \qquad\qquad \frac{AB}{XY} = \frac{BC}{YZ} \qquad\qquad \frac{AC}{XZ} = \frac{BC}{YZ}$$

Notice that in each proportion, the numerators come from one figure and the denominators are corresponding lengths in the other.

The Means and the Extremes in a Proportion

The four numbers in a proportion are called the **terms of the proportion.** The four terms in $\frac{a}{b} = \frac{c}{d}$ or $a{:}b = c{:}d$ have two sets of names and are numbered in order. The first and fourth terms a and d are the **extremes**, and the second and third terms b and c are the **means**.

Mental Math

Calculate

a. a 20% tip on a $38 bill.

b. a 5% raise on a $7.50 per hour rate.

c. a 30% savings on a $45 sweater.

1st term ⟶ $\dfrac{a}{b} = \dfrac{c}{d}$ ⟵ 3rd term
2nd term ⟶ ⟵ 4th term

extremes

$\dfrac{a}{b} = \dfrac{c}{d}$

means

In previous courses, you have learned that in any proportion, *the product of the means equals the product of the extremes*. This is the *Means-Extremes Property*. You solved proportions in earlier courses based on this property.

> **Theorem (Means-Extremes Property)**
>
> If $\dfrac{a}{b} = \dfrac{c}{d}$, then $ad = bc$.

Proof Suppose $\dfrac{a}{b} = \dfrac{c}{d}$. Then, by the Multiplication Property of Equality, $bd \cdot \dfrac{a}{b} = bd \cdot \dfrac{c}{d}$. Simplifying both sides, $ad = bc$.

One consequence of the Means-Extremes Property is that there are many proportions that produce the same products of means and extremes and therefore are equivalent equations. This is summarized in the following series of equivalent proportions.

$$\dfrac{x}{a} = \dfrac{p}{q} \qquad \dfrac{a}{x} = \dfrac{q}{p} \qquad \dfrac{q}{a} = \dfrac{p}{x} \qquad \dfrac{a}{q} = \dfrac{x}{p}$$

You might notice that all of these proportions imply $xq = ap$. You might also notice that if two ratios are equal, their reciprocals are equal, so taking the reciprocal of both sides of a *true* proportion leads to a true proportion.

🛑 **QY1**

> ▶ **QY1**
>
> If $\dfrac{5}{7} = \dfrac{r}{h}$, write four other true equations involving r and h.

Example 1

$S_k (\triangle BCD) = \triangle B'C'D'$, $BD = 5$, $BC = 6$, and $B'D' = 14$.
Find $B'C'$.

Solution The magnitude of the size change is the ratio of a length on the image to a length on the preimage. So

$$\dfrac{B'D'}{BD} = \dfrac{B'C'}{BC}.$$

Substituting the given information,

$$\dfrac{14}{5} = \dfrac{x}{6}.$$

By the Means-Extremes Property,

$$5x = 14 \cdot 6$$
$$5x = 84$$
$$x = 16.8.$$

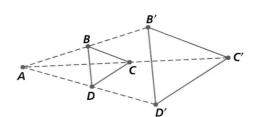

Proportions in Scale Models

Example 2

Pictured at the right is a Fabergé egg presented by Tsar Nicholas II of Russia in 1908 to his wife Alexandra Fedorovna. Shown below is the buildng inside this egg. The building is a scale model of the Alexander Palace and is made of gold. The scale model building is 30 mm by 65 mm. The palace is $\frac{13}{24,600}$ actual size. What is the length of the actual palace?

Solution Because corresponding lengths are proportional,

$$\frac{\text{model length}}{\text{actual length}} = \frac{13}{24,600}.$$

Substitute the known dimensions.

$$\frac{65}{L} = \frac{13}{24,600}$$

Solve using the Means-Extremes Property.

$$24,600 \cdot 65 \text{ mm} = 13L$$
$$1,599,000 \text{ mm} = 13L$$
$$123,000 \text{ mm} = L$$
$$123 \text{ m} = L$$

So the Alexander Palace is about 123 meters long.

 QY2

> ▶ QY2
>
> Find the width of the actual Alexander Palace.

Questions

COVERING THE IDEAS

1. **Multiple Choice** Which is *not* a way of writing the ratio of 3 to 5?

   A 3/5   B 3:5   C $\frac{3}{5}$ D 3.5

2. **Fill in the Blank** A ratio is a ___?___ of two numbers.

3. **Fill in the Blanks** A proportion is a statement that two ___?___ are ___?___.

4. Given $\frac{12}{x} = \frac{y}{11}$, name the

   a. extremes.   b. means.   c. second term.

5. Refer to the proportion of Question 4.

   a. What can you conclude because of the Means-Extremes Property?

   b. Give two possible pairs of values of x and y.

6. In a flag, the *hoist* is the width and the *fly* is the length. In a United States flag, the ratio of the hoist to the fly is typically 1:1.9. What is the fly on the flag whose hoist is 60 inches?

7. a. What equation results if both sides of the equation $ad = bc$ are divided by bd?

   b. What equation results if both sides of the equation $ad = bc$ are divided by ac?

In 8 and 9, *ABCD* at the right is the image of *FGHE* under a size change with center *V*.

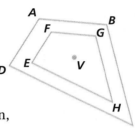

8. **Fill in the Blank** $\dfrac{AB}{FG} = \dfrac{AD}{?}$

9. If $FG = 10$, $AB = 12$ and $BC = 15$, what is GH?

10. Suppose $\dfrac{u}{v} = \dfrac{w}{x} = \dfrac{y}{z}$. From this information, form three true proportions.

11. In Question 10, explain why it is true that $\dfrac{x}{w} = \dfrac{z}{y}$.

APPLYING THE MATHEMATICS

12. **Multiple Choice** A 5-megapixel digital image has dimensions 1944×2592 pixels. The available sizes at a photo printer are given below. If a digital photo were printed on each size photo paper, on which size would the photo fit exactly?

   A 5 in. × 7 in.   **B** 8 in. × 10 in.   **C** 4 in. × 6 in.

   D 2.5 in. × 3.5 in. (wallet size)   **E** none of these

13. Suppose $S_k \circ S_k(\triangle ABC) = \triangle DEF$, as shown below; $AB = 4$; and $DF = 36$. Find $A'B'$. Write three equal ratios involving the sides of $\triangle ABC$ and $\triangle A'B'C'$.

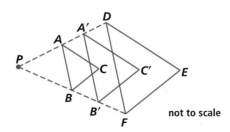

not to scale

14. In the figure at the right, the quadrilaterals are squares. Joining the midpoints of the sides of the outer squares forms the inner squares. What is the value of the ratio $\frac{FG}{BA}$?

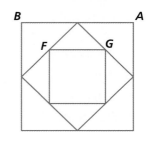

15. Solve for x: $\frac{x+1}{1} = \frac{1}{x}$

16. If $wx = yz$, write three true proportions using w, x, y, and z.

REVIEW

17. Let S be a size transformation with magnitude k and center O. Let T be a size transformation with magnitude k and center P. Suppose $S(\triangle ABC) = \triangle A'B'C'$ and $T(\triangle ABC) = \triangle A''B''C''$. Explain why $\triangle A'B'C' \cong \triangle A''B''C''$. (**Lesson 12-1**)

18. Show that the transformation R with equation $R(x, y) = (12x, 2y)$ does not preserve distance. (**Lesson 11-5**)

19. Use an indirect proof to prove the following fact: There are infinitely many numbers that are divisible by 2. (Hint: What could you do if there was a greatest number that was divisible by 2?) (**Lesson 11-3**)

20. A blacksmith takes a cylinder of metal with height h cm and base with area 8 cm². He hammers the cylinder into a cube with equal volume. What is the surface area of the cube? (**Lessons 10-3, 9-9**)

21. a. How does the area of a rectangle change when all of its sides are multiplied by k?
    b. How does the area of a circle change when its radius is multiplied by k? (**Lessons 8-9, 8-2**)

22. a. According to the definition of congruence, when are two figures congruent? (**Lesson 5-1**)
    b. What is an isometry? (**Lesson 4-7**)

EXPLORATION

23. Look up the *harmonic mean* of two numbers, and find one way in which it is used.

Lesson

12-3 Similar Figures

Vocabulary

similar

similarity transformation

ratio of similitude,
  scale factor

▶ **BIG IDEA** By composing size transformations and reflections, similar figures are created.

The word "congruent" is a precise term that refers to figures with the same size and shape. The words "size" and "shape" do not have precise definitions. So, in defining congruent figures, we used ideas of distance and transformations. In Lesson 5-1, three terms were precisely defined:

- congruent figures
- ≅ (is congruent to)
- congruence transformation (isometry).

There are corresponding terms for figures that have the same shape, but not necessarily the same size:

- similar figures
- ~ (is similar to)
- similarity transformation.

Mental Math

Jorge has a box that is 2 ft wide, 2 ft long, and 3 ft high; another box of the same base that is three times as high; and a third box four times as long and wide as the first box, but the same height.

a. What is the volume of the first box?

b. What is the ratio of the volume of the second box to that of the first?

c. What is the ratio of the volume of the third box to that of the first?

A Precise Definition for Similar Figures

Under a size transformation, figures and their images seem to have the same shape. So under any definition, these figures should be similar. Performing an isometry (reflection, rotation, translation, or glide reflection) on a figure also does not change its shape. The precise definition of *similar figures* encompasses all of these possibilities.

Definition of Similar

Two figures F and G are **similar**, written as $F \sim G$, if and only if there is a composite of size transformations and reflections mapping F onto G.

Triangles $\triangle ABC$, $\triangle PQR$, $\triangle SQT$, and $\triangle XYZ$ on the next page are similar.

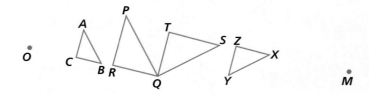

$\triangle PQR$ is a size-change image of $\triangle ABC$ with center O and magnitude 1.8. $\triangle SQT$ is a rotation image of $\triangle PQR$ with center Q and magnitude $-90°$. $\triangle XYZ$ is a size-change image of $\triangle SQT$ with center M and magnitude $\frac{2}{3}$.

The symbol "~" is read "is similar to." In the figure at the right, $S_{1.7}(MELON) = FRUIT$, where the center of $S_{1.7}$ is point A. You can write

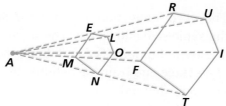

$$MELON \sim FRUIT$$

and say "pentagon $MELON$ is similar to pentagon $FRUIT$." As with congruent figures, corresponding vertices of similar figures are written in corresponding order.

Which Transformations Are Similarity Transformations?

Transformations that give rise to similar figures are called *similarity transformations*.

Definition of Similarity Transformation

A transformation is a **similarity transformation** if and only if it is the composite of size transformations and reflections.

According to this definition, a composite of reflections is a similarity transformation, so all congruent figures are similar figures.

Here is a hierarchy of transformations you have studied, with similarity transformations, size transformations, and the identity transformation included.

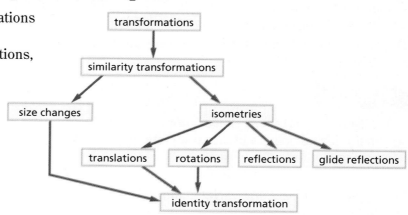

Properties of Similar Figures

The basic properties of similar figures come from preservation properties of similarity transformations. Compare these lists.

Preserved under reflections	Preserved under size transformations
Angle measure	Angle measure
Betweenness	Betweenness
Collinearity	Collinearity
Distance	

The properties common to both columns are preserved by similarity transformations. Thus similarity transformations have the A-B-C preservation properties, but distance is usually *not* preserved. Still, because reflections preserve distance and size transformations multiply distance by a certain constant amount, similarity transformations multiply distance by a constant amount. Thus, in a similarity transformation, the ratios of image lengths to preimage lengths are equal. This causes corresponding lengths in similar figures to be proportional.

Similar Figures Theorem

If two figures are similar, then

1. corresponding angles are congruent.
2. corresponding lengths are proportional.

The Similar Figures Theorem allows lengths and angle measures in similar figures to be found. From this theorem and the other definitions in this lesson, it is clear that all regular polygons with the same number of sides are similar, because we can always find a similarity transformation that will map one such polygon onto another. This means that any equilateral triangle is similar to any other equilateral triangle, any square is similar to any other square, etc. Extending this further, you can say that any circle is similar to any other circle.

GUIDED

Example 1

In the figure at the right, $r_m \circ S_{1.8}(CARP) = FISH$, and the center of $S_{1.8}$ is O.

a. What is the justification for the statement $CARP \sim FISH$?

b. If $m\angle P = 82$, what angle in $FISH$ has measure 82?

c. If $RP = 6$, what length in $FISH$ can be determined, and what is it?

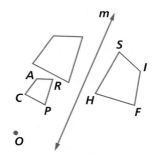

Solution

a. $CARP \sim FISH$ by the definition of ___?___ because $r_m \circ S_{1.8}$ is a composite of ___?___ and ___?___.

b. By the Similar Figures Theorem, corresponding angles are congruent. so $\angle P \cong \angle H$. Because $m\angle P = 82$, ___?___ $= 82$.

c. By the Similar Figures Theorem, corresponding sides are proportional. The ratio of these sides is the magnitude of the size transformation.

$$\frac{SH}{RP} = \underline{\quad ? \quad}$$

Substitute the given length for RP.

$$\frac{SH}{?} = \underline{\quad ? \quad}$$

$$SH = \underline{\quad ? \quad}$$

🛑 **QY**

▶ **QY**

Refer to Example 1. Write three other ratios of lengths equal to 1.8.

The Ratio of Similitude

The ratio of a length in an image to the corresponding length in a similar preimage is called the **ratio of similitude**. In many applications, the ratio of similitude is called the **scale factor.** Unless otherwise specified, when $F \sim G$ with ratio of similitude k, lengths in G (the second figure mentioned) divided by corresponding lengths in F equal k. The ratio of similitude is the product of the size-change factors of all the size transformations used in the similarity transformation. In Example 1, the ratio of similitude is 1.8.

When examining similar figures, always look first for corresponding vertices. These give pairs of congruent angles. Then look at corresponding sides.

▶ **READING MATH**

One of the episodes in the third season of the television series *Star Trek: Enterprise* was titled "Similitude." This episode first aired in 2003. *Similitude* is also the name of a concept that is used in the testing of engineering models.

Example 2

$\triangle ANT \sim \triangle BUG$ with angle measure and lengths as indicated in the figure at the right.

a. Find as many other angle measures in these triangles as possible.

b. Find as many other lengths of sides in these triangles as possible.

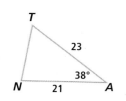

Solution

a. ∠A corresponds to ∠B. In similar figures, corresponding angles are congruent. So m∠A = m∠B = 38.
Because you do not know the measures of any other angles, the measures of the other four angles cannot be found using the Similar Figures Theorem.

b. Because corresponding sides are proportional,
$$\frac{AN}{BU} = \frac{NT}{UG} = \frac{TA}{GB}.$$
Any one of these ratios equals the ratio of similitude.

Substitute the three known lengths.
$$\frac{21}{BU} = \frac{NT}{UG} = \frac{23}{15}$$
Use the equality of the first and the last ratios.
$$\frac{21}{BU} = \frac{23}{15}$$
By the Means-Extremes Property,
$$21 \cdot 15 = 23 \cdot BU$$
$$BU = \frac{315}{23} = 13\frac{16}{23} \approx 13.696.$$
Because you do not know the length of either of the corresponding sides $\overline{NT}$ or $\overline{UG}$, neither NT nor UG can be found.

In Example 2, measures of two sides and their included angle of △ANT are given. This is the SAS condition, so the lengths of the other side and the measures of the other two angles are determined. These measures can be found using the trigonometry that you will learn in later mathematics courses.

Questions

COVERING THE IDEAS

1. Shown at the right are the letters b and d in different font sizes, as indicated.
   a. Explain why the letters are similar.
   b. What is the ratio of similitude of the similarity transformation mapping the letter b onto the letter d?
   c. What is the ratio of similitude of the similarity transformation mapping the letter d onto the letter b?

48-point   36-point

2. Define *similar figures*.

3. **True or False**
   a. Similar figures must be congruent.
   b. Congruent figures must be similar.

In 4–8, use the drawing at the right. Figure III is the image of Figure I under $S \circ r_m$ and S is the size transformation with center C and magnitude 2.3.

4. Are Figure I and Figure III similar? Why or why not?

5. If $AB = 6.8$ inches, find the lengths of two other segments in the drawing.

6. If $MN = 230$ mm, find the lengths of two other segments in the drawing.

7. Which angles in the drawing have the same measure as $\angle GFJ$?

8. Let $CH = 4$. Use this fact to find the distance between another pair of points in the drawing.

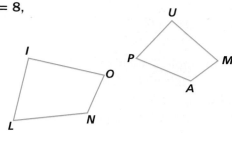

In 9–12, tell whether the property is preserved by every similarity transformation. If your answer is no, then give a counterexample.

9. angle measure

10. collinearity

11. distance

12. orientation

In 13 and 14, *LION* ~ *PUMA*. The ratio of similitude is $\frac{3}{5}$. $LN = 8$, $m\angle M = 80$, and $m\angle N = 119$.

13. Find the measures of as many other sides as you can.

14. Find the measures of as many other angles as you can.

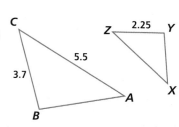

15. $\triangle ABC \sim \triangle XYZ$ in the figure at the right. Find as many lengths as you can.

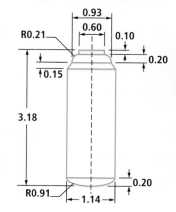

APPLYING THE MATHEMATICS

16. The figure at the right is a part of NASA's blueprints for making a scale model of the service module for the International Space Station. This scale model is similar to the actual space station with a ratio of similitude of $\frac{1}{100}$. The width of the top portion of this part of the module is marked as 0.93, meaning 0.93 inch. What is the actual width of this part of the module?

Continuous human occupation of the International Space Station began on November 2, 2000.

17. The distance from Houston, Texas, to Dallas, Texas, on a map is 6 inches. The actual distance from Houston to Dallas is 240 miles.

   a. If San Antonio, Texas, appears on the map 5 inches from Houston, how far is it from Houston to San Antonio?

   b. What is the ratio of similitude of the map to the actual state?

18. When a new house is built, builders use blueprints that are similar to cross-sections of the actual house. One of the considerations in the construction is the steepness of the roof. This steepness is known as the *pitch* of the roof, and is given in blueprints as the *x* pitch, where *x* is the number of inches the roof rises for every 12 inches across. If the blueprints for a particular house states that the roof is to have a 7 pitch and the height of the roof is to be 10 feet, what is the horizontal length of this part of the roof?

19. One rectangle has dimensions 4 feet and 6 feet. Another rectangle has dimensions 3 cm and 2 cm. Must the two rectangles be similar? Why or why not?

REVIEW

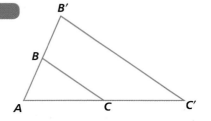

20. Suppose B' and C' are the images of B and C under a size transformation with center A. $AB = 15$, $BB' = 9$, and $BC = 20$. What is $B'C'$? (**Lesson 12-2**)

21. Solve $\frac{10}{x} = \frac{14}{x - 2}$. (**Lesson 12-2, Previous Course**)

22. Let $A = (0, 0)$, $B = (0, 7)$, and $C = (3, 0)$. Find an equation for the circle containing points A, B, and C. (**Lesson 11-6**)

23. An inflated basketball has a diameter 23.9 cm. It is made up of air, and a layer of leather that is 1.5 mm thick. What percent of the volume of the ball is air? What percent is leather? (**Lesson 10-6**)

24. You have four congruent 30-60-90 triangular regions and wish to fit them together, with no overlap, to form a quadrilateral region. How many non-congruent regions can be formed? Draw and describe each region. (**Lessons 8-7, 6-4**)

25. **True or False** Any transformation that is a composite of ten reflections is also a composite of four or less reflections. (**Lesson 4-7**)

EXPLORATION

26. In Lesson 5-1 you saw proofs of the Reflexive, Symmetric, and Transitive Properties of Congruence. Are there comparable properties for similarity? If so, state them, and explain why they are true. If not, explain why not.

QY ANSWER

$\frac{FI}{CA}, \frac{IS}{AR}, \frac{HF}{PC}$

Lesson

12-4

The Fundamental Theorem of Similarity

▶ **BIG IDEA**   In similar figures that are not congruent, ratios of corresponding lengths, areas, and volumes are all determined by the ratio of similitude and are not equal to each other.

Similar figures may be 2-dimensional or 3-dimensional because reflections and size transformations can be done in three dimensions, as well as in two. Any plane can be the reflecting plane for a 3-dimensional reflection, as you saw in Lesson 9-7. Any point can be the center of a size transformation and any nonzero real number can be the magnitude of the transformation. The Similar Figures Theorem holds for space figures. Ratios of corresponding distances are equal to the ratio of similitude and corresponding angles are congruent.

Because ratios of corresponding distances are equal in similar figures, ratios of perimeters, areas, and volumes also can be determined.

Mental Math

State the negation of each statement.

a. $x + y = 14$

b. Quadrilateral *HIJK* has no more than two sides of length 11.5.

c. $m > \frac{1}{6}$

Activity

Step 1 Draw a box with dimensions 3 units by 5 units by 6 units.

Step 2 Apply a size change of magnitude 4 and draw the image box.

Step 3 Calculate:

   a. the perimeter of a face of the larger box.

   b. the perimeter of the corresponding face of the smaller box.

   c. the ratio of the larger perimeter to the smaller.

Step 4 Calculate:

   a. the surface area of the larger box.

   b. the surface area of the smaller box.

   c. the ratio of the larger surface area to the smaller.

Step 5 Calculate:

   a. the volume of the larger box.

   b. the volume of the smaller box.

   c. the ratio of the larger volume to the smaller.

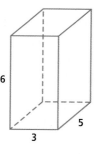

In Activity 1, you should have obtained different answers for Steps 3c, 4c, and 5c. You could predict these results without actually calculating any perimeters, surface areas, or volumes. The scale factor (or ratio of similitude) of the boxes is 4. The ratio of corresponding lengths is 4 or 4^1. The ratio of corresponding surface areas is $4 \cdot 4 = 4^2 = 16$, because area involves the multiplication of 2 lengths and each of these has been multiplied by 4. The ratio of corresponding volumes is $4 \cdot 4 \cdot 4 = 4^3 = 64$, because volume involves the multiplication of 3 lengths, each of which has been multiplied by 4.

 QY1

▶ **QY1**

If you apply a size transformation of magnitude 7 to the 3-by-5-by-6 box in the Activity, what are the ratios of the perimeters, the areas, and the volumes of the image to the preimage?

The General Theorem

The results you should have found in Activity 1 are instances of the following theorem, which applies to all similar figures in two or three dimensions.

Fundamental Theorem of Similarity (first statement)

If $F \sim F'$ with a ratio of similitude k, then:

1. Length in $F' = k \cdot$ Corresponding Length in F,
   or $\dfrac{\text{Length in } F'}{\text{Corresponding Length in } F} = k.$

2. Surface Area $(F') = k^2 \cdot$ Surface Area (F), or
   $\dfrac{\text{Surface Area } (F')}{\text{Surface Area } (F)} = k^2.$

3. Volume $(F') = k^3 \cdot$ Volume (F), or $\dfrac{\text{Volume } (F')}{\text{Volume } (F)} = k^3.$

Proof of 1 Suppose that one of the linear dimensions of F is a. Then the length of the corresponding linear dimension of F' is ka.
So, $\dfrac{\text{Length in } F'}{\text{Corresponding Length in } F} = \dfrac{ka}{a} = k.$

Proof of 2 Let $A =$ Area(F). Then you could think of the area of F as the sum of the areas of A squares with sides of length 1 (unit squares). Then the area of F' is the sum of the areas of A squares with sides of length k units. Since each square in F' has area k^2,
Area $(F') = A \cdot k^2 = k^2 \cdot$ Area (F).

Proof of 3 This argument is much like that for the area. Let $V =$ Volume (F). Then the volume of F equals that of V cubes with edges of length 1 (unit cubes). The volume of F' is the sum of the volumes of V cubes each with edges of length k. Since each cube in F' has volume k^3,
Volume $(F') = V \cdot k^3 = k^3 \cdot$ Volume (F).

Notice that the Fundamental Theorem of Similarity is not restricted to polygons, circles, pyramids, cylinders, or other figures with names. The theorem and the proof apply to *all figures*.

Here is an alternate statement of the Fundamental Theorem of Similarity. It includes the fact that because both reflections and size transformations preserve angle measure, the measures of corresponding angles in similar figures are equal.

> ### Fundamental Theorem of Similarity (alternate statement)
>
> If two figures are similar with ratio of similitude k, then
>
> 1. corresponding angle measures are equal.
> 2. corresponding lengths and perimeters are in the ratio k.
> 3. corresponding areas and surface areas are in the ratio k^2.
> 4. corresponding volumes are in the ratio k^3.

Example 1

In the figure at the right, $\triangle ABC \sim \triangle DEF$. Give the ratio of the perimeters and areas of these triangles.

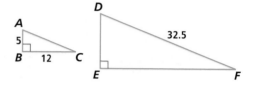

Solution 1 Although no two corresponding lengths are given, the problem can still be solved because the triangles are right triangles. By the Pythagorean Theorem, $AC = 13$. From this, the ratio of similitude k between these triangles is $k = \frac{32.5}{13} = 2.5$.

Consequently, using the Fundamental Theorem of Similarity, the ratio of perimeters is 2.5 and the ratio of areas is 2.5^2, or 6.25.

Solution 2 Suppose you did not remember the Fundamental Theorem of Similarity. Then find the ratio of similitude as in Solution 1. From this,

$$\frac{DE}{AB} = \frac{DF}{AC} = \frac{EF}{BC} = 2.5.$$

Substitute the lengths that you know.

$$\frac{DE}{5} = \frac{32.5}{AC} = \frac{EF}{12} = 2.5.$$

$$\frac{DE}{5} = 2.5, \text{ which means } DE = 5 \cdot 2.5 = 12.5.$$

$$\frac{EF}{12} = 2.5, \text{ which means } EF = 12 \cdot 2.5 = 30.$$

Now, the perimeters and areas of these two triangles can be determined directly.

Perimeter of $\triangle ABC = 5 + 12 + 13 = 30$ units

Perimeter of $\triangle DEF = 12.5 + 30 + 32.5 = 75$ units

So, the ratio of perimeters $= \frac{75}{30} = 2.5$.

Area of $\triangle ABC = \frac{1}{2} \cdot 5 \cdot 12 = 30$ square units

Area of $\triangle DEF = \frac{1}{2} \cdot 12.5 \cdot 30 = 187.5$ square units

So, the ratio of areas $= \frac{187.5}{30} = 6.25$.

GUIDED

Example 2

In the figure at the right, *PINK* ~ *ROSE*. If *PINK* has an area of 20 square units, what is the area of *ROSE*?

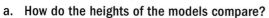

Solution The ratio of the areas is the square of the ratio of similitude.

$$\frac{\text{Area}(ROSE)}{\text{Area}(PINK)} = (\underline{\ ?\ })^2 = \underline{\ ?\ }$$

Substituting the given information,

$$\frac{\text{Area}(ROSE)}{20} = \underline{\ ?\ }.$$

Solving this equation for Area(*ROSE*),

$$\text{Area}(ROSE) = \underline{\ ?\ } \text{ square units.}$$

STOP QY2

> **QY2**
>
> In Example 2, if you know $m\angle N$, what other angle measure can you find, and how?

GUIDED

Example 3

Two scale models of the same car are shown at the right. Each model is similar to the actual car, so they are similar to one another. The smaller model is 15 cm in length, while the larger model is 75 cm in length.

a. How do the heights of the models compare?

b. How do the surface areas of the models compare?

c. How do the volumes of the cars compare?

Solution The ratio of similitude k, can be determined.

$$k = \frac{\text{length of larger model}}{\text{length of smaller model}} = \underline{\ ?\ }$$

a. The height is a length, so the ratio of the heights is k.

$$\frac{\text{height of larger model}}{\text{height of smaller model}} = k = \underline{\ ?\ }$$

So, the height of the larger model is $\underline{\ ?\ }$ times the height of the smaller model.

(continued on next page)

b. The ratio of the surface areas of the two model cars is k^2.

$$\frac{\text{surface area of larger model}}{\text{surface area of smaller model}} = k^2 = \underline{\ ?\ }$$

So, the surface area of the larger model is __?__ times the surface area of the smaller model.

c. The ratio of the volumes of the two model cars is k^3.

$$\frac{\text{volume of larger model}}{\text{volume of smaller model}} = k^3 = \underline{\ ?\ }.$$

So, the volume of the larger model is __?__ times the volume of the smaller model.

Questions

COVERING THE IDEAS

1. Give an example to show what happens to the area and perimeter of a rectangle if you multiply each of its dimensions by 10.

2. Give an example to show what happens to the surface area and volume of a cube if you multiply each of its dimensions by 0.3.

3. Consider two similar regular triangular pyramids. Are all the corresponding angles congruent?

4. At the right, $\triangle ABC \sim \triangle DEF$.
   a. Give the ratio of similitude.
   b. Give the ratio of the perimeters.
   c. Give the ratio of the areas.

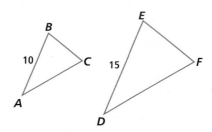

5. If the ratio of similitude of two polyhedra is $\frac{7}{4}$, then find the ratios of
   a. the perimeters of corresponding faces.
   b. their surface areas.
   c. their volumes.
   d. lengths of corresponding edges.
   e. measures of corresponding angles.

6. The handball and the marble pictured at the right are spheres and hence are similar figures. An official handball has a diameter of $1\frac{7}{8}$ inches and an official marble has a diameter of $\frac{5}{8}$ inches.
   a. What is the ratio of their volumes?
   b. What is the ratio of their surface areas?

7. Two equilateral triangles have sides with lengths in the ratio $\frac{5}{9}$. Find the ratio of their
   a. altitudes. b. perimeters. c. areas.

8. Use the formula $V = \ell wh$ to show that if you multiply the dimensions of a box by k, then the volume is multiplied by k^3.

9. Use the formula $A = \pi r^2$ to show that if you multiply the radius of a circle by 15, then the circle's area is multiplied by 225.

10. At the right, $ABCDE \sim JKLMN$ with ratio of similitude $\frac{8}{5}$.
    a. If the area of $JKLMN$ is 192 cm^2, then what is the area of $ABCDE$?
    b. If $AE = 6$ cm, find all other lengths that you can.

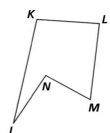

APPLYING THE MATHEMATICS

11. Two circles have areas of 49π cm^2 and 25π cm^2. Find the ratio of their diameters.

12. Two cylindrical glasses are similar with ratio of similitude 3. The larger glass is too big to fit under a faucet, so the smaller glass will be used to fill the larger glass. How many times must you fill the smaller glass and empty its contents into the larger glass in order to fill the larger cylinder?

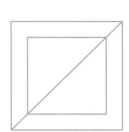

13. The Rollins family had an addition built onto their existing house. With this addition, the house was 40% longer and 20% wider than before. If the house had 2100 ft^2 of floor area before, what is its new floor area?

14. In the figure at the right, the two squares have the same center. The diagonal of the larger square has length 8. Find the length of a diagonal of the smaller square such that the area of the smaller square is exactly half the area of the larger square.

15. Consider two triangular pyramids that are similar. Suppose the ratio of the volumes of the two pyramids is $\frac{4013}{512}$.
    a. Find the ratio of the surface areas of the pyramids.
    b. Find the ratio of the perimeters of corresponding faces.

16. The figure at the right is a sector of a circle of radius 10. If m$\angle AOB = 45$, find OC such that the area of the sector COD is half the area of sector AOB.

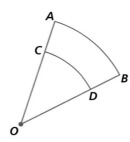

REVIEW

17. Let $T(x, y) = (3x, y)$. Is T a similarity transformation? Why, or why not? (**Lesson 12-3**)

18. A goal post and its shadow are shown. The total length of its shadow is 50 ft. The crossbar is 10 ft high and its shadow falls 16 ft from the base of the goal post. What is the height of the goal post? (**Lesson 12 2**)

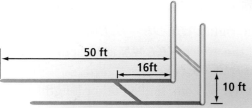

19. What kind of a figure is the set of all points in 3-dimensional space whose z-coordinate is equal to 1? (**Lessons 11-8**)

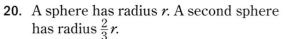

50 ft

16ft

10 ft

20. A sphere has radius r. A second sphere has radius $\frac{2}{3}r$.
    a. How do their volumes compare? (**Lesson 10-6**)
    b. How do their surface areas compare? (**Lesson 10-7**)

21. Define each term. (**Lesson 5-1**)
    a. congruent
    b. oppositely congruent

22. **True or False** (Lesson 2-5)
    a. The intersection of two overlapping rectangular regions whose sides are vertical and horizontal is another rectangular region.
    b. The intersection of any two rectangular regions is a rectangular region.

EXPLORATION

23. Find a model of a car, plane, building, or other object.
    a. Determine the ratio of similitude between the model and the actual object.
    b. Use the ratio of similitude and the weight of the model to estimate the weight of the actual object.
    c. Does your answer to Part b seem to be correct?

QY ANSWERS

1. The ratio of the perimeters is 7; the ratio of the areas is 49; the ratio of the volumes is 343.

2. m∠S, using the Fundamental Theorem of Similarity

Lesson
12-5
Can There Be Giants?

> ▶ **BIG IDEA** There cannot be people many times the height of a typical person.

Giants are common characters in children's stories. Saturday morning cartoons often have giant creatures with human shape. But, according to the *Guinness Book of World Records*, the tallest man on record was Robert Wadlow from Alton, Illinois. On June 27, 1940 at age 22, he was measured at 8 feet, 11.1 inches. Wadlow was about 1.5 times the height of a typical male. Can humans be much taller? The Fundamental Theorem of Similarity provides the answer.

Giants in *Gulliver's Travels*

Let's look at what the theorem reveals, using an example from a famous novel. In *Gulliver's Travels*, Jonathan Swift writes about Gulliver, a young man who visits the land of Brobdingnag, where the Brobdingnagians are similar to us, but 12 times as tall. Volume is proportional to weight, so their volume, and thus their weight, would be 12^3 times ours. If you weigh 140 pounds, a similar Brobdingnagian would weigh 1728 times that, or 241,920 pounds! The problem is that the weight that can be supported by the legs of a person is proportional to the areas of cross sections of his bones and muscles. The Brobdignagian giant's bones and muscles would be about 12^2, or 144 times as strong as Gulliver's. They could support a person weighing about 20,160 pounds, but Brobdignagians weigh 12 times that! If a Brobdingnagian were similar to a human, his skeletal structure would not support his weight, and his body would just collapse.

Even champion weight lifters seldom lift more than twice their body weight, and when they do, it is only for a few seconds. Lifting 12 times your weight would quickly break your bones!

You might think that a giant body would find some way of dealing with the extra weight. But it can't. Wadlow had to wear a leg brace to support his weight. While getting out of a car (which was hard for him to do), the brace cut deep into his leg. The wound became infected and cellulitis set in. Eighteen days after his height was measured for the last time in 1940, he died in Manistee, Michigan.

Mental Math

a. If half a pound of cheese costs $2.75, at this price, what does 3 pounds of cheese cost?

b. A model of a building is $\frac{1}{1000}$ actual size. If the model is 24 cm tall, how tall is the building, in meters?

c. If you walked 2.5 miles in 40 minutes at a steady rate, how long did it take you to walk three-quarters of a mile?

In 2007, Leonid Stadnyk of Ukraine was named the world's tallest living man at 8 feet 5.5 inches. He wore size 64 shoes.

The Fundamental Theorem of Similarity in Nature

Animals in nature have developed within the constraints imposed by the Fundamental Theorem of Similarity. Elephants have legs with large horizontal cross-sectional areas to support their great weight. Thoroughbred race horses have skinny legs, which enable them to run fast, but the legs are small for their bodies and break easily. Draught horses, which pull wagons, have thicker, stronger legs, but they are slow. A mosquito can walk on the surface of water because it is so light that it will not break the surface tension. Its thin legs are sufficient to support its light body, but that body has a relatively large surface area. Should a raindrop force the body into the water, the surface tension acts like glue on the body's surface and the thin legs cannot pull the mosquito from the water.

Thoroughbred mare and foal

Large flying birds like an eagle or a buzzard have fairly small bodies compared to nonflying animals. Consider a small bird, like a sparrow. If the sparrow were twice as long from beak to tail, its volume would be 2^3 times as large. That means it would be 8 times as heavy and its wings would have to be 8 times as strong. But its wings would be only 4 times as strong, so they would need to be twice as long to get enough strength to fly. Large birds have much longer wings compared to the length of their bodies and tend to fly differently than small birds. They soar and glide and tend to fly much higher.

Draught horse

The amount of food needed by an animal is proportional to its volume. The Brobdingnagians would need to consume 1728 times the food needed by Gulliver. For a person like Gulliver, about 19 calories per day per pound of body weight (perhaps 3000 calories) are needed to maintain body weight. The Brobdingnagians would require 1728 times 3000 calories daily to maintain their body weights. That's a lot of food.

Tiny People in *Gulliver's Travels*

Gulliver also visited the land of Lilliput, where people were $\frac{1}{12}$ his height. For Gulliver, as for us, a new coat would require about 2 square yards of material. Clothing is proportional to surface area and is multiplied by k^2, the square of the ratio of similitude. So the Brobdingnagians would require 12^2 or 144 times as much material. The Lilliputians would require only $\left(\frac{1}{12}\right)^2$ or $\frac{1}{144}$ times the 2 square yards needed by Gulliver. Thus, geometry answers questions about clothing and food needs, as well as the properties of giants.

Example 1

Suppose a restaurant sells 12"-diameter pizzas with cheese and one topping for $12.99 (plus tax). Also suppose that the restaurant bases its prices on the amount of ingredients used to make the pizza, and that the pizzas have the same thickness. What should the restaurant charge for a 14"-diameter pizza with cheese and one topping?

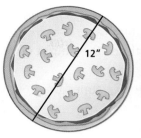 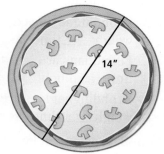

Solution Because the tops of both pizzas are circles, they are similar, with $k = \frac{14}{12} = \frac{7}{6}$. An immediate response is to charge $\frac{7}{6} \cdot \$12.99 \approx \$15.16$. However, this is not correct because the crust and toppings are proportional to the area, not the linear dimensions. So you must multiply by $\left(\frac{7}{6}\right)^2$, not $\frac{7}{6}$. $\left(\frac{7}{6}\right)^2 = \frac{49}{36}$. The price, based only on amount of ingredients, would be $\$12.99 \cdot \frac{49}{36} \approx \$17.68$.

Usually the prices of pizzas are not based only on ingredients. Other factors, such as salaries, equipment, and overhead must be taken into account.

Example 2

Suppose a solid clay figurine weighs 5 pounds. A similar figurine, twice as tall, stands next to it. Could a 4-year-old child pick up the taller figurine?

Solution Weight is dependent upon volume. Thus, the multiplying factor is the cube of the ratio of similitude, in this case 2^3. The taller figurine weighs $2^3 \cdot 5 = 40$ lb. Most 4-year- olds would not be able to pick up the figurine, and many adults would be surprised at the weight of the figurine.

The Fundamental Theorem of Similarity was known to Euclid, but the structural applications were not recognized until over 1800 years later by the Italian scientist Galileo. He considered this discovery as important as his more famous discovery that when heavier-than-air objects of different weights are dropped from the same height, they fall to the ground at the same time. The two discoveries were announced in the same book of Galileo's, *On Two New Sciences*.

Questions

COVERING THE IDEAS

1. Who was Robert Wadlow?

Fill in the Blanks In 2–4, use the information about *Gulliver's Travels* given in this lesson and the Fundamental Theorem of Similarity.

2. Brobdingnagians are __?__ times the height of Gulliver, and weigh __?__ times as much.

3. Lilliputians are __?__ times the height of Gulliver and weigh __?__ times as much.

4. Brobdingnagians are __?__ times the height of Lilliputians and weigh __?__ times as much.

In 5–8, consider an imaginary giantess 27 feet tall, which is about 5 times the height of an average woman. If the giantess and woman had similar shapes, how would each quantity compare?

5. weight

6. length of an index finger

7. area of a footprint

8. waist circumference

9. Explain why it is impossible for a 120-pound weightlifter to lift three times her weight for more than a few seconds.

10. Why does an elephant need thicker legs for its height than a mosquito?

11. **Fill in the Blanks** A scale model is $\frac{1}{10}$ actual size. If it is made from the same materials as the original object, then its weight will be __?__ times the weight of the object. The amount of paint to cover the exterior will be __?__ times the paint used to cover the original.

12. **True or False** Prices of pizzas are proportional to their diameters.

Yao Defen, believed to be Asia's tallest woman, is 7 feet 9 inches tall. (She is not related to Yao Ming, mentioned in Question 16.)

APPLYING THE MATHEMATICS

13. Two similar solid statues are 20 cm and 30 cm tall. If the shorter one weighs 6 kg, how much will the taller one weigh?

14. Suppose that the manager of the pizza restaurant in Example 1 changes his pricing strategy. Instead of charging solely based upon ingredients, he sells each pizza at $8.00 over the cost of the ingredients, no matter what the size. A 12″ pizza with one topping still costs $12.99. Using his new pricing plan, how much should he charge for a 14″ pizza with one topping?

15. If the price of a TV set is proportional to the area of its screen, and a 42-inch plasma TV can be purchased for $1500, estimate the cost of a 58-inch plasma set. Justify your answer.

16. In 2007, NBA player LeBron James was 6 feet 8 inches tall and weighed about 250 pounds. Yao Ming was 7 feet 6 inches tall. If Yao was the same shape as James, about how much should he weigh?

17. Consider both Earth and Mars to be spheres. The surface area of Mars is about 28% that of Earth. What is the ratio of
    a. their radii?
    b. their volumes?

18. Many movies have been made in which people are shrunk down to a fraction of their size. Suppose Armando is now strong enough to lift 60 pounds. If he were shrunk down using a ratio of similitude of $\frac{1}{16}$ (about the size of a toy action figure), how many pounds could he lift?

REVIEW

19. A hexagon has area 115.5. What is the area of its image under a size change of magnitude $\frac{2}{5}$? (Lesson 12-4)

20. a. Define what it means for two figures *not* to be similar.
    b. Draw two nonsimilar triangles. (Lesson 12-3)

21. What kind(s) of size transformations can be composites of reflections? (Lesson 12-1)

22. Draw the image of $\triangle ABC$ at the right under a size change with center P and magnitude 3. (Lesson 12-1)

23. Write a full statement of the SAS Congruence Theorem in words. (Lesson 7-2)

24. X, Y, and Z are three points on a circle, $m\overset{\frown}{XY} = 160°$, and $m\overset{\frown}{YZ} = 38°$. What are the measures of the three angles of $\triangle XYZ$? (Lesson 6-3)

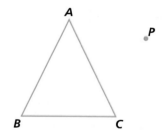

EXPLORATION

25. Below are the world record amounts of weight lifted in women's weightlifting competition, as of 2007.
    a. Calculate the ratio of weight lifted to each weightlifter's maximum weight. (Ignore the 75+ kg class.)
    b. What trends do you see?
    c. Give an explanation for any trends you find.

Svetlana Podobedova's lift of 75 kg won the gold medal in the 2012 Olympic Games.

Weight Class	Record Holder	Country	Total Weight (2 lifts)
48 kg	Yang Lian	China	217 kg
53 kg	Qiu Hongxia	China	226 kg
58 kg	Chen Yanqing	China	251 kg
63 kg	Liu Haixia	China	257 kg
69 kg	Oxana Slivenko	Russia	276 kg
75 kg	Svetlana Podobedova	Russia	286 kg
75+ kg	Mu Shuangshuang	China	319 kg

Lesson

12-6

The SSS Similarity Theorem

▶ **BIG IDEA**   If ratios of the lengths of all three pairs of corresponding sides of two triangles are equal, then the triangles are similar.

Because similar figures possess important relationships to each other, it is useful to know when figures are similar. So far in this chapter, you have always been told that two figures are similar or that one figure is the image of the other under a similarity transformation. In this lesson and the next, you will learn conditions that tell when triangles are similar. These conditions are quite a bit like those for congruent triangles.

Imagine that you are the head of a design team. You need two triangular supports with dimensions 5 cm, 8 cm, and 10 cm. You tell your team members to construct two triangles with side lengths 5, 8, and 10. One member of the team constructed such a triangle. A second member of the team constructed a triangle with dimensions 5 in., 8 in., and 10 in. You can see at the right that even though both triangles have sides 5, 8, and 10, they are not congruent because the lengths are not in the same units.

In centimeters, the sides of $\triangle DEF$ have lengths 12.7, 20.32, and 25.4. The lengths of the sides of $\triangle DEF$ are 2.54 times the length of corresponding sides of $\triangle ABC$. The side lengths are proportional.

$$\frac{12.7}{5} = \frac{20.32}{8} = \frac{25.4}{10} = 2.54, \text{ the ratio of similitude}$$

Suppose you applied a size transformation of magnitude 2.54 to $\triangle ABC$. The resulting image $\triangle A'B'C'$ (not drawn) has side lengths equal to those of $\triangle DEF$. This means that $\triangle A'B'C' \cong \triangle DEF$ by the SSS Congruence Theorem. Since $\triangle DEF$ is the image of $\triangle ABC$ under a similarity transformation, $\triangle ABC \sim \triangle DEF$.

In general, if three pairs of corresponding sides of any two triangles are proportional, then there must exist a similarity transformation mapping one triangle to the other. This means that the triangles are similar. This result is called the SSS Similarity Theorem. It is proved by generalizing the process described in showing $\triangle ABC \sim \triangle DEF$ above.

Mental Math

$\triangle VEB \sim \triangle JAN$, $VE = 6$, and $JA = 4.5$.

a. What is the ratio of similitude?

b. What is the ratio of the perimeters of the triangles?

c. What is the ratio of the areas of the triangles?

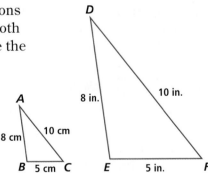

SSS Similarity Theorem

If the three sides of one triangle are proportional to three sides of a second triangle, then the triangles are similar.

Given $\dfrac{XY}{AB} = \dfrac{YZ}{BC} = \dfrac{XZ}{AC}$

Prove $\triangle ABC \sim \triangle XYZ$

Proof First we determine the magnitude of a size change needed to transform $\triangle ABC$ into a triangle that is congruent to $\triangle XYZ$. That magnitude is $\dfrac{XY}{AB}$, so let $k = \dfrac{XY}{AB}$. Then, by the Transitive Property of Equality, $k = \dfrac{YZ}{BC}$ and $k = \dfrac{XZ}{AC}$.

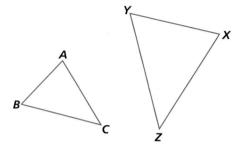

Apply any size transformation with magnitude k to $\triangle ABC$.

In the image $\triangle A'B'C'$, $A'B' = k \cdot AB$, $B'C' = k \cdot BC$, and $A'C' = k \cdot AC$.

Then $A'B' = k \cdot AB = \dfrac{XY}{AB} \cdot AB = XY$,

$B'C' = k \cdot BC = \dfrac{YZ}{BC} \cdot BC = YZ$,

and $A'C' = k \cdot AC = \dfrac{XZ}{AC} \cdot AC = XZ$.

Thus, the three sides of $\triangle A'B'C'$ have the same lengths as the corresponding sides of $\triangle XYZ$. By the SSS Congruence Theorem, $\triangle A'B'C' \cong \triangle XYZ$. The definition of congruence tells us there is an isometry mapping $\triangle A'B'C'$ onto $\triangle XYZ$. So there is a composite of a size change (the one with magnitude k) and an isometry mapping $\triangle ABC$ onto $\triangle XYZ$. Therefore, by the definition of similarity, $\triangle ABC \sim \triangle XYZ$.

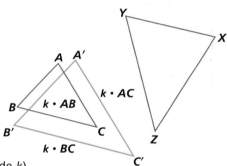

Applying the SSS Similarity Theorem

One way to tell whether two triangles are similar is to order the sides of each triangle by their lengths. Then compare the ratios formed by their corresponding lengths.

Example 1

Determine if the triangles shown at the right are similar.

Solution Put the sides in order from shortest to longest and take the ratios: $\dfrac{3}{8}, \dfrac{6}{16}, \dfrac{8}{\frac{64}{3}}$.

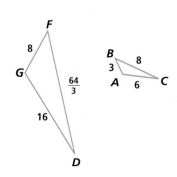

Simplifying: $\dfrac{6}{16} = \dfrac{3}{8}$

$\dfrac{8}{\frac{64}{3}} = 8 \cdot \dfrac{3}{64} = \dfrac{3}{8}$

Because all the ratios are equal, the triangles are similar due to the SSS Similarity Theorem.

In Example 1, the ratio of similitude is either $\frac{3}{8}$ or $\frac{8}{3}$, depending on which triangle is viewed as the preimage, and which as the image. If we write $\triangle FGD \sim \triangle BAC$, then the ratio of similitude is $\frac{3}{8}$. The corresponding sides tell you which vertices correspond.

STOP QY

QY

One triangle has side lengths of 20, 30, and 40. A second triangle has side lengths of 80, 60, and 40. Are the triangles similar? Why or why not?

GUIDED

Example 2

At the right, two triangles and some side lengths are given. The angle measures are approximations to the nearest tenth of a degree.

a. If $\triangle FED \sim \triangle GOV$, what is the ratio of similitude?

b. What length of $\overline{GO}$ would make $\triangle FED \sim \triangle GOV$?

c. Estimate the measure of each angle of $\triangle GOV$ to the nearest tenth of a degree.

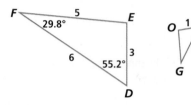

Solution

a. The ratio of similitude is the ratio of a side of $\triangle GOV$ to the corresponding side of $\triangle FED$. Because we have a value for both *ED* and *OV* and these two sides correspond, we use these lengths to find the ratio of similitude. **The ratio of similitude is $\frac{OV}{ED}$ or __?__.**

b. $\triangle FED \sim \triangle GOV$ by the SSS Similarity Theorem if the corresponding sides are proportional. We need to find *GO* such that: $\frac{OV}{ED} = \frac{GV}{?} = \frac{GO}{?} = $ __?__. Solving the proportion, $GO = $ __?__.

c. Because $\triangle FED \sim \triangle GOV$, $\angle F \cong$ __?__, $\angle E \cong$ __?__, and $\angle D \cong$ __?__.

In $\triangle GOV$, $m\angle$__?__ ≈ 29.8, and $m\angle$__?__ ≈ 55.2. The measure of the third angle can be found using the Triangle-Sum Theorem, so $m\angle$__?__ ≈ 95.

Families of Similar Triangles

Because of the SSS Similarity Theorem, if you multiply or divide the lengths of the three sides of any triangle by the same positive number, the triangle with the new side lengths will be similar to the orginal triangle. That is, if a, b, and c are lengths of sides of a triangle, then ka, kb, and kc (where $k > 0$) are the side lengths of a similar triangle. This is especially useful when it comes to Pythagorean triples.

You know that a 3-4-5 triangle is a right triangle by the Converse of the Pythagorean Theorem ($3^2 + 4^2 = 5^2$). By the SSS Similarity Theorem, we know that multiplying the side lengths by a positive constant will create side lengths of a similar right triangle. Thus, a triangle with sides of length $3k$, $4k$, and $5k$ ($k > 0$) is a right triangle similar to the first. Likewise, we know that a right triangle is uniquely determined by either its two legs or a hypotenuse and a leg. Thus, if we know that two of the side lengths of a right triangle are multiples of two corresponding parts of a Pythagorean triple, then we know the third part will be a multiple of the other part of the triple.

Example 3

Solve for x in the figure at the right.

Solution You could calculate the length x using the Pythagorean Theorem. However, you should also recognize that 5-12-13 is a Pythagorean triple and that the numbers 50 and 130 are 10 times 5 and 10 times 13, respectively. So, x must be 10 times 12. Thus, $x = 120$.

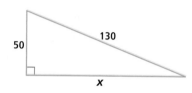

GUIDED

Example 4

Use a Pythagorean triple to find the missing length in the right triangle at the right.

Solution These lengths are $\frac{1}{7}$ of two lengths in a __?__-__?__-__?__ triple. So, the third side is $\frac{1}{7}$ of __?__. So $x =$ __?__.

Questions

COVERING THE IDEAS

1. a. If a member of the design team at the beginning of this lesson had constructed his triangle in feet, would his triangle be similar to either the triangle measured in centimeters or the triangle measured in inches? Why or why not?

   b. Would his triangle be congruent to any drawn triangle?

2. State the SSS Similarity Theorem.

3. **True or False**

   a. If two triangles are similar, then there exists a size transformation that maps one to the other.

   b. The SSS Similarity Theorem can be used to prove triangles congruent.

4. If $\triangle YES \sim \triangle MAD$, state two ratios equal to $\frac{YS}{MD}$.

In 5 and 6, determine if the following triangles are similar. If they are similar, provide the ratio of similitude and write a similarity statement indicating which vertices correspond.

5.

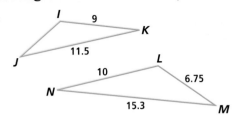

6.

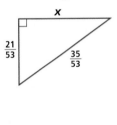
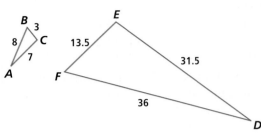

7. In $\triangle MNO$, $MN = 4$, $NO = 13$, and $OM = 15$. To the nearest hundredth of a degree, m$\angle O = 14.25$ and m$\angle M = 53.13$. If $\triangle MNO \sim \triangle PQR$ and $PR = 20$, find the lengths of all the other sides of $\triangle PQR$ and the measures of all of its angles to the nearest hundredth of a degree.

In 8–10, use Pythagorean triples to solve for the unknown length.

8.

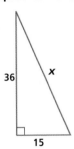

9.

10.

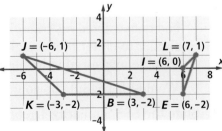

11. **True or False** A triangle with sides of length $\frac{1}{4}, \frac{1}{3}$, and $\frac{1}{6}$ is similar to a triangle that has sides of length 4, 3, and 6.

12. Triangles BJK and EIL are graphed at the right on the coordinate plane. Prove that $\triangle BJK \sim \triangle ELI$.

13. In Inman Square in Cambridge, Massachusetts, there is a triangular piece of land created by the intersections of Webster Street, Newton Street, and Prospect Street. The lengths of the sides of the triangle are: 240' on Webster Street, 600' on Prospect Street and 450' on Newton Street. How could you use this information, a protractor, and a ruler to determine the measures of the angles of the triangle formed by the streets?

14. A teacher wants to enlarge the fact triangle at the right as a demonstration tool. If the length of each side of the triangle is 3.4 cm, what will the dimensions and angles of the triangle be after the teacher uses a 450% magnification on a copy machine?

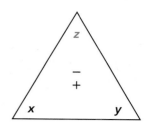

15. Write a paragraph proof.

    Given D is the midpoint of $\overline{BC}$;
    E is the midpoint of $\overline{AC}$.

    Prove $\triangle BCA \sim \triangle ECD$

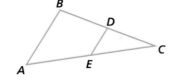

16. Suppose the sides of one quadrilateral have lengths 3, 5, 3, and 5 in that order, and the sides of a second quadrilateral have lengths 6, 10, 6, and 10 in that order. Explain why the two quadrilaterals do not have to be similar.

17. If D, E, and F are midpoints of the sides of $\triangle ABC$ at the right, explain why $\dfrac{\text{area}(\triangle DEF)}{\text{area}(\triangle ABC)} = \dfrac{1}{4}$.

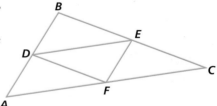

<hr/>

REVIEW

18. A company produces two sizes of bottled water. The two types of bottles are similar. Suppose that the smaller size holds 0.5 liter, and the larger size holds 750 milliliters. Approximately how many times as much paper is required for the label on the larger bottle than on the smaller bottle? (**Lesson 12-4**)

19. **True or False** If all of the lengths of the edges of a cube are increased by 5 cm, then the area of each face of the cube increases by 25 cm² and the cube's volume increases by 125 cm³. (**Lesson 12-4**)

20. Consider the transformation S_k, where $k \neq 0$. (**Lesson 12-1**)
    a. What kind of a transformation is $S_k \circ S_k$?
    b. Are there any values of k for which $S_k \circ S_k$ reverses orientation? Explain your answer.

21. $\triangle ABC$ at the right is an isosceles right triangle with m$\angle B = 90$. $EFGD$ is a square with sides of length a. Express the area of $\triangle ABC$ in terms of a. (**Lesson 8-7**)

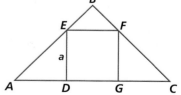

22. Name all of the Triangle Congruence Theorems that you know. (**Lessons 7-5, 7-2**)

<hr/>

EXPLORATION

23. Two sides of one triangle have lengths 6 and 7. Two sides of another triangle have lengths 8 and 9. Can the triangles be similar? If so, how? If not, why not?

QY ANSWER

Yes; If you match up corresponding sides, the ratio of similitude is 2.

▶ **BIG IDEA** The conditions that guarantee similar triangles are analogous to the conditions that guarantee congruent triangles.

A variety of conditions can lead to similar triangles.

Activity

MATERIALS DGS (optional)

Do this activity either by hand or with a DGS.

Step 1 Draw an angle A such that m∠A = 40.

Step 2 On one side of ∠A, identify a point B. Draw an angle ABX such that m∠ABX = 75. A triangle is formed by the sides of ∠A and ∠B. Name the third vertex of this triangle C.

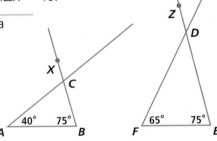

Step 3 Measure the lengths of each of the sides of △ABC.

Step 4 On the same screen or sheet of paper, draw an angle F such that m∠F = 65.

Step 5 On one side of ∠F, identify a point E. Draw an angle FEZ such that m∠FEZ = 75. A triangle is formed by the sides of ∠E and ∠F. Name the third vertex of this triangle D.

Step 6 Measure or use the calculator feature on your DGS to have your DGS display the following ratios: $\frac{AB}{DE} = \frac{?}{?} = \underline{\quad?\quad}; \frac{BC}{EF} = \frac{?}{?} = \underline{\quad?\quad};$ $\frac{AC}{DF} = \frac{?}{?} = \underline{\quad?\quad}.$

Step 7 Are the two triangles similar? If so, which vertices correspond?

The AA Similarity Condition

In the Activity, you should have found that corresponding sides of the two triangles are in the same ratio. Then they are similar by the SSS Similarity Theorem. But no side lengths were given, so what is the given information that led to the similarity of the triangles?

You may not have realized that the triangles were chosen to have angles of the same measure. This is not obvious, because the measures of noncorresponding angles were given, but you only need the measures of two angles to determine the measure of the third angle.

 QY1

In fact, for any two triangles, if two pairs of corresponding angles are congruent, then the triangles must be similar. The argument used in proving this statement is very much like the one used in proving the SSS Similarity Theorem. A size change is applied to one triangle so that its image is congruent to the other triangle. The key decision in the proof is the choice of the magnitude k of the size change. Once k is chosen, the only other thing to do is to identify which triangle congruence theorem to use.

> ## AA Similarity Theorem
>
> If two angles of one triangle are congruent to two angles of another, then the triangles are similar.

Given Triangles ABC and XYZ with $\angle A \cong \angle X$ and $\angle B \cong \angle Y$

Prove $\triangle ABC \sim \triangle XYZ$

Drawing The congruent angles signal the corresponding vertices. This indicates the corresponding sides and enables a picture to be drawn and marked.

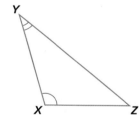

Proof The included sides $\overline{XY}$ and $\overline{AB}$ correspond, so $\frac{XY}{AB}$ is the magnitude of a size transformation applied to $\triangle ABC$ that will cause its image to be congruent to $\triangle XYZ$. Let $k = \frac{XY}{AB}$.

Then
$$A'B' = k \cdot AB$$
$$= \frac{XY}{AB} \cdot AB$$
$$= XY.$$

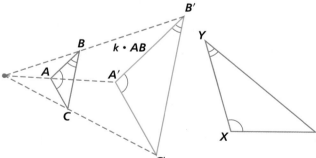

Also, because size transformations preserve angle measure, $\angle A \cong \angle A'$ and $\angle B \cong \angle B'$. Thus, because of the transitivity of congruence, $\angle A' \cong \angle X$ and $\angle B' \cong \angle Y$. So $\triangle A'B'C' \cong \triangle XYZ$ by the ASA Congruence Theorem. Thus, $\triangle ABC$ can be mapped onto $\triangle XYZ$ by a composite of size changes and reflections, so $\triangle ABC \sim \triangle XYZ$.

STOP QY2

If triangles are known to be similar, then ratios of corresponding sides are equal. From the equal ratios, lengths of unknown sides can be found.

▶ **QY2**

By identifying the angles with the same measure, write a similarity statement for the triangles below.

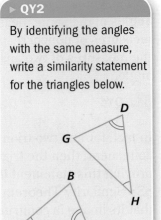

GUIDED

Example 1

If a meter stick casts a 80 cm long shadow at the same time that a cactus casts a shadow 15 m long, how tall is the cactus?

not to scale

Solution We assume that the cactus and the meter stick are perpendicular to the ground, forming right angles. The key to the solution is that the Sun is so far away that its rays can be considered to be parallel. Consequently, right triangles are formed with congruent acute angles.

The triangles are similar by the AA Similarity Theorem. Thus, the sides are proportional:

$$\frac{x}{?\ m} = \frac{?\ m}{?\ cm}$$

Convert to the same units. We choose to convert to centimeters.

$$\frac{x}{?\ cm} = \frac{?\ cm}{?\ cm}$$

1 m

80 cm

15 m

The product of the means in the proportion equals the product of the extremes, so

$$\underline{\quad ?\quad} \cdot x = \underline{\quad ?\quad}$$
$$x = 1875\ cm.$$

The cactus is 1875 cm or about 19 m high.

The Cardón cactus, found primarily in the Sonoran Desert in Baja California, Mexico, is the tallest cactus in the world.

The SAS Similarity Condition

A third condition that leads to similar triangles is called the SAS condition. Notice that this condition is not the same as the SAS congruence condition. In the SAS congruence condition, the sides are *congruent*, while in the SAS similarity condition, the sides are *proportional*. The proof of the SAS Similarity Theorem is very similar to the proof of the AA Similarity Theorem. You are asked to prove it in Question 8.

> ### SAS Similarity Theorem
>
> If, in two triangles, the ratios of two pairs of corresponding sides are equal and the included angles are congruent, then the triangles are similar.

Specifically, if in $\triangle ABC$ and $\triangle XYZ$, $\angle B \cong \angle Y$, and $\frac{XY}{AB} = \frac{YZ}{BC}$, then $\triangle ABC \sim \triangle XYZ$.

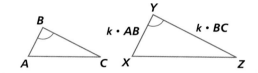

GUIDED

Example 2

Given: $\overline{WT}$ and $\overline{VU}$ intersect at X with right angles as indicated. Name the theorem that proves that the pair of triangles are similar and provide a similarity statement.

Solution __?__ $\cong$ __?__ because vertical angles are congruent. __?__ $\cong$ __?__ because right angles are congruent. So $\triangle TUX \sim \triangle$ __?__ by the __?__ Similarity Theorem.

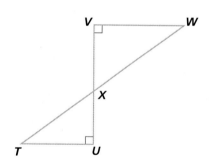

Summary

Each triangle congruence theorem has a counterpart triangle similarity theorem. In the triangle similarity theorems, "A" still denotes a pair of congruent angles, but "S" denotes a *ratio* of corresponding sides.

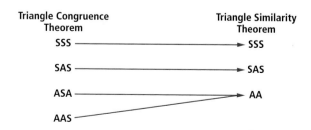

Questions

COVERING THE IDEAS

1. Describe the general strategy used to prove the three triangle similarity theorems mentioned in this lesson.

2. Name and describe three sufficient conditions for triangles to be similar.

For 3–6, each figure contains at least two triangles.

   a. Are there two similar triangles?

   b. If so, what triangle similarity theorem guarantees their similarity? If not, explain why not.

   c. If two triangles are similar, provide a similarity statement indicating which vertices correspond.

3.
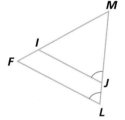

4.
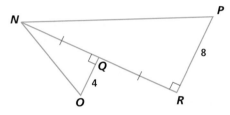

5.
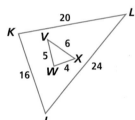

6.
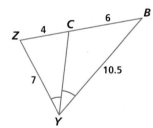

7. Suppose a 5'8"-tall person casts a 3' shadow at the same time that a flagpole casts a 13'4" shadow. To the nearest foot, how tall is the flagpole?

8. Prove the SAS Similarity Theorem using the figure at the right. (Hint: Use the general idea of the proof of the AA Triangle Similarity Theorem.)

   Given $\angle B \cong \angle Y$, $\frac{AB}{XY} = \frac{BC}{YZ}$

   Prove $\triangle ABC \sim \triangle XYZ$

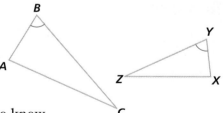

9. Kenya Ketchit, a softball player, is 5 feet tall. She wanted to know the height of the foul pole. To estimate the height, she stood so that the end of her shadow and the end of the shadow of the foul pole coincided. At this time, she was 16 feet from the pole and her shadow was 8 feet long. How tall is the foul pole?

10. A tree in Marisa's yard grew so tall that it was hitting her house. The tree removal company asked her to estimate the height of the tree. She stands so that the edge of her shadow coincides with the end of the shadow of the tree as shown in the diagram below.

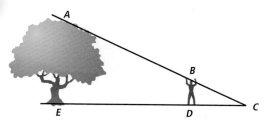

a. Why is $\triangle AEC \sim \triangle BDC$?

b. Complete the following: $\dfrac{AE}{?} = \dfrac{EC}{?}$

c. Marisa is 5'6". If her shadow is 4' and she is standing 11' from the tree, how tall is the tree?

APPLYING THE MATHEMATICS

11. You can determine the height of a tree even if the Sun is not out and there is no shadow. (This drawing is not to scale.)

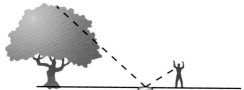

André places a mirror on the ground so that when he looks down into the mirror he sees the top of the tree. Suppose the middle of the mirror is 21 feet from the base of the tree and 19 inches from André. André's eyes are 68 inches above the ground. How tall is the tree?

12. The following method can be used to estimate the distance to an object. Hold one arm straight in front of you with your thumb pointing upward. Close one eye, and align your thumb up with a particular spot on the object. Without moving your arm, change eyes. Your thumb will appear to jump. The distance your thumb jumped can be used to estimate the distance to the object. In the diagram at the right, L and R represent your eyes, T represents your thumb and CD represents the distance the object moved.

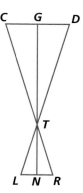

a. Why is $\triangle LRT \sim \triangle DCT$?

b. Corresponding altitudes of similar triangles are in the same ratio as corresponding sides. If the distance between your eyes is 3 inches, your thumb is 16 inches from your face, and your thumb seems to move 2 feet against the background of the object, how far are you from the object?

Chapter 12

13. Russian balalaika orchestras consist of as many as six different-sized balalaikas, long-necked stringed instruments with triangular-faced soundboxes. The shapes of two balalaika soundboxes are shown below.

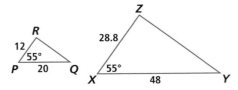

a. What is the magnitude of a size change applied to $\triangle PQR$ which would produce an image congruent to $\triangle XYZ$?

b. Which triangle congruence theorem is the justification that the image of the smaller triangle is congruent to the larger triangle?

14. $ABCD$ is a parallelogram.

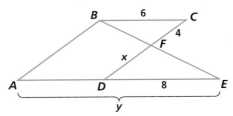

a. Name all pairs of similar triangles.

b. Find x.   c. Find y.   d. Find AB.

15. Prove that two isosceles triangles are similar if their vertex angles are congruent.

16. Given: $\odot A$,
    BD bisects $\angle CBE$.
    Prove: $\triangle BFE \sim \triangle BCD$

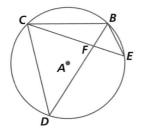

17. Given: $m\angle STU = 90$,
    $m\angle SVT = 90$
    Prove: $\triangle STU \sim \triangle SVT$

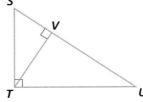

REVIEW

18. The sides of one 30-60-90 triangle have lengths 3, $3\sqrt{3}$, and 6. The longest side of a second 30-60-90 triangle has length 15. What are the lengths of the other two sides of the second triangle? (**Lesson 12-6**)

19. An ant can carry 50 times its weight. The ant's strength is proportional to the area of a cross section of its muscles. **(Lesson 12-5)**

a. If there were a creature similar to an ant and 500 times as long, how many times the ant's weight would it have?

b. Do you think this larger ant could carry 50 times its weight? Explain.

20. a. **True or False** Any two segments are similar figures.

b. **True or False** Any two angles are similar figures. **(Lesson 12-3)**

Leaf-cutter ants do not eat the leaves they cut. They use the leaves to grow a fungus, and it is the fungus that they eat.

21. **Fill in the Blanks** In the figure at the right, $\overline{AB} \parallel \overline{CD}$. Complete the indirect proof below to show that there is no size transformation S such that $S(A) = B$ and $S(C) = D$.

Suppose that there is such a transformation S. Let O be the center of the size transformation. Since $S(A) = B$, O is on the line ___?___. Because $S(C) = D$, O is on the line ___?___. This is impossible because ___?___. Therefore, ___?___.
(Lessons 12-1, 11-3)

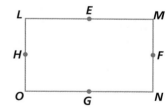

22. Provide a coordinate proof, using the convenient location for a rectangle, **(Lessons 11-7, 11-4)**

Given *LMNO* is a rectangle; *E, F, G,* and *H* are the midpoints of its sides as shown in the figure.

Prove $\overline{GE} \perp \overline{FH}$

23. A prism has the same base as a pyramid. If the height of the prism is h, what would the height of the pyramid have to be in order for their volumes to be the same? **(Lessons 10-4, 10-3)**

EXPLORATION

24. *LMNK* is a parallelogram. *P* is the midpoint of $\overline{LM}$ and *O* is the midpoint of $\overline{MN}$. $\overline{KO}$ and $\overline{KP}$ intersect $\overline{LN}$ at *Q* and *R*. Is it true that $\overline{LQ} \cong \overline{QR} \cong \overline{RN}$? If so, prove this. If not, explain why not. (You may want to explore this situation using a DGS.)

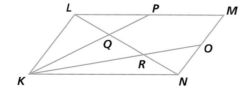

Chapter 12 Projects

1 Finish This Story

A pharaoh in ancient Egypt asked a goldsmith to make two statues, identical except for scale, of the pharaoh and his son. The statue of the pharaoh was to be six times the height of the statue of his son. The statue of the pharaoh required 60 pieces of gold, which the pharaoh supplied to the goldsmith. After that statue was completed, the goldsmith asked for 10 pieces of gold to make the statue for the son. The pharaoh gave the goldsmith the 10 pieces, the goldsmith made the statue, and the pharaoh was so pleased with the result that he rewarded the goldsmith with 5 gold pieces. Some time later, one of the pharaoh's advisors told him that the goldsmith had cheated him. Write a conclusion to this story, which will answer the following questions: Was the pharaoh really cheated? If so, by how much? If not, why is this advisor wrong? If the goldsmith hired a lawyer to defend him, what might the lawyer say? If the advisor hired a lawyer to defend him, what would that lawyer say?

2 Folding Paper

Take a piece of paper and fold it in half. Then turn it 90° and fold it again. Do this as many times as you can. Conventional wisdom says you will not be able to do more than seven folds. It is actually possible to do more folds if you have a much larger and thinner piece of paper. Using tape, or any other method you like, put together smaller sheets of paper to create a piece of paper as big as you can. How many times can you fold this?

3 Models of Real Objects

Make a 3-dimensional scale model of a real object. Calculate the surface areas and volumes of the original object and your scale model. Write a report on how you built your model, including calculations.

4 The Shrinking Person

Suppose a 3-dimensional size change of magnitude $\frac{1}{2}$ were applied to you and everything where you live. Discuss how each of the following would compare to now. There may not be a clear "correct" answer to some of these, so provide a reason for your response.

- the amount of paint required to paint your place
- the cost to heat and cool your place
- the time to clean the place
- your food expenses

5 What Is the Right Size?

The article "On Being the Right Size" by scientist J. B. S. Haldane was published in 1928 and reprinted in *The World of Mathematics* many years later. Find a copy of this article to read and write a short report summarizing Haldane's key points. Discuss the ways in which the mathematics from this chapter supports the ideas set forth by Haldane.

Chapter 12 Summary and Vocabulary

○ In the coordinate plane, a **size change** centered at the origin can be achieved by multiplying coordinates of points by a given **size-change factor.** Because a coordinate system can be created with any point as the origin, size transformations can be centered at any point. Size transformations can occur in two or three dimensions.

○ Two figures are **similar** if and only if one can be mapped onto the other by a composite of reflections and size transformations. The ability to draw or construct similar figures is necessary in the making of scale drawings, toys or scale models, maps, blueprints, and other diagrams. In similar figures with **ratio of similitude** k, angles and their images are congruent. Lengths of image segments are k times the lengths of preimage segments, where k is a positive number called the ratio of similitude. Areas of images are k^2 times the areas of their preimages. Volumes of images are k^3 times the volumes of their preimages. These relationships between two similar figures help explain why large animals need relatively thicker legs than small animals, and why there cannot be giants.

○ When one quantity is k times another, then the ratio of the quantities equals k. An equality of two ratios is called a proportion. Whenever there are similar figures, lengths are proportional. Solving proportions can help you determine unknown measurements.

○ For triangles, the conditions guaranteeing similarity correspond to those for congruence. The most commonly used triangle similarity conditions are SSS, AA, and SAS where A indicates equal angle measures and S indicates equal ratios of sides.

Vocabulary

Lesson 12-1
*size change, size transformation with any center
size-change factor
dilation, dilatation
*expansion
*contraction
*identity transformation

Lesson 12-2
ratio
proportion
terms of a proportion
extremes
means

Lesson 12-3
*similar
*similarity transformation
*ratio of similitude, scale factor

Postulates, Theorems, and Properties

Size-Change Preservation Properties
   Theorem (p. 720)
Size-Change Distance Theorem
   (p. 721)
Figure Size-Change Theorem (p. 723)
Means-Extremes Property (p. 727)

Similar Figures Theorem (p. 733)
Fundamental Theorem of Similarity
   (pp. 739 and 740)
SSS Similarity Theorem (p. 751)
AA Similarity Theorem (p. 757)
SAS Similarity Theorem (p. 759)

Chapter 12 Self-Test

Take this test as you would take a test in class. You will need tracing paper. Then use the Selected Answers section in the back of the book to check your work.

1. In the figure at the right, $\overline{ZW}$ is the image of $\overline{XY}$ under a size change with center M. Use a ruler to find the magnitude of this size change.

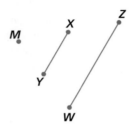

2. In the figure below, m$\angle B = 90$. S_{-3} is a size transformation with center O. $S_{-3}(\triangle ABC) = \triangle PQR$. Find the perimeter of $\triangle PQR$.

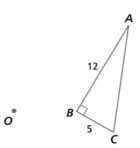

3. Suppose M is the midpoint of $\overline{AB}$ and S is a size transformation. Explain why $S(M)$ is the midpoint of $S(\overline{AB})$.

4. Tetrahedron $PSMR$ is the image of $FLAX$ under the size transformation S_4 about point A. $PS = 8$ cm, the area of $\triangle FLX$ is 10 cm^2, and the volume of $FLAX$ is 16 cm^3. Determine FL, the area of $\triangle PSR$, and the volume of $PSMR$.

5. In the figure below, $\triangle QRP$ is the size change image of $\triangle TSU$. Trace the figure, and find the center of the size transformation.

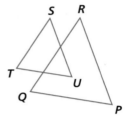

6. Suppose $\frac{x}{y} = \frac{z}{w}$. Write three other true proportions using x, y, z, and w.

7. In the figure below, $PENTA \sim FIGUR$. Find as many lengths and angle measures as you can.

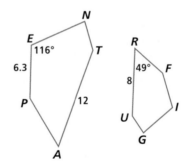

8. Two statues of Archimedes are similar and made of the same material. The larger of the two weighs eight times as much as the smaller one. The length of Archimedes' left pinky finger in the smaller one is 3". What is the length of his left pinky finger in the larger statue?

9. A book has pages with dimensions $6\frac{1}{2}''$ by $9\frac{1}{2}''$. An enlargement is made of a page in the book. The smaller dimension of the enlargement is $9''$. What is the larger dimension?

10. Figure A is similar to figure B. The ratio of their corresponding sides is 3:1. Figure B is similar to figure C. The ratio of their corresponding sides is 2:1. How does the surface area of figure A compare to the surface area of figure C?

11. Explain why larger animals need fatter legs to support their weight.

12. **Multiple Choice** $\triangle ABC$ has sides of length 7, 10, and 13. Which of the following cannot be lengths of the sides of a triangle similar to $\triangle ABC$?

    A 7, 13, and 10

    B 14, 20, and 26

    C 8, 11, and 14

    D 6.5, 3.5, and 5

13. Prove that any two isosceles right triangles are similar.

14. Claudia drew the chart below for a food pyramid. $\overleftrightarrow{DE}$, $\overrightarrow{FG}$, and $\overleftrightarrow{HI}$ are all parallel to $\overleftrightarrow{AB}$.

    a. Prove $\triangle CDE \sim \triangle CAB$.

    b. If $AB = 16$, $CD = 5$ and $CA = 20$, find DE.

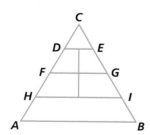

Chapter 12 Chapter Review

SKILLS Procedures used to get answers

OBJECTIVE A Draw the transformation images of figures. (Lesson 12-1)

In 1–3 trace the figure. Draw the image of *ABCD* under each transformation.

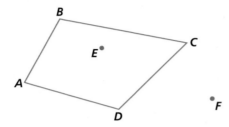

1. center *E*, magnitude 0.25
2. center *B*, magnitude 3
3. center *F*, magnitude –1

4. Trace the figure at the right. Draw its image under a size change with center *Z* and magnitude $\frac{2}{3}$.

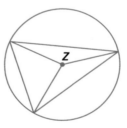

In 5 and 6, the larger figure is a size-change image of the smaller figure. Use a ruler to locate the center and find the magnitude of this size change.

5.

6.

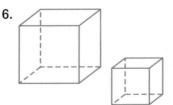

7. △*JKL* is the image of △*DEF* (not drawn) under a size change with center *O* and magnitude $\frac{3}{2}$. Trace the figure and draw △*DEF*.

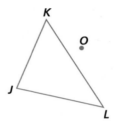

OBJECTIVE B Find angle measures, lengths, perimeters, areas, and volumes in similar figures. (Lessons 12-3, 12-4)

8. In the figure below △*BDC* ~ △*ONM*. Find the following lengths.
   a. *ON*
   b. *CD*

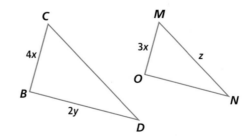

In 9 and 10, refer to the figure below, in which $\triangle FGE \sim \triangle HIE$.

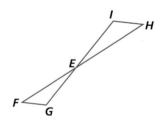

9. Identify three pairs of congruent angles.

10. Suppose $FE = 10$, $EH = 12.5$. What is $\dfrac{\text{perimeter } (\triangle FEG)}{\text{perimeter } (\triangle HEI)}$?

11. Two pyramids are similar. The ratio of the areas of their bases is 9.

   a. Find the ratio of similitude.

   b. Find the ratio of their volumes.

12. In the figure below, $IJKLMN \sim UVWXYZ$. Find as many angle measures and side lengths as you can.

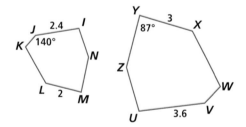

Principles behind the mathematics

OBJECTIVE C Recognize and apply properties of size transformations and similar figures. (Lessons 12-1, 12-3, 12-5)

13. Suppose figure F is the image of figure G under a size transformation S, and that F and G have the same perimeter.

   a. Find the ratio of similitude.

   b. **True or False** F and G are congruent.

14. Suppose $\ell \parallel m$, and S is a size transformation. Explain why $S(\ell) \parallel S(m)$.

15. The dashed figure below is the image of the solid figure under a size transformation with center O. Measure to find the magnitude of this transformation.

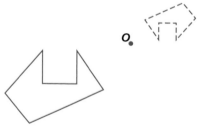

16. **Multiple Choice** Which of the following properties does a size change with magnitude 0.01 preserve?

   A perimeter   B diameter of circles

   C area   D collinearity

OBJECTIVE D Given a true proportion, find another true proportion with the same terms. (Lesson 12-2)

In 17–19, a true proportion is given in which none of the terms is equal to 0. Give another true proportion with the same terms.

17. $\dfrac{x}{y} = \dfrac{a}{b}$ 18. $\dfrac{2}{3} = \dfrac{w}{3}$

19. $\dfrac{484}{300} = \dfrac{363}{225}$

20. Consider the true proportion $\dfrac{u}{v} = \dfrac{s}{t}$. Write another true proportion with these terms in which both s and t are in the denominator.

OBJECTIVE E Determine whether triangles are similar using the AA, SAS, or SSS Similarity Theorems. (Lessons 12-6, 12-7)

21. One triangle has sides of length 7, 9, and 13. Another triangle has sides of length 107, 109, and 113. Are these two triangles similar? Why, or why not?

In 22–24, are the triangles similar? If so, why? If not, why not?

22.

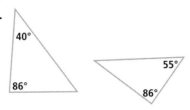

23.

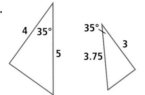

24.

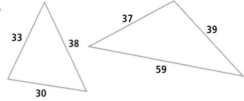

25. Given: D is the midpoint of $\overline{AB}$ and E is the midpoint of $\overline{AC}$.

Prove: $\triangle ADE \sim \triangle ABC$

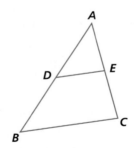

26. Given: $\overline{XV} \perp \overline{YZ}, \overline{YW} \perp \overline{XZ}$

Prove: $\triangle YWZ \sim \triangle XVZ$

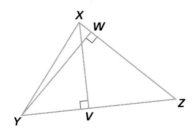

USES Applications of mathematics in real-world situations

OBJECTIVE F Identify and determine proportional lengths and distances in real situations. (Lessons 12-2, 12-3)

27. On a map of Europe, the distance between London and Paris is 2.5 cm. The actual distance between the two cities is 343 km. The distance between London and Rome is 1444 km. How far apart are they on the map?

28. As it left the solar system, the space probe Voyager 1 was traveling at a speed of approximately 5.2×10^8 km/year. On a computer screen showing the Voyager's route, the Voyager is shown to travel 7 inches in 5 years. What is the scale of the image on the screen?

29. An artist makes a sculpture of a man with scale $x{:}y$, where x is a sculpture length and y is the actual length. If the man's eyes are 4 cm apart, how far apart are the eyes on the sculpture?

30. In a scale model of a house, the height is 15 cm and the width of the front window is 6 cm. If the height of the actual house is 8 meters, what is the width of its front window?

31. Two computer screens are similar. One is a 19″ screen (measured along the diagonal), which is 16″ wide. The other is a 21″ screen. What is the width of the second screen?

OBJECTIVE G Apply the Fundamental Theorem of Similarity in real situations. (Lessons 12-4, 12-5)

32. The Statue of Liberty in New York weighs approximately 450,000 pounds. A 1:1000 model of the statue weighs 2 oz. Could the statue and the model be made out of the same material? Explain your answer.

33. In the 1990 report "Estimation of the Total Surface Area in Indian Elephants," K. P. Sreekumar and G. Nirmalan measured the height and surface areas of elephants. One elephant, weighing approximately 5300 kg, was found to have a surface area of approximately 21 m². What would you expect the surface area of a similar, 4800 kg elephant to be?

34. Dolls are often $\frac{1}{12}$ actual size. The same cloth used for a real dress can be used to make how many doll dresses?

35. The amount of heat an animal radiates is proportional to its surface area. Given two similar animals: one smaller and one larger, which will radiate more heat relative to its body weight?

36. Suppose there was a giant who was similar to a person, but 100 times as tall.

    a. How many times would the giant's weight be compared to the person?

    b. How many times would the surface area of the giant be compared to the person?

OBJECTIVE H Use the Triangle Similarity Theorems to find lengths and distances in real situations. (Lessons 12-6, 12-7)

37. A tree casts a shadow 21 feet long. At the same time a yardstick casts a shadow 2 feet long. How tall is the tree?

38. In the beach chair pictured below, $\overline{AE}$ and $\overline{BD}$ intersect at C, and $\overleftrightarrow{AB} \parallel \overleftrightarrow{DE}$.

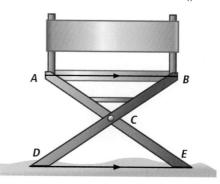

    a. Prove that $\triangle ABC$ and $\triangle EDC$ are similar.

    b. If $AB = 30$ cm, $AC = 20$ cm, and $CE = 25$ cm, find DE.

39. Kiara read that in ancient times, before erosion set it, each of the faces of the Great Pyramid at Giza (which are in the shape of isosceles triangles) had a base of 231 meters, and congruent legs of length 220 meters. Suppose Kiara uses a ruler to construct a triangle with sides of lengths 22 cm, 22 cm, and 23.1 cm. Explain how she could use a protractor to calculate the angles of the faces of the Great Pyramid at Giza.

40. Lyric is cross-stitching two right triangles. She already stitched one triangle with legs of length 2.5 cm, and 3.5 cm. She wants one of the legs of her second triangle to be 1.5 cm. long. What are the possible lengths of the other leg, if Lyric's triangles are to be similar?

REPRESENTATIONS Pictures, graphs, or objects that illustrate concepts.

There are no objectives for representations in this chapter.

Chapter

13 Similar Triangles and Trigonometry

Contents

13-1 The Side-Splitting Theorems

13-2 The Angle Bisector Theorem

13-3 Geometric Means in Right Triangles

13-4 The Golden Ratio

13-5 The Tangent of an Angle

13-6 The Sine and Cosine Ratios

Surveyors, pilots, navigators, sailors, builders, engineers, and many other people estimate lengths or distances as part of their job or their recreation. It is often interesting or useful to determine the height of tall objects or distances that cannot be directly measured, either because they are too large or too small or because there is an obstruction in the way. To do this, similar triangles or theorems that can be derived from similar triangles are often used.

One consequence of the AA Triangle Similarity Theorem is that if you know that a right triangle also has an 82° angle, then you know that this triangle is quite thin for its length. If you also know the length of one side, the lengths of the other sides can be determined. For example, if you know that the right triangle has an 82° angle and the leg between the right angle and the 82° angle has length 10", then the triangle is determined (because of ASA Congruence).

But what are the lengths of the other sides in this case? That is the subject of *trigonometry*. Trigonometry is based on properties of similar right triangles and enables people to determine lengths and angle measures that otherwise could only be estimated.

Here is an example of a problem using trigonometry: An airport is built on the outskirts of a growing city near a mountain range, and the airport needs another runway. The only direction available is one that points directly to the mountain range.

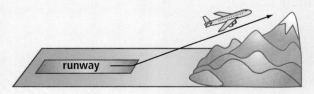

The engineers designing the airport know how tall the range is and how far away it is from the airport and how fast various airplanes can climb. They need to make sure it is safe to build the new runway. Before they proceed, they need to know that the altitude of any airplane taking off or landing on the new runway will always be greater than the height of the mountain range as it passes. Trial and error is not a reasonable option here. Trigonometry will help.

In this chapter, you will see how to use similar triangles and trigonometry to solve problems involving missing lengths and angle measures.

Lesson

13-1

The Side-Splitting Theorems

▶ **BIG IDEA** When two transversals intersect parallel lines, proportional lengths are created.

A variety of situations lead to figures in which there are two parallel lines and two nonparallel transversals. Pictured at the right is part of the front view of a house. The sides of the roof are two transversals and the siding creates parallel horizontal lines.

Mental Math

If *r, s, t,* and *v* are nonzero numbers such that $\frac{r}{s} = \frac{t}{v}$, tell whether the statement is *always, sometimes but not always,* or *never* true.

a. $rv = st$

b. $sv = rt$

c. $\frac{s}{v} = \frac{r}{t}$

Activity

MATERIALS DGS

Step 1 Use a DGS to construct a triangle, $\triangle ABC$.

Step 2 Construct a point D on $\overline{AB}$ and construct line ℓ through D parallel to $\overline{CB}$.

Step 3 Name E the point of intersection of ℓ and $\overline{AC}$.

Step 4 Display the ratios $\frac{AE}{EC}$ and $\frac{AD}{DB}$. What do you notice?

Step 5 Drag point D along $\overline{AB}$. What happens to the two ratios in Step 4?

Step 6 Drag point B anywhere. What happens to the two ratios in Step 4?

The Activity illustrates a result we call the *Triangle Side-Splitting Theorem.* Its proof depends on the AA Similarity Theorem and properties of algebra.

Triangle Side-Splitting Theorem

If a line is parallel to a side of a triangle and intersects the other two sides in distinct points, it splits these sides into segments of proportional lengths.

Given $\triangle ABC$ with point P on $\overline{AB}$, Q on $\overline{AC}$, and $\overleftrightarrow{PQ} \parallel \overleftrightarrow{BC}$

Prove $\dfrac{AP}{PB} = \dfrac{AQ}{QC}$

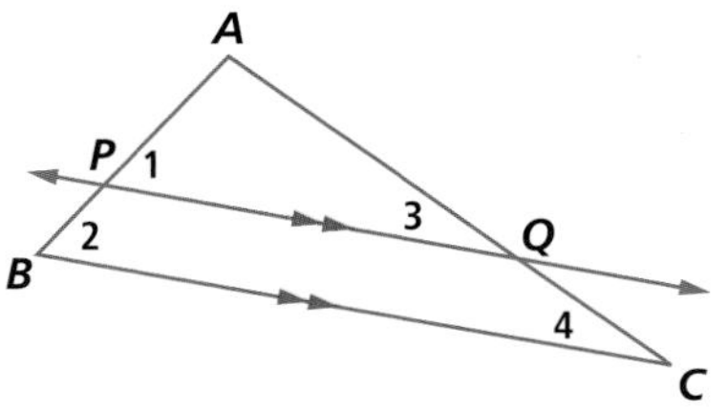

Proof $\angle 1 \cong \angle 2$ and $\angle 3 \cong \angle 4$ by the Corresponding Angles Postulate. Thus, by the AA Similarity Theorem, $\triangle APQ \sim \triangle ABC$. Because these triangles are similar, corresponding sides are proportional.

$$\frac{AB}{AP} = \frac{AC}{AQ}$$

To make the algebra easier to follow, let $AP = w$, $PB = x$, $AQ = y$, and $QC = z$. Substitute for these variables and use the Additive Property of the Distance Postulate to obtain AB and AC.

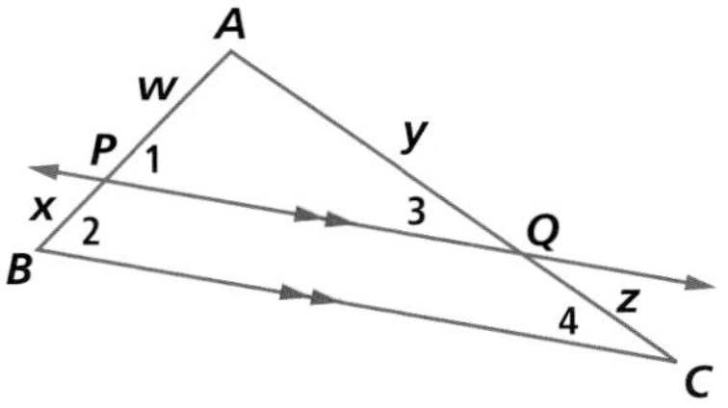

$$\frac{w+x}{w} = \frac{y+z}{y}$$

Separating the fractions, $\quad \frac{w}{w} + \frac{x}{w} = \frac{y}{y} + \frac{z}{y}$

$$1 + \frac{x}{w} = 1 + \frac{z}{y}$$

Subtract 1 from both sides. $\quad \frac{x}{w} = \frac{z}{y}$

Take the reciprocal of each side. $\quad \frac{w}{x} = \frac{y}{z}$

So by substitution, $\quad \dfrac{AP}{PB} = \dfrac{AQ}{QC}$.

The usefulness of the Triangle Side-Splitting Theorem is that it involves the lengths PB and QC that are not sides of the similar triangles, so you do not have to find similar triangles. This is the case in Example 1a below. But for Example 1b, similar triangles are still needed because the sides involved are not split.

GUIDED

Example 1

Use lengths as shown in the diagram at the right, in which $\overline{DE} \parallel \overline{BC}$.

a. Find AE. b. Find DE.

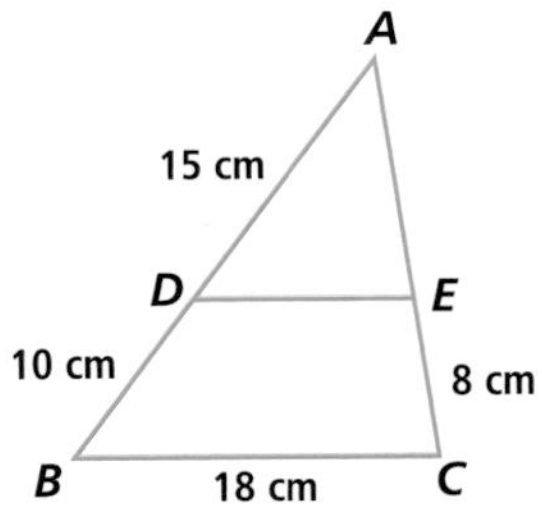

Solution

a. Since $\overline{DE} \parallel \overline{BC}$, you can use the Side-Splitting Theorem to calculate AE.

$$\frac{AE}{EC} = \frac{?}{?}$$
$$\frac{AE}{?} = \frac{?}{?}$$
$$\underline{\quad?\quad} \cdot AE = \underline{\quad?\quad}$$
$$AE = \underline{\quad?\quad}$$

b. The ratio of DE to BC is not the same as the ratio of AD to DB or of AE to EC because $\overline{DB}$ and $\overline{EC}$ are not sides of triangles. Similar triangles must be used. Since $\overline{DE} \parallel \overline{BC}$, $\underline{\quad?\quad} \sim \underline{\quad?\quad}$ by AA Similarity. As a result, the ratios of corresponding lengths are equal.

(continued on next page)

$$\frac{AD}{AB} = \frac{DE}{?}$$

$$\frac{?}{?} = \frac{DE}{?}$$

$$\underline{\quad ? \quad} \cdot DE = \underline{\quad ? \quad}$$

$$DE = \underline{\quad ? \quad}$$

The Converse of the Triangle Side-Splitting Theorem

Notice in the Triangle Side-Splitting Theorem that the four lengths w, x, y, and z of the proportion have the same relative positions both in the fractions and in the diagram.

The converse of the Triangle Side-Splitting Theorem is also true: When w, x, y, and z have these positions in a true proportion, then the line splitting the two sides is parallel to the third side of the triangle. The proof reverses the steps in the proof of the Side-Splitting Theorem.

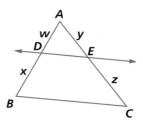

Triangle Side-Splitting Converse Theorem

If a line intersects $\overline{AB}$ and $\overline{AC}$

in distinct points D and E so that

$\frac{AD}{DB} = \frac{AE}{EC}$, then $\overleftrightarrow{DE} \parallel \overleftrightarrow{BC}$.

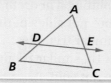

Given $\frac{AD}{DB} = \frac{AE}{EC}$

Prove $\overleftrightarrow{DE} \parallel \overleftrightarrow{BC}$

Proof Let $AD = w$, $DB = x$, $AE = y$, and $EC = z$. Then

Conclusions	Justifications
1. $\frac{w}{x} = \frac{y}{z}$	1. Given; Substitution
2. $\frac{x}{w} = \frac{z}{y}$	2. Take the reciprocal of each side.
3. $1 + \frac{x}{w} = 1 + \frac{z}{y}$	3. Addition Property of Equality
4. $\frac{w + x}{w} = \frac{y + z}{y}$	4. Addition of Fractions Property
5. $\frac{AB}{AD} = \frac{AC}{AE}$	5. Substitution
6. $\angle BAC \cong \angle DAE$	6. Reflexive Property of Congruence
7. $\triangle ABC \sim \triangle ADE$	7. SAS Similarity Theorem
8. $\angle ABC \cong \angle ADE$	8. Corresponding angles in similar figures are congruent. (Similar Figures Theorem)
9. $\overleftrightarrow{DE} \parallel \overleftrightarrow{BC}$	9. Corresponding Angles Postulate

You can use the Side-Splitting Converse Theorem to conclude that lines are parallel.

Example 2

When putting siding on a building, it is important that all of the pieces are parallel. Suppose the siding installer has a triangular section of wall and wants to make sure that piece *F* is parallel to piece *A*. The installer has measured *ZP*, *PQ*, *ZR*, and *RS* as shown. Is piece *F* parallel to piece *A*?

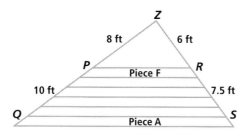

Solution The pieces are parallel if $\frac{ZP}{PQ} = \frac{ZR}{RS}$. Substitute for the lengths *ZP*, *PQ*, *ZR*, and *RS*. $\frac{8}{10} = \frac{6}{7.5}$ is true, so by the Triangle Side-Splitting Converse Theorem, $\overline{PR} \parallel \overline{QS}$.

From Triangles to Trapezoids

The Triangle Side-Splitting Theorem can be used to create a more general theorem about trapezoids. A proof of this generalization takes only a few steps.

Trapezoid Side-Splitting Theorem

If a line parallel to the bases intersects the legs of a trapezoid and divides those legs into two segments, then the lengths of those segments are proportional.

Given $\overline{AD} \parallel \overline{BE} \parallel \overline{CF}$

Prove $\frac{a}{b} = \frac{p}{q}$

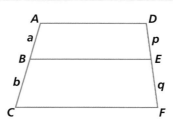

(continued on next page)

Proof Draw the diagonal $\overline{AF}$ of the trapezoid, as shown at the right.

By the Triangle Side-Splitting Theorem applied to $\triangle ACF$,

$$\frac{a}{b} = \frac{x}{y}.$$

By the Triangle Side-Splitting Theorem applied to $\triangle ADF$,

$$\frac{y}{x} = \frac{q}{p}$$

So, taking the reciprocal of each side,

$$\frac{x}{y} = \frac{p}{q}$$

Thus, by the Transitive Property of Equality, $\frac{a}{b} = \frac{p}{q}$.

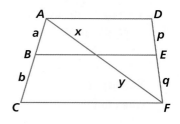

Questions

COVERING THE IDEAS

Fill in the Blank In 1–4, use the figure at the right to find a different fraction equal to the given fraction.

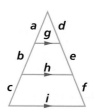

1. $\frac{a}{b} = $ __?__

2. $\frac{b}{c} = $ __?__

3. $\frac{g}{h} = $ __?__

4. $\frac{h}{i} = $ __?__

5. In the triangle at the right, is $\overline{GH} \parallel \overline{IJ}$? Why or why not?

6. In the diagram below, the line segments are parallel to the side of the triangle. Solve for the lengths x, y, and z.

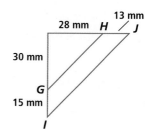

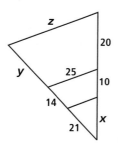

APPLYING THE MATHEMATICS

7. A student said, "I have an easy proof of the Trapezoid Side-Splitting Theorem. Use the figure on the bottom of page 777. Start by translating $\overline{DF}$ so that the translation image of D is A." Finish this student's proof.

8. In the diagram at the right, $\ell \parallel m \parallel n \parallel p$. Prove that $\frac{a}{c} = \frac{x}{z}$.

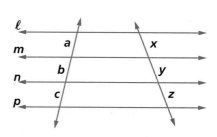

9. In the figure at the right, $\overline{B'C'}$ is the image of $\overline{BC}$ under a size transformation with center A. $AB' = 16$, $BB' = 24$.

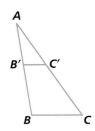

  a. What is the magnitude of the size transformation?
  b. If $CC' = 30$, how long is $\overline{C'A}$?
  c. What guarantees that $\overline{B'C'} \parallel \overline{BC}$?

10. You can use the Side-Splitting Theorems to split a segment into any rational ratio of parts using only a compass and straightedge. For instance, suppose you wished to divide $\overline{PZ}$ into 6 parts of equal length.

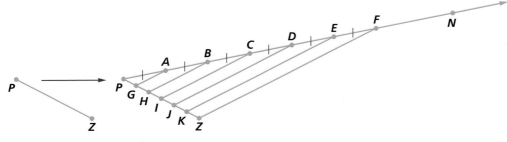

Draw any ray $\overrightarrow{PN}$ that is not collinear with $\overline{PZ}$, and mark off six segments of equal length on $\overrightarrow{PN}$ starting at P. Call the sixth endpoint F. Draw $\overline{FZ}$. Finally, construct segments parallel to $\overline{FZ}$ at each of the other endpoints that you marked off.

  a. Why is $\dfrac{DE}{EF} = \dfrac{JK}{KZ}$?
  b. Use Part a to explain why $JK = KZ$. This argument could be made for any two adjacent segments, proving that $\overline{PZ}$ has been divided into 6 parts of equal length.
  c. What is $\dfrac{PG}{GZ}$?

11. Draw a segment $\overline{RU}$. Use the method described in Question 10 to divide $\overline{RU}$ into 3 parts of equal length using only a compass and straightedge.

12. Copy the segment below. Use the method described in Question 10 to divide $\overline{AB}$ into a 2:3 ratio.

13. Explain how you could use the lines on a piece of notebook paper to find 5 points that divide a toothpick into 6 congruent pieces.

14. A child is sliding down a straight playground slide. The top of the slide is 9 feet off the ground and the bottom of the slide is 1 foot off the ground. The slide is 12 feet long. The child begins sliding, but stops after 3 feet. How high off of the ground is the child?

15. A right cone with a height of 12 inches and radius of 4 inches is cut by a plane parallel to the base. A frustum is created with a height of 3 inches.
    a. What is the radius of the smaller circular base of the frustum?
    b. What is the volume of the frustum in terms of π?

REVIEW

16. A cake is cut in half. (**Lesson 12-4**)
    a. How does the weight of each half relate to the weight of the whole cake?
    b. Why doesn't your answer from Part a contradict the Fundamental Theorem of Similarity?

17. In the figure at the right, $\overline{DE} \parallel \overline{GF}$. Find x.
    (**Lesson 12-3**)

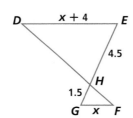

18. In the figure for Question 17, name three pairs of congruent angles. (**Lesson 5-4, 3-3**)

19. a. Find an equation in standard form for the line through the points $(7, -18)$ and $(-32, 5)$.
    b. Find the slope of the line in Part a. (**Lesson 1-2**)

20. Simplify the expression $\sqrt{3y^2}\sqrt{27y^4}$. (**Previous Course**)

EXPLORATION

21. Prove that if $\frac{a}{b} = \frac{c}{d}$, then it is also true that $\frac{a}{b} = \frac{a+c}{b+d}$.

Lesson

13-2 The Angle Bisector Theorem

> **BIG IDEA** The bisector of any angle of a triangle splits the opposite side into two segments whose lengths are in the same ratio as the sides including the original angle.

In Lesson 6-2, you learned that the bisector of the vertex angle of an isosceles triangle is a symmetry line for the triangle. This means that the angle bisector also divides the opposite side into congruent parts. A natural question that might follow is, "What if the triangle is scalene?" That is, what is the relationship between p and q in the diagram below at the right?

Mental Math

In the diagram, $\overleftrightarrow{ST} \parallel \overleftrightarrow{MW}$. Justify the conclusion.

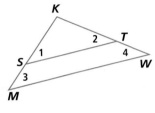

a. $\angle 1 \cong \angle 3$

b. $\angle 2 \cong \angle 4$

c. $\triangle KST \sim \triangle KMW$

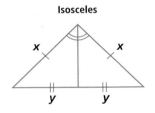

Isosceles

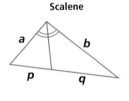

Scalene

Activity 1

MATERIALS DGS

Step 1 Create $\triangle ABC$ and the bisector of $\angle A$.

Step 2 Let F be the point where the angle bisector of $\angle A$ intersects $\overline{BC}$.

Step 3 Display the ratios $\frac{AB}{AC}$ and $\frac{BF}{FC}$.

Step 4 Drag one of the vertices and make a conjecture about these ratios.

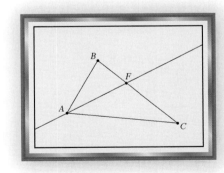

In Activity 1, you probably noticed that the point F divided the side it was on into segments whose lengths were proportional to the other two side lengths of the triangle. That is, $\frac{BF}{FC} = \frac{AB}{AC}$. This generalization can be proved using similar triangles.

The key to proving this theorem is in knowing where to draw an auxiliary segment to create the similar triangles. In Activity 2, we provide the statements in the proof and ask you to provide the justifications.

Angle Bisector Theorem

The bisector of any angle of a triangle divides the opposite side into segments whose lengths are proportional to the adjacent sides of the triangle.

Activity 2

Work in a group to fill in the missing justifications for the following proof of the Angle Bisector Theorem.

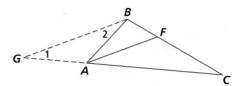

Given $\triangle ABC$; angle bisector $\overrightarrow{AF}$, where F is on $\overline{BC}$

Prove $\dfrac{AB}{AC} = \dfrac{BF}{FC}$

Proof Let G be the point on $\overrightarrow{CA}$ such that $GA = AB$.

Conclusions	Justifications
1. $m\angle 1 = m\angle 2$	1. _____?_____
2. $m\angle BAC = m\angle 1 + m\angle 2$	2. _____?_____
3. $m\angle BAC = 2m\angle 1$	3. _____?_____
4. $\frac{1}{2}m\angle BAC = m\angle 1$	4. _____?_____
5. $\frac{1}{2}m\angle BAC = m\angle FAC$	5. definition of angle bisector
6. $m\angle 1 = m\angle FAC$	6. _____?_____
7. $\overline{AF} \parallel \overline{GB}$	7. _____?_____
8. $\dfrac{GA}{AC} = \dfrac{BF}{FC}$	8. _____?_____
9. $\dfrac{AB}{AC} = \dfrac{BF}{FC}$	9. _____?_____

 QY

> **QY**
>
> In the scalene triangle at the beginning of the lesson, how are a, b, p, and q related?

Example 1

Suppose $\overline{BD}$ bisects $\angle ABC$ in the figure at the right. Find x.

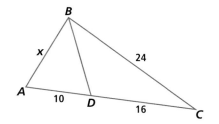

Solution Apply the Angle Bisector Theorem.

$$\frac{BA}{BC} = \frac{DA}{DC}$$

$$\frac{x}{24} = \frac{10}{16}$$

$$x = 15$$

GUIDED

Example 2

In the figure at the right, $\overrightarrow{LN}$ bisects $\angle KLM$ and $KM = 15$. Find KN.

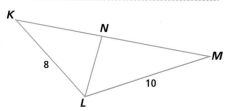

Solution Let $KN = x$. Then $MN = \underline{\ ?\ }$.

From the Angle Bisector Theorem, $\dfrac{LK}{LM} = \dfrac{?}{?}$.

Substitute. $\dfrac{?}{?} = \dfrac{?}{?}$

Solve the proportion for x. $KN = \underline{\ ?\ }$

Dividing Segments Into a Given Ratio Using a Compass and Straightedge

Like the Triangle Side-Splitting Theorem, the Angle Bisector Theorem also allows you to construct a point that divides a given segment exactly into two lengths of a given ratio. This is shown in Example 3.

Example 3

Use a compass and straightedge to divide $\overline{AB}$ into two lengths that are in a 4:5 ratio.

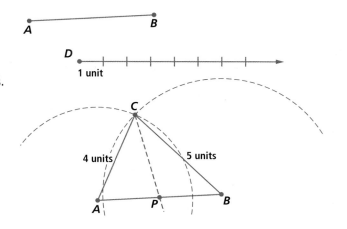

Solution Create a "ruler" (the ray with endpoint D) to obtain 4-unit and 5-unit lengths. Then construct an arc with a 4-unit radius with center A, and an arc with a 5-unit radius with center B. Let C be one of the points where these arcs intersect, and connect $\overline{AC}$ and $\overline{BC}$. Finally, bisect $\angle C$ to find point P on $\overline{AB}$. By the Angle Bisector Theorem, we know that

$$\frac{PA}{PB} = \frac{4}{5}.$$

Questions

COVERING THE IDEAS

Fill in the Blank In 1–4, use the diagram at the right. $\overrightarrow{VT}$ bisects $\angle SVW$ and $\overrightarrow{WR}$ bisects $\angle SWV$. Write a fraction that is equal to the given fraction but uses different lengths from the diagram.

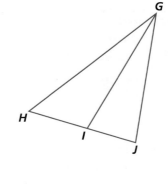

1. $\dfrac{VR}{SR} = \underline{\ ?\ }$

2. $\dfrac{VW}{TW} = \underline{\ ?\ }$

3. $\dfrac{VW}{VS} = \underline{\ ?\ }$

4. $\dfrac{RZ}{ZW} = \underline{\ ?\ }$

5. Suppose you were trying to reproduce the proof of the Angle Bisector Theorem using the diagram at the right (where $\overrightarrow{GI}$ bisects $\angle HGJ$). Copy the diagram and draw the auxiliary lines that you would use for your proof.

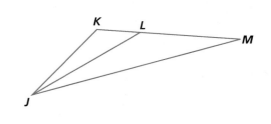

6. In the figure at the right $\overrightarrow{SQ}$ bisects $\angle PSR$. Find x.

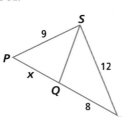

7. In the figure below, suppose $\overrightarrow{DB}$ is an angle bisector. Find $\overline{DC}$.

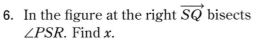

8. In the figure at the right, $\overrightarrow{JL}$ bisects $\angle J$. If $JM = 14$, $JK = 6$, and $KM = 10$, find LM.

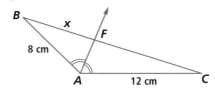

9. In the figure below, if $BC = 15$ cm, find x.

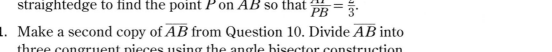

10. Copy $\overline{AB}$ onto another piece of paper. Use a compass and straightedge to find the point P on $\overline{AB}$ so that $\dfrac{AP}{PB} = \dfrac{2}{3}$.

11. Make a second copy of $\overline{AB}$ from Question 10. Divide $\overline{AB}$ into three congruent pieces using the angle bisector construction method from this lesson.

APPLYING THE MATHEMATICS

Fill in the Blank In 12 and 13, use the diagram at the right, in which $\overrightarrow{CD}$ and $\overrightarrow{BE}$ are angle bisectors.

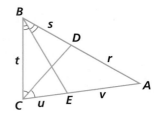

12. $\dfrac{v + u}{t} = \underline{\ ?\ }$

13. $\dfrac{v}{u} = \underline{\ ?\ }$

14. In the figure at the right, $\overrightarrow{AD}$ is an angle bisector.

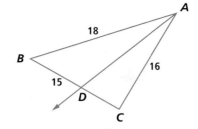

   a. What is $\dfrac{\text{Area}(\triangle ABD)}{\text{Area}(\triangle ACD)}$?
   (Hint: You do not have to calculate any areas to find the ratio.)

   b. Generalize Part a.

15. In the 30-60-90 right triangle shown at the right, the 30° angle is bisected. Find x to the nearest hundredth.

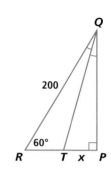

16. In the diagram at the right, $\overrightarrow{ED} \parallel \overline{AC}$, $\overrightarrow{AD}$ bisects $\angle A$, $AB = 9$, and $AC = 12$. Find ED.

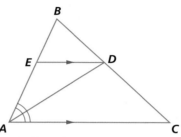

REVIEW

17. The surface of the pill pictured at the right consists of two hemispheres and a cylinder. The radius of each hemisphere is r and the height of the cylindrical portion is $2r$.

   a. Write an expression for the volume of the pill.
   (**Lessons 10-6, 10-3**)

   b. Write an expression for the surface area of the pill.
   (**Lessons 10-7, 9-9**)

18. Suppose a circle is inscribed in an equilateral triangle with sides of length 16. What is the exact area of the circle? (**Lesson 8-9, 8-7**)

19. Find the exact area of the triangle in Question 18. (**Lessons 8-7, 8-4**)

20. Draw a figure that has 4-fold rotation symmetry but no reflection symmetry. (**Lessons 6-7, 6-1**)

EXPLORATION

21. Suppose one angle of an equilateral triangle is trisected by two rays, as shown at the right. Do those rays divide the opposite side into 3 equal lengths? Justify your answer.

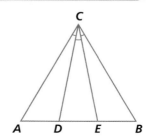

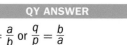

Lesson

13-3

Geometric Means in Right Triangles

Vocabulary

geometric mean

> ▶ **BIG IDEA** The altitude to the hypotenuse of a right triangle creates two triangles similar to the original triangle, from which relationships between the lengths of their sides can be derived.

In this lesson, you will see applications of a number called the *geometric mean* of two numbers.

What Is the Geometric Mean of Two Numbers?

In 1990 there were about 13,000,000 people living in Florida. By the year 2000, that number had grown to about 16,000,000. If the growth rate was constant during the decade from 1900 to 2000, what is a good estimate for the population in 1995?

The answer to this question is *not* 14,500,000 because then the growth factor from 1990 to 1995 would have been

$\frac{14,500,000}{13,000,000}$ or about 1.115, meaning a growth rate of about 11.5%,

while the growth factor from 1995 to 2000 would have been

$\frac{16,000,000}{14,500,000}$ or about 1.103, or a growth rate of about 10.3%.

We are looking for a population P such that the two growth factors are equal.

$$\frac{16,000,000}{P} = \frac{P}{13,000,0000}$$

To find P, we solve the proportion, Use the Means-Extremes Property.

$$P^2 = 16,000,000 \cdot 13,000,000$$
$$P^2 = 208,000,000,000,000$$
$$P = \sqrt{208,000,000,000,000}$$
$$\approx 14,422,000$$

 QY1

> **Mental Math**
>
> **Consider the circle with equation $(x - 2)^2 + (y - 4)^2 = 144$.**
>
> **a.** What is the center of the circle?
>
> **b.** What is the radius of the circle?
>
> **c.** Identify the coordinates of 4 points on the circle.

> ▶ **QY1**
>
> **a.** If the population of Florida was 14,422,000 in 1995, what was the population growth rate from 1990 to 1995?
>
> **b.** If the population of Florida was 14,422,000 in 1995, what was the population growth rate from 1995 to 2000?

In the Florida population situation, the two means in the proportion $\frac{16,000,000}{P} = \frac{P}{13,000,000}$ are equal. When this happens, P is called a *geometric mean* of the other two numbers.

Definition of Geometric Mean

Let a, b, and g be positive numbers. g is the **geometric mean** of a and b if and only if

$$\frac{a}{g} = \frac{g}{b}$$

The definition provides a way of calculating the geometric mean.

Example 1

Find the geometric mean of 4 and 64.

Solution Let g be the geometric mean. From the definition of geometric mean,

$$\frac{4}{g} = \frac{g}{64}.$$

Using the Means-Extremes Property,

$$g^2 = 256.$$

From the definition of geometric mean, g is positive; so

$$g = \sqrt{256} = 16.$$

The geometric mean of 4 and 64 is 16.

Check Does $\frac{4}{16} = \frac{16}{64}$? Yes, 16 is the geometric mean of 4 and 64.

From the definition of geometric mean, if $\frac{a}{g} = \frac{g}{b}$, then $g^2 = ab$; thus $g = \sqrt{ab}$. This yields the following theorem.

Geometric Mean Theorem

The geometric mean of the positive numbers a and b is $\sqrt{ab}$.

The name "geometric mean" comes from relationships among lengths in any right triangle, relationships that were discovered by the ancient Greeks and are in Euclid's *Elements*. The proofs of these relationships depend mainly on the AA Similarity Theorem.

 QY2

▸ QY2

What is the geometric mean of 2 and 18?

Similar Triangles Within a Right Triangle

$\triangle ABC$ at the right is split into two triangles
CBD and ACD by drawing the altitude to its
hypotenuse. As usual, we use the small letters
a and b for the lengths of the legs of a right
triangle, and c for the length of the hypotenuse.
The letter h is used for the length of the altitude
to the hypotenuse. The point where the altitude
intersects the hypotenuse splits it into two
lengths, x and y, so $x + y = c$.

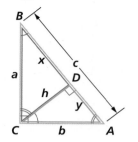
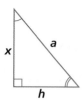

Notice that m$\angle A$ = m$\angle BCD$, because both angles are complements
of $\angle B$. So $\triangle CBD$, $\triangle ACD$, and $\triangle ABC$ are all right triangles, and
each includes an angle congruent to $\angle A$. Thus, by the AA Similarity
Theorem, all three triangles are similar: $\triangle ABC \sim \triangle CBD \sim \triangle ACD$.
Because the triangles are similar, corresponding sides are proportional.

Activity

Step 1 Consider the similar triangles CBD and ACD above. These two
triangles share the side $\overline{CD}$ with length h. Write a true proportion
involving lengths of their sides in which h appears twice.

Step 2 Use $\triangle CBD$ and the original (biggest) triangle. $\overline{BC}$ is a side in both.
Write a true proportion involving lengths of their sides in which
BC appears twice.

Step 3 Use $\triangle ACD$ and the original triangle. $\overline{AC}$ is a side in both. Write a true
proportion involving lengths of their sides in which AC appears twice.

From the Activity, we see that in a right triangle the altitude h to the
hypotenuse and the legs a and b are geometric means of other lengths.

Right-Triangle Altitude Theorem

In every right triangle:
1. The altitude to the hypotenuse is the geometric mean of the
segments into which it divides the hypotenuse.
2. Each leg is the geometric mean of the hypotenuse and the
segment of the hypotenuse adjacent to the leg.

> **QY3**
>
> Write an expression for k,
> the length of one of the
> legs of $\triangle ABC$.

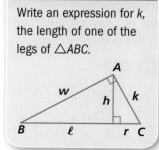

In the figure at the right, $h = \sqrt{xy}$, $a = \sqrt{cx}$, and
$b = \sqrt{cy}$. Most people are quite surprised to learn
that these lengths are related in so many ways, yet
all the relationships come from similar triangles.

STOP QY3

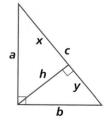

GUIDED

Example 2

$\overrightarrow{CD}$ is the altitude to the hypotenuse of right triangle *ABC*, as shown at the right. If *AD* = 10, and *BD* = 16, find *CD, CA,* and *CB.*

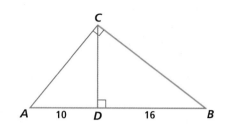

Solution *CD* is the geometric mean of *AD* and __?__.

$$\text{So } CD = \sqrt{AD \cdot \underline{\ ?\ }}$$
$$= \sqrt{10 \cdot \underline{\ ?\ }}$$
$$= \sqrt{\underline{\ ?\ }}$$
$$= 4\sqrt{10} \approx 12.65.$$

CA is the geometric mean of *AD* and __?__.

$$\text{So } CA = \sqrt{AD \cdot \underline{\ ?\ }}$$
$$= \sqrt{10 \cdot \underline{\ ?\ }}$$
$$= \sqrt{\underline{\ ?\ }}$$
$$= 2\sqrt{65} \approx 16.12.$$

__?__ is the geometric mean of *DB* and *AB.*

$$\text{So } \underline{\ ?\ } = \sqrt{BD \cdot BA}$$
$$= \sqrt{\underline{\ ?\ } \cdot 16}$$
$$= \sqrt{\underline{\ ?\ }}$$
$$= 4\sqrt{26} \approx 20.40.$$

Geometric means can be found in semicircles because an angle inscribed in a semicircle is a right angle.

Example 3

Find the diameter of the semicircle pictured at the right if *AP* = 5, *BP* = 10 and $\overline{BP} \perp \overline{PA}$.

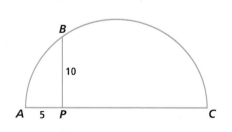

Solution Draw $\overline{BA}$ and $\overline{BC}$ to form $\triangle ABC$. By Thales' Theorem, $\angle ABC$ is a right triangle. Now find *PC* by applying the Right-Triangle Altitude Theorem.

$$BP^2 = AP \cdot PC$$
$$10^2 = 5 \cdot PC$$
$$100 = 5(PC)$$
$$20 = PC$$

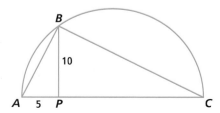

AC = *PC* + *AP* = 5 + 20 = 25, so the diameter of the semicircle is 25.

Questions

COVERING THE IDEAS

In 1 and 2, find the geometric mean of the given numbers to the nearest hundredth.

1. 2 and 50

2. 9 and 12

3. A colony of bacteria grew at a constant rate from 30 bacteria in a culture at 10 A.M. Monday to 12,000 bacteria in the culture two days later at 10 A.M. Wednesday.

   a. What is a good estimate for the number of bacteria in the culture at 10 A.M. Tuesday?

   b. Explain why the arithmetic mean of 30 and 12,000 is not a good estimate for the number of bacteria in the culture at 10 A.M. Tuesday.

Fill in the Blanks In 4–9, use the figure at the right in which $\overline{GH} \perp \overline{EH}$ and $\overline{HF} \perp \overline{GE}$.

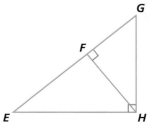

4. $\triangle EGH \sim$ ___?___ $\sim$ ___?___

5. $\dfrac{EG}{EH} = \dfrac{EH}{?}$

6. $\dfrac{EF}{?} = \dfrac{?}{FG}$, where the two blanks represent the same number.

7. FH is the geometric mean of ___?___ and ___?___.

8. EH is the geometric mean of ___?___ and ___?___.

9. GH is the geometric mean of ___?___ and ___?___.

10. In the diagram at the right, $QP = 40$ and $PM = 200$. Find each length.

    a. NP

    b. NQ

    c. NM

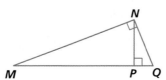

APPLYING THE MATHEMATICS

11. At the right is a famous right triangle with the altitude drawn to its hypotenuse. Find x, y, and h.

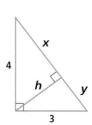

12. In the figure below, $\overline{CG}$ is an altitude of $\triangle ACE$. O is the center of the circle. $AG = 10$ and $GE = 4$. Find CG, AC, EC, and the area of the circle correct to three decimal places.

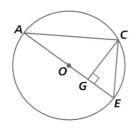

13. The Right-Triangle Altitude Theorem gives us a nice way to prove the Pythagorean Theorem. Provide the missing conclusions or justifications for this proof:

Given $\angle ACB$ is a right angle and $\overline{CP}$ is an altitude of $\triangle ABC$.

Prove $AC^2 + BC^2 = AB^2$

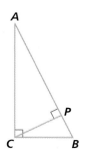

Conclusions	Justifications
a. $\angle ACB$ is right and $\overline{CP}$ is an altitude of $\triangle ABC$.	a. ___?___
b. $AC^2 = (\underline{\ ?\ })(\underline{\ ?\ })$	b. Right-Triangle Altitude Theorem
c. $BC^2 = (\underline{\ ?\ })(\underline{\ ?\ })$	c. Right-Triangle Altitude Theorem
d. $AC^2 + BC^2 = \underline{\ \ ?\ \ }$	d. Addition Property of Equality
e. $AC^2 + BC^2 = (AB)(AP + BP)$	e. ___?___
f. $AC^2 + BC^2 = (AB)(AB)$	f. ___?___
g. $AC^2 + BC^2 = \underline{\ \ ?\ \ }$	g. definition of 2nd power

14. In $\triangle ABD$ below, two right angles are marked. $BD = 41$ and $DC = 16$. Find the perimeter of $\triangle ABC$.

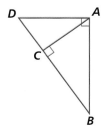

15. Robert showed his geometry teacher how to find the height of a tree using a notebook. He held the corner of his notebook near his eye and moved back from the tree until he could sight both the top and the bottom of the tree along with two adjacent edges of the notebook. His eye was 4 feet off the ground when he was 8 feet from the tree.

a. Which part of the Right-Triangle Altitude Theorem could Robert use to find the height of the tree?

b. How tall is the tree?

16. At the right, $\overline{BA}$ is a diameter of the circle, $\overline{BA} \perp \overline{ST}$, $BW = 2$, and $SW = WT = 4$. Determine the diameter of the circle. (You may find it useful to draw a triangle or two.)

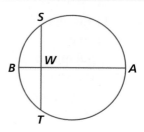

REVIEW

17. Find AM if $\overrightarrow{AT}$ bisects $\angle OAM$ in the figure below. **(Lesson 13-2)**

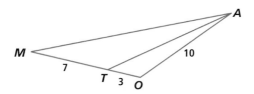

18. In the figure at the right, $\overline{AB} \parallel \overline{CD} \parallel \overline{EF}$. Suppose $AH = HG$. Name as many other pairs of segments that are equal in length as you can. **(Lesson 13-1)**

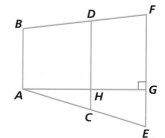

19. Draw an example of two nonsimilar quadrilaterals whose angles have the same measures. **(Lesson 12-3)**

20. In a 30-60-90 triangle, the longer of the two legs has length a and the shorter one has length b. Find the value of $\frac{a}{b}$. **(Lesson 8-7)**

21. Write an expression equal to $\dfrac{2\sqrt{125}}{\sqrt{10}}$ with no radical sign in the denominator. Your answer should be in simplified form. **(Previous Course)**

EXPLORATION

22. Construct a semicircle like the one at the right.
  a. Construct a segment with length equal to the arithmetic mean (average) of AP and PC.
  b. Construct a segment with length equal to the geometric mean of AP and PC.

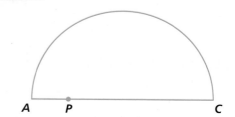

QY ANSWERS

1. a. about 10.9%
   b. about 10.9%
2. 6
3. $k = \sqrt{\ell r + r^2}$

Lesson
13-4   The Golden Ratio

Vocabulary

golden rectangle

golden ratio, phi (φ)

▶ **BIG IDEA**   The golden ratio is a ratio of sides of a rectangle that occurs in nature and is thought to be particularly appealing to the eye.

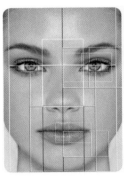

Mental Math

A 3 in. by 4 in. photo is enlarged so that one of its sides is 24 in. long.

a. Find the two possible lengths for the other side of the enlargement.

b. Find the two possible areas of the enlargement.

The face, the ancient Greek building, the shell, and the violin pictured here have something in common. In all of these things there are lengths whose ratio is known as the *golden ratio*. The golden ratio is a number like π. It is one of the most famous numbers in mathematics because it has applications in so many places. It arises from similar rectangles.

Similar Rectangles

We would like a simple criterion to determine when rectangles are similar. For this, we go back to the definition of similarity: Two figures are *similar* if and only if there is a composite of reflections and size transformations that maps one onto the other. Reflections do not change angle measures or side lengths. Size transformations do not change angle measures, but they do multiply side lengths by the size change factor k. So, when two rectangles are similar, the dimensions of one will be a fixed multiple of the dimensions of the other.

If the first rectangle has dimensions a and b, then the second rectangle must have dimensions of the form ka and kb, where k is a fixed positive number.

Suppose the second rectangle has dimensions ℓ and w. Either (1) $\ell = ka$ and $w = kb$, or (2) $w = ka$ and $\ell = kb$. If (1), then $\frac{\ell}{w} = \frac{ka}{kb} = \frac{a}{b}$. If (2), then $\frac{\ell}{w} = \frac{kb}{ka} = \frac{b}{a}$. Thus, if two rectangles are similar, the ratios of their dimensions (in one order or the other) must be the same. This proves the following theorem.

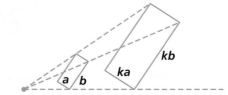

Rectangle Similarity Theorem

Two rectangles are similar if and only if their dimensions are in the same ratio.

Example 1

Which of the rectangles shown below are similar?

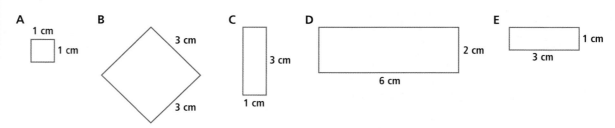

Solution Rectangles A and B are similar because the ratio of their dimensions is 1:1.

Rectangles C, D, and E are all similar because the ratio of their dimensions is 1:3.

Example 1 illustrates that, like triangles, rectangles do not have to be tilted in the same way to be similar.

GUIDED

Example 2

Rectangle I has dimensions 4 cm and 10 cm. A second similar rectangle II has one dimension 6 cm. What are the *two* possible other dimensions of the second rectangle?

Solution First draw a picture. Let x be the unknown dimension.

You can see that there are two possibilities for rectangle II. Call them IIA and IIB.

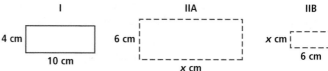

Suppose the 6-cm side of rectangle II corresponds to the 4-cm side of rectangle I. (This is rectangle IIA.) Then $\frac{4}{6} = \frac{10}{x}$, from which $x = \underline{}$ cm.

If the 6-cm side of rectangle II (call it IIB) corresponds to the 10-cm side of rectangle I, then $\frac{6}{10} = \frac{?}{?}$, from which $x = $ __?__ cm.

The Golden Rectangle

Here are pictures of three houses. The front view of each is roughly a rectangle, but the houses are not similar. Which one has the most pleasing look?

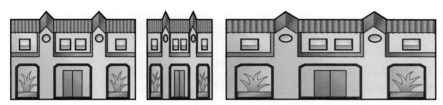

The ancient Greeks had a particular shape of rectangle they felt was more beautiful than others. They used it often in their art and architecture. We call this shape the *golden rectangle*. A **golden rectangle** is one in which, if you take a square away from it, then the result is a rectangle that is similar to the entire figure.

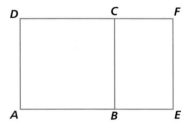

Specifically, in rectangle *DAEF* at the right, *ABCD* is a square. Then *DAEF* is a golden rectangle if it is similar to the smaller rectangle *BEFC*. Notice how alike the shape of *DAEF* is to the house farthest left above.

From the definition of *golden rectangle* given above, you can determine the exact value of the ratio of its dimensions. Suppose $DA = 1$ and let $AE = x$. (Notice that $x > 1$.) Then $BE = x - 1$. Then, by the Rectangle Similarity Theorem,

$\frac{DA}{AE} = \frac{BE}{EF}$ and by substitution $\frac{1}{x} = \frac{x-1}{1}$.

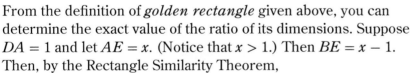

Now solve this proportion for x. The Means-Extremes Property gives the quadratic equation

$$x^2 - x = 1$$
$$x^2 - x - 1 = 0.$$

Use the Quadratic Formula with $a = 1$, $b = -1$, and $c = -1$.

$$x = \frac{-b \pm \sqrt{b^2 - 4ac}}{2a}$$
$$= \frac{-(-1) \pm \sqrt{(-1)^2 - (4)(1)(-1)}}{2(1)}$$
$$= \frac{1 \pm \sqrt{5}}{2}$$

Only one of these two values of x is positive, $\frac{1 + \sqrt{5}}{2} = 1.618 \ldots$.
Thus, a golden rectangle is a rectangle whose dimensions are in the
ratio $\frac{1 + \sqrt{5}}{2}$, and the number $\frac{1 + \sqrt{5}}{2}$ is consequently called the
golden ratio.

At the right is a picture of the Parthenon, a temple to
the goddess Athena built by the architect Phidias in the
5th century BCE on the Acropolis, a hill overlooking the
city of Athens. Overlaid on the picture are many golden
rectangles. The golden ratio $\frac{1 + \sqrt{5}}{2}$ is represented by
the Greek letter φ (phi), in honor of Phidias.

One of the nicest applications of this idea is found in
nature. The shell of a chambered nautilus (pictured
below at the left) has the shape of a sequence of golden
rectangles as drawn below at the right.

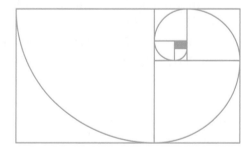

Questions

COVERING THE IDEAS

1. A rectangular room is 9' by 15'. A second rectangular room of
   similar shape has one dimension of 12'. What are the possible
   dimensions of the second room?

2. Follow these directions on paper or with a DGS to construct a
   golden rectangle:
   a. Construct a square. Label it $ABCD$. Extend $\overleftrightarrow{AB}$.
   b. Find the midpoint of $\overline{AB}$. Label it M.
   c. Construct the circle with center M and radius MC.
   d. Let E be one of the points of intersection of the circle
      and $\overleftrightarrow{AB}$.
   e. Complete rectangle $AEFD$.
   f. Suppose $AB = 1$ unit. Find MC.
   g. Show that AE is the golden ratio.

3. Copy the spiral of rectangles that immediately precedes these Questions. Add the next rectangle to your drawing.

4. **a.** The number ϕ is the solution to what quadratic equation with integer coefficients?

   b. What is the exact second solution to that quadratic equation?

   c. Estimate the second solution to the nearest thousandth.

5. Are all rectangles with length twice their width similar to each other? Why or why not?

APPLYING THE MATHEMATICS

6. One kite has sides of lengths 4, 4, 8, and 8. Another kite has sides of lengths 6, 6, 12, and 12. Must the kites be similar?

7. One parallelogram has sides of lengths 4, 8, 4, and 8. Another parallelogram has sides of lengths 6, 12, 6, and 12. Must the parallelograms be similar?

8. One rhombus has diagonals of lengths 30 cm and 35 cm. Another rhombus has diagonals of lengths 7 cm and 6 cm. Must the rhombi be similar?

9. A kite has diagonals of lengths 3 units and 4 units. A second kite has diagonals of lengths 3 units and 4 units.

   a. Must the kites be similar?

   b. Must the kites be congruent?

10. Write down the decimal approximation of ϕ. Now, write down the decimal approximation of the reciprocal of ϕ. (Recall that the reciprocal of ϕ is $\frac{1}{\phi}$.). What do you notice? Write an equation that relates ϕ and $\frac{1}{\phi}$.

11. Prove that $\phi^2 = 1 + \phi$ by squaring $\frac{1 + \sqrt{5}}{2}$.

12. Estimate the first ten integer powers of ϕ to the nearest integer.

13. Consider the regular pentagon $ABCDE$ inscribed in the circle at the right. Let each side of the regular pentagon have length 1. Because of symmetry, each of the diagonals shown has the same length. Call that length d.

    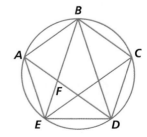

    a. Explain why $m\angle AEB = m\angle BEC = m\angle CED$.

    b. Find the measure of each of the three angles in Part a.

    c. Explain why $\triangle AEF \sim \triangle DBF$.

    d. Explain why $EF = d - 1$.

    e. Use the similar triangles of Part c to find the value of d. (You may be surprised!)

REVIEW

14. Given a 3-cm segment and an 8-cm segment, construct (with compass and straightedge) a segment whose length is the geometric mean of 3 and 8 by following Parts a–f. (**Lesson 13-3**)

    a. Use a ruler to draw a 3-cm segment $\overline{AP}$ and an 8-cm segment $\overline{PB}$ that are collinear and adjacent.

    b. Find M, the midpoint of $\overline{AB}$.

    c. Draw the circle with center M and radius MB.

    d. Construct a perpendicular to $\overleftrightarrow{AB}$ through P. Name C as one of the points of intersection of this perpendicular and the circle.

    e. Explain why CP is the geometric mean between AP and PB.

    f. Measure CP with a ruler to check that it is the geometric mean of 3 and 8.

15. a. What segments in the construction in Question 14 have a length equal to the arithmetic mean of AP and PB?

    b. Explain why the geometric mean will be smaller than the arithmetic mean. (**Lesson 13-3**)

16. In trapezoid $AEFB$ at the right, suppose $EC = 16$, $AE = 40$, and $FD = 20$. (**Lesson 13-1**)

    a. Find FB.   b. Find DB.

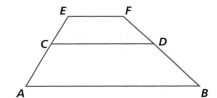

17. **Matching** Match the irrational number to the ratio. (**Lessons 8-8, 8-7**)

    a. π b. $\sqrt{2}$ c. $\sqrt{3}$

    i   the ratio of the length of a diagonal of a square to its side

    ii   the ratio of the longer leg to the shorter leg of a 30-60-90 triangle

    iii the ratio of the circumference of a circle to its diameter

18. Recall the two bars of the song "Row, Row, Row Your Boat" below. The notes are C, C, C, D, and E. (**Lesson 4-8**)

    Row,   row,   row   your   boat

    a. If the melody is transposed so that the first note is an E, write what the other notes would be.

    b. Write what the original melody would be in retrograde.

19. The picture below depicts a film being projected on a screen. The plane of the film is parallel to the plane of the screen. (**Lessons 12-3, 10-2**)

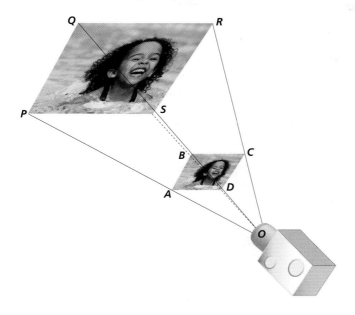

    a. A size transformation is created. If the size change factor is 20, what is the ratio of similitude?

    b. If a linear measure on the film is 0.4 cm, what is the corresponding measure on the screen?

    c. If an area on the film is 2.2 cm², what is the corresponding area on the screen?

    d. If $OA = 12$ cm, find AP.

EXPLORATION

20. Just as there are famous numbers in mathematics (π and ϕ are two of them), there are also famous number patterns. One such pattern is known as the *Fibonacci sequence*. The sequence is named after Leonardo Fibonacci (1170–1250), who was one of the most talented mathematicians of the middle ages. The first 8 terms of the sequence are 1, 1, 2, 3, 5, 8, 13, and 21.

    a. Write the next 4 terms of the Fibonacci sequence.

    b. Describe how the sequence is created.

    c. The Fibonacci sequence and the golden ratio ϕ are related. Do some research to find out how.

Lesson

13-5

The Tangent of an Angle

▶ **BIG IDEA** The tangent of an angle is a trigonometric ratio determined by the angle's measure with which you can determine segment lengths and angle measures in right triangles.

If you know the lengths of two sides of any right triangle, you can always find the length of the third side by using the Pythagorean Theorem. In certain special right triangles, you can find the lengths of any two sides of the triangle given the length of just one side. In particular, you have learned that with a 30-60-90 triangle or a 45-45-90 triangle (an isosceles right triangle), you need only to know one side length to determine the other two. But most triangles are not one of these special triangles. In this lesson and the next, you will learn how you can use the measure of one side and the measure of one acute angle of *any* right triangle to find the lengths of the other two sides.

Activity

MATERIALS DGS or paper, pencil, and protractor

Step 1 Draw a right triangle *ACB* with a right angle at *C* and $m\angle A = 35$.

Step 2 Measure *AC*, *BC*, and calculate $\frac{BC}{AC}$ to the nearest hundredth.

Step 3 Draw a second right triangle similar to *ACB*, but bigger. Call this $\triangle A'C'B'$.

Step 4 Measure *A′C′*, *B′C′*, and calculate $\frac{B'C'}{A'C'}$ to the nearest hundredth. Is $\frac{B'C'}{A'C'}$ greater than, less than, or equal to $\frac{BC}{AC}$?

Step 5 Draw a third right triangle similar to *ACB*, but smaller. Call this $\triangle A''C''B''$.

Step 6 Measure *A″C″*, *B″C″*, and calculate $\frac{B''C''}{A''C''}$ to the nearest hundredth. Do you think $\frac{B''C''}{A''C''}$ will be greater than, less than, or equal to $\frac{BC}{AC}$? Check your prediction with calculations.

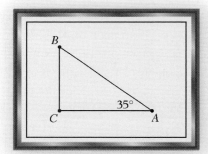

Ratios of Legs in Right Triangles

The activity on the preceding page should convince you that the ratio of legs of a right triangle is the same for a given angle regardless of the size of the triangle. In other words, the ratio of the legs depends only on the measure of the angle selected. This can be proved.

Consider two right triangles $\triangle ABC$ and $\triangle XYZ$ with congruent acute angles A and X. The drawing at the right below follows the custom that a represents the length of the side opposite angle A, b represents the length of the side opposite angle B, and so on. $\triangle ABC \sim \triangle XYZ$ by the AA Similarity Theorem.

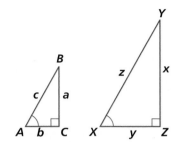

Because corresponding sides in similar figures are proportional, $\frac{a}{x} = \frac{b}{y}$.

Multiplying both sides by $\frac{x}{b}$ yields a ratio of lengths of legs in each triangle.

$$\frac{a}{x} \cdot \frac{x}{b} = \frac{b}{y} \cdot \frac{x}{b}$$
$$\frac{a}{b} = \frac{x}{y}$$

The ratios of the lengths of the legs are equal. So, the ratio depends only on the measure of $\angle A$.

The Tangent Ratio

The legs with lengths a and x are **opposite** the congruent angles $\angle A$ and $\angle X$. The legs with lengths b and y are **adjacent** to angles $\angle A$ and $\angle X$. The previous argument proves that in a right triangle, the ratio

$$\frac{\text{the length of the leg opposite } \angle A}{\text{the length of the leg adjacent to } \angle A}, \text{ or } \frac{a}{b},$$

is the same for any angle congruent to $\angle A$. This ratio is called the *tangent of $\angle A$* and written tan A or tan (A).

Definition of Tangent

Let $\triangle ABC$ be a right triangle with right angle C. The **tangent of $\angle A$,** written **tan A,** is

$$\frac{\text{length of leg opposite } \angle A}{\text{length of leg adjacent to } \angle A}$$

In the triangle at the right, tan $A = \frac{a}{b}$.

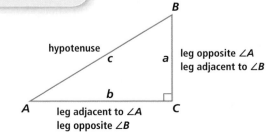

In the Activity, you estimated tan 35° by measuring. The exact value of tan 35° is an irrational number so it cannot be represented by a finite or infinite repeating decimal. Calculators give approximations. Find the TAN key on your calculator. To use this key, make sure your calculator is measuring angles in DEGREE mode. (There are other modes, but we do not discuss them in this book,) Your calculator may display .7002075382. This means that tan(35°) ≈ 0.7002075382.

Notice that as m∠A gets larger, $\overline{BC}$ will be longer compared to $\overline{AC}$, and the tangent ratio will increase.

 QY1

▶ **QY1**

Find tan(36°) rounded to the nearest thousandth.

Using Tangents to Find Lengths

When the measure of an angle and one leg in a right triangle are known, the tangent ratio can help to find the length of the other leg.

Example 1

At a location 30 m from the base of a flagpole, the angle to the top of the flagpole is 34°. Determine the height of the flagpole to the nearest meter.

Solution

$$\tan(34°) = \frac{\text{length of leg opposite } \angle A}{\text{length of leg adjacent } \angle A} = \frac{BC}{AC}$$

$$\tan(34°) = \frac{BC}{30}$$

$$30 \cdot \tan(34°) = BC$$

$$30 \cdot (0.6745085168) \approx BC$$

$$20.235 \approx BC$$

To the nearest meter, the height of the flagpole is 20 meters.

The exact value of the tangent of an acute angle can be determined if you know the lengths of the two legs in a right triangle with that angle.

Example 2

In the right triangle at the right, find tan D.

Solution By definition of the tangent of an angle,

$$\tan D = \frac{\text{length of leg opposite } \angle D}{\text{length of leg adjacent to } \angle D}$$

$$= \frac{EF}{DE}.$$

It is given that $EF = 40$. Use the Pythagorean Theorem to find DE.

$$DE^2 + 40^2 = 41^2$$
$$DE^2 + 1600 = 1681$$
$$DE = 9$$

So $\tan D = \dfrac{EF}{DE} = \dfrac{40}{9}$.

Once you know the tangent of an angle, your calculator has the ability to work backward from the value of the tangent to the angle. The key that does this is usually directly above the $\boxed{\text{TAN}}$ key and is identified with a different color, meaning it requires two keystrokes to use. Some calculators call this key TAN^{-1}, some call it ATAN, and some call it ARCTAN. In any case, this key will output an angle that has the given tangent. It is called the "inverse tangent key" or sometimes the "arctan key."

 QY2

▶ **QY2**

Use your calculator to determine the measure of an angle whose tangent is $\dfrac{40}{9}$, to the nearest hundredth.

Exact Values of the Tangent Ratio

Tangents of most angles can only be approximated by decimals. However, you can find the exact values of tangents of angles in special triangles.

Example 3

Give an exact value for $\tan(30°)$.

Solution Draw a 30-60-90 triangle, as in the figure at the right.

$$\tan(30°) = \tan(\angle CTA)$$
$$= \frac{\text{length of leg opposite } \angle T}{\text{length of leg adjacent to } \angle T} = \frac{CA}{AT}$$

$\overline{CA}$ is the shorter leg. Call its length x. Then $AT = x\sqrt{3}$.

Substituting, $\tan(30°) = \dfrac{x}{x\sqrt{3}} = \dfrac{1}{\sqrt{3}} = \dfrac{\sqrt{3}}{3}$.

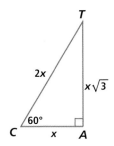

Check For numerical confirmation of this, use a calculator to verify that $\tan(30°) \approx 0.5773502692$ and $\dfrac{\sqrt{3}}{3} \approx 0.5773502692$.

The tangent of an angle is an example of a *trigonometric ratio*. You will study two other trigonometric ratios in the next lesson. With these ratios, you can find all sides and angles of a right triangle whenever you have enough information to satisfy any one of the triangle congruence conditions.

▶ **READING MATH**

The words *trigonometry* and *trigonometric* come from two Greek words: *trigonon,* meaning "triangle" and *metron* meaning "measure."

Questions

COVERING THE IDEAS

1. a. Draw a right triangle with a 25° angle. Measure the sides, and use those measurements to estimate tan(25°).
    b. Use a calculator to estimate tan(25°) to the nearest thousandth.

2. When the Sun is up 50° from the horizon, the wall of a store casts a shadow 28 meters long. To the nearest meter, how high is the wall?

In 3 and 4, use △PQR below. Give the exact value.

3. tan P
4. tan R

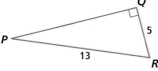

5. Draw your own figure and give the exact value of tan 60°.

6. Draw your own figure and give the exact value of tan 45°.

7. Draw a right triangle with legs of length 3 cm and 8 cm. Use your triangle to estimate answers to the following questions.
    a. What is m∠A if tan $A = \frac{3}{8}$?
    b. What is m∠B if tan $B = \frac{8}{3}$?
    c. Check your answers to Parts a and b with a calculator.

8. a. Draw a right triangle with an acute angle whose tangent is 3.
    b. Use a calculator to determine the measure of that angle to the nearest degree.

9. Can the tangent of an angle be 0.001? If so, what can you say about the angle? If not, why not?

10. Suppose a size change of magnitude 4.5 is applied to △ABC resulting in △A'B'C' (not drawn).
    a. What is tan A?
    b. What is m∠A?

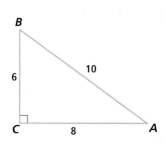

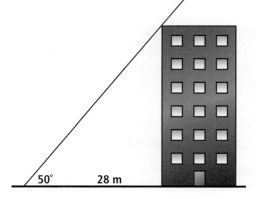

APPLYING THE MATHEMATICS

11. Suppose you are 10 feet from a tall building and your eye level is 5 feet above the ground. If you need to look up 80° to see the top of the building, how tall is the building to the nearest tenth of a foot? (The drawing at the right is to scale.)

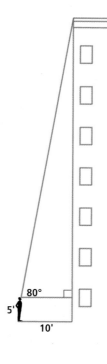

12. In the diagram at the right, O is the center of the circle. $\overline{RS} \perp \overline{PQ}$, $PS = 4$, and $SQ = 9$.
    a. Find RS.
    b. Find $\tan P$.
    c. Find $m\angle P$.

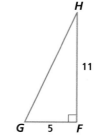

13. Determine the measure of the angle the line with equation $y = \frac{2}{3}x$ makes with the x-axis, to the nearest degree.

14. Refer to right triangle FGH at the right.
    a. Find $\tan G$.
    b. Find $\tan H$.
    c. How are angles G and H related?
    d. How are their tangents related?
    e. Generalize the result of Parts a–d.

15. Use the grid at the right to complete the table below. In this table, $\overrightarrow{PO}$ is the first side of each angle. Find the tangent first and then estimate the measure of the angle using the inverse tangent key on your calculator.

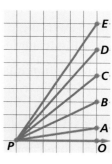

Angle	Tangent	angle measure	slope of 2nd side of angle
∠OPA	?	?	?
∠OPB	?	?	?
∠OPC	?	?	?
∠OPD	?	?	?
∠OPE	?	?	?

16. In the figure at the right, $\overline{CA} \perp \overline{CD}$, $\overline{BC} \perp \overline{DA}$, $AB = 9$ cm, and $BD = 4$ cm. Determine the exact value of

    a. $\tan(D)$.

    b. $\tan(\angle DCB)$.

    c. $\tan(\angle BCA)$.

    d. $\tan(A)$.

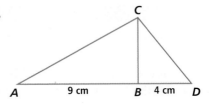

REVIEW

17. a. In the triangle at the right, show that
    $a^2 + b^2 = x^2 + 2xy + y^2$.
    (Lesson 8-6)

    b. Show that $\dfrac{a^2 + b^2 - x^2 - y^2}{2} = h^2$.
    (Lesson 13-3, Previous Course)

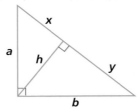

18. State the contrapositive of the Triangle Side-Splitting Theorem. (Lessons 13-1, 11-2)

In 19 and 20, refer to the drawing at the right of a snowman that is built out of three spheres and a cone for a nose. A side view of this snowman is shown in the drawing.

19. If $R' = \frac{2}{3}R$ and $R'' = \frac{2}{3}R'$, find the volume of the snowman in terms of R (without the nose). (Lessons 10-6, 10-2)

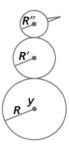

20. Draw a front view and a top view of the snowman. (Lesson 9-5)

21. A 45-45-90 triangle has a hypotenuse of length c. Express the area of the triangle in terms of c. (Lesson 8-7, 8-4)

22. The translation image of the point $(2, -3)$ is $(1, 2)$. Describe the vector for this translation as an ordered pair. (Lesson 4-6)

EXPLORATION

23. Choose three angle measures, other than 90, whose sum is 180. (For example, you could choose 32, 72, and 76.)

    a. Use a calculator to find the sum of the tangents of the measures you have chosen.

    b. Calculate the product of the tangents of the measures you have chosen.

    c. Repeat Parts a and b with a different set of three angle measures.

    d. Make a conjecture based on your results.

QY ANSWERS

1. 0.727

2. 77.32°

Lesson 13-6

The Sine and Cosine Ratios

▶ **BIG IDEA**   The sine and cosine of an angle are trigonometric ratios that, like the tangent, can be utilized to determine segment lengths and angle measures in right triangles.

It is recommended by manufacturers of extension ladders that, when leaning a ladder against a vertical structure, the measure of the angle the ladder makes with the ground not exceed 75°. If a 24-foot ladder is leaned against the side of a building, what is the maximum safe height that can be reached on the building?

In this situation, shown in the drawing at the right, the measure of an acute angle of a right triangle is known and so is the hypotenuse. But neither leg is given, so the tangent of the angle is not helpful. Instead, we need a ratio that compares a leg of a right triangle to its hypotenuse. Two such ratios are commonly used, the sine and the cosine.

Mental Math

GARDN is a regular pentagon.

a. How many diagonals does *GARDN* have?

b. What is the measure of an exterior angle of *GARDN*?

c. What is the measure of an interior angle of *GARDN*?

Definition of Sine, Cosine

Let △*ABC* be a right triangle with right angle *C*.

The **sine of ∠A**, written **sin A** or **sin(A)**, is
$$\frac{\text{length of leg opposite } \angle A}{\text{length of hypotenuse}}.$$

The **cosine of ∠A**, written **cos A** or **cos(A)**, is
$$\frac{\text{length of leg adjacent to } \angle A}{\text{length of hypotenuse}}.$$

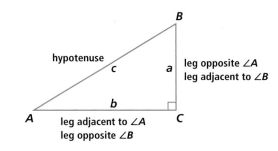

In the triangle at the right, $\sin A = \frac{a}{c}$ and $\cos A = \frac{b}{c}$.

In any other right triangle with an acute angle congruent to ∠*A*, the sine and cosine ratios of that angle will have these same values because of the AA Similarity Theorem.

Example 1

Right triangle *PAL* has side lengths as indicated. Find each value.

a. sin *P*   b. cos *P*   c. sin *L*   d. cos *L*

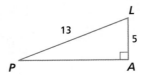

Solution

a. $\sin P = \dfrac{\text{length of leg opposite } \angle A}{\text{length of hypotenuse}} = \dfrac{5}{13}$

b. For the cosine, you need the length of $\overline{PA}$, the side adjacent to $\angle P$.
   Use the Pythagorean Theorem to find the length.

$$PA^2 + 5^2 = 13^2$$
$$PA^2 + 25 = 169$$
$$PA^2 = 144$$
$$PA = \sqrt{144} = 12$$

So $\cos P = \dfrac{\text{length of leg adjacent to } \angle P}{\text{length of hypotenuse}} = \dfrac{12}{13}$

c. $\sin L = \dfrac{\text{length of leg opposite to } \angle L}{\text{length of hypotenuse}} = \dfrac{12}{13}$

d. $\cos L = \dfrac{\text{length of leg adjacent to } \angle L}{\text{length of hypotenuse}} = \dfrac{5}{13}$

In Example 1, notice that sin *P* = cos *L* because the leg opposite $\angle P$ is the leg adjacent to $\angle L$. Also remember that angles *P* and *L* are complementary. So the cosine of an angle is the sine of its complement. This is the origin of the term "cosine"; *cosine* is short for "*co*mplement's *sine.*"

Finding Unknown Lengths Using the Sine and Cosine

The sine and cosine ratios have a great number and variety of applications, and a long history. The ancient Babylonians and Greeks measured the sides and angles of triangles carefully, needing such measurements for navigation, surveying, and astronomy. The first table of trigonometric values was constructed by Claudius Ptolemy in the second century CE. Values like our present-day sine, cosine, and tangent values were first obtained by the German astronomer Regiomontanus in the 1400s. The abbreviations *sin, cos,* and *tan* were first used by Euler in the 1700s.

With the sine, you can answer the question that began this lesson.

Example 2

What is the maximum safe height of the ladder in the first paragraph of the lesson?

Solution The height will be maximized when $m\angle A = 75°$. In the figure at the right, BC is the desired height.

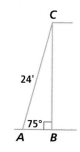

$\overline{BC}$ is the side opposite $\angle A$ and the hypotenuse is known to be 24 feet, so use the sine ratio.

$$\sin(75°) = \frac{\text{length of leg opposite } \angle A}{\text{length of hypotenuse}} = \frac{BC}{24}$$

Thus, $\qquad BC = 24 \sin(75°)$.

Using a calculator, $\qquad BC \approx 23.18221983$.

The ladder can safely reach a height of about 23 feet on the building.

The measure of an acute angle can be found if the value of its sine or cosine is known. The process is similar to that used for the tangent in Lesson 13-5.

Example 3

Suppose a person has a 5-meter ladder and places it so that the foot of the ladder is 1 meter from the base of a vertical wall. What angle does the ladder make with the ground?

Solution Draw a picture. The side adjacent to $\angle A$ has a known length, as does the hypotenuse, which suggests using the cosine ratio.

$$\cos A = \frac{1}{5}$$

Find the inverse cosine command ([cos^{-1}] or an equivalent) on your calculator.

$$m\angle A = \cos^{-1}\left(\frac{1}{5}\right) \approx 78.46°$$

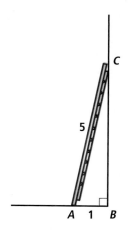

Angles of Depression and Elevation

The **angle of depression** is the angle that the line of sight makes with a horizontal when looking down at something. Likewise, the **angle of elevation** is the angle the line of sight makes with the horizontal when looking up at something.

> ▶ **READING MATH**
>
> To *elevate* literally means to "raise to a higher level." An *elevator* takes one up to a higher level. To *depress* literally means to "push to a lower level." A tongue depressor moves a tongue down.

Notice that the angle of elevation and the angle of depression for the same situation are alternate interior angles, so they will always have the same measure because of the Parallel Lines Theorem.

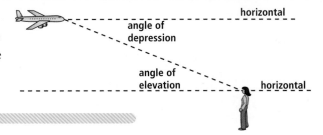

Example 4

A police helicopter is providing protection for an important person. The helicopter needs to be far away so as to remain unnoticed, yet close enough to respond if needed. The helicopter is 1500 feet above ground and sites the person at an angle of depression of 24°. How far is the person from the point on the ground directly below the helicopter?

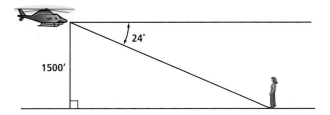

Solution Because the angle of depression has the same measure as the angle of elevation, the triangle formed by the person, the helicopter, and the point on the ground is a right triangle with height 1500 ft and an angle of 24° opposite that side. Because the side we know is opposite the known angle and the side we want is adjacent, the appropriate trigonometric ratio is the tangent.

$$\tan(24°) = \frac{1500}{x}$$

$$x \cdot \tan(24°) = 1500$$

$$x = \frac{1500}{\tan(24°)} \approx 3369 \text{ feet}$$

The person is about 3400 feet away.

To select the appropriate trigonometric ratio to use in a problem, focus on the unknown side, the known acute angle, and the known side. Those determine the appropriate ratio. If you need to find an angle measure, focus on the two known sides and their relationship to the angle, thinking about *opposite, adjacent,* and *hypotenuse.* There is often more than one way to solve these problems.

Exact Values of Sines and Cosines

When the lengths of sides are known exactly, as in special right triangles, you can find exact values for the sine and cosine of an angle.

Example 5

Find the exact values of $\sin(60°)$ and $\cos(60°)$.

Solution Sketch a 30-60-90 triangle as shown at the right.

The legs have lengths x and $x\sqrt{3}$, while the hypotenuse has length $2x$.

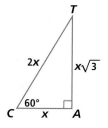

$$\sin(60°) = \frac{\text{length of leg opposite } 60° \text{ angle}}{\text{length of hypotenuse}} = \frac{x\sqrt{3}}{2x} = \frac{\sqrt{3}}{2}$$

$$\cos(60°) = \frac{x}{2x} = \frac{1}{2}$$

Check For numerical confirmation of this, use your calculator to find:

$\sin(60°) \approx 0.8660254038$ and $\frac{\sqrt{3}}{2} \approx 0.8660254038$.

$\cos(60°) = 0.5$ and $\frac{1}{2} = 0.5$.

Questions

COVERING THE IDEAS

1. Use $\triangle NBC$ at the right to find an exact value for
   a. $\sin C$.
   b. $\cos N$.
   c. $\tan N$.

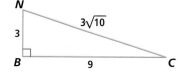

2. Explain why $\sin B = \sin E$ in the triangles at the right.

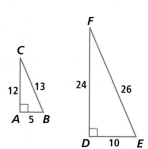

In 3–6, use trigonometry to find x to the nearest hundredth.

3.

4.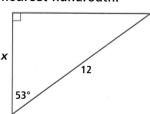

5.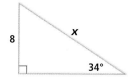

6.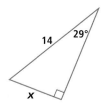

In 7 and 8, consider a 15-foot ladder with the same safety restrictions as in Example 2. Answer to the nearest inch.

7. How far up on the vertical wall can the ladder safely reach?

8. How far, at a minimum, should the bottom of the ladder be from the wall?

9. Explain why the sine of an acute angle of a right triangle is equal to the cosine of its complement.

10. Explain why the sine of an acute angle of a right triangle can never be greater than 1.

In 11−14, find the exact value without a calculator.

11. sin 30°   12. cos 30°   13. sin 45°   14. cos 45°

APPLYING THE MATHEMATICS

15. Suppose an airplane is about to land and will approach an airport at a 3° angle of depression. If the plane must descend from 35,000 ft, find the distance x along the ground between the airport P and the point Q directly below the plane at the time the pilot needs to begin descending. Give your answer to the nearest hundred feet.

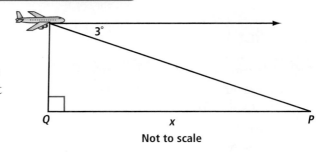

Not to scale

16. An escalator is being built in a new store. If you were standing at the top of the escalator, the angle of depression to the bottom of the escalator is 34°. If the vertical distance between the floors is 5.5 meters, how long is the escalator, to the nearest tenth of a meter?

17. A rhombus has side length 32 cm, and one of its angles has measure 50°.
    a. Find the exact length of the shorter diagonal. (Hint: The exact length will involve a trigonometric ratio.)
    b. Find the exact area of the rhombus.
    c. Estimate the area to the nearest square centimeter.

18. Explain why the cosine of an angle decreases in value as an angle increases its measure between 0° and 90°.

19. In the figure at the right, O and N are midpoints and $\angle L$ is a right angle. Explain why sin K = sin N.

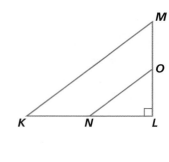

20. Find the measures of the base angles of the isosceles trapezoid at the right, to the nearest degree.

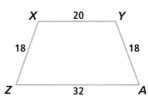

21. **a.** Of angles 1, 2, and 3 pictured at the right, which angle has the largest sine?

    b. Which angle has the largest cosine?

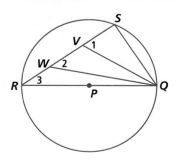

22. Recall from Question 13 of Lesson 13-4 that if a side of a regular pentagon has length 1, then a diagonal of that pentagon has length $\phi = \dfrac{1 + \sqrt{5}}{2}$. Use this fact to explain why $\sin 54° = \dfrac{\phi}{2}$.

REVIEW

23. A plane takes off at an angle of 15°. After traveling a horizontal distance of 300 meters, how high is the plane, to the nearest meter? **(Lesson 13-5)**

24. The Leaning Tower of Pisa is famous for being at a slant, due to the fact that it was built with a poor foundation. The top of the tower is 55.86 meters high. A stone dropped from a height of 47 meters lands 3.5 meters from the base. Find the angle at which the tower leans, to the nearest degree. **(Lesson 13-5)**

25. The circle below has center $(0, 0)$.

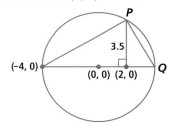

    a. Find the coordinates of P and Q. **(Lesson 11-4)**

    b. Find an equation for the circle. **(Lesson 11-6)**

26. Chantel puts a drawing in a copy machine, and enlarges it by a factor of k. She then takes the enlarged drawing, and enlarges it again by a factor of k. She continues doing this. After five enlargements, the area of the final image is how many times the area of the original image? **(Lesson 12-4)**

27. Define *vector*. **(Lesson 4-6)**

EXPLORATION

28. Choose any 10 angle measures between 0 and 90.

    a. For each measure x, calculate $(\sin x)^2 + (\cos x)^2$. Make a reasonable conjecture based on your findings.

    b. Try to prove your conjecture by using the definitions in this lesson and algebra.

Chapter 13 Projects

1 The Golden Ratio and Music

The great violin makers Stradivari and Baginsky used the golden ratio to construct violins. Composers like Mozart, Beethoven, Bartók, Debussy, Schubert, and Bach used the golden ratio when composing their music. Write a report about how the golden ratio is used in music or compose a short piece of music that uses the golden ratio. Share what you've done with your class.

2 The Geometric Mean and Financial Reporting

It is the geometric mean, not the arithmetic mean, that tells you the average financial rate of return when looking at investments. Research why it is used instead of the arithmetic mean and how is it calculated to find the rate of return. Give some examples of its use.

3 Similar Airplanes?

Boeing's 700 Series airplanes were lined up on the taxiway as shown in the picture. The planes were the 707, 717, 727, 737, 747, 757, 767, and 777. If the company builds a series of planes, does this mean that the planes in that series are similar? Explain and give data to support your conclusion(s).

4 The Golden Ratio in Nature

The golden ratio appears in the arrangement of certain features of many plants. In particular, there is a relationship between the arrangement of seeds in a sunflower and the golden ratio.

a. You have probably seen the Fibonacci numbers in the past. Use the Internet or any other source to write a short description of these numbers, and their relationship to the golden ratio.

b. Use the Internet to find images of several plants whose shape have to do with the golden ratio. Display these images, together with an explanation of how the golden ratio fits into the arrangement.

c. Find out why some scientists believe that the golden ratio appears so often in these plants.

5 Regular Polygons and Unit Circles

Each circle below has radius 1 unit. At the left, the regular hexagon is said to be *circumscribed about* the circle. At the right, the regular octagon is *inscribed in* the circle. There are formulas for the area and perimeter of circumscribed and inscribed regular polygons in terms of the tangent and sine ratios.

circumscribed regular hexagon

Perimeter $= 2n \cdot \tan\left(\frac{180°}{n}\right)$ units

Area $= n \cdot \tan\left(\frac{180°}{n}\right)$ units^2

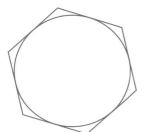 

inscribed regular octagon

Perimeter $= 2n \cdot \sin\left(\frac{180°}{n}\right)$ units

Area $= n \cdot \sin\left(\frac{180°}{n}\right) \cdot \cos\left(\frac{180°}{n}\right)$ units^2

a. Show that these formulas work when the regular n-gon is a square.

b. Prove any two of the formulas.

6 NASA and Big Flags

NASA (the National Aeronautics and Space Administration) painted a flag and a Bicentenntial Emblem on the front of the Vehicle Assembly Building (VAB) in 1976. When you notice the palm trees in front of the building, you can get an idea of how big it is. It took 6000 gallons of paint to paint this flag on the building. How big is this flag? If a flag was 9 feet long (standard size for a 30-foot flagpole), find the scale of the standard flag to the one on the VAB and determine how many gallons of paint it would take to paint a flag 9 feet long.

Chapter 13 Summary and Vocabulary

○ In Chapter 12, you were able to use what you knew about lengths in one of two similar figures to learn about lengths in the other figure. In this chapter, you saw ways in which the similarity can be used to study the relationships among measures of sides within a single figure.

○ The main ideas and theorems in this chapter come out of five situations in which similar figures appear. The five situations are shown below. In situations I and IV, the similar triangles overlap but still are easy to see. In situations II and III, you may find it more difficult to see the similar figures because they are not tilted in the same direction. In situation V, the original figure does not contain any similar triangles, but an auxiliary segment creates a situation like that of situation I.

Vocabulary

13-3
*geometric mean

13-4
golden rectangle
golden ratio phi (ϕ)

13-5
leg opposite an angle
leg adjacent to an angle
*tangent of an angle, tan A

13-6
*sine of $\angle A$, sin A
*cosine of $\angle A$, cos A
angle of depression
angle of elevation

I   line parallel to side of triangle

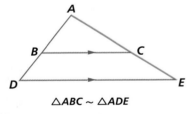

$\triangle ABC \sim \triangle ADE$

II   altitude to hypotenuse of right triangle

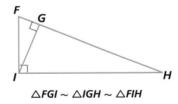

$\triangle FGI \sim \triangle IGH \sim \triangle FIH$

III   rectangle split to form square and similar rectangle

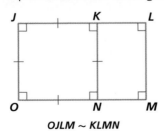

$OJLM \sim KLMN$

IV   line parallel to leg of right triangle

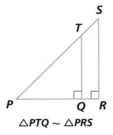

$\triangle PTQ \sim \triangle PRS$

V   angle bisector and line parallel to side

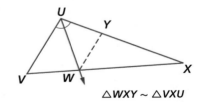

$\triangle WXY \sim \triangle VXU$

◗ Imbedded within these figures are the seeds of important concepts. In situation I are a variety of proportions, one of which we call the **Side-Splitting Theorem** $\frac{AB}{BD} = \frac{AC}{CE}$. That is, a line parallel to a side of a triangle splits the other two sides into segments whose lengths are proportional. The converse of this statement is true, and the same idea applies to trapezoid.

◗ Situation II harbors three examples of **geometric means.** The altitude IG is the geometric mean of the parts FG and GH of the hypotenuse. Either leg is the geometric mean of the hypotenuse and the part of the hypotenuse closest to it.

◗ Situation III leads to the number ϕ, which is called the **golden ratio**. The golden ratio has applications in art, architecture, and biology. Other applications arise from situation IV, in the important area of *trigonometry*. The ratios of sides of a right triangle are called *trigonometric ratios*. These ratios are determined by the measure of an acute angle of the triangle. Three of the ratios are the tangent, sine, and cosine. Using these ratios, you can determine the lengths of sides and measures of angles of any right triangle given sufficient information for a unique triangle. The trigonometric ratios can be roughly estimated using an accurate drawing, estimated to many decimal places with a calculator, or calculated exactly if you know two sides of the right triangle or know one side of a 30-60-90 or 45-45-90 triangle. Situation V shows that lengths of segments can be related in nonobvious ways. Each angle bisector in a triangle splits the opposite side into parts whose lengths are in the same ratio as the other two sides of the triangle.

Postulates, Theorems, and Properties

Triangle Side-Splitting Theorem (p. 774)

Triangle Side-Splitting Converse Theorem (p. 776)

Trapezoid Side-Splitting Theorem (p. 777)

Angle Bisector Theorem (p. 782)

Geometric Mean Theorem (p. 787)

Right-Triangle Altitude Theorem (p. 788)

Rectangle Similarity Theorem (p. 794)

Chapter

13 Self-Test

Take this test as you would take a test in class. You will need a calculator. Then use the Selected Answers section in the back of the book to check your work.

1. In the figure at the right, $\overline{DE} \parallel \overline{AC}$. Find AD.

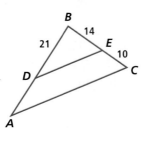

2. Suppose $\triangle ABC$ is a right triangle with a right angle at A, $BC = 18$, and $AC = 5$. Find tan C.

3. In the figure at the right, $XYWZ$ is a rectangle.

   a. Calculate YW.

   b. Calculate XY.

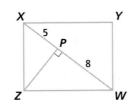

4. **True or False** If $\triangle MNO$ is a right triangle with m$\angle M = 90$ and $\triangle MNO \sim \triangle OPQ$, then sin $P =$ sin N. Explain your answer.

5. In the diagram at the right, $\overline{BD}$ bisects $\angle ABC$ and $AC = 15$. Find DC.

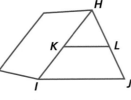

6. To the nearest tenth, find the lengths of the legs of a right triangle that has a $37°$ angle and a hypotenuse 12 in. long.

7. The beam $\overline{KL}$ is parallel to $\overline{IJ}$ and splits the sides of an asymmetric roof. Suppose $HK = 31$, $HL = 28$, and $KI = 23$. Find LJ.

8. A rectangle is 4 cm by 5 cm. A second rectangle has a side length of 12 cm. Give dimensions of the second rectangle so that the two rectangles are similar.

9. **Multiple Choice** Keanu thinks that golden rectangles are the most aesthetic of all rectangles. Which rectangular frame would Keanu find the most visually pleasing?

   A 20 cm by 32 cm   B 20 cm by 30 cm

   C 30 cm by 40 cm   D 40 cm by 50 cm

10. Mahala is standing on an overpass with her eyes 30 feet above the surface of the road. When she first spots a car coming toward the bridge, the angle of depression to the car is $3°$. Fifteen seconds later, the car has passed under her and is traveling away from her, and the angle of depression is again $3°$. To the nearest foot, how far has the car traveled in 15 seconds?

11. To the nearest inch, how far up a vertical wall will a 17-foot ladder reach if the angle it makes with the ground is $70°$?

12. Find the measure of an angle V, to the nearest hundredth of a degree, such that cos $V = 0.32$.

13. In the figure at the right, which numbered angle has the greatest sine, and why?

14. $LMNO$ and $OLPQ$ are similar rectangles. Find x.

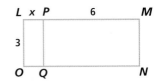

Chapter 13

Chapter Review

SKILLS Procedures used to get answers

OBJECTIVE A Find lengths in figures by applying the Side-Splitting Theorems and Side-Splitting Converse Theorem. (Lesson 13-1)

1. In the figure below $\overline{BE} \parallel \overline{CD}$. Find the following lengths to the nearest tenth.
   a. AB b. BE

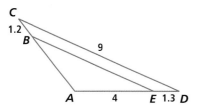

In 2 and 3, is $\overline{ST} \parallel \overline{UV}$? Why or why not?

2.

3.

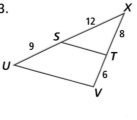

4. Name all pairs of parallel segments in the figure below.

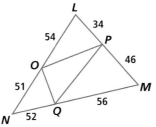

5. $GHJI$ is a trapezoid with $\overline{GH} \parallel \overline{KL}$. Determine GK.

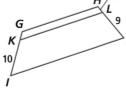

OBJECTIVE B Calculate lengths using the Angle Bisector Theorem. (Lesson 13-2)

In 6–9, $\overline{ZR}$ bisects $\angle PZF$.

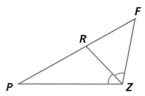

6. If $PZ = 12$, $ZF = 8$, and $RF = 4$, find PR.
7. If $\frac{PR}{RF} = \frac{7}{5}$ and $ZF = 25$, find PZ.
8. The perimeter of $\triangle PZF$ is 50 cm. If $PZ = 18$ cm and $ZF = 12$ cm, find RF.
9. If $PZ = 7$, $ZF = 5$, and $RF = 4$, find PF.

OBJECTIVE C Calculate lengths using the Right-Triangle Altitude Theorem. (Lesson 13-3)

In 10–13, refer to the figure below.

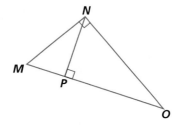

10. Suppose $MP = 4$ and $PO = 13$. Find NP.
11. Suppose $MP = 16$ and $PO = 24$. Find NO.
12. Suppose $MN = 7$ and $MO = 12$. Find PM.
13. Suppose $NP = t$ and $NO = 2t$. Find as many other lengths as you can.

OBJECTIVE D Determine sines, cosines, and tangents of angles. (Lessons 13-5, 13-6)

In 14–16, use a ruler and the figure at the right to determine the given values.

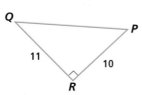

14. $\tan(S)$

15. $\cos(R)$

16. $m\angle R$

17. Use $\triangle PQR$, to find each ratio, to the nearest thousandth.

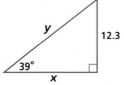

   a. $\sin P$ **b.** $\cos P$ **c.** $\tan P$

In 18–19 give exact values.

18. **a.** $\sin 30°$ 19. **a.** $\cos 45°$

   b. $\tan 30°$ **b.** $\tan 45°$

In 20 and 21, estimate to the nearest thousandth.

20. $\sin(53.2°)$ 21. $\cos(88.9°)$

OBJECTIVE E Use the trigonometric ratios to find lengths and angle measures (Lessons 13-5, 13-6)

22. In the triangle at the right, find x and y to the nearest tenth.

23. Find the measures of the acute angles in a 5-12-13 right triangle, to the nearest hundredth of a degree.

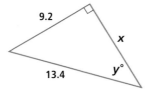

24. In the triangle at the right, find x and y to the nearest hundredth.

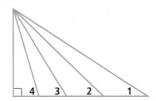

25. An isosceles triangle has congruent sides 12 inches long and a base 9 inches long. To the nearest tenth,

   a. find the altitude to the base.

   b. find the measure of the vertex angle.

26. A regular octagon with sides of length 8 cm is inscribed in a circle. Find the radius of the circle to the nearest centimeter.

PROPERTIES Principles behind the mathematics

OBJECTIVE F Know the definitions of sine, cosine, and tangent. (Lessons 13-5, 13-6)

In 27–29, define the trigonometric ratio if $\triangle XYZ$ is a right triangle with right angle X.

27. $\cos Y$ 28. $\sin Y$ 29. $\tan Z$

30. For $\triangle ABC$ with right angle A, give a numerical value for $\tan B \cdot \tan C$.

In 31 and 32, use the figure below. Choose from the numbered angles.

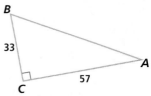

31. Which angle has the largest sine?

32. Which angle has the largest tangent?

33. In $\triangle BCA$, which is the greatest: $\sin A$, $\cos A$, or $\tan A$?

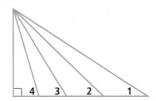

OBJECTIVE G Determine when two rectangles are similar. (Lesson 13-4)

In 34–37, $ABCD$ is a rectangle with $AB = t$ and $BC = 3t$. Determine if the given rectangle is similar to $ABCD$.

34. a rectangle with one side of length 7 and one side of length 21

35. the image of $ABCD$ under a composition of a size transformation and a rotation

36. one of the rectangles created by cutting $ABCD$ into two rectangles by connecting the midpoints of $\overline{AB}$ and $\overline{CD}$

37. a square with the same area as $ABCD$

USES Applications of mathematics in real-world situations

OBJECTIVE H Use the Side-Splitting Theorems to find lengths and distances in real situations. (Lesson 13-1)

38. At a point 1.5 meters from its bottom, a slide is 1.3 meters tall. If the slide runs for 4 meters along the ground, what is its height at its highest point? (Round your answer to the nearest tenth of a meter.)

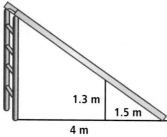

1.3 m
1.5 m
4 m

39. Paris Avenue, Lyon Avenue and Marseilles Avenue are east-to-west streets. They are crossed by Vienna Street and Salzburg Street, as shown below. Find the distance from Marseilles to Lyons Avenues, along Vienna Street.

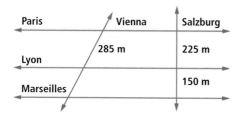

Paris   Vienna   Salzburg
               285 m   225 m
Lyon
                          150 m
Marseilles

OBJECTIVE I Recognize the golden ratio in real-life situations. (Lesson 13-4)

40. This figure is an illustration from Leonardo Da Vinci's *De Divina Proportione*. Several of the rectangles in the figure are golden rectangles. Trace the rectangles (without tracing the face), and mark 3 that appear to be golden rectangles.

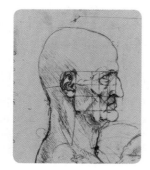

41. Suppose you had 60 1 × 1 square stickers, and wished to arrange them next to each other in the shape of a rectangle that is as close as possible to a golden rectangle. What would the dimensions of this rectangle be?

OBJECTIVE J Use sines, cosines, and tangents to determine unknown lengths in real situations. (Lessons 13-5, 13-6)

42. How far up a vertical wall can a 3.5-meter ladder reach, if the angle it forms with the ground is 80°? Answer to the nearest centimeter.

43. A woman is 25 feet away from a 45-foot tall building. Looking up from 5 feet above the ground, she can just see the top of the building. At what angle above the ground is she looking?

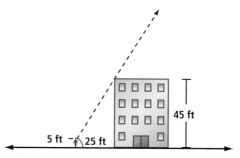

45 ft
5 ft   25 ft

44. An airplane takes off at an angle of 20°, and travels a horizontal distance of 750 meters. How high is it above the ground?

45. When the Sun is at an angle of 80° above the horizon, Shawn casts a shadow 30 cm long. How tall is Shawn?

REPRESENTATIONS Pictures, graphs, or objects that illustrate concepts

There are no representation objectives for this chapter.

Chapter

14 Further Work with Circles

▶ **Contents**

14-1 Chord Length and Arc Measure

14-2 Regular Polygons and Schedules

14-3 Angles Formed by Chords or Secants

14-4 Tangents to Circles and Spheres

14-5 Angles Formed by Tangents and a General Theorem

14-6 Three Circles Associated with a Triangle

14-7 Lengths of Chords, Secants, and Tangents

14-8 The Isoperimetric Inequality

14-9 The Isoperimetric Inequality in Three Dimensions

Why would Galileo mention only triangles and circles from all geometric figures possible? Obviously he felt that there was something very special about them. We have already discussed that triangles are rigid, thus making them important in structures. You have seen how the areas of polygons can be found from the areas of triangles. And you have seen how triangles are basic to the study of congruence and similarity of figures. But why circles?

Circles are basic to the study of distance because all points in a plane at a fixed distance from a point lie on a circle. Circles occur naturally. Drop a stone in water and watch the ripples go out from the point of contact. Because the ripples expand at a constant rate, they form concentric circles.

The great Italian scientist Galileo Galilei wrote in 1623:

> . . . the universe . . . is written in the language of mathematics, and its characters are triangles, circles, and other geometrical figures, without which it is humanly impossible to understand a single word of it; without these, one is wandering about in a dark labyrinth.

Many fruits like grapes, oranges, and apples, naturally resemble spheres, while others like watermelons and cucumbers, resemble cylinders. These all have circular cross sections. Also, the cross sections of flower stems, tree branches, and bones are circular, because they grow at the same rate in every direction perpendicular to their axis. This helps prevent plants and trees from being weak in any direction and makes them less likely to topple due to wind.

Many manufactured objects, such as wheels, hoses, rockets, balls, camera lenses and writing pens, also have circular cross sections. It is generally true that containers for liquids, like water bottles, soda cans, or drinking glasses are cylinders and so have circular cross sections.

In this chapter, some of the more important and beautiful properties of circles are deduced and applied. The results are fascinating and often quite surprising.

Lesson

14-1 Chord Length and Arc Measure

▶ **BIG IDEA** Properties of chords in circles follow from properties of isosceles triangles. Knowing any two of the length of a chord, the measure of its arc, and the radius of the circle, you can determine the third.

The Flower Fountain in Centennial Park in Holland, Michigan, is protected by an iron fence that is in the shape of a regular 16-gon. Recall that the vertices of a regular polygon all lie on a single circle. The sides of that polygon are *chords* of the circle. In general, a segment $\overline{DB}$ is called the **chord of arc** $\overset{\frown}{DB}$. If the center of the circle is C, then $\angle BCD$ is called the **central angle of the chord** $\overline{DB}$.

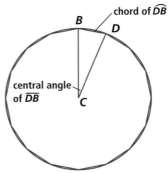

Mental Math

$\overline{BE}$ is a diameter of circle C.

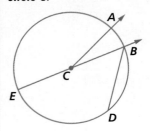

a. What kind of angle is $\angle ACB$?

b. What kind of angle is $\angle EBD$?

c. How does m$\angle ACB$ compare to m$\overset{\frown}{AB}$?

d. How does m$\angle EBD$ compare to m$\overset{\frown}{ED}$?

e. Which of $\overset{\frown}{AB}$ and $\overset{\frown}{BDA}$ is the minor arc? Which the major arc?

Congruent Circles and Congruent Arcs

If two circles $\odot X$ and $\odot Y$ have equal radii, then $\odot X$ can be mapped onto $\odot Y$ by the translation vector $\overrightarrow{XY}$. So $\odot X$ is congruent to $\odot Y$.

If two circles do not have equal radii, no isometry will map one onto the other, since isometries preserve distance. These arguments prove that *circles are congruent if and only if their radii have the same length.*

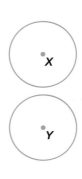

Suppose two arcs $\overset{\frown}{AB}$ and $\overset{\frown}{CD}$ have the same measure, as in the circle at the right. Then you can rotate $\overset{\frown}{AB}$ about O by the measure of $\angle AOC$ to the position of $\overset{\frown}{CD}$. Then the chord $\overline{AB}$ rotates to $\overline{CD}$, so $\overline{AB} \cong \overline{CD}$. Thus, in a circle, arcs of the same measure are congruent and have congruent chords. This proves Part a of the next theorem. The proof of Part b is left for you as Question 17.

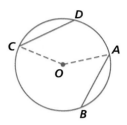

Arc-Chord Congruence Theorem

In a circle or in congruent circles:

a. If two arcs have the same measure, they are congruent and their chords are congruent.

b. If two chords have the same length, their minor arcs have the same measure.

The Arc-Chord Congruence Theorem applies only in a single circle or in circles with the same radii. In circles with different radii, arcs of the same measure cannot be congruent, nor can their chords be congruent. Even so, in those circles, arcs with the same measure are *similar*, because one can be mapped onto the other by a composite of a translation, a size change, and a rotation. *All circular arcs with the same measure are similar.*

Properties of Chords

When an arc of a circle is not a semicircle, the radii drawn to the endpoints of the arc and the arc's corresponding chord form an isosceles triangle. Recall from Lesson 6-2 that, in an isosceles triangle, the bisector of the vertex angle, the perpendicular bisector of the base, the altitude from the vertex, and the median from the vertex all lie on the same line. In the language of circles and chords, this leads to the following theorem.

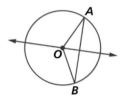

Chord-Center Theorem

1. The line containing the center of a circle perpendicular to a chord bisects the chord.

2. The line containing the center of a circle and the midpoint of a chord bisects the central angle of the chord.

3. The bisector of the central angle of a chord is perpendicular to the chord and bisects the chord.

4. The perpendicular bisector of a chord of a circle contains the center of the circle.

As you know, the distance from a point to a line is the length of the perpendicular segment from the point to the line. By applying the Chord-Center Theorem and the Pythagorean Theorem, if you know any two of the radius of the circle, the length of a chord, or the distance of the chord from the center of the circle, you can find the third.

Example 1

A chord 13 cm long lies in a circle of radius 10 cm. How far is the chord from the center of the circle?

Solution Draw a diagram with the given information: a circle with radius 10 cm and a chord that is 13 cm in length. Draw the full isosceles triangle because it helps in thinking about the question. The goal is to find the distance OL from the center O to the chord.

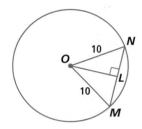

Because $\overline{LO} \perp \overline{NM}$, by Part (1) of the Chord-Center Theorem, $\overline{LO}$ bisects $\overline{MN}$. This means $ML = LN = 6.5$. You now know the leg and the hypotenuse of a right triangle. By the Pythagorean Theorem, $6.5^2 + OL^2 = 10^2$, from which $OL^2 = 57.75$, and so $OL = \sqrt{57.75} \approx 7.59934$. The chord is about 7.60 cm from the center of the circle.

How to Find the Length of a Chord

Given the measure of an arc and the radius of the circle, the length of its corresponding chord can always be found using trigonometry. Now we return to the iron fence around the Flower Fountain.

GUIDED

Example 2

Colleen would like to hang a string of lights on the fence around the Flower Fountain. The fence is a 16-gon inscribed in a circle of radius 15 ft. How long a string of lights will she need to go once around the fountain?

Solution A picture is drawn at the right. Chord $\overline{AB}$ is one side of the fence. Multiplying this length by 16 yields the total length of the string of lights needed.

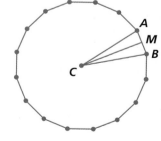

$\triangle ACB$ is isosceles with vertex angle C. $\overset{\frown}{AB}$ is $\frac{1}{16}$ of the circle, so

$m\overset{\frown}{AB} = m\angle ACB = \dfrac{?}{16}^{\circ}$.

Let M be the midpoint of $\overline{AB}$. Then, by Parts 2 and 3 of the Chord-Center Theorem, $\overline{CM}$ bisects ___?___. Also, ___?___ is a right triangle. So $m\angle MCA = \frac{1}{2} \cdot \dfrac{?}{16}^{\circ} = 11.25^{\circ}$.

To find AM, use trigonometry.

$$\sin(11.25)° \approx \frac{AM}{?}$$

$$AM \approx \underline{\ ?\ } \cdot \sin(11.25)°$$

Because $AM = \frac{1}{2}AB$, $AB = 2 \cdot AM \approx 2 \cdot \underline{\ ?\ } = \underline{\ ?\ }$.

She needs $16 \cdot \underline{\ ?\ } = 480 \cdot \sin(11.25)° \approx 93.6$ ft of rope.

If the measure of an arc is a multiple of 30 or 45, it is possible to find the exact length using special right triangles and without trigonometry.

Example 3

In a circle with a radius of 25 centimeters, find the length of a chord of an arc with the given measure.

a. 60°   b. 90°   c. 120°

Solution Always draw a picture.

a. △AOB is an isosceles triangle with a 60° vertex angle, so it is equilateral. Therefore, AB = 25 cm.

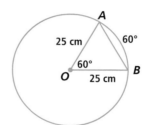

b. Because m∠COD = 90, △COD is an isosceles right triangle. So $CD = 25\sqrt{2}$ cm, or about 35.36 cm.

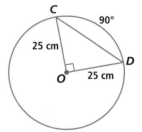

c. Because m∠FOE = 120, m∠F = 30 and m∠E = 30. Now, draw the altitude $\overline{OG}$ to the base of △OEF. The two triangles formed are 30-60-90 triangles.

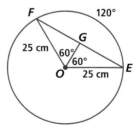

$OE = 2 \cdot OG$, so $OG = \frac{25}{2}$ cm.

$GE = \sqrt{3} \cdot OG$, so $GE = \frac{25}{2}\sqrt{3}$ cm.

Therefore, $FE = 2 \cdot GE = 25\sqrt{3}$ cm ≈ 43.30 cm.

 QY

▶ QY

What is the perimeter of an equilateral triangle whose vertices are on a circle with radius 25 in.?

Questions

COVERING THE IDEAS

In 1–4, use the diagram at the right.

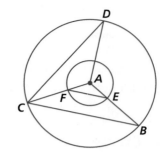

1. Name three chords.
2. **True or False** If $m\overset{\frown}{CB} = m\overset{\frown}{CD}$, then $\overline{CB} \cong \overline{CD}$.
3. **True or False** If $m\overset{\frown}{CB} = m\overset{\frown}{CD}$, then $m\overset{\frown}{FE} = m\overset{\frown}{CD}$.
4. $m\overset{\frown}{CB} = m\overset{\frown}{FE}$, yet $\overline{CB}$ is not congruent to $\overline{FE}$. Explain why.

5. **Multiple Choice** Two circles are congruent if and only if
   A they have the same center.
   B they have radii equal in length.
   C they have arcs of the same length.
   D they have chords of the same length.

Fill in the Blanks In 6–9, use the diagram at the right.

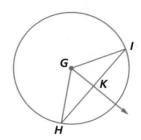

6. If $\overline{GK} \perp \overline{HI}$ and $HK = 6$ cm, then $HI =$ __?__.
7. If K is the midpoint of $\overline{HI}$ and $m\angle HGI = 140$, then
   $m\angle HGK =$ __?__.
8. If $\overrightarrow{GK}$ bisects $\angle HGI$ and $HI = 6$ cm, then $HK =$ __?__ and
   $m\angle GKH =$ __?__.
9. Explain why $\triangle GHI$ is isosceles.

10. A 30-cm chord lies in a circle of radius 17 cm. Find the distance from the center of the circle to the chord.

In 11–14, ⊙O at the right has radius 10 m.

11. Find the exact length of a chord of a 60° arc.
12. Find the exact length of a chord of a 90° arc.
13. Find the exact length of a chord of a 120° arc.
14. Find the approximate length of a chord of a 23° arc to the nearest thousandth.

15. Refer to Example 2. How far off would Colleen have been in her calculation if, to estimate the length of lights needed, she calculated the circumference of the circle rather than the perimeter of the regular polygon?

16. Find the perimeter of a regular 20-gon that is inscribed in a circle of radius 1.

APPLYING THE MATHEMATICS

17. Complete this proof of Part b of the Arc-Chord Congruence Theorem.

    Given $AB = CD$

    Prove $m\overset{\frown}{AB} = m\overset{\frown}{CD}$

    (Hint: The measure of an arc equals the measure of its central angle.)

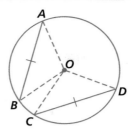

18. Of all the planets, the orbit of Venus is closest to being a circle. Venus makes one full revolution around our Sun every 225 days at a distance of about 67.2 million miles from the Sun. From one Earth day to the next, how far has Venus traveled?

It takes 243 Earth days for Venus to rotate on its axis once, longer than it takes for Venus to orbit the Sun.

19. A circle of radius 10 cm lies on a sphere. Its center is 5 cm from the center of the sphere. Find the radius of the sphere.

20. Find the exact perimeter of an equilateral triangle that is inscribed in the circle with equation $x^2 + y^2 = 100$.

21. To allow access to the sewer level, workmen must climb down a manhole. A workman is concerned about fitting down the hole. If a ladder of width 40 cm is built into a manhole of diameter 61 cm, then what is the maximum distance from the ladder to the edge of the manhole?

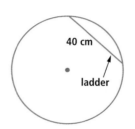

22. Mariana found one of her grandmother's antique plates. The plate is lined with gold border in the shape of a regular decagon. The decagon's vertices are on the edge of the plate. If the diameter of the plate is 8″, find the length of the gold border to the nearest thousandth.

REVIEW

23. The figure below shows a stepladder. If the slant height of the ladder is 13′, what height is the top rung above the ground? (**Lesson 13-6**)

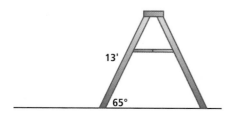

24. Given that $ABCD$ is a rectangle, find m$\angle DBA$ to the nearest tenth of a degree. **(Lesson 13-5)**

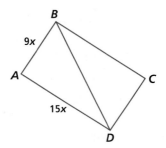

25. A curtain is placed inside a circular auditorium. When viewed from the center of the auditorium, it takes up 64.7° of the viewer's field of vision, as shown in the figure below. What degree of the viewer's field of vision does it take up when viewed from points L, M, and N? **(Lesson 6-3)**

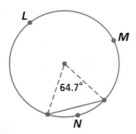

26. **a.** Trace the triangle below, and construct a point that is equidistant from all three of its vertices.

   b. How does the point from Part a relate to the circle in which the triangle can be inscribed? **(Lesson 5-5)**

EXPLORATION

27. Use a DGS. Construct a circle with radius 1. Let $\overline{AC}$ be a diameter of the circle. Place point B on the circle. Move point B from point A to point C. As point B moves, record $x = \text{m}\overarc{AB}$ and $y = $ length of the chord $\overline{AB}$. Graph the points (x, y) and describe the graph.

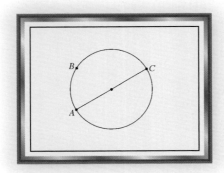

Lesson

14-2

Regular Polygons and Schedules

Vocabulary

round-robin tournament

▶ **BIG IDEA**   An easy, yet not obvious way to schedule teams in a tournament involves chords in a circle.

In this lesson, we show an application of regular polygons and circles to a discrete geometry situation in which points stand for teams.

Mental Math

Name all the types of figures that can be intersections of a right cylinder and a plane.

A Round-Robin Tournament

In many tournaments or leagues, from basketball to chess to bowling, each competitor or team plays all the others. Each time as many teams as possible are playing, the set of matches is called a *round*. When each competitor plays each other competitor exactly once, it is called a **round-robin tournament**. Scheduling a round-robin tournament can be tricky. Suppose there are 7 teams to be scheduled so that each plays the other six. The first thing you might do is number the teams 1 through 7. Because there is an odd number of teams, in each round, one team doesn't play. That team gets a *bye*.

Activity 1

Round-robin schedules can be created in many ways. One example of an incomplete round-robin schedule for seven teams is given. The table lists who each team is playing in each round. For example, since team 1 is paired with team 2 in Round 1, a "2" is placed in the team-1 cell and a "1" is placed in the team-2 cell. Copy the table into your notebook. Complete the table allowing each team to play each other team exactly once.

	Team 1	Team 2	Team 3	Team 4	Team 5	Team 6	Team 7
Round 1	2	1	?	?	?	?	bye
Round 2	3	?	1	?	?	bye	?
Round 3	4	?	?	1	bye	?	?
Round 4	5	?	?	bye	1	?	?
Round 5	6	?	bye	?	?	1	?
Round 6	7	bye	?	?	?	?	1
Round 7	bye	?	?	?	?	?	?

Using Regular Polygons to Schedule an Odd Number of Teams

In Activity 1, you are asked to schedule a round-robin tournament for seven teams. What if there were more teams? It would be nice if there was some algorithm that automatically created the schedule. The algorithm described here is surprising in that it uses properties of regular polygons and circles. The example below is given for 9 teams.

Activity 2

Step 1 Let the 9 teams be represented by vertices of a regular 9-gon (nonagon). By the Center of a Regular Polygon Theorem, we know we can place the vertices equally spaced on a circle for convenience. Copy the figure at the right.

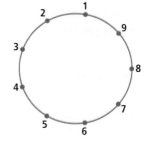

Step 2 (the first round) Draw a chord and all chords parallel to it connecting pairs of numbered points. Because the polygon has an odd number of sides, the minor arcs of each chord pictured have different measures. So no two chords have the same length. The endpoints of these chords are the first round's schedule.

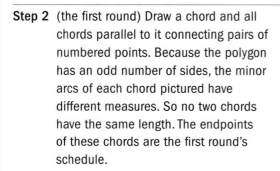

First round: 2-9   3-8   4-7   5-6   1-bye

Because the top chord connects the numbers 2 and 9, team 2 plays team 9 in the first round. This is called a *pairing,* and is written 2-9. Also in the first round, team 3 plays team 8, team 4 plays team 7, team 5 plays team 6, and team 1 gets a bye. The full schedule will be completed when all sides and diagonals of the nonagon have been drawn.

Step 3 (the second round) Rotate the chords $\frac{1}{9}$ of a revolution. For example, the chord pairing 2-9 for the first round rotates into the pairing 1-8 for the second round.

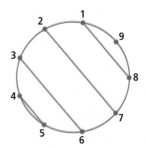

Second round: 1-8   2-7   3-6   4-5   9-bye

Step 4 Continue rotating $\frac{1}{9}$ of a revolution for each round. Because in a round no two chords have the same length, no pairing repeats. (For example, look at the chord that forms the pair 3-8 in round 1. No other chord has the same length, so no other chord will pair team 3 with team 8.)

In a total of nine rounds, the schedule is complete. We leave the 3rd through 9th rounds for you to figure out.

Third round:	9-7	1-6	2-5	3-4	?
Fourth round:	?	?	?	?	?
Fifth round:	?	?	?	?	?
Sixth round:	?	?	?	?	?
Seventh round:	?	?	?	?	?
Eighth round:	?	?	?	?	?
Ninth round:	?	?	?	?	?

In a nonagon, there are 27 diagonals and 9 sides, making a total of 36 segments. Each diagonal can be placed in one of three groups based on its length.

With 9 teams there are 4 games for each of 9 rounds, or 36 pairings. Of these, 9 are sides of the regular nonagon, 9 are congruent shortest diagonals, 9 are congruent middle-length diagonals, and 9 are congruent longest diagonals. Thus, the number of pairings in a round robin with n teams is equal to n plus the number of diagonals.

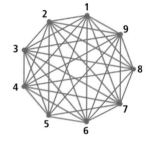

How would you determine the number of diagonals in an n-gon? If n is small, counting the diagonals would be an easy task. But what if n was a large number? Activity 3 leads you to an algebraic expression for the number of diagonals of an n-gon.

Activity 3

The diagram at the right shows that, from a given vertex of a hexagon, it is possible to draw 3 diagonals.

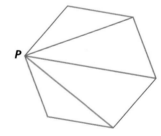

Step 1 Complete the table below by sketching each polygon and counting. We have done the hexagon for you.

number of sides	4	5	6	7	8	...	n
number of vertices	?	?	?	?	?		?
number of diagonals from each vertex	?	?	?	?	?	...	?

Step 2 Why is the number of diagonals from each vertex always 3 fewer than the number of vertices?

Step 3 Create an expression for the *total* number of diagonals that can be drawn in the polygon from all of the vertices.

Step 4 Use your expression to calculate the total number of diagonals in a hexagon. Then count the diagonals and verify that your expression gave you the correct number.

An expression that you might have developed in Activity 3 is stated in the theorem below.

> ### Diagonals in a Polygon Theorem
>
> A polygon with n sides has $\dfrac{n(n-3)}{2}$ diagonals.

Proof In a polygon with n sides, there are n vertices. From each vertex, $(n-3)$ diagonals can be drawn. The product $n(n-3)$ is divided by 2 because the product counts every diagonal twice.

 QY

> ▶ **QY**
>
> How many diagonals does a 27-gon have?

Example

A polygon has 20 diagonals. How many sides does it have?

Solution In an n-gon, there are $\dfrac{n(n-3)}{2}$ diagonals. So,

$$\frac{n(n-3)}{2} = 20$$
$$n(n-3) = 40$$
$$n(n-3) - 40 = 0$$
$$n^2 - 3n - 40 = 0$$
$$(n-8)(n+5) = 0$$
$$n - 8 = 0 \quad \text{or} \quad n + 5 = 0$$
$$n = 8 \quad \text{or} \quad n = -5.$$

Since a polygon cannot have a negative number of sides, the polygon with 20 diagonals has 8 sides. It is an octagon.

Scheduling an Even Number of Teams

The procedure for scheduling an odd number of teams will not work with an even number of teams. If parallel chords are drawn using the vertices of a regular decagon, some will have the same length. You can see that as you rotate, you will repeat pairings.

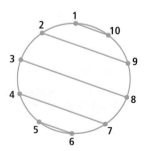

However, instead of putting the tenth team on the circle, it can be placed at the center of the circle. The radius joins team 10 and the team receiving the bye in the schedule for 9 teams. As you rotate the chords to make the schedule, rotate the radius too! This shows the surprising result: it takes as many rounds for a schedule of 9 teams as it does for a schedule of 10 teams. In general, when n is even, it takes as many weeks for a schedule of $n-1$ teams as for a schedule of n teams.

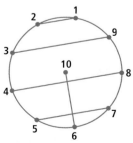

In the Questions, you should assume that all schedules are round-robin.

Questions

1. Refer to Activity 2.
   a. What teams are paired to play in round 6?
   b. What team plays team 7 in the 5th round?
   c. In what round does team 4 play team 5?

2. Refer to the use of the regular nonagon to schedule 9 teams.
   a. How is each team represented?
   b. How is a game between teams 3 and 6 represented?
   c. What is the relationship between all the chords drawn for a given round?
   d. How are the pairings of one week related to those of the next week?
   e. What are the pairings in the 9th round of play?

3. A regular nonagon has diagonals of how many different lengths?

4. When an odd number of teams are scheduled, how many byes are in each round?

5. How many rounds of play are needed to accommodate 32 teams?

6. How is the number of pairings in a round robin of n teams related to the number of diagonals of an n-gon?

7. How many diagonals does a 15-gon have?

8. According to the Diagonals of a Polygon Theorem, how many diagonals does a triangle have? Is that correct?

9. If there is an even number of teams in the tournament, how does the algorithm for creating the schedule change?

10. **True or False** It takes as many weeks for a schedule of 23 teams as it does for a schedule of 24 teams.

11. Make a complete schedule for a tournament with 6 teams.

12. A polygon has 77 diagonals. How many sides does it have?

13. Explain why you cannot begin with the diagram at the right and rotate if you wish to schedule 8 teams.

14. Shreya is scheduling a 20-team round robin.
   a. How many rounds will it take?
   b. Explain why there are no byes.

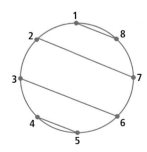

15. **Multiple Choice** Which two of the following descriptions of a point are used in the idea of scheduling?

   A dot   **B** ordered pair

   C location   **D** node of network

16. A round-robin tournament has 66 pairings. How many teams are in the tournament?

17. Each interior angle of a regular polygon measures $135°$. Find the number of diagonals of the polygon.

18. In the figure at the right, each point on the circle represents a person and each chord represents a handshake.

   a. What is the minimum number of handshakes needed for each person in a group of 9 to shake hands with every other person?

   b. What is the minimum number of handshakes needed for each person in a party of n to shake hands with every other person?

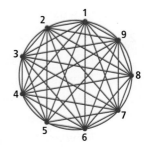

REVIEW

19. Suppose a chord has length 99 in a circle with diameter 100. How far is the chord from the center of the circle? (**Lesson 14-1**)

20. Refer to the figure at the right. (**Lessons 12-7, 6-3**)

   a. Name two angles whose intercepted arc is $\overset{\frown}{AB}$.

   b. What can you say about the measure of these angles?

   c. Name two angles whose intercepted arc is $\overset{\frown}{CD}$. What can you say about the measure of these angles?

   d. Explain why $\triangle CEA \sim \triangle DEB$.

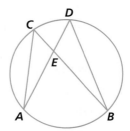

21. The circle at the right has center O, radius r, and $m\angle O = 90$. Write an expression for the area of the circle outside of $\triangle OBC$. (**Lessons 8-9, 8-4**)

22. How many of the diagonals of a regular 19-gon are on lines of symmetry? (**Lesson 6-8**)

23. State the Exterior Angle Theorem. (**Lesson 5-7**)

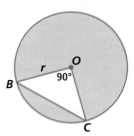

EXPLORATION

24. **a.** Find a schedule for teams in a league involving your school or community.

   b. What factors affect schedules that this lesson does not mention?

25. Explain why, other than a quadrilateral, no polygon has a number of diagonals that is an integer power of 2.

QY ANSWER

324

Lesson

14-3

Angles Formed by Chords or Secants

Vocabulary

secant to a circle

▶ **BIG IDEA** The measure of an angle formed by two chords (secants) is one-half the sum (difference) of the intercepted arcs.

Suppose you draw any two chords, $\overline{AB}$ and $\overline{CD}$, in a circle. If the chords are not parallel, then the lines containing the chords will intersect. The point of intersection E can be inside, on, or outside of the circle as shown below.

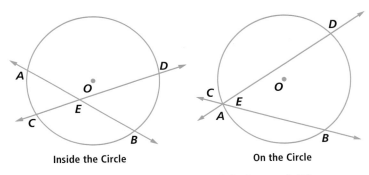

Inside the Circle   On the Circle   Outside the Circle

In each of the above cases, $\angle DEB$ is formed. The measure of $\angle DEB$ is related to the measures of the arcs of the circle. When E is on the circle, $\angle DEB$ is an inscribed angle and its measure is half of the measure of $\overset{\frown}{DB}$. But what about the other two cases? How is m$\angle DEB$ related to the arcs that the chords form on the circle? We examine those cases in this lesson.

Activity

MATERIALS DGS

Step 1 Construct $\odot O$ and points A, B, C, and D on $\odot O$.

Step 2 Draw $\overleftrightarrow{AB}$ and $\overleftrightarrow{CD}$ and drag $\overleftrightarrow{AB}$ and $\overleftrightarrow{CD}$ so that they intersect. Name the point of intersection E. Make your diagram look like the one at the right.

(continued on next page)

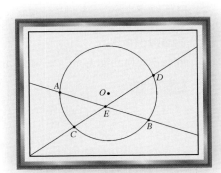

Step 3 Recall that the measure of a minor arc is equal to the measure of its central angle. Thus, the measure of minor arc $\overset{\frown}{AC}$ is the measure of $\angle AOC$. Use this method to find and display the measure of minor arcs $\overset{\frown}{AC}$ and $\overset{\frown}{DB}$ and $m\angle AEC$.

Step 4 Drag point A to different locations on the circle. At each location examine the measures of $\overset{\frown}{AC}$, $\overset{\frown}{DB}$, and $m\angle AEC$. Make a conjecture about the relationship between the angle and the two arcs. Test your conjecture by seeing if it works in several cases. If you get stuck, try adding the measures of the two arcs and comparing the sum to $m\angle AEC$.

The activity might have led you to the conjecture that is stated as the theorem below.

Angle-Chord Theorem

The measure of an angle formed by two intersecting chords is the mean of the measures of the arcs intercepted by it and its vertical angle.

Given Chords $\overline{AB}$ and $\overline{CD}$ in $\odot O$ intersect at point E.

Prove $m\angle AEC = \dfrac{m\overset{\frown}{AC} + m\overset{\frown}{DB}}{2}$

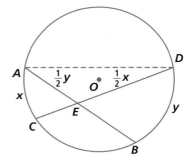

Proof For simplicity, we let $x = m\overset{\frown}{AC}$ and $y = m\overset{\frown}{DB}$. We also draw auxiliary line segment $\overline{AD}$. $\angle DAB$ and $\angle CDA$ are inscribed angles whose measures are half the measures of their intercepted arcs. Thus, $m\angle DAB = \frac{1}{2}y$ and $m\angle CDA = \frac{1}{2}x$. $\angle AEC$ is an exterior angle of $\triangle AED$. By the Exterior Angle Theorem, its measure is the sum of the measures of the interior angles at the other two vertices of the triangle. Thus, $m\angle AEC = \frac{1}{2}x + \frac{1}{2}y = \dfrac{x + y}{2} = \dfrac{m\overset{\frown}{AC} + m\overset{\frown}{DB}}{2}$.

Example 1

In $\odot E$ at the right, $m\overset{\frown}{CA} = 60°$ and $m\overset{\frown}{BD} = 82°$. Find $m\angle CQA$.

Solution

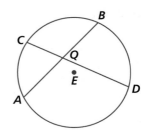

$$m\angle CQA = \tfrac{1}{2}(m\overset{\frown}{CA} + m\overset{\frown}{BD})$$

$$= \tfrac{1}{2}(60° + 82°)$$

$$= \tfrac{1}{2}(142°)$$

$$= 71°$$

QY1

The Inscribed Angle Theorem that you studied in Lesson 6-3 is a special case of the Angle-Chord Theorem. With an inscribed angle, one of the measures of the intercepted arcs is 0 because the vertex of the angle is on the circle. The mean of a measure and 0 is half the measure.

Angles Formed by Secants

A more general case of the Angle-Chord Theorem is the case in which the angle is formed by two intersecting secant lines. A **secant to a circle** is a line that intersects a circle in two different points.

Below at the left, $\angle E$ is formed by two secants and $\angle E$ intercepts the two arcs $\overset{\frown}{AC}$ and $\overset{\frown}{BD}$ with measures 88° and 26°, respectively. To find $m\angle E$, again draw the auxiliary segment $\overline{AD}$, as pictured below at the right.

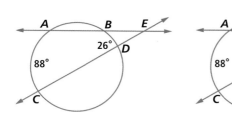

This again forms two inscribed angles, $\angle ADC$ and $\angle BAD$. Of these, $\angle ADC$ is an exterior angle to $\triangle ADE$. By the Exterior Angle Theorem,

$$m\angle DAE + m\angle E = m\angle ADC.$$

Solve for $m\angle E$.	$m\angle E = m\angle ADC - m\angle DAE$
Use the Inscribed Angle Theorem.	$m\angle E = \frac{1}{2}m\overset{\frown}{AC} - \frac{1}{2}m\overset{\frown}{BD}$
Substitute the given measures.	$m\angle E = \frac{1}{2} \cdot 88° - \frac{1}{2} \cdot 26°$
	$= 44° - 13°$

So $m\angle E = 31$.

Notice that $m\angle E$ is half the difference of the measures of $\overset{\frown}{AC}$ and $\overset{\frown}{BD}$. The above argument can be generalized.

Angle-Secant Theorem

The measure of an angle formed by two secants intersecting outside a circle is half the difference of the arcs intercepted by the angle.

 QY2

> **QY1**
>
> In Example 1, what are possible measures of arcs $\overset{\frown}{AD}$ and $\overset{\frown}{BC}$?

> **READING MATH**
>
> The word *secant* comes from the Latin verb *secare*, which means "to cut." This word was chosen to describe a line that cuts through a circle.

> **QY2**
>
> In the figure above at the left, if $m\overset{\frown}{AC}$ were 90° instead of 88°, what would $m\angle E$ be?

Example 2

In circle O at the right, determine each measure.

a. $m\angle D$ b. $m\widehat{BE}$ c. $m\widehat{CD}$

d. $m\widehat{ED}$ e. $m\angle C$ f. $m\angle BED$

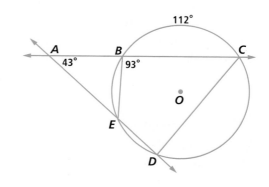

Solution

a. Because $\angle B$ is an inscribed angle, $m\angle B = \frac{1}{2}m\widehat{CDE}$.

So $m\widehat{CDE} = 2 \cdot m\angle B = 2 \cdot \underline{\ ?\ } = \underline{\ ?\ }$.

Thus, $m\widehat{EBC} = 360° - \underline{\ ?\ } = 174°$.

Consequently, $m\angle D = \frac{1}{2}m\widehat{EBC} = \frac{1}{2} \cdot \underline{\ ?\ } = \underline{\ ?\ }$.

b. By the Inscribed Angle Theorem, $m\angle D = \frac{1}{2}m\widehat{EBC}$.

So $\underline{\ ?\ } = \frac{1}{2}(112° + m\widehat{BE})$, which implies that

$m\widehat{BE} = \underline{\ ?\ } - 112° = \underline{\ ?\ }$.

c. By the Angle-Secant Theorem, $m\angle A = \frac{1}{2}(m\widehat{CD} - m\widehat{BE})$,

so $43 = \frac{1}{2}(m\widehat{CD} - \underline{\ ?\ })$. Solving this equation, $m\widehat{CD} = \underline{\ ?\ }$.

d. Because the sum of the measures of the arcs of a circle is $360°$,

$360° = m\widehat{ED} + m\widehat{CD} + 112° + m\widehat{BE}$, so $m\widehat{ED} = 360° -$

$\underline{\ ?\ } = \underline{\ ?\ }$

e. Again, by the Inscribed Angle Theorem, $m\angle C = \frac{1}{2}m\widehat{BED}$, so $m\angle C =$

$\frac{1}{2}(\underline{\ ?\ } + \underline{\ ?\ }) = \underline{\ ?\ }$.

f. Because the sum of the interior angle measures of a quadrilateral is $360°$,

$360 = m\angle BED + 93 + \underline{\ ?\ } + \underline{\ ?\ }$. Solving this equation,

$m\angle BED = \underline{\ ?\ }$.

In Question 14 in Lesson 6-3, a *cyclic quadrilateral* was defined as a quadrilateral with all four vertices on a circle. It was asserted that the opposite angles of a cyclic quadrilateral are supplementary. The measures of the angles in quadrilateral $BCDE$ should confirm that result.

Questions

COVERING THE IDEAS

1. Name a previously proved theorem that is a special case of the Angle-Chord Theorem and the Angle-Secant Theorem.

2. Isosceles trapezoid *BCDE* is inscribed in ⊙*A* and the sides are extended to intersect at *F*. m∠*F* = 40 and m$\widehat{DE}$ = 130°. Determine each indicated measure.

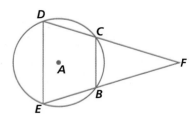

a. m$\widehat{BC}$ b. m$\widehat{BE}$ c. m$\widehat{CD}$

d. m∠*BCD*   e. m∠*CBE*

3. In the figure below, *w* = 78, *x* = 89, and *y* = 130. Find the measure of the indicated angle.

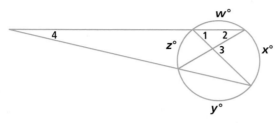

a. ∠1   b. ∠2   c. ∠3   d. ∠4

4. a. In the drawing below, how are *x*, *y*, and 25 related?

   b. Give two different possible pairs of values for *x* and *y*.

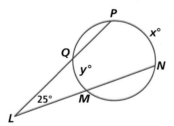

In 5 and 6, use the figure at the right.

5. What additional arc measure(s) do you need in order to find m∠1?

6. Suppose $\widehat{BD}$ is a diameter. Find m∠1.

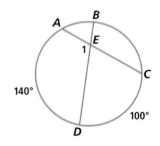

7. a. Using the diagram below, is enough information given to find m∠*F*?

   b. If so, find m∠*F*. If not, what more do you need to know?

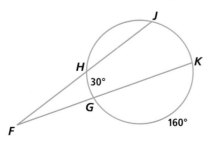

8. In the figure below, $m\overset{\frown}{LP} = 60°$ and $m\overset{\frown}{MO} = 40°$. Determine each indicated measure.

    a. $m\angle MQO$

    b. $m\angle N$

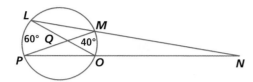

APPLYING THE MATHEMATICS

9. Two intersecting diagonals of a regular pentagon form four angles. What are the measures of those four angles?

10. The *throw* of a speaker in a sound system is the angle of spread from a speaker, much like the picture angle from a camera. Speakers in homes often have throws of up to 120°. Speakers in arenas have a smaller throw so that the sound is more focused and will travel farther. Sports arenas use speakers with 40° throws in clusters to maximize their coverage. The diagram at the right shows the speaker coverage of two loudspeakers mounted back-to-back at point C. If the front speaker covers $\frac{1}{6}$ of the arena, how much does the back speaker cover?

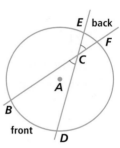

11. A point P is outside $\odot O$. How many secants of $\odot O$ contain P?

12. In the circle below, $m\overset{\frown}{TR} = (x + 10)°$ and $m\overset{\frown}{SU} = (5x)°$. Find $m\overset{\frown}{SU}$.

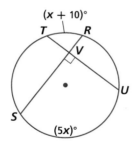

13. Rectangle $VWXY$ is inscribed in the circle at the right. Suppose $m\overset{\frown}{WX} = 50°$.

    a. What is the measure of the acute angle between the diagonals?

    b. What is $m\angle VYW$?

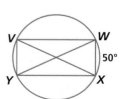

14. In the diagram at the right, the sides of regular hexagon *ABCDEF* have been extended to form a regular six-pointed star. Find:

   a. $m\overset{\frown}{CB}$.

   b. $m\overset{\frown}{DEF}$.

   c. $m\angle G$.

   d. the sum of the measures of the angles in the star points.

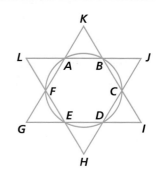

REVIEW

15. **Fill in the Blank** In a round-robin tournament, it takes as many rounds for a schedule of 11 teams as it does for a schedule of __?__ teams. **(Lesson 14-2)**

16. Given: $\overline{AB}$ and $\overline{CD}$ are chords in a circle, $\overline{CD}$ is a diameter of the circle, and $\overline{CD} \perp \overline{AB}$.

   Prove: $\triangle ACB$ is isosceles. **(Lesson 14-1)**

17. Two teams climb a mountain from opposite sides, as pictured at the right. The dashed line shows the altitude of 500 meters. When the first team crossed the dashed line, it had already traveled a distance of 600 meters, and had 720 more meters to travel. When the second team crossed this line, they had traveled 540 meters. How much farther did the second team have to travel? **(Lesson 13-1)**

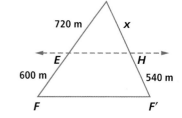

18. **True or False** The statement not-(not-(not-*p*)) is true if and only if not-*p* is true. **(Lesson 11-2)**

EXPLORATION

19. The sides of an inscribed pentagon *ABCDE* at the right are extended to form a five-pointed star.

   a. What is the sum of the measures of angles *F, G, H, I,* and *J* if the pentagon is regular?

   b. What is the largest and smallest this sum can be if the inscribed pentagon is not regular?

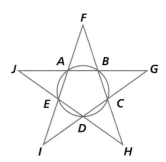

20. Repeat Question 19 if the five-pointed star is inscribed in a circle.

Lesson

14-4

Tangents to Circles and Spheres

Vocabulary

tangent to a circle

point of tangency

tangent to a sphere

common tangents

tangent circles

internally tangent circles

externally tangent circles

▶ **BIG IDEA**   Tangents to circles and spheres are perpendicular to radii drawn to the point of tangency.

If you twirl a weighted object on a string around in a circle and let go, in which direction will it go?

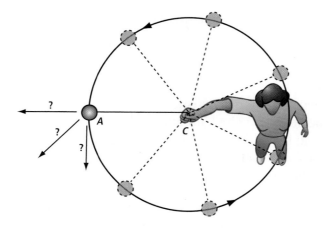

Mental Math

a. Give abbreviations for six sufficient conditions for triangle congruence.

b. Give abbreviations for three sufficient conditions for triangle similarity.

The answer is that the object will travel in the last direction that it was going before you let go. In the diagram above, you can see that the ball constantly changes direction, but if it is let go at A, it will continue in the direction perpendicular to $\overrightarrow{CA}$. The same idea applies if you are throwing a ball or other object.

What Is a Tangent to a Circle?

The line containing the path of the object in the picture is called a *tangent line* to the circle at point A. The word "tangent" comes from the Latin word meaning "touching." The path that the object travels does not intersect the circle again after the object leaves the string.

▶ **READING MATH**

The word *tangent* comes from the Latin verb *tangere*, which means "to touch." This word was chosen to describe a line that just touches a circle in one point.

Definition of Tangent to a Circle, Point of Tangency

A line **tangent to a circle** is a line in the plane of the circle that intersects the circle in exactly one point (the **point of tangency**).

In the picture at the right, you can think of the wheel of the bike (circle) as tangent to the sidewalk (line) as it rolls along.

Activity 1

MATERIALS DGS

Step 1 On a clear DGS screen, construct a circle with center at point *O* and two points *A* and *B* on the circle.

Step 2 Construct the secant line $\overleftrightarrow{AB}$ through these two points.

Step 3 Measure ∠OBA.

Step 4 Drag point *A* toward point *B*. Make a conjecture about the measure of ∠OBA as point *A* gets closer and closer to point *B* (as the secant line approaches a tangent line).

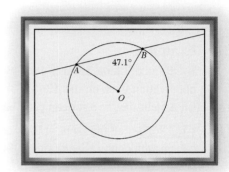

The results of Activity 1 should lead you to the following theorem:

Radius-Tangent Theorem

A line is tangent to a circle if and only if it is perpendicular to the radius at the endpoint of the radius on the circle.

Because this is a biconditional statement, both conditionals need to be proved. The proof of one of these conditionals is given below. You are asked to prove the other in Question 19.

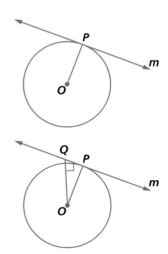

Given *m* is tangent to ⊙*O* at point *P*.

Prove $\overline{OP} \perp m$

Proof We use an indirect proof. Suppose $\overline{OP}$ is not perpendicular to *m*. Then there is a different segment $\overline{OQ}$, with $\overline{OQ} \perp m$ and *Q* on *m*. △*OPQ* is a right triangle with hypotenuse $\overline{OP}$. Consequently $OQ < OP$ because the hypotenuse of a right triangle is longer than either leg. Since $\overline{OP}$ is a radius, *Q* is in the interior of ⊙*O*, so *m* must intersect the circle at a second point. Thus, *m* is not a tangent, which contradicts the given. So, by the Law of Indirect Reasoning, $\overline{OP} \perp m$.

Tangents to Spheres

As you have learned, many ideas studied with 2-dimensional figures extend to 3-dimensional figures. The idea of tangency extends very easily to spheres. A **tangent to a sphere** is a line or plane that intersects the sphere in exactly one point. A common example of a plane tangent to a sphere is a ball in one of its positions as it slides down a ramp.

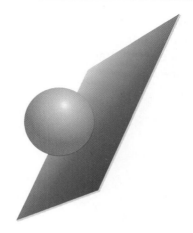

If you have ever traveled in an airplane, then you know that when the sky is clear you can see quite far to the horizon. The Radius-Tangent Theorem enables you to calculate how far you can see from an airplane, or from the top of a building or hill assuming nothing is blocking you and that Earth is a sphere.

Example

The Bank of China Tower in Hong Kong is about 1211 ft tall, and has an observation deck on the 43rd floor that is about 722 ft off of the ground. On a clear day, how far can you see from the observation deck?

Solution Call the point at the observation deck P. We assume Earth is a sphere with center C and radius 3960 miles.

Any tangent line from P to Earth intersects Earth at what we call the horizon. Let T be a point on the horizon and X be the point on Earth directly below P. We wish to calculate PT.

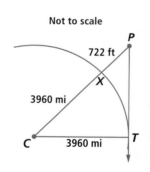

Not to scale

$P, C,$ and T determine a plane. This diagram is a cross section of Earth determined by the plane. This figure is very distorted to fit everything into it. By the Radius-Tangent Theorem, $\angle CTP$ is a right angle. So PT can be found by using the Pythagorean Theorem. However, first 722 ft needs to be changed to miles.

$$722 \text{ ft} = 722 \text{ ft} \cdot \frac{1 \text{ mi}}{5280 \text{ ft}} = \frac{722}{5280} \text{ mi} = \frac{361}{2640} \text{ mi} \approx 0.137 \text{ mi.}$$

This means that CP is about 3960.137 miles. To keep from losing accuracy, use the fraction when you apply the Pythagorean Theorem.

$$PT^2 + 3960^2 = \left(3960 + \frac{361}{2640}\right)^2$$

$$PT^2 = \left(3960 + \frac{361}{2640}\right)^2 - 3960^2$$

$$PT = \frac{19\sqrt{20,909,161}}{300} \approx 33 \text{ miles}$$

From the observation deck of the Bank of China Tower, you can see about 33 miles.

QY1

The numbers in the Example are difficult to deal with. Sailors on the masts of ships could calculate the distance to land and pilots could get a good indication of the location of an object on the horizon using this idea, but it would be more convenient if there were a formula that used any altitude above the ground with the radius of Earth. For these reasons, it is desirable to use variables to obtain a general expression for the distance to the horizon.

But then, why just consider Earth? Letting r be the radius of any sphere and h be the height above the sphere, the distance to the horizon can be shown to be $\sqrt{2rh + h^2}$. (In the Example, $r = 3960$ mi and $h = \frac{361}{2640}$ mi.) You are asked to derive this formula in Question 11.

> **QY1**
>
> The drawing in the Solution to the Example is said to be distorted. Where is the distortion?

Other Tangent Lines

Because there are an infinite number of points on any given circle, there are an infinite number of tangent lines to that circle. The following activity explores one of these situations.

Activity 2

MATERIALS DGS

Step 1 On a clear DGS screen, construct a circle with center O and two points A and B on the circle.

Step 2 Construct the radii $\overline{OA}$ and $\overline{OB}$.

Step 3 Construct lines perpendicular to these radii at points A and B.

Step 4 Drag either point A or point B so that the tangent lines constructed in Step 3 intersect on the screen. Label the point of intersection D.

Step 5 Measure each *tangent segment* $\overline{AD}$ and $\overline{BD}$ (the part of a tangent line with one endpoint on the circle and the other at the point of tangency).

Step 6 Drag points A or B and make a conjecture about the lengths of these tangent segments.

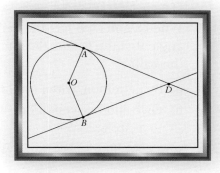

 QY2

The results of Activity 2 lead to the following theorem that you are asked to prove in Question 12.

> **QY2**
>
> In Activity 2, is it possible that the two tangents do not intersect? If so, when does that occur?

Two Tangent Theorem

The two tangent segments from a point to a circle have the same length.

It is also possible for a line to be tangent to more than one circle. Lines that are tangent to two circles are called **common tangents** to those circles. In the figure at the right, $\overleftrightarrow{AB}$ and $\overleftrightarrow{CD}$ are common tangents to both $\odot P$ and $\odot O$.

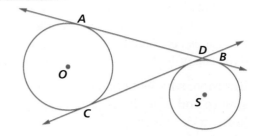

Tangent Circles

Two circles are **tangent circles** if and only if they have exactly one point in common. The figures at the right each show two tangent circles.

On the left, $\odot Q$ and $\odot R$ are **internally tangent circles** because $\odot R$ is inside $\odot Q$. On the right, $\odot T$ and $\odot S$ are **externally tangent circles**.

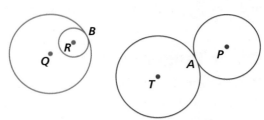

Questions

COVERING THE IDEAS

1. When a golf club hits the ball, what is the measure of the angle that the face of the club will make with the path of the ball?

2. a. By definition, when is a line tangent to a circle?
   b. Give another condition sufficient for a line to be tangent to a circle.

3. When are two circles internally tangent?

4. In the figure below, $\overline{AB}$ and $\overline{AC}$ are tangent to $\odot P$. Find x, if $AB = 5x - 11$ and $AC = 2x + 7$.

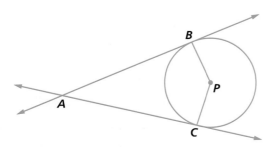

5. Refer to the Example.

   a. Why is $\overline{PT} \perp \overline{TC}$?

   b. Where does the distance 3960 miles come from?

6. In 2003, the Queen Mary 2 was the largest ocean liner ever built. There is a webcam mounted to the bridge of the liner so people can log on and see the captain's view. If the height of the bridge above the water line is 60 m, how far can the webcam see to the horizon?

   a. Give your answer in miles.   b. Give your answer in meters.

In 7–9, give a real-world example of the given mathematical idea.

7. a line tangent to a circle   8. a plane tangent to a sphere

9. a line tangent to a sphere

APPLYING THE MATHEMATICS

10. Determine how far it is to the horizon from a point 200 meters above the surface of the Moon. (The radius of the Moon is about 1750 kilometers.)

11. Given: $\overrightarrow{PT}$ is tangent to circle C, as shown at the right.

   Prove: $d = \sqrt{2rh + h^2}$.

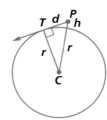

12. Use the figure at the right to prove the Two Tangent Theorem.

   Given Point P outside $\odot N$; $\overline{PX}$ and $\overline{PY}$ are tangent to $\odot N$ at points X and Y.

   Prove $PX = PY$

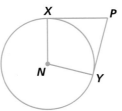

13. a. Copy the figure below and draw the four common tangents to circles O and P.

   b. State at least one property that seems to be true of common tangents.

14. $\overline{PT}$ is tangent to $\odot O$ at T. $PT = 12$ and $PO = 15$.

a. What is the area of the circle?

b. What is the distance from P to N, the nearest point on the circle?

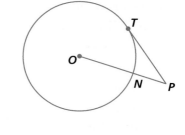

15. The polygon below is circumscribed about circle A. Use this figure and the Two Tangent Theorem to prove the following statement "For any quadrilateral circumscribed about a circle, the sums of the lengths of opposite sides are equal."

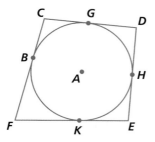

16. An array of congruent tangent circles is shown at the right. Assume that $\overline{RO}$ contains A and is tangent to the circles containing R and O.

a. Prove $\triangle OAJ \cong \triangle RAM$.

b. If the radius of each circle is 1 unit, find AR.

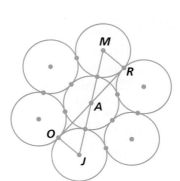

17. a. Extend the Radius-Tangent Theorem of this lesson to apply to spheres.

b. Is the extension true?

18. In the figure at the right, $\overleftrightarrow{ZI}$ is tangent to sphere P at point I. How many other tangents are there from point Z to sphere P?

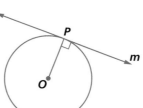

19. Using the information and the diagram provided at the right, prove the following part of the Radius-Tangent Theorem. (Hint: Use an indirect proof.)

Given $\odot O$ with point P on it, $\overline{OP} \perp m$.

Prove m is tangent to the circle.

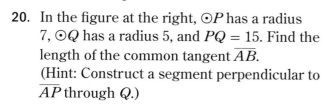

20. In the figure at the right, $\odot P$ has a radius 7, $\odot Q$ has a radius 5, and $PQ = 15$. Find the length of the common tangent $\overline{AB}$. (Hint: Construct a segment perpendicular to $\overline{AP}$ through Q.)

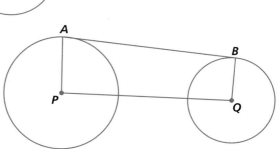

REVIEW

21. In the figure below, $m\angle C = 28$ and $m\widehat{AB} = 34°$. Find $m\widehat{DE}$. (**Lesson 14-3**)

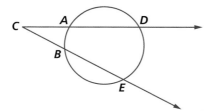

22. Seven people meet at a party. Each wants to exchange telephone numbers with all of the rest. How many exchanges of telephone numbers are required? (**Lesson 14-2**)

23. In a circle of radius 10, what is the length of a chord that intercepts an arc of:
    a. 60°?
    b. 61°? (**Lesson 14-1**)

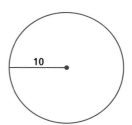

24. Which two of the following statements are contradictory? (**Lessons 11-3, 6-2**)
    A. In $\triangle ABC$, no median is also an altitude.
    B. In $\triangle ABC$, there is a median which is not an altitude.
    C. In $\triangle ABC$ there are two angles which are not congruent.
    D. $\triangle ABC$ has a line of symmetry.

25. Which has the least area: an equilateral triangle with perimeter 1 cm, a square with perimeter 1 cm, or a circle with circumference 1 cm? (**Lessons 8-9, 8-8, 8-7**)

26. What is the sum of the measures of the exterior angles of a convex 168-gon? (**Lesson 5-7**)

EXPLORATION

27. Draw, if possible, the following situations:
    a. two circles with 0 common tangent lines
    b. two circles with only 1 common tangent line
    c. two circles with 2 common tangent lines
    d. two circles with 3 common tangent lines
    e. two circles with 4 common tangent lines
    f. two circles with 5 common tangent lines
    g. two circles with 6 common tangent lines

QY ANSWERS

1. Because 722 ft is about 0.137 mi, $\overline{PX}$ should be significantly shorter than shown.

2. Yes; the two tangents do not intersect if A and B are endpoints of a diameter of $\odot O$. In this case, the tangent lines are parallel because both are perpendicular to the same diameter.

Lesson
14-5 Angles Formed by Tangents and a General Theorem

▶ **BIG IDEA**   The measures of angles formed by tangents to circles are extensions of the measures of angles formed by secants.

So far, you have learned three theorems relating measures of angles and areas.

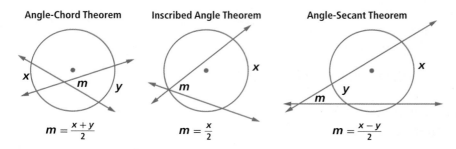

Angle-Chord Theorem

$$m = \frac{x+y}{2}$$

Inscribed Angle Theorem

$$m = \frac{x}{2}$$

Angle-Secant Theorem

$$m = \frac{x-y}{2}$$

Mental Math

Estimate the side length of a cube with volume

a. 508 in³ (to the nearest inch)

b. 0.1252 cm³ (to the nearest tenth of a centimeter)

c. 124.78 ft³ (to the nearest foot)

These three theorems do not cover situations in which one or both of the two intersecting lines is a tangent line. There are three cases in which this occurs, as shown below.

Case 1: tangent-chord

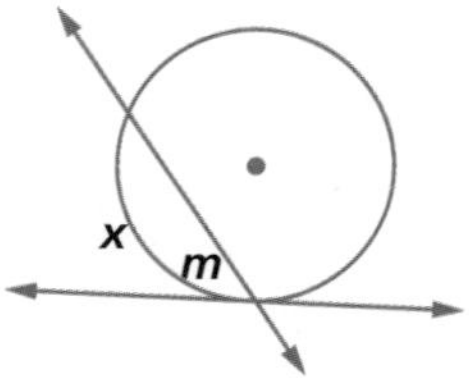

Case 2: tangent-secant

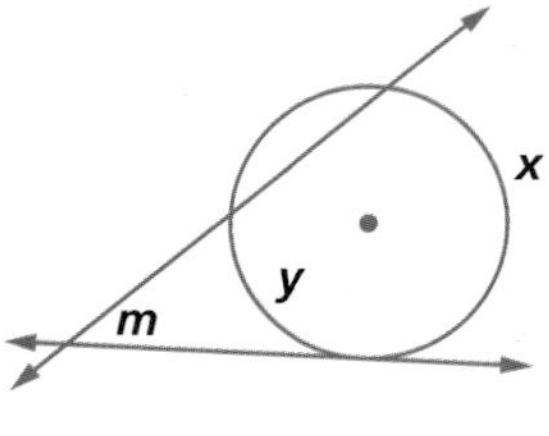

Case 3: tangent-tangent

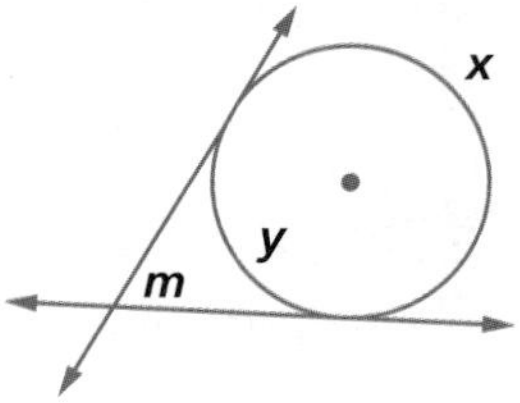

In this lesson, you will explore the relationship between an angle measure *m* and the intercepted arc measures *x* and *y* in these three cases.

Activity

Step 1 On a DGS, construct circle O, a point V on the exterior of the circle, and two secants that go through point V. Label your diagram as shown at the right.

Step 2 Drag point B to point D. What happens to the secant $\overleftrightarrow{VB}$ once point B is the same as point D? This is an example of Case 2.

Step 3 Drag point A to C to illustrate Case 3. What happens to secant $\overleftrightarrow{VA}$?

Step 4 Drag A away from C. Drag point V to point B on the circle to illustrate Case 1. $\overleftrightarrow{VA}$ is still a secant, but what happened to $\overleftrightarrow{VB}$?

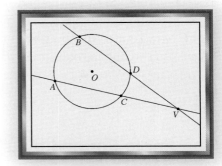

As shown in the Activity, a secant can easily be manipulated to become a tangent, which implies that angles formed by tangents and angles formed by secants are related.

The Angle Formed by a Tangent and a Chord

Consider first the angle formed by a tangent and a chord. Suppose $\overleftrightarrow{BC}$ is tangent to $\odot O$ at B and $m\widehat{AB} = 72°$, as pictured here.

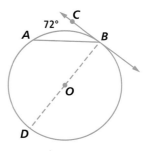

To find $m\angle ABC$, draw the diameter containing B and O. The semicircle $\widehat{BAD}$ has measure $180°$, so $m\widehat{AD} = 108°$. Thus, $m\angle ABD = \frac{1}{2} \cdot 108 = 54$. Now, since $\overleftrightarrow{CB} \perp \overline{BD}$, $\angle CBA$ is complementary to $\angle ABD$. So $m\angle CBA = 36$. In general,

$$m\angle ABC = 90 - m\angle ABD$$
$$= \frac{1}{2} \cdot 180 - \frac{1}{2}m\widehat{AD}$$
$$= \frac{1}{2}(180 - m\widehat{AD})$$
$$= \frac{1}{2}m\widehat{AB}.$$

This proves the following theorem.

Tangent-Chord Theorem

The measure of an angle formed by a tangent and a chord is half the measure of the intercepted arc.

The Tangent-Chord Theorem should look familiar, for it is very much like the Inscribed Angle Theorem. Think of an inscribed angle, like ∠BAC in the first figure at the right. Now rotate $\overleftrightarrow{AC}$ about A until $\overleftrightarrow{AC}$ is tangent to the circle. As the line is rotated, m∠BAC is always half of m$\widehat{BC}$.

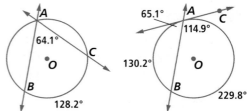

 QY

The Angle Formed by a Tangent and a Secant

Measures of angles between tangents and secants are calculated just like measures of angles between secants.

▶ QY

Sketch a regular octagon inscribed in a circle. If A is a vertex of the octagon, what is the measure of an angle between a side of the octagon and the tangent to the circle at point A?

Tangent-Secant Theorem

The measure of the angle between two tangents, or between a tangent and a secant, is half the difference of the intercepted arcs.

Drawing Secant $\overleftrightarrow{AB}$ and tangent $\overleftrightarrow{EC}$ at point C form ∠E, as shown.

Given m$\widehat{AC}$ = x° and m$\widehat{BC}$ = y°

Prove m∠E = $\frac{1}{2}(x - y)$

Proof Again we draw a segment to form a triangle. Draw $\overline{AC}$. Now m∠1 = $\frac{1}{2}x$ and m∠2 = $\frac{1}{2}y$. But ∠DCA is an exterior angle of △ACE. So

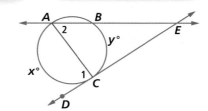

$$m\angle 2 + m\angle E = m\angle 1 \qquad \text{Exterior Angle Theorem}$$
$$m\angle E = m\angle 1 - m\angle 2 \qquad \text{Additive Property of Equality}$$
$$= \tfrac{1}{2}x - \tfrac{1}{2}y \qquad \text{Substitution}$$
$$= \tfrac{1}{2}(x - y). \qquad \text{Distributive Property}$$

The proof for an angle between two tangents is similar. It is left to you as Question 8.

Example 1

Use the figure at the right. $\overrightarrow{AB}$ is tangent to the circle at B. Find m∠A.

Solution The circle measures 360°, so

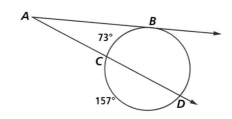

$$m\widehat{BD} = 360° - \underline{\ ?\ } - \underline{\ ?\ }$$
$$= \underline{\ ?\ }$$
$$m\angle A = \tfrac{1}{2}(m\widehat{BD} - m\widehat{BC})$$
$$= \tfrac{1}{2}(\underline{\ ?\ } - \underline{\ ?\ }) = \underline{\ ?\ }$$

Angles Formed by Two Tangents

When two tangents are drawn from the same point to a circle, they split the circle into just two arcs. If the measure of the minor arc is $x°$, then the measure of the major arc is $(360 - x)°$.

Example 2

Refer to the figure at the right. $m\angle P = 40$. What is $m\widehat{QSR}$?

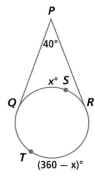

Solution Let $m\widehat{QSR} = x°$. Then $m\widehat{QTR} = (360 - x)°$. By the Tangent-Secant Theorem,

$$m\angle P = \tfrac{1}{2}(m\widehat{QTR} - m\widehat{QSR})$$

$$= \tfrac{1}{2}((360 - x)° - x°).$$

Substitute for $m\angle P$ and solve for x.

$$40 = \tfrac{1}{2}(360 - 2x)$$

$$40 = 180 - x$$

$$x = 140$$

Check When $x = 140$, $360 - x = 220$. Does $m\angle P$ equal half the difference of the two arcs? Yes, 40 is half of $220 - 140$.

A General Theorem

The three theorems from the previous lessons and the two theorems from this lesson are all special cases of a general theorem we call the *General Angle-Arc Measure Theorem*. This theorem uses the idea of arcs facing toward you and away from you.

Consider the two figures drawn at the right. The figure ABC formed by $\overline{AB}$, $\overline{AC}$, and $\widehat{BC}$ at the right is not convex. We say that $\widehat{BC}$ *faces away* from point A. The figure ADE formed by $\overline{AD}$, $\overline{AE}$, and $\widehat{DE}$ at the left is convex. We say that $\widehat{DE}$ *faces toward* point A. When some DGS measure arcs, arcs that face away from the vertex of an angle are given negative measure.

convex

$\widehat{DE}$ faces toward A.

nonconvex

$\widehat{BC}$ faces away from A.

With this terminology, all of the theorems about angle and arc measures in a circle can be interpreted as special cases of one general theorem.

General Angle-Arc Measure Theorem

Suppose line ℓ intersects a circle at points A and C, and line m intersects the same circle at B and D, where we allow the possibilities that as many as three of A, B, C, and D may coincide. If $\ell \cap m = \{P\}$, then $m\angle APB = \frac{1}{2}(m\widehat{AB} + m\widehat{CD})$, provided that arcs that face away from P are viewed as having negative measure.

The five theorems and the definition of the measure of a central angle are related as shown in this hierarchy. Suppose ℓ and m are lines intersecting at P and with both lines intersecting $\odot O$ at at least one point.

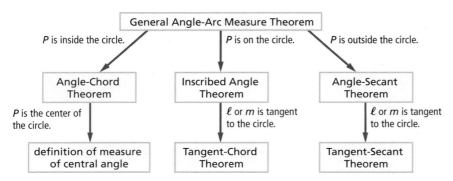

Questions

COVERING THE IDEAS

In 1 and 2, use the figure at the right in which $\overleftrightarrow{BC}$ is tangent to $\odot A$ at B and $m\widehat{BD} = 111°$.

1. Find $m\angle ABC$.

2. Find $m\angle DBC$.

3. Find the measure of $\angle V$ in the circle pictured at the right, given the measures of the arcs it intercepts.

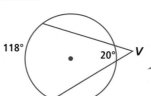

4. $\overline{KJ}$ and $\overline{KL}$ are tangents to $\odot O$, and $m\widehat{JL} = 100.4°$.

    a. Draw this situation.

    b. Find $m\angle JKL$.

In 5–7, use the drawing at the right. $\overline{PT}$ is tangent to circle U at point S.

5. **Fill in the Blank** $m\angle P = $ __?__

6. **Fill in the Blank** $m\angle TSR = $ __?__

7. **True or False** $m\angle RSP = \frac{1}{2}m\widehat{RQS}$

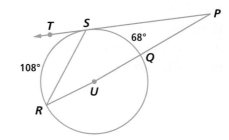

In 8 and 9, use the drawing at the right.

8. Prove that $m\angle P = \frac{1}{2}(m\widehat{QTR} - m\widehat{QSR})$. (Hint: Draw $\overline{QR}$ and use an exterior angle of $\triangle PQR$.)

9. If $m\angle P = 25$, find $m\widehat{QSR}$ and $m\widehat{QTR}$.

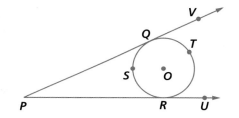

APPLYING THE MATHEMATICS

10. Draw a picture of the special case of the General Angle-Arc Measure Theorem when P is outside the circle and ℓ is tangent to the circle.

11. Explain how the Tangent-Chord Theorem is related to the Inscribed Angle Theorem.

12. $\overleftrightarrow{EF}$ is tangent to circle O at D. $\overline{BC}$ is parallel to $\overleftrightarrow{EF}$. If $m\angle ABC = 100$, and $m\widehat{AD} = 60°$, find the following.
    a. $m\widehat{CD}$
    b. $m\widehat{ABC}$
    c. $m\angle CDF$
    d. $m\angle CDE$
    e. $m\angle BCD$
    f. $m\angle A$

13. Given: $\overrightarrow{UT}$ and $\overrightarrow{UV}$ are tangent to $\odot S$, as shown at the right.
    Prove: $m\widehat{VT} + m\angle U = 180°$.

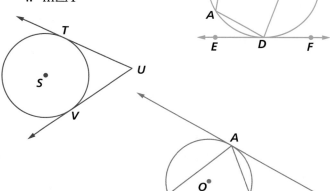

14. $\triangle ABC$ is inscribed in $\odot O$, $\overleftrightarrow{AP}$ is tangent to $\odot O$, $m\angle ABC = 70$, and $m\angle P = 30$. Find the following measures.
    a. $m\angle ABP$
    b. $m\angle ACP$
    c. $m\angle CAB$
    d. What kind of triangle is $\triangle ABC$?

15. $\overrightarrow{PR}$ is tangent to $\odot O$ below at R. If $m\angle P = 4$, what is $m\widehat{QR}$? (The drawing is not to scale.)

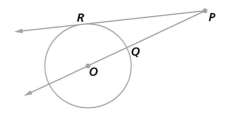

16. At the right, $\overleftrightarrow{ST}$ is tangent to $\odot X$ at T. If $m\angle RTS = 125$, find $m\widehat{TR}$.

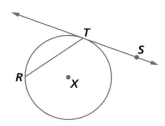

REVIEW

17. The B-2 Spirit stealth bomber is one of the most expensive airplanes ever built (costing up to 2.2 billion dollars for a single plane). It can fly up to a height of 50,000 feet. At that height, how many miles in front of the plane is the horizon? **(Lesson 14-4)**

18. Use the figure below. **(Lesson 14-3)**

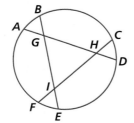

  a. Use the Angle-Chord theorem to find an expression for m∠*IGH* + m∠*GHI* + m∠*HIG* in terms of the arcs of the circle.

  b. Explain how the answer to Part a verifies that the sum of the measures of the angles of △*GHI* is 180°.

19. Suppose $\overline{AB}$ is a chord that intercepts a minor arc with measure 30° in a circle of radius r_1 and $\overline{CD}$ is a chord that intercepts a minor arc with measure 60° in a circle of radius r_2. Write an expression for $\frac{AB}{CD}$. Simplify your expression as much as possible. **(Lesson 14-1)**

20. What is the area of a quadrilateral with vertices (–6, 0), (15, 0), (11, 17), and (–4, 17)? **(Lesson 8-5)**

EXPLORATION

21. For any point *P* outside of ⊙*O*, the *angle of vision* that ⊙*O* covers is the angle between the two tangents to ⊙*O* through *P*. For the circle *O* drawn below, describe the set of all points *P* from which the angle of vision that ⊙*O* covers has measure 40°.

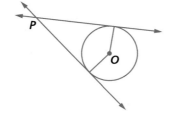

QY ANSWER

22.5°

Lesson 14-6
Three Circles Associated with a Triangle

Vocabulary

circumcircle
circumcenter
circumradius
circumscribed circle
incircle
incenter
inradius
inscribed circle
concurrent lines
point of concurrency
centroid
orthocenter
foot of the altitude
nine-point circle

▶ **BIG IDEA** Every triangle has a circumcircle, incircle, and nine-point circle, each with interesting general properties.

Suppose you bought a circular coffee table and wished to place it on its base so that it balanced properly. To do this you need to find the center of the table. Recall from Lesson 14-1 that the perpendicular bisector of a chord of a circle contains the center of the circle. So, if two segments are chords of the same circle and are not parallel, then the intersection of their perpendicular bisectors is the center of the circle. That method for locating the center of a circle is called the *perpendicular bisector method.*

Activity 1

MATERIALS tracing paper, compass, straightedge.

Step 1 Place a sheet of tracing paper over the circle below. Identify three points *A, B,* and *C* on the circle and trace them onto your sheet of paper. Connect the points to form △*ABC*.

Step 2 Construct the perpendicular bisectors of two of the sides of △*ABC*.

Step 3 Name *P* as the point of intersection of the perpendicular bisectors constructed in Step 2. Draw the circle with center *P* containing one of your points to check that it is the center of the circle containing all three points.

Mental Math

Name all types of quadrilaterals with

a. one symmetry line.

b. two symmetry lines.

c. four symmetry lines.

d. a center of symmetry.

The Circumcircle of a Triangle

Activity 2

MATERIALS DGS

Step 1 On a clear DGS screen, construct an acute triangle, $\triangle ABC$.

Step 2 Construct the perpendicular bisectors of two of the sides of the triangle.

Step 3 Label the point of intersection as point O.

Step 4 Construct $\odot O$ with a radius of either OA, OB, or OC.

Step 5 Drag any of the vertices of the triangle to verify that $\odot O$ always contains points A, B, and C.

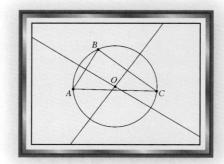

Activities 1 and 2 repeat a construction you learned in Lesson 5-5 to find a point that is equidistant from the three vertices of a triangle. The circle that contains points A, B, and C is called the **circumcircle** of $\triangle ABC$, its center is called the **circumcenter**, and its radius is called the **circumradius**. The circumcenter of a triangle is the unique point that is equidistant from the three vertices of the triangle. We say that the circle is **circumscribed** about $\triangle ABC$.

 QY1

> ▶ **QY1**
>
> In Step 4 of Activity 2, a circle was constructed with radius of either OA, OB, or OC. Why didn't it make a difference which radius you chose?

The Incircle of a Triangle

The circumcircle of a triangle is the smallest circle that contains all the points either on or inside the triangle. In contrast, the *incircle* of a triangle is the largest circle that contains no points inside the triangle.

The center of the incircle is a point that is equidistant from the sides of the triangle. Remember that the distance from a point to a line is measured along the perpendicular from the point to the line.

Activity 3

MATERIALS DGS

Step 1 On a clear DGS screen, construct an acute triangle, $\triangle ABC$.

Step 2 Construct the three angle bisectors of the angles of the triangle.

Step 3 Drag any of the vertices of the triangle to verify that the three angle bisectors intersect at a single point. Label this point I.

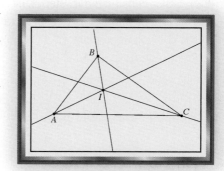

Step 4 Hide the three angle bisectors. Construct a line perpendicular to one of the sides that passes through point *I*.

Step 5 Construct the point of intersection of this perpendicular line and the side of the triangle. Label that point *D*.

Step 6 Construct ⊙*I* with radius *ID*.

Step 7 Hide the perpendicular line. Drag any of the vertices of the triangle to verify that the circle always stays inside the triangle.

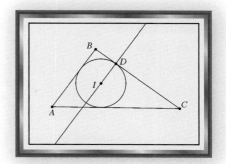

 QY2

Examine closely the circle from Activity 3. Because *I* is at a distance *r* from the three sides of the triangle, ⊙*I* intersects each side of the triangle in exactly one point. Because of the Radius-Tangent Theorem from Lesson 14-5, the radii from *I* to the sides are perpendicular to the sides. The circle whose center is the intersection of the angle bisectors of the triangle and that is tangent to the three sides is called the **incircle** of the triangle. Its center is called the **incenter** of the triangle and *r* is called the **inradius**. We say that the circle is **inscribed** in the triangle.

▶ **QY2**

In Step 4 of Activity 3, you constructed a line through point *I* that was perpendicular to one of the sides. Why didn't it matter which side you chose?

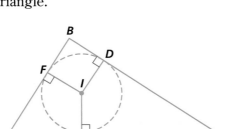

Points of Concurrency

When three or more lines contain the same point, they are called **concurrent lines** and the common point is called the **point of concurrency**. The circumcenter of a triangle is the point of concurrency of the perpendicular bisectors of the sides of the triangle. The incenter of a triangle is the point of concurrency of the bisectors of the angles of the triangle.

The three medians of a triangle are also concurrent, at a point called the **centroid** of the triangle. The centroid has a very nice property that you are asked to find in Activity 4.

Activity 4

MATERIALS DGS

Step 1 Draw any triangle, △ABC.

Step 2 Construct L, the midpoint of $\overline{BC}$, M, the midpoint of $\overline{AC}$, and N, the midpoint of $\overline{AB}$.

Step 3 Construct $\overline{AL}$, $\overline{BM}$, and $\overline{CN}$. If your construction is correct, these segments will be concurrent at the centroid of the triangle. Name the centroid G.

Step 4 Evaluate the three ratios $\frac{AG}{AL}$, $\frac{BG}{BM}$, and $\frac{CG}{CN}$.

Step 5 Make a conjecture based on the results you find in Step 4 and test that conjecture by dragging A, B, and C.

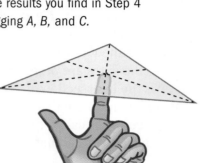

The centroid is the *center of gravity* of the triangular region ABC. It is the point on which the region would balance, it is quite surprising that the angle bisectors, the medians, and the perpendiculars of the sides of a triangle are concurrent. Are the altitudes concurrent as well? Activity 5 investigates this possibility.

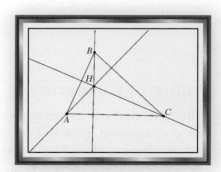

Activity 5

MATERIALS DGS

Step 1 On a clear DGS screen, construct an acute triangle △ABC.

Step 2 Construct lines $\overleftrightarrow{AB}$, $\overleftrightarrow{AC}$, and $\overleftrightarrow{BC}$.

Step 3 Construct perpendicular lines from each of the three vertices to the lines which contain the opposite sides of the triangle.

Step 4 Drag any of the vertices to verify that the three perpendicular lines are concurrent. Label the point of concurrency H.

Step 5 Save your work as you will use this screen in the next activity.

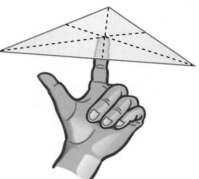

Notice that the perpendicular lines constructed in Activity 5 contain the three altitudes of the triangle. The point of concurrency of the lines containing the altitudes of any triangle is called the **orthocenter** of the triangle. The point of intersection of the altitude and the side of the triangle or the line containing the side is called the **foot of the altitude**.

The Nine-Point Circle

The third circle related to a triangle that we discuss in this lesson has some astounding properties. As you know, there is exactly one circle through three noncollinear points. This last circle contains nine easily identifiable points of a triangle.

Activity 6

MATERIALS DGS

Step 1 Use the last screen from Activity 5. The orthocenter and the three lines containing the altitudes should be shown.

Step 2 Construct the circumcenter of △ABC by constructing perpendicular bisectors as in Activity 2. Label the circumcenter O. Then hide the perpendicular bisectors.

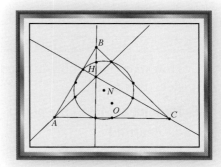

Step 3 Construct the circle that contains the three feet of the altitudes from Step 2. Label the center of this circle N.

Step 4 Construct the three midpoints of the sides of △ABC. What do you notice about these points?

Step 5 Construct segments from the orthocenter H to each of the three vertices of the triangle. Construct the midpoints of these three segments. What do you notice about these midpoints?

Step 6 Drag any of the vertices of △ABC. Are your observations from Steps 5 and 6 still true?

The circle that you constructed in Step 3 of Activity 6 is called the **nine-point circle** of the triangle. The circle gets its name because it contains the midpoints of the three sides, the feet of the three altitudes, and the midpoints of the three segments connecting the orthocenter to the vertices of the triangle.

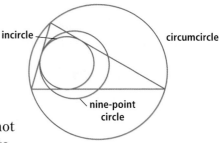

Although the ancient Greek mathematicians knew about the orthocenter, the main properties of the nine-point circle were not discovered until 2000 years after Euclid. Some of these are quite amazing. Here are three.

1. The center of the nine-point circle is the midpoint of the segment connecting the circumcenter and the orthocenter, a segment that also contains the centroid.

2. The radius of the nine-point circle is half the radius of the circumcircle.

3. The nine-point circle is tangent to the incircle.

In this way, the three circles of this lesson are quite closely related to each other.

The discovery in the 1700s of properties like the ones discussed in this lesson encouraged mathematicians to look for more special points and circles associated with triangles, quadrilaterals, and other polygons.

During the 1800s, thousands of theorems were proved about these figures, and new relationships continue to be discovered until the present day. For geometers, these properties demonstrate the extraordinary beauty and richness that can be found in simple figures.

Questions

COVERING THE IDEAS

1. How many lines do you need for them to be concurrent and what does this mean?

2. **Matching** Match the points of concurrency with their origins.
   a. centroid
   b. circumcenter
   c. incenter
   d. orthocenter

   i. perpendicular bisectors
   ii. angle bisectors
   iii. altitudes
   iv. medians

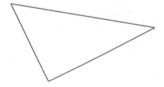

3. Trace the triangle at the right, find its circumcenter, and draw the circumcircle.

4. From what is the incenter of a triangle equidistant?

5. Trace the circle at the right and find its center.

6. Trace the triangle from Question 3. Use a protractor to find its incenter and draw its incircle.

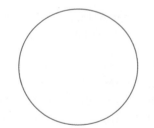

7. With a DGS or by hand, draw a large triangle with an obtuse angle. Draw the nine-point circle of this triangle.

APPLYING THE MATHEMATICS

8. Trace the figure at the right. Is there a point that is equidistant from all of its sides? How do you know?

9. A method for determining the unknown center of a given circle is known as the *right-angle method*. The first step in this process is to choose a point on that circle. Then, draw a right angle at that point. (This is often done with either a T-square or a carpenter's square.)

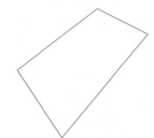

   a. Connecting the points where the sides of the right angle intersect the circle results in a diameter of the circle. Explain why this is true.

   b. By choosing a different point on the circle, you can find a different diameter. The two diameters intersect at the center of the circle. Draw a circle using the bottom of a can or bottle. Find the center of the circle using this method.

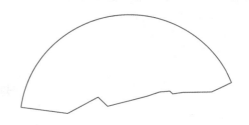

10. At the right is a picture of part of a broken dinner plate.
   a. Trace the figure and find where the center of the unbroken circular plate was.
   b. What was the radius of the unbroken plate?

11. Draw a triangle where the circumcenter and the incenter are the same point.

REVIEW

12. A satellite is placed at point B above Earth. Points A and C lie on the equator, are points of tangency, and represent the borders of the satellite's field of vision. Points A and C are 8562 km apart along the equator. The radius of Earth at the equator is 6378 km. Find $m\angle B$. (**Lesson 14-5**)

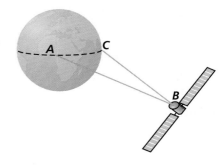

13. **True or False** A line tangent to a circle is tangent to any radius of the circle. (**Lesson 14-4**)

14. The network at the right is made up of a dodecagon, together with all of its diagonals.
   a. How many diagonals and sides are there? (**Lesson 14-2**)
   b. It this network traversable? (**Lesson 1-3**)

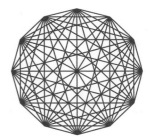

15. Explain why in a circle, the larger a minor arc is, the longer its chord is. (**Lesson 14-1**)

16. Find the area of a regular nonagon that is inscribed in a circle of radius 5. (**Lesson 14-1**)

17. A treasure map reads, "At the palm tree, face west, then turn 35° to the south. Walk 300 paces." How far west and how far south from the palm tree (in paces) will a person be after completing these instructions? (**Lesson 13-6**)

18. The point $P = (3, 4)$ lies on the circle with equation $(x - 2)^2 + (y - 3)^2 = 2$. (**Lessons 11-6, 3-8**)
   a. Let O be the center of the circle. Find the coordinates of O.
   b. Find an equation for the line through P that is perpendicular to $\overleftrightarrow{OP}$.

19. **True or False** Every 2-dimensional cross section of a sphere is a circle. (**Lesson 9-6**)

EXPLORATION

20. Each of the three circles at the right overlaps the other two. The three chords common to each pair of circles are drawn. They seem to have a point in common. Using a DGS or drawings with a compass and ruler, experiment and decide whether this is always true.

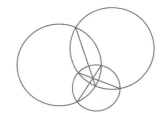

Lesson

14-7

Lengths of Chords, Secants, and Tangents

▶ **BIG IDEA** When two lines intersect each other and also each intersect a circle, the lengths of the segments from the point of intersection to the circle are related in a surprising way.

In Lesson 14-5, you saw that chords, secants, and tangents determine arcs and angles whose measures are related in simple ways. The lengths of chords, secants, and tangent segments in a circle are also related in simple ways.

Mental Math

Ricky completely fills a right cylindrical jar with 104 similar jellybeans. How many jellybeans should he estimate could fit in a jar with

a. twice the dimensions.

b. half the dimensions.

c. twice as tall, with the same base

d. half as tall, with the same base.

Activity

MATERIALS DGS

Step 1 Construct ⊙*O* and place a point *P* in the interior of ⊙*O*.

Step 2 Place a point *A* on ⊙*O*; Construct $\overleftrightarrow{PA}$.

Step 3 Name *B* as the other point where $\overleftrightarrow{PA}$ intersects ⊙*O*. Display the lengths *PA* and *PB*, and *PA* · *PB*.

Step 4 Drag *A*. What do you notice about *PA* · *PB*?

Step 5 Drag *P* to a new point on the interior of the circle. You will notice that *PA* · *PB* changes. However, once you choose a position for *P*, drag *A* again. What do you notice?

Step 6 Now move *P* to the exterior of the circle. Drag point *A* again and observe the product *PA* · *PB*. (Make sure that *A* and *B* are always the two distinct points where the secant line intersects the circle. If they are not, redefine *B* to be the second point.) What do you notice about the product *PA* · *PB* as you drag point *A*?

Step 7 Drag point *A* so that it appears to be a point of tangency. What happens to the location of *B*? Why?

Step 8 Does *PA* · *PB* change when *A* becomes a point of tangency?

Step 9 Move *P* onto the circle itself. Drag point *A*. Does the product remain constant? If so, what is the product in this case and why?

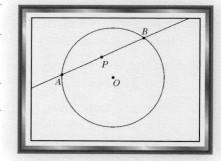

While doing the Activity, you should have noticed that for a given location of P, the product $PA \cdot PB$ is constant. The Swiss geometer Jacob Steiner (1796–1863) called the constant product of these segments of the secant the **power of the point P for the circle O**.

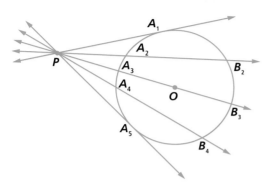
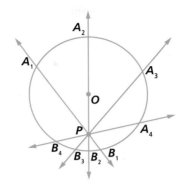

$$(PA_1)^2 = PA_2 \cdot PB_2 = PA_3 \cdot PB_3 = PA_4 \cdot PB_4 = (PA_5)^2 \qquad PA_1 \cdot PB_1 = PA_2 \cdot PB_2 = PA_3 \cdot PB_3 = PA_4 \cdot PB_4$$

This result was known to Euclid, and we still use his proof. A surprise is that the same proof works, letter for letter, whether P is in the interior or the exterior of the circle (with one slight difference in justifications). In both cases we draw auxiliary segments to form similar triangles that give us the ratio that produces the equal products.

Secant Length Theorem

Suppose one secant intersects $\odot O$ at A and B, and a second secant intersects $\odot O$ at C and D. If the secants intersect at P, then $PA \cdot PB = PC \cdot PD$.

Given $\odot O$; secants $\overleftrightarrow{AB}$ and $\overleftrightarrow{CD}$ intersect at P.

Prove $PA \cdot PB = PC \cdot PD$

Proof

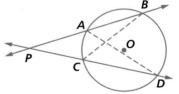

Figure 1

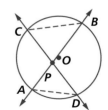

Figure 2

Conclusions	Justifications
1. $\odot O$, secants $\overleftrightarrow{AB}$ and $\overleftrightarrow{CD}$ intersect at P.	1. Given
2. Draw $\overline{AD}$ and $\overline{BC}$.	2. Two points determine a line.
3. $\angle APD \cong \angle CPB$	3. Reflexive Property of Congruence (Figure 1) or Vertical Angles Theorem (Figure 2)
4. $\angle PBC \cong \angle PDA$	4. In a circle, inscribed angles intercepting the same arc are congruent.
5. $\triangle APD \cong \triangle CPB$	5. AA Similarity Theorem
6. $\dfrac{PA}{PC} = \dfrac{PD}{PB}$	6. Corresponding sides of similar figures are proportional.
7. $PA \cdot PB = PC \cdot PD$	7. Means-Extremes Property

Example 1

In the circle at the right, chords $\overline{AB}$ and $\overline{CD}$ intersect at P. Find PB and the power of P.

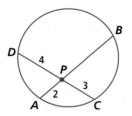

Solution By the Secant Length Theorem, the product of the segments with endpoint P of any chord containing P is the same.

$$PA \cdot PB = PC \cdot PD$$
$$2 \cdot PB = 3 \cdot 4$$
$$2 \cdot PB = 12$$
$$PB = 6$$

The power of P is $3 \cdot 4 = 12$.

If only part of the length of a secant is known, the Secant Length Theorem can lead to needing to solve a quadratic equation.

Example 2

In the figure at the right, find SR.

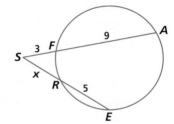

Solution The key to finding this length is to realize that the length of SE is 5 more than the length of SR. Let $SR = x$. Then $SE = x + 5$.

By the Secant Length Theorem, the product of the segments with endpoint S of any secant containing S is the same.

$$SF \cdot SA = SR \cdot SE$$
$$3 \cdot 12 = x(x + 5)$$
$$36 = x(x + 5)$$
$$0 = x^2 + 5x - 36$$
$$0 = (x + 9)(x - 4)$$
$$x = -9 \quad \text{or} \quad x = 4$$

Because x is a length, x is positive, so $x = 4$.

Check $SR \cdot SE = 4(5 + 4) = 36$ and $SF \cdot SA = 3 \cdot 12 = 36$, which is the power of point S.

The Length of a Tangent from a Point to a Circle

If you move the two points A and B of intersection of a secant with a circle so that they coincide, then the secant becomes tangent to the circle. If T is the point of tangency, then the constant product $PA \cdot PB$ becomes $PT \cdot PT$. Your work in the Activity that begins this lesson should have verified this result, which we state and prove in the following theorem.

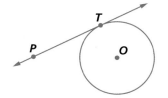

> ### Tangent Length Theorem
>
> Let P be any point in the exterior of $\odot O$, and let $\overleftrightarrow{PT}$ be tangent to $\odot O$ at T. Then the power of P for $\odot O$ is PT^2.

Given   Point P outside $\odot O$, and $\overleftrightarrow{PT}$ tangent to $\odot O$ at T.

Prove   The power of point P for $\odot O$ is PT^2.

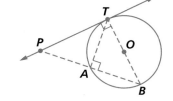

Proof   Draw $\overrightarrow{TO}$, which intersects $\odot O$ at B. Let $\overline{PB}$ intersect $\odot O$ at A and B. Since $\overline{PT} \perp \overline{TB}$ and because $\angle TAB$ is inscribed in a semicircle, $\triangle PTB$ is a right triangle with altitude $\overline{TA}$.

Thus, $PT^2 = PA \cdot PB$ by the Right Triangle Altitude Theorem. But $PA \cdot PB$ is the power of point P, so PT^2 is also the power of point P for $\odot O$.

In other words, the length of the tangent segment from a point to a circle is the square of the power of the point for that point and circle.

Because of the Tangent Length Theorem, you can find the length of a tangent segment given lengths on a secant, or vice versa.

GUIDED

Example 3

In the figure at the right, $\overleftrightarrow{EG}$ is tangent to the circle at E. Find GE and the power of the point G for the circle.

Solution   By the Tangent Length Theorem, $EG^2 = \underline{\quad?\quad} \cdot \underline{\quad?\quad}$. Thus, $EG^2 = \underline{\quad?\quad} \cdot \underline{\quad?\quad} = \underline{\quad?\quad}$. This is the power of the point G for the circle. Taking the square root, $EG = \sqrt{\underline{\quad?\quad}} \approx 5.196$.

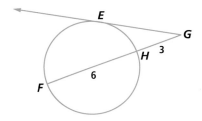

Questions

COVERING THE IDEAS

In 1–3, find the unknown length and the power of B. Each circle's center is O.

1.

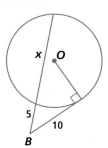

2.

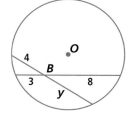

3.

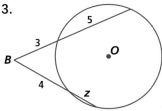

In 4–6, write an expression for the power of the point P.

4.
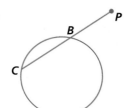

5.
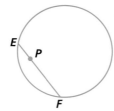

6.
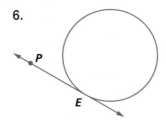

7. If length is measured in centimeters, in what unit is the power of a given point for a given circle measured?

8. In the figure at the right, if $BD = 24$, what is $AP \cdot PC$?

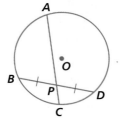

APPLYING THE MATHEMATICS

9. If a point P is on $\odot O$, what is the power of P for $\odot O$?

10. Suppose $\odot O$ has a radius r, P is a point outside the circle, and $OP = d$. Prove that the power of P is $d^2 - r^2$. (Hint: Draw a figure, including the tangent from P to the circle.)

11. Suppose $\odot O$ has a radius r, P is a point inside the circle, and $OP = d$. Prove that the power of the point P is $r^2 - d^2$. (Hint: Draw a figure, including $\overline{OP}$ and the chord perpendicular to $\overline{OP}$ at P.)

12. Suppose a point is P is 3 cm away from the center of a circle with radius 6 cm. What is the power of P?

13. In the diagram at the right, $PA \cdot PB = PC \cdot PD$. Find one other pair of lengths whose product is also equal to these products. (Hint: Use the similar triangles and find another equal ratio.)

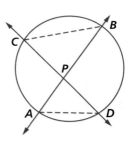

14. A broken plate has partial dimensions as shown in the figure below. What is the diameter of the plate?

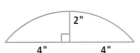

15. A carpenter needs to construct an arch over a doorway that is an arc of a circle. The door is 3 ft wide, 7 ft high on the sides, and 8 ft high in the middle. The dimensions of the doorway are shown at the right.
    a. What is the radius of the circular arc?
    b. What is the total area of the doorway?

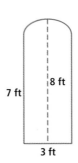

REVIEW

16. The figure below shows arcs in two different tires. Trace the figure and find out if the two tires are the same size. **(Lesson 14-6)**

17. The figure at the right shows a circle of radius r, from which a segment of angle $x°$ was cut out. Find an expression for the perimeter of this figure in terms of r and x. **(Lessons 14-1, 8-8)**

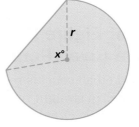

18. Use $\triangle ABC$ below. Give the exact numerical value for each expression. **(Lessons 13-6, 13-5)**

    a. $\sin S$ **b.** $\cos S$ **c.** $(\sin S)^2 + (\cos S)^2$

    d. $\tan S$ **e.** $\dfrac{\sin S}{\cos S}$

19. A sphere of radius 10 has the same surface area as a box with edges x, $2x$, and 8. Find x. **(Lessons 10-7, 9-9)**

20. Suppose the interior of a convex polygon does *not* contain the point $(0, 0)$. What are the possible numbers of quadrants in which the interior of this polygon has points? **(Lesson 2-6)**

EXPLORATION

21. **a.** Suppose a circle has a radius of 5. Accurately describe the set of all points whose power of a point with respect to the circle is 100.

    b. Generalize Part a and prove your generalization.

Lesson 14-8

The Isoperimetric Inequality

Vocabulary

isoperimetric

▶ **BIG IDEA** Figures with the same perimeter may have any area up to the area of a circle with that perimeter (circumference).

What Figure Has the Maximum Area for a Given Perimeter?

A basketball court is rectangular in shape, and it is easier to put benches in straight rows than in circles. So why are large basketball arenas often circular? The reason is because of an important property of circles involving areas and perimeters.

Activity 1

Below are four figures, each with perimeter of 60.

Step 1 Calculate the area of each figure. (Hint: Divide the regular hexagon into six equilateral triangles or use trigonometry.)

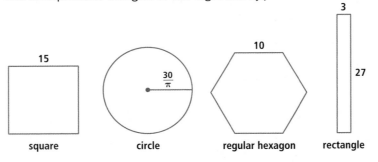

| square | circle | regular hexagon | rectangle |

These figures are drawn to the same scale, so you can compare their areas by sight. It looks as if the rectangle has the least area. Your calculations should bear this out. But which figure has the greatest area?

Step 2 Complete this table for the figures in the picture above.

Figure	Perimeter	Area
square	?	?
circle	?	?
regular hexagon	?	?
rectangle	?	?

Mental Math

In the figure below, ∠A is a right angle.

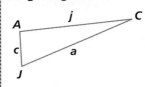

a. Find sin J.

b. sin J is the cosine of which angle?

c. How is sin J affected if m∠J increases?

That the circle has the largest area for a particular perimeter is known as the *Isoperimetric Theorem*. **Isoperimetric** means "having equal perimeters."

> ### Isoperimetric Theorem for the Plane
>
> Of all plane figures with the same perimeter, the circle has the maximum area.

The proof of this theorem requires advanced calculus, a subject usually not studied until college. One reason the proof is difficult is that it requires discussing all sorts of figures, not just polygons and circles. At the near right is an ellipse that is close to circular and thus encloses a substantial area for its perimeter. At the far right is a nonconvex curve with the same perimeter as the ellipse. As you can see, it encloses less area for its perimeter.

When limited to polygons with a given number of sides, regular polygons have the maximum area for their perimeter. As the number of sides increases, these regular polygons look more like circles.

With a Fixed Perimeter, How Large Can the Area of a Figure Be?

Using the Isoperimetric Theorem, the maximum area enclosed by any perimeter p can be found. The result is known as the *Isoperimetric Inequality*.

> ### Isoperimetric Inequality for the Plane
>
> If a plane figure has area A and perimeter p, then $A \le \frac{p^2}{4\pi}$.

Given A plane figure with area A and perimeter p

Prove $A \le \frac{p^2}{4\pi}$.

Proof For any circle with area A, radius r, and perimeter (circumference) p,

$p = 2\pi r$, so $r = \frac{p}{2\pi}$. Now $A = \pi r^2 = \pi\left(\frac{p}{2\pi}\right)^2 = \pi \cdot \frac{p^2}{4\pi^2} = \frac{p^2}{4\pi}$. So another formula for the area of a circle is $A = \frac{p^2}{4\pi}$. By the Isoperimetric Theorem, the area A of any other plane figure with perimeter p must be less than $\frac{p^2}{4\pi}$, so $A \le \frac{p^2}{4\pi}$.

Example

Suppose a figure has perimeter 20 cm.

a. What is its maximum possible area?

b. What is its minimum possible area?

Solution

a. The maximum possible area is given by the Isoperimetric Inequality with $p = 20$ cm.

$$A \leq \frac{p^2}{4\pi} \text{ cm}^2$$

$$A \leq \frac{20^2}{4\pi} \text{ cm}^2$$

$$A \leq \frac{100}{\pi} \text{ cm}^2$$

$$A \leq 31.8 \text{ cm}^2 \text{ (approximately)}$$

The maximum possible area is about 31.8 cm², when the figure is a circle.

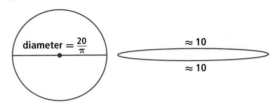

diameter $= \frac{20}{\pi}$ ≈ 10 ≈ 10

b. There is no figure with the smallest area. It is possible to find figures with perimeter 20 cm and area as close to 0 as you like.

 QY

▶ **QY**

Describe a polygon with perimeter 20 units and area 25 square units.

What Is the Least Perimeter Possible for a Given Area?

The inequality $A \leq \frac{p^2}{4\pi}$ gives the range of possible areas for a fixed perimeter p. By solving the inequality for p, you can obtain the range of possible perimeters when the area is fixed. Here is how:

Multiply both sides by 4π. $4\pi A \leq p^2$

Rewrite the inequality to put p^2 on the left. $p^2 \geq 4\pi A$

Take the positive square root of each side. $p \geq \sqrt{4\pi A}$

Thus, of all plane figures with a given area A, the perimeter p is at least $\sqrt{4\pi A}$, the perimeter of a circle whose area is A. This result is another way of stating the Isoperimetric Theorem.

Isoperimetric Theorem for the Plane (alternate statement)

Of all plane figures with the same area, the circle has the least perimeter.

Activity 2

Verify the Isoperimetric Theorem's alternate statement by completing this table.

Figure	Area	Perimeter
a square whose area is 36	?	?
a circle whose area is 36	?	?
a 3- by-12 rectangle	?	?

The restatement of the Isoperimetric Theorem explains why the circle is a popular shape for arenas. In an arena, people like to sit close to the action. Because each person takes up about the same amount of space, each person can be thought of as a unit of area. To minimize the farthest distance from the seats to the center of the court, the perimeter enclosing them should be as small as possible. The smallest perimeter is given by a circle.

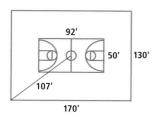
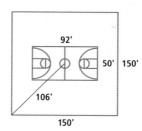
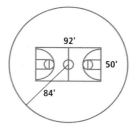

Each of these arenas has an area of about 22,500 ft^2.

Questions

COVERING THE IDEAS

1. Of all figures with perimeter 10, which has the maximum area?

2. Of all figures with perimeter 10, which has the least area?

3. Of all triangles with perimeter 10, which has the maximum area?

4. Draw a nonpolygonal figure whose area is small for its perimeter.

5. Draw a polygonal figure whose area is small for its perimeter.

6. Consider all figures with area 1000 square meters. Which has the least perimeter?

7. **a.** Complete the table at the right.
   b. What theorem asserts that the circle in the table will have the least perimeter?

Figure	Area	Perimeter
circle	12π	?
square	12π	?
rectangle with width 2	12π	?

8. **Fill in the Blanks**
   a. Of all figures with the same area, the ___?___ has the ___?___ perimeter.
   b. Of all the figures with the same perimeter, the ___?___ has the ___?___ area.

9. The Arthur Ashe Stadium at the United States Tennis Association Billie Jean King National Tennis Center in New York City is the largest outdoor tennis-only arena in the world. Although a tennis court is not circular, the tennis center is close to circular. What is an advantage of making a tennis stadium as nearly circular as possible?

Arthur Ashe Stadium

APPLYING THE MATHEMATICS

10. Home plate in baseball is a square foot, with two 45-45-90 triangles added as shown at the right.
    a. Find the perimeter of home plate.
    b. Find the area of home plate.
    c. If the rules were changed so that home plate was a circle but keeping its original area, would the new perimeter be larger or smaller?
    d. Verify your answer to Part c by calculating the new perimeter.

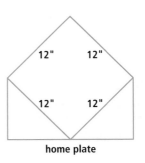

home plate

11. If a rectangle has perimeter $4x$, then its sides can be called $x - y$, $x + y$, $x - y$, and $x + y$.
    a. What is the area of this rectangle?
    b. For what value of y is the area the greatest?

12. Many nomads of Mongolia today live in yurts, cylindrical tents that are made of animal skins and are easy to put up or take down. Some Native Americans used to live in tepees or tipis, cone shaped tents with the same properties as yurts. Explain why the circular base is a wise shape.

Most Mongolian yurts can be set up in about 30 minutes and can be packed on a single draft animal.

13. Convert the nonconvex quadrilateral below into a convex quadrilateral with the same perimeter but greater area.

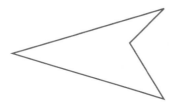

14. Mr. Peano bet his class that he could walk through a hole he cut out of an ordinary sheet of notebook paper. At the right is how he cut the hole. (The hole can be seen by spreading out the cut sheet of paper.)

    a. Could Mr. Peano also have cut a hole that an elephant could walk through?

    b. Although this figure is not exactly a polygon, could a polygon be drawn that approximates this figure well?

    c. How would the perimeter of this polygon compare with its area?

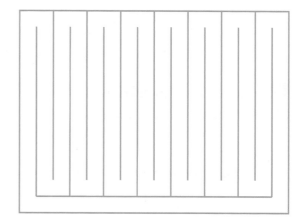

REVIEW

15. Given: Points B and D on $\odot A$, $m\angle A = 90$, and B and D are points of tangency for $\overline{CB}$ and $\overline{CD}$.

    Prove: $ABCD$ is a square. (**Lesson 14-7**)

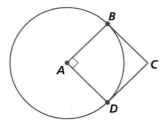

16. Siblings Cassandra, Gloria, and Antonio each started walking on a straight path in a different direction from the same point. All three of them walked at the same speed. After 7 minutes they put marks in the ground. The figure below shows their positions after 10 minutes. (Lessons 14-6, 12-1)

    a. Trace the figure at the right and locate the point they started from.

    b. On your drawing, draw where the siblings' marks were.

C.

G

A

17. a. **True or False** The exterior of circle is a convex figure.

    b. **True or False** If a line is tangent to a circle, then none of the points of the line are in the interior of the circle.
    (Lessons 14-4, 2-4)

18. One cubic centimeter of iron weighs 7.87 grams. One cubic centimeter of aluminum weighs 2.7 grams. Suppose a solid iron statue weighs 14 kg. How much would a similar statue weigh if it were 10 times the linear dimensions, and made of aluminum? (Lesson 12-5)

19. A typical bowling ball has a diameter of 8.5". Three cylindrical holes are bored into the ball. They are $\frac{3}{4}$" in diameter and $1\frac{3}{4}$" deep. Approximate the volume of the bowling ball that is left, to the nearest cubic inch. (Lessons 10-6, 10-3)

EXPLORATION

20. a. Explain why the perimeter of the polygon below right is double the perimeter of the polygon below left, but the area is left the same.

Start with a square with side 3.   Replace each side with this: To get this:

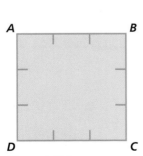

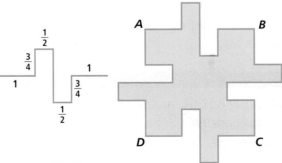

    b. Repeat this process once more on the figure at the right and draw an accurate figure. (You can repeat this process in each $\frac{1}{4}$ unit of border and again double the perimeter and leave the area alone. This demonstrates that for any given area, you can construct a polygon with as large a perimeter as you wish.)

14-9

The Isoperimetric Inequality in Three Dimensions

▶ **BIG IDEA**   Figures with the same surface area may have any volume up to the volume of a sphere with that surface area.

Mental Math

Fill in the Blank

a. The arithmetic mean of 0 and ___?___ is 4.

b. The arithmetic mean of 0 and ___?___ is x.

c. The geometric mean of 1 and ___?___ is 4.

d. The geometric mean of 1 and ___?___ is x.

Above is an engraving of an Inuit village from Martin Frobisher's 1576–1578 expedition in the Canadian Arctic, one of the first expeditions by Central Europeans so far north. Notice the igloos. Ice for the igloo is always available, so an igloo can be built anywhere. Also, a small igloo can be built in about an hour. (How long do you think it took to build the building you live in?) A large igloo, which takes only a little longer to build, usually has an entranceway leading to a smaller part and then to the main part of the igloo. Usually the bases to the smaller and main parts are circles. As you have learned, this shape gives the maximum floor area for a given perimeter.

Both the smaller and main parts of the igloo have the shape of a hemisphere. By using a hemisphere, Inuits can enclose more space with the same amount of ice than with a cylinder, cone, or polyhedron that covers the same base. The sphere plays the same role in 3-dimensional relationships between surface area and volume as the circle does in 2-dimensional relationships between perimeter and area. To emphasize the sameness of the roles, mathematicians use the same term, "isoperimetric," to indicate "same boundary."

Isoperimetric Theorem for Space

Of all solids with the same surface area, the sphere has the largest volume.

Verifying that the Sphere Has the Maximum Volume

Like its counterpart in two dimensions, a proof of the space version of the Isoperimetric Theorem requires advanced mathematics. But the theorem can be verified in different ways. In the Activity, you are asked to do this with the sphere, cylinder, and cube pictured below.

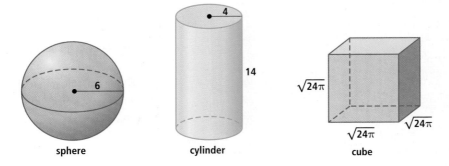

sphere   cylinder   cube

Activity

Complete this table for the three figures shown above.

Figure	Surface Area	Volume
sphere	?	?
cylinder	?	?
cube	?	?

You should have found in the Activity that the sphere has the largest volume for a given surface area. The sphere is the most efficient container. You would see spherical containers everywhere if they could tessellate space and didn't roll.

What Figure Has the Minimum Surface Area for a Given Volume?

Now, rather than a constant boundary, keep the capacity of the figure constant. To do this, consider shapes with the same volume. Soap bubbles consist of some soapy water with a fixed volume of air trapped inside. Because of surface tension, the bubble takes a shape to minimize the surface area surrounding the trapped air. That shape is a sphere. (You never see cubical bubbles.)

> **Isoperimetric Theorem for Space (alternate statement)**
>
> Of all solids with the same volume, the sphere has the least surface area.

For instance, a plastic container that would hold a gallon of milk and use the least amount of plastic would be shaped like a sphere. The following example verifies the alternate statement of this theorem numerically.

Example

The cube, the cylinder, and the sphere below each have volume 1000 cubic meters. Find their surface areas.

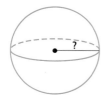

sphere with volume 1000

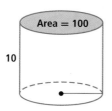

cylinder with volume 1000
(Area of base is 100)

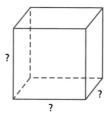

cube with volume 1000

Solution First write down the relevant formulas.

Figure	Volume	Surface Area
cube	$V = s^3$	S.A. $= 6s^2$
sphere	$V = \frac{4}{3}\pi r^3$	S.A. $= 4\pi r^2$
cylinder	$V = \pi r^2 h$	S.A. $= 2\pi r^2 + 2\pi rh$

Now substitute 1000 for the volume of each figure and 10 for the height of the cylinder, determine r and s, and calculate the surface area.

For the cube, $1000 = s^3$, so $s = \sqrt[3]{1000} = 10$ meters.

(continued on next page)

To find the surface area of this cube, substitute 10 for s in the formula for surface area of the cube. S.A. $= 6s^2 = 600$ square meters.

For the sphere, $$1000 = \tfrac{4}{3}\pi r^3$$
$$\frac{750}{\pi} = r^3$$
$$238.7 \approx r^3$$

Take the cube root to find $r \approx 6.2$ meters.

The surface area of the sphere is $4\pi r^2 \approx 4 \cdot \pi \cdot 6.2^2 \approx$ 483 square meters.

The sphere has considerably less surface area than the cube. The cube "wastes" surface near its edges and corners.

For the cylinder, $1000 = \pi r^2 \cdot 10$, so $\pi r^2 = 100$, and $r^2 = \frac{100}{\pi}$.

Thus, $r = \sqrt{\frac{100}{\pi}} \approx 5.65$.

Now S.A. $= 2\pi r^2 + 2\pi rh = 200 + 20\pi r \approx 200 + 20 \cdot \pi \cdot 5.65 \approx$ 555.0 square meters.

You can see that, of these figures, the sphere has the least surface area for its volume.

Figures with Large Surface Areas for a Given Volume

It is possible for an object with small volume to have a large surface area. Start with a 4 × 4 × 4 cube.

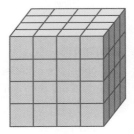

On each of the six faces, take out two opposite cubes from the middle of the face and put them on top of the two other opposite cubes in the middle. Below is the way one face will look after this is done.

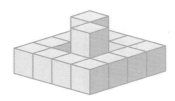

Taking out one cube and reattaching it as is shown adds 8 units to the surface area of the figure. So moving two cubes adds 16 units. Since the original surface area of the face was 16 units, this process doubles the surface area of the figure. When done to all six faces, the result is that the volume of the new polyhedron is the same as that of the original cube, but the surface area has been doubled.

The resulting polyhedron looks a bit like a sponge. Holes give a sponge their large surface area. The large surface area enables a sponge to absorb and hold a lot of water. The water is held in because it clings to the surface.

Concluding Remarks

The Isoperimetric Theorems involve square and cube roots, π, polygons, circles, polyhedra, and spheres. They explain properties of arenas, fences, soap bubbles, and sponges. They demonstrate the broad applicability of geometry and the unity of mathematics.

Many people enjoy the way mathematics connects diverse topics. Others like mathematics for its uses. Still others like the way mathematics fits together and grows. The Isoperimetric Theorems exemplify all of these characteristics of mathematics. We hope that you have found studying them to be an enjoyable way to end your study of geometry using this book.

Questions

COVERING THE IDEAS

1. *Geodesic* houses are made of triangles to produce a close approximation to spherical surfaces. Why might a geodesic house be an efficient dwelling for holding in heat?

2. Consider a rectangular prism with dimensions 5, 10, and 20 meters. Its volume is the same as the volumes of the figures in the Example. How does the surface area of this prism compare to the surface areas of those figures?

3. Suppose the surface area of a solid is 800 square meters. To the nearest 100 cubic meters, what is the maximum possible volume of the solid?

4. **Fill in the Blanks** Of all solids with the same surface area, the ___?___ has the maximum ___?___.

5. **Fill in the Blanks** Of all solids with the same volume, the ___?___ has the least ___?___.

6. A cube and a sphere each have volume 8.
   a. Find the surface area of the cube.
   b. Find the surface area of the sphere.

7. A $1 \times 3 \times 4$ box-shaped pan will hold more water than a $1 \times 3 \times 4$ box-shaped sponge. But, when tipped, the sponge will retain more water than the pan. Why is this?

APPLYING THE MATHEMATICS

8. **a.** A sphere has volume 36π cubic meters. What is its exact surface area?

    b. Give possible dimensions for a cylinder with the same volume as the sphere in Part a.

    c. Will the surface area of the cylinder be more or less than the surface area of the cylinder?

    d. Based on the surface area of the sphere, what percent of surface area is gained by changing to the cylinder? (Give your answer to the nearest tenth of a percent.)

9. **a.** A right circular cone's altitude is 4. The radius of its base is 3. Find the surface area of the cone.

    b. A sphere with the same surface area as the cone is designed. What is the radius of the sphere?

    c. Find the volume of the cone.

    d. Find the volume of the sphere.

    e. Express the increase or decrease in volume from cone to sphere as a percent. (Write your answer as a complete sentence.)

10. A water purifier depends on a cylindrical charcoal filter of diameter 6 cm and height 20 cm. The manufacturer claims that the filter has over 100 acres of surface area.

    a. Can this claim possibly be true?

    b. If so, why is such a large surface area desirable? If not, why can't the claim be true?

REVIEW

11. **a.** Solve the isoperimetric inequality $A \leq \dfrac{p^2}{4\pi}$ for p.

    b. Suppose a region in the plane has area 9 cm^2. Can it have a perimeter of 9 cm? Why or why not? **(Lesson 14-8)**

12. Refer to the figure at the right. **(Lessons 14-8, 4-2)**

    a. Draw the figure resulting from replacing $\overset{\frown}{DE}$ of $\odot H$ by its reflection image across line ℓ, replacing $\overset{\frown}{AF}$ by its reflection image across line k and replacing $\overset{\frown}{BC}$ by its reflection image across line j.

    b. Will the new figure have the same perimeter as the original circle? Why or why not?

    c. Will the figure have the same, more, or less area than the original circle? Give a reason for your answer.

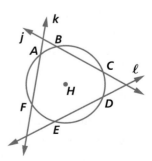

13. Refer to the figure at the right. $\overleftrightarrow{CD}$ and $\overleftrightarrow{CB}$ are tangents to $\odot A$. On a number line, graph the possible measure of angle C. (**Lesson 14-5**)

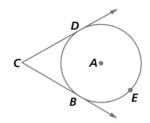

14. The image below shows a front view of a line that is tangent to a sphere. The line is perpendicular to the line of sight. Draw a side and top view of this figure. (**Lessons 14-5, 9-5**)

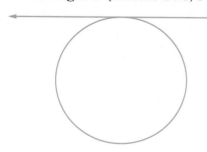

15. $\triangle ABC$ is a right triangle with $m\angle A = 90$ and $\cos B = 0.3$. If $BC = 10$ cm, find AB. (**Lesson 13-6**)

16. **True or False** In this book, several postulates of geometry were proved using theorems. (**Lesson 1-5**)

EXPLORATION

17. According to legend, the first person to use the Isoperimetric Inequality was Dido, the Queen of Carthage.
    a. Where is, or was, Carthage?
    b. How did Dido use this inequality? (Use a book, encyclopedia, or the Internet.)

18. Develop an Isoperimetric Inequality for three dimensions. That is, find a relationship between the volume V and the surface area A of any three-dimensional figure.

Chapter 14 Projects

1 Fractals

In Lesson 14-8, examples are given of how to construct figures that have a large perimeter for their area. The figure below is an example of a *fractal*. Look up what a fractal is, and write an explanation of why such a 2-dimensional figure would have a large perimeter for its area. The particular fractal pictured is called the Heighway dragon. It is also referred to as the Jurassic Park fractal, as its iterations appear on the section title pages of Michael Crichton's novel, *Jurassic Park*. Look up one way to draw the Heighway dragon, either by hand or with a DGS, and show at least enough iterations to see the general shape.

2 Linear and Circular Motion

Many engines work by making turbines move up and down. This is called *linear motion* because the motion of the turbine traces a line. It is often necessary to use this linear motion to cause something, such as the wheel of a locomotive, to move in a circle. The problem of converting linear motion to circular motion baffled engineers in the dawn of the industrial age. How was this problem solved? Find out one way in which this can be done, and make a model explaining this process. This can be either a real-life model or a computer version using the animation features of a DGS.

3 Write a Children's Geometry Book

To get some ideas and inspiration for this project, you might read the book *Sir Cumference and the Great Knight of Angleland: A Math Adventure* by Cindy Neuschwander. Write and illustrate a children's geometry storybook that includes at least 10 "children's book" pages. To prepare:

a. Pick a geometry topic from any chapter of this book, along with at least ten geometry vocabulary words to be used in your book.

b. Choose a title, characters, a setting, and a plot.

4 Geodesic Dome

A *geodesic dome* is a man-made structure that becomes proportionally stronger as it increases in size. It also has the highest ratio of enclosed volume to weight. Create a poster that shows what a geodesic dome looks like, give some history of the geodesic dome, what its advantages and disadvantages are, and where examples of this type of architecture can be found.

5 Making the Most Out of Paper

Start with a standard sheet of paper. Using scissors, cut the sheet of paper so that the resulting shape has a very large perimeter for its area. Compete with your friends to see who can make a shape with the greatest perimeter out of a single piece of paper. Write out an explanation of how you actually measured the perimeter of your shape.

6 Japanese Temple Geometry Theorems

In the Edo period of the history of Japan (1603–1867), when Japan isolated itself from the rest of the world, there was a tradition of posing geometry problems on tablets hung under the roofs of shrines and temples. Many of the theorems involve tangent lines and circles and were unknown outside of Japan until recently. Many of them are astounding. Here are four of these theorems, with the dates of their appearance given in parentheses. In these problems, the radius of circle O_1 is r_1, of circle O_2 is r_2, and so on. Pick at least two of them and draw two different accurate drawings verifying the relationships described in the theorem.

a. (1824) If circle O_1 is tangent to circle O_2, and if circle O_3 is tangent to the two circles O_1 and O_2 and to their common tangent $\overleftrightarrow{AB}$, then $\dfrac{1}{\sqrt{r_3}} = \dfrac{1}{\sqrt{r_1}} + \dfrac{1}{\sqrt{r_2}}$.

b. (1828) Four circles O_1, to O_4 each contain point P and are internally tangent to circle O at consecutive points on the circle. If circles O_1 and O_3 are tangent, and circles O_2 and O_4 are tangent, then $\dfrac{1}{r_1} + \dfrac{1}{r_3} = \dfrac{1}{r_2} + \dfrac{1}{r_4}$.

c. (1782) If circles O_1 to O_4 (of different sizes) are tangent to each other and to circle O at points A, B, C, and D, then $AB \cdot CD = AD \cdot BC$.

d. (1791) Let $\overline{AC}$ and $\overline{BD}$ be any two chords of a circle. Let r_1, r_2, r_3, and r_4 be the radii of the largest circles that can be drawn in the four regions determined by $\overline{AC}$ and $\overline{BD}$. (The circles are tangent to the chords and the larger circle.) Then $\dfrac{1}{r_1} + \dfrac{1}{r_3} = \dfrac{1}{r_2} + \dfrac{1}{r_4}$.

7 The Euler Line

A triangle's orthocenter, incenter, circumcenter, centroid, and the center of its nine-point circle can be studied together. Four of these five points all lie on the same line, called the *Euler line* because Euler discovered some of its properties. The following DGS activity asks you to explore which of these four points are thus related and how these four points are related to each other on that line.

Step 1 Construct a scalene triangle $\triangle ABC$ of reasonable size for your screen.

Step 2 Construct the circumcenter of $\triangle ABC$ and label this point O.

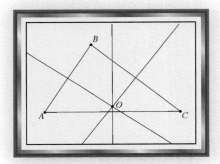

Step 3 Hide the perpendicular bisectors used in Step 2 and then construct the incenter of $\triangle ABC$. Label this point I.

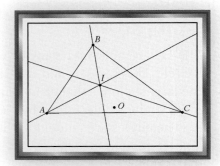

Step 4 Hide the angle bisectors used in Step 3.

Step 5 Construct two medians of $\triangle ABC$ and then construct the centroid, the intersection of the medians of the triangle. Label this point G.

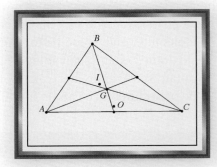

Step 6 Hide the medians and midpoints used in Step 5.

Step 7 Construct two altitudes of $\triangle ABC$ and then construct the orthocenter, the intersection of the altitudes of the triangle. Label this point H.

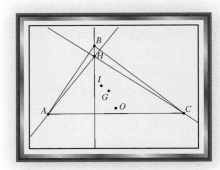

Step 8 Construct the nine-point circle of $\triangle ABC$ and label its center N.

Step 9 Hide the altitudes used in Steps 7 and 8.

Step 10 Drag the vertices of $\triangle ABC$ to see which of points O, I, G, H, and N appear to be aligned. What are the ratios of their distances from each other?

Step 11 Drag any of the vertices of $\triangle ABC$. Which ratios from Step 10 change? Which stay the same?

Chapter 14 Summary and Vocabulary

○ Perpendicular lines are important in circles. A line perpendicular to a chord bisects the chord if and only if it contains the center of the circle. If the sides of an inscribed angle are perpendicular, then the angle intercepts a semicircle. These theorems give ways of finding the center of a circle. A line perpendicular to a radius at its endpoint on a circle is **tangent to the circle**. All other lines through that point are secants to the circle.

○ The chapter also applies congruence and similarity. In a circle, arcs of the same measure are congruent if and only if they have congruent chords. This property of chords enables a round-robin tournament to be scheduled by thinking of the teams as vertices of a regular polygon and each game as a segment joining two teams. An n-gon has $\frac{n(n-3)}{2}$ diagonals.

○ If a **secant** ℓ intersects a circle at points A and B and a secant m intersects the circle at two different points C and D, and $\ell \cap m = \{E\}$, then $\triangle EAC \sim \triangle EDB$. As a result of the similar triangles, $AE \cdot BE = CE \cdot DE$. Either product is the power of point E with respect to the circle. This result holds regardless of the location of E, and even when some of the five points coincide so that the secant becomes a **tangent**. The **power of a point** outside a circle is the square of the length of a tangent from that point to the circle. This property enables you to determine how far you can see from a particular altitude above the ground.

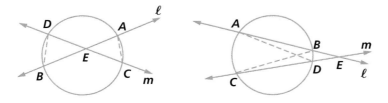

Vocabulary

14-1
*chord of an arc
*central angle of a chord

14-2
round-robin tournament

14-3
*secant to a circle

14-4
*tangent to a circle
point of tangency
*tangent to a sphere
common tangents
tangent circles
internally tangent circles
externally tangent circles

14-6
*circumcircle
circumcenter
circumradius
circumscribed circle
*incircle
incenter
inradius
inscribed circle
*concurrent lines
*point of concurrency
*centroid
*orthocenter
foot of an altitude
nine-point circle

14-7
power of a point P for a circle

14-8
isoperimetric

○ Relationships among measures of angles and arcs are all derived from a theorem you saw in an earlier chapter, that the measure of an inscribed angle is half the intercepted arc. The measure of an angle between two chords is half the sum of the intercepted arcs. (On the previous page at the left, $m\angle AEC = \frac{1}{2}(m\,\widehat{AC} + m\,\widehat{BD})$.) The measure of the angle between secants or tangents is half the difference of the intercepted arcs. (In the figure at the right on the previous page, $m\angle AED = \frac{1}{2}(m\,\widehat{AC} - m\,\widehat{BD})$.)

○ In this chapter, you saw three circles that are associated with triangles: (1) the **circumcircle**, which contains the three vertices of the triangle; (2) the **incircle**, which is tangent to the three sides of the triangle and whose center is in the interior of the triangle; and (3) the **nine-point circle**, which contains the midpoints of the sides of the triangle, the feet of its altitudes, and the midpoints of the segments connecting the orthocenter to the vertices of the triangle. These circles have many surprising properties. Among them is that the nine-point circle is tangent to the incircle and its radius is half the radius of the circumcircle, and that the center of the nine-point circle is on the same line as the **centroid, orthocenter**, and **circumcenter** of the triangle.

○ The Isoperimetric Theorems relate perimeters, areas, and volumes of figures. In a plane, of all figures with the same perimeter, the circle has the most area. Of all figures with the same area, the circle has the least perimeter. In space, of all figures with the same surface area, the sphere has the greatest volume. Of all figures with the same volume, the sphere has the least surface area. Many properties of real objects can be explained by these inequalities.

Postulates, Properties, and Theorems

Arc-Chord Congruence Theorem
  (p. 825)
Chord-Center Theorem (p. 825)
Diagonals of a Polygon Theorem
  (p. 834)
Angle-Chord Theorem (p. 838)
Angle-Secant Theorem (p. 839)
Radius-Tangent Theorem (p. 845)
Two Tangent Theorem (p. 848)
Tangent-Chord Theorem (p. 853)
Tangent-Secant Theorem (p. 854)

General Angle-Arc Measure Theorem
  (p. 856)
Secant Length Theorem (p. 867)
Tangent Length Theorem (p. 869)
Isoperimetric Theorem for the Plane
  (p. 873)
Isoperimetric Inequality for the Plane
  (pp. 873, 875)
Isoperimetric Theorem for Space
  (pp. 880, 881)

Chapter 14   Self-Test

Take this test as you would take a test in class. You will need a straightedge, compass, and protractor. Then check your work with the solutions in the Selected Answers section in the back of the book.

1. In the figure below, m$\widehat{QS}$ = 90°, m$\widehat{SR}$ = (3x)°, m$\widehat{RN}$ = 178°, and m$\widehat{NQ}$ = x°.

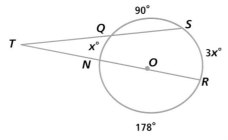

   a. Find x.
   b. Find m∠STR.

2. In the figure below, m$\widehat{AB}$ = 35°. Find m$\widehat{CD}$.

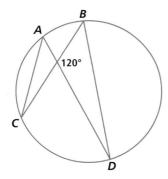

3. In the figure below, $\overline{GH}$ and $\overline{GI}$ are tangent to the circle. Suppose m∠G = x and m$\widehat{HI}$ = (2x)°. Find the value of x.

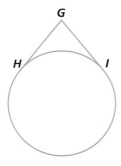

4. Draw an example of a figure with a large perimeter for its area.

5. In the figure below M, O, and N are the midpoints of the sides of △XYZ, respectively. A circle is drawn through these three points. Trace this figure. Then use it to locate the point that is equidistant from the feet of the three altitudes of the triangle.

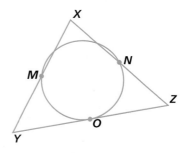

6. In the figure below, in which $\overline{PO}$ is tangent to the circle, find PO.

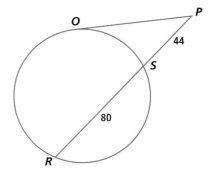

7. In the figure below, $\overline{UV} \perp \overline{UW}$, and $\overline{UV}$ and $\overline{UW}$ are tangent to the circle. Explain why m∠UVW = 45.

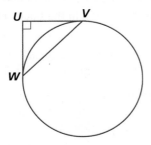

8. Given: $\overleftrightarrow{XY}$ is a tangent to ⊙Z at Y.

Prove: m∠XYV = m∠W

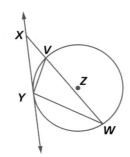

9. **Multiple Choice** In the figure below, which describes a relationship between m∠O and m $\overparen{RS}$?

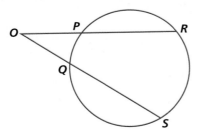

A m∠O > m $\overparen{RS}$

B m∠O = m $\overparen{RS}$

C m∠O < m $\overparen{RS}$

10. **True or False** The three angle bisectors of any triangle are concurrent.

11. Identify the nine points referred to in the nine-point circle.

12. The radius of Mars is about 3400 km. Olympus Mons is a volcano on Mars. At 27 km high, Olympus Mons is the tallest volcano in the solar system. How far could an astronaut see on the surface of Mars, standing on the peak of Olympus Mons?

13. The Super 14 is a union of 14 rugby teams in the southern hemisphere. Each year, they play a round-robin tournament. How many games are played in this tournament?

14. Give an example of a real-life object that has a large surface area for its volume, and an example of an object that has a small surface area for its volume.

15. Chaska wants to buy 3 square kilometers of land. What is the smallest possible length of fence he would need to buy to enclose a region of that area, and what should be the shape of the fence?

16. A regular pentagon is inscribed in a circle of radius 12 cm. Find the lengths of the sides of the pentagon, to the nearest tenth of a centimeter.

Chapter 14 Chapter Review

SKILLS
PROPERTIES
USES
REPRESENTATIONS

SKILLS Procedures used to get answers

OBJECTIVE A Calculate lengths of chords and arcs. (Lesson 14-1)

In 1 and 2, a circle has a radius of 45".

1. Give an exact expression for the length of the chord of a 60° arc.

2. Estimate the length of a chord of a 32° arc to the nearest thousandth of an inch.

3. Find the length of the chord of an $x°$ arc in a circle of radius r.

4. In the figure below, $\triangle ABC$ is an equilateral triangle. The radius of the circle is 20 cm. Find BC.

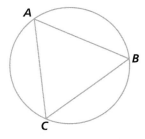

5. In the figure below, X is the center of the circle, $XY = 4$ cm, and $YZ = 7.6$ cm. Find $m\angle X$ to the nearest tenth of a degree.

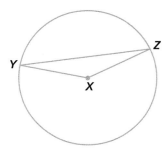

6. A regular 30-gon is inscribed in a circle of radius 14 cm. Find the length of each side of the 30-gon.

OBJECTIVE B Calculate measures of angles between chords, secants, or tangents from measurements of intercepted arcs, and vice versa. (Lessons 14-3, 14-5)

7. In the figure below, all sides of $MKLH$ are tangent to the circle at the indicated points, m$\overarc{FG}$ = 130°, m$\overarc{IJ}$ = 105°, and m$\overarc{IG}$ = 60°. Find the measures of the angles of $MKLH$.

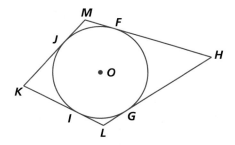

8. In the figure below, m$\overarc{RS}$ = 23° and m$\overarc{QP}$ = 58°. Find the measures of as many angles in the figure as you can.

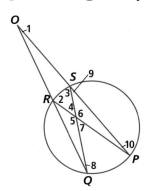

In 9–11, refer to the figure below, in which A is the center of the circle, $\overline{DB}$ and $\overline{DC}$ are tangents. $m\widehat{EF} = 80°$ and $m\widehat{BC} = 120°$. Find the indicated measure.

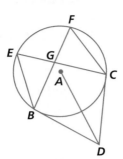

9. $m\angle BGC$

10. $m\angle BDC$

11. $m\angle ADC$

| **OBJECTIVE C** Locate the center of a circle, and find the various centers and circles associated with triangles. (Lesson 14-6)

12. Draw a circle with radius 2″ and locate its center using the Right Angle Method.

13. The figure below shows an arc of a circle. Trace the figure and locate its center.

14. Trace the triangle below and draw its nine-point circle. Locate each of the nine points associated to this circle.

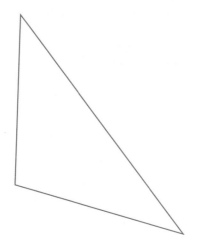

15. Trace the triangle below and find a point that is equidistant from its three sides.

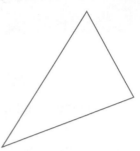

16. Trace the three points below, and find a point that is equidistant from all three of them.

W
•

A
•

G
•

| **OBJECTIVE D** Apply the Secant Length Theorem and Tangent Length Theorem. (Lesson 14-7)

17. **a.** Find the power of L in the figure below.

   b. Find LW.

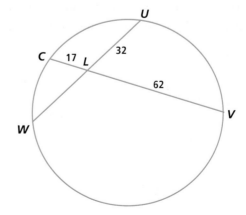

18. In the figure below, $\overrightarrow{MN}$ is tangent to the circle. Find MN.

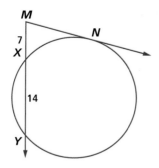

In 19–21, refer to the figure at the right. $\overline{EF}$ is tangent to the circle and $\overline{EG}$ and $\overline{EH}$ are secant lines.

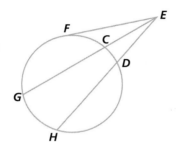

19. Suppose $EF = t$ and $EC = \frac{3}{4}t$. Find CG.

20. Suppose $EF = 12$ cm. What would be the length of any other tangent to the circle through E?

21. Suppose $ED = 14$ cm and $DH = 35$ cm. Find the power of E.

PROPERTIES Principles behind the mathematics

OBJECTIVE E Make deductions from properties of radii, chords, and tangents. (Lessons 14-1, 14-4, 14-5, 14-7)

22. Given: $\odot O$, $\overleftrightarrow{AB}$ is tangent to the circle, and m$\angle ODE = 90$.

   a. Prove: m$\angle A =$ m$\angle E$.

   b. Are $\triangle ODE$ and $\triangle OBA$ congruent? Explain.

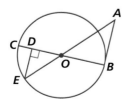

23. Given: A is the center of the smaller circle and $\overline{AC}$ is a diameter of the larger circle.

   Prove: $\overleftrightarrow{CD}$ is tangent to $\odot A$.

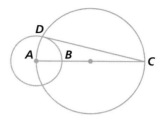

24. Given: $\odot O$, H is the midpoint of $\overline{EF}$.

   Prove: $\triangle EGF$ is isosceles.

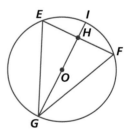

25. Suppose that a tangent to a circle is perpendicular to a chord that passes through the point of tangency. What can you say about that chord?

26. Suppose that two chords in a circle are perpendicular to each other and bisect each other. What can you say about these chords?

OBJECTIVE F Make deductions from properties of angles formed by chords, tangents, or secants. (Lessons 14-3, 14-5)

27. Given: $\overleftrightarrow{QR}$ is tangent to the circle and $QP = QR$.

   Prove: $\triangle QRS$ is isosceles.

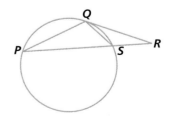

In 28 and 29, use the figure below, in which all of the lines are tangent to the circle, and $EF = EG$.

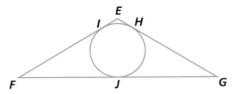

28. Prove that $FJ = JG$.

29. Write an expression for m∠F in terms of m$\widehat{IH}$.

In 30 and 31, refer to the figure at the right. *WXYZ* is a rectangle.

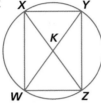

30. a. Explain why $\overline{WY}$ is a diameter of the circle.

   b. Explain why K is the center of the circle.

31. Express m∠XWY in terms of $\widehat{WZ}$.

OBJECTIVE G Apply the Isoperimetric Theorems and the Isoperimetric Inequality to determine which figures have the greatest or the least area, perimeter, or volume. (Lessons 14-8, 14-9)

32. Consider all figures with perimeter 25 cm.
   a. Which has the most area?
   b. What is the area of the figure from Part a?

33. Consider all figures with area 450 mm².
   a. Which one has the smallest perimeter?
   b. What is the perimeter of the figure from Part a?

34. A box has the same volume as a sphere. Which one has less surface area?

35. A figure has an area of 2 cm². Can it have a perimeter of 2 cm? Explain why or why not.

36. Of all the solids with surface area 500 m², is there one with the smallest volume? If so, what is it? If not, why not?

OBJECTIVE H Know if various special points in a triangle all lie on a circle, and if special lines in a triangle are concurrent. (Lesson 14-6)

In 37–39, △*ABC* is a triangle. Are the given lines always concurrent?

37. the three sides of the triangle

38. the three altitudes of the triangle

39. the three angle bisectors and the three medians

In 40–42, △*DEF* is a triangle. Do the given points always lie on one circle?

40. D, E, and F

41. D, E, and the midpoint of $\overline{DE}$

42. The midpoints of the three sides and the feet of the three altitudes

USES Applications of mathematics in real-world situations

OBJECTIVE I Apply the method of scheduling round-robin tournaments to real-life situations. (Lesson 14-2)

43. There are 30 teams in the NBA (National Basketball Association.) How many games would be played if they were all in a round-robin tournament?

44. Write a schedule for a round-robin tournament for 5 teams (numbered 1–5).

45. Write a schedule for a round-robin tournament with 8 teams (numbered 1–8).

46. Kendrick's Little League team plays in a round-robin tournament with every other team in the league. There is one round of play each week. Last year there were seven teams, but this year another team joined the league. Will the number of weeks of the tournament change? Explain your answer.

OBJECTIVE J Use tangent lines and chords in real-life situations. (Lesson 14-1, 14-4, 14-7)

In 47–49, assume that the radius of Earth is 6373 kilometers, and ignore any hills or obstructions that could block your view.

47. How far can you see if your eye level is 160 cm above the ground?

48. How far can you see from a plane flying over an ocean at the height of the top of Mount Everest, which is 8848 meters above sea level?

49. The lighthouse of Alexandria was one of the Seven Wonders of the Ancient World. It was destroyed by earthquakes in the 14th century. Ancient accounts estimate its height as being between 115 and 135 meters. Give upper and lower bounds on how far you could see if you were standing at its top.

50. A square breakfast table has 4 leaves so that a larger, circular table can be formed. If a side of the square is 3 ft long, how much area is gained by using all 4 leaves?

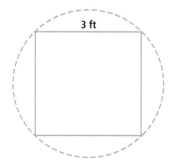

51. A wall of Maggie's room includes a window, as shown in the diagram below. What is the radius of the circular arc?

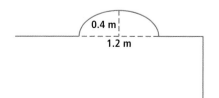

OBJECTIVE K Apply the Isoperimetric Theorems and the Isoperimetric Inequality in real-life situations. (Lessons 14-8, 14-9)

52. Give an example of a figure with a large surface area for its volume.

53. a. A mineral water company wants to manufacture plastic bottles containing 1 liter of water. What shape of bottle will use the least amount of plastic?

    b. Do you think the company will use the shape from Part a? Why, or why not?

54. What is the minimum length of fence necessary to enclose an area of 455 square meters?

55. You have a shopping bag like the one below. Explain how you could increase its volume without changing its surface area.

56. Give an example of a state of the United States that has a large perimeter for its area.

REPRESENTATIONS Pictures, graphs, or objects that illustrate concepts

There are no representation objectives for this chapter.

Postulates

Postulates are statements that are assumed to be true. The postulates listed below may be different from those found in other geometry books.

Chapter 1

Point-Line-Plane Postulate *(Lesson 1-5, pp. 32–33)*

a. Unique Line Assumption
Through any two points there is exactly one line. If the two points are in a plane, the line containing them is in the plane.

b. Number Line Assumption
Every line is a set of points that can be put into a one-to-one correspondence with the real numbers, with any point on it corresponding to 0 and any other point corresponding to 1.

c. Dimension Assumption
1. There are at least two points in space.
2. Given a line in a plane, there is at least one point in the plane that is not on the line.
3. Given a plane in space, there is at least one point in space that is not in the plane.

Distance Postulate *(Lesson 1-6, p. 40)*

a. Uniqueness Property
On a coordinatized line, there is a unique distance between two points.

b. Distance Formula
If two points on a line have coordinates x and y, the distance between them is $|x-y|$.

c. Additive Property
If B is on $\overline{AC}$, then $AB + BC = AC$.

Triangle Inequality Postulate *(Lesson 1-7, p. 47)*
The sum of the lengths of any two sides of a triangle is greater than the length of the third side.

Chapter 3

Angle Measure Postulate *(Lesson 3-1, p. 116)*

a. Unique Measure Assumption
Every angle has a unique measure from 0° to 180°.

b. Unique Angle Assumption
Given any ray $\overrightarrow{VA}$ and a real number r between 0 and 180, there is a unique angle BVA on each side of $\overrightarrow{VA}$ such that m$\angle BVA = r°$.

c. Straight Angle Assumption
If $\overrightarrow{AV}$ and $\overrightarrow{AB}$ are opposite rays, then m$\angle VAB = 180°$.

d. Zero Angle Assumption
If $\overrightarrow{AV}$ and $\overrightarrow{AB}$ are the same ray, then m$\angle VAB = 0°$.

e. Angle Addition Assumption *(Lesson 3-3, p. 127)*
If angles AVC and CVB are adjacent angles, then m$\angle AVC +$ m$\angle CVB =$ m$\angle AVB$.

Postulates of Equality *(Lesson 3-4, p. 133)*
For any real numbers a, b, and c:

Reflexive Property of Equality
$a = a$

Symmetric Property of Equality
If $a = b$, then $b = a$.

Transitive Property of Equality
If $a = b$ and $b = c$, then $a = c$.

Postulates of Equality and Operations
(Lesson 3-4, p. 134)
For any real numbers a, b, and c:

Addition Property of Equality
If $a = b$, then $a + c = b + c$.

Multiplication Property of Equality
If $a = b$, then $ac = bc$.

Postulates of Inequality and Operations
(Lesson 3-4, p. 135)
For any real numbers a, b, and c:

Transitive Property of Inequality
If $a < b$ and $b < c$, then $a < c$.

Addition Property of Inequality
If $a < b$, then $a + c < b + c$.

Multiplication Property of Inequality
If $a < b$ and $c > 0$, then $ac < bc$. If $a < b$ and $c < 0$, then $ac > bc$.

Postulates of Equality and Inequality
(Lesson 3-4, p. 136)
For any real numbers a, b, and c:

Equation to Inequality Property
If a and b are positive numbers and $a + b = c$, then $c > a$ and $c > b$.

Substitution Property
If $a = b$, then a may be substituted for b in any expression.

Corresponding Angles Postulate *(Lesson 3-6, p. 146)*
Suppose two coplanar lines are cut by a transversal.

a. If two corresponding angles have the same measure, then the lines are parallel.
Abbreviation: (measures of corr. ∠s =) ⇒ (∥ lines)

b. If the lines are parallel, then corresponding angles have the same measure.
Abbreviation: (∥ lines) ⇒ (measures of corr. ∠s =)

Chapter 4
Reflection Postulate *(Lesson 4-2, p. 190)*

Under a Reflection	In Other Words
a. There is a one-to-one correspondence between points and their images.	Each preimage has a unique (exactly one) image, and each image has a unique preimage.
b. Collinearity is preserved. If three points are collinear, then their images are collinear.	The image of a line is a line.
c. Betweenness is preserved. If B is between A and C, then the image B' is between the images A' and C'.	The image of a line segment is a line segment.
d. Distance is preserved. If $\overline{A'B'}$ is the image of $\overline{AB}$, then $AB = A'B'$.	The distance between two preimage points equals the distance between their image points.
e. Angle measure is preserved. If $\angle B'A'D'$ is the image of $\angle BAD$, then $m\angle B'A'D' = m\angle BAD$.	The measure of the preimage of an angle is equal to the measure of the image of the angle.
f. Orientation is reversed. A polygon and its image, with vertices taken in corresponding order, have opposite orientations. *(Lesson 4-2, p. 192)*	If a preimage has clockwise orientation, its image has counterclockwise orientation, and vice versa.

Chapter 8
Area Postulate *(Lesson 8-2, p. 458)*
a. Uniqueness Property
Given a unit region, every polygonal region has a unique area.
b. Congruence Property
Congruent figures have the same area.
c. Additive Property
The area of the union of two nonoverlapping regions is the sum of the areas of the regions.
d. Rectangle Formula
The area of a rectangle with dimensions ℓ and w is ℓw. ($A = \ell w$)

Chapter 9
Point-Line-Plane Postulate (Expanded)
(Lesson 9-1, pp. 518, 519)
See Chapter 1 Postulates for parts a–c.
d. Unique Plane Assumption
Through three noncollinear points, there is exactly one plane.
e. Intersection Assumption
Two different planes either do not intersect or intersect in exactly one line.

Chapter 10
Volume Postulate *(Lesson 10-1, p. 597)*
a. Uniqueness Property
Given a unit cube, every polyhedral region has a unique volume.
b. Congruence Property
Congruent figures have the same volume.
c. Additive Property
The volume of the union of two nonoverlapping solids is the sum of the volumes of the solids.
d. Box Formula
The volume of a box with dimensions ℓ, w, and h is ℓwh. ($V = \ell wh$).
e. Cavalieri's Principle *(Lesson 10-3, p. 611)*
Let I and II be two solids included between parallel planes. If every plane P parallel to the given planes intersects I and II in sections with the same area, then Volume(I) = Volume(II).

Chapter 11
Postulates of Logic
Law of Ruling out Possibilities *(Lesson 11-1, p. 648)*
When p or q is true and q is not true, then p is true.
Law of Detachment *(Lesson 11-2, p. 655)*
From a true conditional $p \Rightarrow q$ and a statement or given information p, you may conclude q.
Law of Transitivity *(Lesson 11-2, p. 655)*
If $p \Rightarrow q$ and $q \Rightarrow r$ are true, then $p \Rightarrow r$ is true (also known as Law of Transitivity of Implication).
Double Negative Property *(Lesson 11-2, p. 657)*
The statement not-(not-p) is true exactly when p is true and false exactly when p is false.
Law of the Contrapositive *(Lesson 11-2, p. 658)*
A conditional ($p \Rightarrow q$) and its contrapositive (not-$q \Rightarrow$ not-p) are either both true or both false.
Law of Indirect Reasoning *(Lesson 11-3, p. 663)*
If valid reasoning from a statement p leads to a false conclusion, then p is false.

Theorems

Theorems are statements that have been proved, or can be proved, from the postulates. They are given in order of appearance.

Chapter 1
Line Intersection Theorem *(Lesson 1-5, p. 34)*
Two different lines intersect in at most one point.

Chapter 3
Equal Angle Measures Theorem
(Lesson 3-3, p. 128)
a. If two angles have the same measure, their complements have the same measure.
b. If two angles have the same measure, their supplements have the same measure.

Linear Pair Theorem *(Lesson 3-3, p. 129)*
If two angles form a linear pair, then they are supplementary.

Vertical Angles Theorem *(Lesson 3-3, p. 129)*
If two angles are vertical angles, then their measures are equal.

Parallel Lines and Slopes Theorem
(Lesson 3-6, p. 148)
Two nonvertical lines are parallel if and only if they have the same slope.

Transitivity of Parallelism Theorem
(Lesson 3-6, p. 148)
If line ℓ is parallel to line m and line m is parallel to line n, then line ℓ is parallel to line n.

S_k Theorem 1: Parallel Property *(Lesson 3-7, p. 153)*
Under a size change S_k, the line through any two preimage points is parallel to the line through their images.

S_k Theorem 2: Collinearity Is Preserved
(Lesson 3-7, p. 154)
Under S_k, the images of collinear points are collinear.

S_k Theorem 3: Angle Measure Is Preserved
(Lesson 3-7, p. 154)
Under S_k, an angle and its image have the same measure.

Two Perpendiculars Theorem *(Lesson 3-8, p. 160)*
If two coplanar lines ℓ and m are each perpendicular to the same line, then they are parallel to each other.

Perpendicular to Parallels Theorem
(Lesson 3-8, p. 160)
In a plane, if a line is perpendicular to one of two parallel lines, then it is also perpendicular to the other.

Perpendicular Lines and Slopes Theorem
(Lesson 3-8, p. 161)
Two nonvertical lines are perpendicular if and only if the product of their slopes is –1.

Chapter 4
Figure Transformation Theorem *(Lesson 4-2, p. 191)*
If a figure is determined by certain points, then its transformation image is the corresponding figure determined by the transformation images of those points.

Two-Reflection Theorem for Translations
(Lesson 4-4, p. 206)
If $m \parallel n$, the translation $r_n \circ r_m$ has magnitude two times the distance between m and n in the direction from m perpendicular to n.

Two-Reflection Theorem for Rotations
(Lesson 4-5, p. 212)
If m intersects ℓ, the rotation $r_m \circ r_\ell$ has a center at the point of intersection of m and ℓ and has a magnitude twice the measure of an angle formed by these lines, in the direction from ℓ to m.

Chapter 5
A-B-C-D Theorem *(Lesson 5-1, p. 252)*
Every isometry preserves Angle measure, Betweenness, Collinearity (lines), and Distance (lengths of segments).

Theorem (Equivalence Properties of Congruence) *(Lesson 5-1, p. 254)*
For any figures F, G, and H:

1. $F \cong F$
   (Reflexive Property of Congruence)
2. If $F \cong G$, then $G \cong F$.
   (Symmetric Property of Congruence)
3. If $F \cong G$ and $G \cong H$, then $F \cong H$.
   (Transitive Property of Congruence)

Segment Congruence Theorem *(Lesson 5-2, p. 258)*
Two segments are congruent if and only if they have the same length.

Angle Congruence Theorem *(Lesson 5-2, p. 258)*
Two angles are congruent if and only if they have the same measure.

Corresponding Parts in Congruent Figures (CPCF) Theorem *(Lesson 5-2, p. 259)*
If two figures are congruent, then any pair of corresponding parts are congruent.

Parallel Lines Theorem *(Lesson 5-4, p. 271)*
If two parallel lines are cut by a transversal:

1. Alternate interior angles are congruent.
2. Alternate exterior angles are congruent.
3. Same-side interior angles are supplementary.

Alternate Interior Angles Theorem
(Lesson 5-4, p. 272)
If two lines are cut by a transversal so that a pair of alternate interior angles are congruent, then the two lines are parallel.

Alternate Exterior Angles Theorem
(Lesson 5-4, p. 272)
If two lines are cut by a transversal so that a pair of alternate exterior angles are congruent, then the lines are parallel.

Same-Side Interior Angles Theorem
(Lesson 5-4, p. 272)
If two lines are cut by a transversal so that same-side interior angles are supplementary, then the lines are parallel.

Perpendicular Bisector Theorem *(Lesson 5-5, p. 277)*
If a point is on the perpendicular bisector of a segment, then it is equidistant from the endpoints of the segment.

Unique Circle Theorem *(Lesson 5-6, p. 283)*
There is exactly one circle (a unique circle) through three given noncollinear points.

Uniqueness of Parallels Theorem
(Lesson 5-6, p. 284)
Through a point not on a line, there is exactly one line parallel to the given line.

Triangle-Sum Theorem *(Lesson 5-6, p. 285)*
The sum of the measures of the angles of any triangle is 180°.

Quadrilateral-Sum Theorem *(Lesson 5-7, p. 289)*
The sum of the measures of the angles of a convex quadrilateral is 360.

Polygon-Sum Theorem *(Lesson 5-7, p. 290)*
The sum of the measures of the angles of a convex n-gon is $(n-2) \cdot 180$.

Exterior Angle Theorem for Triangles
(Lesson 5-7, p. 290)
In a triangle, the measure of an exterior angle is equal to the sum of the measures of the interior angles at the other two vertices of the triangle.

Polygon Exterior Angle Theorem *(Lesson 5-7, p. 291)*
The sum of the measures of the exterior angles of a convex n-gon (one per vertex), is 360.

Chapter 6

Flip-Flop Theorem *(Lesson 6-1, p. 309)*
1. If F and G are points and $r_\ell(F) = G$, then $r_\ell(G) = F$.
2. If F and G are figures and $r_\ell(F) = G$, then $r_\ell(G) = F$.

Segment Symmetry Theorem *(Lesson 6-1, p. 309)*
Every segment has exactly two symmetry lines:
1. the line containing the segment, and
2. its perpendicular bisector.

Side-Switching Theorem *(Lesson 6-1, p. 310)*
If one side of an angle is reflected over the line containing the angle bisector, its image is the other side of the angle.

Angle Symmetry Theorem *(Lesson 6-1, p. 311)*
The line containing the bisector of an angle is a symmetry line of the angle.

Circle Symmetry Theorem *(Lesson 6-1, p. 311)*
A circle is reflection-symmetric to any line through its center.

Symmetric Figures Theorem *(Lesson 6-1, p. 312)*
If a figure is symmetric, then any pair of corresponding parts is congruent.

Isosceles Triangle Symmetry Theorem
(Lesson 6-2, p. 317)
The line containing the bisector of the vertex angle of an isosceles triangle is a symmetry line for the triangle.

Isosceles Triangle Coincidence Theorem
(Lesson 6-2, p. 318)
In an isosceles triangle, the bisector of the vertex angle, the perpendicular bisector of the base, and the median to the base determine the same line.

Isosceles Triangle Base Angles Theorem
(Lesson 6-2, p. 318)
If two sides of a triangle are congruent, the angles opposite those sides are congruent.

Unequal Sides Theorem *(Lesson 6-2, p. 320)*
If two sides of a triangle are not congruent, then the angles opposite them are not congruent, and the larger angle is opposite the larger side.

Converse of the Isosceles Triangle Base Angles Theorem *(Lesson 6-2, p. 321)*
If two angles of a triangle are congruent, the sides opposite those angles are congruent.

Unequal Angles Theorem *(Lesson 6-2, p. 321)*
If two angles of a triangle are not congruent, the sides opposite them are not congruent, and the longer side is opposite the larger angle.

Inscribed Angle Theorem *(Lesson 6-3, p. 325)*
The measure of an angle inscribed in a circle is half the measure of its intercepted arc.

Thales' Theorem *(Lesson 6-3, p. 326)*
If an inscribed angle intercepts a semicircle, then the angle is a right angle.

Quadrilateral Hierarchy Theorem *(Lesson 6-4, p. 335)*

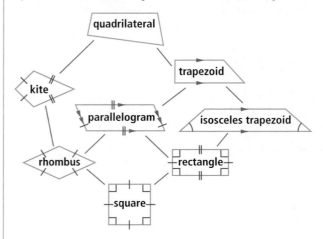

The seven types of quadrilaterals are related as shown in the hierarchy pictured above. Every property true of all figures of one type in the hierarchy is also true of all figures of all types below it to which the first type is connected.

Kite Symmetry Theorem *(Lesson 6-5, p. 340)*
The line containing the ends of a kite is a symmetry line for the kite.

Kite Diagonal Theorem *(Lesson 6-5, p. 341)*
The symmetry line of a kite is the perpendicular bisector of the other diagonal and bisects the two angles at the ends of the kite.

Rhombus Diagonal Theorem *(Lesson 6-5, p. 342)*
Each diagonal of a rhombus is a symmetry line of the rhombus and the perpendicular bisector of the other diagonal.

Trapezoid Angle Theorem *(Lesson 6-6, p. 346)*
In a trapezoid, consecutive angles between a pair of parallel sides are supplementary.

Isosceles Trapezoid Symmetry Theorem
(Lesson 6-6, p. 346)
The perpendicular bisector of one base of an isosceles trapezoid is the perpendicular bisector of the other base and a symmetry line for the trapezoid.

Isosceles Trapezoid Theorem *(Lesson 6-6, p. 347)*
In an isosceles trapezoid, nonbase sides are congruent.

Rectangle Symmetry Theorem *(Lesson 6-6, p. 348)*
The perpendicular bisectors of the sides of a rectangle are symmetry lines for the rectangle.

Theorem *(Lesson 6-7, p. 352)*
If a figure possesses two lines of symmetry intersecting at a point P, then it is rotation-symmetric with a center of symmetry at P.

Center of a Regular Polygon Theorem
(Lesson 6-8, p. 358)
For any regular polygon, there is a unique point (its center) that is equidistant from its vertices.

**Regular Polygon Rotation Symmetry
Theorem** *(Lesson 6-8, p. 359)*
A regular n-gon has n-fold rotation symmetry.

**Regular Polygon Reflection Symmetry
Theorem** *(Lesson 6-8, p. 360)*
Every regular polygon has reflection symmetry about:

1. each line containing its center and a vertex.
2. each perpendicular bisector of its sides.

Chapter 7

SSS Congruence Theorem *(Lesson 7-2, p. 386)*
If, in two triangles, three sides of one are congruent to three sides of the other, then the triangles are congruent.

SAS Congruence Theorem *(Lesson 7-2, p. 387)*
If, in two triangles, two sides and the included angle of one are congruent to two sides and the included angle of the other, then the triangles are congruent.

ASA Congruence Theorem *(Lesson 7-2, p. 388)*
If, in two triangles, two angles and the included side of one are congruent to two angles and the included side of the other, then the two triangles are congruent.

AAS Congruence Theorem *(Lesson 7-2, p. 389)*
If, in two triangles, two angles and a nonincluded side of one are congruent respectively to two angles and the *corresponding* nonincluded side of the other, then triangles are congruent.

Third Angle Theorem *(Lesson 7-2, p. 389)*
If two triangles have two pairs of angles congruent, then their third pair of angles is congruent.

SsA Congruence Theorem *(Lesson 7-5, p. 407)*
If two sides and the angle opposite the longer of the two sides in one triangle are congruent respectively to two sides and the corresponding angle in another triangle, then the triangles are congruent.

HL Congruence Theorem *(Lesson 7-5, p. 408)*
If, in two right triangles, the hypotenuse and a leg of one are congruent to the hypotenuse and a leg of the other, then the two triangles are congruent.

Properties of a Parallelogram Theorem
(Lesson 7-7, p. 420)
In any parallelogram:

a. opposite sides are congruent.
b. opposite angles are congruent.
c. consecutive angles are supplementary.
d. the diagonals intersect at their midpoints.

Distance Between Parallel Lines Theorem
(Lesson 7-7, p. 421)
The distance between two given parallel lines is constant.

Parallelogram Symmetry Theorem
(Lesson 7-7, p. 422)
Every parallelogram has 2-fold rotation symmetry about the intersection of its diagonals.

**Sufficient Conditions for a Parallelogram
Theorem** *(Lesson 7-8, p. 428)*
If, in a quadrilateral,

a. one pair of sides is both parallel and congruent, or

b. both pairs of opposite sides are congruent, or

c. the diagonals bisect each other, or

d. both pairs of opposite angles are congruent,

then the quadrilateral is
a parallelogram.

Chapter 8

Pythagorean Theorem *(Lesson 8-6, p. 481)*
In any right triangle with legs of lengths a and b and hypotenuse of length c, $a^2 + b^2 = c^2$.

Pythagorean Theorem (alternate statement)
(Lesson 8-6, p. 481)
In any right triangle, the sum of the areas of the squares on its legs equals the area of the square on its hypotenuse.

Pythagorean Converse Theorem *(Lesson 8-6, p. 483)*
Suppose a triangle has sides of lengths a, b, and c.
If $a^2 + b^2 = c^2$, then the triangle is a right triangle.

Isosceles Right Triangle Theorem
(Lesson 8-7, p. 488)
In an isosceles right triangle, if a leg has length x, then the hypotenuse has length $x\sqrt{2}$.

30-60-90 Triangle Theorem *(Lesson 8-7, p. 519)*
In a 30-60-90 triangle, if the length of the shorter leg is x, then the length of the longer leg is $x\sqrt{3}$ and the length of the hypotenuse is $2x$.

Chapter 9

Unique Plane Theorem *(Lesson 9-1, p. 520)*
There is exactly one plane through a line and a point not on that line.

Line-Plane Perpendicular Theorem
(Lesson 9-1, p. 521)
If a line is perpendicular to two different lines at their point of intersection, then it is perpendicular to the plane that contains those lines.

Chapter 11

Infinitude of Primes Theorem *(Lesson 11-3, p. 665)*
There are infinitely many prime numbers.

Theorem (Pythagorean Distance Formula on the Coordinate Plane) *(Lesson 11-5, p. 677)*
The distance d between two points (x_1, y_1) and (x_2, y_2) on the coordinate plane is $d = \sqrt{(x_2 - x_1)^2 + (y_2 - y_1)^2}$

Theorem (Equation for a Circle)
(Lesson 11-6, p. 682)

The circle with center (h, k) and radius r is the set of points (x, y) satisfying the equation
$(x - h)^2 + (y - k)^2 = r^2$.

Theorem (Midpoint Formula on a Number Line) *(Lesson 11-7, p. 688)*
On a number line, the midpoint of the segment with endpoints x_1 and x_2 has coordinate $\frac{x_1 + x_2}{2}$.

Theorem (Midpoint Formula on the Plane)
(Lesson 11-7, p. 689)
In the coordinate plane, the midpoint of the segment with endpoints (x_1, y_1) and (x_2, y_2) is $\left(\frac{x_1 + x_2}{2}, \frac{y_1 + y_2}{2}\right)$.

Midsegment of a Trapezoid Theorem
(Lesson 11-8, p. 695)
The midsegment of a trapezoid is parallel to the bases and its length is equal to the average of the lengths of the two bases.

Midsegment of a Triangle Theorem
(Lesson 11-8, p. 695)
A midsegment of a triangle is parallel to and half the length of the base of the triangle.

Theorem (Three Dimension Distance Formula)
(Lesson 11-9, p. 702)
The distance d between two points (x_1, y_1, z_1) and (x_2, y_2, z_2) is given by
$d = \sqrt{(x_1 - x_2)^2 + (y_1 - y_2)^2 + (z_1 - z_2)^2}$.

Theorem (Equation for a Sphere)
(Lesson 11-9, p. 703)
The sphere with center (h, k, j) and radius r is the set of points (x, y, z) satisfying the equation
$(x - h)^2 + (y - k)^2 + (z - j)^2 = r^2$.

Chapter 12

Theorem *(Lesson 12-1, p. 720)*
The transformation S_k that maps (x, y) onto (kx, ky), with $k \neq 0$, is a size transformation S with center $(0, 0)$ and magnitude k.

Size-Change Preservation Properties Theorem *(Lesson 12-1, p. 721)*
Every size transformation preserves:
1. angle measure.
2. betweenness.
3. collinearity.

Size-Change Distance Theorem *(Lesson 12-1, p. 721)*
Under any size change with magnitude $k \neq 0$, the distance between any two image points is $|k|$ times the distance between their preimages.

Figure Size-Change Theorem *(Lesson 12-1, p. 723)*
If a figure is determined by certain points, then its size-change image is the corresponding figure determined by the size-change images of those points.

Theorem (Means-Extremes Property)

(Lesson 12-2, p. 727)

If $\frac{a}{b} = \frac{c}{d}$, then $ad = bc$.

Similar Figures Theorem *(Lesson 12-3, p. 733)*

If two figures are similar, then

1. corresponding angles are congruent.
2. corresponding lengths are proportional.

Fundamental Theorem of Similarity

(Lesson 12-4, p. 739)

If $F \sim F'$ with a ratio of similitude k, then:

1. Length in $F' = k \cdot$ Corresponding Length in F, or $\frac{\text{Length in } F'}{\text{Corresponding Length in } F} = k$.
2. Surface Area $(F') = k^2 \cdot$ Surface Area (F), or $\frac{\text{Surface Area } (F')}{\text{Surface Area } (F)} = k^2$.
3. Volume $(F') = k^3 \cdot$ Volume (F), or $= k^3$.

Fundamental Theorem of Similarity (alternate statement) *(Lesson 12-4, p. 740)*

If two figures are similar with ratio of similitude k, then

1. corresponding angle measures are equal.
2. corresponding lengths and perimeters are in the ratio k.
3. corresponding areas and surface areas are in the ratio k^2.
4. corresponding volumes are in the ratio k^3.

SSS Similarity Theorem *(Lesson 12-6, p. 751)*

If the three sides of one triangle are proportional to three sides of a second triangle, then the triangles are similar.

AA Similarity Theorem *(Lesson 12-7, p. 757)*

If two angles of one triangle are congruent to two angles of another, then the triangles are similar.

SAS Similarity Theorem *(Lesson 12-7, p. 759)*

If, in two triangles, the ratios of two pairs of corresponding sides are equal and the included angles are congruent, then the triangles are similar.

Chapter 13

Triangle Side-Splitting Theorem

(Lesson 13-1, p. 774)

If a line is parallel to a side of a triangle and intersects the other two sides in distinct points, it splits these sides into segments of proportional lengths.

Triangle Side-Splitting Converse Theorem

(Lesson 13-1, p. 776)

If a line intersects $\overleftrightarrow{AB}$ and $\overleftrightarrow{AC}$ in distinct points D and E so that $\frac{AD}{DB} = \frac{AE}{BC}$, then $\overleftrightarrow{DE} \parallel \overleftrightarrow{BC}$.

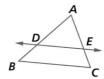

Trapezoid Side-Splitting Theorem

(Lesson 13-1, p. 777)

If a line parallel to the bases intersects the legs of a trapezoid and divides those legs into two segments, then the lengths of those segments are proportional.

Angle Bisector Theorem *(Lesson 13-2, p. 782)*

The bisector of any angle of a triangle divides the opposite side into segments whose lengths are proportional to the adjacent sides of the triangle.

Geometric Mean Theorem *(Lesson 13-3, p. 787)*

The geometric mean of the positive numbers a and b is $\sqrt{ab}$.

Right-Triangle Altitude Theorem

(Lesson 13-3, p. 788)

In every right triangle:

1. The altitude to the hypotenuse is the geometric mean of the segments into which it divides the hypotenuse.
2. Each leg is the geometric mean of the hypotenuse and the segment of the hypotenuse adjacent to the leg.

Rectangle Similarity Theorem *(Lesson 13-4, p. 794)*

Two rectangles are similar if and only if their dimensions are in the same ratio.

Chapter 14

Arc-Chord Congruence Theorem
(Lesson 14-1, p. 825)
In a circle or in congruent circles:

a. If two arcs have the same measure, they are congruent and their chords are congruent.

b. If two chords have the same length, their minor arcs have the same measure.

Chord-Center Theorem *(Lesson 14-1, p. 825)*

1. The line containing the center of a circle perpendicular to a chord bisects the chord.

2. The line containing the center of a circle and the midpoint of a chord bisects the central angle of the chord.

3. The bisector of the central angle of a chord is perpendicular to the chord and bisects the chord.

4. The perpendicular bisector of a chord of a circle contains the center of the circle.

Diagonals of a Polygon Theorem
(Lesson 14-2, p. 833)
A polygon with n sides has $\frac{n(n-3)}{2}$ diagonals.

Angle-Chord Theorem *(Lesson 14-3, p. 837)*
The measure of an angle formed by two intersecting chords is the mean of the measures of the arcs intercepted by it and its vertical angle.

Angle-Secant Theorem *(Lesson 14-3, p. 838)*
The measure of an angle formed by two secants intersecting outside a circle is half the difference of the arcs intercepted by the angle.

Radius-Tangent Theorem *(Lesson 14-4, p. 843)*
A line is tangent to a circle if and only if it is perpendicular to the radius at the endpoint of the radius on the circle.

Two Tangent Theorem *(Lesson 14-4, p. 845)*
The two tangent segments from a point to a circle have the same length.

Tangent-Chord Theorem *(Lesson 14-5, p. 850)*
The measure of an angle formed by a tangent and a chord is half the measure of the intercepted arc.

Tangent-Secant Theorem *(Lesson 14-5, p. 851)*
The measure of the angle between two tangents, or between a tangent and a secant, is half the difference of the intercepted arcs.

General Angle-Arc Measure Theorem
(Lesson 14-5, p. 853)
Suppose line ℓ intersects a circle at points A and C, and line m intersects the same circle at B and D, where we allow the possibilities that as many as three of A, B, C, and D may coincide. If $\ell \cap m = \{P\}$, then $m\angle APB = \frac{1}{2}(m\widehat{AB} + m\widehat{CD})$, provided that arcs that face away from P are viewed as having negative measure.

Secant Length Theorem *(Lesson 14-7, p. 867)*
Suppose one secant intersects $\odot O$ at A and B, and a second secant intersects $\odot O$ at C and D. If the secants intersect at P, then $PA \cdot PB = PC \cdot PD$.

Tangent Length Theorem *(Lesson 14-7, p. 866)*
Let P be any point in the exterior of $\odot O$, and let $\overleftrightarrow{PT}$ be tangent to $\odot O$ at T. Then the power of P for $\odot O$ is PT^2.

Isoperimetric Theorem for the Plane
(Lesson 14-8, p. 870)
Of all plane figures with the same perimeter, the circle has the maximum area.

Isoperimetric Inequality for the Plane
(Lesson 14-8, p. 870)
If a plane figure has area A and perimeter p, then $A \le \frac{p^2}{4\pi}$.

Isoperimetric Theorem for the Plane (alternate statement) *(Lesson 14-8, p. 872)*
Of all plane figures with the same area, the circle has the least perimeter.

Isoperimetric Theorem for Space
(Lesson 14-9, p. 877)
Of all solids with the same surface area, the sphere has the largest volume.

Isoperimetric Theorem for Space (alternate statement) *(Lesson 14-9, p. 878)*
Of all solids with the same volume, the sphere has the least surface area.

Formulas

Right Triangle Area Formula *(Lesson 8-4, p. 468)*
The area of a right triangle is half the product of the lengths of its legs. $A = \frac{1}{2}hb$

Triangle Area Formula *(Lesson 8-4, p. 469)*
The area of a triangle is half the product of the length of a side (the base) and the altitude (height) to that side. $A = \frac{1}{2}hb$

Trapezoid Area Formula *(Lesson 8-5, p. 475)*
The area of a trapezoid equals half the product of its altitude and the sum of the lengths of its bases.
$A = \frac{1}{2}h(b_1 + b_2)$

Parallelogram Area Formula *(Lesson 8-5, p. 475)*
The area of a parallelogram is the product of the length of one of its bases and the altitude to that base. $A = hb$

Perpendicular Diagonals Quadrilateral Area Formula *(Lesson 8-5, p. 477)*
The area of a quadrilateral with perpendicular diagonals equals half the product of its diagonals. When d_1 and d_2 are the lengths of its diagonals.
$A = \frac{1}{2}d_1d_2$

Regular Polygon Area Formula *(Lesson 8-7, p. 491)*
The area of a regular polygon is half the product of its apothem a and its perimeter p. $A = \frac{1}{2}ap$

Circle Circumference Formula *(Lesson 8-8, p. 494)*
If a circle has circumference C and diameter d, and radius r, then $C = \pi d$ or $C = 2\pi r$.

Circle Area Formula *(Lesson 8-9, p. 501)*
The area A of a circle with radius r is πr^2. $A = \pi r^2$

Lateral Area Formula for Right Cylindrical Solids *(Lesson 9-9, p. 571)*
The lateral surface area, L.A., of any right cylindrical surface is the product of its height h and the distance p around its base. L.A. $= ph$

Surface Area Formula for Cylindrical Solids *(Lesson 9-9, p. 572)*
The surface area S.A. of any cylindrical solid is the sum of its lateral area L.A. and twice the area B of a base. S.A. $=$ L.A. $+ 2B$

Surface Area Formula for Pyramids and Cones *(Lesson 9-10, p. 577)*
The surface area, S.A., of any pyramid or cone is the sum of the lateral area L.A. and the area B of its base. S.A. $=$ L.A. $+ B$

Lateral Area Formula for Regular Pyramids and Cones *(Lesson 9-10, p. 579)*
The lateral area L.A. of a regular pyramid or right cone is half the product of its slant height ℓ and the perimeter (circumference) p of the base. L.A. $= \frac{1}{2}\ell p$

Prism-Cylinder Volume Formula *(Lesson 10-3, p. 611)*
The volume V of *any* prism or cylinder is the product of its height h and the area B of its base. $V = Bh$

Pyramid-Cone Volume Formula *(Lesson 10-4, p. 617)*
The volume V of any pyramid or cone equals one-third the product of its height h and its base area B. $V = \frac{1}{3}Bh$

Sphere Volume Formula *(Lesson 10-6, p. 628)*
The volume V of any sphere is $\frac{4}{3}\pi$ times the cube of its radius r. $V = \frac{4}{3}\pi r^3$

Sphere Surface Area Formula *(Lesson 10-7, p. 633)*
The surface area of a sphere with radius r is $4\pi r^2$.
S.A. $= 4\pi r^2$

Box Diagonal Formula *(Lesson 11-9, p. 703)*

The length d of the longest diagonal in a box with dimensions ℓ, w, and h is given by $d = \sqrt{\ell^2 + w^2 + h^2}$.

Three Dimension Midpoint Formula
(Lesson 11-9, p. 704)

The midpoint of the segment with endpoints (x_1, y_1, z_1) and (x_2, y_2, z_2) is $\left(\dfrac{x_1 + x_2}{2}, \dfrac{y_1 + y_2}{2}, \dfrac{z_1 + z_2}{2} \right)$.

Conversion Formulas

Length

Within Customary
1 foot (ft) = 12 inches (in.)
1 yard (yd) = 3 feet
1 mile (mi) = 5280 feet

Within Metric
1 centimeter = 10 millimeters
1 meter = 100 centimeters
1 kilometer = 1000 meters

Between Customary and Metric
1 inch = 2.54 centimeters
1 foot = 0.3048 meter
1 yard = 0.9144 meter
1 mile ≈ 1.609 kilometers

Area

Within Customary
1 square foot = 144 square inches
1 square yard = 9 square feet

Within Metric
1 square centimeter = 100 square millimeters
1 square meter = 10,000 square centimeters

Between Customary and Metric
1 square inch = 6.4516 square centimeters
1 square yard ≈ 0.836 square meter
1 square mile ≈ 2.5889 square kilometers

Land Area

Within Customary
1 square mile = 640 acres

Within Metric
1 hectare = 10,000 square meters

Between Customary and Metric
1 hectare ≈ 2.471 acres

Volume

Within Customary
1 cubic foot = 1728 cubic inches
1 cubic yard = 27 cubic feet

Within Metric
1 cubic centimeter = 1000 cubic millimeters
1 cubic meter = 1,000,000 cubic centimeters

Between Customary and Metric
1 cubic inch ≈ 16.387 cubic centimeters
1 cubic yard ≈ 0.765 cubic meter

Liquid Volume

Within Customary
1 pint = 28.875 cubic inches = 16 ounces
1 quart = 57.75 cubic inches = 32 ounces
1 gallon = 231 cubic inches = 128 ounces

Within Metric
1 milliliter = 1000 cubic millimeters
1 liter = 1000 cubic centimeters

Between Customary and Metric
1 ounce ≈ 29.574 milliliters
1 quart ≈ 0.946 liter
1.057 quarts ≈ 1 liter

Weight (Mass)

Within Customary
1 pound = 16 ounces
1 ton = 2000 pounds

Within Metric
1 gram = 1000 milligrams
1 kilogram = 1000 grams
1 metric ton = 1000 kilograms

Between Customary and Metric
1 ounce ≈ 28.350 grams
1 pound ≈ 0.4536 kilogram
2.2 pounds ≈ 1 kilogram
1 ton ≈ 1.016 metric tons

	Customary Abbreviations	Metric Abbreviations
Length	inch = in., foot = ft, yard = yd, mile = mi	millimeter = mm, centimeter = cm, meter = m, kilometer = km
Liquid Volume	ounce = oz, pint = pt, quart = qt, gallon = gal	milliliter = mL, liter = L
Weight (Mass)	pound = lb, ton = tn.	milligram = mg, gram = g, kilogram = kg, metric ton = t

In this book, the following symbols are used:

a, b, c	sides
A	area
B	area of base
b_1, b_2	bases

C	circumference
d	diameter
d_1, d_2	diagonals
h	height
ℓ	length
ℓ	slant height (in conics)
L.A.	lateral area

n	number of sides
p	perimeter
r	radius
s	side
S.A.	surface area
V	volume
w	width

Two-Dimensional Figures

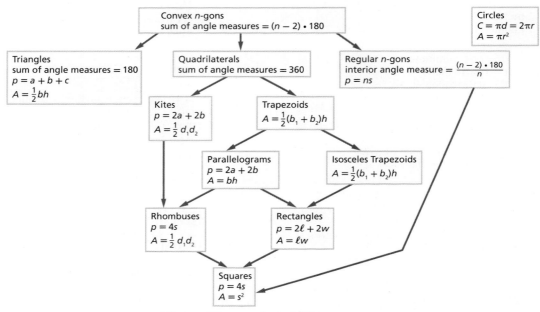

Three-Dimensional Figures

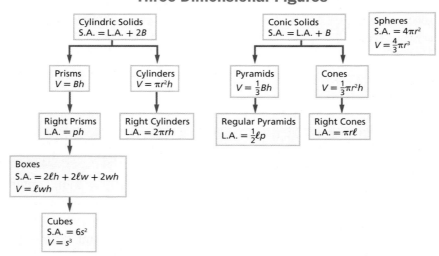

Coordinate Geometry	Distance	Midpoints of a Segment	Slope of a Line
Two Dimensions	$d = \sqrt{(x_2 - x_1)^2 + (y_2 - y_1)^2}$	$M = \left(\dfrac{x_1 + x_2}{2}, \dfrac{y_1 + y_2}{2}\right)$	$m = \dfrac{y_2 - y_1}{x_2 - x_1}$
Three Dimensions	$d = \sqrt{(x_2 - x_1)^2 + (y_2 - y_1)^2 + (z_2 - z_1)^2}$	$M = \left(\dfrac{x_1 + x_2}{2}, \dfrac{y_1 + y_2}{2}, \dfrac{z_1 + z_2}{2}\right)$	

Selected Answers

Chapter 1

Lesson 1-1 (pp. 6–12)

Guided Example: Solution 1: 3; –2; 5; 5   **Solution 2:** 3; –2; 5^2; 5   **Check:** 5

Questions:

1. Greek, "earth-measure"   **3.** Answers vary. Sample: A period has size and shape, but a point does not.   **5.** It is true for all values of c.   **7.** 14   **9.** $|x - y|$ or $|y - x|$ **11.** \$15.37 or \$16.63   **13.** Answers vary. Sample: In the actual situation, the houses would not necessarily be equidistant from the power plant's connection.

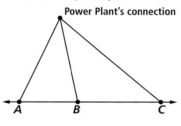
Power Plant's connection

15a. $AB = 175.8$; $BC = 175$; $AC = 350.8$ **b.** true

c.

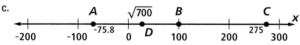

d. $100 - \sqrt{700}$ **e.** 102.3   **17.** –2.3 and 5.7   **19. a.** 15, 21
b. For $n = 5$, $\frac{5(6)}{2} = 15$. For $n = 6$, $\frac{6(7)}{2} = 21$. The formula appears to be correct.   **21.** $x = -1$ and $y = 3$

Lesson 1-2 (pp. 13–18)

Guided Example 3: 9; 4; $\frac{1}{2}$; 9; 4; 7; $\frac{1}{2}$; 7
Guided Example 5: a. 72; 72; 46   **b.** 72; 46; 46; 72; $\frac{1}{2}$; 10; $\frac{1}{2}$; 10
c. 24; 48; 22 ft; 34 ft

Questions:

1. a. A plane is a flat surface without any boundaries.
b. Answers vary. Sample: table tops and tennis courts
3. 3   **5.** vertical   **7.** oblique   **9.** vertical
11.

13.

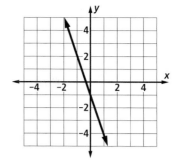

15a.

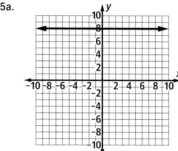

15b.

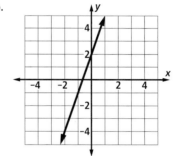

15c. $(2,8)$ **17a.** $(12, 2)$ **b.** $\frac{1}{12}$ **c.** $y = \frac{1}{12}x + 1$ **d.** 1 foot
19. $BC = 7$ **21.** $\sqrt{(-8)^2 + (-15)^2} = \sqrt{64 + 225} = \sqrt{289} = 17$

Lesson 1-3 (pp. 19–26)

Questions:

1a. discrete geometry **b.** Euclidean geometry **c.** graph theory   **3.** one   **5.** no   **7.** that one is able to travel without interruption over all of the arcs in a network exactly once   **9.** discrete geometry (The calculator plots individual points in the graph, but does not plot all of them.)   **11a.** discrete geometry **b.** Euclidean geometry **c.** graph theory   **13.** not traversable
15. Answers vary. Sample: Yes, this network is traversable, because each node has four arcs through it.   **17.** Answers vary. Sample: Any bridge connecting any two points would make the graph traversable, so there are multiple answers to the problem.

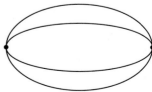

19. Answers vary. Sample: The network is not traversable because there are more than two odd nodes.

21. horizontal

23. vertical

25. Answers vary. Sample: Every distance from zero is labeled with exactly one number.

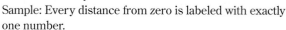

Lesson 1-4 (pp. 27–31)

Guided Example 1: Answers vary. Sample: *B, F, J; C, G, K; D, C, G, H; G, F, J, K* **Guided Example 2:** 2; 3; 1; 3

Questions:

1. Answers vary. Sample: Undefined terms avoid circularity. They provide an initial reference upon which other terms can be defined. 3. Currency is used to define money, and money is used to define currency.
5. Answers vary. Sample: *E, F, J, I; G, H, L, K;* and *B, C, G, F.* 7. Answers vary. Sample (adapted from *Merriam-Webster's Collegiate Dictionary 11th Edition*): "an individual detail" and "an end or object to be achieved"
9a. Answers vary. Sample: No point value is given to an Ace. **b.** Answers vary. Sample: An Ace is worth 5 points. 11. Step 1 is 2-dimensional; Steps 2 and 3 are 3-dimensional. 13. Answers vary. Sample: In Euclidean synthetic geometry, a point is an exact location and a line is a set of points extending in both directions containing the shortest path between any two points on it, while in discrete geometry a point is a dot and a line is a set of dots in a row. 15. $y = -\frac{1}{3}x + \frac{4}{3}$

Lesson 1-5 (pp. 32–37)

Guided Example 1: $9x - 12y = -51$; $x = -3$; -17; 2; -3; 2

Questions:

1. A set of books containing Euclid's organization of theorems and postulates, written around 300 BCE.
3. synthetic and plane coordinate 5. Number Line Assumption (part b of Point-Line-Plane Postulate)
7. Postulates are assumptions that are used to prove theorems. 9. (1, 4) 11. Place your pencil point above or below the paper. 13. The two equations are the same line, so the system has an infinite number of solutions, including (4,0) and (1,2). 15. **a.** true **b.** true **c.** false **d.** true **e.** false **f.** false 17. Answers vary. Sample: By the dimension assumption, there is a line and a point *P* not on the line. By the Number Line Assumption, there are infinitely many points on the line. Connect each of these points to *P* and there are infinitely many lines in the plane. 19. false

21. Answers vary. Sample:

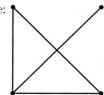

23. 6.75 in.

Lesson 1-6 (pp. 38–43)

Questions:

1. No. 3. ray

5a.

b. a line segment 7. $HG + HK = GK$ 9a. true b. true
c. false d. true

11a. Answers vary. Sample:

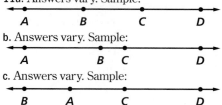

b. Answers vary. Sample:

c. Answers vary. Sample:

13. Answers vary. Sample: It is false because *MQ* and *QM* represent the same length. Rewrite to be *MQ = QM*.
15. the top of his head and the bottom of his foot
17. 4 19. It does not contain the endpoints.
21a. false b. true c. true
23. Answers vary. Sample: There is no unique line between *A* and *B*. 25. $a \leq -\frac{3}{8}$

Lesson 1-7 (pp. 44–50)

Questions:

1. Answers vary. Sample: Sometimes only a simple sketch is needed, but drawing figures by hand is prone to human error, such as slipping rulers, and is time-consuming. A DGS cannot prove theorems. 3. window 5. Answers will vary depending on the DGS used. 7. *AB* 9. *HG + GK > HK, GK + HK > HG, HG + HK > GK* 11a. yes b. no c. no 13. The DGS rounds numbers, so there may be a rounding error so that the values do not add up in this calculation. 15a. *E* is between *D* and *F.* b. *D* is between *E* and *F.* c. *F* is between *D* and *E.* 17. yes 19a. 2 cm b. 2 c. 4 cm 21. B, because the distances given are driving distances, not actual distances. 23. In both cases dragging one will change the shape of the other.
25a., b. Answers vary depending on type of DGS used.
27. –3 and 0 29. a ray 31. neither a ray nor a segment
33. a straight line that contains the two wells

Chapter 1 Self-Test (p. 54)

1. 650 km – 230 km = 420 km 2. Answers vary. Sample: $m = \frac{2-(-7)}{2-(-1)} = \frac{9}{3} = 3$. 3. Use (2, 2) to find *b*; because $y = mx + b$,

$2 = 3 \cdot 2 + b, b = -4$. So $y = 3x - 4$.
3. The Point-Line-Plane Postulate under the Unique Line Assumption states that "through any two points there is exactly one line." Thus, in Question 2, there can only be one line passing through the two points. **4.** Because $AB = AC + BC$, A, B, and C are collinear, with C between A and B **5.** neither, because one endpoint is open **6.** ray, because there is one endpoint **7.** segment, because there are two closed endpoints **8.** Yes, because there are two odd nodes **9.** The network must start and end at V and Z or start at Z and end at V to be traversable. **10a.** No, because $1 + 2 = 3$, which is not greater than the third side of length 3. **b.** Yes. By the Triangle Inequality Postulate, the sum of the lengths of any two sides is greater than the length of the third side. **c.** Yes. By the Triangle Inequality Postulate, the sum of the lengths of any two sides is greater than the length of the third side. **11.** It is discrete geometry, and each band member forms a part of the I, much like the pointillism of Seurat

12a.

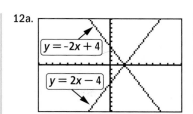

b. Yes, the lines intersect at $(2, 0)$. **13.** $PT = 7 - (-5) = 12$ **14a.** AC does not change because all points on the circle are equidistant from the center. **b.** They change by the same amount.
15. False; a line can use one point for both its endpoints.
16. B, E, D, and C **17.** $2x + 2y = 0$ so $y = -x$, meaning $-x = 2x - 8$, so $3x = 8$, $x = \frac{8}{3}$ and $y = \frac{-8}{3}$. The lines intersect at $(\frac{8}{3}, \frac{-8}{3})$. Graphing $y = -x$ and $y = 2x - 8$ on a DGS will also find the intersection. **18.** $-1 + 3.5 = 2.5$ and $-1 - 3.5 = -4.5$
19. a. $y = -\frac{2}{3}x + -\frac{2}{3}$ **b.**

Self-Test Correlation Chart

The chart below keys the **Self-Test** questions to the objectives in the **Chapter Review** at the end of the chapter. This will enable you to locate those **Chapter Review** questions that correspond to questions missed on the **Self-Test**. The lesson where the material is covered is also indicated on the chart.

Question	1	2	3	4	5	6	7	8	9	10
Objective(s)	G	K	E	F	C	C	C	A, H	A, H	F
Lesson(s)	1-6	1-2	1-4, 1-5	1-6, 1-7	1-5, 1-6	1-5, 1-6	1-5, 1-6	1-1, 1-2, 1-3, 1-4	1-1, 1-2, 1-3, 1-4	1-6, 1-7

Question	11	12	13	14	15	16	17	18	19
Objective(s)	H	J, L	I	B	D	E	L	I	J, K
Lesson(s)	1-1, 1-2, 1-3, 1-4	1-2, 1-5	1-1, 1-6	1-7	1-1, 1-2, 1-3, 1-5	1-4, 1-5	1-5	1-1, 1-6	1-2

Chapter 1 Chapter Review (pp. 55–57)

1a. A, B, E, F **b.** C, D **3.** A and B, B and C **5.** There is always a traversing path in this network starting at A or B and ending at the other. **7.** $\overleftrightarrow{CB}$ and $\overleftrightarrow{CA}$ move, $\overleftrightarrow{AB}$ does not. **9.** outside **11.** No, it is not, because they contain some of the same points (other than the endpoint). **13.** M or Q **15.** false **17.** No, because there may be multiple lines. **19.** Points, plane, line **21.** Dimension Assumption **23.** Point A is between points B and C. **25.** yes **27. a.** $|x - 237|$ **b.** 174 or 300 **29.** 15,248

31a.

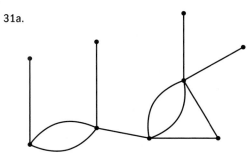

b. No, it cannot be done. The graph that describes the situation is not traversable since it has more than two odd nodes.

33. a.

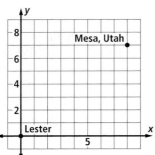

b. $y = \frac{7}{8}x$
35. 20
37. $x + y$
39. 6

41.

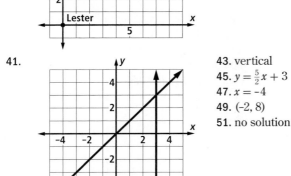

43. vertical
45. $y = \frac{5}{2}x + 3$
47. $x = -4$
49. $(-2, 8)$
51. no solution

Chapter 2

Lesson 2-1 (pp. 60–65)

Guided Example 1: *RESM, RELU; ULSM, ULGM, LGAC, ETNS; RELU, ULSM, ETNS, RESM, RTNM, LGAC* **Guided Example 2:** convex; the line connecting lower left and lower right vertices does not lie in the figure; convex; all segments connecting points in the figure are in the set

Questions:

1. Answers vary. Sample: Junk food is food that doesn't provide many nutritional benefits but is eaten solely on the basis that it may taste good.   3. Answers vary. Sample: It is too small and does not have a regular orbit.   5. false
7. Answers vary. Sample: The angles on the left side of the figure are almost, but not quite, right angles.

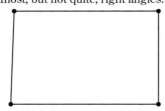

9. Answers vary. Sample: Definitions form one of the basic building blocks of geometry and, along with postulates, are used to prove theorems.   11. Answers vary. Sample: Grant   13. C   15. Answers vary. Sample: Figure A is a filled-in circle. Figure B is a ring in the shape of a circle. Figure C is a "typical" circle. Figure D is a circle as seen from an angle.   17. Answers vary. Sample: (0, 1), (0, 4),

(0, 6), (0, 10), (0, 1)   19. No. Answers vary. Sample: Using a DGS, it is impossible to create the figure with the given sides.   21. a. $x \geq 0$ b. $x \leq 0$ 23. $z < \frac{1}{2}$

Lesson 2-2 (pp. 66–71)

Questions:

1. conditional   3a. If today is Wednesday, then tomorrow is Thursday. b. Today is Wednesday. c. Tomorrow is Thursday.   5. A counterexample is a specific case for which the antecedent is true but the consequent is false.   7. If you are in Guelph, you are in Ontario.
9. If A and B are distinct points, then $\overleftrightarrow{AB}$ exists and it is unique.   11a. Other months with only 30 days are April, September, and November. b. Answers vary. Sample: In August, the day after the 30th is the 31st of August, not the first of September.   13. If something is a filimump, then it is a marklemump; If something is a marklemump, then it is a denifump; If something is a filimump, then it is a denifump.   15a. convex b. nonconvex c. nonconvex
17. No, Answers vary. Sample: A ray does not have an interior and exterior region.   19. $\{n : n \geq -1\}$

Lesson 2-3 (pp. 72–76)

Guided Example 2: a. square; rectangle; true **b.** rectangle; square; false

Questions:

1. Answers vary. Sample: A converse of a conditional is another conditional with the original antecedent as the consequent and the original consequent as the antecedent.   3a. If $x > 2$, then $x > 4$. b. It is false. Answers vary. Sample: If $x = 3$, then $x > 2$, but x is not greater than 4.

5a.

$-3\frac{1}{3}$ $3\frac{1}{3}$

b. Answers vary. Sample: If $|x| \leq 3\frac{1}{3}$, then x is between $-3\frac{1}{3}$ and $3\frac{1}{3}$, $x = 3\frac{1}{3}$, or $x = -3\frac{1}{3}$. c. Answers vary. Sample: If $x = 0$, then $|x| \leq 3\frac{1}{3}$.   7a. no b. If $x = 11$, then $x^2 = 121$. c. yes   9a. Antecedent: Two different lines intersect. Consequent: They intersect at one point. b. If two lines intersect at one point, then they are different.

11a.

b. If something is a fleeper, then it is a beeper. c. If something is a beeper, then it is a fleeper. In order for the converse to be true, everything that is a fleeper or a beeper is also the other, so fleepers and beepers are the same.   13. Answers vary. Sample: No, because C could be an endpoint, which would mean it is not between A and B.

15. Answers vary. Sample: No, the converse need not be true. Nathan could have had wealthy parents who left a good deal of money to him. **17.** Answers vary. Sample: No, because the antecedent is never true. **19a.** no **b.** yes

Lesson 2-4 (pp. 77–82)

Guided Example 1: a. II **b.** Property I **c.** Property III
Questions:
1. A good definition must use only words either commonly understood, defined earlier, or purposely undefined, accurately describe what is being defined, and include no more information than necessary. **3.** ⟺ **5.** 5
7. A diameter is a segment containing two points on the circle and the center of the circle, or the length of that segment. **9a.** Answers vary. Sample: The statement does not specify that it has to be all points in the plane that are the same distance from the center. **b.** Answers vary. Sample: A sphere has not been defined yet.
c. Answers vary. Sample: It does not specify that a circle is a set of points. **d.** Answers vary. Sample: It does not specify that all the points are to be the same distance from the center. **11.** *characteristics* ⇒ *term* **13.** Answers vary. Sample: A set is convex if and only if every segment that connects points of the set lies entirely within the set.
15. no **17a.** If $x + y$ is even, then x and y are both even.
b. no **19.** Antecedent: It is true that if you buy 100 lottery tickets then you are sure to win; consequent: I will buy 100 lottery tickets. **21a.** If a figure is convex, then it is a polygon. **b.** A segment is convex; however, it is not a polygon.

Lesson 2-5 (pp. 83–88)

Questions:
1a. Answers vary. Sample:

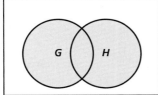

b. Answers vary. Sample:

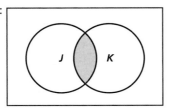

3. $A \cup B = \{(x, y) : (x, y)$ is on the line $y = 2x - 1$ or on the line $y = 3x - 1\}$ $A \cap B = \{(0, -1)\}$ **5.** Answers vary. Sample: The only point that is on both rays is N.
7. $\{S, M\}$ **9.** $\triangle LMF$ **11.** union **13a.** all real numbers **b.** $20 < x < 40$ **15.** $\overline{PQ}, \overline{QR}$

17a.

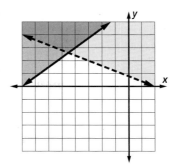

b.

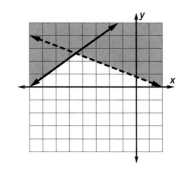

19a. $F \cup M$ is the entire rostered team. **b.** $F \cap M$ is the empty set, because no one can be both a rostered female and a rostered male. **21.** *characteristics* ⇒ *term*
23. The antecedent is "you work more than 40 hours a week" and the consequent is "you will receive bonus pay for overtime." **25.** $x = 120$; $2(180 - 120) = 2 \cdot 60 = 120$, so it checks.

Lesson 2-6 (pp. 89–94)

Questions:
1a. Answers vary. Sample: $\overline{DE}$ does not intersect exactly two other segments. **b.** $\overline{FH}$ and $\overline{IG}$ each intersect three segments instead of two. **c.** Points J and L are not connected by a segment. **3a.** A, B, C, D, E, F
b. $\overline{AB}, \overline{BC}, \overline{CD}, \overline{DE}, \overline{EF}, \overline{FA}$ **c.** Answers vary. Sample: $\overline{CF}$ **d.** Answers vary. Sample: $\overline{AB}, \overline{BC}$ **e.** Answers vary. Sample: A, B **5.** Answers vary. Sample:
7. nonconvex **9.** quadrilateral, nonconvex **11.** hexagon, convex **13.** triangle, polygon, figure **15.** quadrilateral, Colorado **17.** hexagon, Utah
19. a. trianglular region **b.** $(4, 4)$, $(0, 0)$, $(4, 0)$ **21.** pentagon
23. A and B are equivalent sets
25.

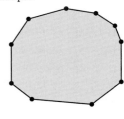

27. $x + 2$

Lesson 2-7 (pp. 95–100)

Questions:

1. an educated guess or opinion   3. a counterexample
5. Answers vary. Sample:
7. Anisa would have to find
every tiger in the world, and
show that each tiger is not
white.   9. Answers vary.
Sample: The sum of two even
whole numbers is an even
whole number.   11. Answers vary. Sample: B
13. Answers vary. Sample: B 15. $\frac{\text{Area } (\triangle ABC)}{\text{Area } (\triangle DEF)} = 4$

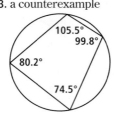

17. Answers vary. Sample: $PENTA$, $PATNE$ 19. Answers
vary. Sample: The intersection of two rays can be a point, a
segment, a ray, or the empty set.

Chapter 2 Self-Test (pp. 104–105)

1. Answers vary. Sample: Words not yet defined are
used.   2a. IV b. II c. V   3. "A triangle is acute" is the
consequent and "a triangle has three acute angles" is the
antecedent.   4. A triangle has three acute angles ⇒ a
triangle is acute.   5. If a figure is a polyhedron, then it
has at least four faces.   6. A triangle has at least two sides
with the same length if and only if it is isosceles.
7a. A is the endpoint in both $\overrightarrow{AB}$ and $\overrightarrow{AC}$, A is between B
and C, so $\overrightarrow{AB}$ and $\overrightarrow{AC}$ are opposite rays, so $\overrightarrow{AB} \cup \overrightarrow{AC}$ is $\overleftrightarrow{BC}$.
b. A is the endpoint in both $\overrightarrow{AB}$ and $\overrightarrow{AD}$, and $\overrightarrow{AB}$ and $\overrightarrow{AD}$
are two rays, so $\overrightarrow{AB} \cup \overrightarrow{AD}$ is $\angle BAD$.   c. Only $\overrightarrow{BC}$ is in both
$\overrightarrow{BC}$ and $\overrightarrow{CB}$ so $\overrightarrow{BC} \cap \overrightarrow{CB}$ is $\overline{BC}$ d. Only A is in both $\overrightarrow{AD}$
and $\overrightarrow{CB}$ so $\overrightarrow{AD} \cap \overleftrightarrow{CB}$ is A.   8. $\triangle MNR$ is equilateral if and
only if $\triangle MNR$ has three angles with equal measure.
9. Answers vary. Sample: She is incorrect, so to prove her
wrong a counterexample is needed. A counterexample
could be found by finding a 16-year-old who does not drive.
10. Answers vary. Sample: No, a student might fail the test
even if he or she uses a pencil. For example, the student
may not have studied for the test and failed but still used
a pencil on the test.   11. a. Answers vary. Sample: $x = 0.5$
gives $0.5 < 1$ and $\frac{1}{0.5} = 2 > 1$.   b. Answers vary. Sample:
$x = -1$ gives $-1 < 1$, but $\frac{1}{-1} = -1 < 1$. Also, $x = -7$ gives
$-7 < 1$, but $\frac{1}{-7} = -\frac{1}{7} < 1$.   c. Answers vary. Sample: If $x < 1$
and $x > 0$, then $\frac{1}{x} > 1$.   12a. Two of the segments

intersect fewer than two other segments. b. Two of the
segments intersect each other at points other than the
endpoints, and the same two segments intersect more
than two other segments. c. The figure does not lie in one
plane.   13. Answers vary. Sample: "Please take one or two
chips."   14. The first statement is true since $-5 \cdot -5 = 25$
but the second statement is false because $5^2 = 25$ as well,
so $x^2 = 25$ does not imply $x = -5$ 15. North America
16. {Presidents of the United States} and {Authors of
Original Proofs of the Pythagorean Theorem}

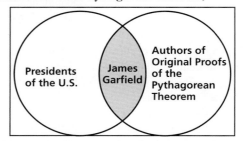

17. We cannot conclude anything because we do not know
if Rachel's refund was because of her score on the test. It
is possible that there were other things that could have
warranted a refund from the company other than a less
than 10-point improvement on the test.
18.

19a.   b.

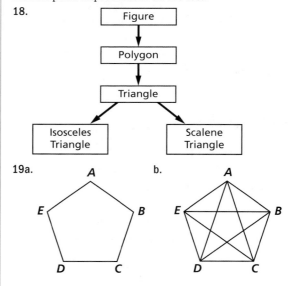

Self-Test Correlation Chart

Question	1	2	3	4	5	6	7	8	9	10
Objective(s)	F	A, D	B	C	C	C	H	C	L	E
Lesson(s)	2-4	2-1, 2-6	2-2	2-2, 2-4	2-2, 2-4	2-2, 2-4	2-5	2-2, 2-4	2-2, 2-7	2-3
Question	11	12	13	14	15	16	17	18	19	
Objective(s)	G	D	K	C	J	J, N	I	M	A, D	
Lesson(s)	2-2, 2-7	2-6	2-1	2-2, 2-4	2-5	2-2, 2-5	2-2, 2-3	2-6	2-1, 2-6	

Chapter 2 Chapter Review (pp. 106-109)

1. nonconvex 3. convex

5. Answers vary. Sample:

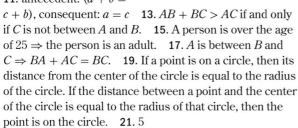

7. antecedent: Two different lines are parallel; consequent: They do not intersect.

9. antecedent: $x = 4$, consequent: $x > -2$

11. antecedent: $(a + b = c + b)$, consequent: $a = c$ 13. $AB + BC > AC$ if and only if C is not between A and B. 15. A person is over the age of 25 $\Rightarrow$ the person is an adult. 17. A is between B and $C \Rightarrow BA + AC = BC$. 19. If a point is on a circle, then its distance from the center of the circle is equal to the radius of the circle. If the distance between a point and the center of the circle is equal to the radius of that circle, then the point is on the circle. 21. 5

23. Answers vary. Sample:

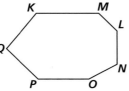

25a. Answers vary. Sample:
b. Answers vary. Sample:
A correct name is $KMLNOPQ$, and an incorrect name is $MQPLONK$. c. Answers vary. $\overline{KM}, \overline{NO}, \overline{PQ}$ 27a. For any real numbers a, b, and c, if $a = b$, then $a + c = b + c$. b. true 29. False, Answers vary. Sample: There are many statements that are true but have false converses, such as "If you are Michael Jordan, then you played for the Bulls." This statement is true. However, the converse, "If you played for the Bulls, then you are Michael Jordan" is not true because many other players, such as Scottie Pippen, also played for the Bulls. 31. False, Answers vary. Questions 29 and 30 disprove both parts of this biconditional statement.

33. Answers vary. Sample: The definition is circular.
35a. Yes, this is a good definition. b. If two circles are intersecting circles, then they intersect in exactly two points. If two circles intersect in exactly two points, then they are intersecting circles.

37a. Answers vary. Sample: b. Answers vary. Sample:

   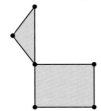

39. Answers vary. Sample: This is an instance of the conjecture, because any letter order will give an order where each letter is connected to the adjacent letter by a segment. 41. $\{M, N, O\}$ 43. $\triangle MNO$ 45. Answers vary. Sample:

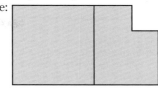

47. No, Julian has not shown the statement to be false. The term "long enough" was not defined, so it is possible that the days Julian spent practicing was not "long enough." 49a. Ø b. Yes. Answers vary. Sample: Franklin Roosevelt was elected President four times, so he is not in either set. 51. Answers vary. Sample: A problem with using the old definition is that the length of the foot of one grown man is not the same as the length of the foot of another. The problem is addressed by using the speed of light and a set time (both are definite quantities) to have a concrete definition. 53a. If you are waiting in a line, then your line will move slower than the other lines. b. Answers vary. Sample: The line you are standing in is moving the fastest. 55. Answers vary. Sample: The tortoise beat the hare; however, slow but steady runners do not typically win a marathon. Instead, the winners of a marathon are typically quite fast and steady.

57.

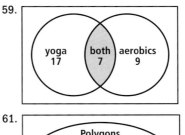

59.

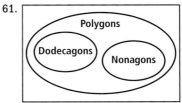

61.

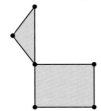

Chapter 3

Lesson 3-1 (pp. 112–119)

Guided Example: a. 32° b. $\overset{\frown}{RS}$; 32°; 328°

Questions:

1.
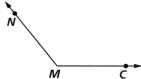

3. Answers vary. Sample:

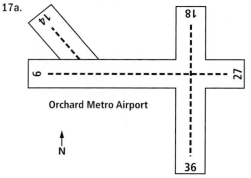

5. G 7. yes 9. yes 11. yes 13a. obtuse b. right
c. obtuse d. acute 15a. ∠CAB, ∠CAD, ∠DAB b. 120°

17a.

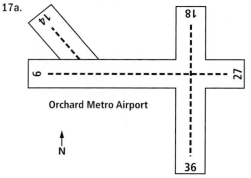

Orchard Metro Airport

b. 32

19a. Answers vary. Sample:

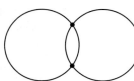

b. Answers vary. Sample:

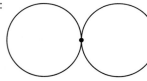

c. Answers vary. Sample:

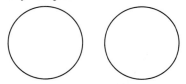

21. Answers vary. Sample: $\overrightarrow{AB} \cap \overrightarrow{BA}$ 23. 7"; 27"
25a. (6, 8) b. $\frac{4}{3}$ c. Answers vary. Sample: $y = \frac{4}{3}x$

Lesson 3-2 (pp. 120–125)

Questions:
1. The object returns to its original position. 3. 55°
5. DF♯A 7. The three angles have the same measure.
9.

11. 1260° 13a. 120°
b. 72° 15. $\frac{1}{8}$
17. a. the set of points
with both x and y
integer coordinates
b. the set of all points
such that at least one
of its coordinates is
an integer

Lesson 3-3 (pp. 126–132)

Guided Example 1: 1. $\overline{FG}, \overline{GD}; \overline{DE}, \overline{EF}$ 2. $\overrightarrow{ZY}; \overrightarrow{ZY}$
Guided Example 2: a. DAC; 90; 125 b. 104, 35, 90; 229; 131

Questions:
1. 180 3. equal 5a. No. ∠AOB and ∠COD do not have
a common side. b. No. There is no common side; in the
interior of the angle formed by the noncommon sides.
c. yes d. No. $\overline{EO}$ is not interior to ∠AOD.
7a. Answers vary. Sample:
b. 25
9. m∠2 = 55; m∠3 = 90;
m∠4 = 35; m∠5 = 145 11. 75 13a. always b. never
15. Answers vary. Sample:

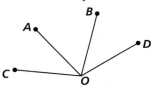

17. Answers vary. Sample:
19. The measure of the
angle is 123°, and its
supplement has
measure 57°.

21.

23. −67.5° 25. Answers vary. Sample: Old McDonald can
measure the obtuse angle from the ground to the grain
pile at the bottom. Because this value forms a linear pair
with the angle of repose, he could subtract the measure of
the obtuse angle from 180 to find the angle of repose.

27. 70°   29. Answers vary. Sample: -45°, 315°
31. the Unique Angle Assumption of the Angle Measure
Postulate   33. Answers vary. Sample: C is the set of odd
positive integers.

Lesson 3-4 (pp. 133-138)

Questions:
1. equal sign   3. $\frac{1}{2} = 50\%$ 5. a. $x < 2$ b. Addition and
Multiplication Properties of Inequality   7. $-4m\angle T <$
$-4m\angle Q$ 9a. The angle has a measure of $90 - 23$.
b. 67°   11. Addition Property of Inequality   13. Equation
to Inequality Property   15. A   17a. Because $PQ = RS$,
$PQ + QR = RS + QR$ by the Addition Property of Equality.
Therefore, $PR = QS$. b. It would be the same except we
would use the Addition Property of Inequality to conclude
that $PR < QS$.   19a. If the measures of two angles are
equal, then they are vertical angles. b. Answers vary.
Sample: Angles do not have to be positioned as vertical
angles to have equal measures. For example, an
equilateral triangle has three angles of equal measure,
no pair of which are vertical angles.   21. $A' = (5, 0)$,
$B' = (5, -5)$, $C' = (0, -5)$ 23. Answers vary. Sample: No,
because they can each see more than half of the circle.
$210° + 210° = 420° > 360°$, which is the whole circle.

Lesson 3-5 (pp. 139-144)

Guided Example 3: 2. Linear Pair   3. supplementary angles
Questions:
1. Angle Addition Property   3. definition of an obtuse
angle   5. postulates, definitions, and previously proved
theorems   7. $m\angle 1 = m\angle 2$ a. Given b. $m\angle 1 = m\angle 2$
9. Answers vary. Sample: It still makes sense for most
people because in most cases DNA, like fingerprints, is
unique to a person. However, unlike fingerprints, identical
twins can have the same DNA.   11. Additive Property
of Distance   13. Substitution Property   15. B
17. $m\angle ABC = 20$ 19. F♯A♯C♯   21a. cannot conclude
b. can conclude c. cannot conclude
23a. 100 b. 11   25. $y = 0$

Lesson 3-6 (pp. 145-150)

Questions:
1. m 3. $\angle 6$ 5. $m \parallel n$ 7. 70°   9. false   11. a. $-\frac{1}{2}$ b. $-\frac{1}{2}$
13. vertical lines   15. Transitivity of Parallelism Theorem
17. Yes, because they all have the same slope (0).
19. The slopes for one pair of opposite, parallel sides
are $\frac{13 - 11}{11 - 4} = \frac{4 - 2}{10 - 3} = \frac{2}{7}$ and the slopes for the other pair
of opposite, parallel sides are $\frac{11 - 2}{4 - 3} = \frac{13 - 4}{11 - 10} = \frac{9}{1} = 9$.
Therefore, the slopes of the opposite lines are equal, so
opposite lines are parallel.   21. ℓ and m 23. Answers
vary. Sample: By the Vertical Angles Theorem, $m\angle AEG =$
$m\angle DEC$, and by the Corresponding Angles Theorem,
$m\angle DEC = m\angle PCS$, so by the Transitive Property of

Equality, we have $m\angle AEG = m\angle PCS$.   25. Addition
Property of Equality   27a.   b. $x = \frac{40}{3}$

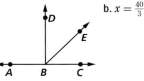

Lesson 3-7 (pp. 151-158)

Guided Example 2: $(3, 15)$; E; $(-9, 0)$; $\frac{5}{4}$; $\frac{5}{4}$; equal
Guided Example 3: 0; 2; 4; 2; 8, 4
Questions:
1. 3   3. 3; 9   5. S_k Theorem 1: Parallel Property   7. $\frac{2}{5}$
9. $(6, 15)$ 11. $\frac{1}{3}$ 13a. $y - b = \frac{kb - b}{ka - a}(x - a) = \frac{b}{a}(x - a)$
b. The y-intercept of this line is $\frac{b}{a}(-a) + b = -b + b = 0$.
Substituting in $(0, 0)$ to the equation yields $-b = -b$, which
is always true; thus, $(0, 0)$ is a solution.
15a.

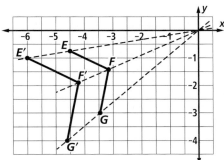

b. 1.5; Answers vary. Sample: By finding the coordinates
for the points.   17. $x = 20, y = 45$ 19. 7; Because $AC > 3$,
$2AC = AB > 6$. Thus, $6 < AB < 8$, and 7 is the only integer
that satisfies these conditions.   21. 60

Lesson 3-8 (pp. 159-164)

Guided Example 2: $-\frac{1}{4}$; $-\frac{1}{4}$; $-\frac{1}{4}$; $-\frac{1}{2}$; $-\frac{5}{2}$; $-\frac{1}{4}x - \frac{5}{2}$
Questions:
1. Two lines are perpendicular if and only if they form a
90° angle.
3.   5. Two Perpendiculars
     Theorem 7. $y = -\frac{3}{2}x$

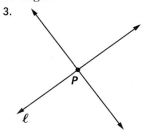

9.

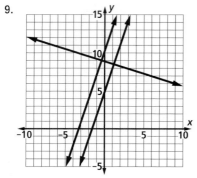

11a. 90° **b.** 40° **c.** 90° **13a. i. b. i. c. iv. d. iv, v.**
15. If two nonvertical lines are perpendicular, then
the product of their slopes is –1. If the product of the
slopes of two nonvertical lines is –1, then the lines are
perpendicular. **17.** Impossible: If two coplanar lines are
each perpendicular to the same line, then they are parallel
to each other. **19.** Answers vary. Sample: No, there is no
k because the figure is not scaled proportionately.
21. The relation ≤ has the reflexive property and the
transitive property, but not the symmetric property.

Lesson 3-9 (pp. 165–171)

Questions:
1.

3. The perpendicular bisector of a segment is any line,
ray, or segment which passes through the midpoint of the
segment at a 90° angle to the segment. **5.** Drag point D
along the line $\overleftrightarrow{AB}$. **7.** Answers vary. Sample:

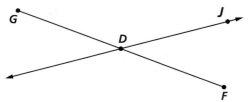

9.

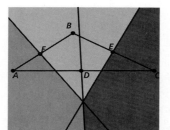

11. Given, Name, and
Construct

13a.

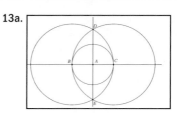

b. Answers vary.
Sample: A is the
midpoint of $\overline{BC}$. B and
C are points on the
circle with radius $\overline{BA}$.
Thus, $\overline{AC}$ is also a
radius of the circle,
so $BA = AC$. **c.** yes $\overleftrightarrow{DE}$ and $\overleftrightarrow{AB}$ are perpendicular

15.

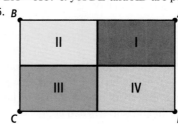

The points closest
to A are all the
points in region I.
The points closest
to B are all the
points in region II.
The points closest
to C are all the

points in region III. The points closest to D are all the
points in region IV. **17a.** all points that lie on the
bisector of $\overleftrightarrow{XY}$ **b.** all points that lie on the bisector
of $\overleftrightarrow{XZ}$ **c.** the intersection point of all three bisectors
19. a. Answers vary. Sample:

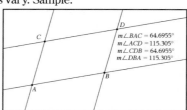

b. Answers vary. Sample:

c. Supplementary pairs: (∠BAC, ∠ACD), (∠BAC, ∠DBA),
(∠ACD, ∠CDB), (∠CDB, ∠DBA); Congruent pairs:
(∠DBA, ∠ACD), (∠BAC, ∠CDB) **d.** Yes; they will always
be true. **e.** The sum is 360°; the sum is always 360°, no
matter how the lines are dragged. **21.** $y = -x + 10$
23. $x = 10$ **25a.** 80° **b.** $(2w)°$

Chapter 3 Self-Test (pp. 176–177)

1. $x + y = 90$ and $x = 81$, so $81 + y = 90$ and $y = 9$; 9
2. $m\angle 3 + m\angle 4 = 180$; $m\angle 4 = 180 - m\angle 3 = 180 - 27 = 153$ 3. $m\angle 3 + m\angle 9 = 180$ $(m\angle 9 - 123) + m\angle 9 = 180$; $2m\angle 9 = 303$; $m\angle 9 = 151.5$, $m\angle 6 = m\angle 3 = 28.5$ 4. $340° - 210° = 130$; $\frac{130°}{2} = 65°$ 5. $m\angle APS = m\angle BPS$; $4q + 8 = 2q + 18$; $2q = 10$; $q = 5$; $m\angle APB = 2(2q + 18) = 56$
6. $m\angle 1 + m\angle 2 = 180$; $(12 + 5m\angle 2) + m\angle 2 = 180$; $6m\angle 2 = 168$; $m\angle 2 = 28$ 7. Transitive Property of Equality   8. Addition Property of Equality   9. $\frac{360°}{12} = 30°$; $30° \cdot 2 = 60°$ 10. $-90 + -45 = -135°$

11.

12a. $m\overset{\frown}{PD} = 360° - 230° = 130°$ b. $m\angle POM = 230° - 180° = 50°$ c. $m\overset{\frown}{WM} = m\overset{\frown}{PD} = 130°$ d. $m\overset{\frown}{MWD} = 180°$
13a. Answers vary. Sample: any three of: $\overset{\frown}{PM}$ and $\overset{\frown}{DW}$, $\overset{\frown}{PMW}$ and $\overset{\frown}{MWD}$, $\overset{\frown}{DP}$ and $\overset{\frown}{WM}$, $\overset{\frown}{PDW}$ and $\overset{\frown}{PMW}$, $\overset{\frown}{MPD}$

and $\overset{\frown}{MWD}$ b. Answers vary. Sample: any two of: $\angle POM$ and $\angle DOW$, $\angle POD$ and $\angle MOW$, $\angle MOD$ and $\angle POW$

14.

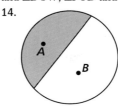

15. Answers vary. Sample: $a \parallel b$ because the supplement to $100°$ is $80°$, which gives an equal corresponding angle to the given $80°$ angle, and $c \parallel d$ because a set of two lines is parallel if and only if a pair of corresponding angles are equal in measure.

16. $2x + 2y = 3x + 57 = 90$; $3x = 33$; $x = 11$; $22 + 2y = 90$; $2y = 68$; $y = 34$ 17a. $3y = -x + 17$; $y = -\frac{x}{3} + \frac{17}{3}$; slope $= -\frac{1}{3}$ b. $-\frac{1}{-\frac{1}{3}} = 3$ 18. Answers vary. Sample: $\overline{SR} = \overline{RW}$ because if R is the midpoint of $\overline{SW}$, it bisects $\overline{SW}$.
19. $m\angle 2 = m\angle 1$ because $\angle 2$ and $\angle 1$ are vertical angles. Therefore, by the Substitution Property of Equality, $m\angle 1 + m\angle 3 = 90$.   20. $A' = (3 \cdot -1, 3 \cdot 2) = (-3, 6)$; $B' = (3 \cdot 0, 3 \cdot -1) = (0, -3)$; $C' = (3 \cdot 2, 3 \cdot 1) = (6, 3)$
21. A   22. a. $-\frac{3}{2}$ b. $\frac{2}{3}$

Self-Test Correlation Chart

Question	1	2	3	4	5	6	7	8	9	10
Objective(s)	B	A	A	H	B	B	F	F	H	D
Lesson(s)	3-3, 3-4	3-1, 3-3	3-1, 3-3	3-1, 3-2	3-3, 3-4	3-3, 3-4	3-4, 3-5	3-4, 3-5	3-1, 3-2	3-2
Question	11	12	13	14	15	16	17	18	19	20
Objective(s)	D	E	E	I	C	B, C	K	G	G	J
Lesson(s)	3-2	3-1	3-1	3-9	3-6, 3-8	3-3, 3-4, 3-6, 3-8	3-6, 3-8	3-3, 3-5, 3-6, 3-8	3-3, 3-5, 3-6, 3-8	3-7
Question	21	22								
Objective(s)	J	K								
Lesson(s)	3-7	3-6, 3-8								

Chapter 3 Chapter Review (pp. 178–181)

1. Answers vary. Sample:

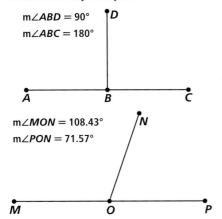

$m\angle ABD = 90°$
$m\angle ABC = 180°$

$m\angle MON = 108.43°$
$m\angle PON = 71.57°$

3. $\angle AEB$ and $\angle CED$ 5. $\angle CEA$ and $\angle DEB$ are obtuse; $\angle AEB$ and $\angle CED$ are acute.   7. $t = \frac{49}{2}$
9.

11. 60   13. $m\angle 2 = m\angle 3 = m\angle 6 = m\angle 7 = 45$; $m\angle 4 = m\angle 5 = m\angle 8 = 135$ 15. yes   17. $m\angle 5 = 90$. Because $v \perp t$, the lines form a $90°$ angle.   19. on the midpoint of $\overline{AC}$ 21. $\triangle CLJ$ 23. $\angle DOE$, $\angle DOC$, $\angle EOC$, $\angle EOA$, $\angle DOA$, $\angle COA$, $\angle DOB$, $\angle EOB$, $\angle COB$, $\angle AOB$
25. true   27. Substitution Property of Equality
29. $LM = NO$ 31a. Multiplication Property of Equality b. Addition Property of Equality c. Multiplication Property of Equality   33. Corresponding Angles Postulate
35. Vertical Angles Theorem   37. $m\angle FOR + m\angle FOP = 3m\angle FOP = 180$, $m\angle FOP = 60$, $m\angle FOR = 180 - m\angle FOP = 120$ 39. $135°$ 41. $135°$

43.

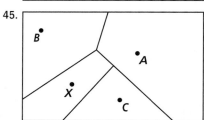

45.

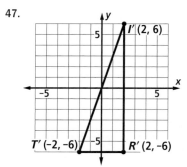

47.

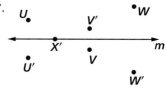

49. 2   **51.** –0.8   **53.** $-\frac{3}{5}$ **55.** $\frac{5}{2}$

Chapter 4

Lesson 4-1 (pp. 184–188)

Guided Example 2: a. 3; 3; 3; 4; 3; 4 **b.** 4; 4; –3; –4
Questions:
1. preimage   **3.** P **5. a.** true **b.** false **c.** false
7.

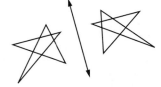

9. a. $(0, -1)$ **b.** $(0, 1)$ **c.** $(1, 0)$ **11. a.** B, ℓ; C, m; D, n.
b. $r_{\ell}(A) = B$; $r_{m}(A) = C$; $r_{n}(A) = D$
13.

15.

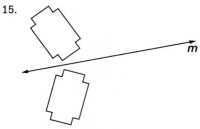

17. a. Yes, the slope of line ℓ is equal to the slope of
line m. **b.** No, the slope of line ℓ is not the opposite
reciprocal of the slope of line n.   **19a.** The measures of
supplementary angles add to 180°. **b.** Corresponding
Angles Postulate   **21.** If a network has no odd nodes, then
it is traversable.

Lesson 4-2 (pp. 189–195)

Guided Example 2: a. right; clockwise **b.** left;
counterclockwise
Questions:
1. No; there is a one to one correspondence between a
point and its image over a single line of reflection.
3. a. 2 cm; distance is preserved. **b.** 39°; angle measure
is preserved. **c.** on $\overline{Y'K'}$; betweenness is preserved. **d.**
Yes; collinearity is preserved. **5.** A figure and its reflection
image have opposite orientations.
7a.

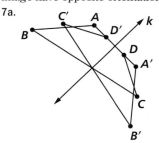

b. *BCDA, CDAB, DABC* **c.** clockwise **d.** Answers vary.
Sample: Measure the lengths of a segment and its image
to see that they are equal.   **9a.** 6 **b.** 107
c.

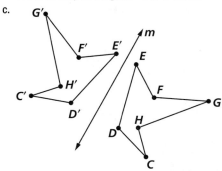

11a. $-\frac{1}{3}$ **b.** $\frac{1}{3}$ **c.** $\frac{1}{3}$ **d.** Answers vary. Sample: The slope of
the reflection image of a line about the x-axis or the y-axis,
is the opposite of the slope of the preimage line.

13.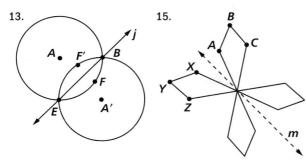

15.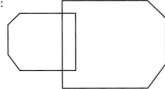

7a.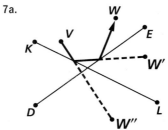

b. 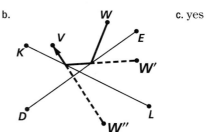 **c.** yes

17. a. $y = \frac{3}{2}x + 2$ **b.** $y = -\frac{3}{2}x + 2$ **19. a.** *DEHIFJCG*
b. *CJFIHEDG* **c.** *I* and *G* **21. a.** Answers vary. Sample:
$y = -\frac{1}{3}x + \frac{25}{3}$ **b.** Answers vary. Sample: $y = 3x - 5$
23. Answers vary. Sample: yes, when $x = 50$, corresponding
angles are congruent, so the lines are parallel.
25. Answers vary. Sample:

9.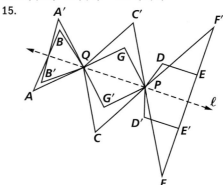

11a. $(1, 10)$ **b.** $(5, 6)$ **13.** $(-2, 0)$

15.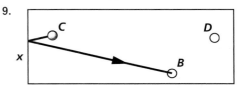

Lesson 4-3 (pp. 196–201)

Questions:

1.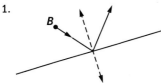

3. Aim at point *A*, which is the intersection of x and $\overline{R'T}$.

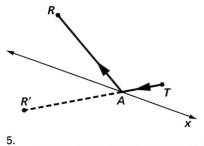

5.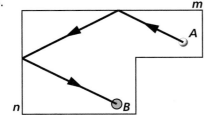

17. 28 **19.** Answers vary. Sample: We know that
$m\angle ABC = 180 - 140 = 3x$ by the definition of
supplementary angles and the Corresponding Angles
Postulate. Therefore, $3x = 40$, so $x = \frac{40}{3}$. Thus, $x < 40$.

21.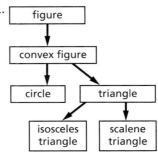

Lesson 4-4 (pp. 202–209)

Guided Example 3: $a + b$; a; $2a + b$; $2a + 2b$

Questions:

1. B 3a. C b. C c. A d. K 5a. C b. I 7a. 13 b. –11
c. –14 d. 20   9a. $r_s(r_t(\triangle XYZ)) = X''Y''Z''$ b. t c. $X''Z''$
d. $\angle Z''X''Y''$ e. 4 cm f. Answers vary. Sample: $\overline{ZZ''}$
11. A is $x + y$ away from m, meaning that A' is also $x + y$
away from m. Therefore, A' is $x + y + x = 2x + y$ away
from n, meaning A'' is also $2x + y$ away from n. Therefore,
$y + AA'' = 2x + y$, so $AA'' = 2x$.   13. $y = -5$
15.

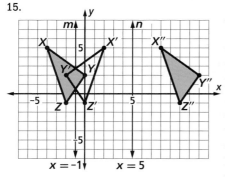

Explanations
vary. Sample:
m and n are
both vertical
lines, so they
are parallel.
So, $r_n \circ r_m$ is a
composite of
two reflections
over parallel
lines and is a
translation.

17. Answers vary. Sample:

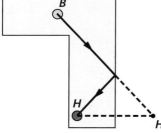

19. counterclockwise   21. $y = 6$ 23. Answers
vary. Sample: If a and b were parallel, then by the
Corresponding Angles Postulate, $3x = x$. We know that
$20 + 3x = 180$, so $3x = 160$, or $x = \frac{160}{3}$. But $3\left(\frac{160}{3}\right) \neq \frac{160}{3}$, so
the lines are not parallel.

Lesson 4-5 (pp. 210–215)

Guided Example 2: M; 114; 228°; 114°; 66°; 132°; –132°

Questions:

1. rotation; center   3. angle measure, betweenness,
collinearity, distance, and orientation   5. 90°
7a. $A'' = (-1, -3)$; $B'' = (-2, -6)$; $C'' = (-4, -6)$;
7a. The rotation is 180° centered about the point (0, 0).
9.

11. a.  b.

c.

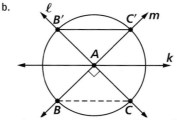

13. Answers vary. Sample: Reflect
over $x = 0$ and then reflect over
$x = -4$.   15. Answers vary. Sample:
Reflect over reflect over $y = x$ and
then reflect over $y = 0$.
17. Answers vary. Sample:
Translations are composites of two
reflections, so because reflections
reverse orientation, the first reflection reverses the
orientation and the second reverses it back to its original
orientation.   19. No, T does not preserve distance.
21. Answers vary. Sample: $-2x + y = 0$ and $-2x + y = 8$

Lesson 4-6 (pp. 216–222)

Guided Example 2: –5; 8; –2; 4

Questions:

1. direction, magnitude   3a. $\langle 0, 4 \rangle$ b. $(a, b + 4)$
5a. $\langle 3.5, -7.1 \rangle$ b. $(a + 3.5, b - 7.1)$ 7. $\langle -5, -3 \rangle$
9.

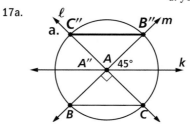

11. (6, 17)   13. $y = 7$
15a. $A' = (7, 7)$;
$B' = (9, 8)$ b. $y = \frac{1}{2}x + \frac{7}{2}$
c. They are parallel
because $\overline{A'B'}$ is the image
of m under a translation.
d. yes

17a.

19. Answers vary. Sample: Yes; reflections preserve
distance, so sides of equal length will have equal lengths
as images.

Lesson 4-7 (pp. 223-229)

Questions:

1. rotation, translation, reflection, glide reflection
3. reflection **5a.** translation **b.** The hands have the same orientation. **7a.** translation **b.** glide reflection **c.** rotation **d.** Answers vary. Sample: A reflection could not have been used. If orientation is reversed then the image of *ABCD* is *FEHG* or *HGFE*. Both of these are glide reflections. **9a.** Answers vary. Sample:

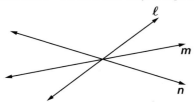

b. Impossible; a composite of reflections over three lines cannot be a translation. **c.** Answers vary. Sample:

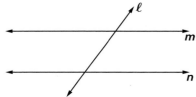

d. Impossible; a composite of reflections over three lines cannot be a rotation. **11.** the stem **13a.** "Even" and "odd" refer to the number of reflecting lines in the isometry. **b.** orientation **15a.** translation **b.** translation **c.** rotation **d.** rotation

17.

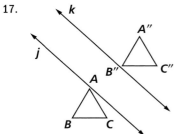

Lesson 4-8 (pp. 230-240)

Questions:

1a. Answers vary. Sample: A change in pitch is the raising or lowering of notes. **b.** C **3a.** Answers vary. Sample: Augmentation is a lengthening of the duration of a group of notes by a fixed scale factor. **b.** D **5a.** Answers vary. Sample: Retrograde is a reversal in the order of a group of notes. **b.** A **7.** $h = 2$ **9a.** change in pitch (transposing) **b.** translation upward 5 notes **c.** Answers vary. Sample: The pitch is higher. **11a.** retrograde **b.** reflection over the vertical line between the middle two

notes **c.** Answers vary. Sample: The melody is played backward. **13a.** rhythmic diminution **b.** size change by a factor of $\frac{1}{2}$ in time **c.** Answers vary. Sample: The music is twice as fast. **15.** A, F♯, D, D, C♯, B, C♯

17a. 1 beat **b.** glide reflection **c.** Answers vary. Sample: Each part is a glide reflection of the part above it. The third and fourth measures of the first part are a horizontal translation of the first and second measures. **19.** inversion and transposition **21.** rhythmic augmentation in 14–16, transposition in 15–16 (by different constants) **23a.** no **b.** Answers vary. Sample: Rotations preserve orientation, so a composite of rotations preserves orientation. A glide reflection reverses orientation.
25. ⟨-3, 4⟩ **27a.** m∠*AOB* = 90 **b.** m∠*ABD* = 270°
c. true

Chapter 4 Self-Test (pp. 244-245)

1.

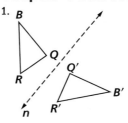

2.

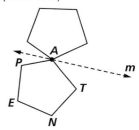

3.

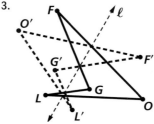

4. $F' = (-3, -6)$; $G' = (4, -4)$; $H' = (-1, 1)$

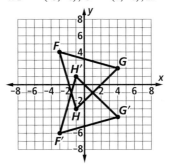

5. Answers vary. Sample:

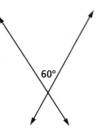

60°

6. Answers vary. Sample:

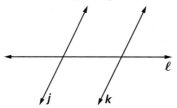

ℓ

j k

7. True; reflections preserve angle measure.   **8.** True; the line connecting the preimage point and the reflection image point will be perpendicular to the line of reflection.
9. False; the rays extend in opposite directions.

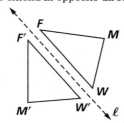

F
F'
M
W
W' ℓ
M'

10. $\overline{AE}$; Going backward, when reflecting $\triangle JWB$ over m, $\overline{SM}$ corresponds to $\overline{BW}$. When reflecting $\triangle AEH$ over n, $\overline{BW}$ corresponds to $\overline{AE}$.

11.

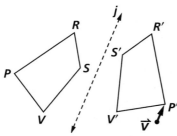

j
R R'
S'
S
P
V V' $\vec{v}$
P'

12. Because lines z and y are parallel and 5 cm apart, $r_z \circ r_y$ is a translation of magnitude 10 cm in the directon from y to z.

13.

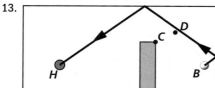

D
C
H B

14.

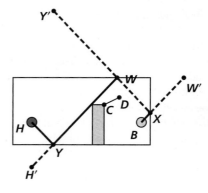

Y'
W
C D W'
H X
B
Y
H'

15. $(2, 0)$

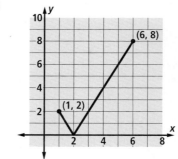

y
10
8 (6, 8)
6
4
2 (1, 2)
x
2 4 6 8

16. Answers vary. Sample: The images in the driveway will reflect to a position straight ahead of the pedestrian.

17. a.

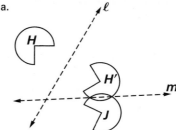

ℓ
H
H'
m
J

b.

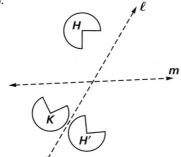

ℓ
H
m
K
H'

c. K and J are both images of rotations of the same magnitude of H and the intersection of ℓ and m. However, the rotations are in opposite directions. **d.** Make ℓ and m perpendicular to each other.   **18.** C   **19.** glide reflection
20. A

Self-Test Correlation Chart

Question	1	2	3	4	5	6	7	8	9	10
Objective(s)	A	A	A	J	C	B	D	D	D	C
Lesson(s)	4-1, 4-2	4-1, 4-2	4-1, 4-2	4-1, 4-2, 4-6	4-4, 4-5	4-6, 4-7	4-1, 4-2	4-1, 4-2	4-1, 4-2	4-4, 4-5

Question	11	12	13	14	15	16	17	18	19	20
Objective(s)	B	F	H	H	H	H	C	E	G	I
Lesson(s)	4-6, 4-7	4-4, 4-5, 4-7	4-3	4-3	4-3	4-3	4-4, 4-5	4-4, 4-5, 4-7	4-7	4-8

Chapter 4 Chapter Review (pp. 246–249)

1. a. 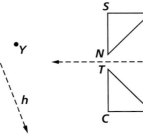 b.

3.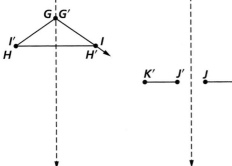

5. Answers vary. Sample:

7.

9.

11.

13.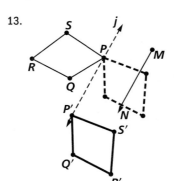

15. a. Two reflections appear as a translation up and to the left. b. Two reflections appear as a translation down and to the right.

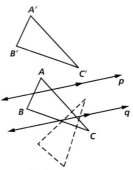

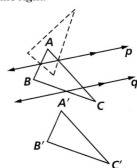

17a. clockwise b. counterclockwise c. clockwise
19. false 21. $2x$ 23. B 25. Any odd number of reflections reverses orientation. 27. It moves every point 2 inches down. 29. $r_n \circ r_k$ or $r_k \circ r_n$ 31. reflection
33. glide reflection 35. translation
37.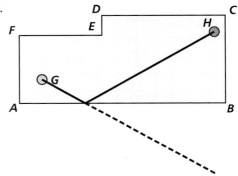

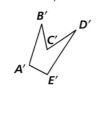

39.

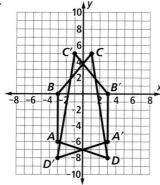

41a. reflection and translation **b.** transposition and retrograde **43.** –2 **45.** (8, 2) **47.** (m, –n) **49.** ⟨–a, –1 – b⟩ **51.**

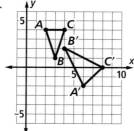

Chapter 5

Lesson 5-1 (pp. 252–256)

Questions:
1. Answers vary. Sample: lightbulbs, batteries, soda cans **3.** composite of reflections; transformation that preserves A-B-C-D; congruence transformation **5. a.** distance **b.** angle measure, distance **7.** △DOG ≅ △CAT **9.** reflections and glide reflections
11. a.

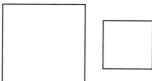

b. No, distance is not preserved. **13. a.** oppositely congruent; reflection **b.** reflection **15.** Answers vary. Sample:

17. no **19.** yes **21.** yes **23. a.** (–2, 1) **b.** (–2, 13)

Lesson 5-2 (pp. 257–262)

Guided Example 1: AR; AR; CA; 36
Guided Example 3: a. ∠CAN; ∠ANC; ∠NCA; $\overline{TO}$; $\overline{OP}$; $\overline{TP}$
Questions:
1. length **3.** angles **5.** $TO \approx 3.4$; $CN \approx 2.7$; $AN \approx 5.2$; m∠C = 118; m∠O = 27; m∠P = 35 **7.** △ABC ≅ △FED
9. Answers vary. Sample: a rotation and then a translation **11a.** $\overrightarrow{LU}$ **b.** m∠C'UP **c.** Sample: By the definition of congruence, if there exists an isometry that maps ∠CLS onto ∠PUN, the angles are congruent.
13a. △ADB ≅ △ADC **b.** $\overline{AB} \cong \overline{AC}$; $\overline{DB} \cong \overline{CD}$; $\overline{AD} \cong \overline{AD}$; ∠ABD ≅ ∠ACD; ∠BDA ≅ ∠CDA; ∠DAB ≅ ∠DAC
15a. △JML ≅ △KML **b.** $\overline{JM} \cong \overline{KM}$; $\overline{ML} \cong \overline{ML}$; $\overline{LJ} \cong \overline{LK}$; ∠JML ≅ ∠KML; ∠MLJ ≅ ∠MLK; ∠LJM ≅ ∠LKM
17. m∠DLI = 140 **19.** translation **21.** If a and b are positive numbers and a + b = c, then c > a and c > b.

Lesson 5-3 (pp. 263–268)

Guided Example 3: Given; CPCF Theorem; Segment
Guided Example 4: M is the midpoint of $\overline{AB}$; $AM = MB$, Definition of midpoint **Guided Example 5:** the Corresponding Angles Postulate; ∠4 ≅ ∠6; Corresponding Angles Postulate **Guided Example 6:** Given; Definition of congruence

Questions:
1. Answers vary. Sample: ∠C ≅ ∠D; A-B-C-D Theorem
3. definitions, postulates, and previously proven facts **5.** Answers vary. Sample: Corresponding Angles Postulate **7.** Segment Congruence Theorem
9. Answers vary. Sample: ∠1 ≅ ∠6, ∠2 ≅ ∠5, ∠3 ≅ ∠8, by the Vertical Angles Theorem, and ∠1 ≅ ∠3, ∠2 ≅ ∠4 by the Corresponding Angles Theorem. **11a.** term ⇒ characteristics **b.** If a point divides a segment into two congruent segments, it is a midpoint of the segment.
13a.

b. Answers vary. Sample: By the given information and the A-B-C-D Theorem, we know that $\overline{AB} \cong \overline{BC}$, $\overline{BC} \cong \overline{CD}$, and $\overline{CD} \cong \overline{DE}$, so by the Transitive Property of Congruence, we know that $\overline{AB} \cong \overline{DE}$.

15. No. Answers vary. Sample: $\overline{AC}$ is not congruent to $\overline{AD}$ because they are radii of noncongruent circles.
17. A **19.**

Lesson 5-4 (pp. 269–276)

Guided Example 2: Vertical Angles Theorem; Transitivity Property of Congruence

Questions:

1. Thales was a Greek mathematician who lived in the 6th century BCE. He is the first person to write proofs like those used today. **3.** All points on a circle are equidistant from the center of the circle by the definition of a circle.

5.

Conclusions	Justifications
1. A with radius AB	1. Given
2. $AD = AB$	2. definition of circle
3. B with radius BA	3. Given
4. $AB = BD$	4. definition of circle
5. $AD = BD$	5. Transitive Property of Equality (with Steps 2 and 4)
6. $\triangle ABD$ is equilateral.	6. definition of equilateral triangle

7. a. true **b.** false **9.** The sides of the handicapped space are parallel to the sides of the regular space. Therefore the angles are corresponding angles formed by parallel lines. So by the Parallel Lines Postulate, they are the same measures.

11. Answers vary. Sample:

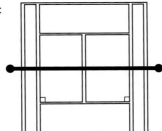

13a. $\angle C$ and $\angle E$; $\angle D$ and $\angle G$ **b.** $\overline{CG}$ and $\overline{DE}$

15.

Conclusions	Justifications
1. B is the midpoint of $\overline{AC}$; C is the midpoint of $\overline{BD}$	1. Given
2. $\overline{AB} \cong \overline{BC}$ and $\overline{BC} \cong \overline{CD}$	2. Def. of midpoint
3. $\overline{AB} \cong \overline{CD}$	3. Trans. Prop. of $\cong$

17. Because $\angle 2$ and $\angle 3$ are a linear pair, they are supplementary angles. So $m\angle 3 + m\angle 2 = 180$. Because $\angle 1$ and $\angle 3$ are corresponding angles, $m\angle 1 = m\angle 3$. By substitution, $m\angle 1 + m\angle 2 = 180$; thus, $\angle 1$ and $\angle 2$ are supplementary angles. **19.** $\overline{AD} \cong \overline{LO}$ **21.** $T(C) = C$ by the A-B-C-D Theorem **23a.** m is the perpendicular bisector of $\overline{PQ}$. **b.** A lies on m. **c.** A is on the perpendicular bisector of $\overline{PQ}$.

Lesson 5-5 (pp. 277–281)

Questions:

1.

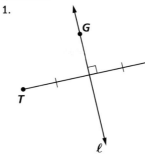

3. Answers vary. Sample: At any given moment, Dexter will be the same distance from A as from B.

5. The pilot is equidistant from three cities at points X and Y.

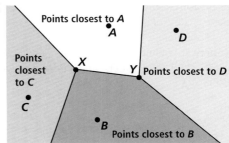

7. a.

b. By the definition of reflection, $r_{\overline{MN}}(M) = M$ and $r_{\overline{MN}}(N) = N$, and since $r_{\overline{MN}}(A) = E$, then $r_{\overline{MN}}(\angle AMN) = \angle EMN$, so, by the definition of congruence, $\angle AMN \cong \angle EMN$. **9.** $m(r_m(\angle ABC)) = 90 = m\angle XYZ$, so $\angle XYZ$ is a right angle.

11a. If a point is equidistant from two points, it lies on the perpendicular bisector of the segment connecting the two given points. **b.** yes **c.** Show that any point equidistant from two given points is on the perpendicular bisector of the segment connecting the given points. **d.** Find a point not on the perpendicular bisector that is equidistant from the given points.

13. By the definition of a circle, the circle centered at E with radius EF contains all points that are at a distance equal to EF from E. Because $EG = EF$, G must lie on the circle. **15.** Because P and Q are on m, $r_m(P) = P$ and $r_m(Q) = Q$ so $r_m(PQBA) = PQCD$. Therefore, by the definition of congruence, $PQBA \cong PQCD$. **17.** Answers vary. Sample: By definition, S_2 and $S_{\frac{1}{2}}$ do not preserve length, and so are not isometries, but because $S_{\frac{1}{2}} \circ S_2 = S_1$, length is preserved, so $S_{\frac{1}{2}} \circ S_2$ is an isometry.

Lesson 5-6 (pp. 282–287)

Questions:

1. No, in Euclidean geometry, there are infinitely many distinct pairs of parallel lines.   **3.** Yes, in Euclidean geometry, a line is uniquely determined by a point and slope.   **5.** The midpoints of segments are uniquely determined, and given two points, the segment connecting these two points is also uniquely determined.   **7.** Two auxiliary lines are used, one parallel to the first line passing through the point P, and another a transversal to the two parallel lines and passing through P.
9. Corresponding Angles Postulate   **11.** A right angle has measure 90 and an obtuse angle measures greater than 90. Their measures sum to over 180, which is impossible in a triangle.   **13.** Not uniquely determined, because both 5 and –5 are solutions.   **15.** Not uniquely determined because there are infinitely many lines perpendicular to a given line.   **17.** Uniquely determined. Answers vary. Sample:

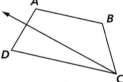

19. Uniquely determined. Answers vary. Sample:

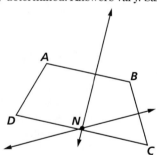

21. $x = 180 + a - b$ **23.** Uniqueness of Parallels Theorem   **25.** The origin is the center; the magnitude is 180.   **27.** no

Lesson 5-7 (pp. 288–295)

Guided Example 1: $\overrightarrow{QA}$; 1; 2; 3; 4; 3; 4; 1; 2; 180; 180; 180; 180; 360

Questions:

1. 32   **3.** 287   **5.** 28   **7.** 360   **9.** 135   **11a.** 180; Triangle-Sum Theorem **b.** 180; Linear Pair Theorem **c.** Substitution **d.** Addition Property of Equality
13. $m\angle 1 + m\angle 2 + m\angle 3 = 180$ by the Triangle-Sum Theorem. $m\angle 1 = m\angle 4$, $m\angle 2 = m\angle 5$, and $m\angle 3 = m\angle 6$ by the Vertical Angles Theorem, $m\angle 4 + m\angle 5 + m\angle 6 = 180$ by substitution.   **15a.** n **b.** $180n$ **c.** 360; The angles form one complete revolution. **d.** $180n - 360$ **e.** $180n - 360 = 180(n - 2)$

17.

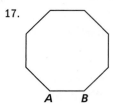

19. B and C **21.** $y = x$

23.

Conclusions	Justifications
1. $\triangle ABC$ and $\triangle DBC$ are equilateral	1. Given
2. $AB = BC = CA$	2. Def. of an equilateral $\triangle$
3. $DB = BC = CD$	3. Def. of an equilateral $\triangle$
4. $AB = BD = DC = CA$	4. Trans. Prop. of $=$
5. $ABCD$ is a rhombus	5. Def. of rhombus

25a. P **b.** P lies on k, ℓ, m, and n; therefore, a reflection of P over these lines does not result in a translation. **c.** identity transformation

Chapter 5 Self-Test (pp. 300–301)

1. $\angle L$ because corresponding parts of congruent figures are congruent.   **2.** $\overline{RB}$ because corresponding parts of congruent figures are congruent.   **3.** $m\angle 2 = 143$. $m\angle 5 = m\angle 3$ by the Vertical Angles Theorem, and $m\angle 2 + m\angle 3 = 180$ by the Parallel Lines Theorem (same-side interior angles).
4.

5a. $AE = EB$ and $AC = BC$ by the Perpendicular Bisector Theorem. Therefore, by the definition of congruence, $\triangle AEC \cong \triangle BEC$, and thus $\angle ECA \cong \angle ECB$ because corresponding angles of congruent figures are congruent. **b.** $2x - 20 = x + 15$; $x = 35$ **6.** By the definition of midpoint, $MN = NB$. Because it is given that $AM = MN$, by transitivity, $AM = NB$. By the Segment Congruence Theorem, $\overline{AM} \cong \overline{NB}$.   **7.** $x + x + 2x = 180$, so $4x = 180$ and $x = 45$; the measure of the third angle is 90.
8a. Given **b.** Vertical Angles Theorem **c.** Given **d.** Transitive Property of Congruence **e.** Corresponding Angles Postulate   **9.** $(10 - 2) \cdot 180 = 1440$, and there are 10 angles, so each angle is 144°.   **10.** $2x$. Because $\angle EAB$ is an exterior angle, its measure is the sum of the measures of the two interior angles not adjacent to it.
11. Answers vary. Sample: Draw auxiliary segments $\overline{EA}$ and $\overline{HG}$. Then $FEAT \cong FHGT$, because both have four right angles and equivalent sides. Now $HO = EX$ by the definition of reflection, as $OG = XA$. Therefore, because $EA = HG$, $\triangle HOG \cong \triangle EXA$. Corresponding angles of congruent figures are congruent, so $\triangle HOG \cong \triangle EXA$.

12.

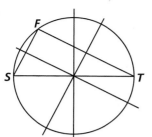

13. *I*, *O*, and *E* are not necessarily collinear, as collinearity cannot be assumed from a picture.

line such that the measure of the alternate interior angle of the given figure is also 40.

16. Answers vary. Sample:

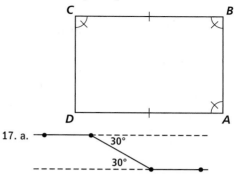

14. a.

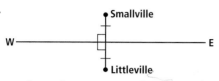

b. The road goes in an east-west direction and is the perpendicular bisector of the segment with endpoints at Smallville and Littleville.   **15.** By the Alternate Interior Angles Theorem, he should make sure that he draws a

17. a.

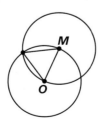

b. She turned 30° to the left. Because the lines of travel west are parallel, the corresponding angles must be congruent, so she turned the same amount to the left.

Self-Test Correlation Chart

Question	1	2	3	4	5	6	7	8	9
Objective(s)	A	A	C	B	C	F	D	G	D
Lesson(s)	5-2	5-2	5-4, 5-5	5-4, 5-5	5-4, 5-5	5-1, 5-2, 5-3	5-6, 5-7	5-4	5-6, 5-7

Question	10	11	12	13	14	15	16	17
Objective(s)	E	H	B	I	J	C	K	J
Lesson(s)	5-7	5-5	5-4, 5-5	5-6	5-4, 5-5	5-4, 5-5	5-6, 5-7	5-4, 5-5

Chapter 5 Chapter Review (pp. 302–305)

1.

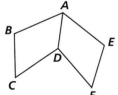

3. $\overline{CD}$ **5.** $DC = 10$ cm, $OA = OB = OC = OD = PK = PL = PM = PN = 5$ cm   **7.** $6x$

9.

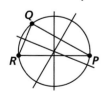

11.

each other, the line of reflection must perpendicularly bisect $\overline{AB}$ and $\overline{CD}$.   **31.** $\angle 1 \cong \angle 2$ is given and $\angle 2 \cong \angle 3$ by the Vertical Angles Theorem. By the Transitive Property of Congruence, $\angle 1 \cong \angle 3$.   **33.** By the A-B-C-D Theorem, angle measure is preserved under an isometry, and the composition of isometries is an isometry. Therefore, because $\triangle DEF \cong \triangle GHI$, $m\angle D = m\angle G$ by the CPCF Theorem.   **35.** Answers vary. Sample: By the A-B-C-D Theorem, distance is preserved, so $QP = XY$, $PS = YZ$, $SR = ZW$, and $RQ = WX$. Therefore, by the definition of congruence, $QPSR$ is congruent to $XYZW$.   **37.** no   **39.** yes   **41.** yes   **43.** $m\angle 1 = m\angle 2$ must so $\angle 1$ and $\angle 2$ are congruent

45.

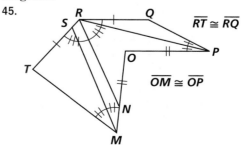

$\overline{RT} \cong \overline{RQ}$

$\overline{OM} \cong \overline{OP}$

13. 120   **15.** 127   **17.** 30, 50, 100   **19. a.** 3780 **b.** $\frac{3780}{23} \approx$ 164.35   **21.** $p = 90$, $q = 90$ **23. a.** 98 **b.** 110 **c.** 12 **d.** 12   **25.** $\overline{QU} \cong \overline{FG}$, $\overline{UA} \cong \overline{GO}$, $\overline{AR} \cong \overline{ON}$, $\overline{RQ} \cong \overline{NF}$, $\angle Q \cong \angle F$, $\angle U \cong \angle G$, $\angle A \cong \angle O$, $\angle R \cong \angle N$ **27.** Angle Congruence Theorem   **29.** $\overline{AB}$ and $\overline{CD}$ have *n* as their perpendicular bisector. Answers vary. Sample: Because *A* and *B* are reflected onto each other and *C* and *D* are reflected onto

Chapter 6

Lesson 6-1 (pp. 308–315)

Guided Example 2: *ONMLKJIHGFEDCBARQP*, *N*, *F*, $\overline{ML}$, $\overline{CB}$, $\overline{LF}$, $\overline{RC}$

Questions:
1. reflection-symmetric; symmetry line   **3.** Group 1: **a, b, d, e, f, g, h, j, k, m, n, p, q, r, s, t, u, y, z**; Group 2: **c, i, v, w**; Group 3: **l, o, x**   **5.** $r_\ell(\overrightarrow{MN}) = \overrightarrow{PQ}$ **7.** $\overrightarrow{AB}$ has one line of symmetry, the line containing the ray

9. 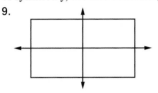   **11.**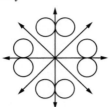

13.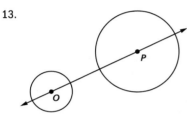

15. TAXI   **17. a.** Answers vary. Sample: basketball and tennis; these sports require that each team has an equal chance of winning, and thus, require symmetric courts.
b. Answers vary. Sample: skiing, sailing. These sports take place in nature, which is not symmetric.   **19.** $x \approx 100.57$
21. m∠*C* = 61   **23a.** *T* cannot be a translation because two applications of the same translation on a point do not return that point to its original location.   **b.** *T* can be a rotation; for instance, a possible rotation magnitude of 180, giving $T \circ T(P) = P$. An impossible rotation magnitude is 90.

Lesson 6-2 (pp. 316–323)

Guided Example 2: 74; 74; 148; 32
Questions: 1. a. $\overline{OX}$ **b.** ∠*S* **c.** ∠*O* and ∠*X* **d.** ∠*O* and ∠*X*
3a. If two angles of a triangle are congruent, the sides opposite those angles are congruent. (Converse of the Isosceles Triangle Base Angles Theorem)   **b.** If two sides of a triangle are congruent, the angles opposite those sides are congruent. (Isosceles Triangle Base Angles Theorem)

5.

Conclusions	Justifications
1. m∠*XYZ* = m∠*WYQ*	**1.** Vertical Angles Theorem
2. m∠*XZY* = m∠*RZT*	**2.** Vertical Angles Theorem
3. m∠*XYZ* = m∠*XZY*	**3.** Isosceles Triangle Base Angles Theorem
4. m∠*WYQ* = m∠*RZT*	**4.** Transitive Property of Equality

7a. 3 **b.** 1 **c.** 0   **9.** G. Wilikers   **11.** 12 cm   **13.** *BC* < *AB* < *AC*   **15.** 40   **17.** 36   **19.** Answers vary. Sample: m∠*R* = m∠*SLR*, m∠*SLR* = m∠*PLW*, Vertical Angles Theorem, m∠*W* = m∠*R*, Transtive Property of Equality
21. *F*   **23.** (0, 2.9)

Lesson 6-3 (pp. 324–330)

Guided Example 2: $\overarc{BPD}$; $\overarc{BPD}$; $\overarc{BPD}$; 80
Questions:
1. a. 168 **b.** 64 **c.** 128   **3. a.** m$\overarc{EF}$ = 64°, m$\overarc{FG}$ = 56°, m$\overarc{EHG}$ = 240° **b.** m$\overarc{EF}$ + m$\overarc{EHG}$ + m$\overarc{FG}$ = 360°
5. a. 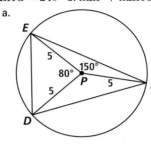 **b.** Answers vary. Sample: △*DPE* and △*EPF* **c.** Answers vary. Sample: ∠*PDE* ≅ ∠*PED* **d.** 75
7a. Angle Addition Property **b.** By the result of Case I of the Inscribed Angle Theorem proof **c.** measure of a central angle = measure of intercepted arc **d.** Distributive Property **e.** Angle Addition Property **f.** measure of a central angle = measure of intercepted arc
9. m∠*HWV* = 39   **11a.** ∠*PBQ* **b.** ∠*PAR* **c.** ∠*APB* and ∠*ARB*   **13a.** 126 **b.** 103.5 **c.** 54 **d.** 76.5 **e.** ∠*B* and ∠*D* are supplementary. **f.** ∠*A* and ∠*C* are supplementary.
15. 40   **17.** *n* = 14   **19.**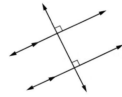

Lesson 6-4 (pp. 331–338)

Questions:
1. Answers vary. Sample:

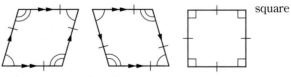 square

3. Answers vary. Sample:

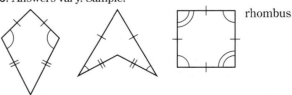 rhombus

5. Answers vary. Sample:

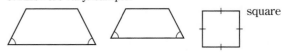 square

7. it is a quadrilateral; it has four right angles   **9.** it is a quadrilateral; it four sides congruent

11.

Quadrilateral	Information marked on figure	Name of figure (most specific)
ABCD	Opposite sides are parallel	parallelogram
EFGH	Four right angles	rectangle
IJKL	Four right angles; all sides are congruent	square
MNOP	All sides are congruent	rhombus

13. rhombus

15. parallelogram

17a. All radii of a circle are congruent. **b.** All radii of a circle are congruent. **c.** kite   **19.** kites and general quadrilaterals

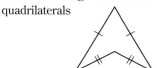

21. 48 and 84 or 66 and 66   **23.** 315

25. a.

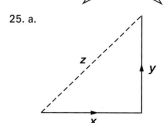

b. The angle formed by the segment of length x and the segment of length y is right. z becomes the length of the hypotenuse of a right triangle, which much be the longest segment in the triangle, because it is opposite the largest angle (90°). Thus, $x < z$ and $y < z$.

Lesson 6-5 (pp. 339–344)

Guided Example 1: 40; 23.5; 80; 8; 16; 66.5; 50; *TEI*; 66.5; *KIE*; 50   **Guided Example 3: 2.** Rhombus Diagonal Theorem **4.** Symmetric Figures Theorem **5.** Transitive Property of Congruence **6.** Parallel Lines Theorem (alternate interior angles) **7.** Definition of parallelogram

Questions:

1. rhombuses and squares   **3a.** B and L **b.** $\overleftrightarrow{BL}$ **c.** $\angle BOL$
5a. m$\angle D$ **b.** $\overleftrightarrow{AC}$ **7.** $\angle ABC$, m$\angle ABC$, angle measure, $\overleftrightarrow{BD}$, $\angle DAB$, m$\angle DAB$ **9a.** The angles that $\overline{CD}$ and $\overline{BC}$ form with $\overline{AC}$ are congruent, as are the angles that $\overline{AB}$ and $\overline{AD}$ form with $\overline{AC}$. Therefore, B is the same distance from $\overleftrightarrow{AC}$ that D is, so D is the reflection image of B. **b.** Because D is the reflection image of B, $AB = AD$ and $CD = CB$. Therefore, by the definition of a kite, $ABCD$ is a kite.
11. squares and rectangles

13. Since $\angle BAD = 63$, m$\widehat{BD} = 126°$, so m$\widehat{DA} = 54°$, so m$\angle BAD = 27$. Therefore, m$\angle ABC = $ m$\angle ABD$.   **15.** Answers vary. Sample: No, the 3 points must also be noncollinear.

Lesson 6-6 (pp. 345–350)

Guided Example 1: 1. Linear Pair Theorem **2.** Corresponding Angles Postulate

Questions:

1. trapezoids, parallelograms, isosceles trapezoids, rhombuses, rectangles, and squares   **3. a.** $\overline{AB}$ and $\overline{DC}$ **b.** $\angle A$ and $\angle B$, or $\angle D$ and $\angle C$ **c.** m$\angle D = 90$, m$\angle B = 150$
5a. $\angle P$ and $\angle E$, $\angle O$ and $\angle Z$ **b.** There are not necessarily any congruent angles.

7.

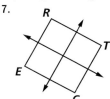

9. $BC = 70$, $FE = 70$
11. a. true **b.** true **c.** true
13. The symmetry lines are the lines that coincide with the two diagonals and the two perpendicular bisectors of the sides.

15. Suppose a kite has diagonals that are symmetry lines. By the definition of kite and the Kite Symmetry Theorem, the four sides of the kite are congruent. Therefore, by definition of rhombus, the kite is a rhombus.
17. m$\angle ABC = $ m$\angle AFC = 35$ **19.** 540

Lesson 6-7 (pp. 351–355)

Guided Example: F, F, rotation
Questions:
1. F   **3a.** 5 **b.** 72°
5. Answers vary. Sample:

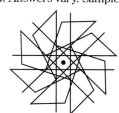

7. Answers vary. Sample:

9. No. According to the theorem given in this lesson, when a figure has 2 or more intersecting symmetry lines, the figure also has rotation symmetry.   **11a.** 180°
b. $DEFABC$ **c.** $\overline{AF}$ and $\overline{DC}$, $\overline{AB}$ and $\overline{DE}$, $\overline{BC}$ and $\overline{EF}$, $\angle A$ and $\angle D$, $\angle B$ and $\angle E$, $\angle C$ and $\angle F$ **13.** This allows the card to be held right-side up or upside down and still be readable.   **15a.** This spiral has 2-fold rotation symmetry. **b.** This spiral does not have any rotation symmetry.   **17a** rhombus   **19a.** Each angle measure is 63.5°, or one is 53 and the other is 74. **b.** 53 and 74

Lesson 6-8 (pp. 356–362)

Guided Example 1: 1080; 8; 135
Questions:
1. Star Ribbed Domes are designed by rotating a pair of parallel arches about some central point so as to create a star pattern.

3.

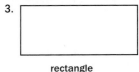

rectangle

5. 8 **7.** Draw a circle. Choose any point on the circle and rotate it $\frac{360}{9}$ or 40° about the center of the circle. Repeat this process eight times and connect the points.

9. 7 **11.** when n is odd **13.** 30° **15.** Regular polygons have rotational symmetry, allowing a wrench to turn the nut from multiple orientations. **17.** 3 **19.** 20 sides
21. $x = 0, y = 3, y = x + 3; y = -x + 3$ **23.** The figure is 9-fold rotation-symmetric. **25.** The length of the base is larger than the lengths of the congruent sides. **27a.** $\angle A$ and $\angle D$, $\angle B$ and $\angle E$, and $\angle C$ and $\angle F$ **b.** $\overline{AB}$ and $\overline{DE}$, $\overline{BC}$ and $\overline{EF}$, $\overline{CA}$ and $\overline{FD}$

Lesson 6-9 (pp. 363–367)

Questions:
1. A frieze pattern is any pattern in which a single fundamental figure is repeated to form a pattern that has a fixed height but infinite length in two opposite directions. **3.** 1 **5.** 7 **7.** 6 **9.** 4 **11.** 6 **13.** 3 **15.** 4
17. 6 **19a.** By definition, a frieze pattern has translation symmetry. Because the reflection of a frieze pattern over a horizontal line is also a frieze pattern, the reflection must also have translation symmetry. Thus, if a frieze pattern has a horizontal symmetry line, it must also have glide reflection symmetry. **b.** no, yes, no, no; no, yes, yes, no; yes, yes, no, no; yes, yes, yes, no **21.** 1 **23.** 7
25a. $(0, \sqrt{3})$ **b.** Yes, the center of the circle in which a regular polygon is inscribed and the center of the regular polygon are always the same point. **27.** sometimes but not always

Chapter 6 Self-Test (pp. 371–372)
1. a. $\overline{GH}, \overline{KJ}, \overline{AB}$ **b.** m is the angle bisector of $\angle BCD$. By the Angle Symmetry Theorem, the symmetry line of an angle contains the bisector of the angle.
2.

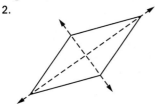

3. Answers vary. Sample:

4. In order for a kite to have rotation symmetry, both its diagonals must be symmetry lines. Any rhombus or square is a kite with rotation symmetry.

5. 1, 2, or 4

isosceles trapezoid   rectangle   square

6. a. Since $\angle RFT$ is a straight angle, $(4x + 50) + (5x + 40) = 180$, so then $9x = 90$, and $x = 10$. **b.** No. $m\angle RBF = 38$ and $m\angle TBF = 43$, so $m\angle R \neq m\angle T$ and $\triangle BRT$ is not isosceles. Thus, $BR \neq BT$. **7.** The largest side is the side opposite the largest angle. $m\angle R = 180 - 40 - 80 = 60$. Because $\angle J$ is the largest angle, $\overline{GR}$ is the longest side.
8. $m\angle P + m\angle X = 180$ by the Trapezoid Angle Theorem. Thus, $m\angle X = 70$. Since $FPXL$ is isosceles, this means that $m\angle L = 70$ as well. **9.** 18 **10a.** 8-fold rotational symmetry **b.** 4-fold rotational symmetry **11.** $\angle W$ intercepts a semicircle, so by Thales' Theorem, it is a right angle, that is, $m\angle W = 90$. **12.** Because $\triangle RAN$ is equilateral, $m\angle R$ and $m\angle RNA$ are both 60. Because the measure of an intercepted arc is twice the measure of the inscribed angle, $m\overset{\frown}{RBA} = 120° = m\overset{\frown}{AN}$. Because $\overset{\frown}{RAN} = \overset{\frown}{RBA} + \overset{\frown}{AN}$, $m\overset{\frown}{RAN} = 240°$. **13a.** Sometimes but not always. All rectangles are trapezoids, but not all trapezoids are rectangles. **b.** Sometimes but not always. In kites, one diagonal is a symmetry line and the perpendicular bisector of the other diagonal. In rhombuses, both diagonals are symmetry lines and perpendicular bisectors of each other. **14a.** The measure of one of the interior angles of a regular n-gon is $\frac{(n-2)180}{n}$, so with $n = 8$, the measure of one of the interior angles is 135. $m\angle MTN = 180 - 135 = 45$. **b.** In $\triangle BFS$, $m\angle B = 135$. The triangle is isosceles, so the base angles are congruent. Thus, $m\angle BFS = \frac{180 - 135}{2} = 22.5$.
15. $\angle R \cong \angle V$ and $\angle T \cong \angle P$ because alternate interior angles formed when parallel lines are cut by a transversal are congruent (Parallel Lines Theorem), and $\angle R \cong \angle P$ is given, so by the Transitive Property of Congruence, $\angle V \cong \angle T$. Thus, $\overline{SV} \cong \overline{ST}$ by the Converse of the Isosceles Triangle Base Angles Theorem. **16.** Diagonals of a rhombus are angle bisectors, so $\triangle FOA$ is equilateral.

Therefore, $FO = OA = FA = 18$ cm, so each side of rhombus $FOAM$ is 18 cm, and the perimeter is 72 cm.

17.

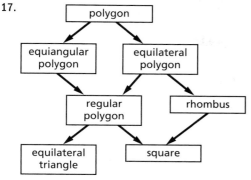

18. Because $MANDC$ is a regular pentagon, m$\angle AND = \frac{(5-2)180}{5} = 108$. Because $AN = ND$, $\triangle AND$ is isosceles with base angles measuring $\frac{180-108}{2} = 36$. So m$\angle ADN = 36$. Because m$\angle CDN = $ m$\angle CDA + $ m$\angle ADN$, $108 = $ m$\angle CDA + 36$, so m$\angle CDA = 72$. $\triangle ACD$ is isosceles, so m$\angle DCA = $ m$\angle CDA = 72$. Thus m$\angle CAD = 180 - 72 - 72 = 36$. Because $MC = CD$, $\triangle MCD$ is isosceles, and by a calculation similar to a previous one, m$\angle MDC = 36$.
19. Quadrilateral II also has congruent diagonals because it is lower in the hierarchy than type G, but nothing can be concluded about Quadrilateral I because it is higher in the hierarchy than type G. **20.** II and V both have type 5 symmetry. III and IV both have type 2 symmetry.

Self-Test Correlation Chart

Question	1	2	3	4	5	6	7	8	9	10
Objective(s)	F	A	B	B	I	G	H	E	F	K
Lesson(s)	6-1, 6-7	6-1, 6-7	6-2, 6-4, 6-7, 6-8	6-2, 6-4, 6-7, 6-8	6-4, 6-5, 6-6	6-2, 6-8	6-2	6-4, 6-5, 6-6, 6-8	6-1, 6-7	6-1, 6-7

Question	11	12	13	14	15	16	17	18	19	20
Objective(s)	D	D	I	E	J	C	M	C	M	L
Lesson(s)	6-3	6-3	6-4, 6-5, 6-6	6-4, 6-5, 6-6, 6-8	6-2, 6-4, 6-5, 6-6	6-2, 6-8	6-4	6-2, 6-8	6-4	6-9

Chapter 6 Chapter Review (pp. 373–377)

1. This figure has 9-fold rotation symmetry.
3. This figure has 12-fold rotation symmetry.
5.

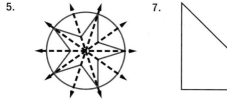

7.

9.

11.

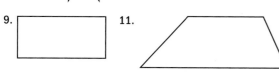

13. 133, 23.5, and 23.5. **15.** 20 **17.** 36 **19.** 134
21. 108 **23.** $\frac{1980}{13} \approx 152.308$ **25.** $OE = 3$ cm, $OD = 6$ cm, $GE = 6$ cm, $FD = 12$ cm
27. a.

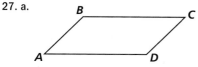

b. m$\angle B = $ m$\angle D = 135$, m$\angle C = 45$ **29.** Answers vary. Sample: $EFLABCD$ and $GFLKJIH$ **31.** squares, rhombuses and kites **33.** true **35.** true **37.** Regular Polygon Reflection Symmetry Theorem. A regular polygon has reflection symmetry about every line containing its center and a vertex and about every perpendicular bisector of its sides. Regular Polygon Rotation Symmetry Theorem: A regular n-gon has n-fold rotation symmetry. **39a.** By the Center of a Regular Polygon Theorem, there is a unique point that is equidistant from the vertices of an equilateral triangle. This point is the center of the equilateral triangle and of the circle in which it is inscribed. The three angle bisectors of an equilateral triangle are radii of this circle, so the three vertices of the triangle are equidistant from their intersection point, which must be the center of the circle. **b.** true **41.** The hypotenuse is always opposite the angle with measure 90 in a right triangle. Because the measures of the interior angles of a triangle add to 180, the other two angles must have measures less than 90. Thus, the hypotenuse must be the longest side of a right triangle. **43.** false **45.** true **47.** parallelogram, rectangle, rhombus, square, kite **49.** isosceles trapezoid, rectangle, square

51.

Conclusions	Justifications
1. $\ell \parallel \overline{MN}$, $\ell \perp \overline{OP}$	**1.** Given
2. $m\angle N = 90 - m\angle 1$, $m\angle M = 90 - m\angle 2$	**2.** definition of complementary angles and congruence of opposite interior angles of parallel lines
3. $\angle N \cong \angle M$	**3.** substitution
4. $\triangle NOP \cong \triangle MOP$	**4.** Angle-Angle-Side
5. $NP = MP$	**8.** corresponding parts of congruent figures are congruent

53.

Conclusions	Justifications
1. $\odot O$ and $\odot P$ intersect at A and B	**1.** Given
2. $OA = OB$ and $AP = BP$	**2.** def. of a circle
3. $\triangle OAB$ and $\triangle PAB$ are isosceles	**3.** def. of isosceles triangles
4. $OAPB$ is a kite	**4.** def. of a kite

55. The figure has 3-fold rotation symmetry.   **57.** 3   **59.** 2

61.

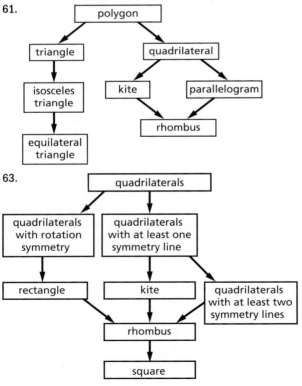

63.

Chapter 7

Lesson 7-1 (pp. 380–385)

Questions:

1. Answers vary. Sample: Triangles are strong, the building blocks of polygons, and simple to describe.

3. SAS   **5.** AAS

7. Answers vary. Sample: **9.** The given angle might be the vertex angle of one triangle and the base angle of the other.

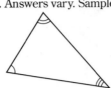

11a. yes

b. Infinitely many triangles are possible. Samples:

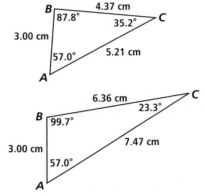

13a. yes **b.** Infinitely many triangles are possible. Samples:

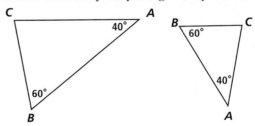

15a. no **b.** Infinitely many triangles are possible. Samples:

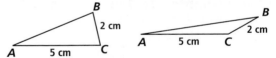

17. Two angles and a non-included side are sufficient to determine a triangle.   **19.** Two sides and a non-included angle are not sufficient to determine a triangle.

21. Yes;

23. The magnitude of R is a multiple of $180°$.

Lesson 7-2 (pp. 386–392)

Guided Example 2: 3. Triangle-Sum Theorem **4.** Transitive
6. Additive **7.** Angle Congruence Theorem
Questions:
1. SSS, SAS, ASA, AAS **3.** C **5.** $\triangle EGF \cong \triangle EGH$; SAS
7. $\triangle XWV \cong \triangle YZA$; SAS **9.** $\triangle RUT \cong \triangle TSR$; ASA
11a. $\angle D \cong \angle X$ **b.** $\angle H \cong \angle R$ **c.** $\overline{PD} \cong \overline{RX}$ **d.** $\overline{HD} \cong \overline{RX}$,
$\overline{PD} \cong \overline{VX}$ **13.** The pattern can be flipped over $\overleftrightarrow{FB}$
because the two triangles are congruent by SAS
congruence. **15.** $m\angle GDF + m\angle FDE = 180$ and
$m\angle GDC' + m\angle C'DE = 180$ because $\angle GDE$ is a straight
angle. It is given that $m\angle FDE = m\angle C'DE$, so by
substitution, $m\angle GDF = m\angle GDC'$. By the definition of
angle bisector, $\overleftrightarrow{EG}$ bisects $\angle FDC'$. **17a.** Center of a
Regular Polygon Theorem **b.** CPCF Theorem
19. The shortest path will
be from E to the point of
intersection of $\overleftrightarrow{EF'}$ and k and
then to F. This uses the
Triangle Inequality Postulate.

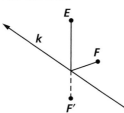

Lesson 7-3 (pp. 393–399)

Guided Example 2: 2. $\angle Y$ **5.** $\overline{PY}$ **6.** $\angle XPY$ **7.** 2, 4, 5
Guided Example 4: Congruent triangles proof:
3. definition of angle bisector **5.** SAS **6.** CPCF Theorem
Transformation proof: **4.** definition of isosceles triangle
6. definition of reflection **8.** definition of congruence
Questions:
1. SAS, ASA, SSS **3.** CPCF Theorem **5.** The justification
of Step 7 would change to the AAS Congruence Theorem
by Steps 2, 5, and 6. **7.** In Step 6, change the conclusion
to $\angle BAO \cong \angle DCO$. The justification is still the CPCF
Theorem. **9b.** $\overline{QM} \cong \overline{MT}$ **c.** $\angle QMR \cong \angle TMS$ **d.** M is the
midpoint of $\overline{RS}$. **e.** $\overline{RM} \cong \overline{MS}$ **f.** $\triangle QMR \cong \triangle TMS$
g. $\angle R \cong \angle S$

11.

Conclusions	Justifications
1. $\overline{RB}$ bisects $\angle ARC$, $\angle A \cong \angle B$, $\overline{AP} \cong \overline{BC}$	1. Given
2. $\angle PRA \cong \angle CRB$	2. def. of $\angle$ bisector
3. $\triangle APR \cong \triangle BCR$	3. AAS $\cong$ Theorem
4. $\overline{AR} \cong \overline{BR}$	4. CPCF Theorem

13a.

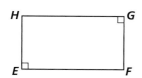

b.

Conclusions	Justifications
1. $\angle E$ and $\angle G$ are right angles.	1. Given
2. $\angle E \cong \angle G$	2. $\angle \cong$ Theorem
3. $\overline{EF} \parallel \overline{HG}$	3. Given
4. $\angle GHF \cong \angle HFE$	4. $\parallel$ Lines Theorem (Alt. Int. Angles)
5. $\overline{HF} \cong \overline{HF}$	5. Refl. Prop. of $\cong$
6. $\triangle EFH \cong \triangle GHF$	6. AAS $\cong$ Theorem

15a. Vertices P, M, and N correspond to vertices N, O, and
P, respectively. **b.** AAS **17a.** true **b.** false **19.** P and Q
must be the same circle. **21a.** Angle Measure Postulate
(Unique Angle Assumption) **b.** reflection

Lesson 7-4 (pp. 400–405)

Guided Example 2: 2. supplementary; Linear Pair Theorem
4. $\angle 1 \cong \angle 2$ **5.** Converse of the Isosceles Triangle
Base Angles Theorem **7.** $\triangle JLN \cong \triangle KLM$
8. CPCF Theorem
Questions:
1. Rectangles are not rigid and could bend easily.
3. $\triangle DCB$ and $\triangle ABC$; $\triangle ABD$ and $\triangle DCA$
5. Answers vary. Sample:

7a. Given **b.** Given **c.** Isosceles Triangle Base Angles
Theorem **d.** Reflexive Property of Congruence **e.** AAS
Congruence Theorem **f.** CPCF Theorem
9.

Conclusions	Justifications
1. $\angle 2 \cong \angle 4$, $KJ \cong KL$	1. Given
2. $\angle 1$ and $\angle 2$ are a linear pair; $\angle 3$ and $\angle 4$ are a linear pair.	2. definition of linear pair
3. $\angle 1$ and $\angle 2$ are supplementary; $\angle 3$ and $\angle 4$ are supplementary.	3. Linear Pair Theorem
4. $\angle 1 \cong \angle 3$	4. Supplements of congruent angles are congruent.
5. $\angle K \cong \angle K$	5. Reflexive Property of congruence.
6. $\triangle KJN \cong \triangle KLM$	6. AAS Congruence Theorem
7. $\angle J \cong \angle L$	7. CPCF Theorem

11a. Given **b.** 1; 2 **c.** Segment Congruence Theorem
d. Additive Property, substitution **e.** Segment Congruence
Theorem **f.** Reflexive Property of Congruence **g.** *ABC*;
DCB **h.** CPCF Theorem

13.

Conclusions	Justifications
1. $\overline{AB} \parallel \overline{CD}$, $AB = CD$	1. Given
2. $\angle ABC \cong \angle DCB$	2. Parallel Lines Theorem (Alt. int. angles)
3. $\overline{BC} \cong \overline{CB}$	3. Reflexive Property of Congruence
4. $\triangle ABC \cong \triangle BCD$	4. SAS Congruence Theorem
5. $BD = CA$	5. CPCF Theorem

15. false **17.** Answers vary. Sample: a movement that
maps one figure to another figure while retaining angle
measure, betweenness, collinearity, and distance; moving,
rotating, or reflecting a figure (or a composite of these)

Lesson 7-5 (pp. 406–412)

Guided Example 2: 1. Given **3.** definition of right triangle
4. Reflexive Property of Congruence **5.** Given **6.** HL
7. CPCF Theorem

Questions:
1. Answers vary. Sample: an angle measure of 45 adjacent
to a side 10 units and opposite a side 8 units in length
3.

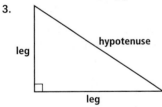

5. SSA leads to congruence when the nonincluded
angle is opposite the larger of the two given sides.
7. $\triangle XUV \cong \triangle XWV$; HL. **9.** $\triangle CDF \cong \triangle BEF$;
Congruence Theorem AAS or ASA

11.

Conclusions	Justifications
1. $\overline{PN} \cong \overline{RN}$	1. Given
2. $\overline{TN} \cong \overline{TN}$	2. Reflexive Property of Congruence
3. $\angle P$ and $\angle R$ are right angles	3. Angles inscribed in a semicircle are right angles. (Thales' Theorem)
4. $\triangle TRN \cong \triangle TPN$	4. HL Congruence Theorem
5. $\overline{TN}$ is a line of reflection-symmetry for *PNRT*	5. definition of reflection symmetry
6. *PNRT* is a kite	6. definition of kite

13.

Conclusions	Justifications
1. $\overline{AB} \cong \overline{DC}$, $\angle F$ is right and $\angle E$ are right angles, $\overline{DE} \cong \overline{AF}$	1. Given
2. $AB = DC$	2. Segment Congruence Theorem
3. $\overline{BC} \cong \overline{CB}$	3. Reflexive Property of Congruence
4. $AC = AB + BC$, $BD = BC + CD$	4. segment addition
5. $AC = BD$	5. substitution
6. $\overline{AC} \cong \overline{BD}$	6. Segment Congruence Theorem
7. $\triangle ACF \cong \triangle DBE$	7. HL Congruence Theorem
8. $\angle B \cong \angle C$	8. CPCF Theorem
9. $\overline{FC} \parallel \overline{BE}$	9. Parallel Lines Theorem (alternate interior angles)

15.

Conclusions	Justifications
1. *FGHI* is an isosceles trapezoid	1. Given
2. $\overline{FI} \cong \overline{GH}$, $\angle IFG \cong \angle HGF$	2. Isosceles Trapezoid Theorem and definition of isosceles trapezoid
3. $\overline{FG} \cong \overline{GF}$	3. Reflexive Property of Congruence
4. $\triangle IFG \cong \triangle HGF$	4. SAS Congruence Theorem
5. $\overline{GI} \cong \overline{FH}$	5. CPCF Theorem
6. $GI = FH$	6. definition of congruence

17.

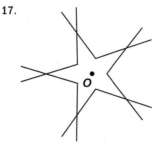

19. $AB = AC = 2$ **21.** size change

Lesson 7-6 (pp. 413–418)

Questions:
1. a covering of a plane with nonoverlapping congruent
regions **3.** If the measure of an interior angle of the
figure evenly divides 360, then the figure tessellates
the plane. **5.** Answers vary. Sample: Which figures
tessellate? **7.**

9. pentagon **11.** *G*, *E*, and *F* are collinear because *F* is
the 180° rotation image of *G* around *E*, and *E*, *F*, and *H* are
collinear because *H* is the 180° rotation image of *E* around

F. Therefore, if *G*, *E*, and *F* are collinear, and *E*, *F*, and *H* are collinear, then *G*, *E*, *F*, and *H* are all collinear.
13. Step 5:

15. Because R_M (*ACA'B*,180°) = *A''C'BA*, $\overleftrightarrow{BC} \parallel \overleftrightarrow{C'A}$ and $\overleftrightarrow{BC'} \parallel \overleftrightarrow{CA}$, so *AC'BC* is a parallelogram by definition. **17.** true **19.** No, because the quadrilateral shown is not convex.

Lesson 7-7 (pp. 419–425)
Guided Example: a. 5 **b.** 8 **c.** 80 **d.** 100
Questions:
1a. 7 **b.** 5 **c.** 3 **3a.** ∠*QSR* **b.** ∠*USR* **c.** ∠*QRU*, ∠*QSU*
d. none **5a.** Corresponding Angles Postulate and Angle Congruence Theorem **b.** Parallel Lines Theorem (alternate interior angles) **c.** Properties of a Parallelogram Theorem, Part a **d.** Reflexive Property of Congruence **e.** SAS Congruence Theorem **f.** CPCF Theorem
7. Answers vary. Sample: Consider parallelogram *ABCD* with diagonals $\overline{AC}$ and $\overline{BD}$ intersecting at *E*. From Part a of the Properties of a Parallelogram Theorem, $\overline{AB} \cong \overline{CD}$. By the Vertical Angles Theorem, m∠*AEB* = m∠*CED*. So ∠*AEB* ≅ ∠*CED* by the Angle Congruence Theorem. Because $\overline{AB} \parallel \overline{CD}$, ∠*BDC* and ∠*DBA* are congruent alternate interior angles (Parallel Lines Theorem). So △*AEB* ≅ △*CED* by the AAS Congruence Theorem. By the CPCF Theorem, $\overline{AE} \cong \overline{CE}$ and $\overline{BE} \cong \overline{DE}$. Thus, by the Segment Congruence Theorem, *AE* = *CE* and *BE* = *DE*.
9a. 64 **b.** 64 **c.** 90 **d.** 52 **e.** 26 **f.** 26 **11.** Distance between Parallel Lines Theorem
13a.

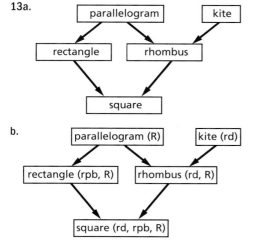

b.

15. Answers vary. Sample:

Conclusions	Justifications
1. *ABCD* is a parallelogram and *M* and *N* are the midpoints of $\overline{AD}$ and $\overline{BC}$	1. Given
2. $\overline{AD} \cong \overline{BC}$	2. Prop. of a Parallelogram Theorem
3. $\overline{AM} \cong \overline{MD}$, $\overline{BN} \cong \overline{NC}$	3. Def. of midpoint
4. 2 · *AM* = 2 · *BN*	4. Substitution
5. *AM* = *BN*	5. Multiplication Prop. of Equality

17. Answers vary. Sample:

19. Answers vary. Sample: Draw the perpendicular bisector for each side of the triangle. The center of the circle in which the triangle is inscribed is the intersection of the perpendicular bisectors. **21.** C, D

Lesson 7-8 (pp. 426–430)
Questions:
1. A parallelogram is a quadrilateral with two pairs of parallel sides. **3.** The diagonals of a quadrilateral bisect each other

Conclusions	Justifications
1. $\overline{AD} \cong \overline{BC}$, $\overline{AB} \cong \overline{DC}$	1. Given
2. $\overline{AC} \cong \overline{AC}$	2. Refl. Prop. of ≅
3. △*ADC* ≅ △*CBA*	3. SSS ≅ Theorem
4. ∠*DCA* ≅ ∠*BAC*	4. CPCF Theorem
5. $\overline{AB} \parallel \overline{DC}$	5. Alt. Int. ∠s Theorem
6. ∠*CAD* ≅ ∠*ACB*	6. CPCF Theorem
7. $\overline{AD} \parallel \overline{BC}$	7. Alt. Int. ∠s Theorem
8. *ABCD* is a parallelogram	8. Def. of parallelogram

5a. no **7a.** yes **b.** Draw the diagonal connecting the vertices that have not been marked with angles, and then there are two similar triangles. Alternate Interior Angles Theorem shows it is a parallelogram.
9a. Given: Quadrilateral *QUAD*, $\overline{QU} \parallel \overline{AD}$, $\overline{QU} \cong \overline{AD}$. Prove: *QUAD* is a parallelogram.

b.

Conclusions	Justifications
1. $\overline{QU} \parallel \overline{AD}$	1. Given
2. $\angle QUD \cong \angle ADU$	2. ∥ Lines Theorem (alt. int. angles)
3. $\overline{QU} \cong \overline{AD}$	3. Given
4. $\overline{UD} \cong \overline{DU}$	4. Refl. Prop. of ≅
5. $\triangle QUD \cong \triangle ADU$	5. SAS ≅ Theorem
6. $\angle UDQ \cong \angle DUA$	6. CPCF Theorem
7. $\overline{QD} \parallel \overline{AU}$	7. Alternate Interior Angles Theorem
8. $QUAD$ is a parallelogram.	8. Def. of parallelogram

11. rectangle, parallelogram, kite **13.** Yes it is possible.
15. Yes it is possible.

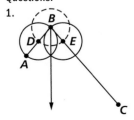

17. $AD = 4$ cm, $DC = 1$ cm, $m\angle C = 72°$, $m\angle B = m\angle D = 108°$ **19.** Answers vary. Sample:

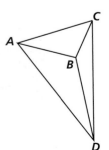

Lesson 7-9 (pp. 431–435)

Questions:
1. kites, rhombuses, and squares **3.** isosceles trapezoids, rectangles, and squares **5.** Draw a line segment and its perpendicular bisector. Place a point randomly on the bisector and reflect it over the original segment. Now, connect the four points. **7.** rectangle **9.** kite **11.** No, these are sufficient conditions for a rectangle. **13.** The diagonals bisect each other. **15.** $ADCE$ is a rectangle because the diagonals bisect each other and are congruent. **17.** No, 7 is not a factor of the area of the rectangle. **19.** no

Lesson 7-10 (pp. 436–439)

Questions:
1.

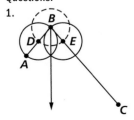

3.

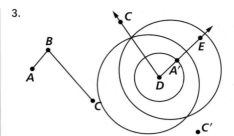

5.

Conclusions	Justifications
1. $\angle AOB$ and circles O, A, and C as described in the construction	1. Given
2. $\overline{OA} \cong \overline{OC}$, $\overline{OA} \cong \overline{AD}$, $\overline{OC} \cong \overline{CD}$	2. Definition of a circle
3. $\overline{OD} \cong \overline{OD}$	3. Reflexive Property of Congruence
4. $\triangle OAD \cong \triangle OCD$ is a rhombus.	4. SSS Congruence Theorem
5. $\angle AOD \cong \angle BOD$	5. CPCF Theorem

7. In Question 6, we constructed the perpendicular of ℓ through P, call it m. Inspecting the construction of the figure, we find that each of the segments of the quadrilateral $PAQB$ is congruent because each is a radius of congruent circles. Thus, $PAQB$ is a rhombus. The diagonals of a rhombus are perpendicular; thus, $\overleftrightarrow{AB}$ is perpendicular to m.

9. **11.**

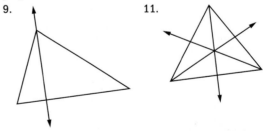

The bisectors intersect at a point inside the triangle, but not necessarily the same point as the point of intersection of the perpendicular bisectors of the sides.
13. A quadrilateral with this property must be concave.
15. True; because the square can tessellate the plane, the pattern used to tessellate the square can also be used to tessellate the plane.
17. less than or equal to

Chapter 7 Self-Test (pp. 443–444)

1. SAS, AAS, ASA, SsA, HL
SSS Congruence Theorem
SAS Congruence Theorem
HL Congruence Theorem
2. $\triangle SPM \cong \triangle JRB$;
3. $\triangle MNT \cong \triangle PTN$;
4. $\triangle OVS \cong \triangle OTP$;
5. Yes, by the SsA
Congruence Theorem, because two sides and the angle opposite the longer of the two sides are known.
6. Below is a piece of the tessellation.

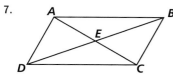

7.

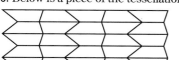

a. The parallelogram must be 2-fold rotation-symmetric about E. b. The figure must have rotation symmetry about E and reflection symmetry about the diagonals.
c. The figure must have rotation symmetry about E and reflection symmetry about the perpendicular bisectors of the sides. d. The square has all of the symmetries listed above. 8. A sufficient condition for a parallelogram is that one pair of sides is both parallel and congruent, so, because $\overline{AB} \cong \overline{CD}$ and $\overline{AB} \parallel \overline{CD}$, $ABCD$ is a parallelogram.

9.

Conclusions	Justifications
1. $\overline{AT} \cong \overline{AR}$, $\angle VTA \cong \angle VRA$	1. Given
2. $\angle A \cong \angle A$	2. Refl. Prop. of $\cong$
3. $\triangle RAS \cong \triangle TAW$	3. ASA $\cong$ Theorem

10.

Conclusions	Justifications
1. $FRWK$ is a parallelogram, $\overline{KT} \cong \overline{MR}$	1. Given
2. $KT = MR$	2. Segment $\cong$ Theorem
3. $\overline{FR} \cong \overline{KW}$, $\overline{FR} \parallel \overline{KW}$	3. Def. of parallelogram
4. $FR = FM + MR$, $KW = KT + TW$	4. Additive Prop.
5. $TW = FM$	5. Substitution
6. $\overline{TW} \cong \overline{FM}$	6. Segment $\cong$ Theorem
7. $\angle FSM \cong \angle WST$	7. Vertical $\angle$s Theorem
8. $\angle RFW \cong \angle KWF$	8. $\parallel$ Lines Theorem (Alt. Int. $\angle$s)
9. $\triangle TWS \cong \triangle MFS$	9. AAS $\cong$ Theorem
10. $\overline{MS} \cong \overline{ST}$	10. CPCF Theorem

11.

Conclusions	Justifications
1. $MRPB$ is an isosceles trapezoid with legs $\overline{RP}$ and $\overline{MB}$.	1. Given
2. $\overline{MB} \cong \overline{RP}$, $\overline{MP} \cong \overline{RB}$, $\overline{MR} \parallel \overline{BP}$	2. Def. of isos. trapezoid
3. $\overline{BP} \cong \overline{PB}$	3. Refl. Prop. of $\cong$
4. $\triangle MBP \cong \triangle RPB$	4. SSS $\cong$ Theorem
5. $\angle PMB \cong \angle BRP$	5. CPCF Theorem

12.

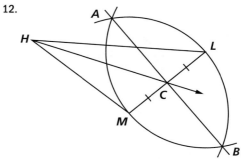

13. Because the circles M and L are congruent by construction, their radii are congruent. Thus, $ALMB$ is a rhombus, implying that the diagonal $\overline{AB}$ bisects $\overline{ML}$. Thus, C is the midpoint of $\overline{ML}$. Because two points determine a unique line, $\overline{HC}$ is the segment from H to the midpoint of $\overline{ML}$. 14. The only quadrilaterals with the property that the diagonals are of equal length are the isosceles trapezoid, rectangle and square; however, because we know each pair of sides is congruent and all four sides are not, the figure must be a rectangle. Because rectangles have right angles at every vertex, the carpenter knows that he has all right angles. 15. Rectangle, because the diagonals are congruent and bisect each other.
16. Rhombus, because one diagonal is the perpendicular bisector of the other, and the diagonals bisect each other. 17. Isosceles trapezoid, because both parts of the diagonals are congruent. 18. Square, rhombus, and kite, according to the hierarchy in Lesson 7-9.

Self-Test Correlation Chart

Question	1	2	3	4	5	6	7	8	9
Objective(s)	C	C	C	C	A	I	F	H	D
Lesson(s)	7-2, 7-5	7-2, 7-5	7-2, 7-5	7-2, 7-5	7-1, 7-2, 7-5	7-6	7-7, 7-8, 7-9	7-2, 7-5, 7-7, 7-8, 7-9	7-3, 7-4, 7-5

Question	10	11	12	13	14	15	16	17	18
Objective(s)	E	E	K	G	H	J	J	J	B
Lesson(s)	7-3, 7-4, 7-5	7-3, 7-4, 7-5	7-10	7-10	7-2, 7-5, 7-7, 7-8, 7-9	7-9	7-9	7-9	7-7, 7-9

Chapter 7 Chapter Review (pp. 445–449)

1. a. No such triangle is possible. b. The lengths do not satisfy the Triangle Inequality Postulate.

3. a.

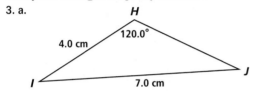

b. Yes, by the SsA Congruence Theorem

5. a.

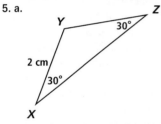

b. Yes, by the AAS Congruence Theorem   7. Yes, by the AAS Congruence Theorem   9. $\overline{AB} \cong \overline{DC}, \overline{AD} \cong \overline{BC},$ $\overline{AO} \cong \overline{OC}, \overline{DO} \cong \overline{OB}$ 11. isosceles trapezoids, rectangles, squares   13. Yes, by AAS; $\triangle MLN \cong \triangle ONL$ 15. Yes, by SAS; $\triangle XYZ \cong \triangle GHI$ 17. a. Answers vary. Sample: $\overline{ZX} \cong \overline{BC}$ b. HL

19.

Conclusions	Justifications
1. $\odot O$ and $\odot P$ intersect at A and B.	1. Given
2. $\overline{OA} \cong \overline{OB}$	2. All radii of a $\odot$ are $\cong$.
3. $\overline{PA} \cong \overline{PB}$	3. All radii of a $\odot$ are $\cong$.
4. $\overline{OP} \cong \overline{OP}$	4. Refl. Prop. of $\cong$
5. $\triangle OAP \cong \triangle OBP$	5. SSS $\cong$ Theorem

21.

Conclusions	Justifications
1. $CD = AB$, $\overline{AC}$ is the longest side in $\triangle CDA$ and $\triangle CAB$.	1. Given
2. $\overline{CA} \cong \overline{AC}$	2. Refl. Prop. of $\cong$
3. $\angle ABC \cong \angle CDA$	3. The inscribed $\angle$s intercept the same arc.
4. $\triangle CAD \cong \triangle ACB$	4. SsA $\cong$ Theorem

23.

Conclusions	Justifications
1. $ABCDEFGH$ is a regular octagon.	1. Given
2. $\overline{FE} \cong \overline{ED}$	2. Regular polygons are equilateral.
3. $\overline{ED} \cong \overline{DC}$	3. Regular polygons are equilateral.
4. $m\angle FED = m\angle EDC$	4. Regular polygons are equiangular.
5. $\triangle FED \cong \triangle EDC$	5. SAS $\cong$ Theorem
6. $\overline{CE} \cong \overline{DF}$	6. CPCF Theorem

25.

Conclusions	Justifications
1. $\overleftrightarrow{VW} \parallel \overleftrightarrow{XY}$, $VW = XY$	1. Given
2. $\angle XZY \cong \angle WZV$	2. Vertical $\angle$s Theorem
3. $\angle XYZ \cong \angle WVZ$	3. $\parallel$ Lines Theorem (alt. int. $\angle$s)
4. $\triangle XYZ \cong \triangle WVZ$	4. AAS $\cong$ Theorem
5. $\overline{XZ} \cong \overline{ZW}$	5. CPCF Theorem
6. $XY = ZW$	6. Segment $\cong$ Theorem

27.

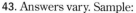

Conclusions	Justifications
1. $\triangle DFE$ is equilateral.	1. Given
2. H, I, and G are the midpoints of $\overline{DF}$, $\overline{DE}$, and $\overline{EF}$.	2. Given
3. $\overline{DF} \cong \overline{DE} \cong \overline{FE}$	3. def. of equilateral $\triangle$s
4. $DH = HF = \frac{1}{2}DF$, $DI = IE = \frac{1}{2}DE$, $FG = GE = \frac{1}{2}FE$	4. def. of midpoint
5. $\overline{DH} \cong \overline{IE} \cong \overline{FG}$, $\overline{DI} \cong \overline{GE} \cong \overline{HF}$	5. Trans. Prop. of $\cong$
6. $\angle HDI \cong \angle HFG$	6. def. of equilateral $\triangle$s
7. $\triangle HFG \cong \triangle HDI \cong \triangle IEG$	7. SAS $\cong$ Theorem
8. $\overline{HG} \cong \overline{IH} \cong \overline{GI}$	8. CPCF Theorem
9. $\triangle HIG$ is equilateral.	9. def. of equilateral $\triangle$s

29. false   **31.** $DG = 4.5$, $GF = 4$, $HF = 3.7$, $HG = 2.1$, $m\angle EFG = 60$, $m\angle DEF = m\angle FGD = 120$ **33.** False. An isosceles trapezoid has congruent diagonals and is not a parallelogram.   **35.** Because all points on a circle are equidistant from its center and B is the center of a circle with diameter $\overline{AC}$, B must be the midpoint of $\overline{AC}$.
37. Because $\overline{AP}$ is perpendicular to $\overline{RB}$, $\triangle RPA$ and $\triangle BPA$ are right triangles. We are given $\overline{RA} \cong \overline{BA}$, and since $\overline{AP} \cong \overline{AP}$ by the Reflexive Property of Congruence, we know $\triangle RAP \cong \triangle BAP$ by the HL Congruence Theorem. Thus $\overline{RP} \cong \overline{PB}$ by the CPCF Theorem, so $RP = PB$ by the Segment Congruence Theorem.   **39.** square

41. If it is possible to tessellate the plane with the part, then the parts can be cut from a single sheet of metal with no wasted material because they all fit together to cover the plane without leaving spaces or overlapping.

43. Answers vary. Sample:

45. true   **47.** parallelograms, rhombuses, rectangles, squares   **49.** Yes;

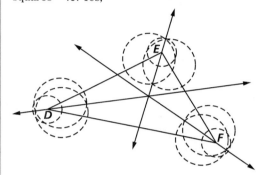

Selected Answers

Chapter 8

Lesson 8-1 (pp. 452–456)
Guided Example 1: a. $4s$ b. $2a + 2b$ c. $6s$ d. ns
Questions:
1. 0.4 mi   3. 1.45 mi   5. $2m + 2n$ 7. 75 yd
9. 160 by 100   11a. No; the route forms a 1.4 mi-by-1.6 mi
rectangle. b. 6 mi c. 1 hr   13. nw; $2h + 2nw$; $n - 1$;
h; $2h + 2nw + h(n - 1)$ 15a. true b. Answers vary.
Sample: equilateral triangle   17. 34
19. Answers vary. Sample:
21a. no b. yes
23. $x > 0$; $y > 0$; $x + y > 17$;
$x + 17 > y$, so $y - x < 17$;
$y + 17 > x$, so $x - y < 17$

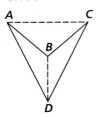

Lesson 8–2 (pp. 457–462)
Questions:
1. Yes; let G be the final image of F after reflection and
rotation. Because a reflection, rotation, and composition
of the two is an isometry, $G \cong F$ by the definition of
congruent figures, and therefore Area$(G) =$ Area(F)
by the Congruence Property of the Area Postulate
3a. about 6.80 cm^2 b. about 680 mm^2 5a. 12 b. 30
c. 22 d. 42   7. 6,300,000   9. $\frac{10}{3}$ 11. $\frac{1}{4}$ mi^2
13a. 61 children b. 42 children   15. 62.8125 mi^2
17a. $50 - \ell$ b. $\ell(50 - \ell) = 50\ell - \ell^2$ 19. $\frac{1}{2}$ 21a. false
b. Yes, because that would imply that all of the sides
are congruent because a pair of congruent sides in two
equilateral triangles gives SSS congruence

23a. $w = \frac{12}{\ell}$

23b.

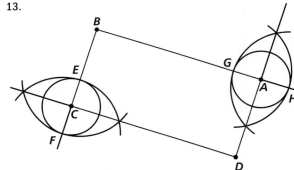

Lesson 8-3 (pp. 463–467)
Questions:
1. Answers vary. Sample: To sell land accurately, to
determine how much can be built on the island, and
to determine what infrastructure is appropriate for the
island.   3a. less than b. greater than   5a. Answers vary.
Sample: 1.25 in^2 b. Answers vary. Sample: 1.28175 in^2
7. Answers vary. Sample: 750 mi^2 9. Answers vary.
Sample: 17.875   11a. $100a$ b. $100b$
13.

15. $1 + 1 + 1 < 4$, so even if the "sides" were arranged
on a straight line there would be no way to connect the
4 segments.
17.

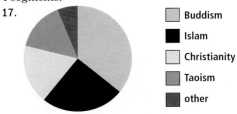

- Buddism
- Islam
- Christianity
- Taoism
- other

Lesson 8-4 (pp. 468–473)
Guided Example 2: a. 6; 7; 21 b. 6; 2; 6 c. DE; 6; 3; 9
d. AC; BE; 6; 10; 30
Questions:
1. $A = \frac{1}{2}hb$

3.

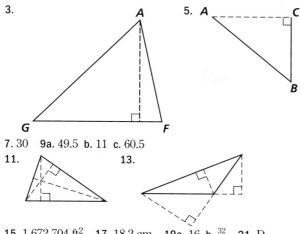

5.

23. 16,320 ft² **25a.** 1 b.

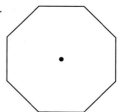

7. 30 **9a.** 49.5 **b.** 11 **c.** 60.5
11.   **13.**

15. 1,672,704 ft² **17.** 18.2 cm **19a.** 16 **b.** $\frac{32}{7}$ **21.** D
23. Straightedges are allowed in constructions, but not
rulers. **25.** No, $AB = \sqrt{58}, BC = \sqrt{37}, AC = \sqrt{17}$

Lesson 8-5 (pp. 474–479)
Guided Example 1: 5; 9; 6; 6; 5; 9; 42
Guided Example 2: 1. Given; **3.** $\frac{1}{2}(DE)(AC)$ Triangle Area
Formula; **5.** $\frac{1}{2}(BE)(AC) + \frac{1}{2}(DE)(AC)$; Substitution
6. Distributive Property
Questions:
1. Divide the polygon into triangles, find the area of
each triangle using the Triangle Area Formula, and sum
the areas. **3.** The bases are parallel, and the distance
between parallel lines is constant. **5.** $\frac{3}{64}$ mi² **7.** 126
9. 69.12 **11.** Answers vary. Sample: The area of any
parallelogram is the length of the base times the height.
The base length of all three parallelograms is FG,
and they all share the height CG, so they all have the
same area, $A = FG \cdot CG$. **13.** 66 **15.** Answers vary.
Sample: If the height of $MILK$ is h, then Area($\triangle MLK$) =
$\frac{1}{2}h \cdot KL$ = Area($\triangle IKL$). Area($\triangle MRK$) = Area($\triangle MLK$) −
Area($\triangle KRL$) = Area($\triangle IKL$) − Area($\triangle KRL$) =
Area($\triangle IRL$). **17.** 75 in² **19a.** $\frac{1}{2}xh$ **b.** $\frac{1}{2}(x + y)h$ **c.** $\frac{1}{2}zh$
21a. 42 **b.** No, there are more than two odd nodes.
23. 60, 120 **25.** $5\sqrt{5}$

Lesson 8-6 (pp. 480–486)
Questions:
1. 25 m **3.** Answers vary. Sample: ancient Greek,
Chinese, Indian, Italian, and American **5.** Instead of
referring to isosceles triangles and square roots, Homer
should have stated that the *sum of the squares* of the legs
of a *right triangle* is equal to the *square* of the remaining
side (hypotenuse). **7.** 15 **9.** $5\sqrt{2}$ **11.** acute
13. obtuse **15.** 40 **17.** $4\sqrt{5}$ **19.** $4\sqrt{2}$
21. $10^2 = 100 = 6^2 + 8^2$; thus, this would assure a right
triangle (and therefore a right angle) by the Pythagorean
Converse Theorem.

Lesson 8-7 (pp. 487–493)
Guided Example 1 a. 6.4; $6.4\sqrt{2}$ **b.** $5\sqrt{2}$; $5\sqrt{2}$; $5\sqrt{2}$
Questions:
1. 45, 45, 90 **3.** If the shorter leg has length x, then the
other leg has length $x\sqrt{3}$ and the hypotenuse has length
$2x$. **5.** $AE = 9.52, EC = 9.52\sqrt{2}$ **7.** Both legs
measure $8\sqrt{2}$. **9.** $A = \frac{1}{2}ap$ **11.** 541.8 m² **13.** 100
15. $\frac{1}{2}(1 + \sqrt{3})h^2$ **17.** 98 **19.** 20 **21.** If the two mirrors
in the periscope are parallel, then by the definition of a
translation (a composite of two reflections over parallel
lines), the image is translated from above the water to the
sailor's eye level.

Lesson 8-8 (pp. 494–499)
Guided Example 1. 26, 29
Guided Example 3. 60°, 60, 60, 1, 10, $\frac{10}{3}$
Questions:
1. circumference **3.** 3.14 is an approximation, so using
this in the calculation can lead to rounding errors.
5a. 8π in. **b.** 25.133 in. **c.** 3π in. **d.** 9.425 in. **7.** The
measure of an arc is the measurement of the central angle
that creates the arc, whereas the length of an arc is the
length of the arc's share of the circumference of the circle.
9. 76.394° **11.** $16\sqrt{2}$ **13.** $\frac{100}{\pi}$ m, or about 31.83 m
15. $\frac{1}{10\pi}$ m, or about 3.18 cm **17.** 2 cm
19. $A = \frac{1}{2}h(b_1 + b_2)$ **21.** $p = 4s$ **23.** 6

Lesson 8-9 (pp. 500–505)
Questions:
1. 100π cm² **3a.** r **b.** $\frac{1}{2}C = \frac{1}{2}(2\pi r) = \pi r$ **c.** πr^2
5. 0.0225 **7.** $\frac{1}{4}$ **9.** about 42.4 mi **11.** 300π ft²
13. approximately \$2.84 **15.** $\frac{25\pi}{2} - 25 \approx 14.27$ cm²
17. 15 and 36 **19.** A **21.** pentagon, hexagon, and
trapezoid

Chapter 8 Self-Test (pp. 509–510)
1a. $RM = PS - WT = 12 - 8 = 4, MW = ST - PR = 12$;
the perimeter of $PSTWMR = PS + ST + TW + WM +$
$MR + RP = 64$. **b.** Area($PSTWMR$) = $8 \cdot 4 + 20 \cdot 8 = 192$
2. Answers vary. Sample: Because the area of a
right triangle is one-half the product of its legs,
Area ($\triangle MQP$) = $\frac{1}{2}QP \cdot MQ$ and Area ($\triangle MQN$) =
$\frac{1}{2}NQ \cdot MQ$. Area ($\triangle MNP$) = Area ($\triangle MQP$) +
Area ($\triangle MQN$) = $\frac{1}{2}QP \cdot MQ + \frac{1}{2}NQ \cdot MQ = \frac{1}{2}(QP + NQ) \cdot$

$MQ = \frac{1}{2}NP \cdot MQ = \frac{1}{2}bh$.   **3a.** By the Pythagorean Theorem, the hypotenuse of each right triangle is 13. Thus the perimeter is $2 \cdot 13 + 2 \cdot 5 = 36$.

b. Area $= 2\left(\frac{1}{2} \cdot 5 \cdot 12\right) = 60$ **4.** $8\sqrt{2}$; The diagonal forms two 45-45-90 triangles. The legs measure 8, so the hypotenuse measures $8\sqrt{2}$.   **5.** 112; The two congruent sides of the trapezoid measure $\frac{1}{2}(48 - 20 - 8) = 10$ and are the hypotenuses of triangles with base measures of $\frac{1}{2}(20 - 8) = 6$. Therefore the height of the trapezoid is 8, so the area is $\frac{1}{2} \cdot 8(20 + 8) = 112$.   **6.** Extend the vertical sides of the stairs to form rectangles, the area is $8(12) + 16(12) + 24(12) + 32(12) + 40(12) = 1440$.

7. $A = 12(28) + \frac{1}{2}\pi \cdot 6^2 = 336 + 18\pi$ ft$^2 \approx 392.55$ ft^2

8. 200 yd; The segment between the runners is made up of the hypotenuses of two right triangles with legs measuring 60 and 80 yards, so each hypotenuse is 100 yd.   **9.** There are 8 squares completely inside, 22 squares partially inside, therefore Hawaii covers about $15^2(8 + (22)) = 4275$ mi^2.   **10.** $A = \frac{1}{2}ap$; $a = 6\sqrt{3}$, so $A = \frac{1}{2} \cdot 6\sqrt{3} \cdot 72 = 216\sqrt{3}$ mm^2 **11.** $A = \frac{80}{360} \cdot \pi r^2 = \frac{2}{9} \cdot \pi(20)^2 \approx 279.25$ ft^2 **12.** $A = x^2 + \pi\left(\frac{x}{2}\right)^2 = \left(1 + \frac{\pi}{4}\right)x^2$ **13.** Form a rectangle and two 30-60-90 triangles; the altitude of the trapezoid is 6 and the base $\overline{NM}$ has length $(2\sqrt{3} + 8\sqrt{3} + 6\sqrt{3})$ or $16\sqrt{3}$. $A = \frac{1}{2} \cdot 6(8\sqrt{3} + 16\sqrt{3}) = 72\sqrt{3}$

14.

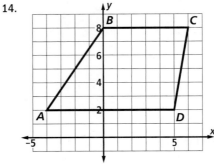

$ABCD$ is a trapezoid. The area of $ABCD$ is $\frac{1}{2}(6)(6 + 9) = 45$.   **15.** $\pi \cdot 10^2 - \pi \cdot 8^2 = 100\pi - 64\pi = 36\pi$ **16.** The cost per square inch is the price of the pizza divided by its area. The radius of the pizza is 4.5". 9" pizza: $\frac{\$12.30}{\pi \cdot 4.5^2} \approx \$0.19$ per square inch   **17.** Four inches are added to both the length and width of the photo. Thus the outer dimensions of the frame are 8 in. and 10 in., and so the perimeter is $2(8) + 2(10) = 36$ in.   **18.** $90\sqrt{2} - 60\sqrt{2} = 30\sqrt{2} \approx 42.4$ ft   **19a.** By the Pythagorean Converse Theorem, the triangle is right, and $\angle Z$ is the right angle. **b.** The triangle is not right, and since the square of the side opposite $\angle Z$ is larger than the sum of the squares of the other two sides, $\angle Z$ is obtuse. **c.** The triangle is not right, and because the square of the side opposite $\angle Z$ is smaller than the sum of the squares of the other two sides, $\angle Z$ is acute.   **20.** about 390 ft; $d = \sqrt{310^2 + 236^2}$

Self-Test Correlation Chart

Question	1	2	3	4	5	6	7	8	9	10
Objective(s)	A, C	H	A, C, E	E	C	K	K, L	J	B	G
Lesson(s)	8-1, 8-2, 8-4, 8-5, 8-7	8-2, 8-3, 8-4, 8-9	8-1, 8-2, 8-4, 8-5, 8-6, 8-7	8-6	8-2, 8-4, 8-5, 8-7	8-2, 8-4, 8-5	8-2, 8-4, 8-5, 8-8, 8-9	8-6	8-3	8-7

Question	11	12	13	14	15	16	17	18	19	20
Objective(s)	L	G, H	G	M	D	L	I	J	F	J
Lesson(s)	8-8, 8-9	8-2, 8-3, 8-4, 8-7, 8-9	8-7	8-2, 8-4, 8-5	8-8, 8-9	8-8, 8-9	8-1	8-6	8-6	8-6

Chapter 8 Chapter Review (pp. 511–515)

1. 76.5 cm   **3.** 48 cm   **5.** 3 ft, 15.5 ft, 15.5 ft   **7.** Answers vary. Sample: 21.5 cm^2 **9.** Answers vary. Sample: 4.5 mi^2 **11.** $13xy$ **13.** 11 cm^2 **15.** 19 cm   **17.** $\frac{7}{2\pi} \approx 1.11$ cm   **19.** 4.71 in.   **21.** 90   **23.** 37 ft   **25.** $5\sqrt{7}$ **27.** yes   **29.** no   **31.** Yes, the sum of the squares of the legs $(10^2 + 7.5^2)$ is equal to the square of the hypotenuse (12.5^2).   **33.** 23.4 cm^2 **35.** 50 cm^2 **37.** The area of a trapezoid with height h and bases b_1 and b_2 is $A = \frac{1}{2}h(b_1 + b_2)$. Using the Trapezoid Area Formula and noting that in a parallelogram the bases have equal length, one can derive the formula for the area of a parallelogram: $A = hb$.   **39.** $\frac{T^2}{n}$; because n copies of figure F tile the square, $n \cdot \text{Area}(F) = \text{Area}(\text{square})$. The square has sides of length T, so its area is T^2. Thus, $n \cdot \text{Area}(F) = T^2$, so $\text{Area}(F) = \frac{T^2}{n}$.   **41.** about 15.34 sec   **43.** 14 min   **45.** about 9.6 ft   **47.** about 283 m   **49.** 144 ft^2 **51a.** 2000 ft^2 **b.** 39.9 in^2 **53a.** 9π mi^2 **b.** $\frac{1}{9}$ **55.** 8π cm^2 **57.** 100   **59.** $3y$

Chapter 9

Lesson 9-1 (pp. 518-524)
Questions:
1. Unique Plane Assumption.
3. Unique Plane Assumption   5. True
7.

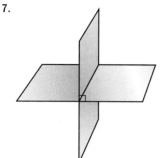

9.

11.

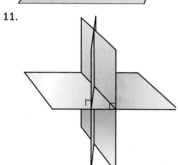

13.

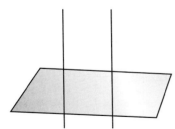

15a. Unique Plane Assumption b. Unique Line
Assumption   17. Consider the lines going from the
front license plate to the back license plate on each
of the bottom two cars. Clearly each of these lines is
perpendicular to the metal spike, yet the lines are not
parallel to one another.   19. 80   21. 2, by the Line-Plane
Perpendicular Theorem   23. 0.96 in.   25. 3 matches,
4 matches, 5 matches

27.

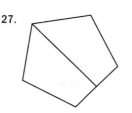

Lesson 9-2 (pp. 525-531)
Questions:
1. A surface separates space into two regions, an interior
and an exterior. A solid is the union of a surface and its
interior.   3. *ABCDEF* and *QRSNOP, FEOP* and *BCSR,
AFPQ* and *CDNS, EDNO* and *ABRQ*   5. true
7. The bases of a cube are squares, which are regular
quadrilaterals. Thus, a cube is a regular prism.
9.

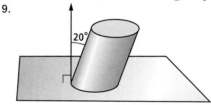

11. The amount of lean plus the measure of the angle
of the lateral edge equals 90°.   13a. 64 b. $4\sqrt{3} \approx 6.9$
15. about 9.6 in.   17a. True; every lateral edge is
produced by equivalent translations of points. b. False; the
lateral edges of an oblique prism are not perpendicular to
the bases. c. False; if the base is not a regular polygon,
the lateral faces may not be congruent. d. True; the
lateral faces of a prism always have four sides; thus,
the bases must be quadrilaterals for every face to be
congruent. The bases must also be regular polygons
and the prism right. These conditions are satisfied only
by a cube.   19. 12*s*   21. The line and the plane are
perpendicular.
23a. no b. Answers may vary. Sample:

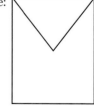

Lesson 9-3 (pp. 532-537)
Questions:
1. a. *EVIR* b. *T* c. 4; $\overline{TE}, \overline{TR}, \overline{TI}, \overline{TV}$
d. 4; *TEV, TVI, TIR, TRE* e. 4; $\overline{RE}, \overline{EV}, \overline{VI}, \overline{IR}$
3. triangular pyramid   5. hexagonal pyramid   7a. 12
b. $4\sqrt{10} \approx 12.65$ 9. The generalization is not true
because the cone has 2 faces, only 1 vertex, and infinitely
many lateral edges.   11. height

13.

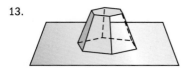

15. about 153 m   17. solid right decagonal prism
19. about 2.9 times   21. The point on $\overline{AC}$ closest to B is
the point D that is the intersection of $\overline{AC}$ and the altitude
to it. The Unequal Angles Theorem guarantees that BD is
less than the distance from B to any other point on $\overline{AC}$.

Lesson 9-4 (pp. 538–543)

Questions:

1. a. yes b. no   3. Artists can, but do not necessarily have
to, use perspective in their drawings

5.

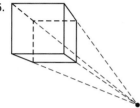

7. 2   9. one-point perspective 11.

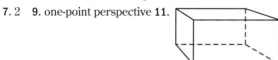

13. a. Answers vary. Sample: the one on the back building
appears longer. b. The segment on the front building is
longer. c. Answers vary. Sample: The perspective makes
the second building seem larger than the first, and hence,
makes the segment on the back building seem larger than
the segment on the front building.   15. On the back of the
U.S. one dollar bill is a solid right quadrilateral pyramid
above a truncated pyramid.   17. circle   19. point
21. For a point P not on a line ℓ, the reflection image of
point P over line ℓ is the point Q if and only if ℓ is the
perpendicular bisector of $\overline{PQ}$. For a point P on ℓ, the
reflection image is P itself.

Lesson 9-5 (pp. 544–549)

Guided Example 2: a. trapezoidal prism b. 15 ft c. 20 ft;
15 ft d. 7, Pythagorean Theorem; 13.0, 260

Questions:

1. view, elevation, isometric drawing   3. B
5a. left view   b.   top view

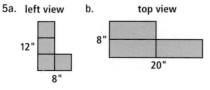

c.   back view

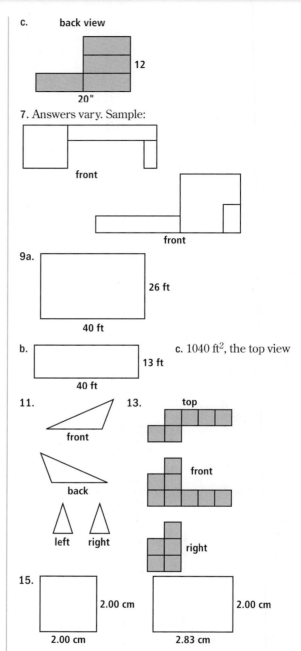

7. Answers vary. Sample:

9a.

b.

c. 1040 ft², the top view

11.   13.

15.

Smallest: 2 cm, 2 cm; Largest: 2 cm, 2.83 cm   17. true
19. Hexagons can tessellate a plane, whereas pentagons
cannot.

Lesson 9-6 (pp. 550–558)

Guided Example: 7.2; 7.2; 7.2; 250,000; 129,250,000; 24,479

Questions:

1.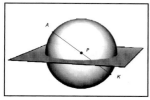

3. A plane section of a 3 dimensional figure is the intersection of that figure with a plane.   **5.** false
7a. circle, ellipse, parabola, hyperbola **b.** Answers vary. Sample: Conic sections are used for telescopes, headlights, and satellites   **9.**

11.

13. **a.**   **b.**

c. pentagon, pentagon
15a.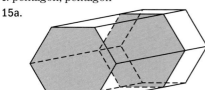

b.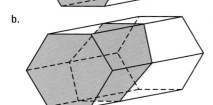

c. hexagon, hexagon   **17.** 4 m
19.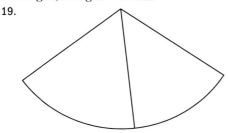

21. the circle centered at *A*, with radius *AB*

Lesson 9-7 (pp. 559–563)

Question:

1a. infinitely many **b.** any plane through the center of both bases; also, the plane parallel to the bases and equidistant from them   **3a.** infinitely many **b.** any plane through the center of the handle base and the apex   **5a.** infinitely many **b.** any plane through the apex and the center of the circular base   **7.** C   **9a.** 4

b.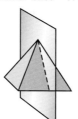

11. yes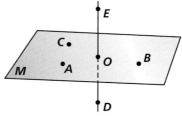

13. 9   **15a.** none of these four views **b.** A corner view of the edge between the top and front sides will reveal the missing cube.   **17.** the back side   **19a.** No, the center, *O*, of the unique circle through *A*, *B*, and *C* is the only point in the plane *M* containing them that is equidistant from them. *D* and *E* must be equidistant from *O* on the line perpendicular to *M* through *O*.

b.

Lesson 9-8 (pp. 564–569)

Guided Example: 7; 5; pentagons; pentagons; bases; pentagonal prism; rectangles; pentagonal prism

Questions:

1.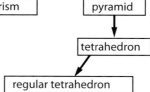

3. polyhedron; edge; faces

5. A cylinder is not the union of polygonal regions.

7a. 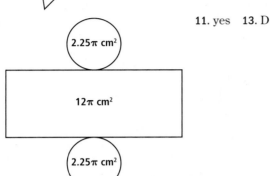 **b.**

9.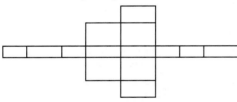

2.25π cm²

12π cm²

2.25π cm²

11. yes   **13.** D

15. Answers vary. Sample:

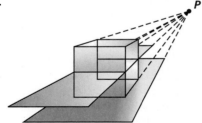

17. The surface that results from slicing a cone with a plane parallel to its base and using the section of the plane as a face is a frustum of a cone.

19.

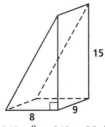

P

21. The creature must be 2-dimensional.
23. m∠EHI = m∠IHG by the definition of a bisector. $\overline{EF} \parallel \overline{HG}$ and $\overline{EH} \parallel \overline{FG}$ by the definition of a parallelogram. m∠IHG = m∠EIH by the Parallel Lines Theorem. m∠EIH = m∠EHI by transitivity. $\overline{EI} \cong \overline{EH}$ by the Converse of the Isosceles Triangle Base Angles Theorem.

Lesson 9-9 (pp. 570–576)

Guided Example 3: πr^2; $2\pi r$; $2\pi rh$; $2\pi rh$; πr^2; $2\pi r(r + h)$
Guided Example 5: 1.5; 0.5; $2.25 - 0.25\pi$; 1.3125; 0.875π; $2.25 - 0.25\pi$; 1.3125; 0.875π; $9.75 + 0.375\pi$
Questions:
1. sum; twice   **3a.** no **b.** no **c.** yes **d.** yes **e.** no
5a. 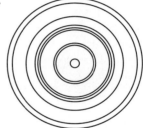   **b.** 342 **c.** 462

15

8   9

7a. 240 m² **b.** $240 + 36\sqrt{3} \approx 302.4$ m²   **9a.** $2xy + 2xz + 2yz$ **b.** $18xy + 18xz + 18yz$ **c.** 9 **d.** If all the dimensions of a box increase by a factor of r, then the surface area of the box increases by r^2.   **11.** $80 - 4x^2$ ft²   **13.** 48π in² ≈ 151 in²
15. The shape would be a rectangle with height 4 in. and length 1350 in. Its area would be 5400 in.² = 37.5 ft².
17. 4 in.   **19.** $\sqrt{3} \approx 1.7$ ft
21.

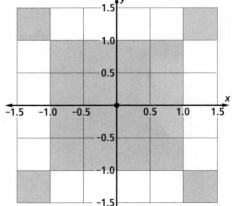

23a. 17.5 **b.** 25.5
c. 19.7 **d.** 95.5
25.

Lesson 9-10 (pp. 577–582)

Guided Example 1: $\frac{1}{2}s\ell$; ns

Questions:

1a. i b. ii c. i d. ii

3a.

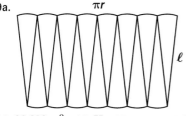

2 in.

2 in.

b. $6\sqrt{3}$ in^2 c. $12\sqrt{3}$ in^2 5. $70\pi \approx 219.9$ cm^2 7. 1621 ft^2

9a.

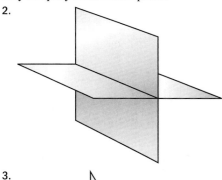

πr

ℓ

b. ℓ and πr

11. 28,899 m^2 13. H 15. true 17. Uniqueness Property, Congruence Property, Additive Property, and Rectangle Formula

Chapter 9 Self-Test (pp. 587–588)

1. One. To determine the plane that the circle lies in, pick any three points on the circle that are not on the same diameter. Because these three points are noncollinear, they uniquely determine a plane.

2.

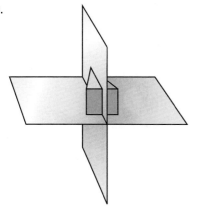

3.

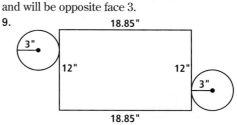

4.

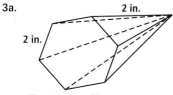

top   side   front

5. cell phone 6. Using the Pythagorean Theorem, the radius r can be found by the equation $r^2 + 16^2 = 20^2$, so that $r^2 = 144$. Therefore, $r = 12$, and because the diameter is twice the radius, the diameter of the base is 24.

7. The area of the base is $12^2 = 144$. The height of each of the triangles, h, is found by solving the equation $6^2 + h^2 = 10^2$, which gives $h^2 = 64$, so $h = 8$. The area of each of the triangles is given by $\frac{1}{2}bh$ so the area of each triangle is 48. Thus, the surface area of the pyramid is $144 + 4 \cdot 48 = 336$. 8. 6. Let face 4 be the base of the cube. Then face 3 will fold up, and face 5 will also fold up and stand adjacent to face 3. Face 6 will then fold around and will be opposite face 3.

9.

18.85"

3"

12"   12"

3"

18.85"

10. The segment connecting the center of the sphere with any point on the intersection of the sphere with the circle will be a radius of that sphere. Therefore, a right triangle is formed with the radius of the circle and the distance between the circle and the center of the sphere as legs of the triangle and the radius of the sphere as the hypotenuse. Because the small circle has area 36π in^2, its radius is 6 in. This gives $6^2 + 6^2 = r^2$, so $r = 6\sqrt{2}$ inches, and because the diameter of the sphere is twice the radius of the sphere, the diameter of the sphere is $12\sqrt{2}$ in.

11. To find the area of one of the lateral faces, first find the height of the triangle that forms a lateral face. Dividing the length of the side of the base in half gives 50, which is the length of one of the legs of the triangle that has a hypotenuse that is the height of the triangle that forms a lateral face. Thus, if h is the height of the lateral face, $h^2 = 50^2 + 125^2$, so $h^2 = 18{,}125$, and $h = 25\sqrt{29}$. Then the area of a lateral face is given by $\frac{1}{2}bh$, where $b = 100$, so the area of a lateral face is $1250\sqrt{29}$ ft^2. Multiplying this number by 4 gives the lateral area, so the lateral area of the pyramid is $5000\sqrt{29}$ ft^2, or about 26,926 ft^2.

12. No; In a prism, the lateral edges must be parallel, and the sides of a triangle are never parallel. 13. The area of each base triangle is given by $\frac{1}{2}bh$, where $b = 5$ and $h = 4$, so the area of each base triangle is 10. To find AC, use the

Pythagorean Theorem, which gives $4^2 + 5^2 = AC^2$, so $AC^2 = 41$, and $AC = \sqrt{41}$. Then the area of $AEDC = 12\sqrt{41}$, the area of $AEFB = 48$, and the area of $BFDC = 60$. Adding the areas gives a surface area of $128 + 12\sqrt{41}$.
14a. ABC, FED **b.** Answers vary. Sample: $AEDC$
c. Answers vary. Sample: $\overline{AE}$ **15.** Converting 10 feet into inches gives 120 inches. The circumference of the base circle is $\pi d = 28\pi$ inches, and multiplying by the height gives approximately 10,556 or 73.3 ft².
16.

| front | side | top |

17a. 8. Seven are planes containing corresponding symmetry lines of the base heptagons and one is parallel to the bases. **b.** 7
18a.

 b.

19. 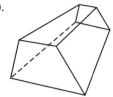 **20.**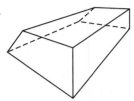

Self-Test Correlation Chart

Question	1	2	3	4	5	6	7	8	9	10
Objective(s)	E	A	A	B	L	C	C, K	K	K	C
Lesson(s)	9-1	9-1, 9-2, 9-3, 9-6	9-1, 9-2, 9-3, 9-6	9-5	9-5	9-2, 9-3, 9-9, 9-10	9-2, 9-3, 9-8, 9-9, 9-10	9-8	9-8	9-2, 9-3, 9-9, 9-10

Question	11	12	13	14	15	16	17	18	19	20
Objective(s)	J	G	C	F	J	B	H	I	D	D
Lesson(s)	9-9, 9-10	9-2, 9-3, 9-6	9-2, 9-3, 9-9, 9-10	9-2, 9-3	9-9, 9-10	9-5	9-7	9-6	9-4	9-4

Chapter 9

Chapter Review (pp. 589–593)

1.

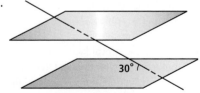

3. Answers may vary. Sample:

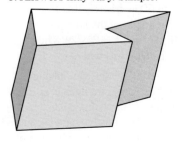

5.

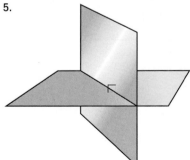

7. 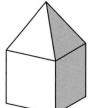 **9a.** **b.** **c.**

11a. $3\sqrt{2}$ **b.** $3\sqrt{3}$ **c.** 54 **13.** 24 **15.** $\frac{40}{3\pi} \approx 4.24$ cm
17. Answers may vary. Sample:

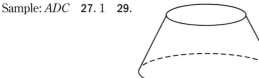

19.

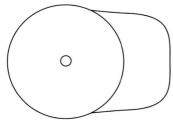

21. The three points are noncollinear. **23.** Plane A contains $\overleftrightarrow{PQ}$. **25a.** $\overline{AO}$ **b.** $DEFBC$ **c.** Answers vary. Sample: ADC **27.** 1 **29.**

31. true **33.** true **35.** 3 **37.** 13 **39.** infinitely many
41. A plane parallel to the ground (on which the fork rests) and intersecting the base of the fork will result in a circle.
43. Answers vary. Sample:

45. 11 cm **47.** $6400\sqrt{2}$ cubits^2 **49a.** no **b.** yes
c. no **d.** no
51.

53.

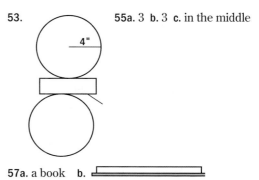

55a. 3 **b.** 3 **c.** in the middle

57a. a book **b.**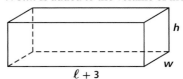

Chapter 10

Lesson 10-1 (pp. 596–602)
Guided Example 2: 6; 3, 2, 10, 6; 10, 6, 22; 22; 5; 5; 4; 26
Questions:
1a. no **b.** no **3.** Answers vary. Sample: Surface area is a 2-dimensional measure of a 2-dimensional figure, while volume is a 3-dimensional measure, of a 3-dimensional figure. Two objects can have the same surface area but different volumes (and vice versa) **5a.** 1428 in^3
b. 730 in^2 **7.** The total volume of two nonoverlapping solids is equal to the sum of the individual volumes
9. 28.875 in^3 **11.** A **13a.** $\sqrt[3]{60}$ cm **b.** 3.91 cm
15. Answers vary. Sample: 1 in. by 1 in. by 72 in.; 1 in. by 8 in. by 9 in.; 2 in. by 4 in. by 9 in. **17a.** the area of the base **b.** $B = \ell w$, so $V = \ell wh = Bh$. **19a.** 16.387064 cm^3
b. 16.39 **21.** 9475 days **23.** about 7.92 ft **25.** No, the sides of the cylinder are still perpendicular to the bases.
27. area of $A = 15$; area of $B = 27$

Lesson 10-2 (pp. 603–608)
Questions:
1. $4 \cdot 5 + 4 \cdot 0.2 + 5 \cdot 0.7 + 0.2 \cdot 0.7$
3a. $(c + d + 5)(a + b + 3)$
b. $ac + ad + bc + bd + 5a + 5b + 3c + 3d + 15$
c. $(4 + 10 + 5)(2 + 3 + 3) = 19 \cdot 8 = 152$,
and $2 \cdot 4 + 2 \cdot 10 + 3 \cdot 4 + 3 \cdot 10 + 5 \cdot 2 + 5 \cdot 3 + 3 \cdot 4 + 3 \cdot 10 + 15 = 152$.
5. The volume of the box is multiplied by a factor of 3.

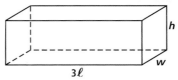

7. $3wh$ is added to the volume of the box.

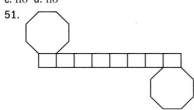

9a. The volume of the original brick is multiplied by xy.
b. The cost will increase as x and y increase. **c.** The cost of the house will increase. **11a.** $x(y + z) = xy + xz$
b. Distributive Property of Multiplication over Addition
13. yes **15.** 15 in. **17.** 204,000 cm^3 **19.** about 270 ft^2
21. $A = hb$; the formula for the area of a trapezoid is $\frac{1}{2}h(b_1 + b_2)$, but in a parallelogram, the two bases, b_1 and b_2 are of equal length, so $\frac{1}{2}(b_1 + b_2) = b$.

Lesson 10-3 (pp. 609–614)

Guided Example 2: 7; 49π; 49π
Questions:
1. about 7.48 gallons **3.** D **5.** about 5.763 m^3
7. $14.205 \approx 14.21$ m^3 **9.** $400\pi \approx 1256.68$ ft^3 **11.** Let I and II be two solids included between parallel planes. If every plane P parallel to the given planes intersects I and II in sections with the same area, then Volume(I) = Volume(II).
13. $404,766.2 **15.** 6 m^2 **17.** The volumes have a ratio of 1:8.
19.

	b	c	
a	ab	ac	
4	4b	4c	

$ab + ac + 4b + 4c$

21a. Answers vary. Sample: Construct plane Z perpendicular to both X and Y. **b.** Answers vary. Sample: Construct plane Z by forming a plane containing one line in plane X and another line in plane Y parallel to the first line. **c.** Answers vary. Sample: Construct plane Z parallel to either X or Y. **23a.** square **b.** parallelogram
c. trapezoid **d.** rectangle

Lesson 10-4 (pp. 615–622)

Guided Example 1: 660, 24; 660, 24, 5280; 380.25; 380.25, 18, 2281.5
Questions:
1. HL Congruence Theorem **3.** It is $\frac{1}{3}$ the volume of the box. **5.** 40 **7.** The volume of the cone is the volume of the cylinder. **9.** $180\sqrt{3}$ **11.** The volume is multiplied by 25. **13.** $\frac{2000\pi}{3} \approx 2094.40$ **15a.** $\frac{610\pi}{3} \approx 638.79$ cm^3
b. $38\pi \approx 119.38$ cm^3 **c.** about 0.19 **d.** For a container like this frustrum, only about 19% of the ice cream is more than 2 cm away from the freezer air. **17a.** 11,088 in^3
b. Cavalieri's Principle tells us that the volume of the stairs is the same as the volume of a rectangular prism with dimensions 42" by 7" by 22". **19.** 9906 **21.** rotation, because it preserves orientation and does not move the figure out of an area if it is repeated

Lesson 10-5 (pp. 623–627)

Guided Example 1: $\frac{1}{3}Bh$; πr^2; $\frac{1}{3}\pi r^2 h$
Questions:
1. The best formulas to remember are the formulas that apply to the most figures. **3.** cones and pyramids
5. S.A. $= 4s^2 + 2s^2 = 6s^2$ **7.** S.A. $= \pi r \ell + \pi r^2$
9a. $A = \frac{1}{2}\pi rh$ **b.** 64π in$^2 \approx 201.1$ in^2
11a. $V = \frac{2}{3}\pi r^3$ **b.** $\sqrt[3]{36}$
13a.

b. $V = \frac{1}{60}\pi d^3$ **c.** about 1362 bushels **15a.** $V = \frac{\sqrt{2}}{2}s^3$
b. about 117.85 cm^3 **17.** $3xyz + 9xz - 9yz - 27z$
19. 6 by 6 by 6 **21.** no

Lesson 10-6 (pp. 628–632)

Guided Example 2: 1221 km; 3480 km; 3480 km; 1221 km; 3480; 1221; 3480; 1221; 40,323,875,140
Questions:
1. He discovered the Sphere Volume Formula.
3a. $20\pi \approx 62.83$ **b.** $20\pi \approx 62.83$ **5.** 8 **7.** $\frac{16}{3}\pi + 192$ m^3
9. less **11.** πr^3 **13.** about 10 m **15a.** If A and B are two solids included between parallel planes such that every plane P parallel to the given planes intersects A and B in sections with the same area, then the volumes of A and B are equal. **b.** If A and B are two solids included between parallel planes and have the same volume, then every plane P parallel to the given planes intersects A and B in sections with the same area. The converse is not true.
17. about 0.5 **19.** Answers vary. Sample: $P(x, y) = (\frac{1}{7}x, \frac{1}{7}y)$

Lesson 10-7 (pp. 633–637)

Guided Example 3: $4\pi r^2$; $\frac{4}{3}\pi r^3$; 2; $2\pi r^2$; $\frac{2}{3}\pi r^3$; 21 in.; $2\pi(21)^2$; 2770.9; $\frac{2}{3}\pi(21)^3$; 19,396.2
Questions:
1. 4 **3a.** 144π cm^2 **b.** 452.4 cm^2 **5.** about 399.20 cm^2
7. S.A. $= \pi d^2$ **9.** They are equal. **11a.** $111\pi \approx 348.72$ in^3
b. $87\pi \approx 273.32$ in^2 **13.** S.A. $= \frac{C^2}{\pi}$ **15.** $\frac{80}{3}$ ft^3 or $26\frac{2}{3}$ ft^3
17. by Cavalieri's Principle **19a.** 141 cm^3 **b.** 78 cm^3

Chapter 10 Self-Test (pp. 640–641)

1a. $V = \pi r^2 h$, $r = \frac{d}{2} = \frac{3}{2}$, so $V = \pi\left(\frac{3}{2}\right)^2(6) = \frac{54\pi}{4} = \frac{27\pi}{2}$ in^3
b. $\frac{0.554 \text{ fl oz}}{1 \text{ in}^3} \cdot \frac{27\pi}{2}$ in$^3 \approx 23.50$ fl oz **2.** $V = s^3 = (0.8)^3 =$
0.512 cm^3 **3.** $V = Bh$, so $40 \cdot h = 300$, $h = \frac{300}{40} = 7.5$ in.
4. S.A. $=$ L.A. $+ 2B$. $MN = \sqrt{15^2 - 12^2} = \sqrt{81} = 9$, so
L.A. $= 18 \cdot 15 + 18 \cdot 12 + 18 \cdot 9 = 648$. $B = \frac{1}{2}(9 \cdot 12) = 54$.
S.A. $= 648 + 2(54) = 756$. The surface area of a cube with side s is $6s^2$, so let $6s^2 = 756$, then $s = \sqrt{126}$.
5. The surface area of the box is $2(10)(50) + 2(10)(2) + 2(50)(2) = 1240$, and the volume of the box is $10 \cdot 50 \cdot 2 = 1000$. The volume of a cube with side s is s^3, so $s^3 = 1000$.

Thus, $s = \sqrt[3]{1000} = 10$. The surface area of the cube is S.A. $= 6s^2 = 6(10)^2 = 600$ cm^2, so the surface area of the box is greater. **6a.** $V = \frac{4}{3}\pi r^3$ and $C = 2\pi r$. If $C = 9$, $9 = 2\pi r$, $r = \frac{9}{2\pi}$, so $V = \frac{4}{3}\pi\left(\frac{9}{2\pi}\right)^3 \approx 12$ in^3. **b.** S.A. $= 4\pi r^2$ and $C = 2\pi r$. If $C = 9.25$, $9.25 = 2\pi r$, $r = \frac{9.25}{2\pi}$ so S.A. $= 4\pi\left(\frac{9.25}{2\pi}\right)^2 \approx .27$ in^2. **7.** $V = \frac{1}{3}\pi r^2 h$, $r = \frac{d}{2} = \frac{10}{2} = 5$, $h = \sqrt{13^2 - 5^2} = \sqrt{144} = 12$, so $V = \frac{1}{3}\pi(5)^2(12) = 100\pi$ in^3. **8a.** $V = (3x)\left(\frac{x}{2}\right)(x) = \frac{3}{2}x^3$ **b.** The new volume is larger than the original volume. **9.** The total volume is the volume of the cylinder plus the volume of the two hemispheres. Since two hemisphere together form a sphere, the total volume is the volume of the cylinder plus the volume of a sphere. $V = \pi r^2 h + \frac{4}{3}\pi r^3$, with $h = w$, $r = \frac{w}{2}$. So $V = \pi\left(\frac{w}{2}\right)^2 w + \frac{4}{3}\pi\left(\frac{w}{2}\right)^3 = \frac{5}{12}\pi w^3$ **10.** $V = (x+3)(x+2)(y+1) = x^2 y + x^2 + 5xy + 5x + 6y + 6$, or $V =$ the sum of the volumes of the smaller boxes $= x^2 y + x^2 + 3xy + 3x + 2xy + 2x + 6 + 6y = x^2 y + x^2 + 5xy + 5x + 6y + 6$ **11.** No, Cavalieri's Principle does not apply to objects of different heights. If the bottoms of the two cups are on the same plane, the tops

will be on different planes, and vice versa. **12.** $V = Bh$, $B = \frac{1}{2}(b_1 + b_2)h_{\text{base}}$, $b_1 = 14$, $b_2 = 14 + \sqrt{10^2 - 8^2} = 14 + 6 = 20$, $h_{\text{base}} = 8$, and $h = 16$. So $V = \frac{1}{2}(8)(14 + 20)(16) = 2176$ in^3 **13. a.** Because 60 is $\frac{1}{6}$ of 360, $V = \frac{1}{6}$ of the volume of the cylinder with the same radius and height as, so the solid $V_{\text{cyl}} = \pi r^2 h = \pi(6)^2 12 = 432\pi$. $V = \frac{1}{6}V_{\text{cyl}} = 72\pi$. **b.** S.A. $=$ Area$(2$ rectangular sides$) + 2B + \frac{1}{6}$S.A.$_{\text{cyl}} = 2(12)(6) + 2\left(\frac{\pi 6^2}{6}\right) + \frac{12(12\pi)}{6} = 144 + 12\pi + 24\pi = 144 + 36\pi$ **14.** The volume of the space between the spheres is the volume of the larger sphere minus the volume of the smaller sphere. The larger sphere has a radius of $15 + \frac{3}{4} = 15.75$ in. So, $V = \frac{4}{3}\pi(15.75)^3 - \frac{4}{3}\pi(15)^3 \approx 2228.4$ in^3. **15.** $V = \frac{1}{3}Bh$, $B = \frac{1}{2}ap = \frac{1}{2} \cdot 30\sqrt{3} \cdot 360 = 5400\sqrt{3}$, $h = \sqrt{100^2 - 60^2} = \sqrt{6400} = 80$. So $V = \frac{1}{3}(5400\sqrt{3} \cdot 80) = 144{,}000\sqrt{3}$ cm^3 **16.** Take the general formula for volume to be Bh. Then, if you know the area formulas for each of the figures' bases, you can easily figure out their volumes. For example, the area of a circle (the base of a cylinder) is πr^2. Therefore, the volume is $\pi r^2 h$.

Self-Test Correlation Chart

Question	1	2	3	4	5	6	7	8
Objective(s)	A, H	J	A	E	E	I, J	B	F
Lesson(s)	10-1, 10-3	10-1, 10-3, 10-4, 10-6	10-3	10-5	10-5	10-1, 10-3, 10-4, 10-6, 10-7	10-4	10-2

Question	9	10	11	12	13	14	15	16
Objective(s)	D	K	G	A	A	C	B	D
Lesson(s)	10-1, 10-5	10-2	10-3, 10-4	10-3	10-3	10-6, 10-7	10-4	10-2, 10-5

Chapter 10 Chapter Review (pp. 642–645)

1. 101.25π **3.** 720 **5.** $139\sqrt{139} \approx 1638.79$ cm^3 **7.** $\frac{4}{3}$ cm^2 **9.** $\frac{17}{3}$ **11.** about 18,000,000 cubits^3 **13.** 36 **15.** $\frac{148}{3}\sqrt{37}\pi$ cm^3 **17.** 11.8 **19.** $V = \frac{1}{3}\pi r^2(r+t)$ **21.** $V = \frac{1}{3}\pi r^2\sqrt{\ell^2 - r^2}$ **23.** $\frac{\sqrt{2}}{12}m^3$ **25.** cylinder **27.** $r\sqrt{\frac{2\pi}{3}}$ **29.** The volume of the cone is $\frac{1}{3}$ the volume of the cylinder. **31a.** It is 52.4 times the original. **b.** It is 4.8 times the original. **33.** The new pizza has four times the volume of the original. **35a.** The volume is multiplied by 5. **b.** no **37.** Yes, because both prisms are right prisms with bases of equal areas, every plane parallel to the given prisms intersects each prism in sections with the same area. So the volumes of the two prisms are equal. **39a.** true **b.** false **41.** 97,500 L **43.** 5.4 L **45.** at least 363,702 pictures **47.** about 2 ft^2

49a. about 44.4% **b.** about 66.7% **51.** $66{,}000\pi$ ft^3 **53.** about 108,678,129 km **55.** $2xy + y^2 - 6xz - 3yz$ **57.** $A = pz + 2wp + 4tz + 8wt$; $A = (p + 4t)(z + 2w)$

Chapter 11

Lesson 11-1 (pp. 648–653)

Questions:
1. Law of Ruling out Possibilities   3a. There is a triangle *ABC*. b. Law of Ruling out Possibilities   5. Answers vary. Sample: Tina is not an actor, and Ms. Willows is not an actor.   7. Aaliyah likes fencing, Hassan likes sailing, Ian likes chess, Jordan likes hockey, and Marissa likes knitting.   9. Point A lies more than 5 units away from *C*.
11. Law of Ruling out Possibilities
13.

2	7	1	10
8	9	2	19
9	8	4	21
19	24	7	

15a. $V = \pi r^2 h + \pi r^3$
b. S. A. $= 2\pi rh + 4\pi r^2$
17. true   19. $x = \sqrt{2}$ or $x = -\sqrt{2}$

Lesson 11-2 (pp. 654–662)

Guided Example 1: rectangle; the diagonals of the figure are congruent; has four right angles; the diagonals of the quadrilateral are congruent; The diagonals of *ABCD* are congruent   **Guided Example 2:** $n = 3$; $n \neq 5$; $n = 5$; $n \neq 3$; $n \neq 3$; $n = 5$; converse; inverse; conditional; contrapositive

Questions:
1. Law of Detachment   3a. If the midpoint of the longest side of a triangle is equidistant from the three vertices of the triangle, then the triangle is a right triangle. b. If a triangle is not a right triangle, then the midpoint of the longest side is not equidistant from the three vertices of the triangle. c. If the midpoint of the longest side of a triangle is not equidistant from the three vertices of the triangle, then the triangle is not a right triangle. d. The original statement and statement c are true.
5. $\triangle ABC$ is isosceles.   7. true   9. false   11a. If you need it, then we have it. b. yes   13. Answers vary. Sample: If a prism is regular, then its bases are rotation-symmetric.   15. We know $p \Rightarrow q$ and $q \Rightarrow r$. Using the Law of Transitivity we conclude $p \Rightarrow r$. The given equation is p, so by the Law of Detachment, we conclude r.
17. Because George Washington is not alive today, no conclusion can be drawn from this statement.
19a. $\frac{32,000}{3\pi}$ cm$^3 \approx 33{,}510.3$ b. about 1.68 cm^3 away
21a. AMBULANCE b. It would make sense to write AMBULANCE from left to right, so that drivers behind it could read the sign.   23. $y = -\frac{1}{3}x + \frac{25}{3}$

Lesson 11-3 (pp. 663–669)

Guided Example 3: *BCD* does not have four right angles. four acute angles; acute angle; m∠*C*; m∠*D*; m∠*A* + m∠*B* + m∠*C* + m∠*D*; the Law of Indirect Reasoning; no convex quadrilateral has four acute angles

Questions:
1. Supposition, Deduction to Contradiction, and Final Conclusion   3. If valid reasoning from a statement

p leads to a false statement, then *p* is false.   5. The two lines intersect.   7a. yes b. If the segments are perpendicular, then by the definition of perpendicular lines, the angle formed by them will have measure 90, not 88.   9. *Supposition:* $2x - y = 5$ and $3y = 6x + 5$ are not parallel. Therefore, the lines will intersect and there will be a solution to the system of their equations. *Deduction to Contradiction:* To find the solution, multiply both sides of the first equation by 3. $\begin{cases} 6x - 3y = 15 \\ -6x + 3y = 5 \end{cases}$ Add the equations to get $0 = 15$. This contradicts the fact that $0 \neq 15$. *Final Conclusion:* Because the two lines do not intersect, they must be parallel.   11. *Supposition:* $\triangle ABC$ is scalene, $\overline{BD}$ is the altitude and the median to $\overline{AC}$. *Deduction to Contradiction:* Because $\overline{BD}$ is an altitude, $\overline{BD} \perp \overline{AC}$. Because $\overline{BD}$ is a median, $AD = DC$. By the Reflexive Property of Congruence, $\overline{BD} \cong \overline{BD}$. So by the SAS Congruence Theorem, $\triangle ADB \cong \triangle CDB$. By the CPCF Theorem, $\overline{AB} \cong \overline{BC}$. But because we assumed that $\triangle ABC$ was scalene, we know $AB \neq BC$. *Final Conclusion:* Therefore, by the Law of Indirect Reasoning, in a scalene triangle, an altitude cannot be a median.   13. *Supposition:* $\triangle ABC$ has two right angles. *Deduction to Contradiction:* If m∠*A* = 90 and m∠*B* = 90, because all angles in a triangle must have measures greater than 0, then m∠*A* + m∠*B* + m∠*C* > 180. But, by the Triangle-Sum Theorem, m∠*A* + m∠*B* + m∠*C* = 180. *Final Conclusion:* Therefore, by the Law of Indirect Reasoning, a triangle cannot have two right angles.   15. Answers vary. Sample: *Supposition:* $\sqrt{2} = \frac{1393}{985}$ *Deduction to Contradiction:* If $\sqrt{2} = \frac{1393}{985}$, then $(\sqrt{2})^2 = (\frac{1393}{985})^2$. But $2 \neq \frac{1{,}940{,}449}{970{,}225} \approx 1.999\ldots$ *Final Conclusion:* Therefore, $\sqrt{2} \neq \frac{1393}{985}$, by the Law of Indirect Reasoning.   17. *Supposition:* There is an integer *N* that is larger than all other integers. *Deduction to Contradiction:* If *N* is greater than all other integers, $N > N + 1$. But we know that for any number *N*, $N < N + 1$. *Final Conclusion:* Therefore, by the Law of Indirect Reasoning, there is no integer larger than all other integers.   19. Converse: If I eat my hat, Jimmy will have shown up on time; Inverse: If Jimmy does not show up on time, I will not eat my hat; Contrapositive: If I do not eat my hat, Jimmy will not have shown up on time.

21.

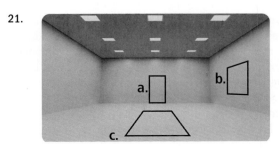

23a. $-\frac{9}{8}$ **b.** Answers vary. Sample: $y = \frac{8}{9}x$ **25.** No. In order for them to be collinear, the slopes between each possible pair of points must be equal. This is not true.

Lesson 11-4 (pp. 670–675)

Questions:

1. variables **3.**

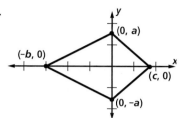

5.

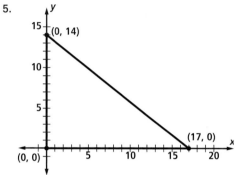

7. The slope of $\overleftrightarrow{RM}$ is $\frac{6-0}{3+3} = 1$. The slope of $\overleftrightarrow{AK}$ is $\frac{5-1}{-2-3} = -1$. Therefore, the slopes are opposite reciprocals, and thus the diagonals are perpendicular. **9.** $(-a, 0)$, $(a, 0)$, $(-c, b)$, (c, b) **11.** $a\sqrt{3}$

13a.

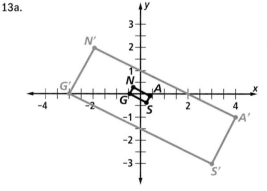

b. $m_{\overleftrightarrow{N'G'}} = \frac{2-0}{-2+3} = 2$, $m_{\overleftrightarrow{S'A'}} = \frac{-3+1}{3-4} = 2$, $m_{\overleftrightarrow{A'N'}} = \frac{2+1}{-2-4} = -\frac{1}{2}$, $m_{\overleftrightarrow{S'G'}} = \frac{0+3}{-3-3} = -\frac{1}{2}$. Opposite sides of the figure are parallel, and thus $S'A'N'G'$ is a parallelogram. **c.** Size changes preserve angle measure, so the angles of $S'A'N'G'$ are congruent to the corresponding angles in $SANG$. Therefore, opposite sides of $SANG$ are parallel and $SANG$ is a parallelogram. **15.** The figure is a parallelogram. **17.** *Converse*: "If you don't copy him, you can't imitate him." *Inverse*: "If you can imitate him, copy him." *Contrapositive*: "If you copy him, you can imitate him."

19. the box with the sphere of radius 2

21a. all points on and inside the boundary shown

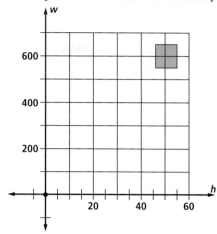

b. It is a rectangle and its interior with vertices at $(46, 550)$, $(46, 650)$, $(55, 550)$, and $(55, 650)$. **23.** $-4xy$

Lesson 11–5 (pp. 676–681)

Guided Example 1: 54, 18; 54, 18, 3240, 3240, 56.9

Guided Example 2: –15; –5; 24; 7; –10; 17

Questions:

1a. $(-5, 3)$ **b.** 16 **c.** $8\sqrt{2} \approx 11.31$ **3.** $8\sqrt{2} \approx 11.31$

5. B **7.** 13 mi

9a.

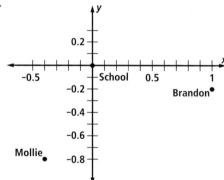

b. They live the same distance. **c.** Mollie lives closer.
11. It is on the circle. **13.** $\sqrt{37} \approx 6.08$ units
15. $2\sqrt{5} \approx 4.47$ **17.** $m_{\overline{AB}} = \frac{50}{9}$, $m_{\overline{BC}} = \frac{-14}{61}$, $m_{\overline{CD}} = \frac{16}{20}$, and $m_{\overline{DA}} = -\frac{1}{2}$. Because none of the sides are parallel, the figure cannot be a trapezoid. **19a.** The intersection of the segments is the midpoint of each segment. **b.** square, rectangle, rhombus, parallelogram. **21.** $(x - 3, y + 2)$

Lesson 11-6 (pp. 682-687)

Guided Example 3: 0; –5; 10; (0, –5); 10

Questions:

1. $(x - h)^2 + (y - k)^2 = r^2$ **3.** $(5, -7); 7$ **5.** $(0, 0); \sqrt{2}$

7a.

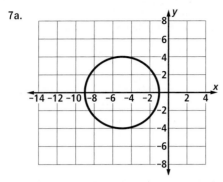

b. Answers vary. Sample: (-1, 0), (-5, 4), (-9, 0), (-5, -4)
9a. $(1, 2 + 2\sqrt{5})$, $(1, 2 - 2\sqrt{5})$ **b.** $(5 + \sqrt{11}, -3)$,
$(5 - \sqrt{11}, -3)$ **c.** no points of intersection
11a. $(x + 7)^2 + (y - 4)^2 = 18$ **b.** $6\sqrt{2}\pi$ **c.** 18π
13. $y = 13; y = -5$ **15a.** $(x + 4)^2 + (y - 2)^2 = 64$
b. $(x + 4)^2 + (y - 2)^2 \leq (8t)^2$ **17.** The eccentric
millionaire implies that he uses the Law of Contrapositives
to absolve Nadia from any blame: Because he did not
see her steal the diamond, and he spent the entire night
with her, she couldn't have been who stole the diamond.
If his logic were correct throughout his argument, then
he would have reached his end conclusion through the
Law of Ruling Out Possibilities: He had seven guests,
and he ruled out six of them. So there was only one
possible culprit left. **19a.** $x^2 + x^2 = 2x^2 = (x\sqrt{2})^2$. The
Pythagorean Converse Theorem justifies that the given
triangle is a right triangle with legs of length x and a
hypotenuse of length $x\sqrt{2}$. **b.** The given triangle is an
isosceles right triangle because it is a right triangle with
two sides at length x. Because the triangle is a right
triangle, there must be a 90° angle. By the Isosceles
Triangle Base Angles Theorem, the remaining angles
must be 45° angles. So the given triangle must be a
45-45-90 triangle. **21a.** $y \geq x; y < 6; x < 0$
b.

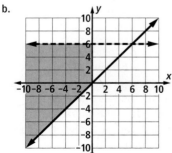

Lesson 11-7 (pp. 688–693)

Guided Example 1: 5; –1; 1; 3
Questions:
1. 4° F **3a.** 98 **b.** $2m - 82$ **5.** (0.105, 4)
7. $y = -\frac{2}{5}x + \frac{24}{5}$ **9a.** (1998, 92.26) **b.** In 1998,
total energy consumption in the United States was
92.26 quadrillions Btu. **11.** no
13a. $\left(\frac{c}{2}, \frac{d}{2}\right)$, $\left(-\frac{c}{2}, \frac{d}{2}\right)$, $\left(-\frac{c}{2}, -\frac{d}{2}\right)$, $\left(\frac{c}{2}, -\frac{d}{2}\right)$ **b.** The figure
formed by connecting the midpoints has two horizontal
sides and two vertical sides, so it has four right angles and
is a rectangle. **c.** $y = 0$ (the x-axis) and $x = 0$ (the y-axis)
d. $y = 0$ (the x-axis) and $x = 0$ (the y-axis) **15a.** 3.5
b. (3.5, 0), (-3.5, 0) **c.** (0, 3.5), (0, -3.5) **d.** $x^2 + y^2 = 12.25$
17. The perpendicular bisector of one base of an isosceles
trapezoid is the perpendicular bisector of the other base
and a symmetry line for the trapezoid.

Lesson 11-8 (pp. 694–699)

Guided Example 1: $\overline{MN} \parallel \overline{QR}$, $MN = \frac{1}{2}QR$; $\frac{0 + 2b}{2}, \frac{0 + 2c}{2}$;
b, c; $\frac{2a + 2b}{2}, \frac{0 + 2c}{2}$; $a + b, c$ **a.** 0; 0; their slopes are equal
b. c; c; 0; 0, 0
Questions:
1. $KL = 4$, $MN = 12$ **3a.** (4, 1) **b.** no **5.** $A = hm$,
where $m = $ length of midsegment **7.** 12.75 inches; the
Midsegment of a Trapezoid Theorem assures that the
support is parallel to the beams and therefore vertical.
9. Use the figure in Question 8 and draw the diagonal HJ.
Because the diagonals of a rectangle are congruent,
$IK = HJ$. Using the Midsegment of a Triangle Theorem
as in Question 8, $GD = DE = EF = FG = \frac{1}{2} \cdot IK$, and so
the quadrilateral $DEFG$ is equilateral. Thus $DEFG$ is, by
definition, a rhombus. **11.** 24,600 m² **13a.** (1, –3)
b. $5\sqrt{2}$ **c.** $(x - 1)^2 + (y + 3)^2 = 50$
15. $-6 < x < 6$, $-4 < y < 4$ **17.** The volume of the large
sphere is $3\sqrt{3}$ times the volume of the small sphere.

Lesson 11-9 (pp. 700–707)

Guided Example 2: $-5 - 3$; $-3 - 7$; $4 - (-2)$; $(-8)^2$; $(-10)^2$; 6^2;
200; 14.1 **Guided Example 3:** -3; 2; 1; 7; -3; 2; 1; 49;
$x + 3$; $y - 2$; $z - 1$; 49
Questions:
1. triple **3.** (800, –620, –290) **5.** $\sqrt{122}$ units
7. $(x + 2)^2 + (y - 4)^2 + (z + 6)^2 = 25$ **9a.** (4, 0, –7) **b.**
$4\sqrt{2}$ **11a.** (2, –5, 7) **b.** (5, –2, 19), (5, –2, –5) **c.** $3\sqrt{17}$
d. $9\sqrt{2}$ **e.** $36\sqrt{17}$ **f.** 144 **13a.** $(x - 3)^2 + (y + 1)^2 + (z - 1)^2 = 45$ **b.** $180\pi\sqrt{5} \approx 1264.47$ **c.** $180\pi \approx 565.49$
15a. (1, 6, 5) **b.** $\sqrt{138} \approx 11.75$ **17a.** $ED = z$; $EF = y$; $DF = x$; $AD = y$; $DC = y$; $CF = z$; $FB = z$; $BE = x$;
$EA = x$ **b.** The triangles are all congruent because of
the Midsegment of a Triangle Theorem and the SSS
Congruence Theorem. **19.** r is true. **21.** $\frac{30}{13}$, or about
2.31 mi

Chapter 11 Self-Test (pp. 711-712)

1a. yes; A hexagon is, by definition, a polygon. **b.** If a figure is not a hexagon, then it is not a polygon. **c.** The figure below is not a hexagon, but it is a polygon; thus, the inverse is false.

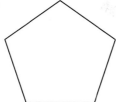

2a. If two angles do not form a linear pair, then they are not adjacent. **b.** no; The original statement is not true, so its contrapositive is not true. **3a.** $\angle A$ is a right angle. **b.** the Law of Ruling Out Possibilities **4.** Indirect proof = (supposition) Assume that $\triangle ABC$ is a triangle with $m\angle A \angle 50$, $m\angle B \angle 50$, and $m\angle C \angle 50$. (Deduction to Contradiction) By the Addition Property of Inequality, $m\angle A + m\angle B + m\angle C \angle 150$. But, by the Triangle-Sum Theorem, $m\angle A + m\angle B, + m\angle C = 180$. (Final Conclusion) The contradiction indicates that the supposition is false, so, by Law of Indirect Reasoning, no triangle can have three angles all with measures less than 30°.

5. The Jack is from the 18th century, the Queen from the 19th, the King from the 20th, and the Ace from the 17th.

	17th	18th	19th	20th
Jack	X	O	X_2	X_2
Queen	X	X	O	X_1
King	X	X	X	O
Ace	O	X_3	X_3	X_3

6. Dante is not a baby and Dante is not teething. To show this is true, assume that Dante is a baby. Then by (1), Dante must be happy. However, by (3), it is known that Dante is sad. This contradicts (1), so the assumption that Dante is a baby must be wrong. Therefore, Dante is not a baby and by (2) we know that Dante cannot be teething. **7.** By definition, $\overline{FD}$ is a midsegment of $\triangle ABC$. By the Midsegment of a Triangle Theorem, $\overline{FD} \parallel \overline{EB}$. Also, $\overline{EF}$ is a midsegment of $\triangle ABC$, so $\overline{EF} \parallel \overline{BD}$. By definition, $BDFE$ is a parallelogram. **8.** Because E and D are midpoints, $AE = EB = 5.5$ and $BD = DC = 11.15$. Also, by the Midsegment of a Triangle Theorem, $EF = BD = 11.15$ and $DF = EB = 5.5$. **9.** $RS = \sqrt{(8-3)^2 + (4-4)^2} = 5$; $ST = \sqrt{(11-8^2) + (8-4)^2} = 5$; $TR = \sqrt{(11-3)^2 + (8-4)^2} = \sqrt{80} = 4\sqrt{5}$ The perimeter is $RS + ST + TR = 10 + 4\sqrt{5}$. **10a.** $(-1, 9)$ **b.** 5 **c.** Answers vary. Sample: $(-1 + 5, 9 + 0) = (4, 9)$

11. A Find the slopes of the sides. Slope of $\overline{AB} = \frac{3}{4}$, slope of $\overline{BC} = -\frac{4}{3}$, slope of $\overline{CD} = \frac{3}{4}$, slope of $\overline{DA} = -\frac{4}{3}$. Opposite sides are parallel, so this is a parallelogram, but because adjacent sides are perpendicular, the most specific name is a rectangle. **12.** $\left(\frac{0+14}{2}, \frac{-4+82}{2}, \frac{8+-16}{2}\right) = (7, 39, -4)$

13. $E = \left(\frac{16+6}{2}, \frac{8+8}{2}\right) = (11, 8); I = \left(\frac{16+10}{2}, \frac{8+0}{2}\right) = (13, 4);$ $M = \left(\frac{10+0}{2}, \frac{0+0}{2}\right) = (5, 0); U = \left(\frac{6+0}{2}, \frac{8+0}{2}\right) = (3, 4)$

14. The slopes of UE and MI are $\frac{1}{2}$; the slopes of UM and EI are -20; thus, we see that $\overline{UE} \perp \overline{UM}, \overline{UM} \perp \overline{MI}, \overline{MI} \perp \overline{EI},$ and $\overline{EI} \perp \overline{UE}$ so that $EIMU$ has four right angles and is a rectangle by definition. **15.** Answers vary. Sample: $(-a, 0), (a, 0), (0, b), (0, -c)$

16a. $\left(\frac{3+4}{2}, \frac{2+9}{2}, \frac{-1+7}{2}\right) = (3.5, 5, 5, 3)$

b. $\sqrt{(3.5-3)^2 + (5.5-2)^2 + 3--1} = \sqrt{28.5} \approx 5.34$

17. $\sqrt{200^2 + 150^2 + 123^2} \approx 279$ yd **18.** The midpoint of $\overline{XZ}$ is $\left(\frac{2a+2b}{2}, \frac{2c+0}{2}\right)$.

The midpoint of $\overline{WY}$ is $\left(\frac{2a+2b+0}{2}, \frac{2c+0}{2}\right) = (a + b, c)$. Because the coordinates are the same, the diagonals have the same midpoint. **19.** *Supposition:* Given $\odot O$ with $\overline{OB}$ not an altitude of $\triangle AOC$. Suppose $\overline{OB}$ bisects $\triangle COA$. *Deduction to Contradiction:* By the definition of a circle, $AO = OC$. So $\triangle AOC$ is isosceles, with vertex angle $\triangle COA$. In an isosceles triangle, the bisector of the vertex angle and the perpendicular bisector of the base determine the same line, so $\overline{OB}$ is an altitude of $\triangle AOC$. *Final Conclusion:* This is a contradiction. By the Law of Indirect Reasoning, $\overline{OB}$ does not bisect $\angle COA$.

20. $d = \sqrt{(-12-39)^2 + (60--36)^2} \approx 108.7$ mi

Self-Test Correlation Chart

Question	1	2	3	4	5	6	7	8	9	10
Objective(s)	E	E	F	G	I	I	C	C	A	K
Lesson(s)	11-2	11-2	11-1, 11-2, 11-3	11-3	11-1, 11-2, 11-3	11-1, 11-2, 11-3	11-8	11-8	11-5	11-6, 11-9

Question	11	12	13	14	15	16	17	18	19	20
Objective(s)	H	D	B	H	L	K	D	H	G	J
Lesson(s)	11-4, 11-5, 11-6, 11-7	11-9	11-7	11-4, 11-5, 11-6, 11-7	11-4	11-6, 11-9	11-9	11-4, 11-5, 11-6, 11-7	11-3	11-5, 11-9

Chapter 11 Chapter Review (pp. 713-713)

1. $\sqrt{10}$ 3. $\sqrt{24.5}$ or $3.5\sqrt{2}$ 5a. 36 b. scalene
7. $\sqrt{x^2 + y^2}$ 9. (83, –98.5)   11. –5   13. Answers vary.
Sample: *AFDC, ABDE, BCEF*   15. 34   17. $\sqrt{6629}$ 19.
(3.5, 5.5, –3.5)
21a. (–1, 9, 9) b. $2\sqrt{11}$ 23a. If $x^2 = 16$, then $x = 4$. b. If
$x \neq 4$, then $x^2 \neq 16$. c. If $x^2 \neq 16$, then $x \neq 4$. d. The
conditional and its contrapositive are true.   25a. If B is the
midpoint of $\overline{AC}$, then $AB = BC$. b. If $AB \neq BC$, then B is
not the midpoint of $\overline{AC}$. c. If B is not the midpoint of $\overline{AC}$,
then $AB \neq BC$. d. The converse and inverse are true.
27a. $j \parallel k$ b. Law of Ruling Out Possibilites
29a. B does not lie on $\odot O$. b. Law of Detachment, Law of
the Contrapositive   31. not-p 33. *Supposition:* Suppose
ABCD is a quadrilateral with no congruent sides, and
that $\overline{AC}$ and $\overline{BD}$ bisect each other at point O. *Deduction
to Contradiction:* Then, because O is the midpoint of
both $\overline{AC}$ and $\overline{BD}$, $AO = OC$ and $BO = OD$. m$\angle AOD =$
m$\angle COB$, because $\angle AOD$ and $\angle COB$ are vertical angles.
$\triangle AOD \cong \triangle COB$, by the SAS Congruence Theorem. AD
$= BC$ by the CPCF Theorem. But this contradicts the fact
that no two sides of *ABCD* have the same length. *Final
Conclusion:* The supposition must be false. By the Law of
Indirect Reasoning, the diagonals of *ABCD* do not bisect
each other.   35. *Supposition:* Suppose that there are only
finitely many integers whose units digit is 0. *Deduction to
Contradiction:* Then, because the set of integers whose
units digit is 0 is finite, by supposition, there must be a
largest such number, call it m. $10m$ has a units digit of
0, because m has a units digit of 0, and $10m > m$. This
contradicts the fact that m is the largest integer whose
units digit is 0. *Final Conclusion:* So the supposition must
be false. Consequently, applying the Law of Indirect
Reasoning, there are infinitely many integers whose units
digit is 0.   37. Suppose an arbitrary rectangle, *ABCD,* has
vertices $A = (x, y)$, $B = (x, -y)$, $C = (-x, -y)$, $D = (-x, y)$.
The midpoint of $\overline{AC}$ is (0, 0). Similarly, the midpoint of
$\overline{BD}$ is (0, 0). Therefore, by the definition of midpoint and
bisects, $\overline{AC}$ and $\overline{BD}$ bisect each other at (0, 0).   39. The
midpoints of *ABCD* are $E = (-3, 15)$, $F = (14, 18)$, $G =$
(18, 5), and $H = (1, 2)$. The slope of $\overleftrightarrow{EF}$ is $\frac{18 - 15}{14 - -3} = \frac{3}{17}$
and the slope of $\overleftrightarrow{GH}$ is $\frac{2 - 5}{1 - 18} = \frac{-3}{-17} = \frac{3}{17}$ so $\overline{EF}$ and $\overline{GH}$ are
parallel. The slope of $\overleftrightarrow{FG}$ is $\frac{5 - 18}{18 - 14} = \frac{-13}{4}$ and the slope
of $\overleftrightarrow{HE}$ is $\frac{2 - 15}{11 - -3} = \frac{-13}{14}$, so $\overline{FG}$ and $\overline{HE}$ are parallel. Thus
EFGH is a parallelogram.   41. That one Friday afternoon,
Sophia fenced.   43. Jacqueline lives in France and cannot
see the Northern Lights.   45. 19,039.67 ft   47. 11.09 ft
49. $(x - 102)^2 + (y + 89)^2 = 56.25$

51.

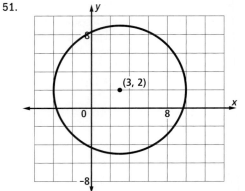

53. (–3, –12, 17), 2.5   55. $(a, 0)$, $(-a, 0)$, $(-b, c)$, (b, c)
57. $(2a, a)$, $(2a, -a)$, $(-2a, -a)$, $(-2a, a)$

Chapter 12

Lesson 12-1 (pp. 718-725)

Guided Example 2: 70: $A'B'$; $B'C'$; $A'C'$; AB; BC; AC; 70; 21
Questions:
1.

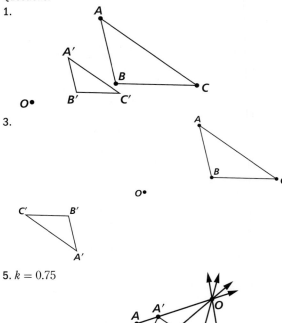

3.

5. $k = 0.75$

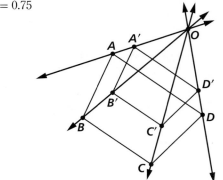

7. angle measure; betweenness; collinearity   9a. First,
apply a size transformation of magnitude 0.9 and center

O, where *O* is a point not on Arim. Then, apply a glide reflection so that the image created is facing Arim and on the line. **b.** no, because that will produce a figure that is the composite of a size transformation of positive magnitude and a rotation, not the reflection required to change orientation **11.** $SS' = 4$ cm, $F'S' = 6$ cm **13.** By the theorem on page 720, $k(AB) = A'B'$ and $k(BC) = B'C'$. Therefore, $\frac{A'B'}{B'C'} = \frac{k(AB)}{k(BC)} = \frac{AB}{BC}$. **15a.** $(x-4)^2 + (y+2)^2 = 25$ **b.** $(x-12)^2 + (y+6)^2 = 225$ **c.** $(x-4)^2 + (y+2)^2 = 225$ **17.** a circle **19.** $8640°$ **21a.** 8 cm **b.** 126 cm **c.** no, because growth typically occurs in short bursts that are not constant throughout life and slows greatly by age 20

Lesson 12-2 (pp. 726–730)

Questions:
1. D **3.** ratios; equal **5a.** $xy = 132$ **b.** Answers vary. Sample: $x = 2$, $y = 66$ and $x = 4$, $y = 33$ **7a.** $\frac{a}{b} = \frac{c}{d}$ **b.** $\frac{d}{c} = \frac{b}{a}$
9. 12.5 **11.** By the Means-Extremes Property, $wz = xy$. Divide both sides by wy to get $\frac{x}{w} = \frac{z}{y}$. **13.** $A'B' = 12$; $\frac{AC}{A'C'} = \frac{BC}{B'C'} = \frac{AB}{A'B'}$ **15.** $x = \frac{-\sqrt{5}-1}{2}$, $x = \frac{\sqrt{5}-1}{2}$
17. The magnitude of both S and T is k, so the side lengths of $\triangle ABC$ are multiplied by k in each size transformation. Thus, by the SSS Congruence Theorem, the two images are congruent. **19.** Assume that there are only finitely many numbers divisible by 2. Let N be the greatest number divisible by 2. Then, $2N$ is also a number that is divisible by 2, and $2N$ is greater than N. This contradicts the assumption that there is a greatest number that is divisible by 2, so there must be infinitely many numbers that are divisible by 2. **21a.** The area is multiplied by k^2.
b. The area is multiplied by k^2.

Lesson 12-3 (pp. 731–737)

Guided Example 1: a. similarity; a size transformation; a reflection b. m$\angle H$ c. 1.8; 6; 1.8; 10.8
Questions:
1. a. The letters are similar because the d is the image of the b under a reflection and size transformation. **b.** $\frac{3}{4}$
c. $\frac{4}{3}$ **3a.** false **b.** true **5.** $GF = 6.8$ in.; $LK = 15.64$ in.
7. $\angle ABE$ and $\angle LKN$ **9.** yes **11.** No, Figure I and Figure III in Question 4 are similar, but $AB \neq LK$.
13. $PA = 4.8$ **15.** $XZ \approx 3.34$ **17a.** 200 mi. **b.** $\frac{1}{2,534,400}$
19. Yes, the rectangles must be similar because corresponding angles are congruent (all angles are right angles) and corresponding lengths are proportional $\left(\frac{2 \text{ cm}}{4 \text{ ft}} = \frac{3 \text{ cm}}{6 \text{ ft}}\right)$. **21.** $x = -5$ **23.** 96.28% air, 3.72% leather **25.** true

Lesson 12-4 (pp. 738–744)

Guided Example 2: $\frac{11}{5}$; $\frac{121}{25}$; $\frac{121}{25}$; 96.8 **Guided Example 3:** 5; 5; 5; 25; 25; 125; 125

Questions:
1. Answers vary. Sample: Let R be a 3 cm by 5 cm rectangle. R has area 15 cm^2 and perimeter 16 cm. Let R' be a 30 cm by 50 cm rectangle. R' has area 1500 cm^2 and perimeter 160 cm. Thus, when each dimension is multiplied by 10, the area of the image is multiplied by 100 and the perimeter is multiplied by 10. **3.** yes
5a. $\frac{7}{4}$ **b.** $\frac{49}{16}$ **c.** $\frac{343}{64}$ **d.** $\frac{7}{4}$ **e.** 1 **7a.** $\frac{5}{9}$ **b.** $\frac{5}{9}$ **c.** $\frac{25}{81}$
9. $A = \pi r^2$, therefore $A' = \pi(r')^2 = \pi(15r)^2 = 225\,\pi r^2 = 225A$ **11.** $\frac{5}{7}$ **13.** 3528 ft^2 **15a.** 3.946 **b.** 1.986
17. No; because T only changes one coordinate, it is not a size transformation. **19.** a plane **21a.** Two figures F and G are congruent figures if and only if G is the image of F under an isometry. **b.** Two figures are oppositely congruent if they are congruent and have opposite orientation.

Lesson 12-5 (pp. 745–749)

Questions:
1. He was the tallest man on record, standing 8 ft, 11.1 in.
3. $\frac{1}{12}$; $\frac{1}{1728}$ **5.** The giantess is 125 times the weight of the woman. **7.** The giantess's footprint is 25 times the area of the woman's footprint. **9.** The amount of weight her legs can support is proportional to the cross-sectional area of her bones and muscles; three times her own weight is just too much for her bones and muscles to support.
11. $\frac{1}{1000}$; $\frac{1}{100}$ **13.** 20.25 kg **15.** $\left(\frac{58}{42}\right)^2 \cdot \$1500 \approx \$2860.54$
17a. about 1.89 **b.** about 6.75 **19.** 18.48 **21.** Size transformations of magnitude $k = 1$ can be composites of reflections. **23.** If, in two triangles, two sides and the included angle of one are congruent to two sides and the included angle of the other, then the triangles are congruent.

Lesson 12-6 (pp. 750–755)

Guided Example 2: a. $\frac{1}{3}$ b. FD, FE, $\frac{1}{3}$, $\frac{5}{3}$ c. $\angle G$; $\angle V$; $\angle O$; G; V; O **Guided Example 4:** 3; 4; 5; 4; $\frac{4}{7}$

Questions:
1a. Yes, it would be similar. The conversion factor to change the measure from feet to inches or centimeters is the ratio of similitude in each case. **b.** No, it would not be congruent to either triangle because the lengths are not in the same units. **3a.** true **b.** false **5.** No, they are not similar. **7.** $PQ = \frac{16}{3}$, $QR = \frac{52}{3}$, m$\angle P = 53.13$, m$\angle R = 14.25$, m$\angle Q = 112.62$ **9.** 39 **11.** false **13.** By the SSS Similarity Theorem, you can construct a triangle using the ruler with side lengths 12", 9", and 4.8" that will be similar to the piece of land. Then the measure of the angles can be determined by measuring them with a protractor, and the corresponding angle measures will be congruent because the triangles are similar. **15.** Because D and E

are midpoints, then $EC = \frac{1}{2}AC$ and $DC = \frac{1}{2}BC$. Further, $DE = \frac{1}{2}AB$ by the Midsegment of a Triangle Theorem, so because the three sides of the triangles are proportional with ratio of similitude. Then, by SSS Similarity Theorem, $\triangle BCA \sim \triangle ECD$. **17.** The ratio of similitude for the sides of the two triangles is $\frac{1}{2}$, so the ratio of the areas is $\frac{1}{4}$.
19. false **21.** $\frac{9a^2}{4}$

Lesson 12-7 (pp. 756–763)

Guided Example 1: 15; 1; 80; 1500; 100; 80; 80; 150,000

Guided Example 2: $\angle VXW$; $\angle UXT$; $\angle WVX$; $\angle TUX$; WVX; AA

Questions:
1. Apply a size change to a triangle to show its image is congruent to its preimage. **3a.** yes **b.** AA Similarity Theorem **c.** $\triangle IJM \sim \triangle FLM$ **5a.** yes **b.** SSS Similarity Theorem **c.** $\triangle JKL \sim \triangle XWV$ **7.** 25 ft **9.** 15 ft **11.** 75 ft 2 in. **13a.** 2.4 **b.** SAS Congruence Theorem **15.** If two isosceles triangles have congruent vertex angles with measure x, then all of their remaining angles have measure $\frac{180 - x}{2}$. By the AA Similarity Theorem, the two triangles are similar.

17.

Conclusions	Justifications
1. m$\angle STU = 90$, m$\angle SVT = 90$	1. Given
2. $\angle STU \cong \angle SVT$	2. Reflexive Property of Equality, Angle Congruence Theorem
3. $\angle S \cong \angle S$	3. Reflexive Property of Congruence
4. $\triangle STU \sim \triangle SVT$	4. AA Similarity Theorem

19a. 125,000,000 times **b.** No, the area of a cross section of its muscles would only be 250,000 times larger, so the larger ant would be able to lift $\frac{1}{10}$ of its own weight. **21.** $\overleftrightarrow{AB}$; $\overleftrightarrow{CD}$; $\overline{AB} \parallel \overline{CD}$ there is no size transformation S such that $S(A) = B$ and $S(C) = D$. **23.** $3h$

Chapter 12 Self-Test (pp. 766–767)

1. The magnitude of the size change is 2 because $WZ = 2(XY)$. **2.** $\triangle ABC$ is a 5-12-13 triangle, so it has perimeter 30. Because a size transformation with magnitude 3 means that the length of each side will be multiplied by 3, we can multiply the perimeter by 3 to find the perimeter of $\triangle PQR$. Thus, the perimeter of $\triangle PQR$ is 90. **3.** If M is the midpoint of $\overline{AB}$, then $AM = MB$ and M is between A and B. Let $S(A) = A'$, $S(B) = B'$, and $S(M) = M'$. Size transformations preserve betweenness, so $S(M)$ is between $S(A)$ and $S(B)$. Further, $A'M' = k \cdot AM$ and $M'B' = k \cdot MB$, so $A'M' = k \cdot AM = k \cdot MB = M'B'$, so $A'M' = M'B'$, and M' is the midpoint. **4.** $FL = 4 \cdot 8 = 32$ cm, because length of a segment of the image is the

magnitude of the size transformation times the size of the preimage. By the Fundamental Theorem of Similarity, the area of $\triangle PSR = 4^2 \cdot 10 = 160$ cm^2 and the volume of $PSMR$ is $4^3 \cdot 16 = 1024$ cm^3. **5.** O is the center of the size transformation.

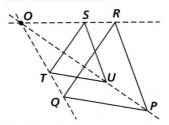

6. First, multiply both sides to get $wx = yz$. Divide both sides of this equation by xz, which gives $\frac{w}{z} = \frac{y}{x}$. Next, divide both sides of $wx = yz$ by wz, which gives $\frac{x}{z} = \frac{y}{w}$. Last, divide $wx = yz$ by xy, which gives $\frac{w}{y} = \frac{z}{x}$. **7.** The corresponding angle to $\angle E$ is $\angle I$, so m$\angle I = 116$. Further, $\angle R$ corresponds to $\angle A$, so m$\angle A = 49$. Because $\overline{RU}$ corresponds to $\overline{AT}$, the magnitude of the size transformation is $\frac{12}{8} = 1.5$. Therefore, since $\overline{PE}$ corresponds to $\overline{FI}$, $FI = \frac{6.3}{1.5} = 42$. **8.** Since weight is a measure of volume, the magnitude of the size transformation is the cube root of the weight proportion. Because $2^3 = 8$, the magnitude of the size change is 2. Therefore, the size of the left pinky finger on the larger statue is $2 \cdot 3 = 6$ in. **9.** The magnitude of the size change is $\frac{9}{6\frac{1}{2}} = \frac{18}{13}$. The larger dimension is $9\frac{1}{2} \cdot \frac{18}{13} = \frac{132}{13}$ in. **10.** The ratio of the corresponding sides of figure A to figure C is 6:1, because $3 \cdot 2 = 6$. Because surface area is a measure of area, the surface-area ratio of figure A to figure C is 36:1 because $6^2 = 36$. **11.** An animal that is twice as tall as another with a similar shape will weigh eight times as much. To support this weight, the bones must be eight times as strong, so must be eight times as thick as the smaller animal's bones. Therefore, much thicker legs are needed to support the great increase in weight. **12.** C, because it is the only set of lengths that are not proportional to 7, 13, and 10. **13.** An isosceles right triangle has one angle with measure 90 and two angles with measure 45. Therefore, since the angles of every isosceles right triangle are congruent, all right isosceles triangles are similar. **14a.** Because $\overleftrightarrow{DE} \parallel \overleftrightarrow{AB}$, $\angle CDE \cong \angle A$ and $\angle CED \cong \angle B$ by the Corresponding Angles Postulate. Therefore, by AA similarity, $\triangle CDE \sim \triangle CAB$. **b.** The magnitude of the size transformation mapping $\triangle CDE$ to $\triangle CAB$ is $\frac{20}{5} = 4$. Therefore, $DE = \frac{AB}{4} = \frac{16}{4} = 4$.

Self-Test Correlation Chart

Question	1	2	3	4	5	6	7
Objective(s)	A	B, C	C	B, C	A	D	B
Lesson(s)	12-1	12-1, 12-3, 12-4, 12-5	12-1, 12-3, 12-5	12-1, 12-3, 12-4, 12-5	12-1	12-2	12-3, 12-4

Question	8	9	10	11	12	13	14
Objective(s)	F, G	F	B, C	G	E	E	E, H
Lesson(s)	12-2, 12-3, 12-4, 12-5	12-2, 12-3, 12-5	12-1, 12-3, 12-4, 12-5	12-4, 12-5	12-6, 12-7	12-6, 12-7	12-6, 12-7

Chapter 12 Chapter Review (pp. 768–771)

1.

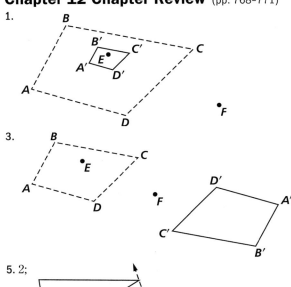

3.

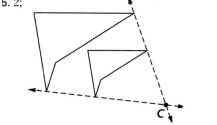

5. 2;

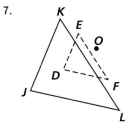

7.

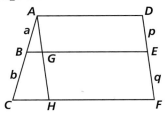

9. $\angle GEF \cong \angle IEH$; $\angle F \cong \angle H$; $\angle G \cong \angle I$ 11a. 3 b. 27
13a. 1 b. true   15. $-\frac{1}{2}$ 17. Answers vary. Sample: $\frac{b}{y} = \frac{a}{x}$
19. Answers vary. Sample: $\frac{300}{484} = \frac{225}{363}$ 21. No, the triangles
are not similar because $\frac{107}{7} = 15.29 \neq 12.11 = \frac{109}{9}$.
23. Yes, the triangles are similar $\frac{4}{3} = \frac{5}{3.75}$, so the triangles
are similar by the SAS Triangle Similarity Theorem.

25.

Conclusions	Justifications
1. D is the midpoint of $\overline{AB}$ and E is the midpoint of $\overline{AC}$.	1. Given
2. $AB = 2 \cdot AD$ and $AC = 2 \cdot AE$	2. Definition of midpoint
3. $\frac{AB}{AD} = \frac{AC}{AE}$	3. Divide by 2.
4. $\angle A \cong \angle A$	4. Reflexive Property of Congruence
5. $\triangle ADE \sim \triangle ABC$	5. SAS Triangle Similarity Theorem

27. 10.52 cm   29. $\frac{4x}{y}$ 31. 17.68 in.   33. 19.66 m²
35. the smaller animal   37. 31.5 ft   39. Kiara could use
a protractor to measure the angles of his triangle, and
because her triangle is similar to the faces of the pyramid
the angles will be congruent.

Chapter 13

Lesson 13-1 (pp. 774–780)
Guided Example 1: $\frac{AD}{DB}, \frac{AE}{8} = \frac{15}{10}$, $10 \cdot AE = 120$; 12 cm;
$\triangle ADE \sim \triangle ABC$; BC; $\frac{15}{25} = \frac{DE}{18}$; $25 \cdot DE = 270$; 10.8 cm

Questions:
1. $\frac{d}{e}$ 3. Answers vary. Sample: $\frac{a}{a+b}$ 5. No; $\frac{30}{15} \neq \frac{28}{13}$.
7. Let G be the translation image of E and let H be the
translation image of F.

Because the translations preserve distance, $AG = DE$, so
$AG = p$, and $GH = EF$, so $GH = q$. By the Triangle Side-
Splitting Theorem, $\frac{AB}{BC} = \frac{AG}{GH}$, or $\frac{a}{b} = \frac{p}{q}$.   9a. $\frac{5}{2}$ b. 20 c. Size
transformations preserve angle measure.

11.

13. Since the lines on a sheet of paper are parallel and equally spaced, a toothpick that is laid such that it intersects 7 of these lines is divided into 6 congruent lengths by these lines.   **15a.** 3 inches **b.** 37π
17. $x = 2$ **19a.** $23x + 39y = -541$ **b.** $-\frac{23}{39}$

Lesson 13-2 (pp. 781–785)

Guided Example 2: $15 - x$; $\frac{NK}{NM}$; $\frac{8}{10} = \frac{x}{15} - x$; $\frac{20}{3}$

Questions:

1. $\frac{WV}{WS}$ **3.** $\frac{WT}{ST}$

5.

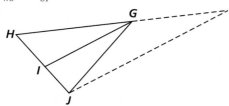

7. $x \approx 9.66$ **9.** $x = 6$ cm

11.

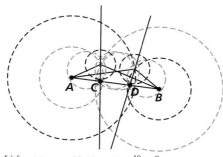

13. $\frac{r+s}{t}$ **15.** $x \approx 46.41$ **17a.** $\frac{10}{3}\pi r^3$
b. $8\pi r^2$ **19.** $64\sqrt{3}$

Lesson 13-3 (pp. 786–792)

Guided Example 2: BD; BD; 16; 160; AB; AB; 26; 260; CB; CB; 26; 416

Questions

1. 10   **3a.** 600 **b.** Answers vary. Sample: It is not a good estimate because bacteria do not grow at a linear rate.   **5.** EF **7.** EF, FG **9.** FG, EG **11.** $x = \frac{16}{5}$, $y = \frac{9}{5}$, $h = \frac{12}{5}$ **13a.** Given **b.** AB; AP **c.** AB; BP
d. $(AB)(AP) + (AB)(BP)$ **e.** Distributive Property
f. Additive Property of Distance Postulate **g.** AB^2
15. a. Part 1 **b.** 20 feet tall   **17.** $AM = \frac{70}{3}$ **19.** Answers vary. Sample:
21. $5\sqrt{2}$

Lesson 13-4 (pp. 793–799)

Guided Example 2: 15; x, 4; 2.4

Questions:

1. 12 ft by 7.2 ft; 12 ft by 20 ft

3.

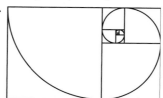

5. Yes. Their dimensions are all in the same ratio of 2:1, so they are similar by the Rectangle Similarity Theorem.
7. no   **9a.** yes **b.** yes   **11.** $\left(\frac{1+\sqrt5}{2}\right) = \left(\frac{1+2\sqrt5+5}{4}\right) = \frac{6+2\sqrt5}{4} = \frac{3+\sqrt5}{2} = \frac{2}{2} + \frac{1+\sqrt5}{2} = 1 + \frac{1+\sqrt5}{2}$ **13a.** $\angle AEB$, $\angle BEC$, $\angle CED$ all intercept congruent arcs. **b.** $36°$ **c.** $\triangle AEF$, $\triangle DBF$ are isosceles triangles with the same base angles, by the argument in Part a, so $\angle FAE = \angle FEA = \angle FBD = \angle FDB$. By the AA Similarity Theorem, $\triangle AEF \sim \triangle DBF$.
d. $m\angle AFB = m\angle FAE + m\angle AEB = 2m\angle AEB$ by the argument in Part a. Also $m\angle ABD = m\angle ABE + m\angle FBD = 2m\angle ABE = 2m\angle AEB$. So, $m\angle AFB = m\angle ABD$. Thus, $m\angle BAD = m\angle ABD = m\angle AFB$, so $\triangle AFB$ is isosceles and thus $AB = BF = 1$ by the Converse of the Isosceles Triangle Base Angles Theorem. Thus $EF = d - 1$. **e.** Since $\triangle AEF \sim \triangle DBF$, we have $\frac{EF}{AE} = \frac{BF}{BD}$. Since diagonals have length d and $AE = 1$, then $\frac{EF}{1} = \frac{d-EF}{d}$. Rearranging and substituting gives $EF = d - 1 = \frac{d}{d+1}$. So $d^2 - d - 1 = 0$. And hence, $d = \frac{1+\sqrt5}{2}$.
15a. $\overline{AM}$ and $\overline{BM}$ **b.** $\left(\sqrt a - \sqrt b\right)^2 \geq 0$, so $a - 2\sqrt{ab} + b \geq 0$, and $\frac{1}{2}\left(a - 2\sqrt{ab} + b\right) \geq 0$. So $\frac{1}{2}a + \frac{1}{2}b - \sqrt{ab} \geq 0$ and thus $\frac{a+b}{2} \geq \sqrt{ab}$.
17a. iii **b.** i **c.** ii   **19a.** 20 **b.** 8 cm **c.** 880 cm^2 **d.** 228 cm

Lesson 13-5 (pp. 800–806)

Questions:

1. a. $\tan 25° = \frac{2.8}{6.0} \approx 0.467$; Triangles vary. Sample:

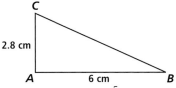

b. $\tan 25° \approx 0.466$ **3.** $\frac{5}{12}$ **5.** $\tan 60° = \sqrt{3}$

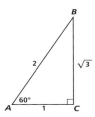

7.

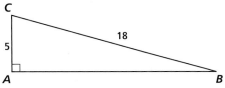

a. $m\angle A \approx 20.6$ **b.** $m\angle B \approx 69.4$ **c.** $\tan^{-1}\left(\frac{3}{8}\right) \approx 20.6°$, $\tan^{-1}\left(\frac{8}{3}\right) \approx 69.4°$ **9.** yes; the angle is very small

11. 61.7 ft   **13.** 34°

15.

Angle	Tangent	angle measure	slope of 2nd side of angle
$\angle OPA$	$\frac{1}{6}$	9.46	$\frac{1}{6}$
$\angle OPB$	$\frac{1}{2}$	26.57	$\frac{1}{2}$
$\angle OPC$	$\frac{5}{6}$	39.81	$\frac{5}{6}$
$\angle OPD$	$\frac{7}{6}$	49.40	$\frac{7}{6}$
$\angle OPE$	$\frac{3}{2}$	56.31	$\frac{3}{2}$

17a. By the Pythagorean Theorem, $a^2 + b^2 = (x + y)^2 = x^2 + 2xy + y^2$. **b.** By the Right-Triangle Altitude Theorem, $h^2 = xy$. From Part a, $xy = \frac{a^2 + b^2 - x^2 - y^2}{2}$. By the Transitive Property of Equality, $h^2 = \frac{a^2 + b^2 - x^2 - y^2}{2}$.
19. $\frac{4036}{2187}\pi R^3$ **21.** $\frac{c^2}{4}$

Lesson 13-6 (pp. 807–813)

Questions:

1. a. $\frac{\sqrt{10}}{10}$ **b.** $\frac{\sqrt{10}}{10}$ **c.** 3   **3.** 14.31   **5.** 6.79   **7.** 174 inches
9. Answers vary. Sample: In a right triangle, the two acute angles are complements. Therefore, since the sine of the first angle is equal to the length of the side opposite it divided by the length of the hypotenuse, the cosine of the second angle is equal to the length of the side adjacent to it divided by the length of the hypotenuse, and since the side opposite the first angle is the same as the side adjacent to the second angle, the sine of the first angle is equal to the cosine of the second angle.   **11.** $\frac{1}{2}$ **13.** $\frac{\sqrt{2}}{2}$
15. 667,800 feet   **17a.** 64sin 25° cm **b.** 2048 sin 25° sin 65° **c.** 784 cm^2 **19.** Answers vary. Sample: Since O and N are midpoints, $MO = OL$ and $KN = NL$. Because $\angle L \cong \angle L$, by the ASA Similarity Theorem, $\triangle NLO \sim \triangle KLM$. Thus the side lengths of each triangle are in the same proportions, and so $\frac{OL}{ON} = \frac{ML}{MK}$, so by the definition of sine, $\sin K = \sin N$.
21a. angle 1 **b.** angle 3   **23.** 80 m   **25a.** $P = (2, 3.5)$; $Q = (4, 0)$ **b.** $x^2 + y^2 = 16$ **27.** A vector is a quantity that can be characterized by its direction and magnitude.

Chapter 13 Self-Test (p. 818)

1. By the Triangle Side-Splitting Theorem, $\frac{BD}{AD} = \frac{BE}{CE}$. Substituting the given lengths, this proportion becomes $\frac{21}{AD} = \frac{14}{10}$, so $14(AD) = 210$, and $AD = \frac{210}{14} = 15$.
2. The tangent ratio involves the lengths of the two legs, but we are given the length of one leg and the hypotenuse.

Use the Pythagorean Theorem to find AB.
$$AB^2 + AC^2 = BC^2$$
$$AB^2 = BC^2 - AC^2$$
$$AB^2 = 18^2 - 5^2$$
$$AB^2 = 299$$
$$AB = \sqrt{299}$$
Then $\tan C = \frac{AB}{AC} = \frac{\sqrt{299}}{5}$.

3a. By the Right Angle Altitude Theorem $ZP = \sqrt{XP \cdot PW} = \sqrt{8 \cdot 5} = \sqrt{40}$, so then by the Pythagorean Theorem, $PZ^2 + XP^2 = XZ^2$, which gives $XZ = \sqrt{65}$. Since $XZ = YW$, $YW = \sqrt{65}$. **b.** By Part a, $ZP = \sqrt{40}$, so by the Pythagorean Theorem, $ZP^2 + PW^2 = ZW^2$, which gives $ZW = \sqrt{104}$. Since $ZW = XY$, $XY = \sqrt{104} = 2\sqrt{26}$.   **4.** True. If the measures of two angles are the same, then the sines of those angles are equal as well. Since the two triangles are similar, corresponding angles are equal, so the sines of corresponding angles are also equal.   **5.** By the Angle Bisector Theorem, $\frac{AB}{BC} = \frac{AD}{DC}$. Let $DC = x$, then $AD = 15 - x$, and $\frac{5}{12} = \frac{15 - x}{x}$. Thus $5x = 180 - 12x$, so $x = \frac{180}{17}$.
6. The lengths of the legs are $12 \cdot \sin 37° \approx 7.2$ in. and $12 \cdot \cos 37° \approx 9.6$ in.   **7.** Because $\overline{KL} \parallel \overline{IJ}$ and $\angle HKL \cong \angle HIJ$, and $\angle H \cong \angle H$, so $\triangle HKL \sim \triangle HIJ$ by the AA Similarity Theorem. Then $\frac{HK}{HI} = \frac{HL}{HJ}$. $HI = HK + KI = 31 + 23 = 54$, so the proportion becomes $\frac{31}{54} = \frac{28}{HJ}$. Then, $31(HJ) = 1512$, and $HJ = \frac{1512}{31}$. Finally, $IJ = HJ - HL = \frac{1512}{31} - 28 = \frac{644}{31} \approx 20.77$.   **8.** There are two possible rectangles represented

by proportions $\frac{4}{5} = \frac{12}{x}$ and $\frac{4}{5} = \frac{x}{12}$. $x = 15$ or $x = 9.6$; so the dimensions of the second rectangle are 12 cm by 15 cm or 9.6 cm by 12 cm.   **9.** A; $\frac{32}{20} = 1.6$ is the closest to the golden ratio of those choices.   **10.** We need to solve $\tan(3°) = \frac{30}{x}$, where x is the distance from the car to the bridge, and then multiply by 2 to find the total distance traveled. $x \approx 572.434$, so the car has traveled about 1145 feet.   **11.** We need to solve $\sin(70°) = \frac{x}{17}$, where x is how far up the vertical wall the ladder goes. This gives about 15.97 feet, and converting to inches and rounding gives 192 inches.   **12.** $\cos^{-1}(0.32) \approx 71.34°$ **13.** The sine of an angle is the length of the opposite side divided by the length of the hypotenuse, $\sin(\theta) = \frac{y}{r}$, so since each angle has the same opposite side, the largest sine will be when the hypotenuse is the shortest. This occurs with angle 3.   **14.** The dimensions of the similar rectangles must be in the same ratio. Thus $\frac{x}{3} = \frac{3}{x+6}$, so $x^2 + 6x = 9$. Thus $x^2 + 6x - 9 = 0$, and solving by the Quadratic Formula gives $x = -3 + 3\sqrt{2} \approx 1.24$ (since x must be positive).

Self-Test Correlation Chart

Question	1	2	3	4	5	6	7
Objective(s)	A	D	C	F	B	E	H
Lesson(s)	13-1	13-5, 13-6	13-3	13-5, 13-6	13-2	13-5, 13-6	13-1
Question	8	9	10	11	12	13	14
Objective(s)	G	I	J	J	E	F	G
Lesson(s)	13-4	13-4	13-5, 13-6	13-5, 13-6	13-5, 13-6	13-5, 13-6	13-4

Chapter 13 Chapter Review (pp. 819–821)

1. a. 3.7 **b.** 6.8   **3.** Yes, since $\frac{12}{9} = \frac{8}{6}$, by the converse of the Side-Splitting Theorem, the segments are parallel.
5. $\frac{20}{9}$ **7.** 35   **9.** $\frac{48}{5}$ **11.** $8\sqrt{15}$ **13.** $MP = \frac{\sqrt{3}}{3}t$, $PO = t\sqrt{3}$, $MN = \frac{2\sqrt{3}}{3}t$ **15.** 0.5   **17a.** 0.740   **b.** 0.673   **c.** 1.1
19a. $\frac{\sqrt{2}}{2}$ **b.** 1   **21.** 0.019   **23.** 22.62, 67.38   **25a.** 11.1 in.
b. 44.0°   **27.** $\frac{XY}{YZ}$ **29.** $\frac{XY}{XZ}$ **31.** Angle 4   **33.** $\cos A$ **35.** yes
37. no   **39.** 190 meters   **41.** 10 by 6   **43.** 58°   **45.** 170 cm

Chapter 14

Lesson 14-1 (pp. 824–830)

Guided Example 2: 360; $\angle ACB$; $\triangle MCA$; 360; 15; 15; $15 \cdot \sin 11.25°$; $30 \cdot \sin 11.25°$; $30 \cdot \sin 11.25°$
Questions:
1. $\overline{CD}, \overline{CB}, \overline{FE}$ **3.** true   **5.** B   **7.** 70   **9.** $\triangle GHI$ is isosceles because $GH = GI$, since they are both radii of circle G.   **11.** 10 m   **13.** $10\sqrt{3}$ m   **15.** $30\pi - 93.6 \approx 0.6$ ft
17. Construct segments $\overline{OA}, \overline{OB}, \overline{OC}$, and $\overline{OD}$. $OA = OB = OC = OD$ since the radii of a circle are congruent. $\triangle BOA = \triangle COD$ by the SSS Triangle Congruence Theorem. Then $\angle BOA = \angle COD$ by the CPCF Theorem. Because the measure of an arc equals the measure of its central angle, $m\widehat{AB} = m\widehat{CD}$.   **19.** $5\sqrt{5}$ cm
21. about 53.5 cm   **23.** 11.78 feet   **25.** L: 32.35°, M: 32.35°, N: 147.65°

Lesson 14-2 (pp. 831–836)

Questions: 1a. 6 and 4, 7 and 3, 8 and 2, 9 and 1 **b.** team 5 **c.** second round   **3.** 4   **5.** 31   **7.** 90   **9.** There are no byes, and one team is placed at the center of the circle, with a radius drawn connecting it to a vertex on the circle.
11. 1-2, 3-4, 5-6; 2-3, 4-5, 6-1; 1-3, 2-5, 4-6; 1-4, 2-6, 3-5; 1-5, 2-4, 3-6   **13.** Some parallel chords of a regular octagon will have the same length. As you rotate you will repeat pairings.   **15.** A and D   **17.** 20   **19.** $\frac{\sqrt{199}}{2}$ **21.** $\pi r^2 - \frac{1}{2}r^2$ **23.** In a triangle, the measure of an exterior angle is equal to the sum of the measures of the interior angles at the other two vertices of the triangle.

Lesson 14-3 (pp. 837–843)

Guided Example 2a. 93; 186°; 186°; 174°; 87   **b.** 87; 174; 62°   **c.** 62°; 148° **d.** 322°; 38° **e.** $m\widehat{BE}$; $m\widehat{ED}$; 50 **f.** 50; 87; 130
Questions:
1. Inscribed Angle Theorem   **3a.** 44.5 **b.** 31.5 **c.** 76 **d.** 13   **5.** $m\widehat{BC}$ **7a.** no   **b.** $m\widehat{JK}$ or $m\widehat{HJ}$ **9.** 108, 72, 108, 72   **11.** infinitely many   **13a.** 50 **b.** 65   **15.** 12
17. 648 m

Lesson 14-4 (pp. 844–851)

Questions:
1. 90   **3.** Two circles in the same plane are internally tangent if they are tangent and one circle is inside the other.   **5a.** $\overline{CT}$ is a radius that contains the point of tangency T of the tangent line $\overleftrightarrow{PT}$. **b.** It is the average radius of Earth.   **7.** Answers vary. Sample: The path of a bicycle is tangent to its wheel.

9. Answers vary. Sample: The path of a ball rolling down a ramp is tangent to the ball.   **11.** Answers vary. Sample: From the Pythagorean Theorem, we have $d^2 = (r + h)^2 - r^2 = 2rh + h^2$. Since lengths can only be positive, $d = \sqrt{2rh + h^2}$.

13a.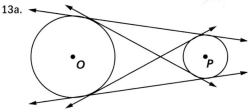

b. Answers vary. Sample: $\overleftrightarrow{OP}$ is a symmetry line for the common tangents.   **15.** By the Two Tangent Theorem, $GC = CB$, $GD = DH$, $BF = FK$, and $KE = EH$. By the Addition Property of Equality and the Substitution Property, $CD + EF = ED + CF$, and thus the sums of the lengths of opposite sides are equal.   **17a.** A line is tangent to a sphere if and only if it is perpendicular to the radius at the radius's endpoint on the sphere. **b.** yes   **19.** It needs to be shown that no point of m other than P is on the circle. Suppose Q is another point on m and on the circle. Since Q is on m, $\triangle OPQ$ is a right triangle with hypotenuse $\overline{OQ}$. So $OQ > OP$. But since Q is on the circle, $OQ = OP$. These two statements are contradictory, so by the Law of Indirect Reasoning, the assumption must be false. So m is tangent to the circle.   **21.** 90°   **23a.** 10 **b.** 20 sin 30.5°   **25.** equilateral triangle with perimeter 1 cm

Lesson 14-5 (pp. 852–858)

Guided Example 1: 73°; 157°; 130°; 130°; 73°; 28.5

Questions:

1. 90   **3.** 49   **5.** 20   **7.** true   **9.** $m\widehat{QSR} = 155°$, $m\widehat{QTR} = 205°$ **11.** If you start with an inscribed angle formed by two chords that meet at a point on the circle, then rotate one of the chords about that point until it is tangent to the circle, the measure of the angle formed by the tangent and the chord always stays the same as it was when the angle was inscribed.   **13.** By the Tangent-Secant Theorem, $m\angle U = \frac{1}{2}\left((360 - m\widehat{VT}) - m\widehat{VT}\right)$, thus $m\angle U = 180 - m\widehat{VT}$, so $m\widehat{VT} + m\angle U = 180°$.   **15.** 86°   **17.** about 274 miles   **19.** $\frac{r_1 \sin 15°}{r_2 \sin 30°} \approx 0.518 \frac{r_1}{r_2}$

Lesson 14-6 (pp. 859–865)

Questions:

1. You need at least 3 lines; concurrent means that 3 or more lines meet at the same point.

3.

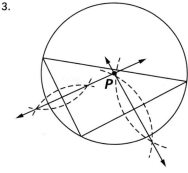

5.

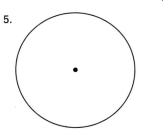

7.

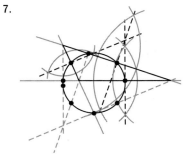

9. a. We know that an inscribed angle intercepts an arc with twice the measure. Therefore, a 90° angle intercepts a 180° arc, i.e. a semicircle, and is therefore a diameter.

b.   **11.** Answers vary. Sample:

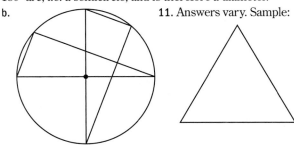

13. false   **15.** In a circle with two different inscribed n-gons, the length of a side of the n-gon with fewer sides is longer than the n-gon with more sides. Each side is a chord of the circle. Thus, the n-gon with fewer sides creates a longer chord and a larger minor arc.   **17.** about 246 paces west and 172 paces south   **19.** true

Lesson 14-7 (pp. 866–871)

Guided Example 3: *GH; GF*; 3; 9; 27; 27

Questions:

1. $x = 15$; power of B is 100.   3. $z = 2$; power of $B = 24$
5. $PE \cdot PF$ 7. square centimeters   9. 0   11. Draw a chord that passes through P perpendicular to PO, and label the points of intersection of the chord with the circle A and B. $PA = PB = \sqrt{r^2 - d^2}$ by the Pythagorean Theorem, and the power of P is $PA \cdot PB = r^2 - d^2$.
13. Answers vary. Sample: $AP \cdot BC = PC \cdot AD$
15a. 1.625 ft b. about 23.17 ft^2 17. $\frac{360 - x}{180} \cdot \pi r + 2r \sin\left(\frac{x}{2}\right)$
19. $\left(2\sqrt{25\pi + 9} - 3\right)$

Lesson 14-8 (pp. 872–878)

Questions:

1. circle with radius $\frac{5}{\pi}$ 3. equilateral triangle with side lengths $\frac{10}{3}$ 5. Answers vary. Sample:

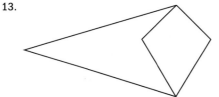

7. a.

Figure	Area	Perimeter
Circle	12π	≈ 21.77
Square	12π	≈ 24.56
Rectangle with width 2	12π	≈ 41.70

b. Isoperimetric Theorem (alternate statement)

9. Answers vary. Sample: Circles maximize area while keeping perimeter small, which allows more fans to fit within the stadium and requires less material to build the stadium.   11a. $x^2 - y^2$ b. $y = 0$
13.

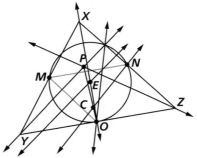

15. Since B is a point of tangency for $\overline{CB}$, we know that $\angle B$ is a right angle and, similarly, that $\angle D$ is a right angle. Therefore, *ABCD* is a rectangle. Since $\overline{AB}$ and $\overline{AD}$ are radii of the same circle, $AB = AD$. *ABCD* has equal length and width, so it is a square.   17a. false b. true   19. 319 in^3

Lesson 14-9 (pp. 879–885)

Questions:

1. Answers may vary. Sample: It would minimize the surface area for a given volume, keeping heating costs low.
3. 2100 m^3 5. sphere; surface area   7. The sponge is actually made up of many sphere-shaped bubbles, which hold water more efficiently than the single box-shaped interior of the pan.   9a. $24\pi \approx 75.40$ b. $\sqrt{6} \approx 2.45$
c. $12\pi \approx 37.70$ d. $8\pi\sqrt{6} \approx 61.56$ e. A sphere with the same surface area as a certain cone will have about 163% of the volume of the cone.   11a. $p \geq 2\sqrt{A\pi}$

---

b. No, $A = p = 9$ does not satisfy the Isoperimetric Inequality.

13.

15. 3 cm

Chapter 14 Self-Test (pp. 891–892)

1. a. Since the total arc measure of a circle is 360°, $x° + 90° + (3x)° + 178° = 360°$. So $(4x)° + 268° = 360°$, $(4x)° = 92°$, $x = 23$.   b. Using the Angle-Secant Theorem, $m\angle STR = \frac{1}{2}(3x - x) = \frac{1}{2}(2x) = x = 23$.   2. Since $m\widehat{AB} = 35°$, by the Inscribed Angle Theorem, $m\angle D = \frac{1}{2} \cdot 35° = 17.5$. Thus, since $m\angle B + 120 + 17.5 = 180$, $m\angle B = 42.5$. By the Inscribed Angle Theorem, $2m\angle B = m\widehat{CD}$ so $m\widehat{CD} = 85°$.   3. Since $m\widehat{HI} = 2x°$, the measure of the major arc connecting H and I is $360° - (2x)°$. Since $\overleftrightarrow{GH}$ and $\overleftrightarrow{GI}$ are tangent to the circle, the Tangent-Secant Theorem applies. Thus, $m\angle G = \frac{1}{2}((360° - 2x°) - 2x°)$. Since $m\angle G = x$, $x = \frac{1}{2}(360 - 4x)$, so $2x = 180 - 4x$, and so $6x = 360$, and $x = 60$.   4. Answers vary. Sample:

5. The circle through the midpoints of the sides of the triangle is the nine-point circle, so it also passes through the feet of the three altitudes. So the point equidistant from the feet of the three altitudes is the center of the nine-point circle, which is the midpoint E of the segment connecting the circumcenter C and the orthocenter O.

6. By the Secant Length Theorem, the power of the point P is $PS \cdot PR = 44 \cdot 124 = 5456$. So by the Tangent Length Theorem, $PO^2 = 5456$ and $PO = \sqrt{5456} \approx 73.86$.
7. Answers vary. Sample: By the Tangent Length Theorem, $UV^2 = UW^2$, and so $UV = UW$. This means that $\triangle UVW$ is an isosceles right triangle, and so $m\angle UVW = 45$.
8. It is given that $\overleftrightarrow{XY}$ is a tangent to $\odot Z$ at Y. By the Tangent-Chord Theorem, $m\angle XYV = \frac{1}{2}m\widehat{VY}$. By the Inscribed Angle Theorem, $m\angle W = \frac{1}{2}m\widehat{VY}$. By the Transitive Property of Equality, $m\angle XYV = m\angle W$.
9. C; By the Secant-Angle Theorem, $m\angle O = \frac{m\widehat{RS} - m\widehat{PQ}}{2}$, so solving for $m\widehat{RS}$ gives $2m\angle O + m\widehat{PQ} = m\widehat{RS}$. Thus, because all the measures are positive, $m\angle O < m\widehat{RS}$.
10. True, the three angle bisectors meet at the incenter.

11. The nine points referred to are the midpoint of each side of the triangle, the foot of each altitude of the triangle, and the midpoint of the segment of each altitude from its vertex to the orthocenter. **12.** Draw a diagram with a circle representing Mars and a point representing the volcano. Draw a tangent to the circle from the point, whose length you will solve for. The length of the segment from the center of Mars to the volcano is $3400 + 27 = 3427$ km. The radius of the circle is 3400 km, so by the Pythagorean Theorem, an astronaut could see $\sqrt{3427^2 - 3400^2} \approx 429$ km. **13.** The number of pairings in a round-robin tournament with n teams is equal to n plus the number of diagonals of an n-gon. $n = 14$ so the total number of games played is $14 + \frac{14(14-3)}{2} = 91$.

14. Answers vary. Example: A sugar cube has a large surface area for its volume, while a basketball has a small surface area for its volume. **15.** The Isoperimetric Inequality states that $A \leq \frac{p^2}{4\pi}$. Since $A = 3$, $3 \leq \frac{p^2}{4\pi}$. Solving for p, $p \geq 2\sqrt{3\pi}$. The smallest possible length of fence is $2\sqrt{3\pi}$ km. The shape of the fence would be a circle. **16.** Draw a diagram of the situation. Form an isosceles triangle by drawing two radii of the circle that intersect consecutive vertices of the pentagon. Then draw the altitude of the isosceles triangle from the center of the circle. The interior angle of the pentagon is 72°, so the altitude has bisected that angle into two 36° angles and created two right triangles. Using trigonometry, the length of a side of the pentagon is $2 \cdot 12 \sin 36° \approx 14.1$ cm.

Self-Test Correlation Chart

Question	1	2	3	4	5	6	7	8
Objective(s)	B	B	B	G	C	D	D	F
Lesson(s)	14-3, 14-5	14-3, 14-5	14-3, 14-5	14-8, 14-9	14-6	14-7	14-7	14-3, 14-5

Question	9	10	11	12	13	14	15	16
Objective(s)	E	H	H	J	I	K	K	A
Lesson(s)	14-1, 14-4, 14-5, 14-7	14-6	14-6	14-1, 14-4, 14-7	14-2	14-8, 14-9	14-8, 14-9	14-1

Chapter 14 Chapter Review (pp. 893–897)

1. 45 in. **3.** $2r \sin \frac{x}{2}$ **5.** 143.6 **7.** $m\angle K = 75$, $m\angle M = 115$, $m\angle H = 50$, $m\angle L = 120$ **9.** 100 **11.** 30

13.

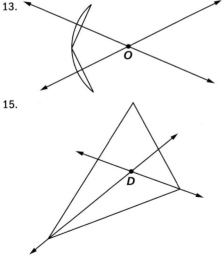

15.

17a. 1054 **b.** $\frac{1054}{32} = 32.9375$ **19.** $\frac{7}{12}t$ **21.** 686
23. Answers vary. Sample: Draw the auxiliary chord $\overline{AD}$. Because it is given that C, D, and A are points on the

circle with diameter $\overline{AC}$, $\angle CDA$ is an inscribed angle with measure 90. Since D is also a point on circle A, $\overline{AD}$ is a radius of the circle. By the Radius-Tangent Theorem, $\overleftrightarrow{CD}$ is tangent to circle A. **25.** The chord is a diameter of the circle. **27.** Since $QP = QR$, $m\angle QPS = m\angle QRS$. The measure of an inscribed angle is half the measure of its intercepted arc, so $m\angle QPS = \frac{1}{2}m\widehat{QS}$. By the Tangent-Chord Theorem, $m\angle RQS = \frac{1}{2}m\widehat{QS}$. Thus, $m\angle QRS = m\angle QPS = m\angle RQS$, and so by definition, $\triangle QRS$ is isosceles. **29.** $\frac{1}{2}m\widehat{IH}$ **31.** $m\angle XWY = \frac{1}{2}m\widehat{WZ}$
33a. a circle with radius $15\sqrt{\frac{2}{\pi}}$ mm **b.** $30\sqrt{2\pi} \approx 75.2$ mm
35. No. The Isoperimetric Inequality, $A \leq \frac{p^2}{4\pi}$, does not hold when $A = p = 2$. **37.** no **39.** no **41.** no
43. 435 games **45.** Answers vary. Sample: Round 1: 1-6, 2-5, 3-4, 7-8; Round 2: 1-5, 2-4, 3-8, 6-7; Round 3: 1-4, 2-3, 6-8, 5-7; Round 4: 1-3, 2-8, 4-7, 5-6; Round 5: 1-2, 3-7, 4-6, 5-8; Round 6: 1-7, 2-6, 3-5, 4-8; Round 7: 1-8, 2-7, 3-6, 4-5
47. about 4.52 km **49.** between 38.3 km and 41.5 km
51. 0.65 m **53a.** a sphere **b.** Answers vary. Sample: No, it will be hard to market to consumers because it is hard to carry around. **55.** Make it more spherical.

Glossary

A

acute angle An angle whose measure is greater than $0°$ and less than $90°$. (**116**)

adjacent angle Two nonstraight and nonzero angles with a common side interior to the angle formed by the noncommon sides. (**126**)

acute triangle A triangle with all three angles acute. (**286**)

adjacent sides See *consecutive sides*. (**90**)

adjacent vertices See *consecutive vertices*. (**90**)

alternate exterior angles Nonadjacent angles formed by two lines and a transversal, with interiors outside the two lines and on different sides of the transversal. (**271**)

alternate interior angles Nonadjacent angles formed by two lines and a transversal, with interiors inside the two lines and on different sides of the transversal. (**271**)

altitude In a triangle or a trapezoid, the segment from a vertex perpendicular to the line containing the opposite side; also, the length of that segment. In a prism or a cylinder, the distance between the planes of its bases. In a pyramid or a cone, the length of the segment from the apex perpendicular to the plane of the base. Also called *height*. (**318, 469, 474, 527, 534**)

analytic geometry See *coordinate geometry*. (**13**)

angle The union of two rays (its sides) that have the same endpoint (its vertex). Symbols: $\angle$, $\angle A$, $\angle ABC$, $\angle 1$. (**83, 114**)

angle bisector The ray in the interior of an angle that divides the angle into two angles of equal measure. (**127**)

angle formed by a line and a plane The angle of smallest measure that the line makes with rays in the plane from the point of intersection. (**520**)

angle of depression An angle measured from the horizontal plane downward from an observer's eye to a given point below the plane. (**809**)

angle of elevation An angle measured from the horizontal plane upward from an observer's eye to a given point above the plane. (**809**)

angle of incidence When an object hits a surface, the angle formed by the path of the object before contact and the line perpendicular to the surface through the point of contact. (**196**)

angle of reflection When an object leaves a hit surface, the angle formed by the path of the object after contact and the line perpendicular to the surface through the point of contact. (**196**)

antecedent The "if" clause of a conditional. Also called *hypothesis*. (**66**)

apex See *conic solid*. (**533**)

apothem The perpendicular segment from the center to any side of a regular *n*-gon. (**491**)

arc (**A**) A path from one point (node) of a network to another point. (**B**) A set of points made up of two points A and B on a circle (its endpoints) and all points on the circle between A and B, denoted $\overset{\frown}{AB}$. (**21, 113**)

area The measure of space covered by a 2-dimensional region. (**457**)

augmentation Changing the tempo in a musical composition so that notes are lengthened, similar to expansion in mathemtics. (**234**)

auxiliary figure A figure that is added to a given figure, often to aid in completing a proof. (**283**)

axis A number line of reference in a coordinate system. Plural: *axes*. (**700**)

axis of a cone The line through a cone's apex and the center of its base. (**533**)

B

base See *cylindrical solid* and *conic solid*. (**526, 533**)

base angles of an isosceles triangle Two angles of an isosceles triangle whose vertices are the endpoints of the base of the triangle. (**316**)

base angles of a trapezoid Two consecutive angles of a trapezoid whose vertices are the endpoints of a base of the trapezoid. (**334**)

base edge of a pyramid In a pyramid, any edge other than a lateral edge. (**533**)

base of a trapezoid Either of two parallel sides of a trapezoid. (**334**)

base of an isosceles triangle The side opposite the vertex angle of the triangle. (**316**)

betweenness of numbers A number is between two other numbers if it is greater than one of them and less than the other. (**38**)

betweenness of points A point is between two other points on the same line if its coordinate is between their coordinates. (**38**)

biconditional statement A statement that includes a conditional and its converse. It may be written in the form "$p \Rightarrow q$ and $q \Rightarrow p$" or "$p \Leftrightarrow q$," read "p if and only if q." Also called *if-and-only-if statement*. (**78**)

bilateral symmetry Symmetry of a space figure such that the reflection image of the figure over a plane is the figure itself. Also called *reflection symmetry*. (**560**)

bisector A point, line, ray, or plane that divides a segment, angle, or figure into two parts of equal measure. (**165**)

bisector of a segment A line, ray, or segment that intersects a segment at its midpoint but does not contain the segment. (**165**)

box A right prism whose faces are rectangles. (**528**)

bye In a tournament, an unpaired status. (**831**)

C

capacity See *volume*. (**596**)

center See *circle, rotation, size change, sphere*. (**79, 120, 153, 550**)

center of a regular polygon The point equidistant from all the vertices of the polygon. (**357**)

center of symmetry For a rotation-symmetric figure, the center of a rotation that maps the figure onto itself. (**351**)

central angle of a chord The central angle of a circle whose rays intersect the endpoints of the chord. (**824**)

central angle of a circle An angle whose vertex is the center of the circle. (**114**)

centroid of a triangle The point at which the medians of a triangle intersect. (**888**)

chord A segment whose endpoints are on a given circle. (**824**)

chord of an arc The chord whose endpoints define the arc. (**824**)

circle The set of points in a plane at a certain positive distance (its radius) from a certain point (its center). (**79**)

circularity The "circling back" that sometimes occurs when one tries to define basic terms; returning to the word that one is trying to define. (**28**)

circumcenter The center of the circumcircle of a triangle. (**857**)

circumcircle The circle that contains the three vertices of a triangle. (**857**)

circumference The perimeter of a circle, which is the limit of the perimeters of inscribed polygons. (**494**)

circumradius The radius of the circumcircle of a triangle. (**857**)

circumscribed circle A circle that contains all the vertices of a given polygon. (**860**)

circumscribed polygon A polygon whose sides are all tangent to a circle. (**860**)

clockwise The direction in which the hands move on an analog clock, designated as negative. (**120**)

clockwise orientation The orientation for which the interior of a figure is on the right as you walk around it. (**191**)

collinear points Points that lie on the same line. (**28**)

common tangent A line that is tangent to two or more distinct circles. (**845**)

compass An instrument for drawing circles. (**166**)

complementary angles Two angles with measures that sum to 90°. (**128**)

complements Two angles that are complementary. (**128**)

composite (of transformations) The transformation $T \circ S$ under which the image of a point or figure F is $T(S(F))$. (**203**)

composition The operation, denoted by the symbol $\circ$, of combining transformation S followed by the transformation T, denoted $T \circ S$, that maps each point P onto $T(S(P))$. (**203**)

compound statement A sentence that combines two or more statements with some type of connective such as *and, or*, or *if . . . then*. (**66**)

concentric circles Two or more circles with the same center and different radii. (**411**)

conclusion See *consequent*. (66)

concurrent Having a point in common. (152)

concurrent lines Three or more lines that contain the same point. Their common point is called the point of concurrency. (152, 858)

conditional statement A statement of the form "If . . ., then . . .." Also called *conditional*. (66)

cone The surface of a conic solid whose base is a circle. (533)

congruence transformation A transformation that is a reflection or composite of reflections. Also called *isometry*. (253)

congruent figures Figures, one of which is the image of the other under a reflection or a composite of reflections (an isometry), denoted by ≅. (253, 559)

conic section The intersection of a plane with the union of two right conical surfaces that have the same apex and whose edges are opposite rays; a circle, an ellipse, a hyperbola, or a parabola. (556)

conic solid The set of points between a given point (its apex) and all points of a given region (its base), together with the apex and the base. (533)

conic surface The boundary of a conic solid. (532)

conjecture An educated guess or opinion. (95)

consecutive angles In a polygon, two angles whose vertices are endpoints of the same side. (334)

consecutive sides In a polygon, two sides with an endpoint in common. Also called *adjacent sides*. (90)

consecutive vertices In a polygon, endpoints of a side. Also called *adjacent vertices*. (90)

consequent The "then" clause of a conditional. Also called *conclusion*. (66)

construction A drawing that is made following certain prescribed rules and using only an unmarked straightedge and a compass. (166)

contraction A size change with absolute value of magnitude between 0 and 1. (719)

contradiction A situation in which there exist contradictory statements. (663)

contradictory statements Two statements that cannot both be true at the same time. (664)

contrapositive of a statement A conditional resulting from negating and switching the antecedent and consequent of the original conditional. The contrapositive of $p \Rightarrow q$ is *not-q* $\Rightarrow$ *not-p*. (657)

convenient location A general location for a figure on a coordinate plane in which its key points are described with the fewest possible variables. (672)

converse The conditional statement formed by switching the antecedent and consequent of a given conditional. The converse of $p \Rightarrow q$ is $q \Rightarrow p$. (72)

convex polygon A polygon whose polygonal region is a convex set. (91)

convex polyhedron A polyhedron in which the segment connecting any two points of it is contained within the polyhedron and its interior. (564)

convex set A set in which every segment that connects points of the set lies entirely in the set. (62)

coordinate The number or numbers associated with the location of a point on a line, a plane, or in space. (8)

coordinate axes (A) A pair of perpendicular coordinatized lines in a plane that intersect at the point with coordinate (0, 0). (B) Three mutually perpendicular coordinatized lines in space that intersect at the point with coordinate (0, 0, 0). (13, 701)

coordinate geometry A geometry in which points are represented by coordinates and in which algebraic methods of reasoning are utilized. (13)

coordinate plane Name given to the plane containing points identified as ordered pairs of real numbers. (13)

coordinatized line A line on which every point is identified with exactly one number and every number is identified with a point on the line. (8)

coplanar points Points lying in the same plane. (28)

corollary A theorem that is easily proved from another theorem. (347)

corresponding angles A pair of angles in similar locations with respect to a transversal and its two intersected lines. (145)

corresponding parts Parts of figures that are images of each other under a transformation. (259)

cosine of an angle In a right triangle, the ratio $\frac{\text{length of leg adjacent to the angle}}{\text{length of hypotenuse}}$. Abbreviated *cos*. (807)

counterclockwise The direction opposite that which the hands move on an analog clock, designated as positive. (**120**)

counterclockwise orientation The orientation for which the interior of a figure as you walk around it is on the left. (**191**)

counterexample An example that shows a conjecture to be false. For a conditional, a specific case for which the antecedent ("if" part) is true but the consequent ("then" part) is false. (**68**)

cross section See *plane section*. (**552**)

cube A box whose faces are all congruent squares. (**528**)

cube root, $\sqrt[3]{\ }$ A real number x is the cube root of a real number y, written $x = \sqrt[3]{y}$, if and only if $x^3 = y$. (**599**)

cubic units The way volume is measured. (**596**)

cyclic quadrilateral A quadrilateral whose vertices all lie on one circle. (**330**)

cylinder The surface of a cylindrical solid whose base is a circle. (**527**)

cylindrical solid The set of points between a region (its base) and its translation image in space, including the region and its image. (**526**)

cylindrical surface The boundary of a cylindrical solid. (**526**)

D

decagon A polygon with 10 sides. (**90**)

definition A description that clearly and uniquely specifies an object or a class of objects. (**77**)

degree (°) A unit of measure used for the measure of an angle, arc, or rotation. (**113**)

diagonal of a polygon A segment connecting nonconsecutive vertices of the polygon. (**90**)

diameter of a circle or a sphere A segment connecting two points on the circle or sphere and containing the center of the circle or sphere; also, the length of that segment. (**79, 550**)

dihedral angle The angle formed by the union of two half-planes with the same edge. (**522**)

dilation, dilatation A size-change transformation. (**719**)

diminution Changing the tempo in a musical composition so that notes are shortened, similar to contraction in mathematics. (**234**)

direct proofs Proofs using direct reasoning. (**663**)

direct reasoning Reasoning that uses the Laws of Detachment and Transitivity to begin with given true information and arrive at a true conclusion. (**663**)

direction of a translation The direction given by the ray from any preimage point through its image point in a translation. (**205**)

directly congruent Figures that are congruent and have the same orientation. (**253**)

discrete geometry The study of discrete points and lines. (**20**)

discrete lines Lines made up of discrete points. (**20**)

discrete points Points described as dots. (**20**)

distance between two parallel planes The length of a segment perpendicular to the planes with an endpoint in each plane. (**521**)

distance between two parallel lines The length of a segment that is perpendicular to the lines with an endpoint on each line. (**421**)

distance between two points The absolute value of the difference of the coordinates of the two points on a coordinatized line. (**9**)

distance to a plane from a point The length of the perpendicular segment from the point to the plane. (**521**)

dodecagon A polygon with 12 sides. (**90**)

dodecahedron A polyhedron with 12 faces. (**564**)

dynamic geometry system (DGS) Any calculator or computer utility that enables the user to draw or construct figures and then dynamically manipulate them. (**44**)

E

edge Any side of a face of a polyhedron. (**526, 532, 564**)

edge of a dihedral angle The boundary line of a dihedral angle. (**522**)

elevations Planar views of 3-dimensional figures given from the top, front, or sides. (**545**)

ellipse The conic section formed by a plane that intersects only one of the right conical surfaces. (556)

empty set See *null set.* (84)

endpoint See *arc, ray, segment.* (21, 39, 113)

ends of a kite The common vertices of the congruent sides of the kite. (340)

equiangular polygon A polygon with all angles congruent. (357)

equidistant At the same distance. (165)

equilateral polygon A polygon with all sides congruent. (357)

equilateral triangle A triangle with all three sides the same length. (91)

Euclidean geometry The collection of propositions about figures that includes or from which can be deduced those given by the mathematician Euclid around 250 BCE. (32)

Euclidean synthetic geometry A geometry studied without the use of coordinates. (6, 8)

even node A node that is the endpoint of an even number of arcs in a network. (23)

expansion A size change with absolute value of magnitude greater than 1. (719)

exterior angle of a polygon An angle that forms a linear pair with an angle of a polygon. (290)

exterior angles Angles formed by two lines and a transversal whose interiors contain no points between the two lines. (270)

exterior of a figure When a figure separates the plane into two parts, one bounded and one not, the unbounded part. (91)

exterior of a surface When a surface separates space into two regions, one bounded and one not, the unbounded region. (525)

exterior of an angle A nonzero angle separates the plane into two sets of points. If the angle is not straight, the nonconvex set is the exterior of the angle. (116)

externally tangent circles Two tangent circles such that one circle lies outside the other. (846)

extremes of a proportion The first and fourth terms of a proportion. (727)

F

face of a polyhedron Any of the polygonal regions that form the surface of the polyhedron. (526, 533)

figure A set of points. (28)

flip See *reflection.* (184)

foot The point of intersection of a segment perpendicular to a plane and the plane. (521)

foot of an altitude The point of intersection of the altitude and the side of a triangle. (862)

45-45-90 triangle An isosceles right triangle whose angles measure 45°, 45°, and 90°. (487)

frieze pattern Any pattern in which a single fundamental figure is repeated to form a pattern that has a fixed height and infinite length in opposite directions. (363)

frustum of a pyramid The polyhedron that results from slicing a pyramid with a plane parallel to its base and using the section of the plane as a face. (569)

fundamental region A region that is used to tessellate a plane. (413)

G

geometric mean Given positive numbers a, b, and g, g is the geometric mean of a and b if and only if $\frac{a}{g} = \frac{g}{b}$. (787)

glide reflection The composite of a reflection and a translation parallel to the reflecting line. (225)

golden ratio The number $\frac{1 + \sqrt{5}}{2} = 1.618\ldots$ represented by φ (phi). (796)

golden rectangle A rectangle whose dimensions are in the golden ratio. (795)

grade The slope of a road, often represented as a percent or a ratio, measured by the angle it makes with a horizontal line. Also called *tilt.* (145)

graph a picture of numbers on a number line or coordinate system. See also *network.* (8, 13, 21, 700)

graph theory The geometry of networks. (21)

great circle of a sphere The intersection of a sphere and a plane that contains that center of the sphere. (551)

H

height See *altitude*. (318, 469, 474, 527, 534)

hemisphere The half of a sphere on one side of a great circle. (551)

heptagon A polygon with 7 sides. Also called *septagon*. (90)

hexagon A polygon with 6 sides. (90)

hexahedron A polyhedron with 6 faces. (564)

hierarchy A diagram that shows how various figures or ideas are related, often with a downward direction that moves from more general to more specific. (91)

horizontal component of a vector The first component in the ordered-pair description of a vector, indicting its magnitude along the *x*-axis of the coordinate plane. (218)

horizontal line A line with equation $y = k$ on the coordinate plane. (13)

hyperbola The conic section formed by a plane that intersects both of the right conical surfaces. (556)

hypotenuse The longest side of a right triangle; the side opposite the right angle of a right triangle. (408)

hypothesis See *antecedent*. (66)

I

icosahedron A polyhedron with 20 faces. (564)

identity transformation A size change with size-change factor $k = 1$. Each point coincides with its image. (719)

if-and-only-if statement See *biconditional*. (78)

if-then statement See *conditional statement*. (66)

image The result of applying a transformation to an original figure, its preimage. (121)

incenter The center of the incircle of a triangle. (858)

incircle The circle whose center is the intersection of the angle bisectors of a triangle and that is tangent to the sides of the triangle. (858)

included angle The angle formed by two consecutive sides of a polygon. (382)

included side The side between two consecutive vertices of a polygon. (388)

indirect proof Proof that uses indirect reasoning. (664)

indirect reasoning Reasoning using the Law of the Contrapositive, Law of Ruling out Possibilities, and Law of Indirect Reasoning to rule out all possibilities other than the one to be proven true. (663)

initial point The beginning point of a vector. (217)

inradius The radius of the incircle of a triangle. (858)

inscribed angle in a circle An angle with its vertex on the circle and with its sides each intersecting the circle at a point other than the vertex. (324)

inscribed circle In a polygon, a circle that is tangent to each side of the polygon. (858)

inscribed polygon In a circle, a polygon whose vertices all lie on the circle. (358)

instance of a conditional A specific case in which both the antecedent ("if" part) and the consequent ("then" part) of the conditional are true. (68)

intercepted arc An arc of a circle in the interior of a central angle. (325)

interior angle of a polygon An angle within a polygon. (290)

interior angles Angles formed by two lines and a transversal whose interiors are partially between the lines. (270)

interior of a figure When a figure separates the plane into two parts, one bounded and one not, the bounded part. (91)

interior of a surface When a surface separates space into two regions, one bounded and one not, the bounded region. (525)

interior of an angle A nonzero angle separates the plane into two sets of points. If the angle is not straight, the convex set is the interior of the angle. (116)

internally tangent circles Two tangent circles such that one circle lies inside the other. (846)

intersecting planes Two planes that contain the same line. (522)

intersection of two sets The set of elements that are in both of the sets, denoted by ∩. (83)

inverse of a statement The conditional statement formed by negating *both* the antecedent and consequent of a given conditional. The converse of $p \Rightarrow q$ is *not-p* $\Rightarrow$ *not-q*. (**657**)

inversion The changing of a musical theme so that it goes down in pitch whenever it went up or vice versa, similar to a horizontal or glide reflection in mathematics. (**234**)

isometric drawing A drawing using isometric graph paper, which is paper consisting of dots located at the vertices of a tessellation of equivalent triangles. (**546**)

isometry A transformation that is a reflection or a composite of reflections. Also called *congruence transformation* or *distance-preserving transformation*. (**223**)

isoperimetric Having equal perimeters. (**870**)

isosceles trapezoid A trapezoid with a pair of base angles congruent. (**334**)

isosceles triangle A triangle with at least two sides equal in length. (**91**)

J

justification A general property applied to a special case. (**139**)

K

kite A quadrilateral with two distinct pairs of consecutive sides congruent. (**333**)

L

lateral area (L.A.) The total area of the lateral surface of a solid. (**571**)

lateral edge of a conic solid In a pyramid, a segment connecting the apex of the pyramid to a vertex of the base. In a cone, any segment whose endpoints are the apex of the cone and a point on the circle of the base. (**533**)

lateral edge of a prism In a prism, the intersection of two lateral faces. (**527**)

lateral face of a conic or cylindrical solid Any face other than a base. (**527, 533**)

lateral surface of a conic or cylindrical solid The surface of the solid not including its base(s). (**527**)

lattice point A point in the coordinate plane or in space with integer coordinates. (**457, 687**)

leg Either side of a right triangle that includes the right angle. (**408**)

leg adjacent to an angle Either of the two legs of a triangle with one endpoint being the vertex of the angle. (**801**)

leg opposite an angle The leg of a triangle with neither endpoint being the vertex of the angle. (**801**)

length (A) The distance between two points measured along the segment or arc joining them. **(B)** A dimension of a rectangle or a rectangular solid. (**40, 483, 526, 596**)

limit The value to which the terms of an infinite sequence get closer and closer in value as one goes farther out in the sequence. (**494**)

line An undefined geometric term. A line containing A and B is denoted $\overleftrightarrow{AB}$. See *Point-Line-Plane Postulate*. (**8, 13**)

line of reflection See *reflecting line*. (**184**)

line segment See *segment*. (**39**)

line tangent to a circle A line that intersects a given circle at exactly one point. (**686**)

linear pair Two adjacent angles whose noncommon sides are opposite rays. (**129**)

logically equivalent Two statements that are both true and both false at the same time. (**658**)

M

magnitude of a size transformation In a size change, the factor by which the length of the preimage is changed, $\frac{A'B'}{AB}$, where A' and B' are the images of A and B under the transformation. Also called a *size-change factor*. (**719**)

magnitude of a translation The distance a figure is translated. (**205**)

magnitude of rotation In a rotation, the amount that the preimage is turned about the center of rotation, measured in degrees from $-180°$ (clockwise) to $180°$ (counterclockwise), m$\angle POP'$, where P' is the image of P under the rotation and O is its center. (**121**)

major arc An arc of a circle with measure greater than $180°$. (**113**)

mapping See *transformation*. **(152)**

means of a proportion The second and third terms in a proportion. **(727)**

measure of an angle The amount of openness of the interior of an angle. **(114)**

measure of an angle between a line and a plane The measure of the smallest angle the line makes with rays in the plane from the point of intersection. **(520)**

measure of an arc The measure of the central angle that includes the endpoints of the arc. **(113)**

medial triangle The union of segments that join the midpoints of the sides of a triangle. **(697)**

median of a triangle A segment from a vertex of the triangle to the midpoint of the opposite side. **(318)**

menu A list of options from which a user of a dynamic geometry system (DGS) can select an operation for the DGS to perform **(45)**

midpoint of a segment The point of the segment equidistant from the segment's endpoints. **(77)**

midsegment of a trapezoid The segment joining the midpoints of the nonparallel sides of a trapezoid. **(694)**

midsegment of a triangle A segment joining the midpoints of any two sides. **(694)**

minor arc An arc of a circle with measure less than 180°. **(113)**

mirror See *reflecting line*. **(559)**

N

n-fold rotation symmetry A figure has *n*-fold rotation symmetry when the smallest magnitude of rotation that will map it onto itself is m where $m = \frac{360°}{n}$. **(352)**

n-gon A polygon with n sides. **(90)**

negation of a statement (not-p) A statement, written "not-p", that is true whenever statement p is false and is false whenever statement p is true. **(656)**

net A 2-dimensional figure that can be folded on its segments or curved on its boundaries into a 3-dimensional surface. **(565)**

network A union of points (its vertices, or nodes) and segments (its arcs) connecting them. Also called *graph*. **(21)**

nine-point circle For a triangle, the circle which contains the midpoints of the sides, the feet of the altitudes, and the midpoints of the segments connecting the orthocenter to the vertices of the triangle. **(863)**

node An endpoint of an arc in a network. Also called *vertex*. **(21)**

non-Euclidean geometry A geometry in which the postulates are not the same as those in Euclidean geometry. **(285)**

nonagon A polygon with 9 sides. **(90)**

nonconvex set A set in which at least one segment that connects points within the set has points that lie outside the set. **(63)**

nonoverlapping figures Figures that do not overlap. **(400)**

null set The set with no elements. Also called *empty set*. **(84)**

number line A line on which points are identified with real numbers. **(8)**

O

oblique cone A cone whose axis is not perpendicular to the plane of the base. **(533)**

oblique cylindrical solid A cylindrical solid formed when the direction of translation of the base is not perpendicular to the plane of the base. **(527)**

oblique line A line that is neither horizontal nor vertical. **(13)**

oblique pyramid A pyramid in which the segment connecting the vertex to the center of symmetry of the base is not perpendicular to the plane of the base. A nonright pyramid. **(533)**

obtuse angle An angle whose measure is between 90° and 180°. **(116)**

obtuse triangle A triangle with an obtuse angle. **(286)**

octagon A polygon with 8 sides. **(90)**

octahedron A polyhedron with 8 faces. **(564)**

odd node A node that is the endpoint of an odd number of arcs in a network. **(23)**

one-dimensional A space, type of geometry, or figure in which all points are collinear. **(29)**

one-step proof A justified conclusion of a conditional requiring a single definition, theorem, or postulate. (263)

opposite rays $\overrightarrow{AB}$ and $\overrightarrow{AC}$ are opposite rays if and only if A is between B and C. (40)

opposite reciprocals Two numbers whose product is −1. (161)

oppositely congruent Figures that are congruent and have opposite orientation. (253)

ordered pair The pair of numbers (a, b) identifying a point in a 2-dimensional coordinate system. (13)

ordered-pair description of a vector The description of a vector as the ordered pair (a, b) where a is the horizontal component and b is the vertical component. (218)

ordered triple The triple of numbers (a, b, c) identifying a point in a 3-dimensional coordinate system. (701)

orientation The order in which the vertices of a polygon are considered, either clockwise or counterclockwise. (191)

orthocenter The point in which the lines containing the altitudes of a triangle are concurrent. (859)

overlapping figures Figures that have interior points in common. (400)

overlapping triangles Triangles that have interior points in common. (400)

P

pairing The assignment of two teams or individuals to compete against one another in a tournament. (831)

parabola The conic section formed by a plane parallel to an edge of the conical surface. (556)

paragraph proof A form of written proof in which conclusions and justifications are incorporated into sentences. (269)

parallel lines Two coplanar lines that have no points in common or are identical. (35)

parallel planes Planes that have no points in common or are identical. (521)

parallelogram A quadrilateral with both pairs of opposite sides parallel. (331)

pentagon A polygon with 5 sides. (90)

perimeter The length of the boundary of a closed region. (483)

perimeter of a polygon The sum of the lengths of the sides of the polygon. (453)

perpendicular Two segments, rays, or lines such that the lines containing them form a 90° angle, denoted by ⊥, ⌐. (159)

perpendicular bisector, ⊥ bisector, of a segment In a plane, the line that bisects and is perpendicular to the segment. In space, the plane that bisects and is perpendicular to the segment. (165, 559)

perpendicular-bisector method A method for finding the center of a circle that involves drawing perpendicular bisectors of two chords. The perpendicular bisectors intersect at the center of the circle. (861)

perpendicular planes Planes that form a 90° dihedral angle. (522)

perpendicular to a plane A line, segment, or ray that intersects the plane and is perpendicular to every line in the plane passing through the point of intersections. (521)

perspective drawing A drawing of a figure made to look as it would in the real world. (538)

pi, π The ratio of the circumference to the diameter of a circle. (494)

pixel A dot on a TV, computer screen, or other monitor. (20)

plane An undefined geometric term. See *Point-Line-Plane Postulate*. (13)

plane coordinate geometry The study of geometric figures using points as ordered pairs of real numbers. (13)

plane figure A set of points that are all in one plane. (28)

plane of reflection See *reflecting plane*. (559)

plane section The intersection of a 3-dimensional figure with a plane. Also called 2-dimensional cross section or section. (552)

point An undefined geometric term. See *Point-Line-Plane Postulate*. (8)

point of concurrency The point at which three or more lines intersect. (152, 858)

point of tangency The point at which a tangent intersects a circle or a sphere. (**842**)

polygon The union of segments in the same plane such that each segment intersects exactly two others, one at each of its endpoints; and if two segments share an endpoint, they are not collinear. (**90**)

polygon inscribed in a circle A polygon whose vertices all lie on a circle. Every regular polygon can be inscribed in a circle. (**358**)

polygonal region The union of a polygon and its interior. (**91**)

polyhedron A 3-dimensional surface that is the union of polygonal regions (its faces) and that has no holes. Plural: *polyhedrons* or *polyhedra*. (**564**)

postulate A statement assumed to be true. (**32**)

power of a point for $\odot O$ For any secant through P intersecting $\odot O$ at A and B, the product $PA \cdot PB$. (**864**)

preimage The original figure in a transformation. (**121**)

preserved property Under a transformation, a property that, if present in a preimage, is present in the image. (**154, 189, 733**)

prime number A number whose only factors are 1 and itself. (**70**)

prism The surface of a cylindrical solid whose base is a polygon. (**527**)

proof A sequence of justified conclusions, leading from what is given or known to a final conclusion. (**139**)

proportion A statement that two ratios are equal. (**726**)

protractor An instrument for measuring angles. (**114**)

pyramid The surface of a conic solid whose base is a polygonal region. (**533**)

Pythagorean triple A set of three whole numbers that can be the lengths of the sides of a right triangle. (**484**)

Q

quadrilateral A polygon with 4 sides. (**90**)

R

radius of a circle or a sphere A segment connecting the center of a circle or a sphere with a point on that circle or sphere; also, the length of that segment. Plural: *radii*. (**79, 550**)

ratio A quotient of two quantities with the same units. (**726**)

ratio of similitude In similar figures, the ratio of a length in an image to the corresponding length in a preimage. (**734**)

ray The set consisting of the distinct points A (its endpoint) and B, all points between A and B, and all points for which B is between the point and A, denoted $\overrightarrow{AB}$. (**39**)

rectangle A quadrilateral with four right angles. (**331**)

reflecting line The line over which a preimage is reflected. Also called *line of reflection*. (**184**)

reflecting plane The plane over which a preimage is reflected. Also called *plane of reflection* or *mirror*. (**559**)

reflection A transformation in which each point is mapped onto its reflection image over a line or a plane, denoted by r. Also called *flip*. (**184**)

reflection image of a figure The set of all reflection images of the points of the figure. (**190**)

reflection image of a point P over a line m If P is not on m, the reflection image of P over line m is the point Q if and only if m is the perpendicular bisector of $\overline{PQ}$. If P is on m, the reflection image is P itself. (**185**)

reflection image of a point P over a plane M If P is not on M, the reflection image of P is the point Q if and only if M is the perpendicular bisector of $\overline{PQ}$. If P is on M, the reflection image is point P itself. (**559**)

reflection-symmetric plane figure A plane figure F for which there is a line m (the symmetry line) such that $r_m(F) = F$. (**308, 560**)

reflection-symmetric space figure A space figure F for which there is a plane M (the symmetry plane) such that $r_M(F) = F$. (**560**)

regular n-gon A regular polygon with n sides. (**356**)

regular polygon A convex polygon whose angles are all congruent and whose sides are all congruent. (**356**)

regular polyhedron A convex polyhedron whose faces are all congruent regular polygons and in which the same number of edges intersect at each of its vertices. (**564**)

regular prism A right prism whose bases are regular polygons. (**527**)

regular pyramid A right pyramid whose base is a regular polygon. (**533**)

retrograde The reversal of a theme in a musical composition, similar to a vertical reflection in mathematics. (**236**)

rhombus A quadrilateral with four congruent sides. (**331**)

right angle An angle whose measure is 90°. (**116**)

right-angle method A method for finding the center of a circle that involves constructing two inscribed right triangles. The intersection of their hypotenuses is the center of the circle. (**861**)

right cone A cone whose axis is perpendicular to the plane of the base. (**533**)

right cylindrical solid A cylindrical solid formed when the direction of translation of the base is perpendicular to the plane of the base. (**527**)

right pyramid A pyramid in which the segment connecting the apex to the center of symmetry of the base is perpendicular to the plane of the base. (**533**)

right triangle A triangle that contains a right angle. (**286**)

rotation The composite of two reflections over intersecting lines; the transformation "turns"the preimage onto the image about a fixed point (its center), denoted by R. Also called *turn*. (**120, 211**)

rotation-symmetric figure A plane figure F such that there is a rotation R with magnitude between 0° and 360° such that $R(F) = F$. The center of R is a center of symmetry for F. (**351**)

round A musical composition in which two or more voices sing the same melody, but at different times, and when each voice finishes the melody the voice goes back to the beginning, similar to a 2-dimensional translation in mathematics. (**232**)

round-robin tournament A tournament in which each competitor plays each other competitor exactly once. (**830**)

S

same-side interior angles Angles formed by two lines and a transversal, with interiors inside the two lines and on the same side of the transversal. (**271**)

scalene triangle A triangle with no two sides the same length. (**91**)

secant to a circle A line that intersects a circle in two different points. (**838**)

section See *plane section*. (**552**)

sector of a circle The figure bounded by two radii and the included arc of the circle. (**501**)

segment The set consisting of the distinct points A and B (its endpoints) and all points between A and B, denoted $\overline{AB}$. Also called *line segment*. (**39**)

semicircle An arc of a circle whose endpoints are the endpoints of a diameter. Its degree measure is 180°. (**113**)

semiregular tessellation A tessellation that combines several different regular polygons. (**418**)

side of an angle See *angle*. (**114**)

side of a polygon One of the segments that make up a polygon; also, the length of that segment. (**90**)

similar figures Figures that have the same shape, denoted by the symbol ~. Figures for which there is a composite of size transformations and reflections mapping one onto the other. (**731**)

similarity transformation A composite of size transformations and reflections. (**732**)

sine of an angle In a right triangle, the ratio $\frac{\text{length of leg opposite the angle}}{\text{length of hypotenuse}}$. Abbreviated *sin*. (**807**)

size change (size transformation) **(A)** The transformation in which the image of (x, y) is (kx, ky) for $k \neq 0$. **(B)** The transformation S such that for a given point P and a nonzero real number k (its magnitude) and any point O (its center), $S(P) = P'$ is the point on $\overrightarrow{OP}$ with $OP' = k \cdot OP$. Also called *size transformation* or *dilation*. (**153, 719**)

size-change factor See *magnitude of a size transformation*. (**719**)

skew lines Lines that do not lie in the same plane. (**521**)

slant height of a regular pyramid. The altitude from the apex of any lateral face of the pyramid. (**534**)

slant height of a right cone The length of a lateral edge of the cone. (**534**)

slide See *translation*. (**204**)

slope In the coordinate plane, the change in y-values divided by the corresponding change in x-values. (**14**)

slope-intercept form of an equation of a line A linear equation of the form $y = mx + b$, where m is the slope and b is the y-intercept. (**14**)

small circle of a sphere The intersection of the sphere and a plane that does not contain the center of the sphere. (**551**)

solid The union of a surface and the region of space enclosed by the surface. (**525**)

space The set of all points in a geometry. (**28**)

sphere The set of points in space at a certain distance (its radius) from a point (its center). (**550**)

square A quadrilateral with four congruent sides and four right angles. (**332**)

square root, $\sqrt{}$ x is a square root of y if and only if $x^2 = y$. If x and y are positive, $x = \sqrt{y}$. (**510**)

square unit The fundamental region used to tessellate a region in order to find its area. (**457**)

standard form of an equation of a line An equation of a line in the form $Ax + By = C$, where A and B are not both zero. (**13**)

statement A sentence that is either true or false and not both. (**66**)

straight angle An angle whose measure is 180°. (**116**)

sufficient condition p is a sufficient condition for q if and only if p implies q. (**381**)

supplementary angles Two angles with measures that sum to 180°. (**128**)

supplements Two angles that are supplementary. (**128**)

surface The boundary of a 3-dimensional figure. (**525**)

surface area (S.A.) The sum of the areas of all the faces or surfaces that enclose a solid. (**570**)

symmetry diagonal of a kite The diagonal of the kite that is on the symmetry line and contains the ends of the kite. (**341**)

symmetry line For a figure, a line m such that the figure coincides with its reflection image over m. (**308**)

symmetry plane For a figure, a plane M such that the figure coincides with its reflection image over M. (**560**)

T

tangent A line, ray, segment, or plane that intersects a curve or curved surface in exactly one point, the point of tangency. (**686, 842**)

tangent circles Two circles that have exactly one point in common. (**846**)

tangent of an angle In a right triangle, the ratio $\dfrac{\text{length of leg opposite the angle}}{\text{length of leg adjacent to the angle}}$. Abbreviated tan. (**801**)

$\tan^{-1}(x)$ The inverse tangent of x. The angle whose tangent has the value of x. Also called $arctan$. (**803**)

taxicab distance A distance between two points found by summing the absolute differences of their coordinates. (**676**)

tempo How fast a piece of music is played or sung. (**233**)

terminal point The endpoint of a vector. (**217**)

terms of a proportion The four numbers in a proportion. (**727**)

tessellate To tile a plane without leaving spaces or overlapping. (**413**)

tessellation A covering of a plane with nonoverlapping congruent regions. (**413**)

tetrahedron A polyhedron with four faces. (**564**)

theorem A statement that follows from postulates, definitions, and other statements already proved. (**32**)

30-60-90 triangle A triangle in which the three angles measure 30°, 60°, and 90°. (**489**)

three-dimensional A space, type of geometry, or figure in which all points do not lie in a single plane. (**29**)

three-dimensional coordinate system A system of coordinates to locate points in space by their distances and directions from three mutually perpendicular lines. (**700**)

3-space 3-dimensional space. Also called $space$. (**701**)

transformation A correspondence between two sets of points such that each point in the preimage set has exactly one image and each point in the image set has exactly one preimage. Also called map. (**152**)

transformation image of a figure The set of all images of the points of the figure under a transformation. (**152**)

translation The composite of two reflections over parallel lines. Also called *slide*. (204)

translation symmetry A figure *F* has translation symmetry when there is a translation *T* with nonzero magnitude such that $T(F) = F$. (363)

translation vector A vector that gives the length and direction of a particular translation. (217)

transposition Raising or lowering all the notes in a musical composition by the same number of notes, similar to a 1-dimensional translation in mathematics. (231)

transversal A line that intersects two or more lines. (145)

trapezoid A quadrilateral with at least one pair of parallel sides. (334)

traversable network A network in which all the arcs may be traced exactly once without picking up the tracing instrument. (22)

triangle A polygon with 3 sides. (90)

triangulate To split a polygon into nonoverlapping triangles. (490)

trigonometric ratio A ratio of the lengths of two sides of a right triangle. (803)

truncated surface A part of a conic surface including its base, the intersection of a plane parallel to the base with the conic solid, and all points of the surface between these. (536)

two-column proof A form of written proof in which the conclusions are written in one column and the justifications beside them in a second column. (269)

two-dimensional A space, type of geometry, or figure in which all points are coplanar. (29)

two-dimensional coordinate system A system of coordinates to locate points in a plane by their distances and directions from two perpendicular lines. (700)

U

undefined term A term used without a specific mathematical definition. (28)

unmarked straightedge An instrument for drawing lines; it has no marks for determining length. (166)

union of two sets The set of elements that are in either or both of the sets, denoted by ∪. (83)

uniquely determined When exactly one thing satisfies some given conditions. (282)

unit circle The circle with center (0, 0) and radius 1. (686)

unit cube A cube in which every edge has length 1 unit. (596)

unit square A square in which each side has length 1 unit. (487)

V

vanishing point A point at which several lines of a drawing appear to meet at a distance from the viewer's eye. (538)

vector A quantity that has both magnitude and direction. (216)

vertex See *angle, network, polygon*. Plural: *vertices*. (90, 114)

vertex angle The angle included by the congruent sides of an isosceles triangle. (316)

vertex of a polyhedron Any of the vertices of the faces of the polyhedron. (526, 564)

vertical angles Two nonstraight angles whose union of their sides is two lines. (129)

vertical component of a vector The second component in the ordered-pair description of a vector, indicating its magnitude along the *y*-axis of the coordinate plane. (218)

vertical line A line with an equation $x = h$ on the coordinate plane. (13)

view A 2-dimensional nonperspective representation of a 3-dimensional figure usually drawn from front, back, top, bottom, left, or right. (544)

volume A measure of how much a 3-dimensional surface will hold. Also called *capacity*. (596)

Voronoi diagram A diagram utilizing the perpendicular bisectors of the segments determined by 2 or more points that outlines the points in the plane closest to each point, named after the Russian mathematician Georgy Voronoi. (167)

W

width of a rectangle A dimension of a rectangle or rectangular solid taken at right angles to the length. (**484**)

window The area in which a dynamic geometry system (DGS) can draw. (**45**)

X

x-**axis** The line in the coordinate plane or in space, usually horizontal, containing those points whose second coordinates (and third, in space) are 0. (**701**)

Y

y-**axis** The line in the coordinate plane or in space, usually vertical, containing those points whose first coordinates (and third, in space) are 0. (**701**)

y-**intercept** In the equation of a line, the value of y when $x = 0$. (**14**)

Z

z-**axis** The line in a 3-dimensional coordinate system containing those points whose first and second coordinates are 0. (**701**)

z-**coordinate** The third number in ordered triple in a 3-dimensional coordinate system. (**700**)

zero angle An angle whose measure is $0°$. (**116**)

Index

3-space, 701
6,4,3,4 tessellation, 418
30-60-90 Triangle Theorem, 489–490
30-60-90 triangles, 489–490, 803, 811
45-45-90 triangles, 487–489

A

A-B-C-D Theorem, 252
AA similarity condition, 756–758
AA Similarity Theorem, 757, 772, 774, 788, 807
AAA condition, 382
AAS condition, 383
AAS Congruence Theorem, 389, 392
Abbott, Edwin A., 551
abbreviations of conditionals, 67
absolute value of a number, 8–10
acre, 537
Activities, 7, 21, 23, 38, 45, 46–47, 62, 63, 89, 90, 95, 96, 97, 98, 117, 120–121, 145–146, 147, 151–152, 161, 165, 166, 167, 168, 184, 185, 189, 198, 202, 204, 205, 210, 211–212, 217, 224, 226–227, 269, 278, 283, 289, 290, 291, 310, 311, 316, 317, 324, 332, 335, 339–340, 357, 359, 380, 381, 382–383, 394, 414, 415, 419, 422, 426, 427, 433, 436, 437, 452, 459, 463–464, 470, 480, 482–483, 490, 491, 500, 522, 528, 535, 540, 541, 546, 552, 554, 555, 565–566, 567, 578, 604, 615–616, 617, 649, 650, 684, 694–695, 718, 722, 738, 756, 774, 781, 782, 788, 800, 831, 832, 833, 837–838, 845, 847, 853, 859, 860–861, 862, 863, 866, 872, 875, 880
acute angles, 116, 143
acute triangles, 286, 483
addition, relationship to intersections, 102
Addition Property of Equality, 134
Addition Property of Inequality, 135
Additive Property
   in Area Postulate, 458, 474

in Distance Postulate, 40
in Volume Postulate, 597
Adenas (American Indians), 620
adjacent angles, 126–129
adjacent leg, 801
adjacent sides of polygons, 90
adjacent vertices of polygons, 90
Alberti, Leon Battista, 538
alchemy, 350
Alexander Palace, 728
Alexandria, 553
algebra properties, 133–136
Alice's Adventures in Wonderland (Carroll), 659, 675
alternate exterior angles, 271
Alternate Exterior Angles Theorem, 272
alternate interior angles, 271, 810
Alternate Interior Angles Theorem, 272
altitude
   of pyramid or cone, 534
   of right triangle, 788
   of solid, 527
   of trapezoid, 474
   of triangle, 318, 469
"America" song, 231
Americans with Disabilities Act (ADA), 17
and, as connective, 66
angle(s), 83, 110–111, 114–116, See also *angle measure; right angles; rotations.*
   acute, 116, 143
   addition of, 126
   adjacent, 126–129
   alternate exterior, 271
   alternate interior, 271, 810
   base, See *base angles.*
   central, See *central angle.*
   complementary, 126, 127–128
   congruent, 258
   corresponding, 145–147, 733, 757–758
   dihedral, 522
   duplicating, 437
   formed by
      chords, 837–839, 853–854
      line and plane, 520–521
      secants, 839–840, 854
      tangents, 852–856
      transversal, 270–273
   inscribed in circles, 324–327
   interior and exterior of polygons, 290–292
   linear pair of, 129
   measuring, 114–115

obtuse, 116, 143
reflection symmetry, 309–310
right, 61, 116 (See also *right angles.*)
straight, 114, 116
supplementary, 126, 127–128, 143, 346
symbol, 114
in trapezoids, 345–346
in triangle-sum theorem, 285–286
in triangles, and sufficient conditions for drawing, 381–383
types of, 116
vertex, 129–130
vertical, 129–130
Angle Addition Assumption, 127
angle bisector, 126–127
   in compass, 131
   of parallelograms, 441
   in triangle, 781–783
   of vertex angle in isosceles triangle, 317
Angle Bisector Theorem, 782–783
Angle-Chord Theorem, 838
Angle Congruence Theorem, 258
angle measure
   congruent, 258
   preserved under reflection, 190, 733
   preserved under size transformation, 154–155, 720, 723, 733
   sums in convex polygons, 288–292
Angle Measure Postulate, 115–116, 127
angle of depression, 809–810
angle of elevation, 809–810
angle of incidence, 196
angle of reflection, 196
angle of repose, 132
angle of vision, 858
Angle-Secant Theorem, 839
Angle Symmetry Theorem, 311
Antarctica, 645
antecedents, 66, 139
apex
   of conic solids, 533
   of triangular pyramid, 615
apothem, 491
applications, See also *mathematics in other disciplines.*
   (Applications are found throughout the text. The following are selected

applications.)
alphabets
    Inuktitut alphabet, 214
    letters of alphabet, 308, 313
animals
    birds, 746
    dodo, 82
    elephants, 746, 771
    foxes and rabbits,
        112–113, 120
    horses, 746
    mosquito, 746
    paw prints, area of, 465
    tigers, 99
    zebra size, 675
apartment complex, 209
art
    Bushongo sewn mat, 281
    Fabergé egg, 728
    Navajo rug, 281
    sidewalk chalk art, 95
    similar statues, 764
    string art with DGS, 52
batteries, AAA, 257
bridge problems, 21–24, 31
carousels, 361
cell phone towers, 167, 168
chess, 66
child care facilities area, 461
clock tower in Rome,
    Georgia, 498
computer disk area, 501
containers
    surface areas, 570
    volumes, 598, 599
diamonds, 357
digital photos, 729
dinner party seating, 456
distances, driving, 43, 49, 51
energy consumption in U.S., 692
Ferris wheels, 158
fingerprints, 140, 173
fire hydrants, 307
flags, 377, 729, 815
food
    container volumes, 594–595,
        598, 599
    pie slices and arc length, 495
    pizza costs, 747
    water consumption, 602
    frieze patterns, 367
games
    antique playing cards, 711
    dice, 251
    handball and marble, 742
    Hopscotch, 507
    proofs as, 297

yo-yo tricks, 180
garden trellis, 323
home improvement
    apartment air
        conditioning, 601
    window fan blades, 504
Japanese citizens, 86
kaleidoscopes, 241
ladder and angle, 807, 809
laser beams, 196, 200
mile markers, 10
music (See also *music and
    transformations*.)
    celesta, 235
    chord rotations, 122–123,
        124, 144
    electronic keyboard, 430
    golden ratio in, 814
    marching bands, 216
    metronome, 233
    piano keyboard, 122, 230–231
    round, as puzzle, 239
    Russian balalaikas, 762
    violin and viola, 240
nature
    ants' proportional
        strength, 763
    basalt prism-shaped
        columns, 575
    cactus height, 758
    crystals and symmetry
        planes, 560
    golden ratio, 793, 796, 814
    honeycomb and
        tessellations, 449
    Sitka willow, 228
    snowflakes, 368
    spider web, 377
    tree height, 761, 791
    tree rings, 576
    width of river, estimating, 449
oil
    barrels of crude, 609
    gasoline transportation and
        storage, 631
    pipelines, 7, 198
    reserves, 612
paint required for coverage, 459,
    572–573
paper
    airplanes, 584
    cutting polygons, 877, 887
    drawing Earth on flat
        paper, 584
    folding, 368, 764
    surface area of bag, 570–571
periscope mirrors, 493

photogrammetry, 708
population in Florida, 786–787
printing press, 296
puzzles (See also *logic puzzles*.)
    diamond theft, 687
    mazes, 172
    musical round, 239
    Rubik's cube, 602
    Sudoku, 646–647, 648, 654
    tangram shapes, 101
    tessellations, 449
    "Truth Tellers," 662
    with unions and
        intersections, 88
    using networks, 26
radio station broadcast
    areas, 506
Rube Goldberg machine, 71
Russian nesting dolls, 716–717
sandcastles, 516–517
scale models, 728, 736, 741, 764
sewing, and yards of cloth, 479
space
    Jupiter diameter, 636
    Mars, 644, 892
    Moon, 119, 636
    planet definition, 60
    Venus, 829
    Voyager 1 space probe, 770
sports
    angles and rotations, 110–111
    baseball, 510, 514, 636, 876
    basketball, 102, 338, 450–451,
        645, 737
    billiards, 197, 199, 631
    bowling, 878
    figure skating, 110, 124
    football, 610
    golf, 195
    hockey, 159
    lacrosse field, 485
    miniature golf, 196–197, 241
    round-robin tournaments,
        831–834
    rugby, 892
    shapes of playing fields and
        courts, 448
    soccer, 362, 434, 482
    T-ball, 27
    tennis court, 274, 876
    track and running, 424, 452,
        498, 582
    weightlifting, 749
Springfield, 68
statues, 520, 764, 770, 878
streets, 18, 65, 472, 498, 754
tape measures, 10

U-turns with chalk, 506
water, 602, 614
arc(s), 112–114
  facing away and toward, 855
  intercepted, by inscribed
    angles, 325
  length of, 496
  in networks, 21–23
Arc-Chord Congruence Theorem,
  825, 829
arches, Roman, 613
Archimedes, 500, 628, 632, 638
arctan key, 803
Arctic National Wildlife Refuge, 7
area, 450–451, 457–460
  of circles, 500–502, 505
  of circumscribed and inscribed
    regular polygons, 815
  fundamental properties of,
    457–459
  of irregular figures, 463–464
  map distortions, 507
  multiplication shown by
    rectangles, 603–604
  of parallelograms, 476
  of quadrilaterals, 474–477, 479
  of regular hexagon, 491
  of regular polygons, 490–491
  relationship to perimeter in
    plane, 872–875
  relationship to volume in space,
    880–883
  of right triangles, 468–469, 473
  of sectors, 502
  square unit measurement, 457
  of trapezoid, 474–475
  of triangles, 468–471, 473
  unit conversion, 459–460
  used to calculate probability, 502
Area Postulate, 458, 468
arithmetic mean, 475
Arlington Cemetery, Washington
  D.C., 520
Art of the Fugue, The (Bach), 235
Arthur Ashe Stadium, 876
ASA condition, 383, 388
ASA Congruence Theorem, 388–389
assumed properties, 133, See also
  Postulates.
  of operations, 136
  of points, lines, and planes,
    518–520
assumptions, 32–33, 40–41,
  116–117, 189–190
augmentation, in music, 234
auxiliary figures, 283
average of two numbers, 475

axes, of 3-dimensional coordinate
  system, 700
axes of rotation, 355
axis of cone, 533
Azrieli towers, Tel-Aviv, Israel, 582

B
Babylonians, 113, 481, 808
Bach, Johann Sebastian, 235, 239
Baldwin, John Jr., 124
Bank of China Tower, Hong
  Kong, 846
barrel, 609
Bartók, Béla, 234–235
base(s)
  of conic solids, 533
  of cylindrical solids, 526
  of isosceles triangles, 316
  of trapezoid, 334
base angles
  of isosceles triangles, 316
  of trapezoid, 334
base edges of pyramids, 533
Beethoven, Ludwig van, 234, 367

Berra, Yogi, 675
between, 38, 79
betweenness, 38, 40
  preserved under reflection,
    190, 733
  preserved under size
    transformation, 720, 722–
    723, 733
Bhaskara (Indian
  mathematician), 481
biconditionals, 78–79, 148
bilateral symmetry, 560
bisect, 81
bisector
  of angle (See angle bisector.)
  of segment, 81, 165 (See also
    perpendicular bisector.)
bounded regions, 525
box, 564, 596, 597, 601, 604–605
Box Diagonal Formula
  Theorem, 703
Box Volume Formula, 597
Brewster, Sir David, 241
"Bridge of Fools," 440
Brunelleschi, Filippo, 538
Bury, Thomas Talbot, 539
bushel, 621
bye, in tournament, 831–833

C
calculator, finding tangent ratio,

802, 803
Carroll, Lewis, 653, 659, 675
Carysfort Reef Lighthouse, 620
Cavalieri, Bonaventura, 611, 614
Cavalieri's Principle, 611, 616, 617,
  628–630
center
  of circle, finding, 859, 864
  of circumcircle, 860
  of gravity, 862
  of incircle, 860
  of regular polygon, 357–358
  of rotation, 121, 212
  of size change, 152–153, 718,
    719–720
  of sphere, 550
  of symmetry, 351
Center of a Regular Polygon
  Theorem, 358
central angle
  of chord, 824
  of circle, 114, 115
centroid of triangle, 861–862, 888
Chapter Review, 55–57, 106–109,
  178–181, 246–249, 302–305,
  373–377, 445–449, 511–515,
  589–593, 642–645, 713–715,
  768–771, 819–821, 893–897,
  See also Review questions.
characteristics, in conditionals,
  79–80
Chasles, Michel, 183
Child Nutrition Promotion and
  School Lunch Protection
  Act, 59
Chinese, ancient, 481
Chord-Center Theorem, 825
chord of arc, 824
  angles formed by, 837–839,
    853–854
  length of, 826–827, 866
  properties of, 825–826
chords, in music, 122–123, 124, 144
Christmas Island, Australia, 463–464
circle(s), 79–80, 822–823, See also
  chord of arc; trigonometric
  ratios.
  angles inscribed in, 324–327
  arc length, 496
  arcs and semicircles, 113–114
  area of, 500–502, 505
  associated with triangles
    circumcircle, 860
    incircle, 860–861
    nine-point circle, 863–864
    points of concurrency,
      861–862

circumference of, 494–495
congruent, 824–825
congruent arcs, 327, 329,
    824–825
conic sections, 556
constructing kites with, 339–340
equations for, 682–685, 703
reflection symmetry of, 311
regular polygons and, 357–360
sections of spheres, 551, 557
sectors of, 501
tangent line, 686
unique circle theorem, 282–283
unit circle, 686
Circle Area Formula, 501
Circle Circumference Formula, 494
Circle Symmetry Theorem, 311
circular arcs, 825
circular motion, 886
circularity, 28, 51
circumcenter, 860, 888
circumcircle of triangle, 860, 861
circumference of circle, 494–495
circumradius, 860
circumscribed about a circle, 815
circumscribed about triangle, 860
clockwise direction, 120
clockwise orientation of polygon,
    191–192
College Board, 692
collinear points, 28–29
collinearity
    preserved by isometry, 223
    preserved under reflection,
        190, 733
    preserved under size
        transformation, 154, 720,
        722–723, 733
Columbo, 621
common tangents, 848
compass
    angle bisectors, 131, 783
    labeling of points of, 118
    permitted in construction, 166
complementary angles, 127–128
complements, 128, 132
complete graph on *n* vertices, 119
compliment, 132
composing transformations, 202
composite, 203
    of reflections, as similarity
        transformation, 732
    of transformation in music, 235
composite numbers, 665
composition
    notation of, 203
    of reflections over intersecting

lines, 210–212
    of reflections over parallel lines,
        202–203
compound statements, 66
concave lens, 62
concentric circles, 411
conclusion, 66, 139–141, 270
Conclusion, Final, 664
concurrent lines, 152, 224, 861
conditional statements, 66–67
    congruent segments and
        length, 257
    instance of, 68
    in programming languages, 70
    proof of, 139, 269
    truth or falsity of, 68–69
    Venn diagrams for analysis, 69
cones, 533, 534
    construction of, 537
    nets for, 567
    plane sections of, 555
    surface area of, 577, 579–580
    volume of, 615–619
congruence, See also *tessellations.*
    counterpart to similarity, 759
    equality and, 260
    equivalence properties of,
        254–255
    genetics and, 296
    justifications for, 265
    literacy and, 296
    trivial, 311
congruence proofs, 250–251. See
    also *proofs*
    congruent figures in (See
        congruent figures.)
    justifications for, 265
    one-step, 263–264
    using reflections, 277–279
    using sums of angle measures in
        polygons, 288–292
    using transitivity, 269–273
    using uniqueness and auxiliary
        figures, 282–286
Congruence Property
    in Area Postulate, 458
    in Volume Postulate, 597
congruence theorems for triangles,
    386–390, 393–396
congruence transformations,
    182–183, 253, See also
    composition; isometries;
    transformations; translations.
    reflecting figures, 189–192
    reflecting points, 184–186
congruent angle measure, 258
congruent angles, 258, 265

congruent arcs, 327, 329, 824–825
congruent circles, 824–825
congruent figures, 250–251, 252–255
    compared to similar figures, 731
    corresponding parts, 257–260
    in space, 559
congruent length, 257–258
congruent segments, 257–258, 265
congruent triangles, 378–379
    compared to similar
        triangles, 750
    congruence of parts, 395–396
    drawing triangles, 380–383
    HL condition, 408–409
    overlapping triangles, 400–402
    parallelograms, 419–422,
        426–429
    proving validity of constructions,
        436–437
    in roof trusses, 380, 381
    SSA condition, 406–408
    sufficient conditions for
        congruence, 381–383 (See
        also *sufficient conditions.*)
    using congruence theorems,
        393–396
conic sections, 556
conic solids, 532, 533
conic surface, 532, 624
conjectures, 95–98
connectives, in statements, 66
consecutive sides of polygons, 90
consecutive vertices of
    polygons, 90
consequents, 66, 139, See also
    conclusion.
Construct, 166
construction, 166
    duplicating an angle, 437
    of perpendicular bisector of
        segment, 166, 167
    of polygon with large perimeter,
        877, 878
    proving validity of, 436–437
contractions, 234, 719
Contradiction, Deduction to, 664
contradictory statements, 664
contrapositive of the original,
    657–658
convenient location for a figure,
    671–673
**Converse of Isosceles Triangle Base
    Angles Theorem**, 320–321
Converse of Pythagorean Theorem,
    482–484, 753
**Converse of Triangle Side-Splitting
    Theorem**, 776

converses, 72
  of the original, 657–658
  truth or falsity of, 73–74
conversions
  of area units, 459–460
  line in standard form to slope-
    intercept form, 15–16
  of units in liquid volumes, 598
convex lens, 62
convex polygons, 91, 288–292
convex polyhedron, 564
convex quadrilateral, 667
convex sets, 62–63, 116
coordinate geometry, 13
  coordinatizing space, 700–702
  proofs in, 670
  unions and intersections, 84
coordinate plane
  convenient locations in, 671–673
  distance on, 676–677
  Pythagorean distance formula
    on, 677–679
  vectors in, 218–219
coordinate proofs, 646, 670–673,
  See also *indirect proofs*.
coordinate system, and box, 608
coordinates, 8
coordinatized line, 8, 9
coordinatizing space, 700–702
coplanar points, 28, 518
Córdoba dome, 356, 357
corollary, 347
corresponding angle measures, 740
corresponding angles, 145–147, 733,
  757–758
Corresponding Angles Postulate,
  146, 160, 439
corresponding areas, 740
corresponding lengths, 733, 740
Corresponding Parts in Congruent
  Figures (CPCF) Theorem,
  259, 395
corresponding parts of congruent
  figures, 257–260
corresponding perimeters, 740
corresponding sides, 758
corresponding surface areas, 740
corresponding vertices, in
  overlapping triangles, 401
corresponding volumes, 740
cosine of an angle, 807–811
counterclockwise direction, 120
counterclockwise orientation of
  polygon, 191–192
counterexample to conditional,
  68–69
CPCF Theorem, 259, 395

Crichton, Michael, 886
cross sections, 551–552
cube(s), 527
  as hexahedron, 564
  net for, 583
  perspective drawing of, 538
  plane sections of, 554–555, 584
  Rubik's cube, 602
  volume of, 599–600
cube root, 599
Cube Volume Formula, 599
cubic units, 596
cubit, length measure, 642
cyclic quadrilateral, 330, 840
cylinders, 525–528
  nets for, 566
  plane sections of, 554–555
  thickness of hollow shapes, 626
  volume of, 609–612
cylindrical solids, 526
  lean in, 530
  surface area of, 571–573
  and surfaces, 526–527
cylindrical surface, 526, 528
  formulas for area and
    volume, 624

D

Da Vinci, Leonardo, 481, 821
De Divina Proportione (Da Vinci), 821
decagon, 90
deduced properties, 133, See also
  Theorems.
deduction of formulas, 623
Deduction to Contradiction, 664
defining properties, 133
definitions
  of geometric terms, 60–62, 77–80
  properties of, 77
  using undefined terms, 28–30
degrees, 113, 116
derivation of formulas, 624–625
Descartes, René, 13, 76, 647
detachment, law of, 654–655
DGS (dynamic geometry system),
  44–45
  animation capabilities of, 51
  construction of equidistant
    points, 166
  construction of parallelogram
    and angle bisectors, 441
  construction of
    quadrilaterals, 433
  drawing triangles, 381–383
  dynamic capabilities of, 46–47
  Euler line relationships, 888
  rotations, 122

star polygons, angle measures
  of, 297
using to test conjectures, 97–98
diagonals
  as bisectors, 431–432
  in box, longest, 703
  of isosceles trapezoid, 678–679
  of kites, 341
  number in *n*-gon, 833–834
  of polygons, 90
  of quadrilaterals, 98, 427,
    431–433, 476–477
  of rhombuses, 342
  of square, 671
Diagonals in a Polygon
  Theorem, 834
diameter, 79, 550
Dido, Queen of Carthage, 885
dihedral angles, 522
Dijkstra's algorithm, 51
dilation/dilatation, 719
Dimension Assumption, 33, 519
diminution, in music, 234
direct proofs, 646, 663
direct reasoning, 663
direction of translations, 205
direction of vector, 216
directly congruent, 253
discrete geometry, 4, 20, 24
discrete lines, 20
discrete points, 20
disproofs, 96
distance, 40–41
  on coordinate plane, 676–677
  estimating distance with
    thumb, 761
  to horizon, using tangent,
    846–847
  multiplication of, under size
    transformation, 721, 723
  on a number line, 8–10
  between parallel planes, 521
  to a plane from a point, 521
  preserved by isometry, 223
  preserved under reflection,
    190, 733
  between two parallel lines, 421
  between two points, 9
Distance between Parallel Lines
  Theorem, 421
Distance Formula, 40
distance formula, in space, 702–703
Distance Formula, Pythagorean,
  676–679
Distance Postulate, 40

Distributive Property of Multiplication over Addition, 136, 605
dodecagon, 90, 361
dodecahedron, 565, 583
domes
    Córdoba dome, 356, 357
    geodesic, 883, 887
    Star Ribbed, 356
dots, 19–20
Double Negative Property, 657, 658
Doyle, Sir Arthur Conan, 648
Drawing, 269
drawings
    assumptions from, 116–117
    figures with rotation symmetry, 353
    isometric, 546
    multiplication shown by area and volume, 603–605
    in perspective, 538–541
    reflection image of a figure, 190–191
    of three-dimensional coordinate system, 701
    translation images given a vector, 218
    triangles, 380–383
    of two-dimensional coordinate system, 701
    views of solids and surfaces, 544–546
dynamic geometry system, See *DGS (dynamic geometry system)*.

E

Earth, See also *applications, space.*
    ancients' beliefs of Earth's shape, 550
    circumference of, 553, 557
    drawing sphere on flat paper, 584
    equator, 499, 552
    as oblate spheroid, 558
    orbit around Sun, 515
    radius, diameter of, 636, 637, 645
    as a sphere, 552–553
    surface area of, 634
    volume of layers of, 630
edge(s)
    of dihedral angle, 522
    of polyhedra, 564
    of surface, 526
Egyptians, ancient measuring system, 484

Einstein, Albert, 653
Elements (Euclid), 32, 269, 284, 408, 440, 665, 787
elevations, 545–546
ellipses, 556
empty set, 84
endpoints, 21, 39, 113
ends of a kite, 340
envelopes of lines, 52
Equal Angle Measures Theorem, 128
equal sign, 134
equality, and congruence, 260
Equality and Inequality Postulates, 136
Equality and Operations Postulates, 134
Equality Postulates, 133–134
equation
    for circle, 682–685, 703
    for planes, 708
    for sphere, 703–704
Equation for a Circle Theorem, 682
Equation for a Sphere Theorem, 704
Equation to Inequality Property, 136
equator of Earth, 449, 552
equiangular polygon, 357
equidistant points, 165
equilateral polygon, 357
equilateral triangle, 91–92, 319
    constructing hyperbolic geometry, 440
    Morley's theorem, 441
    trisected by two rays, 785
Equivalence Properties of Congruence Theorem, 254
equivalent equations, 727
equivalent frieze patterns, 364
Eratosthenes, estimate of Earth's circumference, 553, 557
Escher, Maurits C., 182, 241, 584
Escher kaleidocycles, 440
Escher tessellations, 182, 241, 440
Euclid, 32, 37, 269, 284, 747, 867
Euclidean geometry, 32
Euclidean plane coordinate geometry, 4, 13, 24, 162
Euclidean synthetic geometry, 4, 6, 20, 24, 162, 670
Euler, Leonhard, 21–23, 183
Euler line, 888
Euler's Formula, 568
Europeans in Middle Ages, belief of Earth as flat, 550
even isometries, 228
even nodes, 23
exact location, 8, 28
expansions, 234, 719

Exploration questions, 12, 18, 26, 31, 37, 43, 50, 65, 71, 76, 82, 88, 94, 100, 119, 125, 132, 138, 144, 150, 158, 164, 171, 188, 195, 201, 209, 215, 222, 229, 240, 256, 262, 268, 276, 281, 287, 295, 315, 323, 330, 338, 344, 350, 355, 362, 367, 385, 392, 399, 405, 412, 418, 425, 430, 435, 439, 456, 462, 467, 473, 479, 486, 493, 499, 505, 524, 531, 537, 543, 549, 558, 563, 569, 576, 582, 602, 608, 614, 622, 627, 632, 637, 653, 662, 669, 675, 681, 687, 693, 699, 707, 725, 730, 737, 744, 749, 755, 763, 780, 785, 792, 799, 806, 813, 830, 836, 843, 851, 858, 865, 871, 878, 885
exterior
    of angle, 116
    of polygonal region, 91
    of surface, 525
Exterior Angle Theorem for Triangles, 290, 839
exterior angles, 270–272
exterior angles of polygons, 290–292
externally tangent circles, 848
extremes, in a proportion, 726–727

F

face(s)
    of polyhedra, 564
    of pyramids, 533
    of surface, 526
falsity
    of conditionals, 68–69
    of converses, 73–74
Federal Trade Commission (FTC), 59
Federation of International Basketball Associations (FIBA), 338
Fenway Park, Boston, 510
Fermat, Pierre de, 13
Fibonacci, Leonardo, 799
Fibonacci sequence, 486, 799
fields of vision, 112–113, 120, 138, 201
figure(s), 28, 29
    drawing reflection images, 190–191
    irregular, area of, 463–464
    orientation of, 191–192
    perimeter and area relationship, 872–875
    reflecting, 189–192

surface area and volume
relationship, 880–883
symmetric (See *reflection
symmetry.*)
Figure Size-Change Theorem, 723
Figure Skating Championships,
110, 124
Figure Transformation Theorem,
191, 279
Final Conclusion, 664
fingerprints, 140, 173
Fitzgerald, James, 651
five-pointed star, 843
flat notes, 235
Flatland, 551, 552, 554
Flatland: The Movie, 551
Flatland (Abbott), 551
Flip-Flop Theorem, 309
flip transformation, 183, 220
foot
of the altitude, 862
measurement definition, 109
of segment in a plane, 521
Formulas
box diagonal, 703
box volume, 597
circle area, 501
circle circumference, 494
cube volume, 599
distance, 40
distance, Pythagorean, 676–679
Euler's, 568
lateral area
for regular pyramids and
right cones, 579
for right cylindrical solids, 571
midpoint
on a number line, 688–689
on the plane, 689–690
parallelogram area, 476
perpendicular diagonals
quadrilateral area, 477
prism-cylinder volume, 611
pyramid-cone volume, 617
Pythagorean distance
on the coordinate plane,
677–679
equation for circle, 682
quadratic, 795
rectangle in area postulate, 458,
459–460
regular polygon area, 491
remembering and organizing,
623–625
right triangle area, 468
sphere surface area, 634
sphere volume, 629–630

surface area
for cylindrical solids, 572
for pyramids and cones, 577
three-dimension distance,
702–704
three-dimension midpoint, 704
trapezoid area, 475
triangle area, 469–470
fractals, 886
"Frère Jacques," 238
frieze patterns, 363–367
Frobisher, Martin, 879
frustums
of cones, 595, 619, 620
of polyhedra, 569
of regular pyramid, 698
fundamental region, 413, 415–416
Fundamental Theorem of Similarity
alternate statement, 740
first statement, 739, 745–747

G

Galileo Galilei, 747, 822
gallon, 598
Gardner, Martin, 88
Garfield, James, 105, 481
Gauss, Carl, 368
**General Angle-Arc Measure
Theorem,** 855–856
general quadrilaterals, 336
geodesic houses/domes, 883, 887
Geomagnetic North Pole, 507
geometric construction, 166, See
also *construction.*
geometric mean, 788–789, 814
Geometric Mean Theorem, 787
geometry, See also *logic of geometry;
non-Euclidean geometries.*
algebra properties used in,
133–136
Euclidean, 32
Euclidean plane coordinate, 4,
13, 24, 162
Euclidean synthetic, 4, 6, 20, 24,
162, 670
hyperbolic, 285, 440
spherical (See *spherical
geometry.*)
terms defined, 61–63, 77–80
types of, 4–5, 20–24
Georgia Aquarium, 613
giants, 745–746, 748
**Giant's Causeway, Northern
Ireland,** 575
Given, 166, 269
glide reflections, 183, 225, 228, 229

frieze pattern symmetry,
364–365
inversion in music, 234–235
golden ratio, 793–796, 797, 813, 814
golden rectangle, 795–796
graceful network, 26
grade, in roads, 145
graph theory, 4, 20–24, 119
graphing equations, 708
graphs (networks), 21
great circles, 551, 552
Greeks, ancient
individuals
Archimedes (See *Archimedes.*)
Eratosthenes, 553, 557
Euclid (See *Euclid.*)
Hero (Heron), 473
Herodotus, 537
Hippocrates, 506
Pappus, 100
Proclus, 284
Ptolemy, Claudius, 808
Pythagoras (See *Pythagoras.*)
Thales, 269, 326
knowledge of
belief of Earth as round, 550
constructing a square, given
a cube, 661
Euclidean geometry, 6
finding unknown lengths, 808
golden rectangle, 795–796
on orthocenter, 863
sine and cosine
abbreviations, 808
tools permitted in
construction, 166
transformations and, 183
Greenland, map distortions, 507
guess, educated, See *conjectures.*
*Guinness Book of World
Records,* 745
Gulliver, 745–746
Gulliver's Travels (Swift), 745–746
Gutenberg, Johannes, 296

H

Hagia Sophia, Istanbul, 631
Haldane, J. B. S., 764
half-cylinders, 548
Hamiltonian Circuits, 52
Hamlisch, Marvin, 239
harmonic mean, 730

Haydn, Franz Joseph, 236
Hegel, William, 230

height
of pyramid or cone, 534
of solid, 527
of trapezoid, 474
of triangle, 469
Heighway dragon fractal, 886
hemispheres, 551, 635
Henin, Justine, 124
heptagon, 62, 90
heptagonal triangle, 493
Hero (Heron), 473
Herodotus, Greek historian, 537
Hero's (Heron's) formula, 473
hexagon, 90, 833
regular, 356, 362, 415, 491
hexahedron, 564
hierarchy
of conic surfaces, 534
of cylindrical surfaces, 528
general angle-arc measure
theorem, 856
of quadrilaterals, 332, 335,
342–343, 344
of three-dimensional figures, 585
of transformations, 732
of triangles, 91, 369
Hilbert, David, 28
Hippocrates, 506
Hippocrates' Theorem, 506
HL condition, 408–409
**HL (Hypotenuse-Leg) Congruence
Theorem**, 408–409
Holmes, Sherlock, 263, 648
homonyms, in math, 132
horizontal component of vectors,
218–219
horizontal lines, 13–14
frieze pattern reflection
symmetry over, 364–365
in perspective drawings, 539, 540
slope of, 147
horizontal reflection, See also *glide
reflections.*
inversion in music, 234–235
horizontal translation on piano, 231
hyperbolas, 556
hyperbolic geometry, 285, 440
hypercube, 31
hypotenuse, of right triangle,
375, 408
**Hypotenuse-Leg (HL) Congruence
Theorem**, 408–409
hypothesis, 66, See also *antecedents.*

icosahedron, 565, 583
identity transformation, 719

if ... then, as connective, 66
if and only if, 78–79
image, 121
mapping preimage onto, 151–152
reflections of figures, 189–191
implication, law of transitivity,
655–656
Inca, 345, 348
incenter of triangle, 861
incircle of triangle, 860–861
included angle, 382
included side, 388
India, ancient, 481
Indian mound, 620
indirect proofs, 646, 663–667
infinitude of primes theorem,
665–666
laws of logic, 654–659, 663
line intersection theorem,
664–665
proofs with coordinates, 670–673
Pythagorean distance formula,
676–679
ruling out possibilities, 648–
650, 651
scalene triangles, medians and
altitudes, 666–667
indirect reasoning, 647, 663
**Inequality, and Equality,
Postulates**, 136
**Inequality and Operations
Postulates**, 135
inequality properties, 135, 136
inequality sign, 134
Infinitude of Primes Theorem,
665–666
initial point of vector, 217
**Inman Square, Cambridge,
Massachusetts**, 754
Inoue, Rena, 124
inradius, 861
Inscribed Angle Theorem, 325–236,
327, 839, 854
inscribed angles, in circles, 324–327
inscribed in a circle, 815
inscribed in triangle, 861
inscribed pentagon, 843
inscribed polygon, in circle,
358–359, 815
instance of a conditional, 68
intercepted arcs, 325
interior
of angle, 116
of circle, 81
of polygonal region, 91
of surface, 525

interior angles, 270–272
of polygons, 290, 327
internally tangent circles, 848
**International Astronomical
Union**, 60
International Space Station, 736
international system of units, 109
**intersecting circles in kite
construction**, 339–340
intersecting lines, 34, 210–212
Intersection Assumption, 519
intersection of two sets, 83, 102
**intersection points, in reflections
over three lines**, 223–225
intersections in triangles, 90
intersections of figures, 83–85, 114
Inuit village, 879
Inuktitut alphabet, 214
Inuyasha **cartoon**, 569
inverse of the original, 657–658
inverse tangent key, 803
inversion, in music, 234–235
Inversions **(Kim)**, 368
irregular figures, 463–464
isometric drawings, 546
isometric graph paper, 546
isometries
A-B-C-D property, 252
reflections over three lines,
223–225
types of, 226–227, 228
isoperimetric, 873, 879
isoperimetric inequality
in three dimensions, 879–883
in two dimensions, 872–875
**Isoperimetric Inequality for the
Plane**, 873, 885
**Isoperimetric Theorem for
Space**, 880
alternate statement, 881
**Isoperimetric Theorem for the
Plane**, 873
alternate statement, 875
isosceles right triangle, 487–489
**Isosceles Right Triangle
Theorem**, 488
**Isosceles Trapezoid Symmetry
Theorem**, 346–347
Isosceles Trapezoid Theorem, 347
isosceles trapezoids, 334–335
diagonals of, 678–679
right triangles in, 385
symmetry of, 346–347
used in buildings, 345, 349
isosceles triangles, 91
convenient location for, 672
inscribed in circle, 825

parts of, 316
reflection symmetry of, 317–319
in regular pyramid, 534
Isosceles Triangle Base Angles Converse Theorem, 320–321
Isosceles Triangle Base Angles Theorem, 318–319, 440, 487
Isosceles Triangle Coincidence Theorem, 318
Isosceles Triangle Symmetry Theorem, 317–318
Iwo Jima Memorial statue, 520

J

James, LeBron, 748
James Bond movie, 239
Japan, theorems from, 481, 887
John Hancock Center, Chicago, 345, 349
junk food, 58–59
Jurassic Park (Crichton), 886
Jurassic Park fractal, 886
justifications, 139–142, 263, 265, 270

K

Kagel, Mauricio, 239
Kakeya, Soichi, 506
Kakeya problem, 506
Kaleidocycles, 440
Kepler-Poinsot polyhedra, 583
Kepler's problem, 638
key points, 672
Kim, Scott, 368
Kite Diagonal Theorem, 341, 342, 431
Kite Symmetry Theorem, 340–341, 391
kites, 333–334, 335, 339–341, 672
Königsberg Bridge problem, 21–24
Kreisel tank, 614

L

Lake Okeechobee, area of, 465
Lateral Area Formula for Regular Pyramids and Right Cones, 579
Lateral Area Formula for Right Cylindrical Solids, 571
lateral areas
formulas for, 623–624
of regular pyramids, 577–579
lateral edge(s)
of cones, 533
of prism, 527

of pyramids, 533
lateral face(s)
of prism, 527
of pyramids, 533
lateral surface, 527
of cones, 533
of prism, 527, 571
Latin square, 646
lattice points, 467, 505, 687
Law of Detachment, 654–655
Law of Indirect Reasoning, 663
Law of Ruling Out Possibilities, 648
Law of the Contrapositive, 658
Law of Transitivity (of Implication), 655–656
laws of logic. See *logic, laws of*
lean, in cylindrical solids, 530
Leaning Tower of Pisa, 813
legs, of right triangle, 408, 801
length, 450–451
of arcs, 496
of chord of arc, 826–827
circumference of circle, 494–495
congruent, 257–258
cubits, 642
finding unknown, with sine and cosine, 808–809
perimeter of figures, 452–454 (See also *perimeter.*)
of secants, 867–868
of sides in triangles, 487–490 (See also *Pythagorean Theorem.*)
in similarity theorems, 739–740
in size transformations, 155–156
stade (unit), 553
of tangent from point to circle, 868–869
using tangent to find, 802–803
Liberty Bell, 588
limit in estimating area, 464
limit of area of parallelogram, 501
line(s), 4–5, 28, See also *transversal line(s).*
assumed properties, in space, 518–521
description of, 8, 14, 20, 22
intersecting, 34
as locations, 6–10
parallel, 35, 145–148, 162 (See also *parallel lines.*)
perpendicular, 159–162
perpendicular bisector (See *perpendicular bisector.*)
in Point-Line-Plane Postulate, 32–33
slope-intercept form of equation

of, 14–16
slopes of, 147
in spherical geometry, 296
standard form of equation of, 13–14
tangent to circle, 686
Line Intersection Theorem, 34, 664–665
line of reflection, 184
Line-Plane Perpendicular Theorem, 521
line segment, 39, 296
linear motion, 886
linear pair, 129
Linear Pair Theorem, 129, 141
liquid volume, 598–599
liter, 598
Lobachevski, Nikolai Ivanovich, 652
logic, laws of, 654–659, 663
logic of geometry
conditional statements, 66–69 (See also *conditional statements.*)
conjectures, 95–98
converses, 72–74
definitions, 60–63, 77–80
polygons, 62–63, 89–91 (See also *polygons.*)
unions and intersections of figures, 83–85
logic puzzles, 649–650, 652, 653, 659
logically equivalent, 658
London Eye Ferris wheel, 505
Loomis, Elisha, 481
Louvre art museum, Paris, France, 532–533
Loyd, Sam, 88
lune, quadrature of, 506

M

Machu Picchu, 348

Magellan, Ferdinand, 550
magnitude
of rotation, 121, 212
of size change, 153, 718–719
of translations, 205
of vector, 216, 219
major arcs, 113, 115
mappings, 152
maps
Alberta, Canada, law of contrapositive, 658
contour, 708
drawing Earth on flat paper, 584

Greenland distortions, 507
Nebraska, 64
of preimage onto image, 152
Voronoi diagrams and high
    schools, 172
zoom and size
    transformations, 172
Marino, George, 6
Mathematical Puzzles of Sam Loyd
    (Gardner), 88
mathematics in other disciplines,
    See also *applications.*
    (Examples of math in
    other disciplines are found
    throughout the text. The
    following are selected
    examples.)
agriculture
    farm building shapes, 548,
        581, 602
    grain in conical piles, 132,
        621, 626
business, corporate logos, 368
communications, fiber optic
    cables, 241
construction
    area of octagonal base of
        gazebo, 487, 491, 492
    blueprints, 736, 737
    bricks for building, 607
    bridge supports, 400
    circular saw blades, 502
    concrete measure, 627
    domes and regular polygons,
        356, 357
    fences, 274, 454
    golden rectangle in buildings,
        793, 795, 796
    light fixtures, 305
    lighthouse design, 620
    masons at work, 385
    measuring electric wires, 399
    nuts, hexagonal and square,
        306–307
    patterns in Egyptian tomb, 376
    quadrilateral foundation, 434
    radio tower guy wires, 405
    right angle measures, 486
    roof shapes and pitch, 581, 737
    scale models, 736
    siding on building, 777
    skyscraper shapes, 345,
        349, 582
    temple shapes, in Central
        America, 348, 698
    trusses, in roofs, 380, 381
    yurts and tipis, 877

entertainment
    aquarium tanks, 613, 614
    carousel, 608
    Ferris wheel in London, 505
    speaker coverage, 842
human biology
    blood donation, 601
    giants, 745–746, 748
    human DNA and
        congruence, 296
    human faces, 315
    humans' bilateral
        symmetry, 560
    plane sections, 554
manufacturing
    product packaging, 594–595
    square nut's surface area, 573
transportation
    airplanes, 339, 544, 814, 858
    airports, 118, 165, 167, 277
    bicycle wheel
        circumferences, 494, 495
    long-range navigation
        (LORAN), 556
    parallel roads, 305
    police helicopter, 810
    speed limit, 87
    subway system, 85
    teenage driver, 668
McKellar, Danica, 408
mean, geometric interpretation of,
    688–689
mean temperature (outdoors), 691
means, in a proportion, 726–727
Means-Extremes Property Theorem,
    727, 786, 795
measure of an angle, 114–115
measure of an arc, 113
measures, in music notation, 231
medial triangle, 697
median of a triangle, 318, 693
Meissner, Kimmie, 110
Menger sponge, 576
Mental Math, 6, 13, 19, 27, 32, 38,
    44, 60, 66, 72, 77, 83, 89, 95,
    112, 120, 126, 133, 139, 145,
    151, 159, 165, 184, 189, 196,
    202, 210, 216, 223, 230, 252,
    257, 263, 269, 277, 282, 288,
    308, 316, 324, 331, 339, 345,
    351, 356, 363, 380, 386, 393,
    400, 406, 413, 419, 426, 431,
    436, 452, 457, 463, 468, 474,
    480, 487, 494, 500, 518, 525,
    532, 538, 544, 550, 559, 564,
    570, 577, 596, 603, 609, 615,
    623, 628, 633, 648, 654, 663,

    670, 676, 682, 688, 694, 700,
    718, 726, 731, 738, 745, 750,
    756, 774, 781, 786, 793, 800,
    807, 824, 831, 837, 844, 852,
    859, 866, 872, 879
menu, in DGS, 45
Metric System, of liquid volume, 598
Miamisburg Mound, Ohio, 620
Midpoint Formula on a Number Line
    Theorem, 688–689
Midpoint Formula on the Plane
    Theorem, 689–690
midpoint of a segment, 64, 77–78
    formula for, 704–705
    mean, geometric interpretation
        of, 688
    in two dimensions, 689–691
midsegment of trapezoid, 694–695
Midsegment of Trapezoid
    Theorem, 695
midsegment of triangle, 694
Midsegment of Triangle Theorem,
    695–696
Millennium Biltmore Hotel, Los
    Angeles, 461
Ming, Yao, 42
minor arcs, 113, 115, 825
mirrors
    bisecting a segment, 559
    car, 305
    in periscope, 493
    reflection images, 184, 196,
        201, 207
models, See also *scale models.*
    for multiplication of numbers,
        603, 604
Mongolian nomads, 877
"Moonlight Sonata"
    (Beethoven), 367
Moors, and tessellations, 416
Morley, Frank, 441
Morley's theorem, 441
Mosque of Córdoba, 356, 357
mosque tessellations, 416
Mount Everest, 897
movable drawing, See *DGS*
    (dynamic geometry system).
Mozart, Wolfgang Amadeus, 236,
    239, 367
multiplication
    of lengths in size
        transformations, 155–156
    relationship to unions, 102
    shown by area and volume,
        603–605
Multiplication Property of
    Equality, 134

Multiplication Property of
Inequality, 135
Murphy's Law, 109
music and transformations, 230–237, 241
augmentation and diminution, 234
changing tempo, 233–234
composites of transformations, 235
frieze patterns, 367
inversion, 234–235
notes of Western music, 230
other transformations, 236
retrograde, 235–236
rounds, 232–233
transposing in music, 231
"Music for Strings, Percussion, and Celesta" (Bartók), 235
Muttart Conservatory, Alberta, Canada, 618
"My country 'tis of thee," 234

N

n-fold rotation symmetry, 351–352, 358–359
n-gon, 90, See also polygons.
regular polygons, 356
sums of angle measures in, 288–295
symmetry lines in, 350
Name, 166
National Aeronautics and Space Administration (NASA), 815
National Collegiate Athletic Association Regulations, 450
National Hockey League, 159
natural notes, 235
negation of statement, 656–657
nets, for three-dimensional figures, 565–567, 568, 583, 633
network, 21
graceful, 26
hierarchy of quadrilaterals as, 344
Neuschwander, Cindy, 886
New York City bridge problem, 31
Nicholas II, Tsar of Russia, 728
nine-point circle, 863–864, 888
Nirmalan, G., 771
"Nobody Does It Better," 239
nodes, 21–23
Noether, Emmy, 652
non-Euclidean geometries, 285, 296
hyperbolic, 285, 440
spherical, 146, 285, 296, 416, 507
nonagon, 90, 833

nonconvex polygon, 91
nonconvex set, 63, 116
nonoverlapping figures, 400
North Pole, of Earth, 552–553
not-p, 656–657
notes, in music, 230–233, 235
null set, 84
number, between two other numbers, 38
Number Line Assumption, 33, 519, 520
number lines, 8–10, 688
number patterns, 799

O

oblate spheroid, 558
oblateness of a planet, 558
oblique cones, 533, 580
oblique cylinders, 610
oblique lines, 13–14
oblique pentagonal pyramid, 577
oblique prisms, 610
oblique pyramids, 533
oblique solid, 527
obtuse angles, 116, 143
obtuse triangles, 286, 483, 484
octagon, 90, 487, 491, 492
octahedron, 565, 583
octave, musical, 230
odd isometries, 228
odd nodes, 23
On Two New Sciences (Galileo), 747
one-dimensional figures, 29, 31
one-dimensional lines, 13
one-dimensional size transformations, changing tempo in music, 233–234
one-dimensional space, 29
one-dimensional translation, transposing in music, 231
one-point perspective, 539–540, 542
one-step congruence proofs, 263–264
opinions, See conjectures.
opposite angles, of cyclic quadrilateral, 840
opposite leg, 801
opposite rays, 39–40
opposite reciprocals, 161
oppositely congruent, 253
Optics (Euclid), 37
or, as connective, 66
ordered pairs of real numbers, 13–16
ordered triple, 701
orientation of figures, 191–192
orthocenter of triangle, 862, 888

overlapping figures, 400
overlapping triangles, 400–402

P

pairing, 832–833
palindromes, 235–236
Pappus, Greek geometer, 100
parabolas, 556
parallel lines, 35, 145–148, 162
composing reflections over, 202–203
distance between, 421
slopes of, 147–148
summary of postulates and theorems, 273
transitivity theorem, 148
translation over, 204
Parallel Lines and Slopes Theorem, 148, 670
Parallel Lines Theorem, 271
parallel planes, 521
parallel property, size change theorem, 153–154
parallel segments and rays, 146
Parallelogram Area Formula, 476
Parallelogram Symmetry Theorem, 422
parallelograms, 149, 331–333
area of, 476, 501
convenient location for, 672
diagonals of, 419, 422, 427
in hierarchy of quadrilaterals, 332, 335
from midpoints of quadrilateral, 699
proof, in coordinate geometry, 670–671
properties of, 419–421
as rhombus, 343
rotation symmetry of, 421–422
sufficient conditions for, 426–429
parallels
Playfair's parallel postulate, 284–285
uniqueness of parallels theorem, 283–284
Parthenon, on the Acropolis, 796
patterns, See also frieze patterns.
inscribed quadrilaterals, 412
repeating, rotations of musical chords, 122–123
pentagon, 90, 291–292, 327
regular, 524
tessellations of, 415, 418
Pentagon Building, Washington, D.C., 350
pentagonal prism, 528, 612

perimeter, 452–454
  of circle, 494–495
  of circumscribed and inscribed
    regular polygons, 815
  relationship to area in plane,
    872–875
perpendicular bisector, 165–168
  equation of, 690–691
  in isosceles triangle, 318
  as line of reflection, 184
  proving valid construction, 436
  of quadrilaterals, 341, 342, 431
  in regular polygons, 360
  of segment in space, 559
perpendicular bisector method, 859
Perpendicular Bisector
  Theorem, 277
Perpendicular Diagonals
  Quadrilateral Area
  Formula, 477
Perpendicular Gothic period, 159
perpendicular lines, 159–162
Perpendicular Lines and Slopes
  Theorem, 161
perpendicular planes, 520–522
Perpendicular to Parallels
  Theorem, 160
perspective drawings, 538–541, 584
Peter and the Wolf (Prokofiev), 241
Peterson, Jeret, 110
Petrushka (Stravinsky), 367
phi, golden ratio, 796, 813
photogrammetry, 708
pi, 494, 495
Pick's Theorem, 467
pitch in music, 231, 236
pitch of roof, 737
pixels, 20
plane(s), 13, 28
  equation for, 708
  midpoint formula on, 689
  in Point-Line-Plane Postulate,
    32–33
  in space, 518–522
  symmetry, 560–561
plane coordinate geometry, 13,
  See also Euclidean plane
  coordinate geometry.
plane figure, 28
plane of reflection, 559, See also
  reflecting plane.
plane sections, 552
  conic, 556
  of cube, 554–555, 584
  of prisms and cylinders, 554–555
  of pyramids and cones, 555–556
  of sphere, 551–552

Playfair, John, 284
Playfair's Parallel Postulate,
  284–285, 439
Pluto, 60
point(s), 4–5
  assumed properties of, 518–520
  description of, 8, 14, 20, 22
  as dots, 19–20
  equidistant, 165
  of intersection in reflections over
    three lines, 223–225
  as locations, 6–10
  as nodes in networks, 21–23
  ordered pairs as, 13–16
  in Point-Line-Plane Postulate,
    32–33
  reflecting, 184–186
  between two other points, 38
  as undefined term in
    definitions, 28
Point-Line-Plane Postulate, 32–33
Point-Line-Plane Postulate
  (Expanded), 518–519
point of concurrency, 152, 861–862
point of tangency, 844
polygon(s), 62–63, 89–91, 94, 101,
  356, See also n-gon; regular
  polygons.
  areas of regular polygons,
    490–491
  construction with large
    perimeter, 877, 878
  design of repeating rotation, 125
  equilateral and equiangular, 357
  exterior angles of, 290–292
  inscribed in circle, 358–359
  interior angles of, 290, 327
  lattice points and area, 467
  orientations of, 191–192
  scheduling tournaments,
    831–834
  sums of angle measures in,
    288–292
  symmetry of (See symmetry of
    polygons.)
  unions and intersections, 84
  unit circle and, 815
Polygon Exterior Angle
  Theorem, 291
Polygon-Sum Theorem, 290,
  294, 295
polygonal regions, 91, 525
polyhedra (polyhedrons), 564–565,
  568, 569, See also prisms;
  pyramids.
polynomials, multiplication shown
  by area and volume, 603–605

"Pons Asinorum," 440
positive square root, 480
possibilities, ruling out, 648–
  650, 651
Postulates, 5, 32, 53, 133, 175, 243,
  299, 370, 442, 508, 586, 639,
  710, 765, 817, 890, See also
  Properties; Theorems.
  angle measure, 115–116, 127
  area, 458, 468
  corresponding angles, 146,
    160, 439
  distance, 40
  of equality, 133–134
  of equality and inequality, 136
  of equality and operations, 134
  of inequality and operations, 135
  parallel lines and
    transversals, 273
  Playfair's parallel, 284–285, 439
  point-line-plane, 32–33
  point-line-plane (expanded),
    518–519
  reflection, 190, 192, 196, 198
  triangle inequality, 47, 198
  as type of property, 133
  volume, 597, 611
power of the point for the circle, 867
preimage, 121
  mapped onto an image, 151–152
  reflection images, 184
  reflections of figures, 189–191
preservation properties
  of reflections, 189–190, 277–
    279, 733
  of size transformations, 154–155,
    720, 722–723, 733
prime numbers
  in conditional statements, 70
  infinitude of primes theorem,
    665–666
Prism-Cylinder Volume Formula, 611
prisms, 525–528
  creating from pyramids, 616–617
  plane sections of, 554–555
  surface area of, 570–571
  volume of, 609–612
probability, calculated with
  areas, 502
Proclus, 284
Procter & Gamble, 60
Projects
  Advertising, 368
  "Bridge of Fools, The," 440
  Broadcast Areas, 506
  Circular Definitions in
    English, 51

Combination Locks and Rotations, 172
Congruence and Genetics, 296
Congruence and Literacy, 296
Contour Maps, 708
Convex or Nonconvex?, 102
Creating Tessellations, 440
Equations for Planes, 708
Escher Tessellations, 241
Euler Line, The, 888
Famous Proofs of the Pythagorean Theorem, 507
Finding the Shortest Route, 51
Fingerprint Statistics, 173
Finish This Story, 764
Folding Paper, 764
Folding Planes, 584
Fractals, 886
Frieze Patterns, 368
Geodesic Dome, 887
Geometric Mean and Financial Reporting, The, 814
Golden Ratio and Music, The, 814
Golden Ratio in Nature, The, 814
Graphing Equations, 708
Hamiltonian Circuits, 52
Hopscotch, 507
How Long Can Proofs Get?, 297
Intersection, Union, Addition, and Multiplication, 102
Japanese Temple Geometry Theorems, 887
Kaleidocycles, 440
Kaleidoscopes, 241
Lengths and Areas on the Surface of Earth, 638
Linear and Circular Motion, 886
Making Cones, 638
Making Maps, 584
Making the Most Out of Paper, 887
Maps Distort Area, 507
Mazes, 172
Miniature Golf, 241
Models of Everyday Objects, 764
Morley's Theorem, 441
NASA and Big Flags, 815
Packing Spheres, 638
Photogrammetry, 708
Plane Sections of a Cube, 584
Polygons Are a Matter of Opinion, 101
Proofs as Games, 297
Quadrature of the Lune, 506
Reflections and Communications, 241

Regular Polygons and Unit Circles, 815
Regular Polyhedra, 583
Shrinking Person, The, 764
Similar Airplanes?, 814
Special Regular Polygon, A, 368
Star Polygons, 297
Star Polyhedra, 583
Strange Triangle Tessellation, A, 440
String Art, 52
Structures, 584
Symmetry and Snowflakes, 368
Tangram Shapes, 101
Three-Dimensional Paradoxes, 584
Tiling the Rice Way, 441
Transformations in Music, 241
Triangles on Curved Surfaces, 296
Truth Tables, 708
U-Turns and Area, 506
Upside-Down Names, 368
Use a DGS, 441
Using a DGS to Draw, 51
Using Paper-Folding to Construct a Regular *n*-gon, 368
Venn Problem Solving, 101
Voronoi Diagrams and High Schools, 172
What Is Teddy's Surface Area and Volume?, 638
What Is the Right Size?, 764
What Is the Volume of a Spherical Cap?, 638
Write a Children's Geometry Book, 886
Zoom and Size Transformations, 172
Prokofiev, Sergei, 241
Proof, 270
proofs, 96, 139–142, 269–270, See also *congruence proofs; indirect proofs.*
30-60-90 triangle theorem, 489–490
angle bisector theorem, 782
angle-chord theorem, 838
angle symmetry theorem, 311
arc-chord congruence theorem, 825, 829
circles, 311, 411, 501
congruence properties, 254–255
congruent angles, 437
congruent figures, 279

congruent triangles, 386–390, 392, 393–396, 407–409
definition of proof, 139
diagonals, 671, 678–679, 703
distance formula, in space, 702
equilateral triangle, 270
exterior angles, 291, 293, 294
external angles, 272–273
flip-flop theorem, 309
infinitude of primes theorem, 666
inscribed angle, 325–326, 328, 329
isoperimetric inequality for the plane, 873
isosceles trapezoid, 338, 347
isosceles triangle, 317, 319
law of detachment, 654–655
line intersection theorem, 664–665
means-extremes property, 727
midpoint of segment, 398, 689, 690
midsegment of triangle theorem, 695–696
overlapping triangles, 401–402, 403, 404, 405
parallel lines, 271, 274, 275, 284
parallelograms, 420–422, 428, 670–671
perpendicular bisectors, 278, 279, 397, 436
prism-cylinder volume formula, 611
pyramid-cone volume formula, 617
Pythagorean Theorem, 105, 481–482, 483, 507, 791
quadrilaterals, 288, 476, 667
radius-tangent theorem, 845, 850
regular polygons, 358–360
scalene triangles, medians and altitudes, 666–667
secant length theorem, 867
side-splitting theorems, 775, 776, 777–778
side switching theorem, 310
similarity theorems, 739, 751, 757
size change properties, 153–154, 155, 720, 721
spheres, 557, 628–629, 634, 637, 704
supplementary angles, 143
symmetry theorems, 310, 311, 313, 340, 352
tangent angle theorems, 853, 854
tangent length theorem, 869

trapezoids, 338, 346, 347, 475, 678–679, 777–778
triangle area formula, 469–470
triangle congruence, 407–408, 409
triangle side-splitting theorems, 775, 776
triangle-sum theorem, 285
two-reflection theorems, 206, 212
unequal sides theorem, 320
using distance formula in, 678–679
using reflections, 277–279, 280
using to deduce volume formulas, 625
validity of constructions, 436–437
vertical angles theorem, 130
Properties, 53, 175, 243, 299, 370, 442, 508, 586, 639, 710, 765, 817, 890, See also *Postulates; Theorems.*
addition, of equality, 134
addition, of inequality, 135
additive, 40 (See also *Additive Property.*)
of area, 457–460
congruence (See *Congruence Property.*)
distributive, of multiplication over addition, 136, 605
double negative, 657, 658
of equality, 133–134
equation to inequality, 136
of inequality, 135
means-extremes, 727, 786, 795
multiplication, of equality, 134
multiplication, of inequality, 135
parallel, size change theorem, 153–154
reflexive, of congruence, 254
reflexive, of equality, 133–134
of similar figures, 733–734
size change theorem 1: parallel, 153–154
substitution, 136
symmetric, of congruence, 254
symmetric, of equality, 133–134
transitive, of congruence, 254
transitive, of equality, 133–134, 270, 616
transitive, of inequality, 135
types of, 133

uniqueness, 40 (See also *Uniqueness Property.*)
of volume, 596–600
Properties of a Parallelogram Theorem, 420, 431
proportional sides, 759, 774–778, 781–783
proportions, 726–728
protractors, 114–115
Prove, 269
Ptolemy, Claudius, 808
puzzles, See *applications, puzzles; logic puzzles.*
pyramid(s), 532–533, 534
breaking spheres into, 633–634
frustum of, 569
lateral areas of, 577–579
nets for, 565–566
plane sections of, 555
volume of, 615–619
Pyramid-Cone Volume Formula, 617
Pyramid of Cestius, Rome, 588
Pyramids at Giza, Egypt, 532–533, 537, 579, 642, 645, 771
Pythagoras, 230, 481
Pythagorean Converse Theorem, 482–484, 753
Pythagorean Distance Formula on the Coordinate Plane, 677–679, 682
Pythagorean Theorem, 480–484
alternate statement, 481
finding distance on coordinate plane, 676–677
first statement, 481
generalization to right tetrahedrons, 707
proofs of, 105, 481–482, 507, 791
Pythagorean triples, 484, 486, 752–753

Q

Quadratic Formula, 795
quadrature of a lune, 506
quadrilateral(s), 90
areas of, 474–477, 479
cyclic, 330
diagonals of, 98, 427, 431–433, 476–477
hierarchy of, 332, 335, 342–343, 344, 432
kites (See *kites.*)
lines of symmetry, 350

midpoints forming parallelogram, 699
parallelograms, 149, 426 (See also *parallelograms.*)
with rotation symmetry, 353
tessellations with, 414
trapezoids (See *trapezoids.*)
volume of, 614
Quadrilateral Hierarchy Theorem, 335, 421
Quadrilateral-Sum Theorem, 289

R

radii, in circle, 79
radius
of circle, 79
of pyramid or cone, 534
of sphere, 550
Radius-Tangent Theorem, 845, 846, 850
rate, 726
ratio(s), 726–728
golden, 793–796, 797, 813, 814
ratio of similitude, 734–735, 739–740
rays, 39–40, 41, 146, 159
reciprocals, 727
rectangle(s), 60–61, 331
area formula, 458, 459–460
convenient location for, 672
golden ratio, 793–796, 814
in hierarchy of quadrilaterals, 332, 335
as isosceles trapezoid, 347–348
multiplication shown by, 603–604
similar, 793–794
symmetry of, 348, 352–353
Rectangle Formula, in Area Postulate, 458, 459–460
Rectangle Similarity Theorem, 794
Rectangle Symmetry Theorem, 348, 352
refining a conjecture, 98
reflecting figures, 189–192
reflecting line, 184
reflecting plane, 559
reflecting points, 184–186
reflection(s), 183
applying twice, 309
communications and, 241
composing over intersecting lines, 210–212

composing over parallel lines, 202–203

congruence proofs using, 277–279

horizontal, inversion in music, 234–235

in miniature golf and billiards, 196–197

notation for, 185–186

over three lines, isometries, 223–225

preservation properties of, 189–190, 277–279, 733

and rotation, 211–213

in space, 559–561

as type of transformation, 220, 228

vertical, musical retrograde, 235–236

reflection image of a figure, 190–192

reflection image of point over a plane, 559

reflection images, 184–185, 196

Reflection Postulate, 190, 192, 196, 198

reflection-symmetric figure, 308, 560, 672–673

reflection-symmetric palindromes, 235–236

reflection symmetry, 308

of angles, 310–311

applying same reflection twice, 309

of circles, 311

frieze pattern symmetry, 364–365

of kites, 340–341

and rotation symmetry, 352–353

of segments, 309–310

Reflexive Property of Congruence, 254

Reflexive Property of Equality, 133–134

Regiomontanus, 808

regular hexagon, 356, 362, 415, 491

regular hexagonal prism, 527, 561

regular hexagonal pyramid, 565

regular *n*-gons, 356

regular pentagon, 415, 524

regular pentagonal pyramids, 535, 578

Regular Polygon Area Formula, 491

Regular Polygon Reflection Symmetry Theorem, 360

Regular Polygon Rotation Symmetry Theorem, 359

regular polygons, 356

areas of, 490–491

circles and, 357–360

construction of, 368

reflection symmetry of, 360

rotation symmetry of, 358–359

schedules, for tournaments, 832–834

tessellations with, 414–415, 418

regular polyhedron, 564–565, 583

regular prism, 527

regular pyramids, 533

regular square pyramids, 533, 579, 698

regular tetrahedron, 627

regular triangular pyramid, 560–561

remote sensing, 708

Requiem Mass **(Mozart),** 367

retrograde, in music, 235–236

Review questions, 12, 18, 26, 31, 37, 43, 50, 65, 71, 76, 81–82, 87–88, 94, 100, 118–119, 125, 132, 138, 144, 150, 157–158, 164, 171, 188, 195, 200–201, 209, 215, 221–222, 228–229, 240, 256, 262, 268, 276, 281, 287, 294–295, 315, 323, 330, 338, 344, 349–350, 355, 362, 367, 385, 392, 398–399, 405, 411–412, 417–418, 425, 430, 435, 439, 456, 462, 466–467, 472–473, 478–479, 486, 493, 499, 505, 524, 531, 537, 543, 549, 558, 563, 569, 576, 582, 602, 608, 614, 621–622, 627, 632, 637, 652–653, 661–662, 669, 675, 681, 686–687, 693, 698–699, 706–707, 725, 730, 737, 744, 749, 755, 762–763, 780, 785, 792, 798–799, 806, 813, 829–830, 836, 843, 851, 858, 865, 871, 877–878, 884–885, See also *Chapter Review.*

Rhombus Diagonal Theorem, 342

rhombuses, 295, 331, 332, 335, 342–343

Rice, Marjorie, 441

right-angle method, 864

right angles, 61, 116

inscribed angles in semicircles, 326

measuring in construction, 486

similar triangles within, 788–789

right cones, 533

right cylinders, 560–561, 609–610, 630

right cylindrical solid, 571

right prisms, 609–610

right pyramids, 533, 581

right solid, 527

right square prism, 527, See also *cubes.*

right tetrahedron, 707

right trapezoidal prisms, 575

Right-Triangle Altitude Theorem, 788, 791

Right Triangle Area Formula, 468

right triangles, 286, See also *Pythagorean Theorem.*

ancient Egyptian measuring system, 484

area of, 468–469, 473

convenient location for, 672

lengths of sides, 482–484, 486, 487–490

ratios of legs in, 801

similar triangles, 753

tangent of angle in, 801–802

rigidity of triangles, 380, 822

Roman arch, 613

Rome, Georgia, clock tower, 498

rotation(s), 120–123, 210–211

axes of rotation, 355

in combination locks, 172

and reflections, 211–213

as type of transformation, 220, 228

rotation-symmetric figure, 351, 672–673

rotation-symmetric music, 236

rotation symmetry, 351–353

drawing a figure with, 353

frieze pattern symmetry, 364–365

of parallelograms, 421–422

reflection symmetry and, 352–353

Rotation Symmetry Theorem, 352

round-robin tournaments, 831–834

rounds, in music, 232–233, 238

rounds, in tournaments, 831–833

"Row, Row, Row Your Boat," 232–233, 236

Rubik's cube, 602

rules, 27, 30

ruling out possibilities, 648–650, 651

S

Sager, Carol Bayer, 239

same-side interior angles, 271

Same-Side Interior Angles Theorem, 272

San Francisco area, 472

SAS condition, 382–383, 387, 735
SAS Congruence Theorem,
    387–388, 393
SAS similarity condition, 759
SAS Similarity Theorem, 759
scale factor, 734
scale models, 728, 736, 741, 764
scalene triangles, 91, 316, 413–414,
    666–667
"Scandal in Bohemia, A" (Doyle), 648
schedules, for tournaments, 831–834
Secant Length Theorem, 867–868
secants, to circle, 839–840, 854
sector of a circle, 501, 502
segment(s), 39
    congruent, 257–258
    midpoint of, 77–78
    parallel, 146
    perpendicular, 159
    reflection symmetry, 309–310
    trapezoid side-splitting, 777–778
    triangle angle bisector, 781–783
    triangle side-splitting, 774–777
Segment Congruence Theorem, 258
Segment Symmetry Theorem,
    309–310
Self-Test, 54, 104–105, 176–177,
    244–245, 300–301, 371–372,
    443–444, 509–510, 587–588,
    640–641, 711–712, 766–767,
    818, 891–892
semicircles, 113
    angle inscribed in, 326, 789
    angles that intercept, 326–327
    geometric mean in, 789
semiperimeter of triangle, 473
semiregular tessellations, 418
sentences, rewriting as
    conditionals, 67
sets, 83, 84, 102
Seurat, Georges, 19
Seven Wonders of the Ancient
    World, 532, 897
Seymour Conjecture, 297
Sharapova, Maria, 124
sharp symbol, in music notes, 230
shear, 254, 256
Shuler, Dustin, 523
side(s)
    adjacent an angle, 801
    of an angle, 114
    opposite an angle, 801
    of polygons, 90
    of triangles, 381–383, 487–490
        (See also Pythagorean
        Theorem.)
side-splitting theorems, 774–778, 779

Side-Switching Theorem, 310–311
similar figures, 716–717, 731–735
Similar Figures Theorem, 733, 738
similar rectangles, 793–794
similar triangles, 750–753,
    756–759, 788–789, See also
    trigonometry.
similarity, 716–717, See also size
    transformations.
    in animals and birds, 746
    counterpart to congruence, 759
    fundamental theorems of,
        738–742, 745–747
    human height, 745, 746
    ratios and proportions,
        726–728
    structural applications, 747
similarity transformations, 732
similitude, ratio of, 734–735,
    739–740
Simpsons, The, 484
sine of an angle, 807–811
Sir Cumference and the Great
    Knight of Angleland:
    A Math Adventure
    (Neuschwander), 886
Size-Change Distance Theorem,
    721, 722
size-change factor, 719
Size-Change Preservation
    Properties Theorem, 720
Size Change Theorem 1: Parallel
    Property, 153–154
Size Change Theorem 2: Collinearity
    Is Preserved, 154
Size Change Theorem 3: Angle
    Measure Is Preserved, 154
size changes, 151–152, 254, 716, See
    also size transformations.
Size Transformation Mapping
    Theorem, 720
size transformations, 151–155,
    718–723
    1-dimensional, changing tempo
        in music, 233–234
    with any center, 719–720
    with center (0, 0), 718
    distance in image and preimage,
        721–722
    maps and zoom factors, 172
    multiplication of lengths,
        155–156
    parallel property of images,
        153–154
    preservation properties of,
        154–155, 720, 722–723, 733
    similar figures in, 731–732

skew lines, 521
slant height of pyramid or cone, 534
slide transformation, 183, 204, 220
slope(s), 14
    of lines, 147, 670
    of parallel lines, 147–148
    of perpendicular lines, 160–162
slope-intercept form of equation of
    a line, 14–16
small circles, 551
solid(s), 525
    conic, 532
    view of, 544–546
solid sphere, 551
South Pole, of Earth, 552–553
space, 28
    reflections in, 559–561
    three-dimensional coordinates,
        700–705
special quadrilaterals, 336
sphere(s), 550–552
    breaking into pyramids, 633–634
    comparing with other surfaces,
        628–629
    Earth as, 552–553, 558
    equation for, 703–704
    packing in box, 638
    surface area of, 624, 633–635, 637
    volume of, 624, 628–630, 637
Sphere Surface Area Formula, 634
Sphere Volume Formula, 629–630
spherical cap, 638
spherical geometry, 146, 285, 296,
    416, 507
spherical triangles, 416
"Spindle, The," 523
spirals, 355
"Spy Who Loved Me, The," 239
square, 332
    diagonals of, 671
    in hierarchy of quadrilaterals,
        332, 335
    as quadrilateral, 276, 332
    rotation symmetry of, 353
square pyramids, 618
square root, positive, 480
square root form of distance, 9
square units, 457
squircle, 510
Sreekumar, K. P., 771
SSA condition, 406–408
SsA Congruence Theorem, 406–408
SSS condition, 381–382, 386
SSS Congruence Theorem, 386–387,
    391, 750
SSS Similarity Theorem, 750–753,
    756–757

stade, 553
staffs, in music notation, 231
standard form of equation of a line, 13–14
star polygons, 297
star polyhedra, 583
Star Ribbed Domes, 356
statements, 66
    compound, 66
    conditional, 66–67
    contradictory, 664
    negation of, 656–657
Statue of Liberty, New York, 770
Stein, Rolf, 418
Steiner, Jacob, 867
straight angle, 114, 116
Straight Angle Assumption, 115–116
straightedge, unmarked, 166, 783
Stravinsky, Igor, 367
subsets in Venn diagrams, 69
Substitution Property, 136
Sudoku number puzzles, 646–647, 648, 654
sufficient conditions
    for parallelograms, 426–429
    for quadrilaterals, 432–433
    for triangles, 381–383, 386–389
Sufficient Conditions for a Parallelogram Theorem, 428
Summary/Vocabulary, 53, 103, 174–175, 242–243, 298–299, 369–370, 442, 508, 585–586, 639, 709–710, 765, 816–817, 889–890
Sun City, Arizona, 498
Sunday Afternoon on the Island of Grande Jatte (Seurat), 19
supplementary angles, 127–128, 143, 346
supplements, 128
Supposition, 664
surface(s), 525
    conic, 532
    and cylindrical solids, 526–527
    view of, 544–546
surface area (S.A.), 570
    of complex figures, 573
    of cones, 577, 579–580
    of cylinders, 571–573
    formulas for, 623–624
    of prisms, 570–571
    in product packaging, 595–595
    of pyramids, 577–580
    in similarity theorems, 739–740
    of spheres, 624, 633–635, 637
    volume compare to, 598

Surface Area Formula for Cylindrical Solids, 572
Surface Area Formula for Pyramids and Cones, 577
Swift, Jonathan, 745
Syene, 553
symbols
    alchemy, 350
    angle, 114
    cube root, 599
    degrees, 116
    distance, 41
    equal sign, 135
    golden ratio, phi, 796
    inequality sign, 135
    Inuktitut alphabet, 214
    is less than, 114, 135
    is not equal to, 138
    is similar to, 732
    line, 41
    music notes, 230, 233, 235
    not-*p*, 656
    null set, 84
    parallel lines, 145
    perpendicular bisector of segment, 168
    pi, 494
    positive square root, 480
    ray, 41
    right angles, 61
    segment, 41
    size change, 153
    volume, 597
symmetric figures, 311–312
Symmetric Figures Theorem, 312
Symmetric Property of Congruence, 254
Symmetric Property of Equality, 133–134
symmetry diagonal of a kite, 341
symmetry line, 308, 350
symmetry of polygons
    angles inscribed in circles, 324–327
    frieze patterns, 363–365
    isosceles triangle, 316–321
    kite family, 333–334, 335, 339–341 (See also *kites.*)
    parallelogram family, 331–333, 343, 421–422 (See also *parallelograms.*)
    of reflections (See *reflection symmetry.*)
    regular polygons, 358–359, 360
    rhombuses, 342 (See also *rhombuses.*)
    of rotation, 351–353

trapezoid family, 334–335, 345–348
symmetry planes, 560–561
synthetic geometry, See *Euclidean synthetic geometry.*

T

Tangent-Chord Theorem, 853–854
tangent circles, 848
Tangent Length Theorem, 869
tangent of an angle, 800–803
tangent ratio, 801–803
Tangent-Secant Theorem, 854
tangent segment, 847
tangent to circle, 686, 844–845, 847–848
    angles formed with tangents, 852–855
    length, from point to circle, 868–869
tangent to spheres, 846–847
tangram shapes, 101
taxicab distance, 676, 677
technical drawings, 538
tempo changes, in music, 233–234
terminal point of vector, 217
terms in conditionals, 79–80
terms of a proportion, 726–727
tessellates a plane, 413
tessellations, 182–183, 248
    Escher, 182, 241, 440
    finding area of irregular figures, 463
    with nonpolygons, 415–416
    with polygons, 413–415, 418
    of spherical geometries, 416
    used in finding areas, 457, 463
tetrahedrons (tetrahedra), 440, 564, 565, 583
Texas, area of, 472
Thales, 269, 326
Thales' Theorem, 326
Thebes, tomb in Egypt, 376
"Theme and Variations," 236
Theorems, 32, 53, 133, 175, 243, 299, 370, 442, 508, 586, 639, 710, 765, 817, 890, See also *Postulates; Properties.*
30-60-90 triangle, 489–490
A-B-C-D, 252
AA similarity, 757, 772, 774, 788, 807
AAS congruence, 389, 392
alternate exterior angles, 272
alternate interior angles, 272
angle bisector, 782–783

angle-chord, 838
angle congruence, 258
angle-secant, 839
angle symmetry, 311
arc-chord congruence, 825, 829
ASA congruence, 388–389
box diagonal formula, 703
center of a regular polygon, 358
chord-center, 825
circle symmetry, 311
corresponding parts in congruent figures (CPCF), 259, 395
diagonals in a polygon, 834
distance between parallel lines, 421
equal angle measures, 128
equation for a circle, 682
equation for a sphere, 704
equivalence properties of congruence, 254
exterior angle, for triangles, 290, 839
exterior angle for triangles, 290
figure size-change, 723
figure transformation, 191, 279
flip-flop, 309
fundamental, of similarity (alternate statement), 740
fundamental, of similarity (first statement), 739, 745–747
general angle-arc measure, 855–856
geometric mean, 787
Hippocrates', 506
hypotenuse-leg (HL) congruence, 408–409
infinitude of primes, 665–666
inscribed angle, 325–326, 327, 839, 854
inscribed angles and congruent arcs, 329
isoperimetric, for space, 880
isoperimetric, for space (alternate statement), 881
isoperimetric, for the plane, 873
isoperimetric, for the plane (alternate statement), 875
isosceles right triangle, 488
isosceles trapezoid, 347
isosceles trapezoid symmetry, 346–347
isosceles triangle base angles, 318–319, 440, 487
isosceles triangle base angles converse, 320–321

isosceles triangle coincidence, 318
isosceles triangle symmetry, 317–318
from Japan, 481, 887
kite diagonal, 341, 342, 431
kite symmetry, 340–341, 391
line intersection, 34, 664–665
line-plane perpendicular, 521
linear pair, 129, 141
means-extremes property, 727, 786, 795
midpoint formula on a number line, 688–689
midpoint formula on the plane, 689–690
midsegment of trapezoid, 695
midsegment of triangle, 695–696
Pappus', 100
parallel lines, 271
parallel lines and slopes, 148, 670
parallel lines and transversals, 273
parallelogram symmetry, 422
perpendicular bisector, 277
perpendicular lines and slopes, 161
perpendicular to parallels, 160
Pick's, 467
polygon exterior angle, 291
polygon-sum, 290, 294, 295
properties of a parallelogram, 420, 431
Pythagorean, 481 (See also *Pythagorean Theorem.*)
Pythagorean converse, 483, 753
quadrilateral hierarchy, 335, 421
quadrilateral-sum, 289
radius-tangent, 845, 846, 850
rectangle similarity, 794
rectangle symmetry, 348, 352
regular polygon reflection symmetry, 360
regular polygon rotation symmetry, 359
rhombus diagonal, 342
right-triangle altitude, 788, 791
rotation symmetry, 352
same-side interior angles, 272
SAS congruence, 387–388, 393
SAS similarity, 759
secant length, 867–868
segment congruence, 258
segment symmetry, 309–310
side-splitting, 774–778, 779
side-switching, 310–311
similar figures, 733, 738

size change 1: parallel property, 153–154
size change 2: collinearity is preserved, 154
size change 3: angle measure is preserved, 154
size-change distance, 721, 722
size-change preservation properties, 720
size transformation mapping, 720
SsA congruence, 406–408
SSS congruence, 386–387, 391, 750
SSS similarity, 750–753, 756–757
sufficient conditions for a parallelogram, 428
symmetric figures, 312
tangent-chord, 853–854
tangent length, 869
tangent-secant, 854
Thales', 326
third angle, 389
three-dimension distance formula, 702–704
three-dimension midpoint formula, 704
three squares, 481
transitivity of parallelism, 148
trapezoid angle, 346
trapezoid side-splitting, 777–778
triangle congruence, 759
triangle side-splitting, 774–775
triangle side-splitting converse, 776
triangle similarity, 759
triangle-sum, 285–286, 288, 295, 389
two perpendiculars, 160
two-reflection, for rotations, 212, 352
two-reflection, for translations, 206
two tangent, 848
unequal angles, 321
unequal sides, 320, 408
unique circle, 283
unique plane, 520
uniqueness of parallels, 284, 285
Varignon's, 699
vertical angles, 129–130, 141
Third Angle Theorem, 389
Three-Dimension Distance Formula Theorem, 702–704
Three-Dimension Midpoint Formula Theorem, 704
three-dimensional coordinate system, 700–702

three-dimensional figures, 29, 355, 516–517
  cones, 533, 534, 567 (See also *cones.*)
  cylinders, 525–528, 566 (See also *cylinders.*)
  drawing in perspective, 538–541
  nets, 565–567, 568, 583, 633
  points, lines, and planes in space, 518–522
  polyhedra, 564–565, 568, 569
  prisms, 525–528 (See also *prisms.*)
  pyramids, 532–533, 534, 565–566 (See also *pyramids.*)
  reflections in space, 559–561
  sections of figures, 551–552, 554–556
  spheres, 550–552 (See also *spheres.*)
  surface areas, 570–573, 577–580 (See also *surface area (S.A.).*)
  views of solids and surfaces, 544–546
Three Squares Theorem, 481
Through the Looking Glass (Carroll), 659
tilt, in lines, 145
Tombaugh, Clyde, 60
tools/toolbox, in DGS, 45
tournaments, scheduling teams, 831–834
Transamerica Building, San Francisco, 532
transformation(s), 151–152, See also *size transformations.*
  composites in music, 235
  composition of, 202–203
  isometries, 223 (See also *isometries.*)
  music and (See *music and transformations.*)
  reflections as, 185–186 (See also *reflections.*)
  rotations (See *rotations.*)
  similarity, 732
  that do not yield congruent figures, 253–254
  translations (See *translations.*)
  types of, 219, 220, 223, 228, 243
  using to prove congruence, 396
"Transformation" (Yamaya), 237, 240
Transitive Property of Congruence, 254
Transitive Property of Equality, 133–134, 270, 616

Transitive Property of Inequality, 135
transitivity in proofs, 269–273
transitivity law, 655–656
Transitivity of Parallelism Theorem, 148
translation(s), 183, 204
  1-dimensional, transposing in music, 231
  properties of, 204–207
  2-dimensional, rounds in music, 232
  as type of transformation, 220, 228
  as vectors, 216–219
translation symmetry of frieze patterns, 363–365
translation vectors, 216–217
transposing, in music, 231
transversable network, 22
transversal (line), 145, 270–273
trapezoid(s), 334, 335, See also *isosceles trapezoids.*
  angles in, 345–346
  area of, 474–475
  in buildings, 345, 348, 349
  midsegment of, 694–695
  side-splitting proportional lengths, 777–778
Trapezoid Angle Theorem, 346
Trapezoid Area Formula, 475
Trapezoid Side-Splitting Theorem, 777–778
triangle(s), 89–90, See also *isosceles triangles; right triangles; similar triangles.*
  angle bisector and proportional segments, 781–783
  area of, 468–471, 473
  circles in
    circumcircle, 860
    incircle, 860–861
    nine-point circle, 863–864
    points of concurrency, 861–862
  classification by angles, 286, 331
  drawing, given sides and angles, 381–383
  heptagonal, 493
  hierarchy of, 91, 369
  length of sides, 487–490 (See also *Pythagorean Theorem.*)
  medial, 697
  midsegment of, 694, 695–696
  overlapping, 400–402
  as polygon, 90
  side-splitting proportional lengths, 774–778, 779

in spherical geometry, 296
sufficient conditions for congruence, 381–383, 386–390
types of, 91
with unequal sides and unequal angles, 319–321
Triangle Area Formula, 469–470
Triangle Congruence Theorem, 759
Triangle Inequality Postulate, 47, 198
Triangle Side-Splitting Converse Theorem, 776
Triangle Side-Splitting Theorem, 774–775
Triangle Similarity Theorem, 759
Triangle-Sum Theorem, 285–286, 288, 295, 389
triangular prisms, 527, 616
triangular pyramids, 615–616
"Trias Harmonica" (Bach), 239
Trichotomy Law, 651
trigonometric ratios, 803
  secants, to circle, 839–840, 854
  sine and cosine of an angle, 807–811
  tangents (See *tangent of an angle; tangent to circle.*)
trigonometry, 772–773, 808
  angle bisector in triangle, 781–783
  geometric mean, 788–789, 814
  side-splitting theorems, 774–778
trivial congruences, 311
truncated cone, 536
truncated pyramid, 536
truth
  of conditionals, 68–69
  of conjectures, 96–97
  of converses, 73–74
turn, as transformation, 220
two-dimensional coordinate system, 700
two-dimensional cross sections, 552, See also *plane sections.*
two-dimensional objects, 13
two-dimensional space, 29
two-dimensional translations, rounds in music, 232
Two Perpendiculars Theorem, 160
two-point perspective, 540–541, 543
Two-Reflection Theorem for Rotations, 212, 352
Two-Reflection Theorem for Translations, 206
Two Tangent Theorem, 848

U

unbounded regions, 525
undefined terms, 27–28
Unequal Angles Theorem, 321
Unequal Sides Theorem, 320, 408
union of three segments, 89
union of two sets, 83, 102
unions of figures, 83–85, 90
Unique Angle Assumption, 116
Unique Circle Theorem, 283
Unique Line Assumption, 32–33, 518
Unique Measure Assumption, 116
Unique Plane Assumption, 518
Unique Plane Theorem, 520
uniquely determined, 282
uniqueness in congruence proofs, 282–283, 284–285
Uniqueness of Parallels Theorem, 284, 285
Uniqueness Property
   in Area Postulate, 458
   in Distance Postulate, 40
   in Volume Postulate, 597
unit circle, 686, 815
unit cube, 596
units of area, 457, 459–460
U.S. Customary System, of liquid volume, 598
U.S. Dept. of Agriculture (USDA), 59
U.S. Soccer Federation, 434
U.S. Strategic Petroleum Reserve, 612
U.S. Weather Bureau, 691

V

vanishing points, 538–540
Varignon's Theorem, 699
vectors, 216–219
Venn diagrams
   conditional statement analysis, 69
   intersections and unions, 83
   solving problems, 101
   of squares and rhombuses, 332

vertex (vertices)
   of an angle, 114
   nodes in networks, 21
   of polygon, 90
   of polyhedra, 564
   of surface, 526
vertex angle of isosceles triangle, 316
vertical angles, 129–130
Vertical Angles Theorem, 129–130, 141
vertical component of vectors, 218–219
vertical lines, 13–14
   frieze pattern reflection symmetry over, 364–365
   in perspective drawings, 539, 540
   slope of, 147
vertical reflection, musical retrograde, 235–236
vertical translation, in written music, 231
views of 3-dimensional figures, 544–546
vision
   angle of, 858
   fields of, 112–113, 120, 138, 201
volume, 596
   barrels, 609
   of box, 597, 601
   compared to surface area, 598, 880–883
   of cube, 599–600
   formulas for, 623–625, 628
   liquid measures, 598–599
   multiplication and area, 603–606
   of prisms and cylinders, 609–612
   in product packaging, 594–595
   properties of, 596–600
   of pyramids and cones, 615–619
   in similarity theorems, 739–740
   of spheres, 628–630
   of spherical cap, 638
Volume Postulate, 597, 611
Voronoi, Georgy, 167, 168
Voronoi diagrams, 167, 168, 277

W

Wadlow, Robert, 745
walk transformation, 183, 225
Washington, D. C. subway system, 85
Washington, Denzel, 315
Washington Monument, 309
Webster's Dictionary, 82
"Who Owns the Zebra," 653
Winchester Cathedral, 159
window, in DGS, 45
Wizard of Oz, The, 484
World of Mathematics, The (Haldane), 764

X

x-axis, 701
x-coordinate, 700

Y

y-axis, 701
y-coordinate, 700
y-intercept, 14
Yamaya, Herf, 236–237, 240
Yao Ming, 748

Z

z-axis, 701
z-coordinate, 700
zero angle, 114, 116
Zero Angle Assumption, 115–116
zero-dimensional figure, 31
Zu Chongzhi, 611, 614
Zu Geng, 611

Photo Credits

Photo Credits

Chapters 1–7

Chapters 8–14

We wish to acknowledge the generous support of the **Amoco (now BP) Foundation** and the **Carnegie Corporation of New York** in helping to make it possible for the first edition of these materials to be developed, tested, and distributed, and the additional support of the **Amoco (now BP) Foundation** for the second edition.

Symbols

$>$	is greater than
$<$	is less than
$\geq$	is greater than or equal to
$\leq$	is less than or equal to
$\neq$	is not equal to
$\approx$	is approximately equal to
$\pm$	plus or minus
π	pi
ϕ	phi, the golden ratio
$|n|$	absolute value of n
$\sqrt{}$	positive square root
$\sqrt[3]{}$	cube root
$\|$	is parallel to
$\perp$	is perpendicular to
$\cong$	is congruent to
$\sim$	is similar to
r_m	reflection over line m
$r_m(P)$	reflection image of point P over line m
$r(P)$	reflection image of point P
$R(P)$	rotation image of point P
$T(P)$	transformation image of point P
$T \circ S$, $T(S(P))$	composite of transformation S followed by transformation T
S_k	size change of magnitude k
A'	image of point A
A''	image of point A'
$\Rightarrow$	if-then (implication)
$\Leftrightarrow$	if and only if

$\{\dots\}$	set
$\{\,\}, \varnothing$	empty or null set
$P(E)$	probability of an event E
$\cap$	intersection of sets
$\cup$	union of sets
$\tan A$	tangent of $\angle A$
$\sin A$	sine of $\angle A$
$\cos A$	cosine of $\angle A$
$\overleftrightarrow{AB}$	line through A and B
$\overrightarrow{AB}$	ray with endpoint at A and containing B
$\overline{AB}$	segment with endpoints A and B
AB	distance from A to B
$\angle ABC$	angle ABC
$m\angle ABC$	measure of angle ABC
$\triangle ABC$	triangle with vertices A, B, C
$ABCD\dots$	polygon with vertices A, B, C, D, $\dots$
$\odot O$	circle with center O
$\odot(O, AB)$	circle with center O and radius AB
$\llcorner$	right angle symbol
$n°$	n degrees
$\overset{\frown}{AB}$	minor arc with endpoints A and B
$\overset{\frown}{ADB}$	arc with endpoints A and B containing D
$m\overset{\frown}{AB}$	measure of arc AB in degrees
$\text{Area}(F)$	area of figure F
$\text{Volume}(F)$	volume of figure F
(x, y)	ordered pair
(x, y, z)	ordered triple
$\overrightarrow{AB}$	vector with initial point A and terminal point B
$\mathbf{v}$ or $\vec{v}$	vector v